Dictionary
of Earth Sciences

ENGLISH-FRENCH | FRENCH-ENGLISH
ANGLAIS-FRANÇAIS | FRANÇAIS-ANGLAIS

Dictionnaire
des Sciences de la Terre

J.-P. Michel R.W. Fairbridge

Dictionary of Earth Sciences

ENGLISH-FRENCH | FRENCH-ENGLISH
ANGLAIS-FRANÇAIS | FRANÇAIS-ANGLAIS

Dictionnaire des Sciences de la Terre

Second edition revised and expanded
Deuxième édition revue et augmentée

JOHN WILEY & SONS
Chichester New York
Brisbane Toronto Singapore

1992

MASSON
Paris Milan
Barcelone Bonn

ISBN: 0-471-93535-2

JOHN WILEY AND SONS LTD   Baffins Lane, Chichester, West Sussex PO19 1UD, England

ISBN : 2-225-82395-2

MASSON S.A. 120, boulevard Saint-Germain, 75280 Paris Cedex 06, France

Contents

Table des matières

Preface (1st edition)

Par Rh. W. Fairbridge

Conceived essentially for the use of students and research workers in geology and physical geography, as well as for professionals, this bilingual dictionary of the Earth sciences is addressed to a wide audience. It offers simply word-by-word translations: not explanations. It fills a unique gap in geological literature and in the field of specialized dictionaries. It brings together most of the terms in current use within the following disciplines: physical geography, geomorphology, geodynamics, general and applied geology, petroleum and mining geology, mineralogy, paleontology, sedimentology, tectonics, and so on. Numbers of disused or archaic terms are included, but are so indicated. The terms have mostly been chosen on the basis of frequency of use in classical and recent textbooks and professional publications in geology published in the English and French languages.

This publication is the result of many year's research and collection by the senior author, Jean-Pierre Michel, Docteur ès Sciences, Maître de Conférences à l'Université Pierre et Marie Curie (Paris VI). The associate author, Professor Rhodes W Fairbridge, Department of Geological Sciences, Columbia University, New York, has been able to draw on his long experience in editing numerous volumes in the *Encyclopedia of Earth Science Series* as well as for professional journals and the *Benchmark Series*, adding numbers of terms to the original collection.

We have benefited greatly from the bilingual experience and valuable additions proposed by our colleagues in French Canada at the Université de Québec à Montréal, specifically M. Gilbert Prichonnet, Claude Hilaire-Marcel, and Bernard de Boutteray. Thus both traditional English and North American usages have received close scrutiny. In a work so extensive and desirably comprehensive, it is regrettably inevitable that we have committed some errors and allowed some omissions. We can only ask forbearance.

It is hoped that this multiple review will make the dictionary useful for all geologists, amateur and professional, on a worldwide basis. In particular, it should help people to understand scientific papers and to translate journal articles. In view of the fact that the English language has gradually become the international *lingua franca* of international congresses, even amongst non-English-speaking delegates, it is anticipated that the dictionary should prove an invaluable aid to them.

Avant-propos de la 1ʳᵉ édition

Conçu initialement à l'usage des étudiants et chercheurs en géologie et géographie physique et pour les géologues professionnels, ce dictionnaire bilingue des sciences de la Terre s'adresse en fait à un très vaste public. Ce n'est pas un dictionnaire de définitions mais seulement de traduction. Il comble une lacune dans le domaine des ouvrages de géologie et des dictionnaires spécialisés. Il rassemble les termes les plus employés des disciplines suivantes : géographie physique et géomorphologie, géodynamique, géologie générale et appliquée, géologie pétrolière et minière, minéralogie, paléontologie, sédimentologie, tectonique, etc. De nombreux termes anciens sont inclus et indiqués comme tels. Le choix des termes a été fait en raison de leur fréquence d'utilisation dans les traités classiques récents et autres publications de géologie tant anglais que français.

Cet ouvrage est le fruit de nombreuses années de recherche et de compilation de l'auteur, J. P. Michel, docteur ès sciences, maître de conférences à l'université Pierre et Marie Curie (Paris VI). Le co-auteur, Professeur Rhodes W. Fairbridge, Dept. of Geological Sciences, Columbia University, New York City, a apporté la contribution de sa longue expérience acquise dans la publication de nombreux volumes de l'*Encyclopedia of Earth Science Series* ainsi que dans la collaboration à des revues professionnelles et à la « Benchmark Series », ajoutant un grand nombre de termes à ceux choisis initialement.

Nous avons largement bénéficié également du bilinguisme de nos collègues canadiens-français de l'Université de Québec à Montréal, Messieurs Gilbert Prichonnet, Claude Hilaire-Marcel et Bernard de Boutteray. Ainsi, l'anglais traditionnel et l'anglais spécifiquement nord-américain ont été pris en considération. Bien que nous ayions essayé de faire ce dictionnaire aussi complet que possible, il est certain que des erreurs et des omissions ont pu se glisser.

Nous espérons que le soin apporté à la rédaction de ce dictionnaire le rendra très utile à tous les géologues, amateurs et professionnels. Il devrait notamment permettre de mieux comprendre les travaux scientifiques et articles de périodiques. La langue anglaise devenant de plus en plus la *lingua franca* des congrès internationaux, même pour les délégués des pays non anglophones, nous souhaitons que ce dictionnaire soit une aide précieuse pour tous.

Preface (2nd edition)

This dictionary fills a gap in the documentation of the Earth Sciences; it gathers the most usual geological terms (including Geodynamics, Paleontology, Petrology, and strongly connected specialized disciplines, such as Geomorphology, Mining Geology, Petroleum Geology, Soil Sciences, and Planetology).

This second edition is a quite completelely rehandled version of the Dictionary of Earth Science, published by Editions Masson New York (1980), which included already about 23500 terms in the English-French part, and more than 14000 terms in the French-English part. The authors have added more than 2000 terms in each part. Terminological additions chiefly concern Astronomy (Planetology is now teached with Geology during the first year of the french D.E.U.G.), Geomorphology, Geophysics, Paleontology, and Soil Sciences; in return, many not specifically geological terms have been suppressed.

We have tried to quote the greater number of the terms mentioned in the "Dictionnaire de Géologie", by A. Foucault and J. F. Raoult (Ed. Masson, Paris), with the english equivalent terms (except the names of the paleontological species and of the subsidiaries tectonic events). So, those two dictionaries are complementary to one another. However, the definitions of our "Earth Science dictionary" are very short, without explanations; so, this book is particularly a tool for the geological translation.

In addition to Canadian colleagues, already quoted in the preface of the first edition, we also thank very much Michel Brochu, our geomorphologist colleague at the University of Yaounde (Cameroon) and the Documentation Research assistants of the Earth Sciences Library of P. M. Curie University, for helping.

We hope that this dictionary will be useful for every student and searcher in Earth Sciences, since many papers and oral communications are presented in English during many scientific meetings and congresses.

Préface de la 2ᵉ édition

Ce dictionnaire comble une lacune dans le domaine des sciences de la Terre : il rassemble les termes les plus fréquents, non seulement des différentes branches de la géologie (géodynamique, paléontologie, pétrographie, sédimentologie), mais aussi des disciplines voisines étroitement associées (géomorphologie, géophysique, géologie des mines et du pétrole, minéralogie, pédologie, planétologie).

Cet ouvrage est une version complètement refondue du *Dictionary of Earth Science*, publié par Masson New York en 1980, lequel comprenait déjà environ 23 500 termes pour la partie anglais-français et plus de 14 000 termes pour la partie français-anglais. Les auteurs ont ajouté plus de 2 500 termes à chaque partie. Les additions terminologiques portent principalement sur l'Astronomie (la planétologie fait désormais partie du programme du D.E.U.G.), sur la géomorphologie, la géophysique, la paléontologie et la pédologie. En contrepartie, divers termes, non spécifiquement géologiques, ont été supprimés.

Nous avons essayé de citer la plupart des termes employés dans le *Dictionnaire de géologie* (en français uniquement) de A. Foucault et J. F. Raoult (Éd. Masson, Paris), avec leurs équivalents anglais (sauf les noms de genre ou d'espèce paléontologique, et les noms des phases orogéniques secondaires). Les deux ouvrages sont donc complémentaires, et s'étayent mutuellement. Mais notre ouvrage est lapidaire, et on n'y trouvera pas de longues définitions, puisque c'est avant tout un outil pour la traduction scientifique.

Ce dictionnaire est le fruit de nombreuses années de documentation de l'auteur principal, J. P. Michel, docteur ès sciences, maître de conférences à l'université P. M. Curie (Paris VI). Le professeur Rhodes W. Fairbridge, Department of Geological Sciences, Columbia University, New York City, a apporté la contribution de sa longue expérience acquise dans la publication de nombreux volumes de l'*Encyclopedia of Earth Science Series,* ajoutant de nombreux termes à ceux choisis initialement.

Outre l'aide de collègues canadiens français de l'Université du Québec (voir préface de la première édition), nous avons bénéficié depuis des conseils de Michel Brochu, professeur de géomorphologie à l'université de Yaoundé (Cameroun) et de l'aide des documentalistes de la bibliothèque interuniversitaire de l'université P. M. Curie. Nous les remercions très vivement.

Nous espérons que cet ouvrage sera utile à tous les étudiants et chercheurs des sciences de la Terre, puisque la langue anglaise est devenue pratiquement la langue officielle de communication des congrès scientifiques.

Suggestions for Translators

Decimals and Commas

In French, 0,25 (i. e. one quarter): in English = 0.25
In French, 25.000: in English 25,000
(to readers in U.K. and U.S., 25.000 means 25).

Scientific Abbreviations

Codified by an international commission; the English-language version is published by the Royal Society of London.
Common error in English: "many kms" (plusieurs km), should be "many km" (kms = km per second).
[For comprehensive tables, and physical symbols, see Fairbridge (ed. 1967) "Encyclopedia of Atmospheric Sciences" New York: entry "Units..." p. 1049].

Geographic Names

In French, l'océan Atlantique; in English: Atlantic Ocean (both adjectival name and category noun get a capital letter).
Likewise: le bassin de Paris = the Paris Basin, NOT the Paris basin.

Stratigraphic and Tectonic Names

In French, Tertiaire inférieur = Lower Tertiary (both name and adjective get a capital letter); vieux grès rouge = Old Red Sandstone; Cretaceous System (and Period); Holocene Epoch; Variscan Orogeny. The rules are codified by an international commission of I.U.G.G.

Hyphens

Hyphens are used in English to relate adjectives, example: a grey-green sandstone; but an adjectival modifier eliminates the need, e.g. a greyish green sandstone. When hyphenated adjectives are used very frequently e.g. beach-ridge, ground-water, land-mark, it is usual to simplify them as single words, thus: beachridge, groundwater, landmark (both adj. and noun). But, illogically, some familiar combinations remain separate, e.g. sea level (noun), sea-level rise (2 adjectives).

Common Errors

Accident, Accidenté (Fr.): means (a strong relief, (b) fault line, heavily faulted (terrain); NEVER an "accident" (contresens).
Actuel, Actuellement (Fr.): "at the present", "now". "Presently" is ambiguous in English; in U.S. it means "at the present moment"; in U.K. if means "soon, maybe next week". For clarity, say "At the present time" or "now".
Actual, Actually (Eng.): means "in truth", "real", "genuine".
Actualism, Actualistic (Eng.), *Actualisme* (Fr.): mean the same thing, i.e. the philosophy of interpreting the past by the genuine evidence of the present time.
Altitude (Fr.) means height, elevation, altitude; *Altitude* (Eng.) usually refers to atmospheric heights, not land features ("Elevation"); ex. a jet aircraft flies at 900 m *altitude,* over Mt Everest which has an *elevation* of 8000 m.
Arctic, Antarctic (Eng.): commonly misspelt in U.S. "Artic", "Antartic" (where the middle "c" is not pronounced).
Datation (Fr.): means "Dating" in English; NEVER "datation".
Eustatisme (Fr.), *Eustasy* (Eng.): the -isme suffixes in French indicate philosophic systems and accepted in social sciences (ex. conservatism, Marxism, Existencialism) but rarely in Geology except in a philosophic or historical sense (ex. Neptunism, Plutonism). Eustasy is

based on the Greek word *stasis* (as in isostasy, ecstasy; eustatic, isostatic, ecstatic); these should NEVER be spelled "eustacy", "isostacy", "ecstasy", there being no letter "c" in Greek.

Feldspath, Feldspathoide (Fr.): in U.S., *Feldspar, Feldspathic* (adj.), *Feldspathoid; in U.K. Felspar, Felspathic, Felspathoid.*

Important (Fr.): in Geology means "thick", "considerable", "weighty" (practically NEVER "important" in the English sense).

Vaste (Fr.): in Geology means "broad", "wide", "widespread" (practically NEVER "vast", which is appropriately used for distances measured in light-years, or the size of American national debt).

References

BATES and JACKSON: *Glossary of Geology* (Washington, A.G.I.); U.S. Geological Survey: *Suggestions to Authors* (Washington, U.S. Government Printing Office), many editions *A Guide to Writing, Editing and Printing in Earth Science* (Washington, A.G.I.).

FAIRBRIDGE R. W. (general editor): *The Encyclopedia of Earth Science* series, 16 volumes (*Oceanography, Atmospheric Sciences, Geomorphology, Geochemistry and Environmental Sciences, Structural Geology, Mineralogy, Sedimentology, Paleontology, World Regional Geology, Soil Science, Beaches and Coastal Geology, Climatology, Applied Geology, Field and general Geology, Igneous and Metamorphic Petrology, Solid Earth Geophysics.*

CAILLEUX A. *La Terre*, 591 p., Paris: Bordas (Focus series).

TRICART J. (1964) *Principes et méthodes de la géomorphologie*, Éd. Masson.

JUNG J. (1963) *Précis de pétrographie*, Éd. Masson.

ROUTHIER P. (1963) *Les gisements métallifères*, Éd. Masson.

PIVETEAU (et co-auteurs) *Traité de Paléontologie*, 8 vol., Éd. Masson.

Suggestions aux traducteurs

Décimales et virgules

En français, 0,25 (c'est-à-dire un quart) : en anglais = 0.25
En français, 25.000 : en anglais, 25,000
(pour les lecteurs britanniques et américains, 25.000 signifie 25).

Abréviations scientifiques

Elles sont codifiées par une commission internationale; la version en langue anglaise est publiée par la « Royal Society » de Londres. Citons une erreur fréquente en anglais : « many kms » (plusieurs km), doit s'écrire « many km » (kms = km par seconde).
Pour consulter des tableaux détaillés et les symboles physiques, voir Fairbridge (éd. 1967), *Encyclopedia of Atmospheric Sciences,* New York : paragraphe « unités », p. 1049.

Noms géographiques

En français, l'océan Atlantique; en anglais, Atlantic Ocean (l'adjectif et le nom catégoriel s'écrivent avec une lettre initiale majuscule).
De même : le bassin de Paris = the Paris Basin, et non pas « the Paris basin ».

Noms stratigraphiques et tectoniques

En français, Tertiaire inférieur = Lower Tertiary (le nom et l'adjectif ont leur lettre initiale en capitale);
vieux grès rouge = Old Red Sandstone;
système (et période) Crétacé = System (and Period) Cretaceous;
époque Holocène = Holocen Epoch;
orogenèse Varisque = Variscan Orogeny.
Les règles sont codifiées par une commission internationale de l'I.U.G.G.

Traits d'union

Les traits d'union sont utilisés en anglais pour associer des adjectifs, exemple : un grès gris-vert; mais un qualificatif adjectif peut éviter de placer un trait d'union, par exemple un grès vert grisâtre. Lorsque les adjectifs reliés par un trait d'union sont employés très fréquemment, par exemple, beach-ridge, ground-water, land-mark, on simplifie en les écrivant en un seul mot, tels que : beachridge, groundwater, landmark (à la fois adjectif et nom). Mais, de façon illogique, quelques combinaisons familières restent en deux termes distincts, par ex. sea level (nom), sea-level rise (2 adjectifs).

Erreurs communes

Accident (Fr.) : signifie une ligne de faille, mais jamais un « accident » (contresens).

accidenté (Fr.) : indique un fort relief, un terrain faillé.

actuel, actuellement (Fr.) : « à présent », « maintenant ».

presently est un terme ambigu en anglais; aux U.S.A., il signifie « au moment présent »; en Grande-Bretagne, il signifie « bientôt, peut-être la semaine prochaine ». Pour plus de clarté, disons « à l'instant présent » ou « maintenant ».

actual, actually (Engl.) : signifie « en réalité », réel, authentique.

Actualism, Actualistic (Engl.), *Actualisme* (Fr.) : signifie la même chose, c'est-à-dire la philosophie d'interprétation du passé par des arguments réels de l'époque actuelle.

Altitude (Fr.) : signifie hauteur, altitude; *altitude* (Engl.) se rapporte plutôt aux hauteurs des couches atmosphériques, et non aux caractéristiques terrestres (« altitudes »); par exemple, un avion à réaction vole à 900 m de hauteur au-dessus du Mt Everest, lequel a une altitude de 8 000 m.

Arctic, Antarctic (Engl.) : fréquemment mal orthographié en américain : Artic, Antartic, où le « c » du milieu n'est pas indiqué.

Datation (Fr.) : signifie « dating » en anglais (jamais « datation »).

Eustatisme (Fr.), *Eustasy* (Engl.) : les suffixes -isme en français indiquent les systèmes philosophiques et sont acceptés en sciences sociales (ex : conservatisme, marxisme, existentialisme), mais rarement en géologie, sauf dans un sens philosophique ou historique (par ex. neptunisme, plutonisme). Eustatisme vient du mot grec *stasis* (comme dans isostasie, eustatique, isostatique); il ne faudra jamais les orthographier « eustacy », « isostacy » car il n'y a pas de lettre « c » en Grec.

feldspath, feldspathique, feldspathoïde (Fr.) : en américain, feldspar, feldspathic (adj.), feldspathoid; en anglais, felspar, felspathic, felspathoid.

Important (Fr.) : en géologie, signifie « épais », « considérable » (pratiquement jamais « important » au sens anglais).

vaste (Fr.) : en géologie, signifie « large », « étendu » (pratiquement jamais « vaste », qui est un terme mieux utilisé pour les distances exprimées en années lumières.

DATED GEOLOGICAL TIME SCALE — ÉCHELLE NUMÉRIQUE DES TEMPS GÉOLOGIQUES

Paléozoïque / Paleozoic (French)

ÈRE.SYST.	SÉRIE	SÉRIE	ÉTAGE
CARBONIFÈRE	SILÉSIEN	IV	GZHÉLIEN
			KASIMOVIEN
		V	MOSCOVIEN
			BASHKIRIEN
		VI	SERPUKHOVIEN
	DINANTIEN	VISÉEN	BRIGANTIEN
			ASBIEN
			HOLKÉRIEN
			ARUNDIEN
			CHADIEN
		TOUR-NAISIEN	IVORIEN
			HASTARIEN
DÉVONIEN	SUPÉRIEUR		FAMENNIEN
			FRASNIEN
	MOYEN		GIVÉTIEN
			EIFÉLIEN
	INFÉRIEUR		EMSIEN
			PRAGUIEN
			LOCHKOVIEN
SILURIEN	PRIDOLI		PRIDOLIEN
	LUDLOW		LUDFORDIEN
			GORSTIEN
	WENLOCK		HOMÉRIEN
			SHEINWOODIEN
			TÉLYCHIEN

PALÉOZOÏQUE

Paleozoic (English)

ERA.SYST.	SERIES	SERIES	STAGE	Ma	N.B.
CARBONIFEROUS	SILESIAN	IV	GZHELIAN	295	25
			KASIMOVIAN		26
		V	MOSCOVIAN	305	27
			BASHKIRIAN		28
		VI	SERPUKHOVIAN	315	29
					30
	DINANTIAN	VISEAN	BRIGANTIAN	325	31
			ASBIAN		
			HOLKERIAN		
			ARUNDIAN		
			CHADIAN		
		TOUR-NAISIAN	IVORIAN	350	32
			HASTARIAN		
DEVONIAN	SUPÉRIEUR		FAMENNIEN		33
			FRASNIAN	360	34
	MOYEN		GIVETIAN	365	
			EIFELIAN	375	35
	INFÉRIEUR		EMSIAN	380	
			PRAGUIAN	385	36
			LOCHKOVIAN	390	
SILURIAN	PRIDOLI		PRIDOLIAN	410	37
	LUDLOW		LUDFORDIAN	415	38
			GORSTIAN		
	WENLOCK		HOMERIAN	425	39
			SHEINWOODIAN		
			TELYCHIAN	430	

PALEZOIC

Cénozoïque — Crétacé / Cenozoic — Cretaceous (French)

ÈRE.SYST.	SÉRIE	ÉTAGE
IVre	HOLOCÈNE	
	PLÉISTOCÈNE	CALABRIEN
NÉOGÈNE	PLIOCÈNE	PLAISANCIEN
		ZANCLÉEN
	MIOCÈNE	MESSINIEN
		TORTONIEN
		SERRAVALLIEN
		LANGHIEN
		BURDIGALIEN
		AQUITANIEN
PALÉOGÈNE	OLIGOCÈNE	CHATTIEN
		RUPÉLIEN
	ÉOCÈNE	PRIABONIEN
		BARTONIEN
		LUTÉTIEN
		YPRÉSIEN
	PALÉOCÈNE	THANÉTIEN
		DANIEN
CRÉTACÉ	SUPÉRIEUR	MAASTRICHTIEN
		CAMPANIEN
		SANTONIEN
		CONIACIEN
		TURONIEN
		CÉNOMANIEN
		ALBIEN

CÉNOZOÏQUE

Cenozoic — Cretaceous (English)

ERA.SYST.	SERIES	STAGE	Ma	N.B.
IVry	HOLOCENE		0,01	1
	PLEISTOCENE	CALABRIAN	1,65	2
NEOGENE	PLIOCENE	PLAISANCIAN	3,4	3
		ZANCLEAN	5,3	3
	MIOCENE	MESSINIAN	6,5	
		TORTONIAN	11	
		SERRAVALLIAN	14,5	4
		LANGHIAN	16	5
		BURDIGALIAN	20	
		AQUITANIAN	23,5	
PALEOGENE	OLIGOCENE	CHATTIAN	28	6
		RUPELIAN	34	
	EOCENE	PRIABONIAN	37	
		BARTONIAN	40	
		LUTETIAN	46	7
		YPRESIAN	53	
	PALEOCENE	THANETIAN	59	8
		DANIAN	65	
CRETACEOUS	UPPER	MAASTRICHTIAN	72	
		CAMPANIAN	83	9
		SANTONIAN	87	
		CONIACIAN	88	
		TURONIAN	91	
		CENOMANIAN	96	
		ALBIAN	108	10

CENOZOIC

Mesozoic – Triassic – Permian (English)

Stage	Series	System	Era	Age (Ma)
HAUTERIVIAN	UPPER (MALM)	JURASSIC	MESOZOIC	—
VALANGINIAN				122
BERRIASIAN				130
TITHONIAN				135
KIMMERIDGIAN				141
OXFORDIAN				146
CALLOVIAN	MIDDLE (DOGGER)			154
BATHONIAN				160
BAJOCIAN				167
AALENIAN				176
TOARCIAN	LOWER (LIAS)			180
PLIENSBACHIAN				187
SINEMURIAN				194
HETTANGIAN				201
(RHETIAN)	UPPER	TRIASSIC		205
NORIAN				220
CARNIAN				230
LADINIAN	MIDDLE			235
ANISIAN				240
SCYTHIAN	LOWER			245
TATARIAN	UPPER (I)	PERMIAN	PALEOZOIC	250
KAZANIAN				258
KUNGURIAN	(II)			265
ARTINSKIAN				275
SAKMARIAN	LOWER (III)			285
ASSELIAN				295

Mésozoïque – Trias – Permien (Français)

Étage	Série	Système	Ère
HAUTERIVIEN	SUPÉRIEUR (MALM)	JURASSIQUE	MÉSOZOÏQUE
VALANGINIEN			
BERRIASIEN			
TITHONIEN			
KIMMÉRIDGIEN			
OXFORDIEN			
CALLOVIEN	MOYEN (DOGGER)		
BATHONIEN			
BAJOCIEN			
AALÉNIEN			
TOARCIEN	INFÉRIEUR (LIAS)		
PLIENSBACHIEN			
SINÉMURIEN			
HETTANGIEN			
(RHÉTIEN)	SUPÉRIEUR	TRIAS	
NORIEN			
CARNIEN			
LADINIEN	MOYEN		
ANISIEN			
SCYTHIEN	INFÉRIEUR		
TATARIEN	SUPÉRIEUR (I)	PERMIEN	PALÉOZOÏQUE
KAZANIEN			
KUNGURIEN	(II)		
ARTINSKIEN			
SAKMARIEN	INFÉRIEUR (III)		
ASSÉLIEN			

Ordovician – Cambrian – Proterozoic (English)

Stage	Series	System	Eratem	Eon	Age (Ma)
ASHGILL		ORDOVICIAN			435
CARADOC					445
LLANDEILO					455
LLANVIRN					
ARENIG					470
TREMADOC					485
TREMPEALEAUAN	UPPER	CAMBRIAN			500
FRANCONIAN					
DRESBACHIAN					
MAYAIAN	MIDDLE				
AMGAIAN					
LENIAN	LOWER				
ATDABANIAN					
TOMMOTIAN					530
NEOPROT. III	NEO-PROTEROZOIC	SYSTEM	ERATHEM	PROTEROZOIC	540
CRYOGENIAN					650
TONIAN					850
STENIAN	MESO-PROTEROZOIC				1000
ECTASIAN					1200
CALYMMIAN					1400
STATHERIAN	PALEO-PROTEROZOIC				1600
OROSIRIAN					1800
RHYACIAN					2050
SIDERIAN					2300
			ARCHEAN		2500

Ordovicien – Cambrien – Protérozoïque (Français)

Étage	Série	Système	Érathème	Éon
ASHGILL		ORDOVICIEN		
CARADOC				
LLANDEILO				
LLANVIRN				
ARÉNIG				
TRÉMADOC				
TREMPÉALÉAUIEN	SUPÉRIEUR	CAMBRIEN		
FRANCONIEN				
DRESBACHIEN				
MAYAIEN	MOYEN			
AMGAIEN				
LÉNIEN	INFÉRIEUR			
ATDABANIEN				
TOMMOTIEN				
NÉOPROT. III	NÉO-PROTÉROZOÏQUE	SYSTÈME	ÉRATHÈME	PROTÉROZOÏQUE
CRYOGÉNIEN				
TONIEN				
STÉNIEN	MÉSO-PROTÉROZOÏQUE			
ECTASIEN				
CALYMMIEN				
STATHÉRIEN	PALÉO-PROTÉROZOÏQUE			
OROSIRIEN				
RHYACIEN				
SIDÉRIEN				
			ARCHÉEN	

According to Gilles-Serge ODIN and Chantal ODIN. Géochronique. n° 35, 1990.

D'après Gilles-Serge ODIN et Chantal ODIN, Géochronique, n° 35, 1990.

I THURINGIAN-THURINGIEN
II SAXONIAN-SAXONIEN
III AUTUNIAN-AUTUNIEN
IV STEPHANIAN-STEPHANIEN
V WESTPHALIAN-WESTPHALIEN
VI NAMURIAN-NAMURIEN

Part 1
ENGLISH-FRENCH

1re partie
ANGLAIS-FRANÇAIS

A

A horizon, horizon A (pédol.).

A layer, croûte terrestre (épaisseur : 70-90 km).

aa, aa (laves noires scoriacées).

Aalenian, Aalénien (étage, Lias).

abandoned cliff, falaise morte; **a. shoreline,** ligne de rivage fossile; **a. valley,** vallée morte.

abatement (of pollution), méthode de diminution de la pollution.

abiotic, abiotique.

ablation, fusion, ablation, érosion glaciaire (inusité); formation de sédiments résiduels par lessivage; **a. cone (US),** cône de glace (avec a.); **a. factor,** vitesse de fusion nivale ou glaciaire; **a. till,** moraine d'a.

abnormal (accelerated) erosion, érosion accélérée (d'origine anthropique).

aboral, aboral (pal.).

aboveground, **1.** adj : superficiel, de surface (*mine*); **2.** adv : à la surface.

abrasion, abrasion, érosion; **a. platform,** plate-forme d'a.; **a. coast,** côte d'a.

abrasive rock, roche d'érosion.

abrupt, abrupt.

abruptly, abruptement.

absarokite, absarokite.

absolute, absolu; **a. age,** âge a.; **a. altitude,** altitude a.; **a. datation,** datation a.; **a. dating,** datation a.; **a. temperature,** température a.; **a. zero,** zéro a. ($-273, 18$ °C $= -459,72$ °F.

absorbent, **1.** adj : absorbant; **2.** n : absorbant; **a. of water,** avide d'eau; à forte capacité d'absorption; hydrophile.

absorbing complex (UK), absorption complex (US), complexe absorbant.

absorption, absorption; **a. bands,** bandes d'a.; **a. coefficient,** coefficient d'a.; **a. lines,** raies d'a.; **a. spectrum,** spectre d'a; **atmosphere a.,** a. atmosphérique.

abut (to), buter contre.

abutment, mur de soutènement.

abyss, abysse, abîme.

abyssal, abyssic, abyssal; **a. deposit,** sédiment a.; **a. hill,** colline a.; **a. plain,** plaine a.; **a. rock,** roche plutonique, roche de profondeur; **a. sedimentation,** sédimentation a.; **a. zone,** zone a.

Acadian (N. Am.), Acadien (série, Cambrien moyen); **A. orogeny,** orogenèse a. (Dévonien).

acanthite, acanthite.

Acanthodii, Acanthodiens (pal.).

accelerated erosion, érosion accélérée (anthropique).

acceleration factor, gradient de vitesse; **a. of gravity,** accélération de la pesanteur, intensité de la pesanteur.

accelerogram, accélérogramme.

accelerograph, accélérographe.

accelerometer, accéléromètre.

accessory, accessoire; **a. element,** élément trace; **a. mineral,** minéral a.; **a. plate,** lame a. (pour microscope polarisant).

acclivity, montée, côte.

acclivous, en pente, escarpé.

accordant drainage, réseau hydrographique conséquent, conforme; **a. fold,** pli de même orientation (qu'un ensemble de plis).

accordian folding, accordion folding, plissement en accordéon.

accretion, **1.** accroissement par alluvionnement; **2.** accrétion des continents; **a. coast,** côte d'accumulation; **a. theory,** théorie de l'accroissement de la terre; **a. topography,** relief dû à la sédimentation; **a. vein,** filon minéralisateur.

accretional, en accroissement (tect. plaques).

accretionary prism, prisme d'accrétion; **a. ridge,** levée de plage située à l'arrière du rivage actuel, indiquant un accroissement du continent; **a. lapilli,** nodules volcaniques zonés.

accumulation, **1.** accumulation; **2.** gisement (pétrole); **a. coast,** côte d'a.; **a. zone,** zone d'a.

acerdese, acerdèse, manganite.

achromatic, achromatique.

achromatism, achromatisme.

ACF diagram, diagramme triangulaire (teneur en Al_2O_3, CaO, FeO + MgO).

Acheulian, Acheuléen (Paléolithique ancien).

achirite, dioptase.

achondrite, achondrite (météorite).

achroite (var. of tourmaline), achroïte.

acicular, aciculaire.

aciculate, aciculated, aciculé.

acid, **1.** adj : acide; **2.** n : acide; **a. brown soil,** sol brun a.; **a. clay,** argile a.; **a. lava,** lave a.; **a. rock,** roche a.; **a. soil,** sol a.; **a. treatment,** traitement a.; **weak a.,** a. faible.

acidic rock, roche de type acide.

acidiferous, acidifère.

acidification, acidification.

acidifier, acidifiant.

acidify (to), **1.** acidifier; **2.** s'acidifier.

acidity, acidité.

acidization, acidizing, acidification (introduction d'acide dans un forage).

acidize (to), acidifier.

acidizer, acidifiant.

acidulae, eaux minérales froides chargées d'acide carbonique.

acidulous water, eau acidulée.

aciform, aciculé.

acinase, granuleux, en grappes.

aclinal, aclinic, aclinique.

acline, sans pendage, horizontal.

acmite, acmite, aegyrine; **a. augite,** augite aegyrinique; **a. trachyte,** trachyte à aegyrine.

acoustic, acoustical, acoustique; **a. basement,** socle acoustique; **a. echo-sounding,** écho-sondage a.; **a. frequency,** fréquence a.; **a. horizon,** horizon a.; **a. intensity,** intensité a.; **a. method,** méthode a.; **a. receipter,** récepteur a.; **a. reflection profiling,** méthode de lever par réflection a.; **a. survey,** lever a.; **a. travel-time,** durée de trajet a.; **a. wave,** onde sonore; **a. well logging,** diagraphie a.

acquired character, caractère acquis (*pal.*).

acre, unité de mesure = 0,40468 hectare; **a. foot,** a.-pied : volume de liquide ou de solide nécessaire pour recouvrir 1 acre sur une épaisseur d'1 pied; **a. inch,** a.-pouce : quantité de matériel nécessaire pour recouvrir 1 acre sur une épaisseur d'1 pouce; **a. yield,** rendement à l'a. (pétrole, eau, gaz).

acreage, superficie (en mesures agraires).

acrisol, acrisol (var. de podzol).

Acritarchs, Acritarches (*pal.*).

acrotomous, à clivage parallèle à la base ou au sommet d'un minéral.

Actinaria, Actinaires (*pal.*).

actinic, actinique.

actinium, actinium.

actinoform, de forme radiale, étoilée.

actinolite, actinote, actinolite; **a. facies,** faciès à a.; **a. schist,** schiste a.

actinolitic, actinolithique.

Actinopterygii, Actinoptérygiens (*pal.*).

actinote, actinote.

action zone, zone néritique.

active, actif, en activité; **a. dune,** dune vive; **a. gap,** cluse vive, active; **a. fault,** faille a.; **a. glacier,** glacier a.; **a. layer,** mollisol; **a. margin,** marge active; **a. permafrost,** pergélisol actuel; **a. tectonics,** tectonique a.; **a. volcano,** volcan en activité.

activity, activité; **chemical a.,** a. chimique; **solar a.,** a. solaire; **volcanic a.,** a. volcanique.

actualism, actualisme.

actualistic method, méthode scientifique basée sur les phénomènes réels.

actuapalaeontology, actuapaléontologie (inusité).

acute bisectrix, bissectrice de l'angle aigu formé par les deux axes d'un cristal biaxe.

acyclic, aliphatique.

adamant, diamant.

adamantine, adamantin; **a. luster,** éclat a.; **a. spar,** corindon a.

adamellite (var. of granite), adamellite (*pétro.*).

adaptation, adaptation (au milieu).

adaptive radiation, radiation adaptative; **a. zone,** niche écologique.

adarce, 1. croûte calcaire de source; 2. concrétion salée.

additive, additif.

adductor scar, empreinte des muscles adducteurs (Lamellibranches).

adiabatic, adiabatique; **a. gradient,** gradient a., gradient isothermique.

adiabatically, adiabatiquement.

adinole, adinole (cornéenne métamorphique à albite).

adit, galerie à flanc de coteau, entrée de mine subhorizontale; **a.-cut mining,** exploitation à flanc de coteau.

adjacent sea, mer intérieure.

adjust itself (to), s'adapter (*géogr.*).

adjusted stream, rivière en équilibre.

adjustment, 1. réglage, mise au point; 2. adaptation (*géogr. phys.*); **a. fault,** faille de compensation.

adobe (US), 1. argile loessique (souvent remaniée par les eaux); 2. brique d'argile séchée au soleil.

adolescent river, rivière au stade d'adolescence.

adsorb (to), adsorber.

adsorbed water, eau d'adsorption.

adsorption, adsorption.

adsorptive, adsorbant.

adular, adularia, adulaire (var. de feldspath potassique).

advance, progrès, avancement; **a. heading,** galerie d'a. (*mine*); **a. of a beach,** progression d'une plage vers le large; **a. of geomagnetic field,** déplacement du champ magnétique; **a. of a glacier,** avancée d'un glacier.

advent, arrivée, venue (d'eau).

adventive cone, cône latéral.

aegirine, aegirite, aegyrite, aegyrine.

aegyrine augite, augite aegyrinique; **a. hedenbergite,** hedenbergite aegyrinique.

aeolian, éolien; **a. erosion,** érosion é.; **a. transport,** transport é.

aeolianite (see eol.), sédiment dunaire (éolien) consolidé (Austr.).

aelotropic, anisotropique.

aelotropy, anisotropie.

aeon, durée de 10^9 années (1 milliard d'années).

aerage, ventilation, aération (*mine*).

aeration porosity (US), porosité non capillaire.

aerial, aérien; **a. arch,** voûte anticlinale arasée; **a. fold,** pli dénudé; **a. magnetometer,** magnétomètre aéroporté; **a. mapping,** levé aérien; **a. photography,** photographie aérienne; **a. survey,** photogrammétrie aérienne, levé aérien.

aeriform, gazeux.

aerobian, aerobic, aérobie.

aerodynamic, aérodynamique.

aerolite, aérolite, aérolithe (météorite).

aerologic diagram, diagramme météorologique.

aerology, météorologie.

aeromagnetic map, carte magnétique levée par avion; **a. prospecting,** prospection magnétique aéroportée.

aerometer, aéromètre.

aerometric, aérométrique.

aerometry, aérométrie.

aeronomy, aéronomie.

aerophotography, photographie aérienne.

aerosiderite, aérosidérite (météorite).

aerosite, pyrargyrite.

aerosol, aérosol.

aerugo, vert de gris (oxydation du cuivre).

aff, affinité (*pal.*).

affine deformation, déformation homogène.

affinity, affinité (chimique).

affluent, 1. adj : affluent; 2. n : affluent.

afforestation, boisement, reboisement.

aftershock, réplique séismique.

Aftonian (N. Am.), Aftonien (1er interglaciaire).

agalmatolite, agalmatolite (var. de pyrophyllite massive).

agate, agate; **a. opal,** a. opalisée.

agatiferous, agatifère.

agatize (to), 1. agatiser; 2. s'agatiser.

agatized wood, bois agatisé.

agatoid, agatoïde.

age, âge; **a. dating,** datation; **a. stone,** âge de la pierre (*préhist.*).

agency, agent, action (*géogr.*).

agent, agent (*géogr.*).

agglomerate, 1. adj : aggloméré; 2. n : agglomérat; US; 3) brèche volcanique, tuff dont les fragments sont plus grands que 32 mm.

agglomerate (to), 1. agglomérer; 2. s'agglomérer.

agglomeration, agglomération.

agglutinant, 1. adj : agglutinant; 2. n : liant, agglutinant.

agglutinate, 1. brèche volcanique à ciment vitrifié superficiellement; 2. sol lunaire aggluté.

agglutinate (to), 1. agglutiner; 2. s'agglutiner.

agglutinated, Foraminifères à test formé de particules agglutinées (arénacées ou calcaires).

aggradation, 1. alluvionnement; 2. extension d'une zone à permafrost.

aggrade (to), alluvionner.

aggrading stream, rivière alluvionnante, à l'état d'équilibre.

aggregate, agrégat, groupe; **a. mineral,** a. minéral.

aggregate (to), rassembler, réunir, s'agréger.

aggregation, agglomération.

aggressive, agressif (chimie); **a. magma,** magma intrusif; **a. water,** eau a.

aging cycle of a lake, cycle géomorphologique d'un lac (comblement).

agitate (to), agiter, remuer.

agmatite, agmatite.

Agnatha (*pal.*), Agnathes.

Agoniatitida, Agoniatidés.

agrogeology, agrogéologie.

air, air, vent; **a. blast,** courant d'a., vent; **a. born sand,** sable éolien; **a. chambers,** chambres à a. (des Céphalopodes); **a. course,** voie d'aération (*mine*); **a. crossing,** croisement de voies d'aération; **a. discharge,** sortie d'a. (*mine*); **a. drilling,** forage à l'a. comprimé; **a. exhaust line,** conduite d'évacuation d'a.; **a. fall deposit,** dépôt pyroclastique; **a. fan,** ventilateur (*mine*); **a. flooding,** injection d'a. comprimé; **a. flow meter,** anémomètre; **a. furnace,** foyer d'aérage (*mine*); **a. gap,** buse morte; **a. hammer,** marteau pneumatique; **a. hoisting,** extraction par l'a. comprimé; **a. hole,**

prise d'a.; **a. inlet,** arrivée d'a.; **a. level,** niveau à bulle d'a.; **a. lift,** remontée de pétrole après injection d'a.; **a. pipe,** conduite d'a., buse; **a. saddle,** anticlinal érodé; **a. shaft,** puits d'aération; **a. shooting,** tir en l'air (*sism.*); **a. shrinkage,** rétraction des argiles par dessication à l'a.; **a. stack,** cheminée d'aération; **a. stone,** météorite; **a. stream,** courant aérien, flux d'air; **a. survey camera,** caméra pour photographie aérienne; **a. volcano,** soufflard; **a. wave,** onde acoustique, onde de choc; **a. way,** galerie d'aération (*mine*); **a. surface,** a. interstitiel (dans zone d'aération du sol).

airborne, aéroporté; **a. magnetometer,** magnétomètre a.; **a. scintillation counter,** compteur à scintillation a.; **a. survey,** prospection a.

airing, aération, ventilation (*mine*).

Aitken nucleus, particule d'Aitken (*météo*).

akerite (var. of syenite), akérite.

AKF diagram, diagramme triangulaire :
$$A = Al_2O_3 + Fe_2O_3 + (CaO + Na_2O);$$
$$K = K_2O; \quad F = FeO + MgO + MnO.$$

aklé, aklé.

akmolith, acmolite (var. de laccolite).

alabandite, alabandine, blende manganifère.

alabaster, albâtre; **calcareous a.,** a. calcaire; **gypseous a.,** a. gypseux.

alabasterine, d'albâtre, comme l'albâtre.

alar septum, cloison primaire latérale des Tétracoralliaires.

alaskite (var. of granite), alaskite.

albedo, albédo.

Albertan (N. Am.), série d'Albertan (Cambrien moy.).

albertite, albertite (var. de pyrobitume).

Albian, Albien (étage, Crétacé inf.).

albiclase, albite-oligoclase (série de plagioclases).

Albion US, étage d'Albion (Silurien inf.).

albite, albite; **a. diabase,** dolérite albitisée et altérée; **a. diorite,** diorite à a. (Alaska).

albitisation, albitisation.

albitite, albitite (var. de syénite sodique).

albitized, albitised, albitisé.

albitophyre, roche porphyrique à albite.

alboll, planosol, solonetz à horizon albique (*pédol.*).

alcohol, alcool.

alcrete, croûte pédologique alumineuse.

Alcyonacea, Alcyonidés (*pal.*).

Alcyonaria, Alcyonaires (*pal.*).

Aldanian, Aldanien (Cambrien inf.).

aldehyde, aldéhyde.

Alexandrian, Alexandrien (série, Silurien inf.); désuet.

alexandrite, chrysobéryl.

alfisol, alfisol (*pédol.*).

Alga (Algae pl.), Algue; **fossil a.,** a. fossile; **incrusting a.,** a. incrustante; **lime-secreting a.,** a. calcaire; **red a.,** a. rouge; **algal,** alguaire; **a. bank,** récif d'algues fossiles; **a. biscuit,** biscuit al.; **a. coal,** charbon d'a.; **a. head,** laminite; **a. lime-**

stone, calcaire d'a.; **a. mat,** tapis d'a.; **a. ridge,** crête a. corallienne.

Algoman, Algoman (âge : archéen sup.).

Algonkian, Algonquian, Algonkien (désuet, voir Protérozoïque).

alidade, alidade.

alignment, aline, alinement, 1. alignement, direction; 2. alignement de mégalites (*préhist.*).

aliphatic, aliphatique.

alkali, alkaline, 1. adj : alcalin; 2. n : a) carbonate de sodium ou de potassium, sel superficiel de régions arides; **b)** base forte; **c)** métal alcalin; **a. basalt,** roche basaltique à feldspathoïde; **a. earth,** alcalino-terreux; **a. felspar,** feldspath alcalin; **a. flat,** plaine salée, playa; **a. granite,** granite alcalin; **a. lake,** lac natroné, salé; **a. metal,** métal alcalin; **a. olivine basalt,** basalte à olivine; **a. rock,** roche basique; **a. soil,** sol basique; **a. syenite,** syénite à feldspathoïde, syénite néphélinique; **a. waters,** eaux alcalines.

alkalic, basique, alcalin; **a. calcic,** calco-a.

alkalify, 1. alcaliniser; 2. s'alcaniser.

alkalinity, alcanité.

alkalinous, alkalous, alcalin.

alkalization, alcalinisation.

alkalize (to), 1. alcaliniser; 2. s'alcaliniser.

alkanes, alkanes.

allanite, allanite.

Allegheny orogeny, orogénèse alleghanienne (fin du Paléozoïque).

Allerod, Alleröd (interstade Tardiglaciaire; pays scandinaves).

allite, roche allitique.

allitic, allitique.

allitization, allitisation.

allivalite, allivalite (gabbro à anorthite et olivine).

allochem, allochem (débris chimique et bioch.), allochème.

allochemical, allochimique.

allochromatic, allochromatique.

allochthon, allochthonous, allochtone.

allogeneous, allogenic, allogeneous, allogène, allogénique.

allometry, allométrie (croissance différentielle) (*pal.*).

allomorph, allomorphic, allomorphe.

allomorphism, allomorphie, allomorphisme.

allophane, allophane.

allothigeneous, allothigenous, allogène.

allotriomorphic, allotriomorphe, xénomorphe.

allotrope, allotrope.

allotropic, allotropique.

allotropically, allotropiquement.

allotropy, allotropie, allotropisme.

alloy, alliage.

alloy (to), 1. allier; 2. s'allier.

alloyage, alliage.

alluvial, alluvial, alluvionnaire; **a. apron,** nappe alluviale; **a. claim,** concession de gîte alluvionnaire; **a. cone,** cône torrentiel, de déjection; **a. deposit,** alluvions, gîte alluvionnaire; **a. digger,** orpailleur, chercheur d'or; **a. fan,** cône de déjec-

tion; **a. fill,** remblaiement fluviatile; **a. gold,** or alluvionnaire; **a. ground,** terrain d'alluvion; **a piedmont plain,** glacis de piémont; **a. plain,** plaine alluviale; **a. prospecting,** prospection des alluvions à la batée; **a. sheet,** nappe alluviale; **a. soil,** sol alluvial; **a. slope,** pente alluviale; **a. terrace,** terrasse alluviale; **a. tin-mining,** exploitation de l'étain alluvionnaire; **a. workings,** exploitation d'alluvions.

alluviating stream, fleuve à fort alluvionnement.

alluviation, dépôt d'alluvions, alluvionnement.

alluvion, alluvions.

alluvium, alluvia (pl.), alluvions.

almandine, almandite, almondine, almandine, grenat almandin.

along the dip, suivant le pendage.

alongshore current, courant de dérive littorale.

alpha quartz, quartz alpha, de basse température; **a. radiation,** rayonnement a..

alpine, alpin; **a. glacier,** glacier de type a.; **a. meadow soil,** sol a. humifère; **A. Orogeny (see Laramidian),** orogénèse a. (du Trias à l'actuel).

alstonite, alstonite (carbonate de calcium et de barium).

alter (to), 1. altérer, modifier; 2. s'altérer, se transformer.

alterable, altérable.

alteration, altération, transformation.

altimeter, altimètre.

altimetric, altimetrical, altimétrique.

altimetry, altimétrie.

altiplanation, altiplanation.

altitude, altitude, hauteur; **absolute a.,** a. absolue; **relative a.,** a. relative.

altitudinal, altitudinal.

altocumulus cloud, altocumulus (*météo.*).

altostratus cloud, altostratus (*météo.*).

alum, alun; **a. glass,** a. cristallisé; **a. mine,** alunière, aluminière; **a. salt,** sels naturels d'a.; **a. schist, shale, slate,** schiste aluneux alumineux; **a. works,** alunière, aluminière.

alumina, alumine.

aluminiferous, aluminifère, alunifère.

aluminite, aluminite.

aluminum, aluminium, aluminium; **a. epidote,** épidote alumineuse; **a. garnet,** grenat alumineux; **a. oxide,** oxyde d'a., alumine.

aluminization, alunation.

aluminosilicate, aluminosilicate.

aluminous, alumineux, aluneux.

alumogel, gel d'alumine.

aluniferous, alunifère, aluminifère.

alunite, alunite, alun.

alunitization, alunitisation.

alunitized, alunitisé.

alunogel, gel d'alumine.

alunogen, alunogène.

alveolar, alvéolaire (roche).

Alveolinellidae, Alvéolinidés (*pal.*).

alveolus, alvéole.

amalgam, 1. amalgame; 2. alliage de mercure et d'argent, ou de mercure avec un autre métal.
amalgamable, amalgamable.
amalgamate (to), 1. amalgamer; 2. s'amalgamer.
amazonite, amazonstone, amazonite (microcline).
amber, ambre, succinite.
amblygonite, amblygonite.
ambrite, ambrite.
ambulacral, ambulacraire (*pal.*); **a. area,** zone a.; **a. groove,** sillon a.; **a. plate,** plaque a.; **a. system,** système a.
ambulacrar field, aire a.
ambulacrum, ambulacre.
amethyst, améthyste.
amethystine quartz, quartz améthyste.
Amganian, Amganien (Cambrien moy.).
amiant, amianth, amianthinite, amianthus, amiantus, amiante.
amiantoid, amiantoïde.
ammeter, ampèremètre.
ammite, oolite (désuet).
ammonia, 1. adj : ammoniacal; 2. n : ammoniaque.
ammonification, ammonisation.
Ammonite, Ammonite; **age of A.,** Mésozoïque.
Ammononitidea, Ammonitidés.
Ammonoidea, ammonoids, Ammonitidés.
amorphic, amorphous, amorphose, amorphe.
amosite, amosite (var. d'amiante sud-africaine).
amount, 1. teneur, proportion; 2. quantité.
ampelite, ampélite, schiste noir bitumineux.
ampelitic, ampélitique.
amperemeter, ampèremètre.
Amphibia, Amphibiens.
amphibole, amphibole; **a. schist,** amphiboloschiste.
amphibolic, amphibolique.
amphiboliferous, amphibolifère.
amphibolite, amphibolite; **a. facies,** faciès à a.; **a. schist,** amphiboloschiste.
amphibolitic, à amphiboles.
amphibolitization, amphibolitisation.
amphiboloid, amphiboloïde.
amphibololite, amphibololite.
amphidromic, amphidromique.
amphigene (unusual), leucite.
Amphineura, Amphineures (Mollusques marins).
amphoteric, amphotère.
amphoterite, amphotérite.
amplification, 1. grossissement; 2. amplification.
amplifier, amplificateur.
amplify (to), amplifier, grossir (opt.).
amplifying power, pouvoir grossissant (d'un microscope).
amplitude, amplitude.
amygdule, amygdale, géode, petite cavité.
amygdaloid, amygdaloidal, amygdaloïde, amygdalaire; **a. basalt,** basalte a.
amygdaloids, laves caverneuses, vacuolaires.
anabatic wind, vent anabatique.
anabranch, bras effluent.
anabranching channel, chenal fluviatile effluent.
anaclinal, anaclinal (syn. obséquent).

anaerobic, anaérobie; **a. sediment,** sédiment a..
anafront, front ascendant.
anagenesis, anagenèse.
anaglacial, anaglaciaire.
anal, anal (*pal.*).
analcite, analcime, analcime; **a. basalt,** basalte à a.; **a. dolerite,** dolérite a.; **a. syenite,** syénite à a..
analcitite, analcitite (var. de basalte à feldspathoïdes).
analog computer, ordinateur a.; **a. simulation,** simulation sur ordinateur.
analyser, analyseur (d'un microscope polarisant).
analysis, analyse; **float and sink a.,** a. densimétrique; **gravimetric a.,** a. gravimétrique; **microscopic a.,** a. microscopique; **plug type a.,** a. sur petits échantillons; **well-core a.,** a. d'une carotte.
anamorphic, anamorphique; **a. zone,** zone d'orogenèse.
anamorphism, anamorphisme.
Anapsida, Anapsidés (Reptiles).
anastomosed rivers, anastomosing rivers, rivières anastomosées.
anatase, anatase.
anatexis, anatexie.
anatexite, anatexite.
anchimetamorphism, anchimétamorphisme.
anchor, ancrage (*forage*); **a. ice (ground ice),** glace de fond.
anchored dune, dune fixée.
anchoring, ancrage (*forage*).
ancient river-bed, ancien lit de rivière; **a. ice-age,** âge glaciaire ancien.
andalusite, andalousite; **a. hornstone,** cornéenne à a.
andeclase, série de plagioclases andésine-oligoclase.
andept, andosol (sol sur cendres volcaniques).
andesine, andésine.
andesinite, andésinite.
andesite, andesyte, andésite; **a. line,** ligne andésitique.
andesitic, andésitique.
andosol, andosol (*pédol.*).
andradite, andradite, mélanite (var. de grenat).
anemograph, anémomètre enregistreur.
anemographic, anémographique.
anemography, anémographie.
anemometer, anénomètre.
anemometric, anémométrique.
anemometrograph, anémométrographe.
anemometry, anémométrie.
aneroid, anéroïde; **a. barometer,** baromètre a.
Angara shield (USSR), bouclier, continent de l'Angara : désuet (Précamb. ancien).
Angiospermae, Angiospermes.
angle, angle; **critical a.,** a. limite; **c. a. of bedding,** a. de stratification; **c. a. of declination,** a. de déclinaison; **c. a. of dip,** a. 1. de pendage; 2. inclinaison (magnétique); **c. a. of emergence,** a. d'émergence; **c. a. of hade,** a. d'inclinaison d'une faille; **c. a. of incidence,** a. d'incidence (par rapport à la verticale); **c. a. of reflection,** a. de réflexion; **c. a. of refraction,** a. de réfraction; **c. a. of repose,** a. d'équilibre, de talus; **c. a. of rest,**

pente limite; **c. a. of rotation,** a. de rotation; **c. a. of shear,** a. de cisaillement; **c. a. of strike,** a. de direction.

anglesite, anglésite.

angrite, angrite (météorite achondrite).

angstrom, angström, $Å = 10^{-8}$ cm.

angular, angulaire; **a. distance,** distance angulaire; **a. frequency,** pulsation; **a. unconformity,** discordance a.; **a. velocity,** vitesse a.

angularity, angularité, caractère anguleux (d'un galet, etc.).

angulate drainage, réseau hydrographique à deux directions formant un angle aigu entre elles.

anhedral, allotriomorphe, xénomorphe; **a. crystal,** minéral xénomorphe.

anhydrate (to), déshydrater.

anhydration, déshydratation.

anhydric, anhydre.

anhydrit, anhydrite, anhydrite.

anhydritic, à anhydrite.

anhydrokaolin, kaolin artificiellement déshydraté.

anhydrous, anhydre, sec.

anion, anion.

Anisian, Anisic, Anisien (= Virglorien, Trias moy.).

anisomerous, anisomère.

anisometric, anisométrique.

Anisopleura, Anisopleures (Gastéropodes).

Anisotropic, anisotropal, anisotrope, anisotropical, anisotropous, anisotrope; **a. fabric,** structure orientée.

anisotropically, anisotropiquement.

anisotropy, anisotropie.

ankaramite, ankaramite (basalte à augite et olivine).

ankaramitic, ankaramitique.

ankerite, ankérite.

annabergite, annabergite, nickelocre.

Annelida, Annélides (*pal.*).

annual, annuel; **a. layer,** couche a., varve glacio-lacustre; **a. ring,** cerne a.

annular reef, récif annulaire.

annulus (Mars), anneau de projection autour d'un cratère martien.

anode, anode.

anodont dentition, charnière anodonte (*pal.*).

anomaly, anomalie (*géoph.*); **electromagnetic a.,** a. électromagnétique; **free air a.,** correction à l'air libre; **isostatic a.,** a. isostatique; **magnetic a.,** a. magnétique; **self-potential a.,** a. de potentiel spontané; **tidal a.,** a. de marée terrestre.

anorogenic, atectonique.

anorthic, triclinique.

anorthite, anorthite (plagioclase calcique); **a. basalt,** basalte à a.; **a. rock,** variété d'anorthosite.

anorthitite, anorthitite.

anorthoclase, anorthose, anorthose.

anorthosite, anorthosite.

anoxic water, eau dépourvue d'oxygène (anoxique).

Antarctic, Antarctique.

antecedence, antécédence.

antecedent, antécédent; **a. stream,** fleuve a.

anteclise, antéclise.

antediluvian, antédiluvien.

antenna, antenne (*pal.*).

anterior, antérieur, frontal (*pal.*).

anthophyllite, anthophyllite.

Anthozoa, Anthozoaires.

anthraciferous, anthracifère.

anthracite, anthracite.

anthracitic, anthraciteux.

Anthracolithic, Permo-Carbonifère.

anthraconite, anthracolite, calcaire ou marbre bitumineux et noir.

anthraxolite, anthracite fortement graphitique.

anthraxylon, fusinite.

Anthropogene (unusual), Quaternaire (ère) : inusité.

anthropogeny, anthropologie.

anthropogeography, géographie humaine.

anthropologist, anthropologiste.

Anthropopithecus, Anthropopithèque.

Anthropozoic, Anthropozoïque.

anticathode, anticathode.

anticentre, antiépicentre (*sism.*).

anticlinal, anticline, anticlinal; **a. axis,** axe a.; **a. bulge,** bombement a.; **a. closure,** fermeture d'un a.; **a. composite,** complexe a.; **a. core,** noyau a.; **a. crest,** charnière a.; **a. flank,** flanc d'un a.; **a. fold,** plissement a.; **a. limb,** flanc d'un a.; **a. line,** axe a.; **asymetric a.,** a. asymétrique; **brachya.,** brachy a.; **closed a.,** a. fermé; **faulted a.,** a. faillé; **overturned a.,** a. déversé.

anticlinorium, anticlinorium.

anticyclone, anticyclone.

anticyclonic, anticyclonique.

antiduna, antidune, dune hydraulique.

antiform (US), structure anticlinale, antiforme.

antiformal syneclise, faux anticlinal.

antigorite, antigorite.

antimonate, antimoniate.

antimonial, antimonial; **a. silver,** argent a.

antimoniated, antimonique, stibique.

antimonide, antimoniure.

antimoniferous, antimonifère; **a. arsenic,** allemontite.

antimonious, antimonieux.

antimonite, antimoine sulfuré, stibine.

antimony, antimoine; **a. blende,** kermésite, valentinite; **a. glance,** a. sulfuré, stibine; **a. ochre,** antimoinocre, stibiconite; **a. ore,** a. natif; **a. white,** trioxyde d'a.

antiperthite, antiperthite.

antipodal, antipodal.

antipode, antipode.

antiripplet, ride éolienne transversale.

antiroot, antiracine (matière dense sous les océans).

antistress minerals, minéraux métamorphiques formés à forte température et faible pression.

antithetic faults, failles associées antithétiques.

antitrade, contralizé.

antlerite, antlérite.

apatite, apatite (phosphate).

aperiodic, apériodique (*géoph.*).

aperture, ouverture, orifice.

apex, apex, extrémité, sommet, point le plus élevé d'une couche (pli).

aphanite, aphanite.

aphanitic, aphanitique; **a. limestone,** calcaire lithographique.

Aphebian (Can.), Aphébien (Protérozoïque inf.).

aphotic, aphotique (pas de lumière des profondeurs marines > 200 m).

aphyric, aphanitique (sans phénocristaux).

apical, apical (*pal.*).

aplite, aplite (*pétro.*).

aplitic, aplitique.

apogee, apogée (*astro.*).

apogranite, apogranite.

apomagmatic deposit, gîte apomagmatique (hydrothermal).

apophyllite, apophyllite.

apophysis, apophyse, apophyse.

aporhyolite, rhyolite dévitrifiée.

Appalachian Orogeny, orogenèse appalachienne (Pennsylvanien et Permien, ss).

apparent, apparent; **a. dip,** pendage a.; **a. heave,** rejet horizontal; **a. movement of faults,** mouvement a. de failles; **a. polarity,** polarité a.; **a. slip,** rejet incliné a.; **a. resistivity,** résistivité a.; **a. throw,** rejet vertical a.; **a. velocity,** vitesse a.

appinite, appinite (var. de diorite mélanocrate).

applanation, aplanissement.

applied, appliqué; **a. geophysics,** prospection géophysique; **a. geology,** géologie a.; **a. hydrogeology,** hydrogéologie a.

appraisal drilling, forage d'évaluation; **a. well,** puits d'é.

approach gone, 1. zone d'approche du littoral; 2. travaux dans la zone du littoral.

apron, plaine d'épandage (fluvio-glaciaire), étendue plate; **alluvial a.,** plaine alluviale.

Aptian, Aptien (étage, Crétacé inf.).

aptychus, aptychus (*pal.*).

aquagena tuffs, hyaloclastite.

aqualf, pseudogley lessivé, planosol (*pédol.*).

aquagene tuffs, hyaloclastite.

aquamarine, béryl, aigue-marine.

aquapulse, aquapulse (*géoph.*).

aqua regia, eau régale.

aquaseis, aquaseis.

aqueduct, aqueduc.

aquent, sol hydromorphe, sol alluvial à gley (*pédol.*).

aqueoglacial (unusual), fluvio-glaciaire.

aqueo-igneous, aquo-igné.

aqueous, 1. aqueux; 2. d'origine aquatique; **a. lava,** lave boueuse; **a. ripple-mark,** ride aquatique; **a. rock,** roche d'origine aqueuse, roche sédimentaire..

aquept, gley, pseudogley (*pédol.*).

aquiclude, aquiclude.

aquifer, aquifère; **a. test,** essai de débit.

aquiferous, aquifère.

aquifuge, aquifuge, imperméable.

Aquitanian, Aquitanien (étage, Miocène).

aquitard, couche semi-perméable, aquod (var. de gley).

aquoll, sol humique à gley (*pédol.*).

aquox, sol ferralitique hydromorphe.

aquult, ultisol hydromorphe.

araeometer, aréomètre.

araeometric, aréométrique.

araeometry, aréométrie.

aragonite, aragonite.

arborescent, arborescent, dentritique.

arborized, dendritique.

arc, arc (*géomorph.*); **a. shooting,** tir en a. (*géoph.*); **island a.,** a. insulaire.

arch, arc, voûte; **a. bend,** charnière anticlinale; **a. core,** noyau du pli; **a. dam,** barrage, voûte; **a. limb,** flanc supérieur, flanc normal (d'un pli); **a. pillar,** pilier de voûte (mine); **cave a.,** voûte.

Archaeocyatha, Archéocyathe.

Archaeocyathid, Archéocyathe, Archaéocyathidé.

Archean, Archéen (série).

Archaeopteryges, Archéoptérygiens.

Archeids, Archéides (orogénie anté-cambrienne).

archeological, archéologique.

archeology, archéologie.

archeomagnetism, archéomagnétisme.

Archeozoic, Archaeozoic, Archéozoïque (Précambrien ancien) : ère.

arching, arc-boutement en voûte.

archipelago, archipel.

Archosauria, Archosauriens.

arctic, arctique; **a. pack,** 1. banquise a.; 2. glace polaire.

area, 1. région, zone; 2. surface, superficie; **a. in advance,** avant-pays; **felt a.,** zone où on ressent un séisme; **mining a.,** district minier.

areal, 1. de surface; 2. aréolaire; **a. degradation,** érosion a.; **a. eruption,** éruption avec formation de caldeira.

arenaceo-calcareous, calcaréo-sableux, calcarénitique.

arenaceous, arénacé; **a. deposit,** arénite.

Arenigian, Arenig, Arénigien (étage, Ordovicien inf.).

arenilitic, gréseux.

arenite, arénite.

arenites, roches arénacées.

arenolutite, arénolutite.

arenorudite, arénorudite.

arenose, arenous, arenulous, sablonneux.

arenosol, arénosol.

areographic, aréographique (*astro.*).

areola, aréole (*pal.*).

areometer, aréomètre.

areometric, aréométrique.

areometry, aréométrie.

arete, arête, crête aiguë.

argentiferous, argentifère; **a. lead,** plomb a.

argentite, argentite, argyrose.

argentopyrite, pyrite argentifère (minerai).

argentum, argent.

argillaceous, argileux; **a. schist,** schiste a.

argillaminite, laminite argileuse.

argillic horizon, horizon argillique, à accumulation d'argile.
argilliferous, argilifère.
argillite, argilite, argilite, pélite.
argillitic, pélitique.
argillization, 1. argilisation; 2. formation d'un revêtement argileux dans les forages.
argilloarenaceous, argilo-sableux.
argillocalcareous, argilo-calcaire.
argilloferruginous, argiloferrugineux.
argilloid, argiloïde.
argillomagnesian, argilo-magnésien.
argillous (unusual), argileux.
argon, argon.
Argovian, Argovien (sous-étage, Lusit., Juras. sup.).
argyropyrite, argyropyrite (*miner.*).
argyrose, argyrose, argentite (*miner.*).
argyrythrose, argyrythrose, pyrargyrite (*miner.*).
arheic, aréique (cf. réseau hydrographique du désert).
arheism, aréisme.
arid, aride; **a. index**, indice d'aridité.
aridic soil, sol à encroûtement calcaire.
aridification, aridification.
aridisoil, aridisol, aridosol.
aridity, aridness, aridité.
Arkansan (US), Arkansien (= Carbonifère inférieur).
Arikareean (US), Arikaréen (= Chattien + Aquitanien).
arkose, arkose.
arkosic sandstone, grès feldspathique.
arm of sea, bras de mer.
armenite, arménite, lazulite.
Armorican orogeny, orogenèse Armoricaine.
armoured clayballs, galets d'argile armés, enrobés (débris); **a. concrete**, béton armé.
armored fish, poisson cuirassé; **a. mud ball**, galet d'argile indurée.
aromatic, aromatique. **a. hydrocarbon**, hydrocarbure a.; **a. compounds**, composés a.
aromaticity, caractère aromatique (d'un charbon).
aromatics, hydrocarbures aromatiques.
arrangement, disposition, arrangement.
array, rangée (d'atomes, etc.), alignement (d'appareils), réseau sismique.
arrival, arrivée (d'une onde); **a. time**, temps d'a.
arrowhead, pointe de flèche (*préhist.*).
arrow-head twin, mâcle en fer de lance.
arroyo, arroyo, oued.
arsenate, arseniate, arséniate.
arsenic, 1. adj : arsénique; 2. n : arsenic; **arsenical**, arsénical; **a. nickel**, arsenickel, nickéline (*miner.*); **a. pyrite**, arsénopyrite; **a. silver blende**, argent a.
arsenide, arséniure.
arseniferous, arsénifère.
arsenious, arsénieux.
arsenite, arsenolite, arsénite, arsénolite.
arsenomarcasite, mispickel (*miner.*).
arsenopyrite, arsénopyrite, mispickel.
arteric migmatite, artérites, épibolites.

arterite, artérite.
artesian, artésien; **a. aquifer**, aquifère a.; **a. basin**, bassin sédimentaire a.; **a. spring**, source a.; **a. structure**, structure a.; **a. waters**, eaux a.; **a. well**, puits a.
Arthrodian, Arthrodires.
Arthropoda, Arthropodes.
Articulata, articulate Brachiopoda, Brachiopodes articulés.
artifact, artefact.
artificial satellite, satellite artificiel.
Artinskian, Artinskien (étage, Permien inf.).
Artiodactyla, Artiodactyles.
âs, asar (pl.), esker, remblais fluvio-glaciaires.
asbestoid, asbestoïde.
asbestos, asbeste, amiante.
ascend (to), monter, remonter.
ascensional theory, théorie per ascensum.
aseismic design, ouvrage asismique; **aseismic margin**, marge passive.
ascent, 1. ascension; 2. montée.
ash, 1. cendres (volcaniques); 2. carbonate de sodium anhydre; **a. cone**, cône de cendres, scories; **a. content**, teneur en cendres; **ash-cloud surge**, nuée ardente basale; **a. fall**, chute de cendres; **a. layer**, cinérite; **a. shower**, pluie de cendres.
Ashgillian (US), Ashgillien (sous-étage, Ordovicien sup.).
ashlar, pierre de taille.
ashore, à terre, échoué.
asphalt, asphalte, bitume; **a. base crude**, pétrole brut asphaltique; **a. base petroleum**, pétrole brut asphaltique; **a. bitumen**, bitume asphaltique; **a. bottom**, résidu asphaltique de distillation; **a. deposits**, gisements d'a; **a. residue**, résidu asphaltique; **a. rock**, roche asphaltique; **a. seepage**, suintement d'a; **a. tar**, goudron a.; **native a.**, a. naturel; **natural a.**, a. naturel; **petroleum a.**, a. de pétrole.
asphaltene, asphaltène.
asphaltic, asphaltique; **a. base crude**, pétrole brut a.; **a. coal**, charbon a., albertite; **a. limestone**, calcaire a.; **a. pyrobituminous shale**, schiste pyrobitumineux a.; **a. rock**, roche a.; **a. sand**, sable a.
asphaltite, asphaltite.
asphaltoid, asphaltoïde.
asphaltum, asphalte.
assay, essai, analyse; **a. value**, teneur en métal (d'un minerai).
assay (to), 1. analyser; 2. titrer.
assaying, essai, analyse; **ore assaying 1 % of gold**, minerai titrant 1 % d'or.
assemblage zone, zone d'assemblage, d'association (syn. ~ faunizone, florizone).
assembly, montage, disposition des éléments d'un forage (trépan, tiges, etc.).
assimilation, assimilation (magmatique).
association, association; **a. of igneous rocks**, a. de roches éruptives; **a. of minerals**, cortège minéralogique; **a. of ores**, cortège de minerais.

assortment, 1. classement, classification, triage (granulométrique); **2.** assortiment.
Assynthian Orogeny, orogénie Assyntienne.
Astartian, Astartien (sous-étage, Lusit. Juras. sup.).
astatic, astatique.
asteriated opal, opale à irisations.
asterism, astérisme.
asteroid, astéroïde.
Asteroidea, Astéridés.
asthenolith, asthénolite.
asthenosphere, asthénosphère.
Astian, Astien (étage, Pliocène).
Astreoids, Astréidés.
astrobleme, astroblème.
astrogeology, géologie des planètes et de leurs satellites, astrogéologie.
astrolabe, astrolabe.
astronomical, astronomique.
astronomy, astronomie.
asymmetric, asymmetrical, asymétrique, dissymétrique; **a. crystal,** cristal sans élément de symétrie, cristal triclinique; **a. fold,** pli dissymétrique; **a. ripple-mark,** ride de courant, ride dissymétrique; **a. system,** système triclinique.
atacamite, atacamite.
ataxic, ataxique, mal classé, mal stratifié.
ataxite, ataxite (météorite).
atectonic, non tectonique.
Atlantic period, période Atlantique (intervalle postgl.); **a. series,** province a. (*pétro.*); **a. type coast,** côte discordante; **a. suite,** série a.
atmoclastic rock, roche détritique formée par altération (due aux agents atmosphériques).
atmogenic rock, 1. roche formée par altération (due aux agents atmosphériques); **2.** roche formée par sédimentation éolienne.
atmosphere, atmosphère.
atmospheric, atmosphérique; **a. agent,** agent a.; **a. pollution,** pollution a.; **a. pressure,** pression a.; **a. radiation,** radiation a.; **a. rock,** roche éolienne a.; **a water,** eau a., de pluie.
atmospherics, perturbations a.
atoll, atoll (récif).
atom, atome.
atomic, atomique; **a. bond,** liaison a.; **a. energy,** énergie a.; **a. fission,** fission a.; **a. mass,** masse a.; **a. number,** nombre a.; **a. packing,** empilement d'atomes; **a. radius,** rayon a.; **a. structure,** structure a.; **a. value,** valence a.; **a. weight,** poids a.
atomicity, atomicité, valence atomique.
attack (to), attaquer (un filon) etc.
attacking, corrosion.
attapulgite, attapulgite.
Atterberg limits, limites d'Atterberg (*géotechn.*).
attitude, disposition (d'une strate, c'est-à-dire son orientation et son pendage).
attle, 1. gangue; **2.** déblais, remblais (*mine*), déchets.
attraction, attraction; **gravitational a.,** a. de la gravité.
attrition, attrition, usure.
Aturian, Aturien.

aubrite, aubrite (météorite).
augen gneiss, gneiss œillé; **a. structure,** structure lenticulaire, œillée.
auger, tarière, sonde; **a. drilling,** sondage à la t.; **a. stem,** maîtresse tige (*forage*); **earth a.,** t. à sol.
augite, augite (pyroxène); **a. andesite,** andésite à a.; **a. diorite,** diorite à a.; **a. gneiss,** gneiss à pyroxènes; **a. rock,** pyroxénolite; **a. syenite,** syénite à a.
augitic, augitique.
augitite, augitite (basalte à augite).
augitophyre, basalte à phénocristaux d'augite.
augitophyric, à phénocristaux d'augite.
aulacogen, aulacogène.
aureole, auréole (métamorphique).
auri-argentiferous, auro-argentifère.
auric chloride, chlorure d'or.
auriferous, aurifère.
auro-argentiferous, auro-argentifère.
auroferriferous, auroferrifère.
auroplumbiferous, auroplombifère.
aurora australis, aurore australe; **a. borealis,** a. boréale.
auroral, auroral.
austral, austral.
australite, australite (météorite).
Australopithecus, Australopithèque.
authigene, authigenic, authigenous, authigène.
authigenesis, authigenèse.
autobrecciated lava, coulée de lave formant une brèche superficielle par refroidissement.
autochthon, autochthonous, autochtone.
autoclastic, bréchique (brèche d'orogenèse, brèche de friction, brèche détritique); **a. process,** processus de bréchification.
autodiastrophism, autodiastrophisme (diastrophisme résultant de causes internes).
autogenetic, authigénique; **a. drainage,** réseau fluviatile authigène.
autolith, enclave syngénétique.
autolyse, autolyse (*pétro.*).
automated cartography, cartographie automatique.
automatic picture transmission (APT), transmission automatique des images par satellite.
automation, informatisation.
autometamorphism, automorphism, autometasomatism, autométamorphisme.
automorphic, automorphous, automorphe, idiomorphe.
autopneumatolysis, autopneumatolyse.
autotrophic, autotrophe (organisme).
Autunian, Autunien (étage, Permien).
autunite, autunite, uranite.
Auversian, Auversien = Lédien (étage, Eocène).
auxiliary, auxiliaire, secondaire; **a. fault,** faille ramifiée, secondaire; **a. minerals,** minéraux accessoires.
avalanche, avalanche (de neige); **a. breccia,** brèche de pente; **a. chute,** couloir d'a.; **a. debris,** chaos d'a.; **a. tongue,** langue d'a.; **drift a.,** a. de fond;

dry a., a. sèche; **mud a.,** coulée de boue; **powdery a.,** a. poudreuse.

aven, aven (*karstol.*).

aventurine feldspath, feldspath aventurine; **a. quartz,** quartz a.

average, **1.** adj : moyen; **2.** n : moyenne; **a. annual rainfall,** moyenne pluviométrique annuelle; **a. elevation,** altitude moyenne.

aves, oiseaux (*pal.*).

avoirdupois pound, livre avoir-du-pois = 0,4536 kg.

Avonian, Avonien, = Dinantien (étage, Carbon.).

avulsion, avulsion.

awl, perçoir (*préhist.*).

axe, axis (pl. axes), axe; **a. of elasticity,** axes des indices de réfraction dans un cristal; **crystallographic a.,** a. cristallographique; **fabric axes,** a. cinématiques; **optic a.,** a. optique; **tectonic a., 1.** a. tectonique; **2.** a. cinématique, géométrique (d'un pli).

axial, axial; **a. canal,** canal a. (crinoïdes); **a. compression,** compression a.; **a. elements,** paramètres cristallographiques; **a. lobe,** lobe a. (tribolite); **a. plane,** plan a.; **a. plane-cleavage,** plan de clivage, schistosité; **a. plane foliation,** schistosité de frac-turation; **a. plane separation,** distance des axes de synclinal et anticlinal voisins; **a. ratios,** paramètres cristallographiques; **a. rift zone,** fossé médian d'une ride océanique; **a. section,** coupe longitudinale; **a. symmetry,** symétrie a. des déformations; **a. trace,** trace de l'axe du pli en surface; **a. through,** fléchissement de l'axe d'un pli.

axinite, axinite (cyclosilicate).

axis, axe; **a. of a fold,** a. d'un pli; **a. of rotation,** a. de rotation; **a. of symmetry,** a. de symétrie (cristallographique).

axotropus, à clivage perpendiculaire à un axe cristallographique.

azimuth, azimut; **a. circle,** cercle de relèvement; **a. compass,** boussole azimutale.

azimuthal, azimutal; **a. projection,** projection a.; **a. equal. area projection,** projection a. équivalente; **a. equidistant projection,** projection a. équidistante.

azurmalachite, azurmalachite.

azoic, azoïque.

azonal soil, sol azonal.

azurchalcedony, calcédoine bleue.

azure stone, **1.** lapis lazuli; **2.** azurite.

azurite, azurite (minerai de cuivre).

B

bacalite, bacalite (var. d'ambre).

back, 1. arrière; **2.** amont-pendage (*mine*); **3.** couronne, sommet, toit (US) (d'un passage souterrain); **4.** partie la plus superficielle d'un filon; **b. arc bassin,** bassin arrière-arc, bassin marginal interne; **b. deep,** arrière-fosse; **b. draught,** refoulement; **b. entry,** passage d'aération (*mine*); **b. faulting,** rejeu de faille; **b. fill,** remblai; **b. filling,** remblayage, remplissage; **b. flow, 1.** reflux, courant de retour; **2.** refoulement; **b. folding,** plissement en retour; **b. ground, 1.** soubassement; **2. arrière-plan; 3.** bruit de fond en radioactivité; **4.** fond continu; **b. hole,** trou de toit, trou de voûte; **b. jet current,** courant de retour; **b. land,** arrière-pays; **b. limb,** flanc normal (d'un pli); **b. reef channel,** chenal d'arrière récif; **b. reef,** zone en arrière du récif; **b. reef channel,** chenal post-récifal; **b. set beds, 1.** lits de sable dunaire « sous le vent »; **2.** lit à lamines, incliné en sens inverse (delta); **b. shore,** arrière-plage, gradins de plage; **b. slope,** surface (plaine) structurale; **b. stope,** gradin renversé; **b. stoping,** abatage en gradins renversé; **b. to back,** adossé; **b. slough,** dépression latérale humide; **b. swamp,** dépression latérale humide; **b. thrusting,** rétrocharriage (vers la zone interne); **b. titration,** titrage en retour; **b. tracking,** restitution des paléoprofondeurs; **b. wash, 1.** retour du courant de vagues; **2.** remous; **b. wasting,** recul du front glaciaire; **b. water, 1.** lagune; **2.** eaux dormantes; **3. ressac.**

back (to), 1. adosser; **2.** reculer; **3.** jeter au rebut.

backward erosion, érosion régressive.

backwards, en arrière, vers l'arrière.

bacteriologic origin, origine bactériologique.

bacteriology, bactériologie.

Bacterium, Bactérie; **nodule Bacteria,** bactéries des nodules.

baddeleyite, baddeleyite.

badland, terrain raviné, mauvaises terres.

bagger, drague.

bahada, bahada, plaine de remblaiement, glacis de piedmont.

bahamite, bahamite (roche calcaire cimentée).

bail, anse.

bailer, 1. puisatier, écopeur; **2.** écope; **3.** tube à clapet, cuiller de curage.

bail out (to), curer, puiser (un sondage).

baikerite, baïkerite.

bajada, voir bahada.

Bajocian, Bajocien (étage, Jur. moy.).

balance, 1. balance; **2.** équilibre, bilan; **b. beam,** fléau de b.; **b. bob,** balancier à contre-poids; **b. truck,** chariot contrepoids (*mine*); **b. weight,** contre-poids.

balas ruby, spinelle rosâtre (vient de Badakhshan, Afghanistan).

bald (US), sommet dénudé Sud (USA); **b. headed anticline,** anticlinal à charnière érodée.

ball, 1. boule, nodule, concrétion; **2.** crête sous-marine parallèle au rivage, crête prélittorale; **b. clay,** argile plastique utilisée dans la porcelaine; **b. crusher,** broyeur à boulets; **b. ironstone,** fer carbonaté en rognons; **b. vein,** filon en nodules.

ball (to), 1. agglomérer; **2.** s'agglomérer.

ballast, ballast, empierrement.

balled-up structure, structure noduleuse à concrétion d'argile.

balloon sounding, ballon sonde (*météo.*).

ballstone, calcaire contrétionné, concrétion calcaire (fossilifère) dans matrice argileuse.

banatite, banatite (diorite quartzique).

banco (US), bayou, lac de bras mort (Texas).

band, bande, zone; **b. pass,** passe-bande (*géophys.*); **b. silicate,** inosilicate; **b. spectrum,** spectre de bande.

banded, rubanné, lité, zoné; **b. agate,** agate; **b. ore,** minerai lité; **b. structure,** structure r., z.; **b. vein,** filon z.

banding, zonation; **b. structure,** structure zonée, rubanée.

bank, 1. banc; **2.** berge, rive; **3.** talus, rampe; **4.** carreau d'une mine; **5.** front de taille (*mine*); **6.** bas-fond; **b. claim,** concession riveraine (*océano.*); **b. full stage,** niveau de haut étiage; **b. full stream,** fleuve prêt à déborder; **b. head,** carreau d'une mine; **b. reef,** platier corallien; **b. subsidence,** effondrement des berges; **alluviated b.,** rive alluvionnée; **banner b.,** épi en traîne; **concave b.,** rive concave; **convexe b.,** rive convexe; **inner b.,** rive interne; **outer b.,** rive externe; **raised b.,** levée de rive; **sand b.,** placage de sable; **snow b.,** tache de neige; **undercut b.,** berge sapée

bank (to), amonceler, amasser, entasser, mettre en tas.

banked formation, formation agglomérée.

banket, 1. conglomérat, formation cimentée; **2.** conglomérat aurifère et à pyrite (Afr. du Sud).

banks (the), rivage.

bar, 1. barre (de roche); **2.** banc de sable (fluviatile, etc.); **3.** cordon littoral; **4.** barre (d'un estuaire); **5.** filon-croiseur (mine); **6.** bar (unité de pression); **b. finger sand,** banc de sable alluvial allongé suivant le cours de l'affluent; **b. head,** amont d'un banc; **b. silver,** argent en barres; **b. tail,** aval d'un banc; **b. tin,** étain en saumon; **bay-mouth b.,** flèche barrante à l'embouchure d'une baie; **bay-head b.,** flèche de fond de baie; **bay-side b.,** flèche à l'intérieur d'une baie; **connecting b.,** flèche de jonction; **crest b.,** barre élevée; **cuspate**

b., cordon littoral en V; **gravel b.,** banc de gravier; **headland b.,** flèche avancée, cordon littoral appuyé.; **miner's b.,** barre à mine; **offshore b.,** cordon littoral; **sand b.,** banc de sable; **shingle b.,** cordon de galets; **submarine b.,** remblai sous-marin, barre.

barbados earth, dépôt à Radiolaires.

barbed drainage, réseau hydrographique en barbelé.

barchan, barkhan, barkhane; **b. arm,** aile de b.; **b. swarm,** essaim de b.; **inset b.,** b. ˋemboîtées; **star-like b.,** b. en étoile.

bare, dénudé, sans couvert végétal, sans formations superficielles.

baring, décapage, enlèvement des morts-terrains.

barite, baryte, barytine; **b. dollar,** b. lenticulaire; **b. rose, rosette,** « rose » de b. (de désert).

barium, baryum.

barkevicite, barkévicite (amphibole monoclinique).

barograph, baromètre enregistreur.

barometer, baromètre.

barometric, barometrical, barométrique; **b. elevation,** altitude donnée par mesures b.; **b. levelling,** nivellement; **b. pressure,** pression b.

barometry, barométrie.

barothermograph, barothermographe.

barothermometer, barothermomètre.

barrage, barrage.

barranco, barranca, barranco, ravin abrupt.

barrel, **1.** barril, bbl : 0,15876 m³ (42 US gal); **2.** baril, fût.

barreler, puits de pétrole productif.

Barremian, Barrémien (étage, Crétacé inf.).

barren, stérile; **b. coal-mesures,** terrain houiller s.; **b. ground,** terrain s.; **b. lode,** filon s.; **b. rock,** roche s..

barrens, terrains dénudés.

barrier, barrière; **b. beach, cordon littoral; b. flat,** platier corallien; **b. ice,** glace littorale; **b. iceberg,** iceberg détaché, vêlé; **b. island,** crête d'avant-plage émergée, île barrière; **b. lake,** lac de barrage; **b. pillar,** massif de protection; **b. reef,** récif annulaire, récif barrière.

barring down, abatage, abattage.

barrow, **1.** tumulus, colline; **2.** halde de déblais (*mine*).

Barstovian (US), Barstovien (Burdigalien sup.-Vindobonien).

Bartonian, Bartonien (étage, Eocène sup.).

barylite, barylite.

barysphere, barysphère.

baryta, baryte, barytine, baryte, barytine; **b. feldspar,** feldspath barytique.

barytic, barytifère, barytique.

barytocalcite, barytocalcite.

barytocelestite, barytocelestine.

basal, **1.** basal, de base; **2.** fondamental; **b. cleavage,** clivage parallèle à certaines faces cristallographiques; **b. conglomerate,** conglomérat de b.; **b. moraine,** moraine de fond; **b. sapping,** dégagement d'éboulis; **b. till,** moraine (ou till) de fond.

basals, plaques basales (Crinoïdes) (*pal.*).

basalt, basalte; **b. glass,** tachylite; **b. slate,** b. lié; **b. tuff,** tuf b.; **alkali b.,** b. à feldspathoïde; **alkali olivine b.,** b. à olivine.

basaltic, basaltique; **b. column,** orgues b.; **b. glass,** tachylite; **b. jointing,** prismation b.; **b. lava,** lave b.; **b. layer,** couche b.; **b. rock,** roche b.; **b. structure,** structure prismatique; **b. tuff,** tuf b.

basaltiform, basaltiforme.

basanite, basanite, lydienne.

base, **1.** base, socle; **2.** base (chimie); **3.** bas, pauvre (minerai); **b. conglomerate,** conglomérat de base; **b. exchange,** échange de base; **b. level,** seuil; **b. level karst,** niveau de karstification; **b. level of erosion,** niveau de base d'érosion; **b. line,** ligne de référence (sur la Terre); **b. map,** fond de carte; **b. net,** réseau triangulaire (triangulation); **b. number,** basicité; **b. ore,** minerai pauvre; **b. station,** station de triangulation.

baselap, recouvrement de base sur discordance.

baselevelling, pénéplanation.

basement, soubassement; **b. complex,** socle métamorphique (souvent précambien), gneiss; **b. rock,** roche du socle.

bases, constituants alcalins, basiques du sol.

bash (to), remblayer.

basic, basique; **b. border,** contact b. (de batholites); **b. front,** front chimique de granitisation; **b. lava,** lave b.; **b. rock,** roche b. (peu précis) (relativement pauvre en silice).

basicity, basicité.

basification, alcalinisation.

basin, **1.** dépression, cuvette, bassin sédimentaire; **2.** bassin hydrographique; **3.** synclinal à contour circulaire ou elliptique (USA); **4.** cirque glaciaire (Ouest USA); **b. facies,** faciès sédimentaire profond; **b. and range structure,** structure à horsts et dépressions tectoniques; **b. fold,** pli synclinal; **barred b.,** sillon prélittoral, dépression prélittorale; **catchment b.,** aire d'alimentation fluviatile; **deflation b.,** bassin de déflation; **drainage b.,** bassin hydrographique; **ocean b.,** bassin océanique; **sedimentary b.,** bassin sédimentaire; **structural b.,** cuvette structurale, synclinale; **terminal b.,** bassin glaciaire terminal.

basinal facies, faciès de bassin sédimentaire.

basset, affleurement.

basset (to), affleurer.

bastite, schillerspath, bastite.

bat, gore, gord (argile stratifiée dans charbon).

batea, batée.

batholith, batholite, bathylith, bathylithe, batholite.

batholithic, batholithique.

bathometer, bathymètre.

Bathonian, Bathonien (étage, Jurassique moy.).

bathyal, bathyal (fond océanique de -200 à $-2\,000$ m).

bathygraphic, bathygraphique.

bathymeter, bathymètre.

bathymetric, bathymetrical, bathometric, bathometrical, bathymétrique; **b. map, chart,** carte b.

bathymetry, bathometry, bathymétrie.
bathyorographical map, carte bathyorographique.
bathypelagic, bathypélagique (eaux océanique entre
 − 200 et − 2 000 m); voir bathyal.
bathyscaphe, bathyscaphe.
bathythermograph, bathythermographe.
Batracosauria, Batracosaurien.
batter, talus, angle de glissement (d'un remblai).
batter (to), taluter, damer.
battery ore, minerai d'oxyde de manganèse.
batteryman, bocardeur (mine).
batture, batture, estran (Canada).
bauxite, bauxite.
bauxitic, bauxitique; **b. latosols,** latosols à bauxite.
bauxitization, bauxitisation.
Baveno law (see twin law), loi des mâcles de
 Baveno.
bay, baie; **b. bar,** flèche littorale à l'entrée d'une
 b.; **b. entrance,** entrée de b.; **b. head, 1.** fond de b.
 2. marécage au fond de b. (USA); **b. mouth,**
 embouchure de b.; **b. mouth bar,** flèche littorale
 à l'embouchure d'une b.
bayou, bayou, lac de bras mort, chenal deltaïque
 abandonné.
BC fracture, cassure parallèle au plan structural
 BC et perpendiculaire à l'axe structural A.
beach, rivage, plage; **b. accretion,** accroissement,
 engraissement de p.; **b. bar bedding,** stratification
 de cordon littoral; **b. berm,** gradin de p.; **b.
 combing,** exploitation de sables littoraux; **b.
 cusp,** croissant de p.; **b. deposits,** dépôts côtiers;
 b. drifting, dérive littorale; **b. face,** zone infra-
 littorale à haute énergie; **b. furrow,** sillon de p.;
 b. line, ligne de rivage; **b. placers,** gisements de
 p.; **b. plain,** plaine littorale; **b. profile,** profil de
 p.; **b. rampart,** levée de p.; **b. renourishment,**
 réalimentation en sable des plages; **b. ridge,** levée
 de p.; **b. rock,** dépôt de p. induré, grès de p.;
 b. scarp, talus de p.; **b. swale,** sillon de p.; **b.
 terrace,** terrasse de p.; **barrier b.,** cordon littoral;
 prograding b., p. en progression; **raised b.,** p.
 soulevée; **retrograding b.,** p. en recul; **shingle b.,**
 p. de galets grossiers; **storm b.,** levée de p.
beachy, côtier.
bead test, essai à la perle, essai au chalumeau.
beaded, en chapelet; **b. drainage,** réseau hydrographique en c.; **b. esker,** esker en c.; **b. texture,**
 structure en c.; **b. vein,** filon en c.
beak, 1. cap, pointe, bec; **2.** crochet (Brachiop. et
 Lamellib.).
beam, 1. fléau de balance, balancier; **2.** rayon
 lumineux; **3.** poutre, travée (*mine*); **b. balance,**
 balance à fléau; **b. caliper,** pied à coulisse; **b. of
 light,** faisceau lumineux; **b. well,** faisceau lumineux.
bean ore (or pea ore), 1. limonite en aggrégats
 lenticulaires; **2.** minerai de fer pisolithique
 grossier.
bearing, 1. part : contenant, montrant; **2.** n : apport,
 contribution; **3.** n : relevé de mesures, orientation

cardinale; **b. capacity,** portance; **b. of a lode,**
 orientation du filon; **water b.,** aquifère.
beating, battage, emboutissage, pilage; **b. of waves,**
 action de déferlement mécanique des vagues.
Beaufort wind scale, échelle de Beaufort (vitesse du
 vent).
Becke line, frange de Becke (*minér.*); **B. test,** essai
 de réfringence.
bed, 1. couche (banc : déconseillé), strate; **2.** lit (de
 rivière, de glacier); **3.** lit (repère); **b. claim,**
 concession dans le lit d'une rivière; **b. key,** niveau
 repère; **b. load,** charge de fond; **b. mining,**
 exploitation des couches; **b. roughness,** rugosité
 du lit; **b. surface,** surface de strate; **b. vein,** filon
 couche; **coal b.,** couche de charbon; **glacier b.,**
 lit de glacier; **high-water b.,** lit majeur (d'un
 fleuve); **mean-water b.,** lit mineur; **petroliferous b.,**
 couche pétrolifère; **reservoir b.,** couche réservoir;
 river b., lit fluviatile; **transition b.,** couche de
 passage.
bedded, stratifié, lité; **b. rock,** roche s.; **b. vein,** filon
 couche.
bedding, stratification, litage; **b. angle,** angle de s.;
 b. cleavage, clivage parallèle à la s.; **b. fault,** faille
 parallèle au plan de s.; **b. fissility,** schistosité
 parallèle à la s.; **b. joint,** joint de s.; **b. plane,** plan
 de s.; **b. plane cleavage,** clivage parallèle (stratification); **b. plane slip,** glissement intercouche; **b.
 schistosity,** foliation parallèle à la s.; **conformable b.,** s. concordante; **(criss-)cross b.,** s. entrecroisée; **discordant b.,** s. discordante; **horizontal b.,** horizontale; **large-scale b.,** s. à grande
 échelle; **lenticular b.,** s. en lentilles; **megaripple
 b.,** s. en grandes rides; **oblique b.,** s. oblique;
 rythmic b., s. rythmée; **ripple b.,** s. entrecroisée;
 sand dune b., s. dunaire entrecroisée; **small-
 scale b.,** s. entrecroisée à petite échelle.
Bedoulian, Bédoulien (sous-étage, aptien, Crétacé
 inf.).
bedrock, substratum rocheux, socle, soubassement
 (des dépôts meubles).
bedsole, semelle d'une couche.
beef (US), filon de calcite fibreux.
beekite, beekite 1. variété de calcédoine de substitution; 2. forme concrétionnée de calcite.
beerbachite, beerbachite (var. de gabbro aplitique).
behead (to), décapiter, capturer (le cours d'une
 rivière).
beheaded river, rivière captée.
beidellite, beidellite (min. argileux).
Belemnite, Bélemnite (*pal.*).
Belemnoidea, Belemnoïdés.
bell-metal ore, stannite, étain pyriteux.
belt, zone, ceinture, aire; **b. conveyor,** convoyeur à
 bande (mine); **fold b.,** faisceau de plis; **orogenic b.,**
 zone orogénique; **sorting b.,** bande de triage;
 volcanic b., zone volcanique.
belted plain, plaine de bas de cuesta.
Beltian orogeny, orogenèse Beltienne.
bench, 1. gradin, banc, banquette (littorale); **2.**
 paillasse de laboratoire; **3. b. gravel,** gravier de

terrasse; **b. land,** cuesta; **b. mark,** repère géodésique; **b. placer,** gravier aurifère de terrasse, gisement en terrasse; **b. stoping,** exploitation par gradin; **b. wave-cut,** banquette d'érosion littorale.

bend, 1. courbe, courbure, inflexion (d'une vallée, d'un pli); **2.** argile indurée (Cornouailles); **b. of a fold,** charnière d'un pli; **b. of a seam,** crochon (*mine*); **b. test,** essai à la flexion.

bend (to), arquer, courber, déjeter, fléchir, plier.

bending, courbure, flexion, gauchissement; **b. stress,** effort à la flexion; **b. strength,** résistance à la flexion; **b. test,** essai à la flexion; **b. down,** flexure.

beneficiate (to), enrichir (la teneur d'un minerai).

Benioff plane, plan de Bénioff.

Bennettitales, Bennettitales (*pal.*).

bent, déformé, gauchi, plié.

benthic, benthique.

benthonic, benthique (opposé à pélagique).

benthos, benthos.

bentonite, bentonite (min. argileux).

benzene, benzène.

berg, 1. montagne, colline, (Can.); **2.** iceberg; **b. crystal,** cristal de roche; **b. till,** till déposé par les icebergs.

bergschrund, rimaye.

berm, 1. gradin, **2.** terrasse fluviatile étagée; **b. crest,** bord extérieur d'un gradin; **b. beach,** terrasse de plage.

Berriasian, Berriasien (sous-étage, Valanginien, Crétacé inf.).

Bertrand lens, lentille de Bertrand.

beryl, béryl, béril (cyclosilicate).

beryllium, béryllium.

beta, bêta; **b. particle,** électron; **b. quartz,** quartz de haute température; **b. radiation,** rayonnement b.; **b. rays,** rayon b.

bevel, méplat, biseau.

bevel (to), aplanir, biseauter.

beveled, tronqué, biseauté.

beveling, biseautage.

bevelment, biseau.

B horizon, horizon B (*pédol.*).

bias, 1. tension de polarisation; **2.** biais.

bias (to), polariser.

biatomic, diatomique.

biaxial, biaxe; **b. crystal,** cristal b.; **b. indicatrix,** ellipsoïde des indices.

bicarbonate, bicarbonate.

bichloride, bichlorure.

bichromate, bichromate.

bigh, grande baie ouverte, inflexion, courbure.

bilateral symmetry, symétrie bilatérale.

bill, bec, promontoire.

billabong (Austr.), lac de bras mort, bayou.

billion, 1. USA 10^9 Un. (1 milliard); **2.** Europe, Australie 10^{12} Un. (1 trillion).

billow, grande vague, lame (d'eau).

billowy, houleux.

bimineralic, biminéral.

bimodal, bimodal.

bimodality, caractère bimodal.

bin, silo, trémie.

binary, binaire; **b. granite, 1.** granite à deux micas; **2.** granite à quartz et feldspath prépondérants.

bind, schiste argileux.

bind (to), lier, cimenter.

binder, lien, ciment, substance agglutinante.

bindheimite, bindheimite.

binding, agglutination; **b. agent** agglutinant; **b. coal,** charbon collant; **b. material,** matrice, liant; **b. stone,** parpaing.

bing, minerai de plomb; **b. hole,** cheminée à minerai.

binocular, binoculaire.

binomial system, système binomial (de Linné).

biochemical, biochimique; **b. deposit,** sédiment b.

biochore, biochore.

biochron, biochrone (durée d'une biozone).

biochronology, stratigraphie paléontologique.

bioclast, roche bioclastique.

bioclastic, bioclastique (débris organique).

biocoenose, biocoenosis, biocénose.

biodegradable, biodégradable.

biodegradation, biodégradation.

biofacies, biofaciès.

biogenesis, biogénèse.

biogenetic law, ontogénie.

biogenic, d'origine biologique.

biogeochemical cycling, cycle biogéochimique.

biogeochemistry, biogéochimie.

biogeography, biogéographie.

bioherm, bioherme, récif (formant relief) voir : biostrome.

biolith, biolithite, calcaire construit, roche d'origine biologique.

biolithite, 1. calcaire récifal (construit); **2.** roche formée par des organismes non déplacés.

biologic, biologique; **b. facies,** biofaciès; **b. species,** espèce biologique, interféconde.

biological magnification, concentration biologique (d'éléments géochimiques).

biomass, biomasse.

biome, biome.

biomechanical deposit, sédiment biodétritique.

biometrics, biométrie.

biometry, biométrie.

biomicrite, biomicrite (boue carbonatée à débris de fossiles).

biomineral, biominéral.

biophile, biophile.

biorhexistasy, biorhexistasie.

biosparite, biosparite (débris fossiles à ciment calcaire).

biospecies, 1. espèce vivant actuellement; **2.** espèce biologique.

biosphere, biosphère.

biostratigraphic, biostratigraphique; **b. unit,** unité b.; **b. zone,** zone b.

biostratigraphy, biostratigraphie.

biostratonomy, biostratigraphie.

biostrome, biostrome (accumulation stratiforme d'organismes).

biota, biota, bios.
biotic, biotique.
biotite, biotite.
biotope, biotope.
bioturbate, bioturbé.
bioturbation, bioturbation (dérangement causé par des organismes).
biotype, génotype.
biozone, biozone.
bipyramid, bipyramide.
bipyramidal, bipyramidal, bipyramidé.
bird, oiseau; **b. foot delta,** delta en patte d'oie; **b. foot drainage,** réseau fluviatile digité.
birdseye, yeux de dégazage; **b. limestone,** calcaire lithographique à plages et veinules de calcite cristalline.
birefringence, biréfringence.
birefringent, biréfringent.
biscuit-board topography, topographie accidentée d'origine glaciaire.
bisilicate, silicate double.
bismuth, bismuth; **b. glance,** bismuthinite; **b. ochre,** bismite.
bismuthic, bismuthique.
bismuthiferous, bismuthifère.
bismuthinite, bismuthine, bismuthinite, bismuthine.
bit, 1. fragment, morceau; 2. trépan, foret, mèche; 3. « bit » (informatique); **blade b.,** trépan à lames; **cone b.,** trépan à cônes; **core b.,** carottier; **diamond b.,** trépan à diamant; **drilling b.,** trépan de forage; **mud b.,** trépan à boues; **rock b.,** trépan à molettes; **three-way b.,** trépan à trois lames.
bite (to), mordre, corroder, attaquer (*chimie*).
bitter, amer; **b. earth,** magnésie; **b. lake,** lac salé; **b. salt,** epsomite; **b. spar,** chaux carbonatée, dolomie; **b. water,** eau mère manganésifère (pétrole).
bittern, eau mère, saumure.
bitumen, bitume; **b. residual,** brai; **natural b.,** b. naturel; **petroleum b.,** b. provenant de la distillation du pétrole.
bituminiferous, bituminifère.
bituminization, bituminisation.
bituminous, bitumineux; **b. coal,** houille grasse; **b. limestone,** calcaire b.; **b. sand,** sable b.; **b. schist,** schiste b.; **b. shale,** schiste b.; **b. slate,** schiste ardoisier b.
Bivalvia, Bivalves.
black, noir; **b. alcali soil,** solonetz; **b. band ironstone,** fer carbonaté; **b. band ore,** fer carbonaté; **b. body radiation,** rayonnement du corps n.; **b. dam,** grisou; **b. diamond,** 1. carbonado; 2. charbon; **b. earth,** chernozem; **b. hole,** trou n. (*astro.*); **b. jack,** blende sphalérite; **b. lava,** Aa, basalte; **b. lead,** graphite; **b. lung,** anthracosis; **b. Jura,** Jurassique inférieur; **b. oxide of iron,** magnétite; **b. sands,** sables n. à minéraux lourds; **b. shale,** ampélite; **b. tellurium,** tellure auroplombifère.
blackish, noirâtre.

Blackriverian (US), Backrivérien (étage, Ordovicien inf.).
blade, lame (*préhist.*).
bladelike, laminaire.
Blancan (US), Villafranchien.
blanket, couverture, nappe, filon-couche (inusité); **b. bog,** sol tourbeux; **b. deposit,** 1. placage minéralisé, filon stratiforme; 2. dépôt sédimentaire étendu, peu épais; **sand b.,** couverture de sable (mince); **vegetal b.,** couverture végétale.
blanketing, couverture, manteau (formations superficielles).
blast, 1. déflagration, explosion, coup de mine; 2. vent, courant d'air; **b. area,** zone de tir, de dynamitage; **b. furnace,** haut fourneau; **b. hole,** trou de mine; **b. pressure,** pression du vent; **air b.,** 1. coup de charge; 2. jet d'air; **volcanic b.,** explosion volcanique.
blast (to), faire exploser, faire sauter, dynamiter.
blasted ore, minerai abattu; **b. sand,** sable soufflé.
blaster, 1. dynamiteur; 2. détonateur.
blasthole stoping, abattage par trous de mine.
blastic deformation, recristallisation sous pression métamorphique de minéraux perpendiculairement à la pression.
blasting, 1. abattage aux explosifs, tir, dynamitage; 2. usure par fluide en mouvement; **b. chute,** tir dans une cheminée, dans un couloir; **b. oil,** nitroglycérine; **dry b.,** usure par le vent ou par soufflerie; **electric b.,** tir électrique; **longhole b.,** abattage par trous profonds; **method of b.,** plan de tir.
Blastoidea, Blastoïdes.
blastopelitic rock, pélite métamorphisée.
blastophitic rock, roche métamorphique concernant quelques caractères ophitiques.
blastoporphyritic rock, ancienne roche porphyrique recristallisée par métamorphisme; **b. structure,** structure porphyroblastique.
blastopsammite, enclave de grès dans conglomérat métamorphique.
blastopsammitic rock, grès métamorphisé.
blastopsephitic rock, conglomérat métamorphisé.
B layer, partie supérieure du manteau terrestre.
bleached sand, sable lessivé.
bleach spot, tache de réduction.
bleaching, blanchiment; **b. clay,** argile adsorbante, décolorante; **thermal b.,** b. thermique.
bleb, 1. inclusion poeciclitique; 2. bulle.
bleeding, 1. fuite d'eau, de gaz, exsudation; 2. vidange, purge; **b. core,** carotte à suintement de pétrole.
blend, blende, blende, sphalérite.
blind, sans affleurement, sans ouverture, aveugle; **b. arm,** faux bras; **b. coal,** anthracite; **b. deposit,** gîte caché; **b. drift,** galerie en cul-de-sac; **b. lode,** filon aveugle; **b. roller,** houle; **b. valley,** vallée morte, vallée sèche.
blinding of a sieve, engorgement d'un tamis.
blister cone, blister, pustule volcanique (sur coulée

de lave); **b. water,** eau saumâtre flottant sur de l'eau salée plus dense.

blob, paquet de projections volcaniques.

block, **1.** bloc (L > 256 mm); **2.** bloc, massif (de roche éruptives); **3.** bloc rocheux limité par des failles, compartiment tectonique; **4.** ensemble; **b. caving,** foudroiement, foudroyage; **b. cluster,** amas de b.; **b. diagram,** b. diagramme; **b. embankment,** rempart de b. de lave; **b. field,** champ, chaos de b.; **b. hole,** trou de mine; **b. lava,** Aa, lave blocailleuse; **b. mountain,** b. soulevé et faillé, horst; **b. packing,** amas de b.; **b. rampart,** rempart de b.; **b. strain,** alignement de b.; **b. stream,** coulée de pierres; **b. stripe,** traînée de b.; ***erratic b.,*** b. erratique; **fault b.,** b. faillé; **stray b.,** b. erratique; **sunken b.,** fossé tectonique; **volcanic b.,** projection volcanique (L > 32 mm).

block out a level (to), découper un niveau en bloc d'abattage.

blocky lava, lave scoriacée avec blocs, aa.

bloodstone, héliotrope (jaspe sanguin).

bloom (to), former des efflorescences (minérales).

blossom, **1.** affleurement oxydé; **2.** efflorescence (d'un minéral).

blow, coup de vent, passage de grisou; **b. hole,** trou souffleur, soufflard.

blow dune, dune parallèle.

blow out, blowout, **1.** éruption incontrôlée, jaillissement de gaz ou de pétrole; **2.** creux de déflation (par le vent).

blow up (to), éclater.

blow up, explosion.

blown sands, sables éoliens.

blower, **1.** soufflard; **2.** irruption de grisou (mine); **3.** ventilateur soufflant.

blowing of an ore, vannage d'un minerai.

blowover fan, sable soufflé en avant des dunes.

blowpipe, chalumeau; **b. test,** essai au chalumeau.

blowpiping, analyse au chalumeau.

blue, bleu; **b. asbestos,** crocidolite; **b. band,** **1.** bande de glace compacte d'un glacier (dépourvue de bulles d'air); **2.** un niveau d'argile repère dans charbons d'Illinois-Indiana; **b. copper,** covellite; **b. copper carbonate,** azurite; **b. diamond-bearing ground,** brèche diamantifère (Afrique du Sud), kimberlite; **b.-green algae,** Cyanophycées (*pal.*); **b. ground elvan (GB),** roche filonienne; **b. iron earth,** vivianite; **b. john,** fluorite; **b. malachite,** azurite, chessylite; **b. mud,** vase bleue; **b. ocher,** vivianite; **b. schist,** schiste bleuté à glaucophane; **b. stone,** chalcanthite; **b. stuff,** kimberlite; **b. vitriol,** chalcanthite.

bluff, **1.** adj : escarpé, à pic; **2.** n : escarpement, falaise de rivière (Austr., A du N.); **3.** falaise de sédiments non consolidés.

blunt, émoussé (galet).

board coal (GB), charbon d'aspect ligneux.

boaster, burin.

boasting, taille de la pierre.

boat-level, galerie de navigation (*mine*).

bocca (Italy), bouche éruptive.

bod, balancier.

bodily tide (see earth tide), marée terrestre.

body, corps, masse, massif; **b. centred cubic,** cristal cubique à face centrée; **b. chamber,** chambre d'habitation (*pal.*); **b. of water,** nappe d'eau; **b. tide,** marée terrestre; **b. waves,** ondes sismiques (de volume); **b. whorl,** enroulement spiralé.

boehmite, boehmite.

bog, tourbière, fondrière, marécage, marais; **b. burst,** coulée de boue et de tourbe; **b. iron,** limonite (des marais); **b. iron,** minerai de fer des marais; **b. ore,** fer des marais; **b. manganese mine, mine ore,** manganèse des marais; **low, flat b.,** tourbière basse; **up, high land b.,** tourbière haute; **string b.,** tourbière cordée; **quaking b.,** branloire cordée.

bogaz, bogaz (doline allongée).

boggy, tourbeux, marécageux.

boghead, boghead (var. de charbon); **b. coal,** **1.** charbon bitumeux; **2.** charbon d'algues (non lité).

bohemian garnet, pyrope; **b. ruby,** rubis.

boil, remontée turbulente d'eau.

boiling, **1.** ébullition; **2.** exsolution.

bojite, bojite (gabbro à amphibole primaire) ou diorite à hornblende.

bold, élevé, escarpé.

bole (clay), terre bolaire, argile cuite latéritisée, interstratifiée dans coulée de lave.

bolson (US, Mex.), cuvette endoréique, bassin à drainage centripète.

bolter sieve, tamis.

bolthead, matras, ballon à long col (*labo.*).

bomb (volcanic), bombe (volcanique).

bond, lien, liaison, aggloméré; **b. stone,** parpaing; **b. type,** valence atomique.

bond (to), réunir, connecter, lier, cimenter.

bonding, **1.** liaison; **2.** appareillage (*constr.*); **b. electron,** électron de valence; **b. intermolecular,** liaison intermoléculaire.

bone, os; **b. bed,** couche à débris d'o., de dents, écailles, etc.; **b. breccia,** brèche d'ossements; **b. phosphate,** phosphate d'o.; **b. turquoise,** odontolite, fausse turquoise.

boning, nivellement.

Bononian, Bononien (= Portlandien inf.).

bony, osseux; **b. coal,** charbon schisteux.

book structure, structure en feuillets (d'une roche schistoquartzique).

boomer, «boomeur», dispositif explosif utilisé en sismique marine, générateur sonique.

boortz, diamant noir, bort.

booster, **1.** survolteur, accélérateur; **2.** primaire (explosif).

booze, minerai de plomb.

boracite, boracite.

boralf, sol lessivé boréal.

borate, borate.

borax, borax; **b. bead,** perle de b. (méthode au chalumeau).

border, bordure, lisière, marge; **b. facies of igneous rocks**, faciès de front de roches ignées; **b. land**, cordon de terres bordant un géosynclinal; isolant une mer épicontinentale; **b. line**, limite; **b. moraine**, moraine latérale; **b. sea**, mer bordière; **b. zone**, bordure figée (*pétrol.*).

bore, 1. sondage, forage, trou de sonde; 2. mascaret (*géog. phys.*); 3. banc de sable sous-marin intertidal; **b. bit**, trépan; **b. core**, carotte; **b. hole**, trou de sonde, sondage, trou de forage; **b. holing**, sondage, forage; **b. holing journal**, carnet de sondage; **b. rod**, tige de sonde; **b. well**, sondage.

bore (to), percer, forer, sonder, faire des sondages.

boreal, boréal; **b. climate**, climat boréal; **B. period**, période Boréale (intervalle, Holocène : 9 500-7 200 BP).

borer, 1. sondeur, foreur; 2. sonde, mèche sondeuse, foreuse; 3. perçoir (*préhis.*); **b. organisms**, organisme perforant.

boric acid, acide borique.

boride, borure.

boring, forage, sondage, percement; **b. bar**, tige de sonde, barre de sonde (*mine*); **b. bit**, trépan; **b. by percussion**, s. par battage; **b. chisel**, trépan (*mine*); **b. journal**, rapport de s.; **b. plant**, installation de f.; **b. rig**, appareil de s.; **b. rod**, tige de sonde; **b. sample**, échantillon de f., carotte; **b. site**, emplacement de f.; **b. tool**, outil de f.; **b. tower**, tour de s.; **shot-drills b.**, s. à la grenaille.

bornhardt, relief résiduel (inselberg de grande dimension).

bornite, bornite, érubescite (minerai de cuivre).

borolanite, *borolonite*, borolonite (var. de syénite).

boroll, chernozem.

boron, bore.

borosilicate, borosilicate.

borrow pit, ballastière.

bort, diamant noir.

boss (see stock), 1. protubérance, chicot d'érosion; 2. petit massif intrusif (à contour circulaire).

bostonite, bostonite (var. de microsyénite).

Bothnian, Bothnien (Précambrien balte).

botryogen, botryogène.

botryoidal, aggloméré en grappes.

bottle-neck, étranglement, goulet.

bottom, 1. fond, lit, base, pied, mur, soubassement, socle; 2. aval-pendage (*mine*); 3. régions (US) alluviales; **b. current**, courant de fond; **b. flow**, courant dense de fond; **b. land**, plaine d'inondation, basses terres; **b. level**, galerie de fond; **b. load (bed load)**, charge de fond; **b. of a coal seam**, mur d'une couche de houille; **b. set beds**, couches deltaïques de fond; **b. stopes**, gradins droits; **b. stoping**, abattage descendant; **b. workings**, chantier de fond (*mine*); **b. hill**, pied d'une colline; **b. sandy**, fond sableux; **b. water**, eau océanique profonde.

bottom (to), atteindre le substratum.

bottom up (to), ramener la boue du forage de fond vers la surface.

boudinage, boudinage.

Bouguer anomaly, anomalie de Bouguer; **B. correction**, correction de Bouguer; **B. reduction**, correction de Bouguer.

boulangerite, boulangérite.

boulder, 1. gros bloc (transporté); 2. gros bloc erratique; **b. barricade**, cordon, blanc intertidal de galets; **b. bed**, couche morainique à blocaux; **b. clay**, argile à blocaux; **b. field**, champ de pierres; **b. pavement**, dallage de pierres; **b. period**, période glaciaire; **b. pile**, chaos; **b. rampart**, rempart de galets et de blocs poussés par la glace littorale; **b. stream**, coulée de pierres; **b. train**, alignement de blocs glaciaires; **b. wall**, rempart morainique.

bounce cast, moulage de rebond; **b. mark**, figure (sédimentaire) de rebond.

bound, 1. limite; 2. zone exploitée pour l'étain (Cornouailles).

bound (to), limiter, borner; **b. folding**, plissement entravé; **b. water**, eau de rétention.

boundary, limite, frontière; **b. current**, courant de contour; **b. layer**, couche l.; **b. post**, poteau de bornage; **b. stone**, 1. pierre de bornage; 2. composants organiques liés par un ciment (dans des calcaires construits); **b. waves**, ondes superficielles.

bounding, faille bordière.

bourne, source intermittente.

bournonite, bournonite.

box fold,, pli coffré (Jura).

bow shaped dune, dune arquée.

Bowen's reaction series, séries réactionnelles de Bowen.

bowenite, bowénite (var. de serpentine).

bowl, cuvette, bas-fond; **b. cirque**, cirque peu profond.

bowlder (US), gros bloc.

bowse, minerai de plomb.

box stone, concrétion creuse, grès phosphaté ou ferrugineux (Suffolk).

boxwork, structure cloisonnée.

brachials, plaques brachiales (*pal.*).

brachial valve, valve dorsale.

brachidium, *brachidia (pl.)*, brachidium, appareil brachial.

Brachiopod, *Brachiopoda*, Brachiopode.

brackish, saumâtre.

brachium, *brachia (pl.)*, bras.

brachy-anticline, brachy-anticlinal.

brachydome, brachydôme.

brachyfold, brachypli.

bradygenesis, bradygenèse.

bradyseism, bradyséisme.

bradyseismal, *bradyseismic*, *bradyseismical*, bradysismique.

brae, côte, colline (écossais).

Bragg's angle, angle de diffraction.

Bragg's law, loi de Bragg (diffraction) de rayons X.

braided, anastomosé; **b. river**, rivière a. en tresse.

braiding pattern, réseau anastomosé.

branch, ramification, bifurcation; **b. of lode**, rameau d'un filon; **b. vein**, filon ramifié.

branched, ramifié.
branchiae, branchies (*pal.*).
branchial, branchial; **b. slit**, fente branchiale.
branching, ramification (d'un filon).
branchwork, réseau fluviatile.
brash, éboulis, blocaille; **b. ice**, débâcle des glaces.
brashy, fragmenté, blocailleux.
brass, laiton; **b. balls**, nodules de pyrite; **b. ore**, aurichalcite.
brasses, pyrite (*mine*).
brassil, 1. pyrite; 2. charbon contenant de la pyrite.
brassy, pyriteux.
brattice, cloison d'aérage (*mine*).
braunite, braunite.
Bravais lattice, réseau de Bravais.
brazil, pyrite.
brazilian, brésilien; **b. emerald**, tourmaline verte; **b. peridot**, tourmaline jaune-vert; **b. ruby**, spinelle rougeâtre; **b. sapphire**, tourmaline bleue.
brea, 1. goudron minéral, malthe; 2. sol imprégné de goudron.
breach, brèche, trouée.
breached anticline, anticlinal à cœur érodé; **b. crater**, cratère égueulé.
bread-crust bomb, bombe volcanique en croûte de pain.
break, 1. cassure, faille, rupture; 2. fissure, petite cavité (de couche de charbon); 3. début d'un séisme; **b. in the succession**, lacune stratigraphique; **b. in lode**, faille; **b. in slope, of slope**, rupture de pente; **b. up**, débâcle glaciaire; **sedimentary b.**, lacune stratigraphique.
break (to), 1. casser, concasser, briser, fracturer, fragmenter; 2. se fragmenter, se morceler.
break down (to) ore, abattre le minerai.
breakage, cassure, rupture, fragmentation.
breaks, 1. petites failles; 2. intercalations de roches tendres; 3. ruptures de pente, variations brusques de la topographie, pente disséquée.
breakdown, 1. effondrement, émiettement, décomposition d'une roche; 2. panne.
breaker, brisant (d'une vague); **b. zone**, zone de déferlement, zone à haute énergie.
breakeven point, point mort, point de rupture, point critique.
breaking, 1. abattage, cassage, cassure, concassage; 2. déferlement; **b. down**, abattage (de minerai); **b. ground**, abattage; **b. load**, charge de rupture; **b. point**, point de rupture; **b. strength**, résistance à la rupture; **b. stress**, charge à la rupture, contrainte, tension de rupture; **b. thrust**, chevauchement cassant; **b. up**, 1. fragmentation; 2. débâcle glaciaire.
break off, interruption d'un filon (Derbyshire).
breakpoint, zone de déferlement.
breakthrough, 1. percée (*mine*); 2. boyau de mine.
breakthrust, poussée cassante, faille de chevauchement coupant un flanc de pli.
breakwater, 1. brisant; 2. brise-lames.
breast, front d'abattage, front de taille, taille.
breastopping, abattage de front.

breccia, brèche; **b. marble**, marbre bréchique; **b. pipe**, conduit bréchique métallifère (formé par dissolution de calcaires et affaissement. Ex. : Colorado); **crumbling b.**, b. tectonique; **friction b.**, b. de friction.
brecciated, bréchoïde, bréchique, bréchiforme; **b. marble**, marbre bréchique; **b. vein**, filon bréchique.
brecciation, formation de brèches.
breccio-conglomerate, conglomérat de brèches.
breeder reactor, surrégénérateur.
brenstone, soufre brut.
brick, brique; **b. clay**, argile à b.; **b. earth**, terre à b., loess évolué.
brickwork, maçonnerie.
bridge, liaison atomique.
Bridgerian (US), Bridgérien (âge : Éocène moy.).
bridging oxygen, liaison oxygène.
bright, 1. adj : brillant; 2. n : diamant; **b. coal**, anthracite.
brilliancy, brillance, éclat.
brimstone, soufre (brut).
brine, saumure, eau salée, solution hypersaline; **b. spring**, source salée.
bring (to) an oil-water into production, mettre en production un puits de pétrole.
bring (to) ore to grass, remonter le minerai à la surface.
bring (to) to the surface, remonter à la surface.
Bringewoodian, Bringewoodien (sous-étage, Ludlovien, moy., Sil.).
brink, bord (d'un précipice).
brinkpoint, point de rupture de pente (sur un versant).
british thermal unit (BTU), BTU 252 calories.
brittle, cassant, fragile; **b. failure**, rupture de la roche; **b. ice**, glace cassante; **b. mica**, margarite; **b. silver ore**, stéphanite; **b. star**, Ophiure (*pal.*); **b. strength**, résistance à la fracture.
broaching, entaillage (*mine*); **b. bit**, alésoir.
broadstone bind (GB), schiste argileux se débitant en dalles.
brochantite, brochantite.
broken country, région accidentée.
broken ground, 1. couches fragmentées; 2. couches faillées stériles (Angleterre).
bromide, bromure.
bromine, brome.
bromite, bromyrite.
bromoform, bromoforme.
bromyrite, bromargyrite, bromyrite (bromure d'argent).
brontolith, météorite.
bronze mica, phlogopite.
bronzite, bronzite.
bronzitite, bronzitite, pyroxénite à bronzite.
brook, ruisseau.
brookite, brookite (TiO_2).
brooklet, ruisselet.
brotocrystal, cristal à contour corrodé.
brow, sommet (de colline), front, bord (de falaise).

brown, brun; **b. calcareous soil,** sol b. calcaire; **b. clay ironstone,** limonite argileuse; **b. coal,** lignite; **b. earth,** sol b. forestier; **b. forest soil,** sol b. forestier; **b. hematite,** limonite; **b. iron ore,** limonite; **b. lead ore,** pyromorphite; **b. ocher,** limonite; **b. podzolic soil,** sol b. podzolique; **b. soil,** sol b.; **b. spar,** carbonate teinté en b. par de l'oxyde de fer; **b. steppe soil,** sol b. sub-aride; **b. stone,** grès ferrugineux, pyrite décomposée (Australie).

brownian, brownien.

browsers, organismes marins « brouteurs » (ex. : Gastéropodes).

brucite, brucite [$Mg(OH)_2$].

brunizem, sol brun de prairie.

brush (to) to roof and floor, recouper le toit et le mur, abattre les épontes.

brushite, brushite.

brush ore, hématite dendriforme, stalactiforme.

Bruxellian, Bruxellien (étage, Éocène moy.).

Bryophyta, Bryophytes.

Bryozoa, Bryozoan, Bryozoaires.

bubble, bulle; **b. level,** niveau à b.

bubble (to), bouillonner, dégager des bulles.

bubbly, bulleux.

Bubnoff unit, unité de mouvement géologique (1 µm par année ou 1 m par million d'années).

buck, concasseur de minerai, scheider; **b. quartz,** quartz non aurifère (Australie); **b. stone,** roche non aurifère.

buck (to), broyer, concasser, trier.

bucked ore, minerai de scheidage, minerai scheidé.

bucker, 1. pers. : scheideur; 2. marteau de scheidage.

bucket, godet, benne; **b. dredge,** drague à g.; **b. excavator,** excavateur à g.

bucking ore, scheidage du minerai.

buckle (to), 1. déformer, déjeter, gauchir, voiler; 2. se déformer, se déjeter, se gauchir, se voiler.

buckled, ondulé.

buckling, plissotement, déformation, gauchissement..

bug hole, géode, druse.

building, construction; **b. materials,** matériaux de c.; **b. stone,** pierre à bâtir.

built-up terrace, terrasse construite, terrasse alluviale, terrasse d'accumulation.

bulge, bombement, renflement.

bulk density, densité apparente.

bulk in, en vrac.

bulky, volumineux.

bull (to) a hole, bourrer un trou de mine.

bull quartz, quartz enfumé.

bullion, 1. or, argent, (en lingots); 2. nodules de fer argileux, de schiste pyriteux, etc., renfermant souvent un fossile; 3. concrétion dans les charbons.

bump, choc, secousse, bosse.

bumper, carottier.

bumping screen, tamis à secousses; **b. table,** table à secousses.

bunch of ore, poche de minerai.

bunchy, minéralisé en poches, en nids.

bund, digue.

Bunsen burner, bec Bunsen.

buoyancy, flottabilité, légèreté.

buoyant, flottant.

buoying power, pouvoir de sustentation.

burden, morts-terrains de recouvrement, terrains de couverture.

Burdigalian, Burdigalien (étage, Miocène).

burial, enfouissement; **b. metamorphism,** métamorphisme d'e.

buried, enfoui, enterré; **b. ice,** glace e.; **b. outcrop,** affleurement caché; **b. placer,** gisement caché; **b. topography,** paléorelief enfoui; **b. valley,** ancienne vallée, profonde, entaillant le substratum et colmatée par des alluvions fluvio-glaciaires (inlandsis nord-américain).

burin, burin (*préhist.*).

burmite, burmite (résine fossile).

burn (scot.), ruisseau.

burn (to) plaster, cuire du plâtre.

burning, 1. adj : brûlant, enflammé; 2. n : combustion, cuisson, calcination, grillage; **b. mountain,** volcan; **b. oil,** kérosène; **b. point,** point d'inflammation; **b. test,** essai de combustion.

burnt, cuit; **b. brass,** chalcantite; **b. coal,** charbon altéré par une intrusion de roches éruptives; **b. copper,** oxyde de cuivre; **b. stuff (Australie),** croûte indurée sous-jacente.

burr, bur, roche dure; **b. stone,** meulière.

burrow, 1. halde de déchets; 2. terrier.

burrowing clams, lamellibranches fouisseurs.

burst, 1. explosion (*mine*); 2. essort, explosion (*pal.*); **rock b.,** coup de charge, secousse (*mine*).

burying, enfouissement.

bush, 1. brousse; 2. buisson, arbuste; 3. broussailles; **coral b.,** buisson coralien.

butane, butane.

butte (US), butte-témoin.

butter of tin, chlorure stannique.

buttress, contrefort, éperon, pilier; **b. sand,** lentille sableuse discordante.

buttress (to), étayer, arc-bouter.

by pass, déviation, dérivation.

by product, sous-produit.

bysmalith, intrusion discordante, dôme intrusif, culot intrusif.

byssus, byssus (*pal.*).

bytownite, bytownite.

C

cable, 1. câble, corde; **2.** flûte; **c. drilling,** sondage au c. (*géophys.*); **c. tool drilling,** forage au c.; **blasting c.,** ligne de tir; **hoisting c.,** c. d'extraction.

cacholong, cacholong.

cadastral map, carte du cadastre, c. cadastrale; **c. survey, 1.** service du c.; **2.** levé c.

cadmiferous, cadmifère.

cadmium, cadmium; **c. blende,** c. sulfuré; **c. ochre,** greenokite.

caesium, caesium, césium.

cafemic, cafémique (abrév. : calcium-ferreux-ferrique-magnésium, cf. classification CIPW).

Caenozoic, Tertiaire et Cénozoïque.

cage, cage; **hoisting c.,** c. d'extraction; **mine c.,** c. de mine.

Cainozoic, Cénozoïque, Tertiaire.

cairn, cairn.

cairngorm, quartz jaune fumé.

cake (to), se concrétionner, former une croûte, coller; **c. capacity,** pouvoir agglutinant.

caking coal, charbon collant.

cal (Cornwall), wolframite.

cala, calanque.

Calabrian, Calabrien.

calaite, turquoise.

calamine, calamine; **c. bearing deposit,** gisement de c.

calamite, calamite (*minér.*, $ZnCO_3$).

Calamites, Calamites (*paléobot.*).

calanque, calanque.

calaverite, calavérite (*minér.*).

calc, préfixe indiquant la présence de carbonate de calcium; **c. alkali rock,** roche calco-alcaline; **c. alkaline rock,** roche calco-alcaline; **c. flinta,** cornéenne à silicate calcique; **c. sinter,** travertin calcaire; **c. spar,** carbonate de calcium, calcite; **c. tufa,** tuf calcaire.

Calcarea, éponges calcaires.

calcarenite, calcarénite.

calcareo-argillaceous, calcaréo-argileux.

calcareoferruginous, calcaréo-ferrugineux.

calcareomagnesian, calcaréo-magnésien.

calcareosilicious, calcaréo-siliceux.

calcareous, calcaire; **c. algae,** algues c.; **c. breccia,** brèche c.; **c. concretion,** concrétion c.; **c. grits,** sables calcaréo-siliceux; **c. hardpan,** carapace c.; **c. lithosol,** rendzine embryonnaire; **c. ooze,** vase c., boue c.; **c. sand,** sable c.; **c. sandstone,** grès c.; **c. sinter,** travertin c.; **c. spar,** calcite; **c. tufa,** tuf, travertin c.

calcariferous, calcarifère, calcareux.

calcedonic, chalcedonic, calcédonieux.

calcedonite, calcédonite, calcédoine (*minér.*).

calcic, calcique; **c. mull,** « mull » c.; **c. series,** roches éruptives fortement calco-alcalines.

calciclase, plagioclase calcique (anorthite); **c. syenite,** syénite à anorthite.

calcicrete, encroûtement calcaire.

calciferous, calcifère.

calcification, calcification, transformation en calcaire, encroûtement calcaire.

calcify (to), calcifier.

calcified, enrichi en calcium.

calcilutite, calcilutite.

calcimeter, calcimètre.

calcimorphic soil, sol calcimorphe, sol calcicole.

calcin, calcin.

calcinable, calcinable.

calcination, calcination, grillage.

calcine (to), 1. calciner, griller (un minerai); **2.** se calciner.

calciobiotite, biotite calcique.

calciocelestite, calciocelestine.

calcioscheelite, scheelite.

calciphyre, calcaire cristallin.

calcirudite, calcirudite.

calcisiltite, calcisiltite, limon induré calcaire, aleurolite calcaire, microgrès calcaire.

Calcispongiae, Éponges calcaires.

calcite, calcite; **c. cleavage,** clivage rhomboèdrique, clivage de la c.; **c. trachyte,** trachyte calcique.

calcitic dolomite, dolomie calcaire (10 à 50 % de calcite, 50 à 90 % de dolomie).

calcium, calcium; **c. carbonate,** carbonate de c.; **c. chloride,** chlorure de c.; **c. fluoride,** fluorure de c.; **c. hydrate,** chaux vive; **c. hydroxide,** chaux éteinte; **c. phosphate,** phosphate de c. (cf. apatite); **c. sulfate,** sulfate de c. (cf. gypse).

calcoferrite, calcoferrite.

calcomalachite, calcomalachite.

calcosilicate, silicate calcique.

calcrete, encroûtement calcaire, graviers cimentés, travertin du sol.

calcreted, encroûté, cimenté.

calcrust, croûte concrétionnée.

calcshist, calcshistes.

calc-silicate hornfels, cornéenne à silicates calciques.

calc-sinter, travertin calcaire.

calcspar, calcite en gros cristaux, calcite spathique.

calctufa, calc tufa, tuf calcaire.

calculiform, en forme de galet.

caldera, caldeira.

calderite, caldérite (variété de grossulaire).

Caledonian, Calédonien; **C. Orogeny,** orogenèse calédonienne.

Caledonides, Calédonides.
caledonite, calédonite.
calf, glace flottante.
calibration, 1. datation, âge stratigraphique (d'une formation), « calage »; 2. calibrage, réglage.
caliche (cf. calcrete), 1. caliche ($NaNO_3$); 2. croûte calcaire des sols arides; 3. horizon pédologique d'accumulation calcaire; 4. aggloméat calcaire à graviers, sables et divers (S-O USA, Mexique).
californite, californite (variété verte d'idocrase).
caliper, calibre; **c. gauge,** pied à coulisse; **c. log,** diagramme, log de diamétrage; **c. logging,** diamétrage (forage); **slide c.,** pied à coulisse.
calk (to), étanchéifier.
callainite, callaïnite.
Callovian, Callovien (Jur. moyen).
callow, 1. fondrière; 2. terrains du sommet d'une carrière; 3. terrains superficiels, morts-terrains.
calomel, calomel (chlorure mercureux).
calms, calmes « équatoriaux ».
calorie, calory, calorie; **great c.,** grande c., kilo-c.; **gram c.,** petite c., microthermie; **large c.,** grande c.; **small c.,** petite c., microthermie.
calorific, calorifique, thermique; **c. power,** pouvoir c.
calorimeter, calorimètre.
calorimetric, calorimetrical, calorimétrique.
calorimetry, calorimétrie.
calve (to), vêler (dune, glacier).
calving, 1. vêlage (périgl.); 2. éboulement d'une falaise.
calyx, calice.
cambisol, cambisol (var. de sol brun).
Cambrian, Cambrien.
camera, 1. appareil photographique; 2. chambre, loge (*pal.*); **air c.,** appareil de prise de vues aériennes; **mapping c.,** appareil de prise de vues aériennes; **surveying c.,** appareil de prise de vues aériennes.
camerate, possédant des chambres, des loges (*pal.*).
Campanian, Campanien.
camping site, site de campement préhistorique (*archéol.*).
can, boîte de conserve, boîte, bidon.
Canada balsam, baume de Canada.
Canadian stage, Canadien (= Ordovicien inférieur).
canal, 1. canal (artificiel); 2. chenal; 3. bras de mer; 4. canal, perforation du test des foraminifères; 5. fleuve côtier lent (US).
canalization, canalisation (d'une rivière).
canalize (to), canaliser (une rivière).
cancellated (unusual), réticulé.
canch, entaille.
cancrinite, cancrinite feldspathoïde.
cand, fluorine, fluorite (*minér.*).
candelit, charbon flambant.
candescent, incandescent.
candle coal, charbon gras à spores et pollens.
canga, brèche ferrugineuse (Brésil).
canker (to), corroder, altérer.
cannel coal, charbon gras à spores et pollens.

cannel shale, schiste sapropélique.
cannon hole, souflard.
canoe fold, pli synclinal en forme de bateau.
canon, canyon.
cant, 1. inclinaison, dévers; 2. biseau.
cant (to), 1. incliner; 2. biseauter; 3. s'incliner.
cantilever, encorbellement.
canyon, canyon.
cap, 1. chapeau; 2. détonateur, amorce; **c. rock,** roche couverture, chapeau (d'un gisement); **blasting c.,** amorce, détonateur (*mine*); **casing head c.,** tête de tubage.
cap (to), couronner, coiffer.
cap (to) a well, obturer un puits.
capacity, 1. capacité, contenance, volume; 2. rendement, débit; 3. capacité de transport (fluviatile, éolienne), compétence; **maximum water holding c.,** capacité maximum de rétention en eau.
cape, cap, promontoire; **c. chisel,** burin; **c. ruby,** grenat pyrope.
Cape diamond, diamant du Cap.
capillarity, capillarité.
capillary, capillaire; **c. attraction,** attraction c.; **c. capacity,** capacité c., **c. conductivity,** porosité fine, c.; **c. fringe,** frange c.; **c. interstice,** interstice c.; **c. layer,** couche c.; **c. migration,** migration par capillarité; **c. percolation,** succion; **c. porosity,** porosité c.; **c. potential,** potentiel c.; **c. pressure,** pression c.; **c. pyrite,** millerite; **c. rising,** ascension c.; **c. tube,** tube c.; **c. water,** eau de capillarité.
capping, mort terrain, terrain de recouvrement.
capstan, cabestan, treuil.
captor, 1. cours d'eau capteur; 2. détecteur, capteur (*télédétection*).
capture, capture; **point of c.,** point de c.; **self c.,** autoc.; **spontaneous c.,** c. par déversement.
captured stream, cours d'eau capturé.
car, 1. voiture, chariot; 2. berline, wagonnet (*mine*); **dump c.,** wagon à benne basculante.
Caradocian, Caradocien.
carapace, carapace.
carat, carat.
carbide, carbure; **tungsten c.,** c. de tungstène.
carbohydrates, glucides et hydrates de carbone.
carboid, carboïde.
carbon-14 dating, datation au radiocarbone ^{14}C.
carbon, carbone; **c. bisulfide,** sulfure de c.; **c. disulfide,** sulfure de c.; **c. dioxide,** gaz carbonique; **c. fixed,** matière carbonée solide obtenue après combustion (autre que les cendres); **c. monoxide,** monoxyde de c.; **c. nitrogen ratio,** rapport c. azote; **c. number,** nombre d'atomes de c. dans un hydrocarbure; **c. ratio,** teneur en c.; **c. steel,** acier au c.; **c. tetrachloride,** tétrachlorure de c.; **c. total,** c. total (matière carbonée libre et sèche, y compris les constituants volatils).
carbonaceous, 1. carboné; 2. charbonneux.
carbonado, carbonado, diamant noir industriel.
carbonatation, carbonatation.
carbonate, carbonate; **c. apatite,** apatite calcique; **c. of barium,** witherite; **c. of lime,** carbonate de

chaux; **c. ore**, minerai carbonaté; **c. profile of a soil**, situation des horizons calcaires dans une coupe pédologique; **orthorhombic c.**, carbonates orthorhombiques.

carbonated spring, source carbonatée.

carbonation, 1. carbonatation; 2. dissolution de minéraux et remplacement par des carbonates.

carbonatite, carbonatite.

carbonic, carbonique; **c. acid**, acide c.; **c. oxide**, oxyde de carbone.

Carboniferous, 1. adj : carbonifère, houiller, contenant du charbon; 2. n : Carbonifère (période); **c. formation**, couche contenant du charbon; **C. limestone**, Carbonifère inférieur (GB); **C. period**, période c.; **c. rock**, roche contenant du charbon ou d'âge c.

carbonization, houillification.

carbonizable, carbonisable.

carbonize (to), 1. carboniser; 2. se carboniser; 3. houillifier.

carbonized, 1. carbonisé; 2. transformé en charbon.

carbonous, charbonneux.

carborundum, carborundum, carbure de silicium artificiel (*minér.*).

carburize (to), 1. cémenter; 2. carburer.

cardinal, cardinal, appartenant à la charnière; **c. area**, appareil c.; **c. points**, points cardinaux; **c. process**, appareil c. (Brachiopodes); **c. septum**, cloison, septum principal (des Tétracoralliaires); **c. teeth**, dents c.

carina, carène.

carinate fold, pli isoclinal.

Carlsbad twin, mâcle de Carlsbad.

carnallite, carnallite (*minér.*).

carnelian, cornaline (quartz calcédonieux rouge).

Carnian, Carnien.

carnieule, carnieule.

Carnivora, Carnivores.

Carnosaur, Dinosaure carnivore (*pal.*).

carnotite, carnotite.

Carpoidea, Carpoïdés (*pal.*).

carr, sol marécageux tourbeux.

carried soil, sol allochtone, sol transporté.

carrier bed, lit réservoir (*pétrole*).

carry away (to), enlever, emporter, transporter au loin.

carrying power, puissance de transport.

carse, 1. plaine alluviale (Écosse); 2. vallon (Écosse).

carstone, grès ferrugineux (limonitique).

cartogram, carte ombrée.

cartographer, cartographe.

cartographic, cartographical, cartographique.

cartography, cartographie.

cartometric testing, complètement au sol, vérification au sol.

cartouche, cartouche (cartographique).

carving, 1. modèle du relief; 2. gravure (*préhist.*).

cascade, cascade; **c. folds**, plissotements secondaires formés par glissement sur pente.

case hardening, cimentation superficielle par vernis désertique.

case (to) a well, tuber un puits.

casing, tubage, cuvelage; **c. collar**, joint de t.; **c. elevator**, élévateur à tubes; **c. grab**, accroche-tube; **c. head**, tête de sonde; **c. head gas**, gaz de pétrole; **c. head pressure**, pression en tête de t.; **c. line**, colonne de tubes; **c. perforations**, perforations du t.; **c. pipe**, tube de c.; **c. potential**, potentiel enregistré d'un forage; **c. string**, colonne de t.

cassiterite, cassitérite (SnO_2, minerai d'étain).

cast, 1. adj : moulé, fondu; 2. n : moule, moulage (de fossile, etc.); 3. n : coulée (métal).

castanozem, sol châtain.

cast iron, 1. adj : de fonte, en fonte; 2. fonte de fer.

cast (load), figure de charge.

cast (open), à ciel ouvert.

cast (to), 1. mouler, se mouler; 2. fondre, couler (métal).

castaways, filon stérile.

castellated rock, roche ruiniforme.

castings, déjections fécales, coprolites.

castorite, castorite (var. de feldspathoïde).

cat (GB), argile réfractaire dure; **c. dirt**, a. r.; **c. face**, nodules de pyrite dans front de taille de charbon; **c. gold**, mica doré (désuet); **c. silver**, mica argenté; **c. step**, gradin, terrassette.

cataclasis, cataclase.

cataclasite, roche cataclastique, mylonite.

cataclastic, cataclasting, cataclastique; **c. rock**, roche c.

cataclinal, cataclinal.

cataclysm, cataclysme.

cataclysmal, cataclysmatic, cataclysmic, cataclysmique.

catagenesis, catagenèse.

cataglacial, cataglaciaire.

catalysis, catalyse.

catalyst, catalyseur.

catalytic, catalytique.

cataphorèse, électrophorèse.

cataract, cataracte.

cataspilite, cataspilite (var. altérée de cordiérite).

catastrophism, catastrophisme.

catazone, catazone.

catch fire (to), prendre feu, s'enflammer.

catch water (to), capter l'eau.

catchment, catching, captage, captation; **catchment area**, bassin d'alimentation; **catchment basin**, bassin hydrographique.

catena of soils, chaîne de sols.

catenary complex, chaîne de sols.

catsbrain, concrétion (ferrugineuse).

cathead, 1. cabestan (*for.*); 2. tête de chat (*géol.*).

cathode, cathode; **c. rays**, rayons cathodiques.

cathodic, cathodique.

cathodoluminescence, cathodoluminescence.

cation, cation; **c. exchange**, échanges cationiques, échange de bases.

catlinite, catlinite.

catoctin, relief résiduel, monadnock.

catogene rock, roche sédimentaire.

cat's eye, œil de chat, chrysobéryl; **c. quartz,** quartz œil de chat.

cauk, **1.** barytine (GB); **2.** craie, calcaire (Écosse).

cauldron subsidence, affaissement, effondrement circulaire, subsidence en chaudron.

caulk, barytine (*minér.*).

caustic, caustique; **c. lime,** chaux éteinte; **c. metamorphism,** métamorphisme de contact; **c. potash,** potasse c.; **c. silver,** nitrate d'argent; **c. soda,** soude c.

cave, grotte, caverne; **c. breccia,** brèche de c.; **c. earth,** remplissage sédimentaire de c.; **c. marble,** calcite ou aragonite cryptocristalline zonée; **ice c.,** g. creusée dans la glace; **lava c.,** g. creusée dans des laves; **sea c.,** g. littorale.

cave in (to), s'affaisser, s'effondrer, s'ébouler.

caved, éboulé, effondré.

cavern, caverne.

cavernous, caverneux; **c. porosity,** macroporosité; **c. structure,** structure vésiculaire, structure caverneuse.

caving, **1.** éboulement, effondrement; **2.** exploitation par foudroyage; **block c.,** foudroiement, foudroyage; **undercut c.,** foudroyage après sous-cavage.

cavitation, cavitation.

cavitation erosion, érosion par cavitation.

cavity, cavité, creux.

cawk, **1.** barytine (GB); **2.** craie, calcaire (Écosse).

cay, caye, key, quay, **1.** îlot plat de sable; **2.** île côtière de sable ou de corail; **3.** banc de sable (ou de corail) côtier.

Caytoniales, Caytoniales (*pal.*).

Cazenovian (US), Cazenovien (Dévonien moy.).

cedarite, cédarite (var. de résine).

celadonite, céladonite (*minér.*).

celestial body, corps céleste.

celestine, celestite, célestine (*minér.*, $SrSO_4$).

cell of convection, cellule de convection.

cellular, **1.** cellulaire, alvéolaire; **2.** scoriacé, vésiculaire; **c. dolomite,** carnieule; **c. porosity,** macroporosité; **c. pyrite,** marcasite; **c. soil,** sol polygonal.

Celsius degree, degré centigrade ou degré Celsius; **C. scale,** échelle centigrade.

cement, ciment; **c. formation,** formation cimentée; **c. rock,** calcaire argileux, marne; **c. stone,** calcaire à c.; **c. texture,** substitution du c. de grès par des minerais.

cement (to), cimenter.

cemented gravel, conglomérat.

cementation, **1.** cémentation (métall.); **2.** cimentation, diagenèse.

cementing, cimentation.

Cenomanian, Cénomanien.

cenotypal, néovolcanique (tertiaire ou quaternaire).

Cenozoic, Tertiaire, Cénozoïque; **C. era,** ère Tertiaire.

cenozone, cénozone.

centigrade, centigrade; **c. scale,** échelle c.

centigram, centigramme.

centiliter, centilitre, centilitre.

centimeter, centimètre.

centimeter-gramme-second system, système CGS.

centipoise, centipoise.

central, central; **c. eruption,** éruption c.; **c. vent volcanoe,** volcan à orifice c.; **c. volcanoe,** volcan c., volcan punctiforme.

centre, center, centre, milieu; **c. cut,** fossé c. (*mine*); **c. force,** force centrifuge; **c. line,** axe; **c. of gravity,** c. de gravité; **c. of symmetry,** c. de symétrie.

centrifuger, centrifugeuse.

centripetal force, force centripète.

centroclinal, périclinal; **c. fold,** cuvette synclinale, brachysynclinal.

centrocline, cuvette synclinale.

centrosphere, barysphère.

cephalic, céphalique; **c. limb,** joue (Trilobite).

cephalon, céphalon, tête (Trilobite).

Cephalopoda, Cephalopods, Céphalopodes (*pal.*).

ceramic, céramique.

cerargyrite, cérargyrite.

Ceratite, Cératite; **C. limestone,** Trias moyen (Californie).

Ceratitida, Cératitidés.

cerine, cérine (var. d'allanite).

cerepidote, allanite.

Cerianthera, Cérianthaires.

cerium, cérium.

cerusite, cerussite, cérusite, carbonate de plomb.

cesium, césium.

cesspool, puits absorbant, puits perdu.

Cetacea, Cétacés.

ceylonite, ceylonite.

chabasite, chabazite, chabasie, chabasite (zéolite).

Chadronian (US), Chadronien (= Sannoisien).

chain, **1.** chaîne montagneuse; **2.** chaîne (*mécan.*); **3.** chaîne (*chimie*); **4.** unité de longueur = 20,13 m; **c. conveyor,** transporteur à c.; **c. coral,** Halysitidé (*pal.*); **c. structure silicate,** silicate en c. ou inosilicate; **island c.,** c. insulaire; **mountain c.,** c. insulaire.

chaining, arpentage, chaînage.

chalcanthite, chalcanthite (*minér.*).

chalcedonious, calcédonieux.

chalcedony, calcédoine.

chalco (prefix), cuivre.

chalcocite, chalcosine, chalcocite.

chalcolite, chalcolite, torbernite (*minér.*).

chalcophile, chalcophile (élément ayant des affinités avec le soufre).

chalcopyrite, chalcopyrite.

chalcosiderite, chalcosidérite.

chalcosine, chalcosine.

chalcosphere, chalcosphère (manteau supérieur).

chalcostibite, chalcostibine, chalcostibite.

chalcotrichite, chalcotrichite.

chalk, craie; **c. crust soil,** sol à croûte calcaire; **c. deficient soil,** sol non carbonaté; **c. humus soil,** sol carbonaté humique; **c. marl,** c. glauconieuse; **c. moder,** humus intermédiaire de sol; **c. period,** période crétacée; **c. pit,** crayère, carrière de c.

chalky, crayeux; **c. soil,** sol crayeux (cf. rendzine).

chalybeate, ferrugineux; **c. spring,** source minérale.

chalybite, chalybite, sidérite, sidérose, giobertite ($FeCO_3$).

chamber, **1.** chambre; **2.** alvéole, espace intercloisons, loge (*pal.*); **3.** taille, trou de mine; **magmatic c.,** chambre magmatique; **volcanic c.,** réservoir volcanique.

chambered, cloisonné, à plusieurs loges.

chamberlet, petite loge.

chamosite, chamosite, chamoïsite (var. de chlorite *minér.*).

champion lode, filon principal (Cornouaille).

Champlainian (US), Champlainien (= Ordovicien moyen sup.).

change of level, variation du niveau marin.

channel, **1.** chenal, passe; **2.** détroit; **3.** canal; **4.** canal siphonal (*pal.*); **5.** filon de roche; **c. bed,** couche de gravier; **c. capacity,** débit maximum d'un chenal; **c. deposit,** gîte linéaire; **c. fill deposit,** alluvions sédimentées dans le chenal; **c. flow,** écoulement canalisé dans le lit fluviatile; **c. lag deposit,** dépôt grossier résiduel du chenal fluviatile; **c. sands,** grès fluviatile (souvent minéralisé); **c. storage,** débit d'un chenal; **c. wave,** onde guidée; **c. width,** largeur maximum d'un chenal; **dispersal c.,** bras effluent; **interlacing c.,** réseau de rigoles; **intertwining c.,** lacis de rigoles; **mean water c.,** lit mineur; **subaqueous c.,** ravin sous-aquatique; **subglacial c.,** chenal sous-glaciaire; **sublacustrine c.,** ravin sous-lacustre; **submarine c.,** chenal sous-marin; **tangled c.,** chenaux anastomosés; **tidal c.,** chenal de marée.

channel (to), raviner, canneler.

channeling, **1.** renardage; **2.** existence de chenaux, ravinement; **3.** cannelure; **c. out,** évidage.

channelization, canalisation (d'un cours d'eau).

chaos, chaos; **c. structure,** écailles de chevauchement.

chaotic, chaotique (terrain).

chap, crevasse.

chap (to), se crevasser.

char, résidu carboné de combustion incomplète; **wood c.,** charbon de bois.

char (to), **1.** carboniser; **2.** charbonner.

Characeae, Characées (*paléobot.*); **C. chalk,** craie lacustre.

characteristic radiation, spectre électromagnétique d'un élément chimique.

charcoal, charbon de bois, charbon impur; **c. black test,** essai au chalumeau.

charge, charge de mine (*mine*), charge explosive.

charging, enfournement (*mine*), chargement, facturation.

chark, charbon de bois.

chark (to), carboniser.

Charmouthian, Charmouthien.

charnockite, charnockite (*pétrogr.*).

Charophyta, Charophytes (*paléobot.*).

charred, carbonisé.

charring, carbonisation.

chart, **1.** abaque, graphique, diagramme; **2.** carte; **3.** schéma arborescent d'évolution; **nautical c.,** carte nautique; **rain c.,** carte pluviométrique.

chase, rainure.

chasm, abîme, gouffre.

chassignite, chassignite (var. d'achondrite).

chattemark, coup de gouge (glaciaire).

chatter (to), s'entrechoquer, cogner, vibrer.

Chattian, Chattien.

check, contrôle; **c. basin,** bassin d'irrigation; **c. dam,** barrage submersible; **c. flooding,** irrigation par petits bassins; **c. irrigation,** irrigation régularisée par petits bassins; **c. sample,** échantillon de c.

cheek, joue (*pal.*); **c. of a lode,** parois d'un filon; **c. of a Trilobite,** j. d'un Trilobite (*pal.*).

cheire, cheire.

chelation, désagrégation, décomposition (de roches).

Chelicerates, Chélicérates (*pal.*).

Chelonia, Chéloniens (*pal.*).

chemical, chimique; **c. adsorption,** adsorption c.; **c. analyse,** analyse c.; **c. balance,** balance de laboratoire; **c. bounding,** liaison c.; **c. change,** altération c.; **c. compound,** composé c.; **c. controls,** contrôle géochimique; **c. deposition,** sédimentation c.; **c. equilibrium,** équilibre c.; **c. erosion,** lessivage c.; **c. limestone,** calcaire de précipitation c.; **c. precipitate,** précipité c.; **c. property,** propriété c.; **c. reduction,** réduction c.; **c. remanent magnetization,** aimantation rémanente; **c. valency,** valence c.; **c. weathering,** altération c.

chemically, chimiquement.

chemicomineralogical, chimico-minéralogique.

chemicophysical, chimico-physique.

chemiostratigraphy, chimiostratigraphie.

chemist, chimiste.

chemistry, chimie.

chemochimic origin, d'origine chimique.

Chemungian (US), Chemungien (= Dévonien sup. moyen).

chenier (US), cordon littoral sableux, chênier.

chernozem, chernozem; **c. like alluvial soil,** sol alluvial humifère.

cherry coal, charbon bitumineux.

chert, chert, chaille, phtanite, roche siliceuse (sauf grès, et silex); **c. limestone,** calcaire siliceux.

chertification, silicification.

cherty, siliceux; **c. loam,** limon caillouteux.

chesnut soil, sol châtain.

chessy copper, chessylite, azurite.

chessylite, azurite, chessylite.

Chesterian (US), Chestérien (= Mississipien sup.).

chevron fold, pli en accordéon.

chiastolite, **1.** chiastolite (var. d'andalousite); **2.** mâcle (caractéristique de l'andalousite); **c. slate,** schiste à andalousite.

Chile salpeter, nitrate de soude, salpêtre du Chili.

chill (to), **1.** refroidir, figer (lave); **2.** tremper (métall.).

chilled cast-iron, fonte trempée; **c. contact,** zone de contact à grain fin (roches éruptives); **c. effect,** refroidissement terrestre dû à l'écran atmosphérique; **c. margin,** bordure figée.

chiller, cristallisoir.
chilling, 1. refroidissement; 2. trempe.
chimney, 1. cheminée; 2. fendue; 3. colonne de minerais.
chimney rock, pilier d'érosion, pyramide d'érosion, cheminée de fées, bloc perché.
China clay, China stone, terre à porcelaine, kaolin.
chine, ravin.
chink, crevasse, lézarde, fente, fissure.
chink (to), se fissurer, se lézarder.
chiolite, chiolite (variété de cryolite).
chip, fragment, éclat, écaille, copeau.
chip (to), tailler, buriner.
chippage, fragmentation.
chipped stone age, âge de la pierre taillée, Paléolithique.
chipper, 1. burin; 2. mineur, burineur.
chipping, 1. burinage; 2. gravillonage; **c. chisel,** burin.
chippings, caillasses.
chisel, burin, ciseau, fleuret, trépan; **c. bit,** trépan tranchant.
chiselling, 1. adj : ciselant; 2. n : ciselure, burinage.
chitin, chitine.
chitinous, chitineux.
Chitinozoa, Chitinozoaires.
chloantite, chloantite, smaltite (*minér.*).
chlorargyrite, variété de cérargyrite (*minér.*).
chlorhydric, chlorhydrique.
chloride, chlorure.
chlorinate (to), chlorurer.
chlorinated, chloré; **c. hydrocarbon,** hydrocarbure c. (pesticide).
chlorination, chloration.
chlorine, chlore.
chlorinity, teneur en chlore.
chlorite, chlorite; **c. gneiss,** gneiss chloriteux; **c. schist,** chlorito-schiste, schiste chloriteux; **c. slate,** chlorito-schiste.
chloritic, chloriteux; **c. sand,** sable vert; **c. schist,** chloritoschiste.
chloritization, chloritisation.
chloritoid, chloritoïde.
chloritous, chloriteux.
chloromelanite, chloromélanite (var. de jadéite).
chlorophane, chlorophane.
chlorophyll, chlorophyle.
chlorophyllite, chlorophyllite.
Chlorophyta, Chlorophytes.
chlorospinel, chlorospinelle.
chlorous, chloreux.
Chlorozoan assemblage, association biologique « algues vertes et coraux ».
choke (to), 1. engorger, obstruer; 2. s'engorger, s'obstruer.
Chondrichthyes, Chondrichtyens.
chondrite, chondrite (var. de météorite).
chondrodite, chondrodite.
chondrules, chondrules (nodules d'olivine dans une météorite).
chonolith, chonolite (var. d'intrusion ignée).

chop, clapotis.
chopper, couperet, hachoir (*préhist.*).
chopping, 1. coupe; 2. clapotis; **c. bit,** trépan tranchant.
choppy sea, mer dure.
Chordata, Chordés (*biol.*).
C. horizon, horizon C. (*pédol.*).
chorology, chorologie (*pal.*).
christianite, christianite.
chromatic aberration, aberration chromatique.
chromatography, chromatographie.
chrome, chrome; **c. garnet,** ouvarovite, ou grenat chromifère; **c. iron ore,** fer chromé, chromite; **c. spinel,** spinelle chromifère; **c. steel,** acier au c.
chromiferous, chromifère.
chromite, chromite.
chromium, chrome.
chromopicotite, chromopicotite.
chromosphere, chromosphère.
chronologic unit, unité chronolithologique.
chronostratigraphic scale, échelle chronostratigraphique.
chronostratigraphic unit, unité chronostratigraphique.
chronostratigraphy, chronostratigraphie.
chronotaxial rock unit, formation synchrone.
chronotaxic, chronotaxique (de même âge).
chronotaxis, similitude d'âge.
chronozone, chronozone.
chrysoberyl, chrysobéryl.
chrysocolla, chrysocolle (*minér.*).
chrysolite, 1. chrysolite; 2. olivine, péridot.
chrysolitic, chrysolitique.
chrysoprase, chrysoprase (var. de calcédoine verte).
chrysotile, chrysotile (*minér.*).
churn drilling, sondage percutant, sondage par battage; **c. hole,** marmite de géant.
chute, 1. cheminée (*mine*); 2. couloir, galerie (*mine*); 3. trémie; 4. chute (US), chute d'eau; 5. détroit; 6. masse de minerai allongée (filon).
chute cutoff, chute.
Ciliata, Ciliés (*pal.*).
Cimmerian Orogeny, Orogenèse cimérienne.
cimolite, cimolite (var. de kaolin) (*minér.*).
cinabar, cinabre (*minér.*).
Cincinnatian (US), Cincinnatien (= Ordovicien sup.).
cinder, 1. cendre volcanique, scorie volcanique; 2. laitier (*métallo,*); **c. coal,** 1. charbon altéré par des laves (G.B.); 2. coke naturel de mauvaise qualité (Austr.); **c. cone,** cône de scories; **volcanic c.,** scorie volcanique.
cinereous, cinéritique.
cinerite, cinérite.
cingle, boucle de méandre.
cinnabar, cinabre (*minér.*).
cinnabaric, cinnabrifère.
cipolin (rare), cipolin.
circ (rare), cirque.
circalittoral, circalittoral.

circular section, section circulaire (de l'ellipsoïde des indices); **c. polarization,** polarisation isotrope.

circulating-head, tête de circulation (*for.*).

circulating-water, eaux courantes.

circulation of the mud, injection des boues de forage.

circulation-shaft, puits de circulation du personnel (*mine*).

circumferential wave, onde sismique parallèle à la surface terrestre.

circumlunar, circumlunaire.

circumpacific belt, zone, ceinture péripacifique.

circumpolar, circumpolaire.

circumterrestrial, circumterrestre.

cirque, cirque (glaciaire).

Cirripedia, Cirripèdes.

cirrocumulus cloud, cirrocumulus.

cirrostratus cloud, cirrostratus.

cirrus cloud, cirrus.

citrine, citrine; **c. quartz,** citrine.

civil engineering, génie civil.

clade, phylum (*pal.*).

cladding, revêtement métallique.

Cladocera, Cladocères.

cladogenesis, cladogenèse.

cladogram, arbre évolutif (*pal.*).

claim, concession minière; **c. holder,** concessionnaire, détenteur de concession; **c. licence,** titre de concession.

clam, 1. coquille; 2. coquillage; 3. mollusque lamellibranche.

clamber (to), grimper, gravir.

clamp (to), fixer, attacher, bloquer.

clapotis, vagues stationnaires.

clarain, clarain (*minér.*).

Clarendonian (US), Clarendonien (= Sarmatien).

clark concentration, indice de concentration d'un minerai (par rapport à la teneur moyenne).

clarification, clarification, épuration des eaux.

clarite, clarite = énargite (*minér.*).

class, 1. classe (*pal.*); 2. catégorie.

clast, constituant, fragment de roche détritique; **c. rich breccia,** brèche riche en fragments.

clastation, désagrégation des roches et formation de sédiments détritiques.

clastic, détritique, clastique; **c. dike,** intrusion clastique sédimentaire (ex. : moulage d'une fente de gel).

clastics, roches détritiques.

clastogene, brèche, conglomérat.

clay, argile; **c. auger,** tarière à glaise; **c. band,** intercalation d'a.; **c. bed,** couche argileuse; **c. course,** salbande; **c. iron ore,** minerai de fer argileux; **c. ironstone,** minerai de fer argileux; **c. marl,** marne argileuse; **c. mineral,** minéral argileux; **c. particle,** particule argileuse; **c. parting,** intercalation d'a.; **c. pan,** croûte argileuse dans le sol; **c. pit,** glaisière, carrière d'a.; **c. rock,** pélite; **c. schist,** schiste argileux; **c. slate,** schiste ardoisier argileux; **c. vein,** fissure argileuse dans charbon;

c. with flints, a. à silex; **bedded c.,** a. litée; **boulder c.,** a. à blocaux (désuet), cf. till; **fire c.,** a. réfractaire; **iron c.,** a. ferrugineuse; **laminated c.,** a. feuilletée; **marl c.,** a. marneuse; **mixed-layer c.,** interstratifié (*minér.*); **peat c.,** vase de marais; **potter's c.,** a. plastique; **residual c.,** a. résiduelle d'altération; **sandy c.,** a. sableuse; **scaly c.,** a. écailleuse; **silty c.,** a. limoneuse; **till c.,** a. téguline.

clay (to), glaiser.

claying, glaisage.

clayed podsol, podzol gleyiforme; **c. soil,** sol argileux.

C layer, partie inférieure du manteau terrestre (entre − 370 et − 720 km).

clayey, argileux, glaiseux.

clayiness, teneur en argile.

clayish, glaiseux, argileux.

claystone, 1. argilite, argile indurée; 2. arène feldspathique (désuet).

cleaner, curette (*mine*).

cleap, fissure transversale (dans une couche de charbon).

clear, limpide, clair; **c. crystal,** cristal l.; **c. image,** image nette; **c. of water,** débarrassé d'eau; **c. water,** eau c., l.

clear away (to), déblayer, dégager, désobstruer.

clearer, 1. mineur, piqueur (*mine*); 2. curette.

clearing, 1. déblaiement, enlèvement des débris; 2. défrichement; **c. fails,** déblaiement des éboulements.

clearness, limpidité.

clear out (to) the working face, déblayer le front de taille.

cleat, diaclases, fissures de couches de charbon.

cleavable, clivable.

cleavage, 1. schistosité; 2. clivage; **c. plane,** plan de c.; **axial plane c.,** c. ardoisier axial; **flow c.,** schistosité de pression; **fracture c.,** c. de fracture; **mineral c.,** c. minéralogique; **slip c.,** microfaille; **shear c.,** c. de fracture; **staty c.,** c. ardoisier; **strain-slip c.,** c. par pli-fracture.

cleave (to), 1. cliver, fendre, refendre; 2. se cliver.

cleavelandite, cleavelandite.

cleavibility, clivabilité, fissurabilité.

cleaving, clivage.

cleft, fente, fissure, crevasse.

Clerici solution, liqueur de Clérici.

cleve, falaise (GB).

cliachite, alumine hydratée colloïdale (dans bauxite).

cliff, falaise, escarpement; **c. face,** front de f.; **c. glacier,** glacier de cirque; **c. sapping,** sapement de f.; **c. stoping,** éboulement de f.; **abandoned c.,** f. abandonnée; **ancient c.,** f. morte; **cross c.,** verrou glaciaire; **plunging c.,** f. plongeante; **sea c.,** f. marine; **shore c.,** f. littorale.

cliffed, escarpé; **c. shore,** rivage à falaises; **c. valley,** vallée glaciaire.

climate, climat; **continental c.,** c. continental; **oceanic c.,** c. océanique.

climatic, climatique; **c. change,** variation c.; **c. chart,** carte c.; **c. classification,** classification c.; **c. cycle,** cycle c.; **c. eruption,** maximum d'une phase

éruptive; **c. factor,** facteur c.; **c. geomorphology,** géomorphologique c.; **c. optimum,** optimum c.; **c. province,** province c.; **c. zone,** zone c.
climatogenic, climatique.
climatologic, climatologique.
climatology, climatologie.
climatostratigraphy, climatostratigraphie.
climb, **1.** montée, côté; **2.** ascension.
climb (to), gravir, grimper, monter.
climbing bog (raised bog), tourbière bombée.
climbing dune, dune mouvante.
climosequence, séquence climatique, climoséquence.
clinging, collant.
clink, scorie.
clinker, clinker, laitier, mâchefer, scorie; **c. field,** cheire; **volcanic c.,** lave scoriacée.
clinkstone, phonolite.
clinoamphibole, amphibole monoclinique.
clinochlore, clinochlore (var. de chlorite).
clinoclase, clinoclase.
clinodome, clinodôme.
clinoform, talus subaquatique; en forme de talus.
clinoenstatite, enstatite monoclinique.
clinograph, clinomètre.
clinometer, clinomètre.
clinometric, clinometrical, clinométrique.
clinopyroxene, pyroxène monoclinique.
clinorhombic, monoclinique; **c. system,** système m.
clino-unconformity, discordance angulaire.
clinothem, sédiments déposés sur talus continental et pentes sous-aquatiques.
clinozoïzite, zoïsite monoclinique (var. d'épidote).
clintonite, clintonite (mica).
clints, lapiés (crêtes de).
clip, **1.** pince (pour tubes); **2.** valet (pour platine de microscope).
clod, **1.** motte; **2.** paquet (de laves); **3.** schiste faiblement consolidé; **c. structure,** microstructure en grumeux; **peat c.,** motte de tourbe.
cloddy, motteux (qui se casse en mottes).
clogged channel, bras engorgé, envasé.
close (to) down a mine, fermer une mine.
close, fermeture, clôture; **c. fault,** faille fermée; **c. fold,** pli fermé; **c. grained,** à grain fin, finement cristallisé; **c. jointed,** fortement fissuré, diaclasé; **c. sand,** sable fin peu perméable; **c. timbering,** boisage jointif.
closed, fermé; **c. anticline,** anticlinal f.; **c. basin,** dépression f.; **c. depression,** dépression f.; **c. fold,** pli f.; **c. packing,** fort tassement (des sédiments); **c. stope,** taille remblayée; **c. structure, 1.** structure compacte; **2.** structure f.; **c. work,** exploitation souterraine.
closure, fermeture; **structural c.,** f. structurale; **synclinal c.,** f. d'un synclinal.
clot, **1.** grumeau; **2.** bombe volcanique.
clot (to), floculer, coaguler.
clotted, grumeleux.
clotty, grumeleux.

cloud, nuage, nuée; **c. pattern,** type de n., réseau nuageux; **glowing c.,** nuée ardente.
cloudiness, **1.** turbidité; **2.** caractère nuageux.
cloudy, **1.** nuageux; **2.** trouble (liquide).
clough, ravin, gorge.
clue, indice.
clump, morceau, bloc, masse.
clunch, **1.** argile schisteuse; **2.** bande argileuse dans une couche de houille.
cluse, cluse.
cluster, groupe, amas; **c. analysis,** analyse d'ensemble; **c. of galaxies,** a. de galaxies; **c. of geophones,** « grappe » de géophones.
Clymenida, Clyménidés (*pal.*).
coagulate (to), **1.** figer, coaguler; **2.** se figer.
coal, charbon, houille; **c. ball,** concrétion de débris végétaux dans c.; **c. basin,** bassin houiller; **c. bearing,** houiller, carbonifère; **c. bed,** couche de houille; **c. belt,** sillon houiller; **c. brass,** pyrite; **c. clay,** argile réfractaire; **c. cutter,** haveuse (machine); **c. cutting,** abatage, havage; **c. deposit,** gisement de houille; **c. deposit map,** carte houillère; **c. drift,** fendue; **c. dust,** poussier de houille; **c. face,** front de taille; **c. field,** bassin houiller; **c. formation,** formation houillère; **c. gas,** gaz de houille; **c. horizon,** horizon carbonifère; **c. measures (GB),** carbonifère supérieur; **c. measures,** couches productrices; **c. mine,** houillère; **c. miner,** mineur; **c. mining,** exploitation de charbon; **c. oil,** naphte minéral, pétrole; **c. pipe,** souche d'arbre fossile; **c. plants,** végétaux carbonifères; **c. rank,** classe, catégorie de charbon; **c. seam,** filon houiller; **c. seat,** argile réfractaire; **c. tar,** goudron de houille; **c. types,** catégories de charbons; **c. wall,** front de taille du charbon; **c. washing,** lavage du charbon; **c. winning,** abatage; **anthracite c.,** anthracite; **bituminous c.,** houille grasse; **brown c.,** lignite; **cherry c.,** houille grasse; **flame c.,** charbon flambant; **soft c.,** houille grasse; **steam c.,** charbon demi-gras; **stone c.,** anthracite; **subbituminous c.,** houille maigre.
coalification, houillification.
coaling, charbonnage.
coaly, charbonneux.
coarse, gros, grossier (par la granulométrie); **c. crusher,** broyeur des gros; **c. grain,** à gros grain; **c. grained,** à gros grain; **c. grained fraction,** fraction grossière (en granulométrie); **c. ore,** partie grossière du minerai; **c. sand,** sable grossier; **c. silt,** limon grossier; **c. texture,** granulométrie grossière; **c. topography,** topographie irrégulière; **c. waste,** matériel détritique grossier.
coarseness, grossièreté (d'un sédiment).
coast, côte, rivage, littoral; **c. line,** ligne de c.; **c. of emergence,** r. d'émersion; **c. of submersion,** c. de submersion; **c. shell,** plateau continental; **accretion c.,** c. d'accumulation; **bold c.,** c. élevée; **constructional c.,** c. construite; **depressed c.,** affaissée; **discordant c.,** c. discordante; **embayed c.,** c. découpée; **fault c.,** c. de faille; **fjord c.,** c. à fjord; **flat c.,** c. plate; **glaciated c.,** c. à modelé

glaciaire; **high c.,** c. élevée; **lagoon c.,** c. à lagunes; **longitudinal c.,** c. longitudinale; **raised c.,** c. soulevée; **revealed c.,** l. fossile; **steep c.,** c. abrupte; **tectonic c.,** c. tectonique; **transverse c.,** c. transversale; **volcano c.,** c. volcanique.

coast (to), suivre la côte.

coastal, côtier; **c. bay,** estuaire; **c. comb,** glace de haut estran; **c. current,** courant c.; **c. deposits,** dépôts c. littoraux; **c. dune,** dune littorale; **c. erosion,** érosion littorale; **c. grading,** régularisation du littoral; **c. line,** littoral; **c. marsh,** marais maritime; **c. plain,** plaine littorale; **c. waters,** eaux littorales.

coat, revêtement, enduit.

coating, enduit, revêtement.

cob, sheidage *(mine).*

cob (to), cobb (to), scheider.

cobalt, cobalt; **c. bloom,** érythrite, c. arséniaté; **c. glance,** cobaltite, c. gris; **c. melanterite,** biébérite; **c. pyrite,** linnéite, marcasite (désuet); **c. vitriol,** biébérite; **gray c.,** c. arsenical, smaltite.

cobaltiferous, cobaltifère.

cobaltine, cobaltite, cobaltite.

cobbed ore, minerai de scheidage.

cobber, scheideur.

cobbing, scheidage.

cobble, cobblestone, petit bloc, galet, caillou (64 mm ⩽ L ⩽ 256 mm).

cobbles, 1. galets; 2. gaillette *(mine).*

cobbly, pierreux; **c. soil,** sol pierreux.

Coblentzian, Coblencien (Dévonien).

coccolith, coccolite.

Coccolithophoridae, Coccolithophoridés *(pal.).*

cockade ore, minerai en cocarde; **c. structure,** structure en cocarde.

cockpit, doline; **c. country,** paysage karstique.

cockscomb pyrite, pyrite crêtée, marcasite.

codeclination, codéclinaison.

Codiaceans, Codiacées (algues).

coefficient, coefficient, module; **c. of elasticity,** c. d'élasticité; **c. of expansion,** c. de dilatation; **c. of run-off,** c. d'écoulement; **permeability c.,** c. de perméabilité; **viscosity c.,** c. de viscosité.

Coelenterata, Coelentérés.

coelestine, célestine.

coelome, coelome *(pal.).*

coesite, coésite.

coffer, coffering, coffrage de puits.

coffer (to), coffrer *(mine).*

cognate inclusion, enclave syngénétique.

cohade, pendage.

coherence, cohésion (d'un sédiment).

cohesion less soil, sol non cohérent, qui a une certaine consistance.

coiled shell, coquille enroulée, coquille spiralée.

coke, coke; **c. coal,** c. naturel; **c. deposit,** résidu de c.; **c. dust,** poussier de c.; **c. iron,** fonte au c.; **c. oven,** four à c.; **c. pig iron,** fonte au c.; **c. tar,** goudron de c.; **native c.,** c. naturel.

coke (to), cokéfier, transformer en coke.

coking, 1. adj : cokéfiant; 2. n : cokéfaction; **c. coal,** charbon à coke; **c. plant,** cokerie.

col, 1. col *(géogr.);* 2. col *(météor.).*

colatitude, colatitude.

cold, 1. adj : froid; 2. n : froid; **c. blast,** vent froid, soufflage d'air froid; **c. desert,** désert froid; **c. front,** front froid; **c. glacier,** glacier polaire; **c. snap,** coup de froid; **c. wave,** vague de froid.

Coleoidea, Coléidés *(pal.).*

colemanite, colemanite.

collapse, affaissement, effondrement; **c. breccia,** brèche d'e.; **c. caldera,** caldeira d'e.; **c. sink,** e., doline, puisard; **c. structure,** déformation par glissement; **roof c.,** a. du toit.

collapse (to), s'affaisser, s'ébouler, s'effondrer.

collapsing, éboulement, effondrement.

collect (to), 1. collectionner, rassembler, recueillir, réunir; 2. se rassembler, etc.

collect (to) water, capter les eaux.

collecting minerals, collecte de minéraux; **c. stream,** collecteur fluvial; **c. pit,** puisard.

collection of water, captage d'eau.

collective bed transport, transport en masse de matériaux fluviatiles du fond du lit.

collector, 1. collectionneur; 2. collecteur (d'eau).

colliding of plates, collision de plaques lithosphériques.

collier, 1. mineur de charbon; 2. bateau charbonnier.

colliery, mine de houille, charbonnage.

collimate (to), collimater, viser *(astro.).*

collimation line, axe de visée; **c. axis,** axe de visée.

collimator, collimateur.

collision (of plates), collision de plaques.

colloid, colloïde; **c. clay,** argile colloïdale.

colloidal, colloïdal; **c. clay,** argile c.; **c. complex (Can.),** complexe absorbant; **c. dispersion,** dispersion c.

collophane, collophanite, collophane (phosphate cryptocristallin).

colluvial, colluvial; **c. deposits,** colluvions; **c. soil,** sol colluvial.

colluvium, colluvions.

colonial coral, corail colonial.

colony, colonie (d'organismes).

color, colour, couleur; **c. index,** 1. indice des c.; 2. pourcentage de minéraux foncés d'une roche; **c. ratio,** pourcentage de minéraux mélanocrates; **c. scale,** échelle des c; **c. tracer,** traceur coloré; **rock c. chart,** code des c.

colorimetre, colorimètre.

columbite, columbite.

columella, columelle.

columellar, columellaire; **c. fold,** repli columellaire; **c. lip,** rebord de la columelle.

column, colonne, pilier; **c. like structure,** structure prismatique, structure polyédrique (d'une coulée de basalte); **c. rock,** roche champignon; **c. basalt** prisme basaltique, orgue basaltique; **erosion c.,** pyramide de fée; **fractionating c.,** colonne de fractionnement; **ore c.,** filon vertical minéralisé; **stratigraphic c.,** log stratigraphique.

columnals, Encrines (*pal.*).

columnar, prismatique, en forme de colonne; **c. basalt,** orgue basaltique; **c. jointing,** prismation (basaltique), fissuration prismatique; **c. section,** profil stratigraphique, log stratigraphique; **c. structure, 1.** structure prismatique; **2.** structure polyédrique.

coma, couche gazeuse autour d'une comète.

comagmatic, comagmatique.

Comanchean (US), Comanchéen (= Crétacé inférieur et moyen).

comb, 1. crête d'une colline; **2.** US : espace libre entre deux éponts minéralisées presque jointives d'un filon; **c. rock,** gélifract, matériel de gélivation quaternaire soliflué (GB).

combed vein, filon à éponts fortement minéralisées et presque jointives.

comber, vague déferlante.

combining number, valence atomique.

combustibility, combustibilité.

combustible, combustible; **c. shale,** schiste c., tasmanite.

combustion, combustion.

come water, venue d'eau (*mine*).

comet, comète.

commensalism, commensalisme (*zool.*).

commercial deposit, gisement rentable.

commingle (to), mélanger.

comminute (to), broyer finement, pulvériser.

comminution, pulvérisation, microdésintégration.

commissure, commissure (*pal.*).

common, commun, normal; **c. depht-point method,** couverture multiple (*sism.*); **c. garnet,** grenat c.; **c. lead,** plomb; **c. mica,** muscovite; **c. salt,** sel gemme.

community, communauté (*pal.*).

compact (to), tasser.

compactibility, compactibilité.

compactible, susceptible de tassement.

compaction, tassement, compaction, compactage.

compactness, compacité.

compartment, compartiment (*tecto.*).

compass, 1. boussole; **2.** compas, lecture de la b., relevé à la b.; **c. dial,** b. à cadran; **c. needle,** aiguille de la b.; **c. survey,** levé à la b.; **pocket c.,** b. de poche; **solar c.,** théodolite solaire (isostatique).

compensation, compensation; **c. depht,** profondeur de compensation isostatique; **c. level,** niveau de compensation; **c. point,** point de compensation (*pétro.*); **c. surface,** niveau de compensation isostatique.

competence, competency, compétence (fluviatile).

competent bed, couche compétente.

compile (to), rassembler (des données).

compiled map, carte de compilation, à partir de l'assemblage de données.

complementary dykes, filons intrusifs de natures pétrographiques complémentaires.

completion, achèvement, complètement, conditionnement d'un puits de pétrole.

complex fault, faille composée.

complexing, complexation (*chimie*).

complexometry, complexométrie.

component, 1. constituant, composant; **2.** composante (d'une force).

composite, composite, composé; **c. anticline,** anticlinorium; **c. coast,** côte composite; **c. cone,** stratovolcan; **c. dike,** filon intrusif composite; **c. fault,** faille composée; **c. scarp,** escarpement tectonique complexe (érosion et faille); **c. fold,** pli composé; **c. gneiss,** gneiss d'injection, migmatite; **c. log,** colonne stratigraphique d'un forage (géologique et géophysique); **c. sample,** échantillon composé; **c. sill,** filon couche composé; **c. syncline,** synclinorium; **c. topography,** topographie composite; **c. vein,** filon composite; **c. volcano,** strato-volcan.

composition, composition; **c. face,** plan d'accolement (mâcle); **c. formula,** formule brute; **c. plane,** plan de mâcle; **c. triangle,** diagramme triangulaire.

compositional layering, foliation d'une roche métamorphique (d'un gneiss rubané).

compound, 1. adj : composé; **2.** n : corps composé; **c. coral,** squelette de corail colonial; **c. crystal,** cristal mâclé; **c. eye,** œil composé (*pal.*); **c. meander,** méandre composé; **c. spit,** flèche littorale composite; **c. twin,** mâcle complexe; **c. valley,** vallée composée de deux parties à évolution différente; **c. vein,** filon composé; **c. volcano,** volcan composé, à plusieurs cônes.

compress (to), comprimer.

compressibility, compressibilité.

compressible, compressible, comprimable.

compression, compression, écrasement; **c. fault,** faille de compression; **c. joint,** piézoclase (diaclase de compression); **c. tectonics,** tectonique tangentielle; **c. test,** essai d'écrasement; **c. wave,** ondes de compression; **adiabatic c.,** compression adiabatique; **axial c.,** compression axiale.

compressional fault, faille de compression; **c. margin,** marge active; **c. wave,** onde de compression, onde longitudinale, onde P.

compressive strain, déformation causée par la compression; **c. strength,** résistance à l'écrasement; **c. stress,** effort de compression, contrainte.

computer, ordinateur.

computerized continental drift, dérive des continents reconstituée par ordinateur.

computing machine, calculateur, ordinateur.

concealed fault, faille masquée; **c. podzol,** cryptopodzol.

concentrate, 1. adj : concentré; **2.** n : concentré (de minerai).

concentrate (to), 1. concentrer; **2.** se concentrer.

concentrated acid, acide concentré.

concentrating, concentration, enrichissement des minerais, concentration; **c. plant,** installation de c.

concentric, concentrique; **c. dike,** filon intrusif annulaire; **c. faults,** failles concentriques; **c. folds,**

plis parallèles; **c. fractures,** fractures concentriques; **c. weathering,** désagrégation en boules.
concession, concession.
conch, coquille (Mollusques).
conchiferous, conchitic, coquillier.
conchiolin, conchyoline.
conchoidal, conchoïdal.
Conchostraca, Conchostracés.
concordance, concordance.
concordant, concordant; **c. contact,** contact c.; **c. injection,** intrusion c. interstratifiée; **c. pluton,** massif intrusif c; **c. stratification,** stratification c. (en lits parallèles).
concrete, béton; **c. work,** bétonnage; **reinforced c.,** béton armé.
concrete (to), 1. bétonner; 2. concréfier, se concréfier.
concretion, concrétion; **nodular c.,** c. nodulaire.
concretionary, concrétionné, concrétionnaire.
concussion, secousse, choc; **c. table,** table à s.
condensability, condensabilité.
condensable, condensable.
condensate, produit de condensation.
condensation, condensation.
condense (to), 1. condenser; 2. se condenser.
condensed sequence, série stratigraphique complète, mais d'épaisseur réduite.
condenser, condenseur (*opt.*), condensateur (*électr.*).
condition weather, conditions atmosphériques.
conductibility, conductivity, conductibilité, conductivité; **hydraulic conductivity,** conductivité hydraulique; **thermal conductivity,** conductivité thermique.
conduit, 1. cheminée volcanique, conduit volcanique; 2. conduit aquifère (sous pression hydrostatique).
Condylarthra, Condylarthres.
cone, cône; **c. bit,** trépan à c.; **c. delta,** c. de déjections; **c. in cone structure,** structure intrusive à c. emboîtés; **c. sheet,** complexe annulaire; **adventive c.,** c. adventif; **alluvial c.,** c. alluvial; **avalanche c.,** c. d'avalanche; **cinder c.,** c. de cendres, c. de scories; **composite c.,** stratovolcan; **dribblet c.,** pustule de laves; **ice c.,** c. de glaces; **nested c.,** c. emboîtés; **pyroclastic c.,** c. pyroclastique; **spatter c.,** c. formé de laves projetées.
confined aquifer, nappe captive.
confined ground water, eau artésienne.
confining layer, confining bed, toit imperméable d'une structure souterraine.
confining pressure, pression géostatique, pression hydrostatique.
confining water, eau artésienne.
confluence, confluence; **c. plain,** plaine de c.; **c. step,** gradin de c. glaciaire.
confluent, confluent.
conformability, concordance, conformité.
conformable, concordant, conforme; **c. bedding,** stratification concordante; **c. fault,** faille conforme; **c. relief,** relief conforme; **c. stratification,** stratification concordante; **c. structure,** structure conforme.
conformal map projection, projection cartographique conforme.
conformity, concordance.
congelifluction, congélifluxion.
congelifraction, congélifraction.
congeliturbated, cryoturbé.
congeliturbation, cryoturbation.
congeneric, appartenant au même genre (*pal.*).
conglomerate, 1. adj : congloméré; 2. n : conglomérat; **c. formation,** formation cimentée; **intraformational c.,** conglomérat intraformationnel.
conglomeratic, conglomératique.
congress, congrès.
congruent melting point, point de fusion congruent.
Coniacian, Coniacien.
conical map projection, projection cartographique conique.
Coniferous, Conifères.
conjugate fold, pli synchrone (mais de direction différente).
conjugated fractures, 1. fractures, diaclases, de même direction, mais à pendage opposé; 2. deux ensembles de fractures perpendiculaires.
conjugate joint system, système de deux ensembles de diaclases symétriques; **c. joints,** réseau de diaclases; **c. liquids,** liquides immiscibles, mais en équilibre.
connate water, eau de constitution, eau fossile.
Conodont, Conodonte.
Conrad discontinuity, discontinuité de Conrad.
consanguinity, consanguinité magmatique.
consecutive calderas, caldeiras emboîtées.
consequent stream, cours d'eau conséquent.
consolidation, diagenèse, cimentation et compaction diagénétiques.
conspecific, appartenant à la même espèce (*pal.*).
constancy of winds, indice de constance de direction des vents.
constant, 1. adj : constant; 2. n : constante (*phys.*); **c. of gravitation,** intensité de la pesanteur; **c. slope,** glacis, talus d'éboulis.
constellation, constellation (*astro.*).
constitutional undercooling, surfusion.
constructional fossil, fossile constructeur.
constructive plate margin, marge de plaque lithosphérique en accrétion.
consultant, ingénieur conseil.
consulting geologist, géologue conseil.
contact, 1. contact; 2. surface de contact; **c. aureole,** auréole de c.; **c. bed,** couche adjacente; **c. breccia,** brèche d'intrusion; **c. deposit,** dépôt minéral, filon de c.; **c. lode,** filon de c.; **c. metamorphism,** métamorphisme de c.; **c. metasomatism,** métasomatose de c.; **c. mineral,** minéral métamorphique de c.; **c. twin,** mâcle par accolement; **c. vein,** filon de c.; **c. zone,** auréole de c.
contaminant, polluant.
contaminated rock, roche éruptive contaminée.
contaminate (to), contaminer, polluer.

contamination, 1. pollution; 2. contamination (d'un magma).

contemporaneous fault, faille syngénétique; **c. rocks,** roches de même âge.

content, teneur, titre, proportion; **U content,** teneur en uranium.

contexture, texture.

continent, continent.

continental, continental; **c. accretion,** accrétion c.; **c. apron,** glacis c.; **c. basin,** dépression fermée; **c. block,** bloc c.; **c. climate,** climat c.; **c. crust,** croûte c.; **c. divide,** ligne de partage des eaux; **c. drift,** dérive des continents; **c. glacier,** calotte glaciaire; **c. ice sheet,** calotte glaciaire; **c. island,** île c.; **c. margin,** marge c.; **c. nuclei,** cratons; **c. plate,** plaque lithosphérique; **c. platform,** plate-forme c.; **c. rise,** glacis c.; **c. rock,** roche d'origine c.; **c. shelf,** plate-forme c., plateau c.; **c. shield,** bouclier c.; **c. slope,** talus c.; **c. terrace,** plateau c.

continentality, caractère continental (par opposition à maritime).

continuity, continuité (d'une couche).

continuous, continu; **c. corring,** carottage c.; **c. logging,** enregistrement c.; **c. permafrost zone,** zone à pergélisol c.; **c. profiling,** dispositifs continus (géoph.), profilage c.; **c. seismic profiling,** profilage sismique c.; **c. velocity log,** diagraphie de vitesse; **c. waves,** ondes entretenues.

continuum, fond continu (astro.).

contorted stratum, couche déformée, plissée.

contour, 1. courbe de niveau; 2. contour, profil, tracé; **c. farming,** culture suivant courbes de niveau; **c. interval,** équidistance; **c. line,** courbe de niveau, isobathe; **c. map,** carte en courbes de niveau; **c. strip cropping,** labour suivant courbes de niveau; **c. tillage,** labour suivant courbes de niveau; **structural c.,** courbe structurale; **topographic c.,** courbe topographique.

contour (to), lever les courbes de niveau.

contourite, sédiment marin déposé par courant de contour (de profondeur constante), contourite.

contraction, rétrécissement, retrait, contraction; **c. crack,** fissure de retrait; **c. hypothesis,** hypothèse de la contraction terrestre; **c. vein,** filon minéral occupant une fente de contraction.

contractor, entrepreneur.

contraposed shoreline, côte contraposée.

contribution, 1. note, article; 2. contribution.

contributor, auteur, collaborateur scientifique.

controlled mosaic, mosaïque de photographies aériennes redressées.

controlling of the outflow, contrôle du débit.

control point, point côté, repéré.

Conularida, Conularidés.

convection, convection; **c. cell,** cellule de c.; **c. current,** courant de c.

convectional rain, pluie de convection.

convergence, 1. rapprochement de deux strates; 2. ligne de démarcation entre une eau fluviatile boueuse et une eau lacustre pure; 3. convergence (pal.); **c. zone,** zone de convergence (tecto.).

convergent evolution, convergence, évolution convergente.

conveyer chain, chaîne à godets.

convolute bedding, litage à convolutions.

convolute lamination, lamination contournée, convolutions.

convolute shell, coquille involute (à spires jointives).

convolute spires, spires jointives.

convolution, circonvolution.

convulsion, cataclysme, bouleversement.

cooling, refroidissement; **c. cracks,** fentes de refroidissement, de rétraction thermique.

coomb, combe, cirque.

coomb-rock (GB), roche gélifractée (silex et bouillie crayeuse).

cooperite, coopérite.

coordinates, coordonnées; **astronomical c.,** c. astronomiques.

coordination, coordination, coordinence.

coordination number, nombre de coordination; **octahedral c.,** coordination octahédrique.

coordinence, indice de coordination.

copal, copal (var. de résine).

Copepoda, Copépodes.

Copernican stage, copernicienne (période tardive de formation lunaire).

copper, cuivre; **c. bearing,** cuprifère; **c. glance,** chalcosite; **c. melanterite,** boothite; **c. mica,** chalcophyllite; **c. nickel,** nickéline; **c. ore,** minerai de c.; **c. pyrite,** chalcopyrite; **c. sulfate,** sulfate de c.; **c. uranite,** torbernite, chalcolite; **c. vitriol,** sulfate de c.

copperas, mélantérite.

coppered, copperish, cuivré.

coppery, cuivreux.

coprolith, coprolite.

copropel, vase noire sapropélique.

coquina, calcaire coquillier, lumachelle.

coquinoid, coquillier, lumachellique.

coral, corail (coraux, pl.); **c. bank,** banc corallien; **c. head,** formation corallienne; **c. knoll,** tête de c.; **c. limestone,** calcaire corallien; **c. mud,** vase corallienne; **c. reef,** récif corallien; **c. rock,** roche corallienne; **colonial c.,** c. colonial; **compound c.,** squelette de c. colonial; **fasciculate c.,** c. branchu.

corallian, corallien.

coralliferous, corallifère.

Corallinaceae, Corallinacées (Algues rouges).

coralline, 1. adj : corallien; 2. n : algue calcaire (rouge); **c. oolite,** oolithe corallienne; **c. platform,** plate-forme corallienne.

corallite, squelette coralliaire, individu corallien.

corallum, exosquelette calcaire des coraux.

co-range lines, lignes d'alignement, lignes d'égale amplitude de marée.

Cordaitales, Cordaïtales.

Cordaites, Cordaïtes.

cordierite, cordiérite (cyclosilicate).

cordillera, cordillère.

core, 1. carotte (forage); 2. cœur (d'un pli); 3. noyau de la terre; **c. barrel,** tube carottier; **c. bit,** trépan carottier; **c. cutter,** trépan; **c. drill, 1.** sondage peu profond; **2.** carotteuse, sondeuse; **c. drilling,** carottage mécanique; **c. jet,** jet de plasma (*astro.*); **c. record,** enregistrement d'un carottage; **c. sampling,** collection de carottes; **c. test,** essai de carottage; **c. tube,** tube carottier; **inner c.,** noyau interne, graine; **outer c.,** noyau externe.

core (to), carrotter (forage).

corer, échantillonneur.

corestone (of granite), boule de granite (dans sa matrice d'altération).

coring, carottage.

Coriolis force, force de Coriolis.

cornelian, cornaline.

corneous lead, plomb corné, phosgénite.

corneous silver, cérargyrite.

cornubianitic schist, cornéenne.

cornubianite, schiste tacheté.

corona, **1.** halo (*météo.*); **2.** couronne solaire.

corona structure, structure concentrique, structure orbiculaire.

coronal gas, gaz émis par la couronne solaire.

corona jet (astro.), jet coronal.

cornice, corniche, encorbellement.

cornstones, calcaires gréseux concrétionnés (souvent permo-triasiques).

corrade (to), corroder, éroder.

corrading stream, fleuve érosif.

corrasion, corrasion.

correlation, corrélation (*stratigr.*).

corrie, cirque glaciaire, creux, entonnoir (Écosse); **c. lake,** lac de cirque.

corrode (to), **1.** corroder, ronger; **2.** se corroder.

corrodible, corrosible, oxydable.

corrosion, corrosion, altération; **c. embayment,** golfe de corrosion; **c. rim,** liseré d'altération.

corrosive, corrosif.

corrugate (to), onduler, strier.

corrugated, plissoté, ondulé, ridé, rugueux; **c. soil,** sol mamelonné.

corrugation, ondulation, cannelure; **c. infiltration,** irrigation par infiltration; **c. irrigation,** irrigation par rigoles d'infiltration.

corry, cirque glaciaire.

corsican granite, gabbro orbiculaire.

corsite, corsite (diorite orbiculaire).

cortex, cortex, écorce.

cortice, cortex, partie extérieure.

corundolite, roche corindifère.

corundum, corindon; **c. syenite,** syénite corindifère.

coseismal, coseismic, **1.** adj : cosismal; **2.** n : cosisme.

coset deposits, sédiments à stratification de même direction de courant.

cosmic, cosmique; **c. dust,** particules cosmiques; **c. loop,** boucle c.; **c. nebula,** nébuleuse; **c. radiation,** rayonnement c.

cosmochemistry, cosmochimie.

cosmoid scale, écaille osseuse (Dipneustes).

cosmologic constant, constante cosmologique.

cosmology, cosmologie.

cosmopolita species, espèce cosmopolite, espèce ubiquiste.

costate, avec des côtes (*pal.*).

cotectic, cotectique; **c. boundary line,** ligne c.

cotidal lines, courbes cotidales.

cotunnite, cotunnite.

Cotylosauria, Cotylosauriens.

couloir, ravin, gorge.

coulter counter, compteur de particules (*sédim.*).

counter, **1.** compteur; **2.** filon croiseur; **3.** adj : contre, opposé; **c. current,** contre-courant; **c. flow,** contre-courant; **c. Geiger,** compteur Geiger; **c. level,** galerie costresse; **c. lode,** filon croiseur; **c. scale,** balance Roberval; **c. slope,** contre-pente; **c. vein,** filon croiseur; **scintillation c.,** compteur à scintillation.

countertrade, contralizé.

country rock, roche encaissante.

course, cours (d'un fleuve), direction; **c. of vein,** direction d'un filon; **lower c.,** cours inférieur; **middle c.,** cours moyen; **upper c.,** cours supérieur; **water c.,** cours d'eau.

Couvinian, Couvinien.

covalent bond, liaison covalente.

cove, **1.** anse, petite baie; **2.** cirque d'érosion; **3.** anticlinal érodé.

covelline, covellite, covellite.

cover, couverture, morts-terrains.

cover-glass, lamelle couvre-objets.

covered karst, karst couvert.

coverage, couverture (photographie aérienne).

coversand, sable de couverture, sable superficiel.

crab holey, petite dépression.

crack, fente, fissure, crevasse; **cooling c.,** fissure de refroidissement; **dessication c.,** fissure de dessication.

crack (to), **1.** fendre, fissurer, fêler; **2.** se fendre.

cracked, **1.** fissuré, fendu, craqué, crevassé; **2.** de craquage (*raf.*).

cracking, **1.** cracking; **2.** fendillement, fissuration.

cracky, fissuré.

crag, **1.** rocher escarpé, chicot, verrou glaciaire; **2.** marne sableuse fossilifère marine, falun.

crag and tail, structure due à un verrou glaciaire, avec striage du côté amont, et sédimentation du côté aval.

crag and tail ridge, drumlin.

cragged, rocailleux.

cragginess, anfractuosité.

craggy, anfractueux.

cranch, massif de protection (*mine*).

cranidium, cranidium (*pal.* Tribolites).

cranny, fente, lézarde, niche.

crater, cratère; **c. island,** île-c.; **c. lake,** lac de c.; **c. lip,** bord du c.; **c. rim,** bord du c.; **c. wall,** paroi du c.; **explosion c.,** c. d'explosion; **impact c.,** cratère d'impact (météoritique); **lunar c.,** cratère lunaire; **meteoritic c.,** c. météorique; **mud c.,** c. de boue; **nested c.,** c. emboîté; **ring c.,** c. emboîté; **sand c.,** c. de sable; **volcanic c.,** c. volcanique.

cratered terrain, terrain à cratères (sur la Lune).

crateriform, cratériforme.

cratering, formation de cratères (sur la Lune).

craterlet, **1.** petit cratère; **2.** cratère de boue, de sable.

cratogenic, cratogénique.

craton, craton.

cratonic, cratonique.

crawfoot pattern, réseau en pattes d'oie.

crease, creasing, plissement, pli.

creek, **1.** crique, anse; **2.** ruisseau (US); **tidal c.,** chenal de schorre.

creep, **1.** cheminement, fluage, reptation (des sols), lent glissement; **2.** boursouflement.

creeping, reptation, glissement modéré sur pentes, creeping.

crenate, crénelé.

crenulate coast, côte irrégulière (avec baies et promontoires).

crenulated phyllite, phyllite à crénulations.

crenulation, micropli, ride, plissement microscopique.

crescent, croissant; **c. beach,** c. de plage; **c. dune,** dune en c., barkhane; **c. lake,** lac en c.; **c. shape dune,** barkhane.

crescentic, en forme de croissant.

crescentic gouge, marque d'arrachement (en croissant).

crest, arête, crête, sommet; **c. of an anticline,** charnière anticlinale; **c. line,** ligne de c.; **c. plane,** plan axial (d'un anticlinal); **c. wave,** c. de vague; **oceanic c.,** dorsale océanique.

crest flood, niveau maximum des eaux (lors d'une crue).

crestal, sommital; **c. line,** ligne de crête; **c. plane,** plan axial d'un anticlinal.

crested barite, barytine crêtée.

Cretaceous, Crétacé; **c. period,** période Crétacée.

crevasse, crevasse; **c. channel,** chenal fluviatile provoqué par une fissure d'une berge; **gaping c.,** fissure béante; **lateral c.,** c. latérale; **longitudinal c.,** c. longitudinale; **transverse c.,** c. transversale.

crevasse (to), crevasser.

crevassing, fissuration.

crevice, **1.** fissure, crevasse; **2.** fissure minéralisée; **3.** diaclase.

crevice (to), crevasser, fissurer.

crib, pilier de bois (mine), encoffrement; **c. ring,** cadre de puits (mine).

crib (to), boiser (un puits).

cribbing, boisage d'un puits.

cribble, crible.

cribble (to), cribler.

crinkle (to), froisser, par extension, plissoter.

Crinoid, Crinoidea, Crinoïde (pal.).

crinoidal limestone, calcaire à Crinoïdes.

Crinozoa, Crinozoaires (pal.).

cripple, terrain marécageux, tourbière.

criss-cross, entrecroisé; **c. bedding,** stratification entrecroisée.

cristobalite, cristobalite.

critical, critique; **c. angle,** angle limite d'incidence; **c. point,** point c.; **c. pressure,** pression c.; **c. reflection,** réflection limite; **c. slope,** pente d'équilibre; **c. temperature,** température c.; **c. velocity,** vitesse c.

crocidolite, crocidolite (var. de glaucophane).

Crocodilia, Crocodiliens.

crocoite, crocoïte.

crogball, concrétion.

Croixian (US), Croixien (Cambrien supérieur).

Cromerian (US), Cromérien (interglaciaire Gunz-Mindel).

crop, **1.** récolte; **2.** affleurement.

crop out (to), affleurer.

cropping, affleurement.

cross, **1.** adj : transversal; **2.** n : croix, croisement; **c. bar,** dune transversale; **c. bed,** couche oblique, filon croiseur; **c. bedding,** stratification entrecroisée; **c. cliff,** verrou glaciaire; **c. course lode,** filon croiseur; **c. cut,** travers-blanc, coupe en travers (mine); **c. cutter,** haveuse (mine); **c. cutting,** **1.** travers-blanc, **2.** contact discordant d'une intrusion; **c. dip,** pendage latéral; **c. fault,** faille transversale; **c. folding,** pli transverse; **c. gangway,** galerie transversale; **c. heading,** galerie transversale; **c. lamination,** stratification entrecroisée; **c. profile,** profil transversal de vallée; **c. section,** coupe transversale; **c. shaped twin,** mâcle en croix; **c. strata,** stratification entrecroisée; **c. stratification,** stratification entrecroisée; **c. valley,** cluse; **c. vein,** filon croiseur.

cross cut (to), percer en travers-banc.

cross drive (to), recouper.

crossed, croisé; **c. nicols,** nicols croisés; **c. polars,** polariseurs croisés; **c. twinning,** mâcle quadrillée (microcline).

crosshair, réticule.

crossing, changement des axes d'écoulements fluviatiles.

crossing lode, filon croiseur.

crossite, crossite.

Crossopterigyi, Crossoptérigiens.

crosswork, recoupe (mine).

crowbar, barre de mine, levier, pince.

crown, crête, sommet, bombement.

crucible, creuset.

cruciform twin, mâcle en croix.

crude, brut; **c. mineral oil,** pétrole b.; **c. oil,** pétrole b; **sour c.,** brut corrosif (sulfuré); **topped c.,** brut étêté.

crumb, granule; **c. structure,** structure granuleuse.

crumble (to), **1.** émietter, désagréger; **2.** se déliter, s'effriter.

crumbling, désagrégation, effritement; **c. folding,** plis de couverture.

crumbly, friable; **c. structure,** structure granuleuse.

crumby, grumeleux.

crumpling, plissotement.

crura, crura (pal.).

crush, écrasement; **c. belt,** zone de broyage tectonique; **c. breccia,** brèche de friction; **c. conglome-**

rate, conglomérat de friction, mylonite; **c. movement,** compression, charriage; **c. structure,** structure cataclastique.

crush (to), broyer, écraser, concasser.

crushed, écrasé; **c. ore,** minerai broyé; **c. stone,** cailloutis, ballast; **c. zone,** zone de broyage.

crushing, **1.** broyage, concassage; **2.** écrasement, compression; **c. rolls,** broyeur à cylindres; **c. strength,** résistance à l'écrasement.

crust, croûte, écorce; **c. fracture,** fracture de l'écorce terrestre; **c. soil,** sol à encroûtement; **zoned c.,** croûte zonaire; **earth c.,** c. terrestre; **glass c.,** c. vitreuse.

Crustacea, Crustacés.

crustal, crustal; **c. fold,** pli de couverture; **c. stoping,** digestion par fusion de la croûte *(pétro.);* **c. plate,** plaque lithosphérique.

crustification, formation d'une croûte.

cryergy, cryergie.

cryoaeolian deposit, sédiment éolien déposé sous climat periglaciaire.

cryoclastic, cryoclastique.

cryoclastim, cryoclastisme.

cryoconite, cryoconite.

cryogenic lake, lac de thermokarst.

cryogenic period, période glaciaire.

cryogenics, cryogénie.

cryokarst, cryokarst.

cryolite, cryolite.

cryology, **1.** glaciologie; **2.** étude sur le froid (US).

cryonival, cryonival.

cryopediment, cryopédiment.

cryopedology, cryopédologie.

cryoplanation, cryoplanation, géliplanation.

cryoplanation terrace, terrasse de cryoplanation.

cryosphere, cryosphère.

cryostatic pressure, pression cryostatique.

cryotectonics, cryotectonique.

cryoturbation, cryoturbation.

crypto (prefix), caché, détectable aux rayons X *(pétro.).*

cryptoclastic, cryptodétritique, microdétritique.

cryptocrystalline, cryptocristallin.

cryptodepression, région située en dessous du niveau moyen de la mer.

cryptohalite, cryptohalite.

cryptomerous, microcristallin.

cryptoperthite, cryptoperthite.

cryptovolcanic, cryptovolcanique.

crystal, cristal; **c. axis,** axe cristallin; **c. chemistry,** cristallochimie; **c. class,** classe de symétrie cristalline; **c. defect,** défaut cristallin; **c. form,** forme cristalline; **c. flotation,** flottaison épigmagmatique; **c. fractionation,** différenciation magmatique; **c. growth,** croissance cristalline; **c. habit,** faciès cristallographique; **c. lattice,** réseau cristallin; **c. liquid fractionation,** cristallisation fractionnée; **c. optics,** optique cristalline; **c. rock,** c. de roche, quartz; **c. seeding,** nucléation; **c. settling,** sédimentation des cristaux dans une chambre magmatique; **c. structure,** structure atomique; **c.**

symmetry, symétrie cristalline; **c. system,** système cristallin; **c. tuff,** roche pyroclastique; **c. zoning,** zonation cristalline.

crystalliferous, cristallifère.

crystalline, cristallin; **c. aggregate,** aggrégat c., roche grenue; **c. basement,** socle éruptif; **c. granular,** granitique; **c. limestone,** calcaire c.; **c. rock,** roche c.; **c. schist,** schiste c.

crystallinity, cristallinité.

crystallite, crystallite.

crystallizable, cristallisable.

crystallization, cristallisation; **c. heat,** température de c.; **c. nuclei,** germes de cristaux de glace; **c. system,** système de c.

crystallize (to), cristalliser.

crystallizer, **1.** cristalliser; **2.** cristalliseur.

crystallizing force, force de cristallisation.

crystalloblast, cristalloblaste.

crystalloblastesis, déformation cristalloblastique.

crystalloblastic, cristalloblastique.

crystallogeny, cristallogénie.

crystallographer, cristallographe.

crystallographic, cristallographique; **c. indices,** indices cristallographiques.

crystallography, cristallographie.

crystalloid, cristalloïde.

crystallology, cristallographie.

crystallometry, cristallométrie.

crystallophysics, cristallographie physique.

Ctenodonta, Cténodontes *(pal.).*

cubanite, cubanite.

cube, cube; **c. ore,** pharmacosidérite; **c. spar,** anhydrite.

cubic, cubique; **c. centimeter,** centimètre cube; **c. cleavage,** clivage c.; **c. decimeter,** décimètre cube; **c. foot,** pied cube = 0,028317 mètre cube; **c. inch,** pouce cube = 16, 387 centimètre cube; **c. measurement,** cubage c.; **c. system,** système c.; **c. yard,** yard cube = 0,7645 mètre cube; **c. zeolite,** chabasite, analcime.

cuesta, cuesta, côte; **c. backslope,** revers de cuesta; **c. inface,** front de cuesta.

Cuisian, Cuisien.

culm, **1.** Culm (faciès du Carbonifère); **2.** poussière de charbon.

culmination, **1.** passage au méridien, culmination; **2.** culminaison.

cummingtonite, cummingtonite.

cumulate rocks, cumulates, cumulats, roches éruptives, formées de phénocristaux redéposés ensuite au fond de la chambre magmatique.

cumulative, cumulatif; **c. courbe,** courbe cumulative.

cumulodome, cumulo-dôme.

cumulonimbus cloud, cumulonimbus.

cumulophyric, gloméroporphyrique.

cumulus cloud, cumulus.

cumulus mineral grains, minéraux formés en premier et ayant sédimentés au fond du magma.

cumulo volcano, cumulo-volcan.

cup, **1.** cuvette; **2.** calice *(pal.);* **c. barometer,** baromètre à c.; **c. coral,** coralliaire isolé.

cupola, dôme, coupole (volcanique).
cupellation, coupellation (métallogénie).
cupreous, cuprous, cuivreux.
cupric, cuivrique.
cupriferous, cuprifère.
cuprite, cuprite (CuO_2).
cuprous, cuivreux.
cuprum, cuivre.
Curie point, point de Curie (cf. magnétisme).
curium, curium.
current, courant; **c. bedding,** stratification de c. fluviatile; **c. marks,** marques de c.; **c. ripple,** ride de c.; **back set c.,** c. de retour; **hydraulic c.,** c. marin de décharge; **longshore c.,** c. de dérive littorale; **littoral c.,** c. de dérive littorale; **shore-drift c.,** c. de dérive littorale; **stray c.,** c. vagabond; **wave c.,** c. de vagues.
cursorial animal, animal adapté à la course (*pal.*).
curvature, courbure, inflexion.
cusecs, pied cubique par seconde.
cusp, banc de sable en croissant; **beach cusp,** croissant de plage.
cuspate bar, cordon littoral en V.
cuspate beach, plage à croissants de plage.
cuspate delta, delta lobé.
cuspate foreland, cordon littoral en V.
cut, **1.** coupe, entaille, excavation, saignée, havage; **2.** taille (d'une pierre précieuse); **c. bank,** berge érodée, concave; **c. diamond,** diamant taillé; **c. gem,** pierre précieuse taillée; **c. over,** déboisé; **c. platform,** plate-forme littorale d'abrasion; **c. stone,** pierre de taille; **c. stone quarry,** carrière de pierre de taille; **c. terrace,** terrasse d'érosion.
cut (to), **1.** couper, tailler, trancher; **2.** haver (*mine*); **3.** se tailler, se découper.
cut (to) a lode, recouper un filon.
cut-and-filling, abatage, exploitation par chambre remblayée.
cut-and-fill stoping, abatage, dépilage par chambre remblayée.
cuts and fills, déblais et remblais.
cut and fill process, alternance de creusement et de remblaiement.

cutan, cutane (*pédol.*).
cutoff, **1.** recoupement d'un méandre; **2.** lac de bras mort; **c. grade,** minerai très pauvre, à très faible teneur; **c. lake,** lac de bras mort, lac en croissant.
cutter, **1.** haveuse (*mine*), houilleuse; **2.** diaclase transversale; **3.** tailleur (de pierres); **stonecutter,** tailleur de pierres.
cutting (n.), **1.** débris de forage, déblais, détritus; **2.** tranchée, coupe, taille, entaillage; **3.** havage, sous-cavage (*mine*); **c. across,** percement, rencontre de deux galeries; **c. back,** érosion régressive; **c. bit,** trépan tranchant; **c. gems,** taille des pierres précieuses.
cwm (Wales), cirque glaciaire.
cyanide, cyanure.
cyanite, cyanite, disthène.
Cyanobacteria, Cyanophycées.
Cyanophyta, Cyanophycées.
Cycadales, Cycadales (*pal. vég.*).
cycle, cycle, période; **c. of denudation,** c. d'érosion; **c. of erosion,** c. d'érosion; **c. of sedimentation,** c. sédimentaire; **c. of topographic development,** c. géomorphologique; **geomorphic c.,** c. géomorphologique; **landform c.,** évolution du relief; **marine c.,** c. marin littoral; **physiographic c.,** c. géomorphologique; **river c.,** c. fluvial; **valley c.,** évolution d'une vallée.
cyclic sedimentation, séquence sédimentaire, sédimentation cyclique.
cyclogenesis, formation d'un cyclone.
cyclolysis, déclin d'un cyclone.
cyclonal, cyclonal, cyclonique.
cyclone, cyclone.
cyclosilicate, cyclosilicate.
cyclostrophic wind, vent cyclonique.
cyclothem, cyclothème, rythme sédimentaire.
cylindric, equal area (Lambert's) projection, projection cylindrique de Lambert.
cylindroidal fold, pli cylindrique.
cymophane, cymophane.
cyprine, cyprine.
Cystoidea, Cystoïdés (Echinodermes).

D

dachbank cycle (cf. sohlbank), cycle à banc majeur en tête.

Dacian, Dacien.

dacite, dacite (rhyolite).

dacitic, dacitique.

dahamite, dahamite (*pétro*).

dahlite, dahlite.

daily output, production journalière (d'un puits).

dale, vallon, vallée élargie (Écosse).

Dalradian, Dalradien.

dam, barrage; arch d., b.-voûte; earth d., b. en terre; ice d., embâcle; retention d., b. de retenue; rockfill d., b. en enrochement.

dam (to), barrer, endiguer; d. lake, lac de barrage.

damming, barrage, endigage, endiguement.

damourite, damourite.

damouritization, damouritisation.

damp, 1. adj : humide; 2. n : humidité, mofette (*mine*).

damped, 1. humidifié; 2. amorti (oscillation, onde).

dampening, mouillage, humidification.

damping, 1. humidification; 2. amortissement (d'une oscillation).

dampness, humidité.

damposcope, indicateur de grisou.

dampy, mouillé.

Danian, Danien.

dank, argile sableuse compactée.

danks, schiste houiller.

Darcy, Darcy (unité pratique de perméabilité).

Darcy's law, loi de Darcy.

dark, foncé, sombre; d. forest soil, sol brun forestier lessivé; d. lines, raies s. du spectre; d. matter, matière sombre (*astro*.); d. mineral, minéral foncé; d. position, position d'extinction; d. red silver ore, pyrargyrite; d. ruby silver, pyrargyrite.

Darwinism, Darwinisme.

dashed line, ligne en tireté, en tirets.

Dasycladaceae, Dasycladacées.

data, données; d. acquisition, saisie de d.; d. bank, banque de d.; d. procession, traitement de d.; d. retrieval, restitution de d.

date, âge absolu, à ^{14}C, à U/Th, etc.

dating, datation.

datolite, datholite, datolite, var. de zéolite.

datum, data (pl.), 1. donnée; 2. niveau.

datum elevation, niveau altimétrique de référence; d. horizon, niveau repère; d. level, plan de référence; d. line, niveau de base; d. plane (D.P.), surface de référence (*géophys*.); d. point, point de repère; cartographic vertical d., niveau de base cartographique (niveau marin); geographic d., coordonnées géographiques.

daughter element, élément fils.

davyne, davyne (*minér*.).

day, jour, surface (*mine*); d. coal, couche de charbon la plus proche de la s.; d. colliery, houillère à ciel ouvert; d. drift, galerie débouchant au j.; d. level, houillère à ciel ouvert; d. light mine, mine à ciel ouvert; d. shift, équipe de j.; d. stone, affleurement; d. water, eaux superficielles.

daze (rare), mica.

dead roast (to), griller à mort.

dead, mort, stérile; d. center, point m.; d. cliff, falaise morte, ancienne falaise; d. dune, dune morte, dune fixée; d. glacier, glacier inactif; d. ground, m. terrain; d. ice, glace morte; d. lime, chaux éteinte; d. litter, litière de feuilles mortes; d. lode, filon épuisé; d. oil, huile, pétrole m. (sans gaz); d. quartz, quartz s.; d. rock, roche s.; d. soil, sol s.; d. valley, vallée sèche; d. well, puits s.; d. workings, chantier en m. terrain.

deads, roche stérile, déblais; déchets, stériles.

death assemblage, thanatocœnose.

debacle, débâcle (glaciaire).

debouch (to), déboucher, confluer.

debouchure, 1. embouchure (fluviatile); 2. émergence d'une source; 3. débouché de galeries.

debris, éboulis, déblais, débris; d. avalanche, glissement de terrain; d. cone, cône de déjection; d. fall, chute de fragments; d. flow, coulée boueuse; d. line, laisses de mer de tempête; d. load, charge solide (d'un cours d'eau); d. slide, glissement de terrain.

decalcification, décalcification; d. residue, résidu de d.

decalcified soil, sol décalcifié.

decalcify (to), décalcifier.

decant (to), décanter, transvaser.

decantation, décantation, transvasement.

decanting, décantation.

decapitation, décapitation (*fluv*.).

Decapoda, Décapodes (*pal*.).

decarbonate (to), décarbonater.

decarbonatation, décarbonatation.

decarbonize (to), détartrer, désencrasser, décarburer.

decay, décomposition, désintégration; d. constant, 1. constante de désintégration radioactive; 2. constante de temps (*géophys*.); d. of rocks, décomposition des roches.

decay (to), se décomposer, pourrir.

deck, 1. pont; 2. plancher (d'une cage d'extraction, mine).

decke, nappe (de charriage).

decken structure, structure à nappes de charriage.

decking, encagement (*mine*).

declination, déclinaison (magnétique); d. compass, boussole de d.

decline, diminution, déclin, baisse; **d. of water level,** abaissement du niveau phréatique.

declinometer, déclinomètre.

declivity, déclivité, pente.

decollement, décollement.

decomposable, décomposable.

decompose (to), décomposer, se décomposer.

decomposition, décomposition, désagrégation.

decontamination, décontamination.

decorative stone, pierre ornementale.

decrease, décroissance, diminution, amoindrissement.

decrease (to), diminuer, décroître, amoindrir.

decrepitate (to), décrépiter, calciner.

decussate, disposé en croix, croisé; **d. structure,** structure enchevêtrée (cristaux entrecroisés).

dedolomitization, dédolomitisation.

deep, 1. adj : profond; **2.** n : fosse sous-marine. **d. drilling,** forage profond; **d. focus earthquake,** séisme à foyer profond; **d. folding,** plis de fond; **d. level,** niveau profond (mine); **d. plain,** plaine abyssale; **d. sea,** haute mer, mer profonde; **d. sea deposits,** dépôts abyssaux; **d. sea fan,** cone alluvial de canyon sous-marin; **d. seated,** profond; **d. seated rocks,** roches magmatiques; **d. seated structures,** tectonique de fond; **d. seismic sounding,** sondage sismique profond; **d. spring,** source juvénile; **d. water channel,** passe profonde; **back d.,** arrière fosse; **fore d.,** avant-fosse.

deepen (to), 1. approfondir, creuser, recreuser; **2.** s'approfondir.

deepening, 1. approfondissement (d'un puits); **2.** surcreusement (glaciaire, etc.).

deeply, profondément.

Deerparkian (US), Deerparkien (Dévonien inf. moyen).

deficiency, déficit, insuffisance.

defile, défilé (géogr.).

deflation, déflation, érosion éolienne; **d. basin,** creux de déflation; **d. hole,** creux de déflation; **d. hollow,** creux de déflation.

deflect (to), 1. dévier (une onde, un puits), détourner; **2.** se dévier.

deflecting, déviation.

deflection, 1. déflexion, déviation; **2.** flexion, fléchissement, affaissement (mécan.). **d. strength,** résistance à la flexion; **d. stress,** effort de flexion; **d. test,** essai de flexion; **magnetic d.,** déviation magnétique.

deflocculate (to), défloculer.

deflocculating agent, agent de défloculation.

deflocculation, défloculation.

deformation, déformation (tectonique); **d. ellipsoid,** ellipsoïde des déformations; **d. monitoring,** prévention des éruptions volcaniques par l'étude des déformations du dôme volcanique.

degasification, dégazéification.

degasify (to), dégazer.

degassing, dégazage.

degaussing, effacement (magnétique).

degenerated soil, sol sénile.

deglaciation, déglaciation, recul des glaciers (lors d'un interglaciaire).

degradation, dégradation, décomposition, érosion; **areal d.,** é. en surface, é. aréolaire.

degraded, dégradé (pédol.); **d. alkali soil,** sol salin; **d. chernozem,** sol noir lessivé; **d. humus carbonate,** sol rendziniforme; **d. rendzina,** rendzine d.

degree, degré; **d. of hardness,** d. de dureté; **d. of longitude,** d. de longitude; **d. of the thermometer,** d. de thermomètre.

dehydrate (to), déshydrater.

dehydration water, eau de déshydratation (de réactions chimiques).

de-ironized soil, sol déferruginisé.

delay, retard (ondes sismiques).

delayed run-off, écoulement, ruissellement retardé (par infiltration souterraine).

deleterious, délétère.

delevelled, dénivellé (adj.).

delicate indicator, détecteur de grisou, grisoumètre (mine).

deliquescent, déliquescent.

dell (dry valley), vallée sèche.

dell (Scot.), vallon (boisé).

delta, delta; **d. bedding,** stratification deltaïque; **d. deposit,** sédiment deltaïque; **d. distributory,** bras du d.; **d. fan,** d. sous-marin; **d. interior,** d. intérieur; **d. kame,** dépôt de d. glaciaire lors du recul d'un glacier d'inlandsis; **d. lake,** lac de d.; **d. plain,** plaine deltaïque; **ebb d.,** d. de jusant; **fan d.,** cône de déjections; **flow d.,** d. de flot; **river d.,** d. fluvial; **storm d.,** d. de tempête; **tidal d.,** d. de marée.

deltaïc, deltaïque.

deltoid, triangulaire, en forme de delta.

delthyrium, delthyrium.

deltidial plates, plaques deltidiales.

deltidium, deltidium.

deluvial soil, diluvium.

delve (to), creuser (le sol).

demagnetization, désaimantation.

demagnetise (to), démagnétiser, désaimanter.

demantoid, démantoïde (var. de grenat andradite).

demersal, vivant au fond de la mer, benthique.

Demospongea, Démosponges.

dendriform, arborescent.

dendrite, dendrite.

dendritic drainage, réseau fluviatile dendritique; **d. pattern,** réseau fluviatile dendritique.

dendrochronology, dendrochronologie.

dendrogeomorphic evidence, arguments dendro-géomorphologiques.

dense, dense, compact; **d.-medium separation,** séparation par liquides denses.

denseness, densité.

densilog, diagraphie de densité.

densimeter, densimètre.

densimetry, densimétrie.

densitometer, 1. densimètre; **2.** photomètre.

density, 1. densité, poids spécifique, masse volumique; 2. opacité; **d. current,** courant de turbidité, courant de densité; **absolute d.,** densité absolue.

dent, creux.

dental socket, fossette dentaire (*pal.* lamellibs.).

dentine, dentine (*pal.*).

dentition, dentition (*pal.*).

denudation, érosion, dénudation; **d. chronology,** reconstitution paléogéographique, reconstitution paléogéomorphologique, évolution du paysage; **d. plain,** pénéplaine.

denude (to), déboiser, dénuder.

denuded soil, sol érodé.

denuding agent, agent d'érosion.

deoxidation, désoxydation.

departure, écart, déviation (*phys., math.*).

depauperate fauna, faune appauvrie.

depergelation, processus de dégel du pergélisol.

dephasing, déphasage (*géoph.*).

deplanation, processus géomorphologiques d'aplanissement.

depleted, épuisée (*mine,* etc.); **d. bed,** couche é.; **d. melt,** magma appauvri; **d. well,** puits é.

depletion, épuisement, appauvrissement; **d. area,** région d'ablation (nivale); **d. of bases,** désaturation; **d. layer,** horizon lessivé.

depocenter, point de sédimentation maximum; zone d'accumulation deltaïque exceptionnelle.

depolarization, dépolarisation.

depolarize (to), dépolariser.

deposit, sédiment, gisement, dépôt; **alluvial d.,** alluvions; **eolian d.,** sédiments éoliens; **glacial d.,** dépôts glaciaires; **marine d.,** sédiments marins.

deposit (to), 1. sédimenter, déposer, mise en place; 2. se déposer.

deposit evaluation, évaluation d'un gîte.

deposition, dépôt, accumulation, sédimentation.

depositional fault, faille syngénétique, intraformationnelle, contemporaine de la sédimentation.

depositional magnetization, aimantation rémanente de sédiments.

depositional trap, piège stratigraphique.

depressed coast, côte affaissée.

depressed nappe (hydro), nappe déprimée.

depression, 1. dépression, creux (topo); 2. dépression atmosphérique; **structural d.,** fossé tectonique, dépression tectonique.

depth, profondeur, hauteur; **d. contour,** isobathes; **d. curve,** courbe bathymétrique; **d. finder,** sondeur; **d. of hole,** p. d'un forage; **d. of compensation,** p. de compensation des carbonates; **d. of water,** h. d'eau; **d. point,** côte (*géophys.*); **d. range,** répartition bathymétrique; **d. recorder,** échosondeur; **d. shooting,** tir de p. (*géophys.*); **d. time curve,** courbe p.-temps; **abyssal d.,** p. abyssale.

deputy, « porion », contremaître (*mine*).

deranged drainage, réseau hydrographique désordonné.

Derbyshire spar, fluorine.

derivate rocks, roches sédimentaires.

derivative magma, magma secondaire.

derivative rocks, roches détritiques.

derived fossil, fossile remanié et trouvé dans des couches plus récentes que celles où on le trouve normalement.

derived product, produit dérivé, dérivé (*chimie*).

deroofing, 1. mise à nu d'un massif intrusif (par érosion); 2. fusion du toit de roches intrusives, d'où expansion du magma.

derrick, derrick, tour de forage, chevalement (*mine*); **d. drill,** forage « rotary »; **d. floor,** plancher de manœuvre; **d. leg,** montant de d.; **d. platform,** plate-forme de forage; **d. post,** montant de d.

desalination, dessalage (de l'eau de mer).

desalinization, désalinisation.

desalting, dessalage.

descending, descendant; **d. water,** eau de percolation; **d. spring,** source descendante.

descloizite, descloïzite.

descriptive mineralogy, minéralogie descriptive.

desert, 1. adj : désertique; 2. n : désert; **d. crust,** croûte d.; **d. pavement,** reg, pavement d.; **d. polish,** poli d.; **d. soil,** sol d.; **d. topography,** topographie d.; **d. varnish,** croûte d.; **d. zone,** zone d.; **rock d.,** hamada; **salt d.,** d. salé; **sand d.,** erg.

deserted loop, méandre abandonné.

desertification, désertification.

design earthquake, séisme prévisionnel.

desilication, desilification, désilicification (par lessivage dans les sols tropicaux).

desilverization, désargentation (d'un minerai).

desilverized lead, plomb désargenté.

desilverizing, désargentation.

Desmoinesian, Desmoinésien (Carbonifère sup.).

desolate country, région désolée.

desorganized drainage, réseau hydrographique désorganisé.

desquamation, desquamation.

dessicate (to), déshydrater, dessécher.

dessiccation, dessication; **d. breccia,** brèche de d.; **d. crack,** fente, fissure de d.; **d. fissure,** fente de d.; **d. polygons,** polygones formés par fentes de d.

destruction, destruction; **d. test,** essai de rupture.

destructive, destructeur; **d. plate margin,** marge de plaque en subduction; **d. process,** processus d.; **d. wave,** vagues à fort pouvoir d'érosion littorale.

destructiveness, caractère destructeur (d'un séisme).

desulfurate (to), désulfurer, désoufrer.

desulfuration, désulfuration.

desulfurize (to), désulfurer, désoufrer.

desulfurizer, désulfurizeur.

desulfurizing furnace, four à pyrites.

detailed survey, levé détaillé.

detector, 1. détecteur; 2. séismographe; 3. capteur; **d. of gas,** détecteur de grisou (*mine*); **scintillation d.,** détecteur (compteur) à scintillation.

determinism, déterminisme.

detonate (to), faire détoner, détoner.

detonating, détonant, explosif.

detonation wave, onde d'explosion.

detrital, détritique; **d. remanent magnetism,** aimantation rémanente d.; **d. rock,** roche détritique.

detritism, détritisme.

detritus, **1.** matériel détritique; **2.** détritus, débris pierreux; **d. feeder,** alimentation en matériel d.

deuteric, deutérique (*pétro.*); **d. effects,** effets d'altération et de métasomatose dans les roches ignées avant leur mise en place.

deuterium, deutérium.

deuterogene, deuterogenous rock, roche formée par altération de roches éruptives (dès les derniers stades de cristallisation).

developed, développé, en exploitation (*mine*).

development, **1.** développement (d'une activité, etc.); **2.** développement (*photog.*); **3.** traçage (*mine*); **d. well,** puits de d.; **d. works,** travaux de traçage.

deviated well, sondage dévié.

deviation, déviation, écart; **mean d.,** écart moyen.

device, dispositif, système.

devitrification, dévitrification (de verres volcaniques).

devitrify (to), se dévitrifier.

Devonian, Dévonien.

dew, rosée; **d. point,** point de r., de condensation; **d. ponds,** mares artificielles, d'origine anthropique.

dewater (to), assécher.

dewatering, assèchement, deshydratation, exhaure (*mine*).

dextral, dextre; **d. coiling,** enroulement d.; **d. fault,** faille d.; **d. fold,** pli dissymétrique à flanc décalé vers la droite.

diabase, **1.** diabase (*pétro*); **2.** dolérite (US); **d. amphibolite,** amphibolite formée par dynamométamorphisme.

diabasic, ophitique, diabasique.

diablastic texture, texture diablastique.

diachronism, diachronisme.

diachronous, diachrone (à âge variable, de durée inégale).

diaclase, diaclase.

diaclinal, diaclinal, transversal à un pli.

diadochite, diadochite.

diadochy, diadochie (substitution ionique dans le réseau atomique).

diagenesis, diagénèse.

diagenetic, diagénétique.

diagnostic minerals, minéraux symptomatiques.

diagonal, **1.** adj : diagonal; **2.** n : diagonale; **d. bedding,** stratification oblique; **d. fault,** faille oblique; **d. lamination,** stratification entrecroisée; **d. stratification,** stratification entrecroisée.

diagram, diagramme, figure, schéma; **triaxial d.,** diagramme triangulaire.

diagrammatic section, coupe schématique.

dial, **1.** cadran; **2.** boussole (*mine*).

dial (to), faire le levé au moyen de la boussole (*mine*).

dialing, levé à la boussole.

diallage, diallage (augite).

dialogite, diallogite, dialogite, diallogite, rhodochrosite.

dialysis, dialyse.

diamagnetic, diamagnétique.

diamagnetism, diamagnétisme.

diamantiferous, diamantifère.

diamantine, diamantin.

diamictite, tilloïde, paraconglomérat, dépôt de solifluction, diamictite.

diamond, diamant; **d. bearing,** diamantifère; **d. bit,** trépan à couronne diamantifère; **d. boring crown,** couronne à pointes de d.; **d. coring,** carottage au d.; **d. cutter,** diamantaire; **d. cutting,** taille du d.; **d. drill,** foreuse à pointes de d.; **d. drilling,** forage au d.; **d. field,** terrain diamantifère; **d. mine,** mine de d.; **d. mining industry,** industrie minière diamantifère; **d. pipe,** cheminée diamantifère; **d. producing,** diamantifère; **d. saw,** scie diamantée; **d. spar,** corindon; **d. tin,** grands cristaux de cassitérite.

diamondiferous, diamantifère.

diaphtoresis, rétrométamorphisme, diaphtorèse, rétromorphose.

diapir, diapir; **d. fold,** pli d.

diapiric core, masse interne du diapir; **d. fold,** pli diapir.

diapirism, diapirisme.

diaspore, diaspore.

diastem, interruption stratigraphique, lacune de sédimentation (de courte durée).

diastrophism, diastrophisme (déformation de la croûte terrestre).

diatom, diatomee, diatomée; **d. mud,** vase à d.; **d. ooze,** boue à d.

diatomaceous earth, terre à d., diatomite.

diatomic, diatomique.

diatomite, diatomite.

diatreme, cheminée volcanique, diatrème.

Dibranchiata, Céphalopodes dibranchiaux (*pal.*).

dichotomous, dichotomique.

dichroism, pléochroïsme.

dichroite, dichroïte, cordiérite.

dichromate, bichromate.

diclinic, diclinique.

Dicots, Dicotylédone (abbrév. de).

dicyclic, dicyclique (*pal.* Échino.).

die out (to), se terminer en biseau.

diedral (adj.), dièdre.

differential, différentiel; **d. compaction,** tassement d.; **d. erosion,** érosion d.; **d. melting,** fusion d.; **d. thermal analysis,** analyse thermique d.; **d. weathering,** altération d., a. sélective.

differentiated dike (dyke), filon intrusif composé de plusieurs roches (formées par différenciation magmatique).

differentiation, différenciation (magmatique).

diffluence, diffluence.

diffract (to), diffracter.

diffraction, diffraction; **d. grating,** réseau de d.;
d. line, raie de d.; **d. pattern,** diagramme de d.;
d. spectrum, spectre de d.; **d. spot,** tache de d.;
X ray d., d. aux R.X (rayons X, diffractométrie X).
diffractometer, diffractomètre.
diffractometer trace, diffractogramme.
diffuse (to), diffuser.
diffusion, diffusion.
diffusivity, diffusivité.
dig, salbande.
dig (to), creuser, fouiller.
dig out (to), extraire.
digger, 1. excavateur (*mécha.*); 2. mineur (personne),
orpailleur; **gold d.,** chercheur d'or.
digging, fouille, creusement, excavation, piquage
(*mine*); **d. coal,** piquage de la houille.
diggings, 1. exploitation, gisements alluvionnaires;
2. exploitation de placer (US); **gold d.,** exploitation aurifère.
digital, numérique; **d. computer,** ordinateur; **d.
filter,** filtre n.; **d. recording,** enregistrement n.
digitation, digitation, ramification (d'un anticlinal
secondaire).
digonal, à deux angles.
dihedral (adj.), dièdre.
dihexagonal, bihexagonal.
dike (US), dyke (G.B.), 1. filon intrusif oblique,
penté, dyke; 2. digue; **d. rock,** roche intrusive;
d. set, groupe de filons parallèles; **d. swarm,**
groupe de filons; **dilation d.,** filon intrusif; **res-
training d.,** digue de retenue; **ring-d. structure,**
structure intrusive annulaire.
dikelet, ramification, apophyse d'un filon.
dilapidation, dénudation, dégradation.
dilatability, dilatabilité.
dilatable, dilatable, expansible.
dilatation, dilatation.
dilatational wave, onde séismique P.
dilate (to), 1. dilater; 2. se dilater.
dilatency, dilatation.
dilation, 1. dilatation (de l'eau en glace, etc.); 2.
élargissement d'une fissure volcanique lors de
l'injection du magma; **d. dike,** filon intrusif ayant
écarté les parois de la fissure.
dilute (to), diluer, étendre.
dilution, dilution.
diluvial, diluvien, diluvial.
diluvium, diluvium.
dimension, dimension.
dimensional analysis, analyse dimensionnelle.
dimetric system, système quadratique.
dimictic (lake), lac dimictique.
diminish (to), diminuer, amoindrir, décroître.
diminution, diminution, amoindrissement.
dimorphism, dimorphisme.
dimorphous, dimorphic, dimorphe.
dimple, lentille.
dimyarian, dimyaire.
Dinantian, Dinantien.
dingle, vallon (boisé).
Dinoflagellates, Dinoflagellés.

Dinosaurs, Dinosaures.
diogenite, diogénite (météorite).
diopside, diopside (pyroxène).
dioptase, dioptase.
diorite, diorite.
dioritic, dioritique.
dioxide, bioxyde.
dip, 1. pendage; 2. inclinaison magnétique; **d.
closed,** fermeture par p.; **d. compass,** boussole d'inclinaison; **d. fault,** faille perpendiculaire
à la direction de la couche; **d. fold,** pli plongeant;
d. heading, descenderie; **d. joint,** diaclase perpendiculaire à la direction de la couche; **d. line,**
profil de p.; **d. logging,** pendagemétrie; **d. resolution,** calcul du p. vrai; **d. slip fault,** faille d'effondrement; **d. slope,** revers de cuesta (surface
structurale); **d. vector,** flèche de p.; **down the d.,**
aval-p.; **magnetic d.,** inclinaison magnétique.
dip (to), 1. s'incliner, plonger; 2. immerger, plonger,
tremper.
diphase, diphasé.
dipmeter, pendagemètre.
dipmeter logging, pendagemétrie.
dipolar, bipolaire.
dipole, dipôle.
dipole field, champ dipolaire.
dipping, plongement, inclinaison; **d. bed,** lit incliné;
d. compass, boussole d'inclinaison; **d. needle,**
aiguille d'inclinaison.
Diptera, Diptères (*pal.*).
dipyre, dipyre.
diramation, bifurcation.
direct current, courant continu.
direct run off, ruissellement superficiel (sans infil-
trations).
direct wave, onde directe.
direction, direction, sens; **d. of flow,** d. d'écoulement.
directional, orienté, directionnel; **d. drilling,** forage
o.; **d. structure,** structure sédimentaire à signifi-
cation directionnelle.
directionality, anisotropie.
dirt, 1. saleté, par extension, déblai, remblai, boue;
2. alluvion aurifère; **d. band,** bande boueuse
(*glaciol.*); **d. bed,** 1. mince bande de matériaux
terreux intercalés dans filons de charbon; 2.
paléosol à fragments végétaux; **d. fault,** zone à
charbon broyé.
disaggregate (to), désagréger.
disaggregation, désagrégation.
disaggregated rock, roche décomposée, désagrégée.
disappearing stream, perte karstique.
discard (to), mettre au rebut (un échantillon).
discharge, 1. débit (fluv.), écoulement; 2. décharge,
déversement; **annual d.,** débit annuel; **mean d.,**
débit moyen; **river d.,** débit fluviatile; **spring d.,**
débit d'une source; **solid d.,** débit solide, charge
(fluviatile).
discoidal segregations of a gneiss, «yeux» d'un
gneiss granoblastique «œillé».
disconformity, discordance stratigraphique; **erosional d.,** discordance d'érosion.

discontinuity, discontinuité.

discontinuous, discontinu; **d. deformation**, déformation cassante; **d. permafrost zone**, zone à pergélisol d.

discordance, discordance.

discordant, discordant (*stratig.*); **d. drainage**, réseau fluviatile non adapté, non conséquent; **d. intrusion**, batholite, dyke.

discoverer (of a placer), inventeur.

discovery, découverte (d'un gisement); **d. shaft**, puits de recherches; **d. well**, puits exploitant un gisement jusque là inexploité et inconnu.

disedged, émoussé.

disengage (to), dégager (un gaz).

disengagement, dégagement (d'un gaz).

disgorging spring, dégorgeoir.

dish, **1.** batée (*mine*); **2.** cuvette.

disharmonic fold, pli disharmonique.

disintegrable, désagrégeable.

disintegrate (to), **1.** désagréger, désintégrer, effriter; **2.** se décomposer, se désagréger.

disintegrated, désagrégé, désintégré.

disintegration, désagrégation, effritement;**granular d.**, émiettement.

disjunct apical system, système apical dissocié (*pal.*, Échino.).

disjunctive fold, pli-faille; **d. movement**, faille.

dislevelment, dénivellation, dénivellement.

dislocate (to), disloquer.

dislocation, dislocation, fracture, faille; **d. breccia**, brèche de faille.

dismembered river system, réseau fluvial démembré.

dispersal channel, effluent.

dispersed phase, phase solide dispersée (sous forme colloïdale).

dispersion, dispersion.

dispersive power, pouvoir dispersif.

disphotic zone, zone disphotique (entre − 80 m et − 200 m) sous la surface de la mer.

displaced mass, masse charriée.

displacement, **1.** rejet, décalage (de part et d'autre d'une faille); **2.** remplacement (chimique); **horizontal d.**, rejet horizontal.

displacement theory, théorie de la dérive des continents.

disposal, stockage de matériaux radioactifs.

disrupted, interrompu, disloqué; **d. fold**, pli faille; **d. gouge**, effet de rabotage glaciaire (avec fragmentation).

disruption, rupture, dislocation.

disruptive explosive, explosif brisant.

dissect (to), raviner, éroder par ruissellement.

dissected plain, plaine disséquée par le réseau hydrographique après un soulèvement.

dissection, dissection (du relief) par érosion fluviatile.

disseminate (to), disséminer, éparpiller.

disseminated ore, minerai disséminé (à l'état de fines particules).

dissemination, dissémination.

dissepiment, dissépiment.

dissociation, dissociation (chimique), décomposition; **d. point**, température de dissociation chimique; **d. temperature**, température de dissociation.

dissolubility, dissolubilité.

dissoluble, dissoluble.

dissolution, dissolution; **d. basin**, dépression de d., doline.

dissolve (to), **1.** dissoudre, fondre; **2.** se dissoudre.

dissolved river load, charge fluviatile dissoute; **d. solids**, quantité totale de matériaux dissous.

dissolving, dissolution.

distentional, de distension; **d. fault**, faille de distension.

disthene, disthène.

distil (to), distiller.

distillate, distillation, distillation.

distortional wave, onde transversale.

distributary, effluent (fluv.).

distribution scatter, diagramme de répartition (en fonction de variables).

distributive fault, faille en gradins; **d. province**, province d'alimentation, d'origine (pour des matériaux sédimentaires).

disturbance, **1.** perturbation (*atmosph.*); **2.** mouvements tectoniques ou orogéniques affectant une région, ou une couche géologique; **3.** ébranlement sismique.

disulphide, bisulfure.

ditch, fossé, rigole, tranchée.

ditching, creusement de tranchées; **d. machine**, excavatrice.

Dittonian (GB), Dittonien (Dévonien inf.).

diurnal range, variation diurne (de température), amplitude diurne.

divalent, bivalent.

divaricating channel, diverticule, effluent.

dive (to), plonger (*oceano.*).

diver, plongeur.

divergent plate boundary, limite de plaques lithosphériques divergentes (en accrétion).

diversion, dérivation, changement de cours fluviatile.

diverted stream, fleuve capté.

divide, ligne de partage des eaux (*hydro.*).

divided dial, cadran gradué.

dividing ridge, ligne de partage des eaux.

diviner, sourcier.

divining rod, baguette de sourcier.

diving bell, tourelle de plongée.

diving saucer, soucoupe plongeante (*océano.*).

diving suit, scaphandre.

division, division, séparation, compartimentage.

divisional plane, surface de discontinuité.

D layer, manteau inférieur (− 720 km à − 2 886 km).

do, préfixe indiquant (en pétrographie) qu'un facteur domine les autres dans les proportions de 7/1 et 5/3.

dodecahedron, dodécaèdre.

dog, accrocheur, taquet, tenailles; **d. and chain**, arrache étais (*mine*); **d. hole**, passage (*mine*); **d. house dope**, informations provenant du sondage; **d. legs**, « pattes de chien » (brutal changement de direction d'un forage); **d. tooth spar,** calcite à cristaux pointus («en dents de chien» = en dents de cochon).

Dogger, Dogger (Jurassique moyen).

dolarenite, dolomite à petits cristaux (de la taille de grains de sable).

doldrum, zone de basse pression équatoriale.

dolerite, dolérite, roche à texture doléritique.

doleritic, doléritique.

dolerophanite, dolérophanite (*minér.*).

doline, dolina, dolinen, doline.

dolomicrite, dolomicrite.

dolomite, 1. dolomite (minéral); 2. dolomie (roche); **d. limestone,** calcaire dolomitique; **d. marble,** marbre dolomitique; **ferroan d.,** ankérite.

dolomitic, dolomitique; **d. cemented sandstone,** grès à ciment dolomitique; **d. limestone,** calcaire dolomitique; **d. marl,** marne dolomitique.

dolomitization, dolomitisation.

dolomold, cavité rhomboédrique occupant l'emplacement d'un cristal de dolomite (après dissolution).

dolostone, roche sédimentaire formée de dolomie détritique, concrétionnée ou déposée par précipitation.

domal structure, structure en dôme.

dome, 1. brachyanticlinal; 2. dôme (intrusif); **d. collapse,** effondrement d'un volcan; **lava d.,** dôme de lave; **plug d.,** neck volcanique, dôme formé d'un culot de lave; **salt d.,** dôme de sel, diapir.

Domerian, Domérien.

domical body, massif en forme de dôme.

dominants, espèces dominantes (*pal.*).

domite, domite, trachyte du Puy-de-Dôme (Auvergne).

donga (South Afr.), ravin encaissé.

dopplerite, dopplérite.

Dordonian, Dordonien.

dormant volcano, volcan inactif.

dornick (US), bloc (de minerai de fer).

dose (to), doser.

dosimeter, dosimètre.

dosing, dosage.

dotted (line), en pointillé (courbe).

double refracting spar, cristal de calcite (spath d'Islande) à double réfraction.

double refraction, double réfraction.

doublet, doublet (*minér.*).

douse (to), 1. éteindre, combattre un incendie (de puits de pétrole); 2. chercher des gisements minéraux à l'aide d'une baguette de sourcier.

down bending, flexure.

downbuckle, fossé, affaissement tectonique.

downcast-shaft, puits d'entrée d'air.

downcast side, lèvre affaissée (d'une faille).

downcreep, glissement.

downcutting, érosion.

downcutting stream, fleuve à fort pouvoir de creusement.

downdip, aval pendage; **d. block,** compartiment affaissé.

downdrift, direction de la dérive littorale.

downfall, chute.

downfan, vers l'aval d'un cône alluvial sous-marin.

downfault, faille normale.

downfold, pli synclinal.

downgrade, pente.

downhill, vers le bas.

downlap, couches inclinées sur discordance (*géophys.*).

downpour, averse.

downs, 1. collines dénudées (GB); 2. dunes; 3. prairies (Australie).

downsand, sable dunaire.

downside, lèvre affaissée (d'une faille).

downslip fault, faille normale.

down-stepping erosion surface, surface d'érosion étagée.

downstream, 1. adj : d'aval, aval; 2. adv : en aval.

downthrow, rejet.

downthrown side, compartiment affaissé (faille).

downtime, temps mort (temps perdu aux réparations dans un forage).

Downtonian, Downtonien.

downvalley migration, migration vers l'aval des méandres.

downward, 1. vers le bas; 2. en aval.

downwarp, fléchissement.

downwearing, érosion, aplanissement.

downwarping, affaissement, pli synclinal.

downwasting, fonte glaciaire.

down welling, 1. subduction (*tect. plaques*); 2. enfoncement des eaux océaniques.

down-wind side, face sous le vent.

dowser, 1. sourcier; 2. hydroscope.

dowsing rod, baguette de sourcier.

drag, 1. rebroussement des lèvres d'une faille, crochon; 2. drague; **d. fold,** pli d'étirement, pli d'entraînement, pli de frottement; **d. mark,** figure sédimentaire de frottement; **d. line,** pelle à benne; **d. ore,** minerai broyé.

drag (to), draguer (*océano.*).

Dragonian (US), Dragonien (= Danomontien).

drain (to), drainer, assécher, évacuer.

drain off (to) the water, évacuer les eaux.

drainage, 1. écoulement de l'eau par le réseau hydrographique et par l'écoulement souterrain; 2. drainage, assèchement; **d. area soil,** aire de drainage du pétrole vers le puits; **d. basin,** bassin versant, bassin hydrographique; **d. channel,** 1. chenal d'écoulement; 2. canal de drainage; **d. characteristics,** caractéristiques hydrographiques; **d. density,** densité du réseau hydrographique; **d. level,** galerie de drainage, galerie d'écoulement; **d. line,** thalweg; **d. pattern,** disposition du réseau hydrographique; **d. ratio,** coefficient d'écoulement; **d. system,** réseau hydrographique; **d. wind,** vent catabatique; **exterior d.,** drainage exoréique; **impeded d.,** drainage endoréique; **surface d.,** drainage superficiel.

draining, drainage, assèchement, évacuation des eaux; **d. ditch,** fossé de drainage.

drain shaft, puits de drainage.

drainway, galerie d'écoulement.

drain well, puits absorbant.

drapery, draperie, stalactites.

draught, appel d'air, entrée d'air, tirage, aérage.

dravite, dravite (var. brune de tourmaline).

draw (US), ravin, souvent à sec; oued.

draw (to), 1. dessiner, tracer, tirer (un trait), lever; 2. tirer, traîner; 3. extraire, remonter, arracher; 4. puiser (de l'eau); 5. étirer, filer, tréfiler; **d. off (to),** soutirer un liquide; **d. out (to),** 1. extraire; 2. tirer, étirer.

draw (to) the pillars, déhouiller, dépiler (*mine*).

drawdown, abaissement de la nappe phréatique.

drawer, 1. dessinateur; 2. rouleur, traîneur (*mine*).

drawing, 1. extraction, remontée; 2. dessin; **d. back,** abatage (*mine*); **d. board,** planche à dessin; **d. cage,** cage d'extraction; **d. casing,** arrachage du tubage (forage), étirement; **d. out,** extraction; **d. shaft,** galerie d'extraction; **d. timber,** déboisage (*mine*); **d. table,** table à dessin; **d. water from a well,** puisage de l'eau d'un puits.

drawn, 1. étiré; 2. exploité, épuisé.

drawn-out fold, pli étiré.

dredge, drague; **d. boat,** bateau drague; **d. mining,** exploitation des alluvions par drague.

dredge (to), draguer.

dredger, 1. drague (machine); 2. ouvrier dragueur.

dredging, dragage; **d. depth,** profondeur du d.; **d. ground,** terrains de d.; **d. machine,** drague; **d. pump,** pompe de d.

dreikanter, caillou à facettes.

Dresbachian (US), Dresbachien (Cambrien sup.).

dress (to), tailler, parer, préparer; **d. ore,** préparer mécaniquement du minerai.

dress up (to) the ore, enrichir le minerai.

dressed rocks, roches moutonnées; **d. stone,** pierre taillée.

dressing, triage, préparation, taille; **ore d.,** traitement du minerai; **ore d. works,** installation de traitement des minerais.

driblet cone, cône en pustule (forme adventive).

dried, desséché.

drift, 1. galerie horizontale (*mine*); 2. matériel détritique transporté par le glacier, et déposé à l'état de moraines, surtout quaternaires; 3. formations superficielles; 4. remplissage détritique de cavités (karstiques); 5. matériel abandonné sur une plage; 6. mouvement de la mer par les courants, direction et sens des courants; 7. déviation, dérive; **d. beds,** moraines; **d. barrier lake,** lac morainique; **d. boulder,** bloc erratique; **d. breccia,** brèche glaciaire; **d. clay,** argile à blocaux; **d. covered bedrock,** substratum recouvert de formations morainiques; **d. deposit,** dépôt glaciaire ou fluvioglaciaire, moraine; **d. epoch,** époque glaciaire; **d. glaciers,** petits glaciers alimentés par neige soufflée; **d. ice,** 1. glace flottante; 2. glace déplacée; **d. map,** carte des formations glaciaires et fluvioglaciaires; **d. mine,** mine exploitée par galerie; **d. sand,** sable mouvant; **d. sheet,** couche morainique attribuable à une glaciation; **d. structure,** stratification entrecroisée; **d. theory,** théorie allochtone (de la formation du charbon); **d. tunnel,** galerie d'exploitation; **beach d.,** dérive littorale; **continental d.,** dérive des continents; **cross d.,** galerie de recoupe (*mine*); **glacial d.,** moraine; **river d.,** alluvions; **shore d.,** dérive littorale; **snow d.,** avalanche (poudreuse); **stratified d.,** dépôt fluvio-glaciaire stratifié; **washed d.,** dépôt fluvial-glaciaire stratifié; **wind d.,** dépôt éolien.

drifter, marqueur de courant océanographique.

drifting, 1. percement des galeries; 2. charriage.

drill, 1. foret, mèche, burin; 2. foreuse, sondeuse, sonde perforatrice; **d. bit,** trépan, fleuret; **d. core,** carotte de sondage; **d. cuttings,** déblais de forage; **d. foreman,** contremaître de forage; **d. hole,** forage, trou de sonde; **d. log,** coupe de forage; **d. pipe,** tige de forage (pétrole); **d. pipe coupling,** raccord de tiges; **d. pipe string,** train de tige; **d. rods,** tiges de forage; **d. rope,** câble de sondage; **d. ship,** navire de forage; **d. stem,** maîtresse-tige; **rock-core sampling d.,** carotteuse, pour prélèvement d'échantillons rocheux; **stuck d. pipe string,** tige de forage coincée.

drillability, forabilité.

drillable, forable.

driller, foreur, sondeur.

driller's log, coupe de forage.

drilling, forage, perforation, percement, sondage; **d. barge,** ponton de forage; **d. bit,** trépan; **d. break,** accroissement brusque de la vitesse; **d. by percussion,** sondage par battage du forage; **d. cable,** câble de forage; **d. collars,** manchons, colliers de tubes; **d. core,** carotte de forage; **d. crew,** équipe de forage; **d. fluid,** boue de forage; **d. log,** rapport de forage; **d. machine,** perceuse, foreuse, sondeuse (mine), perforatrice; **d. mud,** boue de forage; **d. pattern,** espacement des puits; **d. pipe,** tige de forage; **d. plant,** installation de forage; **d. reamer,** trépan aléseur; **d. record,** rapport de sondage; **d. rig,** système de forage; **d. sludge,** boue de forage; **d. string,** train de tiges; **d. vessel,** navire de forage; **d. winch,** treuil de forage; **cable-tool d.,** forage au câble; **core d.,** carottage; **directional d.,** forage dévié; **offshore d.,** forage en mer.

drillings, débris de forage.

drillman, ouvrier sondeur, foreur.

drill (to), 1. forer, percer, perforer; 2. trier, classer.

driphole, trou de goutte de pluie.

dripstone, stalactite.

drive, 1. commande, transmission (d'un appareillage); 2. galerie (*mine*); **d. pipe,** colonne de tubage (*for.*); **d. shoe,** sabot de tube (*for.*).

driving, 1. avancement de galeries (*mine*); 2. enfoncement (d'un tubage); 3. commande (*méch.*).

driving mechanism, mécanisme de la dérive des plaques.

drizzle, bruine, crachin.

drop, 1. goutte; 2. chute, baisse, abaissement, dénivellation.

drop off, brusque dénivellation.

drop (to), 1. laisser tomber, descendre; 2. tomber, baisser; 3. goutter, s'égoutter; 4. sédimenter; **d. test**, essai au choc.

droplet, gouttelette.

dropped side, compartiment affaissé d'une faille.

dropstone, 1. bloc isolé d'origine glaciaire dans sédiments glacio-lacustres ou glacio-marins (transportés par les icebergs ou les plaques de glace).; 2. bombe volcanique qui tombe dans l'eau et s'enfonce dans un sédiment.

dross, scories, laitiers.

drossy, 1. sans valeur, de rebut; 2. plein de scories.

drought, sécheresse.

droughty, sec, aride.

drown (to), inonder, submerger.

drowned, submergé; **d. coast**, littoral s.; **d. glacial erosion coast**, côte à modelé glaciaire submergée (côte à fjords).; **d. river mouth**, estuaire s.; **d. topography**, relief s.; **d. valley**, vallée submergée par la remontée du niveau marin dans des zones non glaciaires (= rias).

drum, drumlin, drumlin.

druse, géode, cavité.

drusy, drusique, à géodes.

dry, sec, aride; **d. basin**, dépression intérieure sèche; **d. bone ore**, smithsonite; **d. bulk density**, densité réelle; **d. coal**, charbon maigre; **d. crusher**, broyeur sec; **d. delta**, cône de déjection; **d. essay**, essai par voie sèche; **d. farming**, culture sèche; **d. gap**, cluse morte; **d. hole**, puits stérile, improductif; **d. ice**, neige carbonique; **d. monsoon**, mousson d'hiver; **d. ore**, minerai argentifère pauvre en plomb; **d. pipe**, crépine; **d. process**, voie sèche; **d. river**, oued; **d. sorting**, triage de minerai par voie sèche; **d. valley**, vallée sèche (karstique); **d. wash**, oued (US); **d. well**, puits à sec, puits tari.

dry (to), 1. assécher, dessécher, sécher; 2. s'assécher.

drying, séchage, assèchement; **d. up**, tarissement.

dryness, sécheresse.

Duchesnian (US), Duchesnien (= Ludien).

ductibility, 1. plasticité (d'une roche); 2. malléabilité (d'un métal).

ductile, déformable.

ductile flow, déformation plastique.

duff (US), matte (humus peu décomposé); **d. mull**, humus intermédiaire.

dug earth, déblais.

dull, 1. mat, terne; 2. émoussé; 3. sombre; **d. luster**, éclat mat; **d. weather**, temps sombre.

dull (to), 1. ternir; 2. émousser; 3. s'émousser.

dumb drift, galerie en cul de sac.

dumortierite, dumortiérite.

dump (to), 1. déverser, basculer, jeter; 2. se déverser.

dump, halde, déblais, tas de déblais; **d. car**, wagon à bascule; **ore d.**, halde de minerais.

dumper, camion à bascule.

dumping, 1. déversement; 2. à bascule, basculant; 3. rejets de déblais en mer; **d. ground**, halde, déblais.

dune, dune; **d. bedding**, stratification dunaire; **d. chain**, chaîne de d.; **d. cliff**, falaise de d.; **d. crest**, crête dunaire; **d. field**, champ de d.; **d. lake**, lagune, étang littoral; **d. of sand**, d. de sable; **d. range**, chaîne de d.; **d. ridge**, chaîne de d.; **d. rock**, éolianite; **d. wing**, aile d'une d.; **active d.**, d. vive; **back d.**, arrière d.; **beach d.**, d. d'estran; **bow d.**, d. arquée; **coastal d.**, d. littorale; **crescent d.**, d. en croissant, barkhane; **cross-bar d.**, d. transversal; **fixed d.**, d. fixée; **fore d.**, avant-d.; **inland d.**, d. intérieure; **longitudinal d.**, d. longitudinale; **migrating d.**, d. mouvante; **moving d.**, d. mouvante; **parabolic d.**, d. parabolique; **shifting d.**, d. mouvante; **shore d.**, d. littorale; **snow d.**, d. de neige; **stabilized d.**, d. fixée; **stationary d.**, d. stationnaire; **transverse d.**, d. transversale; **U-shaped**, d. en U; **wandering d.**, d. mouvante.

dunite, dunite.

durability index, indice de résistance à l'usure (de matériaux transportés).

durain, durain.

duration, durée.

duricrust (cf. calcrete, ferricrete, silcrete), croûte pédologique concrétionnée (sous climat semiaride).

duripan, horizon pédologique concrétionné.

dust, poussière, poussier; **d. avalanche**, avalanche de poudreuse; **d. bowl**, désert anthropique; **d. coal**, poussier; **d. collector**, collecteur de poussières; **d. counter**, compteur de particules; **d. devil**, tourbillon de poussière; **d. storm**, tempête de sable, de poussière; **d. tuff**, cinérite; **d. whirl**, tourbillon de poussière; **volcanic d.**, fine cendre volcanique.

duster, puits sec.

dusty, poussiéreux.

dwarf fauna, faune rabougrie.

Dwyka, étage de Dwyka (Karroo inf. d'Afrique du Sud).

dy (Scand.), sapropel.

dying out, terminaison en biseau.

dyke (GB), dyke, filon intrusif.

dynamic, dynamique; **d. breccia**, brèche tectonique; **d. geology**, géologie dynamique; **d. metamorphism**, dynamométamorphisme.

dynamometamorphism, dynamométamorphisme.

dysodont, dysodonte (Lamellibranche).

dystomic, à clivage imparfait, à cassure imparfaite.

dystrophic lake, lac distrophe.

E

ea (GB), ruisseau.

eager, eagre, 1. barre, mascaret; 2. raz de marée.

Eaglefordian (N. Am.), Eaglefordien (étage, Crétacé sup. Turonien).

early, ancien, inférieur (au sens stratigraphique), début; e. diagenesis, syndiagenèse.

earth, 1. globe terrestre, terre; 2. terres, sols; e. auger, tarière à glaise; e. borer, tarière de pédologue; e. coal, lignite; e. column, pyramide de fée; e. constants, constantes terrestres; e. creep, glissement de terrain; e. crust, écorce terrestre; e. current, courant tellurique; e. dam, barrage en terre; e. din, tremblement de terre; e. fall, éboulement; e. flax, amiante; e. flow, glissement de terrain; e. hummock, butte gazonnée, butte à lentille de glace; e. magnetism, magnétisme terrestre; e. mound, butte gazonnée; e. mull, humus doux (à gros grains); e. oil, pétrole; e. orbit, orbite terrestre; e. pillar, demoiselle, cheminée de fée; e. pitch, var. d'asphalte; e. radius, rayon terrestre; e. ring, polygone de terre; (périgl); e. science, géologie; e. shell, croûte terrestre; e. slide, glissement de terrains; e. slope, talus naturel; e. stripes, sol strié terreux (périgl); e. tide, marée terrestre; e. tremor, faible secousse sismique; e. wave, onde sismique; e. wax, ozokérite; e. work, terrassement; e. worm, ver de terre; rare e., terres rares, lanthanides.

earthen, en terre.

earthenware, faïence.

earthquake, tremblement de terre, séisme; e. focus, foyer d'un séisme; e. intensity, intensité d'un séisme; e. magnitude, magnitudes des séismes; e. prediction, prévision des séismes; e. record, séismogramme; e. recorder, séismographe; e. shaking hazards, risques sismiques; e. shock, secousse sismique; e. swarm, série de tremblements de terre; e. wave, onde sismique; e. wave travelling, propagation des ondes sismiques; e. zone, zone sismique.

earthworks, terrassement, travaux de terrassement.

earthworks embankment, terrassement en remblai.

earthworm map, carte géologique du sous-sol.

earthy, terreux; e. calamine, hydrozincite; e. ore, minerai terreux; e. lead ore, var. de cérusite.

easterlies, vents venant de l'est.

easterly (adj.), est, de l'est, vers l'est.

easting, route vers l'est.

eat away (to), ronger, éroder.

ebb, reflux, jusant, marée basse; e. current, courant de jusant; e. tidal delta, delta de marée descendante, delta de jusant; e. tide, marée descendante; e. and flow structure, stratification oblique.

ebb (to), baisser, refluer.

ebulition, ébullition.

Eburnean, orogénèse Éburnéenne (Antécambien).

Eburonian, Éburonien (= Quaternaire inf. moyen).

eccentric, excentrique (astro.).

echelon faults, failles en échelon.

Echinoderm, Echinodermate, Échinoderme.

Echinoidea, Échinoïdes.

Echinozoa, Échinozoaires.

echo, écho; e. ranging sonar, échosondeur; e. sounder, échosondeur (vertical); e. sounding profile, profil d'échosondage.

echogram, échogramme (océano.), profil bathymétrique obtenu par réflexion.

echosounding, échosondage.

eclipse, éclipse.

ecliptic, écliptique.

eclogite, éclogite (pétro.); e. facies, faciès à é.

ecological niche, niche écologique.

ecology, écologie.

economic geology, géologie des ressources naturelles (minerais, eau).

economic mineral, minéral à valeur commerciale.

ecostratigraphy, écostratigraphie.

ecosystem, écosystème.

ecotope, écotope.

ecozone, unité écostratigraphique.

ectinite, ectinite (pétro.).

ectoderm, ectoderme.

edaphic, édaphique.

edaphologist, pédologue.

edaphology, pédologie.

eddy, tourbillon, remous; e. current, courant tourbillonnaire; e. flow, écoulement tourbillonnaire.

Edenian (N. Am.), Édénien (étage, base de l'Ordovicien supérieur).

edenite, édénite (minéral.), variété d'amphibole hornblende.

edge of a crystal, arête d'un cristal; e. water, eau de bordure de gisement; e. wave, vague transversale au rivage.

edgewise conglomerate, conglomérat intraformationnel.

edged, tranchant, anguleux.

Eemian, Eemien (interglaciaire Riss-Wurm : formation marine, mer du Nord).

effective, efficace, réel; e. permeability, perméabilité réelle; e. porosity, porosité réelle; e. size, dimension réelle; e. wind, vent efficace (à compétence suffisante).

effervescence (to), faire effervescence, mousser.

effervescence, effervescency, effervescence.

effervescent, effervescent.

effloresce (to), faire des efflorescences.

efflorescence, efflorescence.

efflorescent, efflorescent.

effluent, 1. cours d'eau dérivé; 2. eau traitée, effluent; **e. glacier,** langue glaciaire de décharge, langue émissaire.

effluvium, effluve.

efflux, écoulement, flux.

effuse (to), faire effusion, s'épancher (roches magmatiques).

effusion, effusion, épanchement; **volcanic e.,** épanchement volcanique.

effusive, effusif, extrusif; **e. eruption,** éruption avec effusion de laves; **e. rock,** roche d'épanchement volcanique, roche effusive.

eggstone, oolithe (inusité).

Eifelian, Eifélien (étage, Dévonien moyen européen).

einkanter, galet éolien à une face éolisée.

eject (to), expulser, émettre, éjecter.

ejecta, ejectamenta, projections volcaniques, rejets.

ejected materials, projections volcaniques.

ejection, jet (de flammes), projection (de laves), expulsion.

eklogite, éclogite (pétro.).

E layer, noyau externe (− 2 886 km à − 5 156 km).

elaeolite, élaeolite, éléolite (var. de néphéline); **e. syenite,** syénite à néphéline.

elaelitic, éléolitique.

Elasmobranchi, Élasmobranches.

elastic, élastique, flexible; **e. bitumen,** élatérite; **e. behavior,** comportement élastique des roches; **e. coefficient,** coefficient d'élasticité; **e. constant,** constante d'élasticité; **e. déformation,** déformation élastique; **e. discontinuity,** discontinuité élastique; **e. limit,** limite d'élasticité; **e. medium,** matériel, milieu élastique; **e. mineral pitch,** élatérite; **e. modulus,** module d'élasticité; **e. rebound,** relaxation de contrainte; **e. rebound theory,** modèle de détente élastique; **e. strength,** limite d'élasticité; **e. waves,** ondes élastiques.

elasticity, élasticité; **e. modulus,** module d'élasticité.

elaterite, élatérite (var. de pyrobitume).

elbaite, elbaïte (minér.) (var. verte de tourmaline).

Elbe (N. Eur.), Elbe (1re glaciation).

elbow, coude; **e. of capture,** coude de capture; **e. shaped twin,** mâcle en genoux (minér.).

electric, electrical, électrique; **e. blasting,** tir électrique; **e. calamine,** calamine; **e. coring,** carottage é.; **e. exploration,** prospection é.; **e. field,** champ é.; **e. log,** diagramme é.; **e. logging,** diagraphie é.; **e. potential,** potentiel é.; **e. polarization,** polarisation; **e. prospecting,** prospection é.; **e. sounding,** sondage é.; e, **welding,** soudure é.; **e. well logging,** diagraphie é.

electroanalysis, électroanalyse.

electrochemical, électrochimique.

electrochemistry, électrochimie.

electrode, électrode.

electrodialysis, électrodialyse.

electrolog, électrolog.

electrolyse, electrolysis, électrolyse.

electrolyte, électrolyte.

electrolytic, electrolytical, électrolytique.

electrolitic refining, affinage électrolytique.

electrolyzable, électrolysable.

electrolyze (to), électrolyser.

electromagnet, électro-aimant.

electromagnetic, électromagnétique; **e. damping,** amortissement é.; **e. field,** champ é.; **e. logging,** diagraphie é.; **e. prospecting,** prospection é.; **e. survey,** étude é.

electromagnetism, électromagnétisme.

electron, électron; **e. diffraction,** analyse par diffraction électronique; **e. microprobe,** sonde électronique, microsonde électronique; **e. probe microanalyser,** microsonde électronique; **e. microscope,** microscope électronique; **e. shell,** couche électronique.

electrophoresis, électrophorèse.

electrorefining, affinage par l'électricité.

electrosilver (to), argenter par électrolyse.

electrostatic, électrostatique; **e. separation,** séparation é.

electrum, 1. électrum, ambre; 2. or argentifère.

element, élément (atomique, chimique), corps simple; **linear e.,** élément structural linéaire; **trace e.,** élément-trace.

elementary body, corps simple.

eleolite, éléolite (var. de néphéline).

Elephantoidea, Éléphantidés.

elevated beach, plage soulevée.

elevation, 1. altitude, hauteur; 2. élévation; **e. correction,** correction à l'air libre (gravimétrie), correction de Bouguer; **e. of a well,** cote d'un sondage.

ellipsoid (adj.), ellipsoidal, ellipsoïdal; **e. lavas,** laves en coussins; **e. structure,** structure en coussins (laves).

ellipsoid (n.), ellipsoïde; **e. of revolution,** e. de révolution.

Elster (N. Eur.), Elster (2e glaciation).

Eltonian (GB), Eltonien (sous-étage, base du Ludovicien).

elutriate (to), décanter, séparer par lavage et filtrage.

elutriation, décantation, séparation, extraction.

elutriator, séparateur, appareil à décantation.

eluvial, éluvial formé par altération; **e. deposit,** matériel altéré, gisement altéré; **e. horizon,** horizon éluvial, horizon lessivé; **e. hydromorphic soil,** sol lessivé à gley; **e. soil,** sol éluvial.

eluviation, éluviation (pédol.).

eluvium, éluvion.

elvan (GB), filon intrusif de granite, microgranite (etc).

emanation, émanation, effluve; **magmatic e.,** e. magmatique; **volcanic e.,** e. volcanique.

embank (to), endiguer, remblayer, terrasser.

embanking, endiguement.

embankment, 1. digue, talus berge; 2. remblai, terrassement en remblai.

embayed quartz grain, cristal de quartz à golfes de corrosion; **e. shore,** littoral découpé, côte à anses.

embayment, baie, golfe.
embed (to), entourer, incorporer, enrober.
embedded, intercalé, entouré.
embolite, embolite (*minér.*).
embouchure, embouchure.
embrechite, embréchite.
emerald, émeraude; **e. copper,** dioptase; **e. nickel,** zaratite.
emerge (to), émerger, déboucher.
emergence, 1. résurgence, émergence (d'un fleuve); **2.** émersion (d'un fond de mer); **coast of e.,** côte d'émersion.
emergent, émergent.
emery, émeri; **e. rock,** roche à corindon; **e. wheel,** meule en é.
eminence, hauteur, éminence, point haut.
emissary, émissaire (d'un lac).
emission, émission, dégagement.
emit (to), émettre.
emplectite, emplectite, euprobismuthite (*minér.*).
empty (to) into, 1. vider, épuiser; **2.** se déverser dans, se décharger dans (fleuve).
Emscherian (Eur.), Emschérien (≅ Coniacien Santonien).
Emsian, Emsien (étage, Dévonien inf. européen).
emulate (to), simuler.
emulator, programme de simulation.
enamel, émail (*pal.*).
enantiomorphic, enantiomorphous, énantiomorphe.
enargite, énargite (*minér.*).
encased valley, vallée encaissée.
enclosed meander, méandre encaissé.
enclosing beds, couches encaissantes.
enclosure, enclave, inclusion (minérale).
encrinital, à entroques.
encrinite, calcaire à entroques.
encrinitic limestone, calcaire à entroques, encrinite.
encroachment, empiétement, envahissement (par de l'eau, dans un forage).
encrust (to), incruster.
encrustation, incrustation.
encrusters, organismes marins « encroûtants » (cf. Bryozoaires, etc).
end, extrémité, bout, fin; **e. moraine,** moraine frontale, terminale; **e. product,** élément radioactif stable.
endemic, endémique.
endemism, endémisme.
ending, terminaison (d'un pli).
endo (prefix), à l'intérieur.
Endoceratidae, Endocératidés.
endocyclic, endocyclique (*pal.*).
endogen, endogenous, endogeneous, endogenetic, endogène; **e. dome,** cryptodôme.
endometamorphic, endomorphic, endomorphous, endomorphe.
endometamorphism, endomorphism, endomorphisme.
endoreic, endorheic, endoréique (drainage intérieur).
endorheism, endoréisme.

endoskeleton, endosquelette.
endothermic, endothermique.
endothermic peak, crochet endothermique (ATD).
endurance limit, limite d'endurance.
en echelon arrangement, disposition en échelons.
energetic, énergétique.
energy, énergie; **e. resource,** ressource énergétique; **geothermal e.,** é. géothermique;**solar e.,** é. solaire; **thermal e.,** é. thermique.
engineering geology, géologie de l'ingénieur.
englacial, intraglaciaire; **e. drift,** débris, ou transport intraglaciaire.
engulfment, engouffrement (de la lave dans une caldeira).
engyscope, microscope à réflexion.
enmeshed streamlets, rigoles de ruissellement enchevêtrées, filets enchevêtrés.
enrichment by flotation, enrichissement par flottage.
enroch (to), enrocher.
enrockment, enrochement.
enstatite, enstatite (*minér.*) variété de pyroxène.
enterolithic structure, fold, pli intraformationnel.
enthalpy, enthalpie.
entisol, cryosol.
entrainment, érosion.
entrapment, piégeage (du pétrole).
entrapped, emprisonné, piégé, enfermé.
entrenched meander, méandre encaissé (par surimposition); **e. stream,** fleuve encaissé.
entrenchment, encaissement.
entrochal limestone, calcaire à entroques.
entropy, entropie.
entry, entrée de galerie, galerie; **e. timbering,** boisage d'une galerie (*mine*).
envelope, 1. roche encaissante métamorphisée par contact de l'intrusion; **2.** enveloppe (d'un pli couché), « capuchon ».
environment, milieu; **high-energy e.,** m. à haute énergie (ex. agité par les vagues); **low-energy e.,** m. à basse énergie (calme); **sedimentary e.,** m. de sédimentation.
environmental geology, géologie de l'environnement; **e. analysis,** étude de milieu.
Eocambrian, Éocambrien (≅ Riphéen; moins 570 à moins 600 MA).
Eocene, Éocène (Europe : système ou période; Am. du N : série ou époque).
Eogene, Éogène (= Paléogène).
eolian, éolien; **e. deposit,** sédiment é.; **e. erosion,** érosion é.; **e. formation,** formation é.; **e. rocks,** roche é.; **e. sand,** sable é., sable dunaire; **e. sand ripple,** ride de sable é.
eolianite, aeolianite, sédiment éolien consolidé.
eolith, éolithe.
Eolithic, Éolithique (rare, avant le Paléolithique).
eon (aeon), éon; **1.** division supérieure à l'ère; ex : Cryptozoïque; **2.** 1 MA.
Eosuchia, Eosuchiens (*pal.*).
Eozoic, Éozoïque ère, (Précambrien).
epeiric, épicontinental; **e. sea,** mer épicontinentale.

epeirocratic paleogeography, paléogéographie dominée par l'émergence des continents.

epeirogenesis, épirogénèse (mouvements verticaux).

epeirogenic, epeirogenetic, épirogénique; **e. movement,** mouvement é.

epeirogeny, épirogénie, épirogénèse.

ephemeral stream, cours d'eau temporaire.

epibiotism, épibiotisme.

epibole, épibole (pal).

epibolite, épibolite (pétro.).

epicentral, épicentral; **e. angle,** angle formé par l'épicentre, la station sismique, et l'hypocentre.

epicentre, epicenter (US), épicentre.

epicentrum, épicentre.

epiclastic, épiclastique, détritique; **e. rock,** roche détritique (sauf les r. pyroclastiques).

epicontinental, épicontinental; **e. sea,** mer épicontinentale.

epicratonic, épicratonique.

epidiagenesis, épidiagenèse (par les eaux météoriques descendantes).

epidiorite, épidiorite (pétro.).

epidote, épidote (minér.).

epidodite, epidosite, épidodite (pétro.).

epidotization, épidotisation.

epifaune, organismes épibrontes.

epigene, 1. de surface (général); 2. épigène (cristallogr.).

epigenesis, épigenèse, épigénie.

epigenetic, épigénétique; **e. drainage,** réseau hydrographique surimposé.

epimagmatic origin, origine épimagmatique.

epineritic, épinéritique (entre le niveau de marée basse et − 40 m).

epipedon, épipédon.

epipelagic, épipélagique (de 0 à − 185 m).

epirogenic, epirogenetic, épirogénique, épirogénétique.

epitaxial, épitaxique; **e. growth,** croissance é. (minér.); **e. overgrowth,** croissance é.

epitaxy, épitaxie.

epithermal, épithermal.

epitheca, épithèque (pal.).

epizona, epizone, épizone.

epsomite, epsom salt, epsomite.

equal area projection, projection équivalente.

equator, équateur.

equatorial, équatorial; **e. projection,** projection cylindrique équatoriale; **e. trough,** zone de basses pressions équatoriales.

equiareal projection, projection cartographique équiaréale (conservant les surfaces).

equigranular, isogranulaire, à grains (minéraux) de même dimension.

equilibrium, équilibre (chimique, électrique); **e. profile,** profil d'é.; **e. regime,** régime d'é.; **profil of e.,** profil d'é.

equinoctual, 1. adj. : équinoxial; 2. n : équateur céleste.

equinox, équinoxe.

equiplanation, pénéplanation (sans perte de matériaux).

Equisetales, Équisétales.

equivalent, 1. équivalent; 2. de même âge ou de même niveau stratigraphique; **e. azimuthal projection,** projection Lambert; **e. grade,** dimension moyenne (arithmétique).

equivoluminal waves, ondes S.

equivolumnar waves, ondes S, ondes transversales.

era, ère.

eradiation, radiation, rayonnement.

erathem, érathème.

Erathosthenian, Érathosthénien (période moyenne de formation lunaire).

E ray, rayon extraordinaire (minéralogie optique).

erect (adj.), droit; **e. anticline,** anticlinal d.; **e. fold,** pli d.

erg, 1. unité de travail CGS, dyne par centimètre; 2. erg, désert sableux.

Erian (N. Am.), Érien (série Dévonien moyen); **e. orogeny (syn. Hibernian),** orogenèse érienne (fin du Silurien).

erinite, érinite.

erionite, érionite = zéolite (minér.).

erode (to), 1. éroder, ronger affouiller, corroder; 2. s'éroder.

eroded soil, sol érodé.

erodibility, érodabilité.

erodible, érodable.

erosion, érosion; **e. base level,** niveau de base d'é.; **e. column,** cheminée de fées, bloc perché; **e. cycle,** cycle d'é.; **e. scarp,** talus, escarpement d'é.; **e. surface,** surface d'é., d'aplanissement; **areal e.,** é. aréolaire; **backward e.,** é. régressive; **chemical e.,** é. chimique, corrosion; **of cycle e.,** cycle d'é.; **differential e.,** é. différentielle; **fluviatile e.,** é. fluviatile; **glacial e.,** é. glaciaire; **headward e.,** é. régressive; **lateral e.,** é. latérale; **mechanical e.,** é. mécanique, corrosion; **retrogressive e.,** é. régressive; **river e.,** é. fluviale; **sheet e.,** é. en nappes; **wind e.,** é. éolienne, déflation.

erosional, formé par érosion, d'érosion; **e. surface,** surface d'érosion; **e. unconformity,** discordance d'érosion.

erosive, érosif.

erosivity, érosivité.

erratic, 1. erratique (glaciaire); 2. irrégulier, variable, accidentel; **e. block,** bloc erratique.

erubescite, érubescite, bornite (minér.).

eructation, éruption volcanique violente.

eruption, éruption; **e. cloud,** nuée volcanique, ardente; **e. cone,** cône éruptif, formé de projections; **e. point,** centre éruptif; **e. flank,** é. latérale; **e. volcanic,** é. volcanique.

eruptive, éruptif; **e. vein,** filon de roches é.; **e. rock,** roche é.

erythrite, érythrite, érythrine (minér.).

erythrosiderite, érythrosidérite (minér.).

Erzgebirgian orogeny, orogenèse Erzgebirgienne (début Carbonifère sup.).

escape shaft, puits de secours.

escarp, escarpement, 1. talus, escarpement; 2. front de cuesta.

eschynite, eschynite (*pétrol.*).

eskar, esker, os, esker (= cordon sinueux de matériaux fluvio-glaciaires).

essential ejecta, projections volcaniques récentes (débris, ou liquides); **e. minerals,** minéraux essentiels (à la classification d'une roche).

essexite, essexite (*pétro.*).

essonite, essonite (*minér.*), variété de grenat grossulaire.

esterellite, estérellite (*pétro.*).

estuarian, estuarien.

estuarine, d'estuaire, estuarien.

estuary, estuaire.

etch (to), attaquer, corroder, graver; **e. marks,** figures de corrosion; **e. plain,** plaine de corrosion; **e. planation,** processus de corrosion.

etched rocks, pierres gravées; **e. surface,** surface mate, dépolie (d'un grain de sable).

etching, 1. attaque, corrosion (par dissolution chimique); 2. gravure.

ethane, éthane.

"ethane plus", hydrocarbures paraffiniques de poids moléculaire supérieur à celui du méthane.

ether, éther.

ethylene, éthylène.

Etrountian (= Strunian), Etroeungtien (étage, Dévonien final).

euclase, euclase (*minér.*).

eucrite, eucrite (gabbro).

eucrystalline, holocristallin, (et bien cristallisé).

eugeosyncline, eugéosynclinal.

euhaline, euryhalin.

euhedral, automorphe.

eulittoral zone, zone littorale (comprise entre 0 et − 50 m).

eulysite, 1. eulysite (var. d'olivine ferrifère); 2. eulysite (roche métamorphique ferrifère).

euphotic, euphotique, zone euphotique (de 0 à environ − 60 m).

euphotide, euphotide (*pétro.*).

eurite, eurite (*pétro.*) [désuet].

euritic, euritique (microgranitique).

euryhaline, euryhalin.

eurypterid, eurypterida, euryptères (*pal.*).

eurythermal organism, organisme eurytherme.

eustatic, eustatique (variation mondiale du niveau de la mer).

eutaxitic fabric, structure à «flammes» eutaxique (alignements cristallins dans roches pyroclastiques).

eutectic, eutectique; **e. mixture,** mélange e.; **e. point,** point e.; **e. temperature,** température e.; **e. texture,** texture e.; **e. valley,** puits e.

eutomus, à clivage net.

eutrophication, eutrophisation.

eutrophic lake, lac eutrophe.

euxenite, euxénite (*minér.*).

euxinic, euxinique.

evaporates (obsolete, see evaporite), sédiments évaporitiques.

evaporate (to), 1. évaporer; 2. s'évaporer.

evaporate down, concentrer par évaporation; **e. deposit,** dépôt d'évaporation, «évaporite».

evaporation, évaporation, volatilisation, vaporisation; **e. pan,** évaporimètre.

evaporite, «évaporite» (dépôt évaporitique).

even, 1. égal, lisse, plat, plan, uni, uniforme; 2. pair (nombre); **e. fracture,** cassure lisse; **e. grained,** à grains de même dimension; **e. ground,** terrain uni.

even (to), aplanir, égaliser, araser.

event, événement, phénomène, cas, séisme; **e. deposit,** tempestite.

ever frozen soil, pergélisol.

evergreen (plants), végétaux à feuilles persistantes.

evolute, évolute (*pal.*), à enroulement lâche.

evolution, 1. évolution (*biol.*); 2. déroulement, tracé d'une courbe (*math.*); 3. dégagement (*phys-ch.*).

evolutionary process, processus évolutif.

evolutionary rate, vitesse d'évolution.

exaggeration of a profile, multiplication de l'échelle d'une coupe.

excavate (to), creuser, déblayer, affouiller, excaver.

excavating bucket, drague, benne excavatrice.

excavation, 1. excavation, cavité, creux, fouille; 2. creusement; 3. déblai.

excavator, 1. excavateur, machine à défoncer; 2. terrassier.

exchange, échange; **e. capacity,** capacité d'é.; **base e.,** é. de base; **ion e.,** é. d'ions.

exciting field, champ d'excitation.

excrements, coprolites.

excursion, 1. excursion; 2. migration du pôle; 3. déflexion (*sismol.*).

exert a pressure (to), exercer une pression.

exfoliate (to), 1. exfolier, desquamer; 2. s'exfolier.

exfoliation, desquamation, exfoliation en écailles, en plaques, etc.

exhalation, exhalation, exhalaison.

exhale (to), exhaler, émettre (un gaz).

exhaust, évacuation, échappement (mécanique).

exhaust (to), épuiser, vider.

exhausted, épuisé (*mine*).

exhaustion, 1. épuisement (d'une *mine*); 2. aspiration.

exhumation, exhumation (d'un paléorelief).

exhumed topography, relief exhumé.

exine, exine.

exinite, exinite.

exocyclic Echinoids, Oursins irréguliers.

exogenetic, exogenic, exogenous, exogène, formé par processus externe; **e. spine,** aiguille d'extrusion.

exomorphic, exomorphique.

exomorphism, exomorphisme.

exoreic, exorheic, exoréique (drainage intérieur).

exorheism, exoréisme.

exoskeleton, exosquelette.

exoscopy, exoscopie.

exosphere, haute atmosphère (terrestre).

exothermal, exothermic, exothermique.

exotic, **1.** allochtone; **2.** introduit, exotique (*pal.*).; **e. block,** bloc charrié ou sédimentaire d'origine lointaine, petit klippe; **e. breccia,** brèche exotique.

expand (to), **1.** se dilater; **2.** se détendre (gaz); **3.** s'agrandir, se développer.

expandable clay, argile gonflante.

expanded foot of a glacier, lobe aval d'un glacier.

expanding earth, globe terrestre en expansion.

expansion, **1.** dilatation; **2.** croissance, expansion; **e. coefficient,** coefficient de dilatation; **e. of the universe,** expansion de l'univers (*astro.*); **e. ratio,** taux de dilatation.

expansive soil, sol gonflant.

experimental geology, géologie expérimentale.

expert, expert, spécialiste.

explanatory note, notice explicative.

exploder, exploseur; **gas e.,** détonateur à gaz (sismique).

exploit (to), exploiter, mettre en exploitation.

exploitation, exploitation, **e. drilling,** forage d'e.

exploration, exploration; **e. boring,** sondage de recherches; **e. expenses,** dépenses d'e.; **e. seismology,** prospection sismique; **e. work,** travaux d'e.

exploratory, d'exploration, de reconnaissance; **e. hole,** sondage de recherches; **e. survey,** levé préliminaire de reconnaissance; **e. well,** sondage de reconnaissance; **e. work,** recherches préliminaires.

explore (to), explorer, reconnaître.

explorer, **1.** explorateur; **2.** instrument de recherches; **3.** satellite artificiel.

explosion, explosion, détonation; **e. breccia,** brèche d'explosion volcanique; **e. caldera,** caldeira d'explosion; **e. crater,** cratère d'explosion; **e. point,** stade explosif; **e. tuff,** tuff volcanique.

explosive, explosif, détonant; **e. evolution,** évolution explosive (*pal.*).

expose (to), **1.** découvrir, mettre à nu; **2.** affleurer.

exposure, **1.** affleurement; **2.** exposition; **e. age,** âge d'exposition.

expulsion, expulsion, éjection.

exsication, assèchement de terrains (par drainage).

exsolution, exsolution (*minér.*).

exsolve (to), entrer en exsolution.

exsolved mineral, minéral formé par exsolution.

exsudation, **1.** écaillage de la surface de roches par exsudation des sels; **2.** exsudation, suintement; **e. vein,** filon d'exsudation, de différenciation magmatique.

exsurgence, exutoire (fluviatile).

extension agreement, accord de prolongation d'une concession; **e. well,** puits d'extension d'un gisement.

extensional margin, marge de distension, marge passive.

external magnetic field, champ magnétique externe.

external mold,, moulage externe.

Externides, Externides.

extinct (volcanoe), (volcan) éteint.

extinction, extinction (*pétro., pal.*), angle d'extinction (*pétro.*); **inclined e.,** e. oblique (*optique*); **straight e.,** e. droite (*optique*);**symmetrical e.,** e. symétrique (*optique*); **undulate e.,** e. roulante, ondulante.

extinguished (volcano), éteint (volcan).

extraclast, fragment bréchique plus ancien.

extract (to), extraire, arracher, retirer.

extractable, extractible.

extracting, extraction.

extraction, extraction; **e. drift,** galerie de taille; **e. plant,** usine d'e.; **e. process,** procédé d'e.; **e. shaft,** puits d'e.

extragalactic, extragalactique.

extramagmatic, extramagmatique.

extraordinary ray, rayon extraordinaire (*optique, phys.*).

extraterrestrial geology, géologie extraterrestre.

extrude (to), **1.** faire extrusion, s'épancher; **2.** refouler; **3.** tréfiler (*métallogénie*).

extrusion, extrusion, épanchement.

extrusive rocks, roches d'épanchement, roches effusives.

eye, œil, œillet; **e. lens,** loupe; **e. piece,** oculaire; **e. structure,** structure œillée.

F

fabric, structure et texture, fabrique, orientation; **f. element,** élément structural; **clastic f.,** structure bréchique; **glassy f.,** structure vitreuse; **planar f.,** foliation; **sequential f.,** structure grenue avec ordre de cristallisation.

face, **1.** front d'abattage, front de taille, taille *(mine);* **2.** face, facette *(minér.);* **3.** surface originellement supérieure d'une couche redressée; **f. man,** abatteur *(mine);* **advancing f.,** taille chassante *(mine);* **crystal f.,** face d'un cristal; **glacier f.,** front glaciaire; **retreating f.,** taille rabattante *(mine);* **rock f.,** muraille, paroi.

face (to), **1.** faire face à, être en face de ; **2.** être dirigé vers des couches plus jeunes (pour des couches renversées).

facet, facette; **solution f.,** f. de dissolution.

facet (to), facetter (une pierre précieuse).

facetted pebble, galet à facettes; **f. spur,** éperon tronqué, chaîne tronquée.

facial suture, suture faciale (Trilobite).

facies, faciès; **f. family,** famille de f.; **f. fauna,** faune de f.; **f. fossil,** fossile de f.; **f. map,** carte de f; **biofacies,** biofaciès; **lithofacies,** lithofaciès; **metamorphic f.,** faciès métamorphique.

faecal (fecal) pellet, pelote fécale.

fahlband, fahlbande (imprégnations de sulfure).

fahlerz (german), minerai de cuivre gris tétraédrite, panabase, tennantite.

fahlore, panabase (minerai de cuivre gris), tétraédrite.

Fahrenheit, Fahrenheit; **f. scale,** échelle Fahrenheit.

failing yield, débit en baisse (d'un puits).

failure, **1.** insuccès, échec; **2.** fracture, rupture, fissure; **f. strenght,** résistance à la rupture.

faint slope, pente faible.

fair weather, beau temps *(météo.).*

fake, roche à débit schisteux.

fall, **1.** éboulement; **2.** chute d'eau, cataracte; **3.** décrue, reflux, jusant; **4.** baisse, abaissement; **5.** automne (Am. du N.); **f. line,** ligne de rapides (d'un fleuve); **f. zone, belt,** zone de rapides, ceinture; **f. of rain,** chute de pluie; **f. of snow,** chute de neige; **f. of stones,** chute de pierres; **f. of the tide,** jusant, reflux; **f. of water-level,** abaissement du niveau de l'eau; **water f.,** chute d'eau.

fall in (to), s'ébouler.

fall into (to), déboucher dans.

fallen in, éboulé, effondré.

faller, taquet à abaissement de cage *(mine).*

falling, chute, éboulement, baisse.

falling in of stones, éboulement de pierres; **f. stone,** météorite; **f. tide,** marée descendante.

false, faux; **f. amethyst,** fluorine violette; **f. bedding (obsolete),** stratifications obliques; **f. cleavage,** pseudo-clivage; **f. color image,** image en fausses couleurs; **f. dip,** pendage apparent; **f. galena,** blende; **f. superposition,** superposition inverse, renversement (de couches); **f. topaz,** citrine.

falun (French), falun (sable coquillier).

famatinite, famatinite *(minér.).*

Famennian, Famennien (étage, Dévonien sup.).

families of igneous rocks, familles de roches éruptives.

fan, **1.** cône de déjection; **2.** ventilateur; **f. cleavage,** fracture en éventail; **f. delta,** cône de déjection; **f. fold,** pli en éventail; **f. shaped structure,** structure en éventail; **f. shooting,** tir en éventail *(sismique);* **f. structure,** cône alluvial; **f. talus,** talus d'éboulis; **alluvial f.,** cône alluvial, cône de déjection; **avalanche f.,** cône d'avalanches; **deep sea f.,** delta sous-marin profond; **talus f.,** cône d'éboulis.

fan (to), ventiler *(mine).*

fanglomerate, dépôt de cône alluvial cimenté (ultérieurement).

farewell, roche stérile.

Farlovian, Farlovien (Dévonien sup.).

fasciculate, groupé en faisceaux.

fassaite, fassaïte (variété d'augite).

fasten (to), **1.** attacher, fixer, amarrer, cramponner; **2.** se fixer, s'attacher.

fast ice, pied de glace (glace littorale).

fat clay, argile plastique; **fat coal,** houille grasse.

fathogram, fathogramme (profil bathymétrique).

fathom, **1.** brasse = 6 pieds = 1,829 m; **2.** volume de 216 pieds cubes.

fathom (to), sonder.

fathometer, échosondeur, fathomètre.

fatty lime, chaux grasse; **f. acid,** acide gras.

fault, faille, accident; **f. basin,** bassin d'effondrement; **f. bench,** gradin de f.; **f. block,** bloc f.; **f. block mountain,** horst; **f. boundary,** limite de f.; **f. breccia,** brèche de f.; **f. bundle,** faisceau de f.; **f. clay,** enduit argileux de f.; **f. cliff,** escarpement de f.; **f. conglomerate,** conglomérat de f.; **f. dip,** inclinaison, pendage de la f.; **f. drag,** rebroussement des lèvres de f.; **f. escarpment,** escarpement de f.; **f. fissure,** fente formée par une f.; **f. fold,** pli-faille; **f. line,** ligne de f.; **f. line scarp,** escarpement de f.; **f. gouge,** mylonite pulvérulente; **f. pit,** effondrement circulaire; **f. plane,** plan de f.; **f. polish,** miroir de f.; **f. rock,** brèche de f.; **f. scarp,** escarpement de f.; **f. set,** ensemble, réseau, de f.; **f. spring,** source d'origine tectonique; **f. strike,** direction de f.; **f. surface,** plan de f.; **f. tectonics,** tectonique cassante; **f. trace,** ligne de f.; **f. trap,** piège de f.; **f. throw,** rejet de f.; **f. through,** fossé tectonique; **f. valley,** vallée tectonique; **f. vein,** filon faillé; **f. wall,** lèvre de la f.; **activated f. zone**, zone faillée; **activated f.**, f. rajeunie; **antithetic f.**, f. contraire; **bedding f.**, f. dans le plan de stratification; **boundary f.**, f. limite; **branch f.**, f. secondaire; **branching f.**, f. ramifiée; **closed f.**, f. fermée; **compressional f.**, f. de compression; **diagonal f.**, f. oblique; **dip f.**, f. transversale; **dip slip f.**, f. normale; **distributive f.**, f. en escalier; **downthrow f.**, f. d'effondement; **inclined f.**, f. inclinée; **gravity f.**, f. normale; **growth f.**, f. directe; **lateral f.**, f. latérale; **longitudinal f.**, f. longitudinale; **multithrow f.**, f. à rejets multiples; **normal f.**, f. normale; **normal slip f.**, f. normale; **oblique slip f.**, f. normale avec déplacement latéral; **open f.**, f. ouverte; **overthrust f.**, plan de charriage; **peripheral f.**, f. radiale; **pivot f.**, f. en ciseaux; **repetitive f.**, f. à répétition; **reverse f.**, f. inverse; **rotary f.**, rotation; **shear f.**, f. de cisaillement; **slip f.**, f. d'effondrement; **splitting f.**, f. ramifiée; **step f.**, f. en gradins; **strike f.**, longitudinale, directionnelle; **strike-slip f.**, décrochement; **tear f.**, décrochement; **thrust f.**, f. chevauchante; **transcurrent f.**, f. de décrochement; **transform f.**, f. transformante; **translational f.**, décrochement; **transverse f.**, f. transversale; **upthrow f.**, chevauchement; **wrench f.**, décrochement.

faultage, dislocation, faille.

faulted, faillé; **f. anticline**, pli-faille; **f. down**, abaissé par faille; **f. terrane**, terrain accidenté.

faulting, formation de failles; **recurrent f.**, rejeu de faille.

fauna assemblage, ensemble biostratigraphique.

faunal province, province faunistique; **faunal list**, liste faunistique.

faunizone, faunizone (unité de biostratigraphie).

fauserite, fausérite (*minér.*).

fay (to), affleurer.

fayalite, fayalite, olivine, ferrifère (*minér.*).

feasibility study, étude de faisabilité.

feather alum, halotrichite (*minér.*); **f. ore**, jamesonite (*minér.*), stibnite fibreuse.

fecal pellet, pelote fécale.

feeder, 1. système d'alimentation; 2. affluent, tributaire; 3. filon nourricier; **feeder sand**, sable pour regarnir les plages érodées.

feeding vent, cheminée volcanique, diatrème.

feed pipe, conduite d'alimentation.

feldspar, feldspath, felspar, feldspath.

feldspathic, felspathic, feldspathique; **f. rock**, roche feldspathique; **f. sandstone (cf. arkosic)**, grès feldspathique (10 à 25 % de feldspath).

feldspathization, feldspathisation.

feldspathoid, feldspathoïde.

feldspathose, feldspathique.

fell, fjell, fjäll (Scand), 1. colline ou plateau rocheux dénudé (Écosse); 2. minerai de plomb.

fells shale, schiste bitumeux (Écosse).

felling, abattage (*mine*).

felsenmeer (german) (cf. boulder field), champ de pierres, chaos rocheux.

felsic minerals, minéraux de couleur claire (quartz, feldspaths, feldspathoïdes).

felsite, felsyte, roche éruptive acide à grain fin : **a)** soit cristalline, et émettant des filons, cf aplite; **b)** soit volcanique, à pâte formée d'agrégats cryptocristallins de minéraux clairs.

felsitic rock, roche claire, à fins cristaux non distincts à l'œil nu; **f. texture**, texture à agrégats cryptocristallins formés par dévitrification de verres volcaniques.

felspar (GB), feldspath.

felspathic, feldspathique.

feldspathoidal, feldspathoïdal.

felstone (obsolete), 1. roche éruptive, claire, à grain fin; 2. feldspath compact.

felty, var. de structure microlithique.

femic, ferromagnésien (au point de vue de la norme); **f. rock**, roche mafique.

fen, marécage, tourbière; **f. fire**, feu follet; **f. soil**, tourbière basse.

fence diagram, 1. bloc diagramme (coupes géologiques); 2. diagramme de stabilité (*géochimie*).

fenestral fabric, structure vacuolaire.

fenestrated, perforé.

fenite, fénite (*pétro.*).

fenitization, fénitisation.

fenny, marécageux.

fenster, fenêtre (de nappe de charriage).

ferralitization, ferralitisation.

ferretto paleosol, paléosol interglaciaire à feretto.

fergusonite, fergusonite (*minér.*).

ferralitic, ferralitique, latéritique; **f. soil**, ferralsol, sol ferralitique, sol latéritique.

ferreous, ferreux.

ferric, ferrique.

ferricrete, conglomérat à ciment ferrugineux.

ferricrust, croûte ferrugineuse.

ferricyanic, ferricyanhydrique.

ferricyanide, ferricyanure.

ferriferous, ferrifère, ferreux.

ferrimorphic soil, sol ferrugineux rouge.

ferrinatrite, ferronatrite (*minér.*).

ferrisols, sols ferrugineux.

ferrite, 1. ferrite (*pétro.*); 2. ferrite (*métallogénie*).

ferro-alloy, ferroalliage.

ferroan, contenant du fer, ferrugineux; **f. dolomite (syn. ankerite)**, dolomie ferrifère.

ferrocalcite, ferrocalcite.

ferrocobaltite, ferrocobalte, cobaltine ferrifère.

ferrod, ferrod, podzol ferrugineux.

ferroferite, magnétite.

ferromagnesian, ferromagnésien.

ferromagnetic material, substance ferromagnétique.

ferromagnetism, ferromagnétisme.

ferromanganese, ferromanganèse.

ferrotellurite, ferrotellurite.

ferrous, ferreux.

ferruginate, à ciment ferrugineux.

ferrugineous, ferruginous, ferrugineux; **f. cuirass**, cuirasse ferrugineuse; **f. spring**, source ferrugineuse; **f. water**, eau ferrugineuse.

ferruginisation, ferruginisation.

ferrum, fer.

fersiallitic soil, sol ferrugineux tropical (sol fersialli-tique).

Festiniogian (Eur.), Festiniogien (étage, Cambrien sup.).

festoon, 1. feston, guirlande (périglaciaire); 2. arc insulaire; 3. type de stratification entrecroisée, à dépressions marquées; **stone f.,** guirlande de solifluxion.

fetch, fetch, étendue.

fiard, fjord à versants doux (Suède).

fibroblastic, nématoblastique, texture métamor-phique dans laquelle dominent les minéraux aciculaires.

fibrolite, variété fibreuse de sillimanite.

fibrous, fibreux; **f. duff,** humus brut f.; **f. fracture,** cassure fibreuse; **f. serpentine,** chrysolite.

field, 1. terrain; 2. gisement; 3. champ (sens propre et figuré); 4. champ électrique; 5. champ magné-tique; 6. champ optique; **f. book,** carnet de terrain; **f. capacity,** capacité capillaire d'un sol; **f. completion,** complétement au sol (de carte); **f. current,** courant inducteur; **f. geology,** géologie de terrain; **f. ice,** banquise; **f. intensity,** intensité du champ magnétique; **f. log,** carnet de terrain; **f. magnet,** inducteur; **f. notes,** carnet de terrain; **f. observations,** observations de terrain; **f. reversal,** inversion de champ magnétique; **f. sampling,** échantillonnage de terrains; **f. survey,** étude de terrain; **f. work,** prospection.

fierry, 1. inflammable; 2. grisouteux.

figuline, argile figuline, terre à poterie.

figurestone, agalmatolite.

Filicales, Filicales (pal.).

fill, 1. remplissage naturel de cavité; 2. matériaux de remplissage (sables, etc.); 3. remblai (artificiel); **f. earth,** terre à remblai; **f. in fill terrace,** terrasse emboîtée; **f. up,** remblayage.

fill (to), 1. remplir, emplir; 2. combler, remblayer; 3. s'emplir, se remplir.

filling, comblement, remplissage, remblai, rembla-yage; **f. machine,** remblayeuse; **f. substance,** matériaux de remblai; **f. system,** système d'ex-ploitation avec remblayage; **f. up,** comblement; **back f.,** remblayage (de mine ou chantier).

filltop terrace, terrasse emboîtée.

film, 1. revêtement mince, pellicule; 2. pellicule, film (photog.); **f. library,** cinémathèque; **f. of oil,** pellicule de pétrole; **boundary f.,** couche limite.

filter, 1. filtre; 2. filtre (géophys.); **f. bed,** couche filtrante; **f. cloth,** toile filtrante; **f. clay,** terre filtrante; **f. correction,** correction de filtre; **f. medium,** milieu filtrant; **f. pass-band,** bande de transmission (*géophys.*); **f. pressing,** processus de différenciation magmatique (expulsion de la phase liquide dans un magma en cours de cristallisa-tion); **f. sand,** sable filtrant; **f. sieve,** tamis filtrant à mailles fines; **f. screen,** 1. crible filtrant; 2. écran filtrant; **f. well,** puits filtrant; **sand f.,** filtre à sable; **water f.,** filtre à eau.

filtered gravity map, carte gravimétrique obtenue par traitement mathématique.

filtering, 1. adj : filtrant; 2. n : filtration, filtrage; **f. stone,** pierre poreuse.

filtrate, filtrat.

filtration, filtration, filtrage, suintement; **f. spring,** source d'infiltration.

fin, nageoire (*pal.*).

final boiling point (FBP), point d'ébullition final.

find (to) one's bearings, s'orienter à la boussole.

fine, 1. fin, à grain fin, de faible granulométrie, menu; 2. pur, fin; 3. beau; **f. crushing,** broyage fin; **f. gold,** or fin; **f. grained,** à grain fin; **f. grained sand,** sable à grain fin; **f. soil,** sol limono-argileux; **f. granular,** à grain fin; **f. ore,** minerai fin; **f. pored,** finement poreux; **f. sand,** sable fin, sablon (0,25 mm < L < 0,125 mm); **f. sandstone,** grès fin; **f. silt,** limon fin; **f. structure,** microstruc-ture; **f. textured,** à grain fin; **f. weather,** beau temps.

fine (to), purifier, affiner.

fineness, 1. titre, qualité (d'un métal); 2. finesse.

finery, affiné.

fines, 1. fraction fine, particules fines; 2. minerai fin, « fines ».

Fingerlakesian (N. Am.), Fingerlakésien (étage, base du Dévonien sup.).

fingering, digitation (d'un gisement).

fingerprint, signature, empreinte d'un phénomène.

fining, purification, affinage.

fining-upward alluvial cycle, grano-classement (avec dépôt de particules fines vers le haut).

fining-up, with top missing, cycle à banc majeur à la base (cf. sohlbank).

finite ressource, ressource non-renouvelable.

fiord, fjord, fjord.

fiorite, fiorite (var. d'opale).

fire, incendie, feu; **f. assay,** essai pyrognostique; **f. belt,** pare-feu (forêts); **f. blende,** pyrostilpnite; **f. brick,** brique réfractaire; **f. clay,** argile réfractaire; **f. coat,** revêtement réfractaire; **f. damp,** gaz des marais, grisou, méthane; **f. damp detector,** détec-teur à grisou; **f. damp outburst,** dégagement de grisou; **f. damp pocket,** poche de grisou; **f. explosion,** explosion de grisou; **f. fountain,** fon-taine de laves; **f. loss,** perte au feu; **f. opal,** opale de feu; **f. setting,** abattage au feu (*mine*); **f. sand,** sable réfractaire; **f. tile,** tuile réfractaire; **f. well,** fontaine de lave.

fire (to), incendier, enflammer.

firing, 1. tir de mine, tir, mise à feu; 2. chauffe, chauffage; 3. cuisson; 4. émission sismique artifi-cielle; **shot f.,** tir des coups de mine.

firn, névé; **f. basin,** bassin d'alimentation glaciaire; **f. ice,** glace de n.; **f. limit,** limite d'ablation estivale d'un n.; **f. moraine,** moraine de n; **f. wind,** vent catabatique.

first, premier; **f. arrivals,** premières ondes sismiques enregistrées; **f. bottom,** plaine alluviale; **f. leg of reflection,** première phase de réflection; **f. motion,** direction du mouvement du sol à l'arrivée des ondes P sur un séismogramme; **f. working,** traçage (*mine*).

firth, estuaire, fjord, bras de mer (Écosse).

fish-eye stone, apophyllite = variété de zéolite (*minér.*).

fish up (to), repêcher (un objet dans un forage).

fishing, repêchage (for.); **f. jar,** coulisse de repêchage; **f. tools,** outils de repêchage.

fishtail bit, trépan à queue de poisson.

fissile, fissile.

fissibility, fissilité.

fission, fission; **f. track age,** âge déterminé à partir des traces de f.

Fissipeda, Fissipèdes (*pal.*).

fissuration, fissuration, fendillement.

fissure, fissure, lente; **f. eruption,** éruption fissurale; **f. network soil,** sol réticulé; **f. vein,** fissure filonienne, filon minéralisé; **shallow f.,** fissure superficielle; **tension f.,** fissure d'extension.

fissure (to), fissurer, fendre, crevasser.

fissuring, fissuration.

fix (to), positionner un point sur une carte.

fixed cheek, joue fixe (Trilobite).

fixed dune, dune stabilisée.

fixed wave, onde stationnaire.

fixism, fixisme.

fjord, fiord, fjord.

flag, dalle, pierre plate; **f. ore,** minerai stratifié.

flagstone, 1. dalle; **2.** roche fissurable en dalle.

flaggy, 1. se débitant en dalles; **2.** de faible épaisseur (1 à 10 cm).

Flagellates, Flagellés.

flags, grès durs ou calcaires lités utilisés comme dalles.

flake, 1. éclat; **2.** écaille, paillette; **3.** feuillet; **4.** flocon; **5.** éclat préhistorique.

flaked, en écailles.

flaking off, exfoliation, écaillage, desquamation.

flaky, en écailles.

flame, flamme; **f. photometry,** spectrophotométrie de f.; **f. proof,** résistant au feu; **f. structure,** figure de charge, avec injection (ou remontée) en « flammes » dans le lit sédimentaire sus-jacent.

flaming coal, charbon flambant.

Flandrian, Flandrien (formation marine, Holocène).

flank, flanc; **f. dip,** pendage latéral; **f. eruption,** éruption latérale; **f. moraine,** moraine latérale; **f. well,** puits latéral.

flanking moraine, moraine latérale.

flap structure, décollement par gravité et renversement d'une couche sur les flancs d'un anticlinal.

flare, torchère, torche; **gas f.,** torche; **solar f.,** éruption solaire.

flare up (to), 1. s'enflammer brusquement; **2.** briller (*astro.*).

flared slope, versant évasé.

flaring, brûlage à la torche.

flaser bedding, stratification entrecroisée.

flaser structure, structure lenticulaire.

flash, 1. éclair; **2.** crue très rapide; **f. flood,** inondation, crue rapide.

flask, fiole, flacon (laboratoire).

flat, 1. adj : plat, plan; **2. n :** bas-fond, marécage; **3. n :** couche subhorizontale de charbon (staffordshire); **f. bottomed valley,** vallée à fond plat; **f. lode,** filon subhorizontal; **f. of ore,** gîte minéral subhorizontal dans un plan de stratification; **f. surface,** surface plane; **mud f.,** estran (surtout la slikke); **sand f.,** estran; **tidal f.,** estran; **valley f.,** fond de vallée.

flatness index, indice d'aplatissement.

flatness ratio, indice d'aplatissement (d'un galet).

flatten (to), 1. aplatir, aplanir; **2.** s'aplatir.

flattening, aplanissement, aplatissement; **f. index,** indice d'aplatissement (*sédim.*).

flaxseed ore, oolithes ferrugineuses ovalaires (en « graine de lin »).

fleckshiefer, schiste tacheté.

flexible sandstone, itacolumite, grès micacé.

flexible silver ore, sternbergite (*minér.*).

flexion, flexion, courbure.

flex point, point d'inflexion.

flexural slip fold, pli par flexion et glissement, flexure.

flexure, 1. pli (Am. du N.); **2.** flexure, courbure; **f. fault,** zone de cisaillement.

flint, silex; **f. age,** âge de la pierre; **f. clay,** argile à s.; **f. stone,** s.

flinty, 1. de silex, en silex; **2.** caillouteux; **f. crush-rock,** mylonite partiellement fondue (dynamométamorphisme); **f. fracture,** cassure conchoïdale; **f. ground,** sol siliceux; **f. slate,** schiste siliceux.

float, 1. minéraux d'altération, fragments rocheux détachés par altération; **2.** flotteur (océano.); **f. and sink testing,** essai de séparation densimétrique; **f. copper,** cuivre natif détritique; **f. gold,** fines paillettes d'or; **f. ore, 1.** paillettes de minerai; **2.** fragments de minerai en aval ou en contrebas de l'affleurement; **f. stone,** variété de pierre ponce.

float (to), 1. flotter, nager, surnager; **2.** transporter; **3.** inonder, submerger.

floating, 1. flottant, libre; **2.** en suspension, dispersion; **f. bog,** marais tremblant; **f. gold,** paillettes d'or; **f. ice,** glace flottante; **f. marsh,** marais tremblant; **f. platform,** plate-forme flottante; **f. sand grain,** grain de sable dispersé (dans une matrice calcaire).

floc, floconts, floculats.

flocculating agent, agent de floculation.

flocculation, floculation.

flocculent, flocon.

flock, flocon (formé par précipitation); **f. point,** point de floculation; **f. test,** essai de floculation.

flocky, floconneux.

floe, glaçon flottant, masse de glaces flottantes, banquise; **f. ice,** glace flottante; **f. rock,** éboulis de grès et argiles réfractaires.

flood, 1. crue (fluviatile); **2.** flot (de la marée); **3.** concentration minérale (en milieu sédimentaire);

f. basalt, basalte de plateau; **f. basin,** plaine d'inondation; **f. control,** surveillance, contrôle des crues; **f. control reservoirs,** réservoirs artificiels de retenue; **f. current,** courant de flot; **f. deposit,** dépôt de crue; **f. peak,** maximum de la crue; **f. plain,** plaine d'inondation, lit majeur; **f. plain silt,** limon de débordement; **f. plain terrace,** terrasse alluviale; **f. planning,** planification des crues; **f. tidal delta,** delta de marée montante, delta de flot (US); **f. tide,** marée montante; **f. zone,** zone inondable, lit majeur; **high f.,** crue; **ice f.,** glaciation; **flash f.,** crue soudaine; **lava f.,** épanchement de laves; **sea f.,** raz de marée; **sheet f.,** ruissellement en nappes.

flood (to), 1. inonder, noyer, submerger; 2. déborder, être en crue.

flooded mine, mine noyée.

flooding, 1. inondation, submersion; 2. débordement, crue (d'une rivière); 3. injection de fluide; **f. of a well,** noyage d'un puits; **f. the mine,** noyage de la mine; **air f.,** injection d'air; **gas f.,** injection de gaz; **water f.,** injection d'eau (pour récupération de pétrole).

floor, 1. plancher, plate-forme; 2. socle, fond, mur (*mine*); **f. limb,** flanc inférieur d'un pli-couché; **f. of seam,** mur d'une couche (*mine*); **f. thrust,** chevauchement basal.

floorman, ouvrier travaillant sur un plancher de forage.

flora, flore.

floral province, province botanique.

floss-ferri, flos-feri, floss-feri (var. d'aragonite dans les stalactites).

flotability, flottabilité.

flotable, flottable.

flotation, flottation, flottage; **f. concentrate,** concentré de flottation; **f. test,** essai de flottation; **enrichment by f.,** enrichissement par flottation.

flour, farine; **f. copper,** très fines paillettes de cuivre natif; **f. gold,** très fines paillettes d'or; **glacial f.,** poussière, f. glaciaire.

flow, 1. écoulement; 2. débit; 3. flot, débit; 4. coulée (de laves, etc.); 5. fluage; **f. banded,** lité; **f. banding,** litage de flux; **f. breccia,** brèche de coulée de lave; **f. cast,** trace inverse d'un courant marin; **f. channel,** chenal d'écoulement; **f. cleavage,** schistosité de pression, clivage de flux; **f. earth,** manteau de solifluxion; **f. fold,** pli de fluage; **f. folding,** pli ptygmatique; **f. head,** tête d'écoulement (*for.*); **f. indicator,** débitmètre; **f. layer,** couche litée; **f. layering,** litage d'écoulement magmatique; **f. line,** 1. ligne de flux; 2. direction d'écoulement; 3. conduite d'écoulement; **f. meter,** débitmètre; **f. of rocks,** coulée de pierres; **f. of the tide,** flot de la marée; **f. rate,** débit; **f. recorder,** débitmètre enregistreur; **f. schedule,** programme de production (*for.*); **f. stage,** stade visqueux, stade fluide; **f. stretching,** orientation des minéraux métamorphiques; **f. structure,** structure fluidale; **f. surface,** surface de feuillets (métamor-

phisme); **f. test,** essai d'écoulement; **f. texture,** texture fluidale; **f. units,** unités d'une coulée de laves; **counter f.,** à contre-courant; **free f.,** écoulement libre; **laminar f.,** écoulement laminaire; **lava f.,** coulée de lave; **mud f.,** coulée de boue; **plastic f.,** écoulement visqueux; **shooting f.,** écoulement violent; **soil f.,** solifluxion; **spring f.,** débit d'une source; **steady f.,** écoulement fluviatile régulier; **turbulent f.,** écoulement torrentiel; **underf.,** sous-écoulement; **unsteady f.,** écoulement torrentiel turbulent.

flow (to), couler, s'écouler; **f. back (to),** refluer; **f. by heads,** jaillir par intermittence; **f. into,** affluer, se verser dans; **f. out,** se vider, s'écouler.

flowage, 1. écoulement plastique; 2. fluage.

flower of iron, floss-ferri, aragonite.

flower of sulfur, fleur de soufre.

flowering plant, plante à fleur.

flowing, 1. adj : coulant, s'écoulant; 2. n : écoulement, jaillissement; **f. well,** puits à jaillissement spontané.

flucan, flookan, salbande argileuse, glaise, terre glaiseuse.

fluctuation, fluctuation, oscillation, variation (du niveau marin).

flue, 1. conduite, canal; 2. carneau (*mine*); **f. gas,** gaz de carneau.

fluid, 1. adj : fluide; 2. n : fluide; **f. inclusion,** inclusion fluide; **f. injection,** injection à l'état fluide; **f. mechanics,** mécanique des fluides; **f. unit,** unité de craquage à catalyseur fluide (raffinage); **drilling f.,** fluide de forage.

fluidal, fluidal; **f. structure,** structure fluidale.

fluidify (to), fluidifier.

fluidity, fluidité.

fluidization, fluidisation.

fluidized, fluidisé.

flume (US), 1. ravin, torrent; 2. canal d'amenée; 3. réservoir.

fluocerite, fluocérine, fluocérite (*minér.*).

fluometer, appareil à doser le fluor.

fluor, 1. fluor; 2. fluorine (*minér.*).

fluorapatite, fluorapatite.

fluorescence, fluorescence.

fluorescein, fluorescéine.

fluorescent, fluorescent.

fluoride, fluorure.

fluorimetry, spectrométrie à fluorescence.

fluorinate (to), fluorer.

fluorination, fluoruration.

fluorine, fluor; **f. dating,** datation au fluor.

fluorite, fluorine, fluorite.

fluoritic, fluoritique.

fluorographic method, méthode de fluorescence aux ultraviolets.

fluorspar, fluorine, fluorite.

flush, remblai d'embouage (*mine*); **f. irrigation,** irrigation par les crues; **f. production of a well,** production éruptive non réglée du début d'exploitation.

flushing, 1. injection d'eau, chasse d'eau; 2. balayage du pétrole par de l'eau (dans un piège); 3. remblayage par embouage; **f. of drill bit,** rinçage d'un trépan; **f. shaft,** puits d'embouage; **core f.,** lavage d'une carotte.

flush out (to), jaillir; **f. out (to),** rincer, laver, curer; **f. over (to),** déborder.

flute, 1. rainure (érosion glaciaire); 2. flute (*sédimentol.*).

flute-mark, marque de courant en dos de cuillère.

flute (to), canneler, strier.

fluting, cannelures glaciaires.

fluvent, sol alluvial (*pédol.*).

fluvial, fluvial; **f. geomorphic cycle,** cycle d'érosion fluviatile; **f. terraces,** terrasses fluviatiles.

fluviation, processus fluviatiles.

fluviatic, fluviatile, fluviatile; **f. dam,** barrage d'alluvions déposées par un affluent; **f. deposits,** alluvions.

fluviogenic soil, sol alluvial.

fluvio-glacial, fluvio-glaciaire; **f. drift,** matériaux fluvio-glaciaires.

fluvio-marine, fluvio-marin.

fluviosol, fluvisol (*pédol.*).

fluvioterrestrial, fluviatile-continental.

flux, 1. flux, flot montant; 2. vitesse d'écoulement (d'énergie) à travers une unité de surface, débit; 3. flux (magnétique); 4. (US) substance abaissant le point de fusion d'un mélange; **f. density,** densité du champ magnétique; **f. gate,** vanne de flux magnétique; **f. meter,** fluxmètre; **gold f.,** aventurine.

flux (to), 1. ruisseler, jaillir; 2. fondre, mettre en fusion, ajouter un fondant.

fluxing agent, fondant (*métall.*).

fluxion structure, structure fluidale.

fluxoturbidite, fluxoturbidite.

flying sand, sable éolien.

flysch, flysch.

flyschoid, flyschoïde.

foam, écume, mousse; **f. earth,** aphrite (*minér.*); **f. line,** ligne de déversement des vagues; **f. spar,** aphrite.

foaming, formation de mousse; **f. earth,** aphrite (*minér.*).

foamy, écumeux, mousseux.

focal, focal; **f. depth,** profondeur de l'hypocentre; **f. length,** distance focale (*optique*); **f. mechanism,** mécanisme au foyer; **f. plane,** plan f.

focalize (to), mettre au point.

focus, foci (pl.), foyer, centre; **f. of an earthquake,** foyer d'un séisme.

focus (to), focuse (to), 1. mettre au point (un microscope); 2. concentrer (un rayon lumineux), faire converger; **to f. on infinity,** mettre au point sur l'infini.

focusing, 1. convergence, concentration; 2. mise au point (*optique*); 3. focalisation; **f. a microscope,** mise au point d'un microscope.

fog, 1. brouillard, brume; 2. buée.

föhn, foëhn.

foïdic, feldspathoïdique.

foïdite, foïdolite, roche à feldspathoïdes.

foïd gabbro, théralite (gabbro à néphéline).

foïds, feldspathoïde (abrév.).

fold, pli; **f. axis,** axe d'un p.; **f. belt,** zone orogénique; **f. bundle,** faisceau de plis; **f. fault,** pli-faille; **f. nappe,** nappe de charriage; **f. overlap,** chevauchement; **f. sheaf,** faisceau de p.; **f. tectonics,** tectonique de couverture, tectonique souple; **f. thrust,** chevauchement avec plissement; **acute f.,** p. serré; **angular f.,** p. en chevron; **back f.,** p. en retour; **box f.,** p. coffré; **carinate f.,** p. isoclinal; **compressed f.,** p. serré; **diapiric f.,** p. diapir; **dipping f.,** p. plongeant; **down f.,** p. synclinal; **drag f.,** p. d'entraînement; **fan shaped f.,** p. en éventail; **flap f.,** p. par flexion; **flow f.,** p. de fluage; **inclined f.,** p. déjeté; **isoclinal f.,** p. isoclinaux; **knee f.,** p. en genou; **monoclinal f.,** p. monoclinal; **oblique f.,** p. oblique; **offset f.,** p. décalé; **overturned f.,** p. déversé; **overfolding f.,** p. replissé; **overthrust f.,** p. faille couché; **piercement f.,** p. de percement, p. diapir; **plunging f.,** p. plongeant; **posthumous f.,** p. posthume; **recumbent f.,** p. renversé, pli couché; **shear f.,** p. de cisaillement; **supratenuous f.,** p. syngénétique; **synclinal f.,** p. synclinal; **truncated f.,** p. tronqué; **upf.,** voûte, p. anticlinal; **upright f.,** p. droit.

folded, plissé, plié.

folding, 1. adj : pliant; 2. n : plissement, pli, pliage; **bound f.,** pli. entravé; **cross f.,** pli transversal; **dysharmonic f.,** pli dysharmonique; **incipient f.,** pli naissant; **injection f.,** pli diapir.

folia, feuillets (schiste).

foliated, 1. feuilleté, lamellaire; 2. schisteux; **f. crystalline rocks,** roches cristallophylliennes.

foliate rock, roche feuilletée.

foliation, 1. schistosité; 2. foliation, clivage de flux; **f. cleavage,** schistosité; **f. mullion,** meneau de clivage; **f. plane,** plan de schistosité; **f. structure,** structure feuilletée; **axial-plane f.,** clivage ardoisier; **closely-spaced f.,** clivage rapproché; **curved f.,** clivage à réfraction continue.

food chain, chaîne alimentaire.

food grooves, sillons ambulacraires alimentaires (Crinoïdes).

fool's gold, pyrite.

foot, 1. pied (unité de mesure = 0,3048 m); 2. base, pied, socle; **f. hill,** avant-mont, contrefort, avant-pays; **f. piece,** semelle, sole (*mine*); **f. print,** empreinte de pas; **f. scale,** échelle graduée en pieds; **f. slope,** bas de pente; **f. wall,** 1. mur d'un filon (*mine*), paroi, lèvre inférieure; 2. série chevauchée (sous un charriage); **f. wall-drift,** galerie au mur (*mine*).

footage, avancement du forage, exprimé en pieds.

footing (US), base, socle, mur.

foramen, foramen, perforation, ouverture.

Foraminifera, Foraminifères.

Foraminiferal ooze, boue à Foraminifères.

Foraminiferous, à Foraminifères.
foramol assemblage, association paléontologique de Foraminifères, Mollusques, Échinides, Bryozoaires, Corallinacées.
Forams, Foraminifères (abrév.).
forced draught, air soufflé (*mine*).
forced injection, injection sous-pression.
forced oscillations, oscillations forcées.
ford, gué (d'un fleuve).
fore, avant; **f. breast,** front de taille; **f. cast,** prévision; **f. deep,** avant-fosse; **f. dune,** avant-dune; **f. field,** front des travaux; **f. ground,** premier-plan; **f. head,** front de taille; **f. land,** avant-pays; **f. limb,** flanc antérieur, inférieur d'un pli déversé; **f. man,** contremaître; **f. poling,** soutènement provisoire; **f. reef,** avant récif; **f. runner,** précurseur; **f. set beds,** lits deltaïques frontaux; **f. shaft,** avant-puits; **f. shock,** secousse sismique prémonitoire; **f. shore,** avant-plage; **f. trough,** avant-fosse; **f. sight,** prévision; **f. winning,** traçage.
forest, forêt; **f. area,** région forestière; **f. bed,** dépôt interglaciaire (sol à débris organiques); **f. soil,** sol forestier à restes de végétaux.
forfeiture of lease, déchéance, abandon d'une concession.
forge, forge.
forge (to), forger, étirer.
forgeman, ouvrier forgeur, forgeur.
fork a mine (to), assécher une mine.
forking, 1. bifurcation; 2. assèchement (*mine*).
form, 1. forme, modelé, relief; 2. coffrage (*constr.*); **f. contour,** courbe topographique tracée par photorestituteur.
formation, formation (*stratigraphie*), terrain, couches; **f. lines,** plans de stratification; **f. map,** carte de formation; **f. resistivity factor,** facteur de résistivité d'une formation; **f. testing,** essai d'évaluation des fluides contenus dans une couche; **f. water,** eau de gisement; **oil producing f.,** couche pétrolifère.
forsterite, forstérite, olivine magnésienne (*minér.*).
fossa (planète), dépression tectonique sur une planète.
fosse, fossé.
fossil, fossile; **f. bearing,** fossilifère; **f. fauna,** faune f.; **f. flour,** diatomite; **f. fuel,** combustible f.; **f. imprint,** empreinte f.; **f. man,** homme f.; **f. oil,** pétrole; **f. ore,** hématite; **f. print,** empreinte f.; **f. salt,** sel gemme; **f. soil,** sol f.; **f. water,** eau f.; **facies f.,** f. de faciès; **guide f.,** f. guide; **index f.,** f. stratigraphique; **key f.,** f. stratigraphique; **persistent f.,** f. à grande survie; **zone f.,** f. de zone.
fossilate (to), fossiliser, pétrifier.
fossiliferous, fossilifère.
fossilification, fossilization, fossilisation.
fossilizable, fossilisable.
fossilize (to), fossility (to), 1. fossiliser; 2. se fossiliser.
fossilizing, fossilisateur.
found native (to be), se trouver à l'état natif.
founder's sand, sable de fonderie.

founding, fonte, coulée, moulage.
foundry, fonderie; **f. casting work,** coulée, fusion; **f. coke,** coke métallurgique; **f. iron,** fonte de moulage; **f. man,** ouvrier fondeur; **f. sand,** matériaux siliceux de fonderie.
fountain, fontaine, puits, source, réservoir; **f. head,** source.
fourble, longueur de forage de quatre tiges; **f. board,** plate-forme d'accrochage.
fowlerite, fowlérite, variété zincifère de rhodonite (*minér.*).
foyaite, foyaïte, (pétrol.) syénite néphélinique contenant

$$\frac{\% \text{ feldspathoïdes}}{\% \text{ feldspath}} \geqslant 2.$$

frac (abrév. of fracturated), fracturé.
fraction, 1. fragment; 2. fraction minérale; **heavy f.,** fraction lourde; **light f.,** fraction légère.
fractional, 1. fractionné; **f. analysis,** analyse fractionnée; **f. crystallization,** cristallisation fractionnée; **f. condensation,** condensation fractionnée; **f. distillation,** distillation fractionnée; **f. fusion,** fusion fractionnée; **f. magma,** magma formé par cristallisation fractionnée.
fractionate (to), fractionner.
fractionating, fractionnement; **f. column,** colonne de f.; **f. tower,** tour de f.
fractionation, fractionnement; **f. product,** produit de cristallisation fractionnée; **f. progressive,** cristallisation fractionnée.
fractionator, colonne de fractionnement.
fracture, fracture, cassure; **f. cleavage,** clivage de fracture; **f. plane,** plan de fracture; **f. porosity,** porosité de fracture; **f. strenght,** résistance à la rupture; **f. thrust,** décollement; **f. zone,** zone transformante (tectonique globale); **open gash f.,** fissure d'extension, joint de tension.
fracture (to), 1. fracturer, fissurer; 2. se fracturer.
fracturing, formation de fissures, degré de fissuration; **hydraulic f.,** fracturation hydraulique.
fragipan, fragipan (*pédol.*).
fragmental, clastique, détritique; **f. rocks,** roches détritiques; **f. texture,** structure détritique.
fragmentary, détritique, formé de débris; **f. rock,** brèche, conglomérat bréchique.
fragmentation, fragmentation.
frame, bâti, cadre, charpente, structure; **f. builder organism,** organisme constructeur; **f. set,** cadre de boisage (*mine*); **f. work,** canevas d'un levé, réseau, ossature (d'un ouvrage), plan d'ensemble, grandes lignes (d'une étude ou d'un rapport), masse d'un récif de corail; **head-f.,** tête de puits, chevalement d'extraction (*mine*).
Franconian (US), Franconien (étage, Cambrien sup.).
franklinite, franklinite (*minér.*).
Frasnian (Eur.), Frasnien (étage, Dévonien sup.).
fray out (to), se coincer en biseau, s'effilocher.
frazil ice (US), glace de fond.

free, libre, débarassé de, dépourvu de, exempt de; **f. air anomaly,** anomalie à l'air libre; **f. air correction,** correction à l'air libre; **f. cheek,** joue libre (Tribolite); **f. face,** surface dégagée (*mine*), affleurement dégagé; **f. gold,** or natif; **f. milling ore,** minerai contenant du métal à l'état libre; **f. nappe,** nappe libre (*hydrol.*); **f. sample,** échantillon gratuit; **f. stone,** pierre de taille.

freezable, gelable, congelable.

freeze, gel; **f. degree day,** jour-degré de gel; **f. ice,** glace de congélation; **f. thaw action,** cycle gel-dégel; **f. up,** prise en glace.

freeze (to), 1. geler, glacer, se solidifier (par congélation); 2. se geler, se congeler.

freezing, 1. congélation, gel; 2. réfrigération; **f. interval,** intervalle de congélation; **f. point,** point, température de congélation; **f. point of water,** température de congélation de l'eau; **f. test,** essai de congélation.

freibergite, freibergite, tétraédrite argentifère (*minér.*).

french chalk, 1. talc, stéatite; 2. craie de meudon.

frequency, fréquence; **f. band,** bande de f.; **f. curve,** courbe de f., courbe en cloche; **f. polygon,** polygone de f.; **f. range,** gamme de f.

fresh, 1. non usé, non altéré (*pétro.*); 2. frais, nouveau; 3. doux, non salé; **f. water,** eau douce; **f. formation,** formation d'eau douce; **f. limestone,** calcaire lacustre.

freshening, 1. dessalement, dessalure (de l'eau de mer); 2. rafraîchissement (de l'atmosphère).

freschet, 1. avalaison, crue, inondation; 2. courant d'eau douce dans la mer.

freshness, fraîcheur (absence d'altération d'une roche).

fret (to), ronger, creuser, user, corroder.

fretting (by salt), effrittement, désagrégation granulaire sous l'effet de sels.

friability, friabilité.

friable, friable.

friction, frottement, friction, attrition; **f. angle,** angle de frottement; **f. breccia,** brèche de dislocation.

frictional resistance, résistance de frottement.

friedelite, friedélite (*minér.*).

frigid zone, zone froide, zone polaire.

fringe, frange, bord, bordure.

fringing, de bordure, marginal; **f. reef,** récif frangeant.

frit (to), fondre partiellement.

frith, 1. fjord, bras de mer; 2. haie.

fritted rock, roche vitrifiée, recuite.

fritting, frittage (fusion superficielle de grains).

frond, fronde (Ptéridophytes).

front, 1. adj : frontal, avant, de devant; 2. n : front, face, partie antérieure, devant; 3. front (*météo.*); **f. view,** vue de face; **thrust f.,** front de charriage; **wave f.,** front d'onde.

frontal, frontal de face; **f. moraine,** moraine frontale, moraine terminale.

frontogenesis, formation d'un front atmosphérique.

frost, 1. gel. gelée; **f. action,** gélivation; **f. blasting,** éclatement par le froid; **f. breaking,** gélifraction; **f. crack,** fissure de gel, fente de gel; **f. crack polygon,** polygone de gélicontraction; **f. creep,** reptation des sols due au gel; **f. desert,** désert de gélifraction; **f. disturbation,** cryoturbation; **f. heaving,** soulèvement par le gel; **f. heaved mound,** sol polygonal; **f. line,** 1. seuil du gel; 2. profondeur maximum du gel; **f. pattern,** réseau polygonal; **f. polygon,** polygone de gel; **f. proof,** résistant au gel; **f. prying,** éclatement par le gel; **f. riving,** gélifraction; **f. shattering,** éclatement par le gel; **f. splitting,** gélidisjonction; **f. stirring,** géliturbation; **f. thrust,** poussée de gel, moraine de poussée; **f. table,** pergélisol; **f. weathering,** action climatique du gel; **f. wedging,** fissuration par le gel; 2. givre; **ground f.,** gelée blanche, g.; **hoar f.,** gelée blanche, givre; **white f.,** gelée blanche.

frosted, 1. dépoli, mat; 2. givré.

frostwork, gélivation.

froth, écume, mousse.

frothy, écumeux, spongieux.

frozen, 1. gelé, glacé, congelé; 2. coincé; **f. drill pipe,** tige de forage coincée; **f. ground,** sol gelé; pergélisol.

frustule, frustule (Diatomée).

fuel, combustible, carburant; **f. gas,** gaz combustible; **f. oil,** mazout, huile combustible; **f. ratio, value,** pouvoir calorifique; **compressed f.,** aggloméré, briquettes de charbon; **heavy f.,** fuel lourd; **saving of f.,** économie d'énergie.

fuel (to), obtenir du combustible, se ravitailler en combustible.

fueling, 1. approvisionnement en combustibles; 2. combustibles.

fulgurite, fulgurite (structure de fusion par la foudre).

full, 1. adj : plein, rempli; 2. crête prélittorale; **f. dip,** pendage vrai, pendage réel; **f. hole drilling,** sondage-carottage complet; **f. scale,** à échelle ou grandeur réelle; **f. timbering,** boisage complet (*mine*).

fuller's earth, terre à détacher, marne à foulon, glaise à dégraisser, argile smectique.

Fultonian (Wash. st.), Fultonien (étage floral, Éocène moy.).

fulvurite, lignite.

fumarole, fumerolle.

fumarolic, fumerollien.

fume, fumée, vapeur, exhalaison, exhalation.

fumes of sulfur, vapeurs sulfureuses.

fuming, fumant.

fundamental gneiss, gneiss du socle.

Fungi, Champignons (Thallophytes).

fur, incrustation, tartre, calcin.

fur (to), 1. entartrer; 2. détartrer, décrasser.

furlong, furlong = 201,168 m.

furnace, four, fourneau, chaudière; **f. coke,** coke métallurgique; **f. shaft,** puits à foyer d'aérage (*mine*).

furred, entartré.

furrow, 1. cannelure, rainure, strie; 2. entaille (*mine*); **f. cast,** trace allongée; **glabellar f.,** sillon glabellaire (de Trilobite).

furrow (to), sillonner, creuser des cannelures.

furrowing, 1. formation de cannelures; 2. érosion karstique (lapiès).

fuschite, fuschite (mica chromifère).

fuse (to), 1. réunir, fusionner (deux images aériennes mentalement); 2. fondre (un métal).

fusibility, fusibilité; **f. scale,** échelle thermique de f.

fusible, fusible; **f. clay,** terre f., soluble; **f. quartz,** obsidienne (GB).

fusibleness, fusibilité.

fusing point, point de fusion, température de fusion.

fusinite, fusinite.

fusion, fusion; **f. curve,** courbe de f.; **f. point,** température de f.; **f. welding,** soudage par f.

Fusulina, Fusuline.

Fusulinids, Fusulinidae, Fusulinidés.

G

gabbro, gabbro; **g. syenite,** monzonite; **alkali g.,** g. alcalin (*pétrogr.*).

gabbroic, gabbroïque.

gabbroid, gabbroïde.

gad, 1. coin (*mine*); **2.** pince (*mine*); **3.** pointe (lance, flèche).

gadolinite, gadolinite (*minér.*).

gage (US), indicateur de pression, manomètre, jauge; **rain g.,** pluviomètre; **slide g.,** pied à coulisse; **tide g.,** indicateur de marée, marégraphe.

gahnite, gahnite, spinelle zincifère.

gain, amplification, accroissement; **g. amplifer,** préamplificateur.

gaining stream, fleuve alimenté par infiltration souterraine.

gaize, gaize (roche siliceuse).

galactic, galactique.

galaxy, galaxie; **the milky way g.,** la voie lactée.

gale, 1. coup de vent, grand vent; **2.** tempête.

galena, galenite, galène (minerai de plomb).

galenobismuthite, galénobismuthite.

gallery, galerie (*mine*).

gallium, gallium.

gallon, gallon; **g. (imperial),** = 4,545.963 l; **g. (US),** = 3,785.41 l.

gallows, cadre incomplet (*mine*); **g. frame,** chevalement (*mine*), bâti de machine à balancier (*for.*).

Galt (UK), Gault (formation, Crétacé inf.).

galvanometre, galvanomètre.

gamma, gamma (unité d'intensité de champ magnétique = 10^{-5} gauss).

gamma-gamma log, diagraphie gamma-gamma.

gamma ray, rayon gamma.

gamma ray spectroscopy, spectroscopie gamma; **g. ray well logging,** méthode de diagraphie par rayons gamma.

gang, 1. chemin, passage; **2.** gangue (*mine*).

gangue, gangue.

gangway, galerie principale (*mine*), galerie maîtresse.

gannister, ganister, 1. roche sédimentaire très réfractaire utilisée dans le revêtement de hauts fourneaux; **2.** mélange de grès et d'argiles réfractaires; **3.** sable siliceux pur situé sous les couches de charbon.

ganoid, ganoïde (*pal.*).

gap, 1. trou, vide, brèche; **2.** col de montagne (Am. du N.) défilé; **3.** composante horizontale du rejet parallèlement à la faille; **4.** lacune, interruption; **g. fault,** faille ouverte; **sedimentary g.,** lacune sédimentaire; **dry, wind, air g.,** cluse sèche, cluse morte; **erosional g.,** lacune d'érosion; **stratigraphic g.,** lacune stratigraphique; **water g.,** lacune fonctionnelle, active, cluse vive.

gaping, béant.

Gargasian (Eur.), Gargasien (sous-étage, Aptien sup.).

garland of stones, croissant, guirlande de pierres (cas de solifluxion entravée).

garnet, grenat; **g. blende,** sphalérite, blende; **g. rock,** grenatite.

garnetiferous, grenatifère.

garnetite, cornéenne à grenats.

garnierite, garniérite, nouméite (*minér.*).

gas, 1. gaz; **2.** grisou (*mine*); **g. bearing,** gazéifère; **g. black,** noir de fumée; **g. blow-out,** éruption de gaz; **g. bubble,** bulle de gaz; **g. cap,** chapeau de gaz (*pétrole*); **g. cap drive,** drainage du pétrole par pression du gaz libre; **g.-cap pool,** gisement possédant un chapeau de gaz; **g. chromatography,** chromatographie en phase gazeuse; **g. coal,** houille grasse; **g. coke,** coke à gaz; **g. condensate,** condensat; **g. cut mud,** boue de forage émulsionnée de gaz; **g. cycling,** recyclage de gaz; **g. detector,** détecteur de grisou; **g. distillate,** gaz humide; **g. drive,** poussée de gaz, drainage par gaz; **g. escape,** dégagement de gaz; **g. expansion method,** méthode par expansion de gaz; **g. explosion,** explosion de grisou; **g. factor,** teneur en gaz; **g. field,** champ de gaz naturel; **g. freeing,** dégazage; **g. horizon,** horizon gazéifère; **g. inclusion,** inclusion gazeuse; **g. indicator,** détecteur de gaz; **g. injection,** injection de gaz; **g. lift,** extraction de pétrole par injection de gaz; **g. line,** conduite de gaz; **g. liquids,** liquides extraits du gaz naturel; **g. main,** conduite de gaz; **g. muds,** boue émulsionnée de gaz; **g. oil level,** niveau de contact entre le gaz et le pétrole; **g. oil ratio (GOR),** proportion gaz pétrole; **g. out burst,** dégagement de gaz; **g. phase,** phase gazeuse; **g. pool,** gisement de gaz; **g. pipe,** conduite de gaz; **g. pressure,** pression de gaz; **g. proof,** étanche au gaz; **g. rock,** roche gazéifère; **g. sand,** sable pétrolifère riche en gaz; **g. seeps,** dégagement gazeux; **g. storage,** stockage de gaz; **g. trap,** « piège » contenant du gaz; **g. verifer,** grisoumètre; **g. well,** sondage à gaz, puits de gaz; **g. yield,** rendement en gaz naturel; **coal g.,** gaz de houille; **marsh g.,** gaz de marais; **oil g.,** gaz de pétrole; **wet g.,** gaz naturel « humide ».

gaseous, 1. gazeux; **2.** grisouteux; **g. hydrocarbon,** hydrocarbure gazeux; **g. inclusion,** inclusion gazeuse; **g. mine,** mine grisouteuse.

gash, cassure, entaille, tranchée (remplie de sédiments); **g. vein,** fissure minéralisée à courte extension verticale.

gasifiable, gazéifiable.

gasification, gazéification.

gasify (to), 1. gazéifier; **2.** se gazéifier.

gasoline (US), essence (de pétrole).

gasser, puits conducteur de gaz.
gassing, **1.** dégagement gazeux; **2.** asphyxie par les gaz.
gassy, **1.** gazeux; **2.** grisouteux.
gastrolith (syn. gizzard stone), gastrolithe.
Gastropod, Gastropoda, classe des Gastéropodes.
gate road, galerie de desserte.
gateway, large défilé (fluvial), voie de roulage (*mine*).
gathering, accumulation; **g. channel**, chenal collecteur; **g. ground**, aire d'alimentation (fluviale); **g. line**, conduite de collecte de petit diamètre; **g. system**, réseau collecteur.
gauge, calibre, jauge; **ionic g.**, manomètre (jauge) ionique; **pressure g.**, manomètre; **vaccum g.**, manomètre à vide.
gauge (to), calibrer, cuber, jauger.
gauging, jaugeage, cubage.
Gault (UK), Gault (GB) (formation, Crétacé inf.).
Gaussian curve, courbe de Gauss.
geanticlinal, geanticline, géanticlinal.
Gedinnian (Eur.), Gédinnien (étage, Dévonien inf.).
gedrite, gédrite (var. d'amphibole).
geest, altérite.
Geiger counter, compteur Geiger.
geisothermal, isogéothermique.
gel, colloïde, gel.
gel (to), se coaguler, se gélifier.
gelatin dynamite, nitrogélatine.
gelation, **1.** congélation; **2.** gélification, prise en gelée, solidification.
gelifluction, solifluxion périglaciaire.
gelifract, gélifract.
gelifracted, gélifracté.
gelisol, gélisol.
geliturbation, géliturbation.
gelivation, gélivation; **g. valley**, vallée périglaciaire.
gelivity, gélivité.
gem, pierre précieuse.
gemmation, bourgeonnement (*pal.*).
gemmology, gemmologie.
gemstone, gemme.
gem bearing, gemmifère; **g. cutting**, taille des pierres précieuses; **g. mine**, mine de pierres précieuses; **g. mining**, exploitation de pierres précieuses.
gemmed, gemmy, gemmé.
gemmiferous, gemmifère.
genal, génal (*pal.*).
genera (plur. of genus), genres (*pal.*).
generating surface, fecht.
genetic drift, changement génétique progressif.
geniculating twin, mâcle en genou.
genital plates, plaques génitales (*pal. Échino.*).
genomorph, génomorphe (*pal.*).
genotype, génotype.
gentle dip, pendage faible.
genus, genre (*pal.*).
geobaric gradient, gradient géobarométrique.
geobotanical, géobotanique.
geochemical, géochimique; **g. cycle**, cycle g.; **g.**

indicator, indice g.; **g. profile**, profil g.; **g. prospecting**, prospection g.
geochemistry, géochimie.
geochronologic, géochronologique; **g. overprint**, superposition de datations géochronologiques; **g. sequence**, stratigraphie g.; **g. unit**, unité g. (*stratigr.*).
geochronology, géochronologie.
geochronometry, géochronologie.
geocratic, géocratique.
geodal, géodique.
geode, géode, druse (dans une géode).
geodepression, géodépression, graben.
geodesy, géodésie.
geodesic, géodésique; **g. coordinates**, coordonnées géodésiques.
geodesist, géodésiste.
geodetic, géodésique; **g. line**, plus courte distance entre deux points de la surface d'une sphère; **g. survey**, bureau g. (US).
geodetics, géodésie.
geodynamics, géodynamique.
geodynamo, « dynamo terrestre ».
geofracture, ancienne fracture de l'écorce terrestre.
geognostic, géognostique.
geognosy, géognosie.
geographer, géographe.
geographic, geographical, géographique.
geographic latitude, latitude; **g. longitude**, longitude.
geographically, géographiquement.
geography, géographie.
geohydrology, géohydrologie.
geoid, géoïde.
geologic, geological, géologique; **g. age**, âge g.; **g. clock**, tableau chronologique; **g. column**, log lithostratigraphique, succession, nature des terrains; **g. engineer**, ingénieur géologue; **g. event**, phénomène g.; **g. map**, carte g.; **g. range**, répartition stratigraphique d'un fossile, durée d'existence d'un fossile; **g. section**, coupe g.; **g. serie**, série g.; **g. setting**, cadre g.; **g. survey**, bureau d'études, service g.; **g. thermometer**, thermomètre g.; **g. time scale**, échelle stratigraphique; **g. window**, fenêtre tectonique.
geologically, géologiquement.
geologist, géologue.
geologist's hammer, marteau de géologue.
geologize (to), faire de la géologie.
geology, géologie; **g. pick**, marteau de géologue; **applied g.**, g. appliquée; **dynamic g.**, g. dynamique; **historical g.**, g. historique; **mining g.**, g. minière; **petroleum g.**, g. pétrolière; **structural g.**, g. structurale.
geomagnetic, géomagnétique; **g. equator**, équateur magnétique; **g. field**, champ magnétique; **g. polarity interval**, chrone; **g. poles**, pôles magnétiques de la Terre; **g. reversal**, inversion g.; **g. secular variation**, variation séculaire g.
geomagnetism, géomagnétisme.
geomechanics, mécanique des sols.

geometer, géomètre.

geomorphic, geomorphologic, géomorphologique; **g. cycle,** cycle g.

geomorphogeny, géomorphogénie.

geomorphometry, géomorphométrie.

geomorphy, geomorphology, géomorphologie.

geonomy, géonomie.

geopetality, orientation vers le centre de la Terre.

geophone, géophone, sismographe; **g. array,** dispositif de géophones; **g. cluster,** grappes de g.; **g. interval,** intervalle entre g.; **g. offset,** distance source-trace.

geophysic, geophysical, géophysique; **g. log,** diagraphie g.; **g. map,** carte g.; **g. method,** méthode g.; **g. prospecting,** prospection g.; **g. survey,** prospection g.; **g. surveying,** relevé g.

geophysicist, géophysicien.

geophysics, géophysique.

geopressure, géopression.

Georgian (N. Am.), Géorgien (étage, Cambrien inf.).

geosphere, géosphère.

geostatic, géostatique.

geostatistics, géostatistique.

geostrophic, géostrophique.

geosynclinal, géosynclinal (adj).

geosyncline, géosynclinal (nom); **g. ridge,** ride géosynclinale; **g. cycle,** cycle tectonique; **eugeosyncline,** eugéosynclinal; **marginal g.,** paragéosynclinal; **miogeosyncline,** miogéosynclinal; **monogeosyncline,** monogéosynclinal; **paragéosyncline,** paragéosynclinal; **polygeosyncline,** polygéosynclinal; **taphrogeosyncline,** taphrogéosynclinal, vallée d'effondrement.

geotechnical, géotechnique (adj); **g. map,** carte g.; **g. property,** propriété g.; **g. survey,** levé g.

geotechnics, géotechnique (nom).

geotectocline, géosynclinal.

geotectogene, ride synclinale.

geotectonic, géotectonique (adj).

geothermal, geothermic, géothermique; **g. degree,** degré g.; **g. energy,** énergie g.; **g. field,** champ g.; **g. gradient,** gradient g.; **g. log,** diagramme g.; **g. metamorphism,** métamorphisme g.

geothermics, geothermy, géothermie.

geothermometry, géothermométrie.

geotumor, bombement de la croûte terrestre.

german silver, maillechort.

germanium, germanium.

gersdorffite, gersdorffite (minér.).

get, production, rendement (d'une mine).

get coal (to), exploiter, extraire du charbon.

gettability, exploitabilité (d'une mine).

gettable, exploitable.

getting oil, exploitation du pétrole.

geyser, geyser; **g. pipe,** conduit de g.

geyseric, geysérien.

geyserite, geysérite, tuf siliceux, opale concrétionnée des dépôts geysériens.

giant, géant; **g. causeway,** « chaussée des géants »; **g. granite,** pegmatite; **g. kettle,** « marmite des géants », kettle ou dépression fermée; **g. ripple,** mégaride.

gibber, galet à facettes.

gibbers, résidu de déflation (Austr.).

gibbsite, gibbsite (minér.).

gieseckite, gieseckite, pinite (minéral d'altération de la néphéline).

gigantism, gigantisme (pal.).

gigantolite, gigantolite (variété de cordiérite) (minér.).

Gigantostraca, Gigantostracés (pal.).

gild (to), dorer, recouvrir d'une pellicule d'or.

gilgai, mottureaux (Austr.).

gill, 1. gorge boisée, ravin boisé; **2.** ruisseau, torrent (coulant dans un ravin) Écosse; **3.** unité de mesure : 1 gill Angl = 0,142.061 litre, Amer. = 0,118.29 l; **4.** branchie (pal.).

gilsonite, gilsonite (var. de bitume).

gin, treuil, cabestan (mine).

gin-truck, camion grue.

Ginkgoales, Ginkgoales (pal.).

giobertite, giobertite, magnésite (minér.).

girasol, girasol (var. d'opale).

gismondite, gismondine, gismondite = zéolite (minér.).

Givetian (Eur.), Givétien (étage, Dévonien moy.).

gizzard stone (syn. gastrolith), gastrolite.

glabella, glabelle (pal., Trilobites).

glacial, glacic, glaciaire; **g. advance,** avancée, crue g.; **g. age,** époque g.; **g. balance,** bilan g.; **g. basin,** ombilic g.; **g. boulder,** bloc erratique; **g. breaching,** modification des réseaux hydrographiques par action des glaciers; **g. canyon,** vallée g. en U; **g. carved valley,** auge, vallée g.; **g. cirque,** cirque g.; **g. cycle,** cycle g.; **g. deposit,** dépôt g.; **g. downwasting,** fusion g.; **g. drift, 1.** apports g., drift; **2.** matériel transporté par la glace; **g. environment,** milieu g.; **g. epoch,** époque g.; **g. erosion,** érosion g.; **g. erratic,** bloc erratique; **g. flow,** écoulement g.; **g. front,** front g.; **g. geology,** géologie des formations g.; **g. groove,** cannelure g.; **g. horn,** aiguiile g.; **g. lake,** lac g.; **g. milk,** lait de glacier; **g. lobe,** lobe de glacier; **g. markings,** marques d'abrasion g.; **g. maximum,** maximum g., plén-glaciaire; **g. meal,** farine g.; **g. mill,** moulin g.; **g. ogive,** ogive g.; **g. period,** période g.; **g. phase,** phase g.; **g. plain,** plaine g.; **g. planing,** rabotage g.; **g. plucking,** délogement g.; **g. polish,** poli g.; **g. pot-hole,** marmite g.; **g. readvance,** récurrence g.; **g. recession,** régression g.; **g. retreat,** recul, retrait g.; **g. rill,** bédière; **g. glacial scour lake,** lac d'érosion g.; **g. sheet,** calotte g.; **g. scratches,** stries g.; **g. spillway,** chenal d'écoulement des eaux provenant de la fusion du glacier; **g. stage,** stade g.; **g. stairway,** vallée g. en gradins; **g. stria(ae),** strie(s) g.; **g. striation,** striage g.; **g. terrace,** terrasse g.; **g. till,** till, moraine; **g. tongue,** langue g.; **g. transport,** transport g.; **g. trough,** auge g.; **g. valley,** vallée g.; **g. wastage,** ablation, fusion g.

glacialism, glacialisme.

glaciate (to), recouvrir de glace, soumettre à l'action d'un glacier.

glaciated, 1. à modelé glaciaire et recouvert de till; **2.** recouvert par un glacier; **g. sea floor,** fond marin littoral à modelé glaciaire quaternaire.

glaciation, glaciation; **g. limit,** limite de glaciation.

glacier, glacier; **g. band,** ogive glaciaire, bandes glaciaires; **g. bed,** lit de g.; **g. burst,** libération d'eau glaciaire; **g. circus, cirque,** cirque glaciaire; **g. corrie,** cirque glaciaire; **g. crevasse, crevice,** crevasse glaciaire; **g. face, front,** front d'un g.; **g. fall,** cascade de g.; **g. flow,** écoulement glaciaire; **g. ice,** glace de g.; **g. lake,** lac de g.; **g. mill,** moulin glaciaire; **g. moulin,** moulin glaciaire; **g. mud,** boue glaciaire; **g. outburst,** débâcle glaciaire; **g. outlet,** exutoire glaciaire; **g. snout,** front d'une langue glaciaire; **g. snow,** névé; **g. spillway,** chenal d'eau de fonte; **g. surge,** crue glaciaire; **g. table,** table glaciaire; **g. trough,** auge glaciaire; **g. tongue,** langue glaciaire; **g. well,** moulin glaciaire; **g. wind,** vent catabatique; **alpin g.,** g. alpin; **cirque g.,** g. de cirque; **continental g.,** inlandsis, calotte glaciaire; **hanging g.,** g. suspendu; **intermont g.** g. d'entremont; **piedmont g.,** g. de piedmont; **plateau g.,** g. de plateau; **recemented g.,** g. régénéré; **rock g.,** g. rocheux; **temperate g.,** g. tempéré; **transection g.,** g. transfluent; **valley g.,** g. de vallée; **g. ice thrust,** glacitectonique.

glacieret, glacier suspendu, petit glacier.

glacierization (GB), glaciation, englacement (d'un continent).

glacio-aqueous clay, argile glacio-aquatique; **g.-fluvial,** fluvio-glaciaire; **g.-eustatism,** glacio-eustatisme; **g. terrace,** terrasse fluvio-glaciaire; **glacioisostasy,** glacioisostasie; **glaciolacustrine,** glaciolacustre; **glaciomarine,** glaciomarin; **glaciologist,** glaciologue; **glaciology,** glaciologie; **glacionival,** glacionival; **glaciovolcanic,** glacio-volcanique.

glacis, glacis, pente modérée.

glacitectonics, glacitectonique.

glade, ouvala (karst).

glance, éclat, brillant; **g. coal,** anthracite, houille luisante; **g. cobalt,** cobaltine; **g. copper,** chalcosine; **g. pitch,** var. d'asphaltite.

glass, 1. verre; **2.** loupe; **3.** baromètre; **4.** jumelles; **g. inclusion,** inclusion vitreuse; **g. matrix,** pâte vitreuse; **g. sand,** sable siliceux pur pour verrerie; **g. sponge (syn. Hyalosponge),** éponge siliceuse (Porifère); **g. structure,** structure vitreuse; **g. tiff,** calcite; **g. ware,** verrerie.

glassy, vitreux; **g. feldspar,** sanidine (minér.); **g. rock,** roche vitreuse.

Glauber's salt, sel de Gauber, sulfate de soude, mirabilite.

glauberite, glaubérite (minér.).

glaucodot, glaucodot (minér.).

glauconiferous, glauconifère.

glauconite, glauconie.

glauconitric, glauconieux; **g. sandstone (greensand),** grès g.

glauconitization, glauconitisation.

glaucophane, glaucophane (amphibole), (minér.); **g. schist,** schiste à glaucophane.

G layer, noyau interne de la Terre.

glazed frost, verglas.

glazy, vitreux.

glebe, 1. glèbe; **2.** terrain minéralisé (GB).

glen, ravin, gorge montagneuse (Écosse).

Glenarm (US), série de Glenarm (Précambrien tardif).

gley, glei, gley; **g. alluvial brown soil,** sol alluvial gleyifié; **g. like,** gleyiforme; **g. like soil,** pseudo-g.; **g. podzol,** sol podzolique à g.; **g. soil,** sol à g.

gleyed forest soil, sol forestier à gley.

gleying process, processus de gleyification.

gleyzation (U.S.), formation d'un niveau gleyifié (pédol.).

glide, gliding, glissement, translation; **g. plane,** plan de réorientation minéralogique; **g. twinning, 1.** mâcle par pseudosymétrie; **2.** mâcle par déformation mécanique.

glimmer, 1. mica; **2.** lame auxiliaire de « mica λ/4 » optique.

glimmerite, roche ultrabasique micacée.

glint, escarpement; **glint line,** limite d'érosion de terrains discordants sur un socle.

global, global; **g. tectonic, g. tectonics,** tectonique globale, des plaques.

Globigerinacea, Globigérines.

Globigerine, Globigérine; **g. mud, -ooze,** boue à Globigérines.

globular, sphérolitique; **g. structure,** structure s.

globularite, roche à structure de dévitrification.

glomerate, conglomérat.

glomeration, formation d'un conglomérat.

glomero-blastic structure, structure gloméro-blastique.

glomerophyric, glomeroporphyritic, gloméroporphyrique (à agrégat de phénocristaux dans la pâte de roches éruptives).

glory hole, excavation à ciel ouvert (mine).

gloss coal, lignite luisant.

Glossopteridales, Glossoptéridées.

gloup, creux de déflation, cadoueyre.

glow (to), 1. rougeoyer; **2.** s'embraser, s'allumer; **3.** se consumer, brûler sans flammes.

glow avalanche, nuée ardente en avalanche.

glow curve, courbe de thermoluminescence.

glowing clouds, nuées ardentes (par explosion).

glycolated sample, échantillon glycériné.

glyptogenesis, glyptogénèse (phase d'érosion).

glyptolith, caillou à facettes.

Gnathostoma, Gnathostomes (pal.).

gneiss, gneiss; **banded g.,** g. lité; **composite g.,** g. d'injection; **foliated g.,** g. en feuillets; **fundamental g.,** g. de socle; **high-grade g.,** g. de fort métamorphisme; **leaf g.,** g. en feuillet; **lenticular banded g.,** g. œillé; **orthogneiss,** orthogneiss (roches ignées métamorphisées); **paragneiss,** paragneiss (roches sédimentaires métamorphisées).

gneissic, gneissoid, gneissose, gneissique; **g. granite,** granito-gneiss; **g. structure,** structure g.; **g. gabbro,** gabbro g.

gneissosity, gneissosité.

Gnetales, Gnétales.

gnomonic projection, projection gnomonique.

goaf, vieux travaux, remblai (*mine*); **g. stower,** machine à remblayer.

gob, remblai; **g. fire explosion,** explosion de grisou; **g. road,** galerie dans les remblais (*mine*); **g. stower,** remblayeuse.

gob (to), remblayer.

gobbing, remblayage.

go dead (to), cesser de produire (puits); **to g. up a river,** remonter une rivière.

goe, valleuse, vallée littorale.

goethite, goethite (*minér.*).

gold, or; **g. bearing,** aurifère; **g. beater,** batteur d'or; **g. claim,** concession de terrains aurifères; **g. digger,** chercheur d'or; **g. digging,** exploitation aurifère; **g. dust,** poudre d'or; **g. field,** champ, district aurifère; **g. finder,** chercheur d'or; **g. flour,** poudre d'or; **g. free,** dépourvu d'or; **g. foil,** feuille d'or, or battu; **g. mine,** mine d'or; **g. mining,** exploitation aurifère; **g. ore,** minerai d'or; **g. placer,** gisement d'or sédimentaire; **g. sand,** sable aurifère; **g. washer,** orpailleur; **g. workings,** exploitation aurifère; **native g.,** or natif.

goldstone, aventurine.

Gondwanaland, continent de Gondwana.

Goniatite, Goniatite (*pal.*).

goniometer, goniomètre.

goniometry, goniométrie.

gonotheca, gonothèque (*pal.*).

gooseberry -stone, grenat grossulaire (*minér.*).

gooseneck (Utah), méandre encaissé.

gopher (drift), galerie de prospection minière taillée irrégulièrement et au hasard.

gopher (to), gaspiller un gisement.

gophering, prospection minière par petits trous ou débuts de galeries.

gorge, ravin, gorge.

Gorgonia, Gorgone (Coralliaires).

gossan, gozzan, chapeau ferrugineux (d'une couche métallifère), partie supérieure altérée d'un filon.

gossaniferous, contenant des produits ferrugineux d'altération.

gossany lode, filon altéré, à produits d'altération.

Gotlandian, Gothlandian (obsolete), Gothlandien (= Silurien en Europe : désuet).

gouge, 1. gouge; 2. rainure (Am.); 3. salbande argileuse (*mine*); 4. fine brèche de faille; **jumping-g.,** coup de gouge glaciaire; **g. zone,** zone broyée; **friction g.,** brèche de friction.

gouge (to), exploiter une mine sans méthode.

gowan, granite altéré.

grab, benne, excavateur; **g. sample,** 1. échantillon de fonds marins prélevé avec des pinces; 2. échantillon prélevé au hasard; **hammer g.,** benne foreuse.

graben, fossé (d'effondrement tectonique).

gradation, 1. gradation, progression, degré; 2. aplanissement, régularisation du relief, régularisation du profil fluviatile (*Géogr. Phys.*).

grade, 1. pente d'équilibre (exprimée en pourcentage); 2. teneur, qualité (d'un minerai); **g. control,** contrôle de la teneur (*mine*); **g. level,** niveau d'équilibre, profil d'équilibre (*fluv.*); **g. scale,** échelle granulométrique (par ex, celle d'Atterberg, de Wentworth, etc); **high g.,** 1. à forte teneur; 2. fort gradient métamorphique; **low g.,** 1. à faible teneur; 2. faible gradient métamorphique.

grade (to), 1. classer, trier; 2. graduer, évaluer la qualité de; 3. régulariser une pente, niveler.

graded, classé, gradué; **g. bed,** lit grano-classé; **g. bedding,** granoclassement ou classement vertical progressif; **g. profile,** profil d'équilibre; **g. river,** rivière régularisée; **g. sediments,** sédiments homométriques bien classés; **g. shoreline,** rivage à tracé régularisé; **g. slope,** pente régulière; **g. stream,** rivière régularisée; **g. valleyside,** vallée à versants régularisés.

grader, trieur, classeur (de minerais).

gradient, 1. pente (fluviatile), inclinaison, dénivellation; 2. gradient; **angle of g.,** angle de pente; **fluid potential g.,** gradient de potentiel fluide; **geothermal g.,** gradient géothermique; **hydraulic g.,** gradient hydraulique; **reversed g.,** contrepente.

grading, 1. triage, classement; 2. nivellement, régularisation de pente; **g. analysis,** analyse granulométrique; **g. curve,** courbe granulométrique; **g. factor,** coefficient de triage granulométrique; **g. screen,** crible classeur, tamis.

gradiometer, gradiomètre, niveau de pente; **magnetic g.,** gradiomètre magnétique.

graduate, 1. diplômé; 2. gradué.

grahamite, grahamite (var. de bitume).

grail, particules fines, sables.

grain, grain; **g. growth,** recristallisation (dans une couche monominérale); **g.-size,** granulométrie; **g.-size curve,** courbe granulométrique; **g.-size distribution,** répartition granulométrique; **g. stone,** calcaire à débris jointifs; **g. structure,** structure granulaire; **coarse g., coarse grained,** à gros grain; **fine g., fine grained,** à grain fin, de faible granulométrie.

grained, granuleux, grenu; **fine g. sand,** sable à grain fin; **g. rock,** roche grenue; **medium-g. sand,** sable à grains moyens.

gram, gramme, gramme; **g. calorie,** calorie g., microthermie, petite calorie; **g. molecule,** molécule g.

grammatite, trémolite, néphrite (var. amphibole).

Grampian, = Dalradien (division du Précambrien) [Écosse].

granite, granite; **g. aplite,** aplite granitique; **g. family,** famille du g.; **g. gneiss,** granito-gneiss; **g. greisen,** greisen granitique (quartz, feldspath, muscovite); **g. pegmatite,** pegmatite granitique (pas de plagioclases); g. **porphyry,** var. de microgranite; g. **wash,** arkose.

granitic, granitique; g. **layer,** croûte continentale granitique; g. **sand,** arène g.

granitification, granitisation.

granitization, granitisation.

granitoid, granitoïde.

granoblastic, granoblastique.

granodiorite, granodiorite (*pétrogr.*).

granogabbro, granogabbro.

granophyre, granophyre (microgranite à texture graphique).

granophyric, graniphyric, granophyrique, texture graphique à petite échelle.

granular, granulaire, grenu; g. **disintegration,** désagrégation granulaire; g. **limestone,** calcaire cristallin, marbre; g. **quartz,** quartzite; g. **structure,** structure grenue; g. **texture,** texture grenue.

granularity, granularité, grosseur de grain.

granulated, granulé, grenu, granuleux cristallisé; g. **rock,** roche grenue.

granulation, 1. broyage (en particules); 2. fragmentation de minéraux au delà de leur limite d'élasticité.

granule, 1. granule (2 à 4 mm); 2. gravier.

granulite, granulite, leptynite; g. **facies,** faciès à granulites (métamorphisme).

granulitic, granulitique; g. **structure,** structure g.

granulometric, granulométrique; g. **composition,** composition g.; g. **curve,** courbe g.

granulometry, granulométrie.

granulose, granulous, granuleux; g. **texture,** texture granoblastique.

granulous, granuleux.

grapestone, calcaire à pellets agglomérés.

graph paper, 1. graphique, diagramme, courbe papier quadrillé; 2. abaque.

graphic, graphique; g. **gold,** sylvanite (*minér.*); g. **granite,** pegmatite g.; g. **intergrowth,** texture g.; g. **log,** colonne lithologique d'un forage; g. **ore,** sylvanite (*minér.*); g. **recorder,** enregistreur g.; g. **structure,** structure g.; g. **tellurium,** sylvanite (*minér.*).

graphite, graphite, plombagine (*minér.*).

graphitic, graphitique.

graphitization, graphitisation.

graphophyric, granophyrique.

Graptolite, Graptolite (*pal.*).

Graptolitic shale, schiste argileux à Graptolites.

Graptolithina, Graptoloidea, Graptolithidés (*pal.*).

grass, 1. herbe; 2. surface, jour (*mine*); g. **land soil,** sol de prairie (brun, rouge); g. **work,** travail en surface.

grass (to), remonter (le minerai).

grate, 1. grille; 2. tamis (*mine*).

grating, spectrographe.

graticule, 1. réticule (*optique*); 2. quadrillage cartographique formé par les parallèles et méridiens.

grating, 1. treillis, grillage; 2. réseau (*optique*); 3. frottement, grincement; 4. réseau cristallin.

grau, grau.

grauwacke (see graywacke), grauwacke.

gravel, gravier; g. **filter,** massif filtrant; g. **mine,** 1. carrière de g.; 2. g. aurifère; g. **packing,** filtre à g.; g. **pit,** ballastière, gravière; g. **stone,** gravier; g. **trains,** alluvions fluvio-glaciaires; **fine g.,** g. fin; **lag g.,** résidu de déflation, reg.

gravelling, empierrement; **fine g.,** gravillonnage.

gravelly, pierreux, à graviers, graveleux; g. **loam,** limon caillouteux; g. **soil,** sol graveleux.

gravific lens, lentille gravitationnelle (*Astro.*).

gravimeter, gravimètre; **astatic g.,** astatisé.

gravimetric, gravimetrical, gravimétrique.

gravimetry, gravimétrie.

gravitate (to), graviter.

gravitation, gravitation.

gravitational, gravitationnel, attractif; g. **compaction,** tassement par gravité; g. **constant,** constante de gravitation; g. **differentiation,** différenciation par gravité; g. **field,** champ de pesanteur; g. **flow,** circulation par gravité; g. **image,** mirage gravitationnel (*Astro.*); g. **lens,** lentille gravitationnelle (*Astro.*); g. **pull,** gravitation; g. **separation,** séparation du gaz, du pétrole et de l'eau par densité (respective); g. **tectonics,** tectonique de gravité, glissements sous-marins; g. **water,** eau libre.

gravitative settling, sédimentation par gravité.

gravity, gravité, pesanteur; g. **anomaly,** anomalie gravimétrique; g. **balance,** balance gravimétrique; g. **collapse structure,** structure d'affaissement tectonique; g. **exploration,** prospection gravimétrique; g. **fault,** faille normale; g. **field,** champ de la pesanteur; g. **fold,** pli d'entraînement; g. **force,** pesanteur, gravité; g. **gliding,** glissement de terrain; g. **line,** conduite d'écoulement naturel; g. **map,** carte gravimétrique; g. **meter,** gravimètre; g. **of oils,** densité des pétroles (système API); g. **scree,** éboulis de gravité; g. **sliding,** phénomènes d'affaissement, de subsidence, etc); g. **survey,** levé gravimétrique; g. **tectonics,** décollement; g. **transport,** mouvement de masse; g. **unit (G. unit),** 1/10 de milligal; **bulk specific g.,** gravité densité apparente; **force of g.,** gravité pesanteur; **specific g.,** gravité poids spécifique (à 60° F).

gray, grey, gris; g. **antimony,** stibine, stibnite (*minér.*); g. **cobalt,** cobalt arsenical, smaltine; g. **copper,** cuivre g., tétrahédrite, panabase, (*minér.*); g. **desert soil,** sol g. désertique; g. **hematite,** spécularite (*minér.*); g. **manganese,** manganite, acerdèse (*minér.*); g. **oxide of manganese,** pyrolusite (*minér.*); g. **ore,** chalcosine (*minér.*); g. **podzol,** podzol g.; g. **soil,** sol g. (aride); g. **sulphuret of copper,** tennantite (*minér.*).

graywacke, grauwacke (grès sombre riche en débris de roches basiques, matrice argilo-micacée, souvent indurée).

grazing angle, angle d'incidence rasante.

greasy, luisant, à aspect gras; g. **feel,** toucher gras; g. **gold,** or fin.

great ice age, époque Pléistocène.

green, vert; **g. carbonate of copper,** malachite (*minér.*); **g. copper,** malachite; **g. copperas,** mélantérite (*minér.*); **g. earth, 1.** glauconie (*minér.*); **2.** chlorite (*minér.*); **g. feldspar,** microcline (*minér.*); **g. john,** fluorine v. (*minér.*); **g. lead ore,** pyromorphite (*minér.*); **g. marble,** serpentine (*minér.*); **g. mineral,** malachite (*minér.*); **g. sand,** sable glauconieux v.; **g. schist,** schiste v.; **g. stone,** roche v.; **g. vitriol,** mélantérite (*minér.*).

greenhouse effect, réchauffement climatique par CO_2, CH_4, etc; effet de serre.

greenland spar, cryolite (*minér.*).

greenockite, greenockite (*minér.*).

greisen, greisen.

greisening, formation de greisen, (type de pneumatolyse acide à mica blanc, quartz, etc.).

Grenville (Can.), série de Grenville, province de Grenville; **g. orogeny,** orogénèse Grenvillienne : — 955 M.A. (Can., USA).

grey, gris; **g. brown podzolic soil,** sol podzolique brun-g., sol brun lessivé; **g. cobalt,** smaltite (*minér.*); **g. copper,** cuivre-g. (*minér.*); **g. desert soil,** sol gris désertique, sierozem; **g. forest soil,** sol g. forestier; **g. wooded soil,** sol g. forestier.

greywacke (see graywacke), grauwacke.

grid, 1. grille, réseau, quadrillage; **2.** tamis; **g. coordinates,** coordonnées rectangulaires d'une carte; **g. method,** méthode du quadrillage, (*photogéol.*); **g. micrometer,** micromètre quadrillé; **g. pattern,** quadrillage de plis; **g. twinning of microcline,** mâcle quadrillée du microcline.

griddle, crible.

griddle (to), cribler.

gridiron twinning, mâcle entrecroisée, d'interpénétration.

grike, fente, crevasse, lapiés.

gril, grille, grille.

grind (to), 1. broyer, moudre, pulvériser; **2.** roder, meuler.

grindability, friabilité.

grinder, broyeur.

grinding machine, broyeur.

grinding of a thin section, meulage d'une lame mince.

grindlet, petit fossé, drain.

grinstone, 1. meule à aiguiser, pierre à aiguiser; **2.** grès quartzeux, fin homogène.

grip, 1. pince, griffe (mécanique); **2.** serrage.

grips, mâchoires d'un étau.

gripper, pinces, griffes.

grit, 1. sable grossier, grès grossier à grains anguleux; **2.** particules abrasives.

gritstone, grès; **millstone g.,** grès meulier, grès pour meules.

gritty, gréseux, sablonneux, à graviers.

grizzly, crible à barreaux.

groin, groyne, épi, éperon (pour protéger le littoral).

groove, 1. cannelure (glaciaire); **2.** sillon (*sédimento.*).

groove-mark, groove cast, sillon (base de strate).

groove (to), rayer, creuser des cannelures.

grossular, grenat grossulaire (*minér.*).

grossularite, grossulaire, grossularite (grenat).

ground, sol, terre, terrain; **g. avalanche,** avalanche de fond; **g. auger,** tarière; **g. cover,** couverture végétale; **g. ice,** hydrolaccolithe; **g. level,** niveau du sol; **g. mass,** pâte, matrice (*pétro.*); **g. method,** mesure au sol; **g. moraine,** moraine de fond, till de fond; **g. noise,** bruit de fond; **g. pressure,** pression géostatique; **g. resistivity,** résistivité du sol; **g. roll,** ondes parasites superficielles (*sism.*), ondes de Rayleigh; **g. sea,** lames de fond; **g. shaking,** secousse du sol; **g. swell,** lames de fond, houle; **g. truth,** « vérité terrain »; **g. water,** eau souterraine; **g.-water dam,** barrage souterrain; **g. water discharge,** résurgence d'eau souterraine; **g. work,** fondement, plan; **g.-ice wedge,** fentes de gel; **g.-water exploration,** prospection de l'eau souterraine; **g.-water level,** niveau piézométrique; **g. reservoir,** aquifère; **g. table,** surface piézométrique; **g. vein,** fissure aquifère (hydrothermale); **dead g.,** mort-terrain; **patterned g.,** sol structuré (*périgl.*); **striped g.,** sol strié (*périgl.*).

grounded ice, glace ancrée au sol.

grounding line, ligne de broyage (*glacio.*).

grouping, classification pétrographique.

grout, 1. déchets d'exploitation de carrière; **2.** (G.B.) : mortier fin; **3.** coulis.

grouting, 1. fonçage des puits par cimentation au mortier liquide; **2.** pénétration, injection; **mud g.,** injection de boue; **g. gallery,** galerie d'injection; **g. hole,** trou d'injection; **g. pressure,** pression d'injection.

growan, granite (*mines*).

growth anticline, anticlinal intraformationnel, contemporain de la sédimentation (Amér. du N.); **g. fault,** faille intraformationnelle syngénétique, contemporaine de la sédimentation (Amér. du N.); **g. line,** site d'accroissement; **g. ring,** strie d'accroissement (*pal.*); **g. twinning,** mâcle de croissance.

groyne, épi, brise-lame.

grumous soil, sol grumeleux.

grumusol, sol noir tropical.

grunerite, grunérite (variété d'amphibole) (*minér.*).

grus, var. d'arène granitique (sous climat aride).

grykes, lapiés.

Gshelian, Gzhelian (USSR), Gshélien (étage, Carbonifère sup.).

Guadalupian (N. Am.), Guadalupien (étage Permien moy. terminal).

guard, rostre (Bélemnite).

gubbin, minerai de fer argileux (angl.).

guidance, direction, orientation.

guide, guide, indication; **g. book,** livret-guide; **g. casing shoe,** sabot de guidage (*for.*); **g. formation,** formation témoin; **g. fossil,** fossile caractéristique; **g. mineral,** minéral guide.

gulch, ravin.

gulf, golfe; **g. coast,** zone pétrolière du Golfe du Mexique.

Gulfian (N. Am.), Gulfien (série, Crétacé sup.).

gullied, ravineux.

gully, ravin, gorge de torrent, rigole; **g. erosion,** ravinement.

gullying, ravinement.

gully (to), raviner.

gum content, teneur en gomme.

gumbo (US), sol collant, argileux, argile collante.

gumbotil (US), argile morainique lessivée et réduite (dérivée d'un till).

gummite, gummite (*minér.*).

gun, fusil, charge de perforation; **g. perforating,** perforation à balles d'un puits; **mud g.,** mitrailleuse à boue.

Gunnison River (US, Col. St), série de « Gunnison River » (Précambrien).

Gunz (Eur.), Günz (glaciation de; 2ᵉ glaciation).

Gunzian (Eur.), Günz (glaciation de; 2ᵉ glaciation).

gurgling well, puits à jaillissement intermittent.

gush, 1. jaillissement; **2.** jet, flot.

gush (to), jaillir.

gusher, 1. jaillissement, débordement; **2.** puits « éruptif » (*for.*); **g. sand,** sable pétrolifère (sous pression de gaz naturels).

gushing spring, source vauclusienne.

gust of wind, rafale.

gut, 1. goulet, passage étroit, défilé; **2.** canalisation (pétrole); **3.** ancien chenal de marée.

gut (to), extraire rapidement l'essentiel d'un gisement.

Gutenberg discontinuity, discontinuité de Gutenberg.

gutter, rigole, cannelure, rainure.

gutter (to), raviner, couler en ruisseaux.

guy, hauban (derrick).

guy (to), haubanner, fixer avec un hauban.

guyed tower, tour de forage fixée au fond de la mer.

guyot, guyot, volcan sous-marin.

G. wave, onde de Love.

Gymnosperm, Gymnospermae, Gymnosperme.

gypcrete (cf calcrete), croûte gypseuse, sol cimenté par du gypse.

gyprock, roche gypseuse.

gypseous, gypseux.

gypsic horizon, horizon gypseux (*pédol.*).

gypsiferous, gypsifère; **g. clay,** argile gypsifère.

gypsite, gypsite.

gypsum, gypse; **g. bearing,** gypsifère; **g. crust,** croûte gypseuse; **g. flower,** rose des sables; **g. plate,** lame auxiliaire « gypse teinte sensible » (optique); **g. quarry,** carrière de pierre à plâtre.

gyres, courants océaniques gyratoires (en boucles).

gyro compass, compas gyroscopique.

gyroscope, gyroscope.

gyroscopic, gyroscopique.

gyttja, gyttja (dépôt organo-terrigène lacustre).

H

haar (G.B.), brume de mer.

habit, habitus, habitus, faciès, aspect, forme (*minér.*); **prismatic habit**, forme prismatique; **tabular habit**, forme tabulaire.

habitat, habitat (*pal.*).

hachure, hachure (*cartographie*).

hack, pic, pioche (*mine*).

hack (to), hacher, ébrécher, piocher, tailler.

hackly fracture, cassure esquilleuse (*minér.*).

hadal gone, zone benthique de mer profonde (fosse).

hade (unusual), angle du plan de pendage d'une faille avec la verticale.

hade (to), s'incliner par rapport à la verticale.

Hadrinian (N. Am.), Hadrynien (Protérozoïque sup. : − 955 à − 600 M.A.).

haematite, hematite, hématite, oligiste (*minér.*).

haff, lagune (Baltique).

hail, grêle.

hailstone, grêlon.

hair pyrite, trichopyrite, millérite (*minér.*); **hair salt**, 1. halotrichite (*minér.*); 2. epsomite (*minér.*).

hairstone, quartz à inclusions aciculaires (rutile).

hair zeolite, zéolite fibreuse.

half-life, période (d'un élément radioactif); **half-tide**, mi-marée; **half-timbered**, demi-boisage (*mine*).

halide, halogénure.

halite, halite, sel gemme.

halitic environment, milieu salin.

Hallian (N. Am.), Hallien (étage, fin Pléistocène).

halloysite (U.S.), halloysite (Eur. métahalloysite).

halmyrolysis, halmyrolyse, altération sous-marine, syndiagenèse.

halo, halo, auréole (d'altération).

halocline, halocline.

halogen, halogenous, halogène.

halogenic, halogénique.

halokarst, karst de roches salifères.

halokinesis, tectonique salifère.

halomorphic soil, sol halomorphe.

halophilic organism, organisme halophile.

halophyte, halophyte.

halotolerant, euryhalin.

halotrichite, halotrichite (*minér.*).

hammada (Arabia), hammada, (région dénudée) surface plane rocheuse désertique.

hammer, marteau, massette; **h. drill**, marteau perforateur; **h. grab**, benne foreuse; **h. sismics**, sismique marteau; **air h.**, marteau pneumatique; **sledge h.**, masse.

hammer (to), marteler.

hammock (U.S.), tertre, butte.

hand, main; **h. auger**, tarière à m.; **h. auger work**, perforation au fleuret à m.; **h. churn drill**, tarière; **h. drill**, perforatrice à m.; **h. hole**, trou de visite;

h. lamp, lampe de mine; **h. lens**, loupe; **h. mining**, abattage à la pioche; **h. mucking**, roulage à bras (*mine*); **h. picked coal**, charbon trié à la m.; **h. picking**, triage à la m.; **h. shovelling**, pelletage à la m. (*mine*); **h. sorting**, triage à la m.; **h. stoping**, abattage à la m.; **h. truck**, diable (*mine*).

handaxe, handstone, coup de poing (Acheuléen).

hang, pente, inclinaison.

hanged retention water, eau de rétention suspendue.

hanger (U.S.), 1. système de suspension pour forage; 2. toit d'une couche, compartiment supérieur d'une faille.

hanging, 1. suspension, accrochage; 2. toit (*mine*); **h. bed**, couche sus-jacente; **h. glacier**, glacier suspendu; **h. side**, toit d'une couche; **h. valley**, vallée suspendue « valleuse »; **h. wall**, lèvre supérieure d'une faille, toit d'une souche; série chevauchante; **h. wall block**, compartiment supérieur d'une faille; **h. water**, eau suspendue.

haplite, aplite, granite filonien hololeucorate (*pétro.*).

hard, dur; **h. ash coal**, charbon maigre; **h. coal**, anthracite, charbon bitumeux; **h. digging**, forage en terrain d.; **h. grade**, degré de dureté; **h. ground**, 1. fond marin induré; 2. paléosol (inusité); **h. lead**, minerai de plomb antimonieux; **h. rock**, roche dure à forer, c.a.d. éruptive ou métamorphique; **h. surface**, surface indurée; **h. water**, eau calcaire et magnésienne.

hardened clay, argile durcie.

hardening, 1. durcissement; 2. trempe (*métall.*).

hardness, dureté; **h. number**, indice de d.; **h. scale**, échelle de d.; **h. test**, essai de d.

hardpan (cf. caliche, calcrete), 1. horizon pédologique, induré; 2. carapace, calcin; 3. alios; 4. niveau caillouteux cimenté par de la limonite.

Harlechian (Eur.), Harlechien (étage, Cambrien inf.).

harmonic tremor, secousses sismiques dues à l'arrivée de laves dans une cheminée volcanique.

harpolith, harpolithe (var. de laccolite).

harpoon, harpon; **antler h.**, h. en bois de renne (*préhist.*).

hartin, hartine (résine fossile).

harzburgite, harzburgite (*pétro.*).

hatchettine, hatchettite, hatchettite, adipocérite (*minér.*).

hatching, hachure (*cartogr.*).

haul (to), tirer, traîner, transporter, remorquer.

haulage, 1. transport (par camion); 2. traînage des wagons (*mine*).

haulageway, galerie de roulage.

haulway, voie de roulage (*mine*).

hausmannite, hausmannite (*minér.*).

Hauterivian, Hauterivien (étage, Crétacé inf.).
hauynite, hauyne, haüyne (feldspathoïde calco-sodique des roches volcaniques).
hauynophyre, var. de trachyte (à haüyne).
Hawaiian eruption, éruption hawaïenne.
hawaiite, hawaïte (andésite).
hawk's eye, crocidolite, variété fibreuse de rié-beckite (amphibole).
haystack hill, colline résiduelle (relief karstique).
haystacks, reliefs résiduels (karstiques).
hazard, risque; **natural h.,** r. naturel; **seismic h.,** r. sismique.
haze, brume légère.
hazy, brumeux.
head, 1. pointe littorale, promontoire; **2.** matériel soliflué périglaciaire (U.K.); **3.** pression hydro-statique; **h. erosion,** érosion régressive; **h. frame,** chevalement (*mine*); **h. gear,** 1. chevalement (*mine*); **2.** superstructure (d'un forage); **h. land,** cap, promontoire; **h. of a bar,** amont d'un banc; **h. of a comet,** tête d'une comète; **h. of a cone,** sommet (apex) d'un cône de déjection; **h. of a delta,** sommet (apex) d'un delta; **h. of the tide,** limite de la marée; **h. pipe,** conduite d'amenée; **h. wave,** onde arrivant en premier (*sismique*); **h. way,** descenderie (*mine*); **artesian h.,** pression artésienne; **casing h.,** tête de tubage; **circulating h.,** tête de circulation; **control h.,** tête de tubage; **dynamic h.,** pression dynamique; **escape h.,** trop-plein; **hydraulic h.,** pression hydraulique; **loss of h.,** perte de pression, décharge; **suction h.,** hau-teur d'aspiration; **well h.,** tête de puits.
heading (U.S.), avancement d'une mine; **h. face,** front d'avancement; **h. stope,** chantier d'avan-cement.
headwall recession, érosion régressive; **headward regression,** érosion régressive; **healed fault,** faille refermée.
heap, amas, accumulation, tas.
heap up (to), 1. amonceler, entasser, amasser; **2.** s'entasser, s'amonceler.
heat, chaleur; **h. balance,** bilan thermique; **h. capacity,** pouvoir calorifique; **h. chamber,** creuset, four; **h. conductivity,** conductibilité thermique; **h. content,** enthalpie; **h. crack,** fissure de dessi-cation; **h. efficiency,** rendement thermique; **h. energy,** énergie thermique; **h. equator,** équateur thermique; **h. flow,** flux thermique; **h. gradient,** gradient thermique; **h. of solution,** c. de dissolu-tion; **h. of vaporization,** c. de vaporisation; **h. release,** dégagement de c.; **h. unit,** unité ther-mique; **h. value,** pouvoir calorifique; **h. waste,** perte de c.; **h. wave,** vague de c. (*météo.*).
heating, chauffage, combustion; **h. capacity,** pou-voir calorifique; **h. power,** pouvoir calorifique; **spontaneous h.,** combustion spontanée (*mine*).
heave, 1. gonflement, boursouflement du mur d'une couche; **2.** composante horizontale (du rejet d'une faille); **stratigraphic h.,** rejet horizontal des couches (imprécis).
heaved side, lèvre soulevée (d'une faille).

heavies, minéraux lourds.
heaving, gonflement, soulèvement du mur d'une couche; **h. bottom,** sol gonflant.
heavy, lourd, dense; **h. baryte,** barytine, baryte; **h. clay soil,** sol argileux lourd; **h. earth,** barytine; **h. fluid separation,** séparation par liqueur dense; **h. fraction,** fraction lourde; **h. gradient,** pente raide; **h. isotope,** isotope lourd; **h. liquid,** liqueur dense; **h. loam,** limon argileux lourd; **h. métal,** métal lourd; **h. minerals,** minéraux lourds; **h. pressure,** haute pression; **h. sand soil,** sol sableux hydromorphe; **h. sea,** forte mer; **h. silt,** limon argileux; **h. spar,** barytine; **h. texture,** granulo-métrie fine; **h. water,** eau lourde.
Hebridean (Scotland), Hébridéen (Précambrien; syn. Lewisien).
hectare, hectare $= 10,000$ m^2; 2,471 acres.
hectogramme, hectogramme $= 100$ grammes.
hectolitre, hectolitre $= 100$ litres.
hectometer, hectomètre $= 100$ mètres.
hedenbergite, hédenbergite (pyroxène).
hedyphane, hédyphane (*minér.*).
height, hauteur, élévation, altitude; **h. above sea level,** altitude par rapport au niveau marin; **h. gauge,** altimètre; **relative h.,** altitude relative; **spot h.,** point côté; **wave h.,** hauteur des vagues.
Helderbergian (N. Am.), Helderbergien (étage, Dévonien inf.).
helicitic, hélicitique (texture originale dans roches métamorphiques); **h. roll,** enroulement hélicoï-dal (*pal.*).
Helikian (N. Am.), Hélikien (Protérozoïque moyen : $- 1 735$ à $- 955$ M.A.).
heliopause, héliopause (*Astro.*).
heliosphere, héliosphère (*Astro.*).
heliotrope, héliotrope.
Heliozoa, Héliozoaires.
helium, hélium.
helmet, casque de protection.
hell raiser, outil de repêchage (*for.*).
Helvetian, Helvétien (sous-étage, Miocène).
hematite, haematite, hématitique.
hemera, hemera, temps d'abondance maximum d'un taxon.
hemycrystalline, semi-cristallin (cristaux + verre ou verre dévitrifié).
hemidome (obsolete), hémidôme (*cristal.*).
hemihedral, hemihedric, hémiédrique, hémièdre.
hemihedrism, hemihedry, hémiédrie.
hemilhedron, hémièdre.
hemihyaline, hémihyalin.
hemimorphic, hemimorphous, hémimorphique.
hemimorphism, hémimorphisme.
hemimorphite, hémimorphite, calamine (*minér.*).
hemipelagic, hémipélagique.
hemiprism, hémiprisme (*cristallo.*).
hemipyramid, demi-pyramide.
hemisphere, hémisphère.
hemispheric, hemispherical, hémisphérique.
hemisymmetric, hemisymmetrical, hémièdre, hé-miédrique.

hemisymmetry, hémiédrie (*cristallo.*).

hemitetrahexahedron, dodécaèdre pentagonal tétraédrique (*cristallo.*).

hemitrope, hemitropic, mâclé.

hemitropism, hemitropy, hémitropie.

hepatite, hépatite, var. de barytine.

hepatic pyrite, marcasite (*minér.*).

Hercynian, Hercynien (Eur.); **h. orogeny,** orogénèse hercynienne (Carbonifère et Permien : syn. varisque).

Hercynides, Hercynides.

hercynite, hercynite (= spinelle) (*minér.*).

hermatypic taxon, taxon hermatypique.

herringbone, « en chevrons »; **h. cross lamination,** stratification transverse en c.; **h. texture,** texture en c.; **h. twin,** mâcle en c.

hessonite, essonite, grossulaire (= variété de grenat).

heteroblastic, hétéroblastique (*pétro.*).

heterochronism, hétérochronisme.

heterochronous, hétérochrone.

heterocyclic, hétérocyclique.

heterodont, hétérodonte (*pal.*).

heterogeneity, hétérogénéité.

heteromorphic, hétéromorphe.

heteromorphism, hétéromorphisme.

heteromyarian, anisomyaire (*pal.*).

heterophyletic, hétérophylétique.

heterosporous, hétérospore.

heterotrophic, hétérotrophe.

heterotactic, heterotactous, heterotaxial, hétérotaxique.

heterozygous, hétérozygote.

Hettangian, Hettangien (étage, Lias inf.).

Heulandite, Heulandite (= zéolite calco-sodique) (*minér.*).

hew (to), tailler, couper.

hewer, 1. tailleur de pierres, carrier; 2. piqueur, mineur.

hewing, havage, piquage, sous-cavage (*mine*); **h. stone,** taille des pierres.

hewn stone, pierre taillée.

Hexacoral, Hexacoralla, Hexacoralliaires.

Hexactinaria, Hexactinaridés.

hexagonal, hexagonal; **h. prism,** prism h.; **h. pyramid,** pyramide h.; **h. system,** système h.

hexahedral, hexahédrique.

hexahedrite, hexahédrite (variété de météorite à clivage cubique).

hexahedron, hexaèdre, cube (*cristallo.*).

hexane, hexane (hydrocarbure liquide).

hexaoctahedron, hexaoctaèdre, hexoctaèdre (*cristallo.*).

hexatetrahedron, hexatétraèdre (*cristallo.*).

hiatal episode, lacune de sédimentation; **h. fabric,** structure, texture hétérométrique (taille différente des cristaux); **h. texture,** structure hétérométrique, inéquigranulaire, par extension poreuse.

hiatus, lacune stratigraphique, lacune de sédimentation.

hiddenite, hiddénite, triphane vert, spodumène chromifère (= pyroxène) (*minér.*).

hidden layer, couche indétectable.

high (in), 1. riche en, à haute teneur; 2. élevé, haut; **h. angle fault,** faille à fort pendage; **h. ash coal,** charbon riche en cendre; **h. energy environment,** milieu à forte énergie; **h. grade,** à forte teneur; **h. gradient,** à forte pente; **h. island,** partie centrale non corallienne d'une île corallienne; **h. level water,** sommet de la nappe phréatique; **h. quartz,** quartz de haute température; **h. resolution,** haute résolution (d'un appareil), fort pouvoir séparateur; **h. sea,** haute mer; **h. speed layer,** couche à forte vitesse de propagation d'ondes; **h. tide,** marée haute; **h. water,** marée haute; **h. water line,** niveau de marée haute; **h. water mark,** laisse de haute mer; **h. water platform,** banquette littorale de tempête; **structural h.,** crête d'un anticlinal, sommet d'un dôme.

higher-grade metamorphic rock, roche à fort gradient de métamorphisme.

highland, pays montagneux, haute terre.

highmoor, tourbière de zones élevées.

hill, colline, coteau, côte; **h. diggings,** exploitation à flanc de coteau; **h. shading,** ombrage des courbes de niveau; **h. side,** versant, flanc; **h. work,** travail à flanc de coteau (*mine*).

hillside creep, mouvement de masse; **h. waste,** manteau détritique, colluvium.

hillock, petite colline, butte, monticule, tertre.

hillslope, versant.

hillwash, ruissellement sur pente.

hilly, montagneux, accidenté.

hilt's rule, règle suivant laquelle la qualité du charbon s'élève avec l'âge.

hinge, charnière, articulation; **h. area,** plateau cardinal; **h. fault,** faille normale dont le rejet diminue progressivement; **h. line,** ligne d'articulation (Brachiop.); **h. teeth,** dents cardinales (*pal.*); **anticlinal h.,** charnière anticlinale.

hinged, à charnière, articulé.

hingement, charnière des valves d'un Ostracode.

hinterland, arrière-pays.

Hippurites, Hippurite (Rudiste).

hirst (syn. hurst), 1. monticule boisé; 2. banc de sable fluviatile.

histic epipedon, horizon pédologique humifère superficiel.

histogram, histogramme.

histosol, histosol, sol organique.

hitch, 1. saccade, secousse; 2. nœud, attache; 3. légère faille (*mine*).

hoarfrost, hoar frost, gelée blanche.

hodograph, hodographe.

hoe, pic, pioche, sape.

hogback, hog's back, crêt(e) monoclinal(e) (symétrique).

hoggin, gravier criblé.

hogshead, fût : 52 gallons, 5 = 240 litres.

hog-tooth span, calcite en dents de cochon (scalénoèdres de calcite).

hog wallow, faiblement ondulé (« en bauge de sanglier »).

hoist, treuil, grue, palan, appareil de levage; **h. frame,** chevalement (*mine*).

hoist (to), remonter, extraire (*mine*).

hoisting, 1. levage, hissage, remontée; 2. extraction (du charbon); **h. compartment,** compartiment d'extraction; **h. crab,** treuil; **h. plant,** installation d'extraction; **h. shaft,** puits d'extraction; **h. winch,** treuil de levage.

holder, 1. support, étau, récipient; 2. concessionnaire, titulaire; **gas h.,** gazomètre; **oil h.,** bidon de pétrole.

hold up, 1. retenue liquide (raffinerie); 2. tenue (des roches).

hole, trou, cavité, forage, sondage, puits; **h. blow,** débourrage (d'un puits); **h. deviation,** déviation du sondage; **h. opener,** élargisseur; **h. section,** coupe de sondage; **blow h.,** soufflard; **bore h.,** sondage; **cased h.,** trou tubé; **churn, eddy, pot h.,** marmite de géant; **deflation h.,** creux de déflation; **dry h.,** puits improductif; **looking h.,** trou de regard; **open h.,** trou non tubé; **sink h.,** doline; **swallow h.,** doline.

hole (to), 1. trouer, percer; 2. creuser, haver souscaver (*mine*).

holer, haveur (*mine*).

holing, perforation, percement, havage (*mine*), souscavage.

holing machine, haveuse.

hollow, creux, cavité, excavation, cuvette, dépression, niche, fondrière; **h. ground,** sol creusé; **h. lode,** filon à géodes; **h. sea,** mer creuse; **h. spar,** andalousite (*minér.*); **nivation h.,** creux de nivation.

hollow out (to), excaver, creuser.

hollowing, creusement.

holm, 1. petite île, îlot (de rivière); 2. rive plate, terrain alluvial.

holoaxial, holoaxe.

holoblast, minéral néoformé (*pétro.*).

Holocene, Holocène, Quaternaire récent.

holocrystalline, holocristallin.

holohedral, holohedric, holoédrique (*cristallo.*).

holohedrism, holoédrie.

holohedron, holoèdre (*cristallo.*).

holohyaline rock, roche vitreuse, roche holohyaline.

holohyaline texture, texture vitreuse.

hololeucocratic, hololeucocrate, à dominance de minéraux blancs (*pétro.*).

holomelanocratic, holomélanocrate (*pétrogr.*), ultrabasique.

holomorphic, holomorphique (à symétrie complète) (*cristallo.*).

holophyte, holophyte (*pal.*).

holosiderite, holosidérite (météorite entièrement ferrique).

Holostea, Holostéens (*pal.*).

holosymmetric, holoèdre, holoédrique (*cristallo.*).

holotype, holotype (*pal.*).

Holsteinian, Holsteinien (interglaciaire Mindel-Riss; Nord de l'Europe).

homeoblastic, 1. homéoblastique (métam.); 2. isométrique, isocristallin.

homeogenesis, développement parallèle (*pal.*).

homeometric, homéométrique.

homeomorph, homeomorphous, homéomorphe.

homeomorphism, homeomorphy, homéomorphie.

homewards method, exploitation en retour, rétrograde (*mine*).

Hominoids, Hominidae, Hominidés.

homoaxial foldings, plissements suivant des axes parallèles.

homoclime, homologue climatique.

homoclinal valley, vallée monoclinale.

homocline, 1. pli monoclinal; 2. structure monoclinale.

Homodont, Taxodonte (*pal.*).

homogene deformation, déformation homogène.

homolographic projection, projection équiaréale.

homology, homologie (ressemblance héréditaire) (*pal.*).

homomyarian, isomyaire.

homonym, homonyme (*pal.*).

homophyletic, homophyllétique.

homoseism, isoséisme.

homoseismal, homoséismique, isoséismique.

homotaxial, homotaxique.

homotaxis, homotaxy, homotaxie (*stratigr.*).

hone, honestone, pierre à aiguiser.

honey-comb structure, texture, structure alvéolaire.

honeycombed, à structure alvéolaire.

hoodoo, cheminée de fée.

hook, 1. crochet, grappin; 2. cap, pointe de terre, flèche littorale recourbée; 3. coude (de rivière).

hook up, montage, installation, tête de puits; **hook valley,** vallée à affluents obliques ou recourbés; **dull hook,** crochet de montage.

hopper, 1. wagonnet basculant; 2. trémie; **h. crystal,** cristal en forme de trémie.

horizon, 1. horizon pédologique; 2. horizon (niveau) stratigraphique (repère); **h. line,** ligne d'horizon; **water h.,** horizon aquifère.

horizonation, existence d'horizons pédologiques.

horizontal, horizontal; **h. cut and fill,** tailles en échelon avec remblayage, exploitation par tranches montantes remblayées (*mine*); **h. displacement,** rejet h.; **h. fault,** faille subhorizontale; **h. joint,** diaclase h., bathroclase; **h. separation,** rejet h.; **h. slicing,** exploitation par tranches h.; **h. slip,** composante h. du rejet net; **h. throw,** rejet h. transversal.

horn, 1. corne; 2. pic glaciaire; **h. coral,** coralliaire isolé; **h. lead,** phosgénite (*minér.*); **h. mercury,** calomel; **h. quicksilver,** calomel; **h. silver ore,** cérargyrite (*minér.*); **h. stone,** silex corné, zoné; **h. tiff,** calcite teintée de matières organiques.

hornblei, phosgénite (*minér.*).

hornblende, hornblende; **h. gneiss,** gneiss à amphibole; **h. schist,** schiste à amphibole (hornblende); **h. syenite,** syénite à hornblende.

hornblendite, hornblendite (*pétro.*).

hornfels, cornéenne.

hornito, hornito, cône de laves.

hornschist, schiste à hornblende.

horse, 1. intercalation stérile (*mine*); 2. chevalet; 3. lambeau de poussée; **h. back,** dos d'âne; **h. flesh ore,** bornite, érubescite; **h. head,** contrepoids de pompe (*pétrole*); **h. shoe curve,** méandre; **h. shoe dune,** dune en fer à cheval; **h. shoe lake,** lac de méandre, en croissant; **h. power,** cheval-vapeur = 0,7457 kW (G.B.); **h. tail,** filon en queue de cheval; **h. teeth feldspar,** feldspath en dents de cheval (dans un granite).

horst, horst, massif soulevé, môle.

host, hôte (*minér.*).

hostrock, roche minéralisée par un apport.

hot, chaud; **h. brime,** saumure chaude; **h. cloud,** nuée ardente; **h. dump,** crassier, terril; **h. point,** point chaud (zone où le magma forme des intrusions, voire des volcans); **h. spot,** point chaud; **h. spring,** source chaude hydrothermale.

hour-glass structure, structure en sablier.

hour-glass valley, vallée « en sablier », vallée resserrée.

hourly flow, débit horaire.

howardite, howardite.

Hoxnian (G.B.), Hoxnien (interglaciaire Mindel-Riss).

hub, 1. mire de nivellement, piquet, repère; 2. moyeu.

hudge, benne.

Hudsonian, Hudsonien (Précambrien canadien); **h. orogeny,** orogénèse hudsonienne (− 1 640 à − 1 820 MA) (équivalent US « Penokean »).

hulking, enlèvement des salbandes (*mine*).

hum, tour karstique, mogote, relief résiduel calcaire.

humate, composé humique.

humic, humique; **h. acid,** acide h.; **h. carbonated soil,** rendzine; **h. gley soil,** sol h. à gley; **h. iron pan,** alios h.; **h. latosol,** sol latéritique; **h. layer,** couche d'humification; **h. podzol,** podzol h.

humid, humid.

humidification, humidification.

humidify (to), humidifier.

humidity, humidité.

humification, humification.

humine coal, humin, humine.

humite, 1. charbon humique; 2. humite (= subsilicate) (*minér.*).

humming device, table à secousses.

humod, podzol humique (non hydromorphe).

hummock, tertre, butte.

hummocky, avec creux et bosses; **h. moraine,** moraine mamelonnée.

humo-calcareous soil, rendzine.

humodite, humodite (charbon).

humogelite, humogélite, vitrinite.

humolite (humic coal), humite.

humox, sol ferrallitique humifère (*pédol.*).

hump, bosse.

humult, ultisol humifère (*pédol.*).

humus, humus; **h. calcareous soil,** rendzine; **h. carbonate soil,** rendzine; **h. impoverishing,** appauvrissement en h.; **h. infiltrated lateritic soil,** sol latéritique humifère; **h. layer,** horizon humifère; **h. silicate soil,** sol humo-silicaté; **mild h.,** h. intermédiaire; **raw h.,** matte.

hundred-weight, quintal; Angl : 112 lbs = 50,802 kg; Amér : 100 lbs = 45,359 kg.

hungry, stérile.

Huronian, Huronien (Précambrien : Protérozoïque inf.).

hurricane, ouragan, tempête; **h. surge,** brusque élévation du niveau marin suite à un ouragan.

hush, décapage des terrains superficiels par courant d'eau.

hust (to), dégager les morts terrains.

hutch, benne roulante (*mine*); **h. road,** galerie de roulage (*mine*).

hutment, baraquement.

hyacinth, hyacinthe (= zircon) (*minér.*).

hyaline, hyalin, transparent, vitreux; **h. quartz,** quartz bleuté légèrement calcédonieux.

hyalite, hyalite, (= variété d'opale) (*minér.*).

hyalo (prefix), vitreux.

hyalobasalt, verre basaltique, basalte vitreux (avec très peu de grands cristaux).

hyaloclastic rock, palagonite (*pétro.*).

hyaloclastite, hyaloclastite.

hyalocrystalline, hyaloporphyrique (var. de structure microlitique).

hyalophitic texture, structure intersertale (ophitique) vitreuse.

hyalosiderite, hyalosidérite (var. d'olivine).

Hyalospongea, Hyalosponges.

hybrid, hybride (*pal.*).

hybridism, hybridation magmatique.

hydatogenesis, hydatogénèse.

hydatogenic, hydatogenous, formé en milieu aqueux.

hydatomorphic, cristallisé en milieu aqueux.

hydrargillite, hydrargillite, gibbsite.

hydrargyrum, mercure.

hydratation, hydradation.

hydrate, hydrate; **calcium h.,** chaux hydratée.

hydrate (to), 1. hydrater; 2. s'hydrater.

hydrated, hydraté.

hydration, hydratation.

hydraulic, hydraulique (adj.).

hydraulic, hydraulique; **h. cement,** ciment h.; **h. circulation system,** forage à injection; **h. engineer,** hydraulicien; **h. fracturing,** fracturation h.; **h. gold mining,** exploitation h. de l'or; **h. gradient,** pente h.; **h. head,** hauteur piézométrique; **h. hoisting,** extraction h.; **h. jack,** vérin h.; **h. lime,** chaux h.; **h. map,** carte h.; **h. mining,** abattage h.; **h. pressure,** pression piézométrique; **h. profile,** coupe verticale d'un aquifère; **h. stowage,** remblayage hydraulique (*mine*).

hydraulic (to), abattre par la méthode hydraulique (*mine*).

hydraulician, hydraulicien.

hydraulicing, hydraulicking, abattage hydraulique.

hydraulicity, hydraulicité.

hydraulics, hydraulique (n).

hydric, hydrique.

hydrobiotite, hydrobiotite.

hydrocarbon, hydrocarbure; **aliphatic h.,** h. aliphatique; **aromatic h.,** h. aromatique; **cyclic h.,** h. cyclique.

hydrocarbonaceous, hydrocarbonic, hydrocarbonous, hydrocarboné.

hydrochemical facies, faciès hydrochimique.

hydrochemistry, hydrochimie.

hydrochloric, chlorhydrique; **h. acid,** acide c.

hydroclastic, roche détritique déposée en milieu aquatique.

Hydrocorallines, Hydrocoralliaires.

hydrocracking, hydrocraquage.

hydrocyanic, cyanhydrique.

hydrocyanite (syn. chalcocyanite), hydrocyanite, sulfate de cuivre blanc.

hydrodynamic, hydrodynamique (adj); **h. lift,** remontée h. des particules.

hydrodynamics, hydrodynamique (n).

hydrodynamists, hydraulyciens.

hydroelectric, hydroélectrique; **h. reservoir,** réservoir hydroélectrique.

hydroexplosion, explosion volcanique sous-marine, explosion volcanique phréatique, etc.

hydrofluoric acid, acide fluorhydrique.

hydrofracturing, fracturation hydraulique.

hydrogen, hydrogène; **h. ion concentration,** pH, concentration en ions H; **h. sulfide,** h. sulfuré.

hydrogenate (to), hydrogéner.

hydrogenic, hydrogenetic, hydrogenous, formé par l'eau; **hydrogenic rock,** évaporite.

hydrogenous coal, charbon à forte teneur en eau.

hydrogeochemical, hydrogéochimique.

hydrogeochemistry, géochimie de l'eau.

hydrogeologist, hydrogéologue.

hydrogeology, hydrogéologie.

hydroglacial, hydroglaciaire.

hydrograph, 1. carte hydrographique; 2. graphique indiquant l'écoulement, la vitesse de l'eau en fonction du temps.

hydrographer, hydrographe (ingénieur).

hydraographic, hydrographique; **h. basin,** bassin h., **h. map,** carte h.

hydrography, hydrographie.

hydrohematite, hydrohématite, turgite (*minér.*).

hydrokinetics, cinétique des fluides.

hydrolaccolith, hydrolaccolithe, syn. pingo.

hydrolite (not explicit), 1. géode remplie d'eau; 2. geysérite; 3. var. de zéolite.

hydrolith, roche d'origine aquatique précipitée chimiquement (évaporite).

hydrologic, hydrological, hydrologique; **h. cycle,** cycle de l'eau.

hydrologist, hydrologue.

hydrology, hydrologie.

hydrolysate, hydrolyzate, 1. produit argileux, latéritique, formé par altération (ex : bauxites, argiles, shales); 2. hydroxyde résultant d'une hydrolyse (chimie).

hydrolysis, hydrolyse.

hydrolyze (to), hydrolyser.

hydromagnesite, hydromagnésite (*minér.*).

hydrometamorphism, hydrométamorphisme.

hydrometer, 1. densimètre; 2. aéromètre.

hydrometry, hydrométrie.

hydromica, hydromica (vermiculite, glauconie, serpentine).

hydromorphic, hydromorphe; **h. soil,** sol h.

hydromuscovite, hydromuscovite, illite, hydromica.

hydronepheline, hydronéphéline (*minér.*).

hydrophane, hydrophane, opale translucide.

hydrophilic, hydrophilous, hydrophile.

hydrophilite, hydrophilite (*minér.*).

hydrophobic, hydrophobe.

hydrophone, hydrophone.

hydropore, hydropore (*pal.*).

hydropower, énergie hydraulique.

hydrosol, hydrosol.

hydrosome, hydrosome, colonie d'hydrozoaires.

hydrosphere, hydrophère.

hydrospheric, hydrosphérique.

hydrostatic, hydrostatical, hydrostatique (adj); **h. head,** pression h.; **h. level,** niveau h.; **h. pressure,** pression h.

hydrostatics, hydrostatique (n).

hydrosulfuric acid, hydrogène sulfuré.

hydrothermal, hydrothermal; **h. alteration,** altération h.; **h. deposit,** gisement minéral h.; **h. synthesis,** genèse cristalline h.; **h. water,** eau thermale venant des profondeurs, eau h.

hydrous, hydraté, aqueux; **h. mica,** illite.

hydrovolcanic process, processus phréatomagmatique.

hydroxide, hydroxyde.

hydrozincite, hydrozincite, zinconise (*minér.*).

Hydrozoa, Hydrozoan, Hydrozoaire (*pal.*).

hygrograph, hygromètre enregistreur.

hygrometer, hygromètre.

hygrometry, hygrométrie.

hygroscopic, hygroscopique; **h. moisture,** état h.; **h. coefficient (U.K.); moisture content (U.S.),** coefficient d'hygroscopicité.

hyolithes, hyolithes (*pal.*).

hypabyssal, hypabyssal; **h. rocks,** roches intrusives de semi-profondeur solidifiées en sills et dykes.

hypautomorphic (syn. hypidiomorphic), hypidiomorphe.

hyperite, gabbro (désuet) à hypersthène et augite.

hypermelanic rock, roche holomélanocrate.

hypersolvus, hypersolvus.

hypersthene, hypersthène.

hypersthenite, hypersthénite (*pétro.*).

hypidiomorphic (syn. hypautomorphic), hypidiomorphique, hypidiomorphe.

hypocenter, hypocentre, hypocentre, foyer.

hypocrystalline, hypocristallin (partiellement cristallin), hypohyalin.

hypogene, hypogène, interne.

hypohyaline, partiellement vitreux.

hypolimnion, niveau inférieur d'un lac dimictique.

hypotaxic, superficiel.

hypotaxic deposits, minerais de surface.

hypothermal, 1. hypothermal (entre 300 et 500 °C); **2.** intervalle hypothermal (syn. « petit âge glaciaire »); **h. deposit,** gisement minéral hypothermal; **h. vein,** filon minéral hypothermal.

hypotype, hypotype (*pal*).

hyppuritic, à Hippurites.

hypsithermal, hypsithermal interval, optimum climatique post-glaciaire.

hypsographic curve, courbe hypsographique.

hypsometry, hypsométrie.

hypsograph, hypsographe.

hypsometer, hypsomètre.

hypsometric, hypsometrical, hypsométrique.

Hystrichomorphe, Hystrichomorphes.

Hystrichosphere, Hystrichosphère (Acritarches et Dinoflagellés).

I

Ice, glace; **i. age,** époque glaciaire; **i. barrier,** barrière de g.; **i. berg,** iceberg; **i. boom,** débâcle des g.; **i. boulder,** bloc erratique; **i. breaker,** brise-g.; **i. cake,** glaçon; **i. cap,** calotte glaciaire, inlandsis de dimensions limitées; **i. cascade,** cascade de séracs; **i. cliff,** falaise de g.; **i. cone,** cônes de g.; **i. cored rock glacier,** glacier rocheux avec glace interstitielle; **i. core,** noyau de g.; **i. crack,** fente de gel; **i. dam,** barrage de g.; **i. dammed-lake,** lac de barrage glaciaire; **i. divide,** ligne de partage glaciaire; **i. dome,** dôme de g.; **i. drift,** charriage de g.; **i. fall, 1.** cascade de g., séracs; **2.** débâcle glaciaire; **i. float,** glaçon; **i. floe,** glaçon; **i. flood,** glaciation; **i. foot,** pied de g.; **i. front, 1.** front glaciaire; **2.** barrière de g. (Antarctique); **i. island,** grand iceberg; **i. jam,** embâcle glaciel; **i. laid drift,** moraine, dépôts glaciaires (général); **i. layer,** couche de g.; **i. ledge,** pied de g.; **i. lens,** lentille de g.; **i. lifted landforms,** formes de terrain créées par soulèvement de g. du lac; **i. lobe,** lobe glaciaire; **i. mark,** trace d'usure glaciaire; **i. mound,** pingo; **i. mountain,** iceberg; **i. pack,** banquise; **i. period,** époque glaciaire; **i. plateau,** calotte glaciaire; **i. point,** température de la g. fondante; **i. push,** poussée de gel; **i. push moraine,** moraine de poussée; **i. rafted block,** bloc transporté par radeaux de g.; **i. rafting,** transport par radeaux de g.; **i. rampart,** rempart latéral de g.; **i. river,** glacier; **i. scour,** érosion glaciaire; **i. scoured plain,** plaine de rabotage glaciaire; **i. sheet,** calotte glaciaire de grandes dimensions; **i. shelf,** plate-forme de g. flottante; **i. shove,** poussée de gel; **i. shove ridge,** bourrelet de poussée glacielle; **i. sill,** filon de g.; **i. slab,** lame de g.; **i. slice,** écaille de g.; **i. smoothed rock,** roche moutonnée; **i. spar,** sanidine (*minér.*); **i. stone,** cryolithe (*minér.*); **i. stream,** langue glaciaire; **i. thrust,** poussée glaciaire; **i. thrusting,** glacio-tectonique; **i. vein,** filon de g.; **i. wall, 1.** pied de g.; **2.** falaise de glace reposant sur le sol ferme, mais avec la possibilité de trouver le fond entre 0 et 900 m; **i. wedge,** fente de gel à remplissage de g.; **i. wedge cast,** moulage (remplissage) de fente de gel; **i. wedge polygon,** polygone de fentes de gel; **i. wedge pseudomorph,** pseudomorphose, fente de gel à remplissage terreux; **i. wedging,** fissuration par le gel; **i. worn,** usé par la g.; **brittle i.,** g. cassante; **bubbly i.,** g. bulleuse; **dead i.,** g. morte; **drifting i.,** g. flottante; **feather i.,** aiguille de g.; **firn i.,** g. de névé; **floating i.,** g. flottante; **inland i.,** inlandsis, calotte glaciaire; **shelf i.,** plate-forme de g. flottante; **shore i.,** g. de rive.

iced, glacé (par la gelée).

Iceland agate, obsidienne d'Islande; **i. spar,** spath d'Islande, calcite.

icelandite, islandite (*pétro.*).

ichnite, empreinte de pas fossile.

ichnofossil, trace fossile.

ichnolite, empreinte de pas fossile.

ichnology, étude des empreintes de pas et de pistes animales fossiles.

ichor, émanation magmatique, minéralisateur.

Ichthyosauria, Ichthyosauriens.

Ichthyopterygia, Ichthoptérigiens.

icicle, pendeloque de glace.

icing, formation de lentilles de glace superficielle dans les plaines alluviales inondées (Alaska).

icy, glacial.

iddingsite, iddingsite (altération de l'olivine).

identified resource, gisement reconnu et évalué.

idioblast, idioblaste (grand cristal automorphe : dans une structure métamorphique).

idioblastic mineral, minéral automorphe, à faces bien cristallisées (dans roche métamorphique).

idioblastic rock, roche métamorphique formée de cristaux automorphes.

idiogenous (better : syngenetic), idiogène (de même origine pétrographique).

idiogeosyncline, idiogéosynclinal.

idiomorphic, idiomorphe, automorphe.

idiomorphism, idiomorphisme.

idocrase, idocrase, vésuvianite (variété de grenat).

idrialite, idrialite (subst. bitumineuse).

igneo-aqueous, formé par l'action conjuguée du feu et de l'eau.

igneous, igné, éruptif, magmatique; **i. breccia, 1.** brèche formée de roches éruptives; **2.** brèche pyroclastique, brèche de coulée volcanique; **i. complex,** complexe de roches éruptives; **i. emanations,** émanations éruptives; **i. facies,** faciès éruptif; **i. intrusion,** intrusion éruptive; **i. lamination,** litage de roches éruptives; **i. rocks,** roches éruptives, ignées, magmatiques; **i. rocks series,** séquence magmatique.

ignescent rock, roche qui émet des étincelles quand on la frappe avec de l'acier.

ignimbrite, ignimbrite (= brèche volcanique, tuff volcanique acide).

ignimbritic, ignimbritique.

ignitability, ignitibility, inflammabilité.

ignitible, inflammable.

ijolith, ijolite (*pétro.*).

Illinoian, Illinoisan (North Am.), Illinoien (3ᵉ glaciation de l'Am. du N.).

illite, illite (argile).

illuminator, source lumineuse d'un microscope.

illuvial, illuvial; **i. horizon,** horizon i. (horizon B); **i. material,** matière illuviale; **i. soil,** sol i.

illuviation, illuviation (dépôt dans l'horizon B).

ilmenite, ilménite, fer titané (*minér.*); **i. norite,** norite à ilménite.

ilmenitite, norite essentiellement formée d'ilménite.

ilmenorutile, ilménorutile, rutile columbifère (*minér.*).

ilvaïte, ilvaïte, liévrite (*minér.*).

imaged, photographié par satellite.

imagery, 1. visualisation (photographique); 2. imagerie par satellite.

imbed (to) *cf.* **embed (to),** enfermer, contenir.

imbibe (to), imbiber.

imbibition, imbibition; **i. water,** eau d'i.

Imbrian, Imbrien (période ancienne de formation lunaire).

imbricate slice, écaille (*tecto.*).

imbricated, imbriqué; **i. structure,** 1. structure en écailles, structure imbriquée; 2. chevauchement, nappes de charriage de même direction.

imbrication, imbrication, structure imbriquée.

immature soil, sol immature, sol non mûr, sol non évolué.

immature sandstone, grès immature, peu évolué.

immerge (to), immerger.

immerse (to), immerger, submerger, plonger.

immersed bog, tourbière immergée.

immersible, immersible.

immersion, immersion; **i. lens,** objectif à i.; **i. mount,** frottis.

immiscibility, immiscibilité.

immiscible, non miscible.

impact, choc, impact, collision; **i. breccia,** brèche formée par l'impact d'un météorite; **i. crater,** cratère de météorite; **i. deposit,** matériaux formés par la collision d'un météorite; **i. law,** loi de chute des particules en milieu liquide, formule densimétrique; **i. metamorphism,** dynamométamorphisme de choc; **i. resistance,** résilience; **i. slag,** verre formé par chute de météorite; **i. statement,** étude d'impact sur l'environnement; **i. strength,** résistance aux chocs.

impactite, impactite (brèche de météorite).

imperforate, non perforé (*pal.*).

impermeability, imperméabilité.

impermeable, imperméable; **i. barrier,** zone i.; **i. layer,** couche i.

impervious, 1. impénétrable; 2. imperméable, étanche.

imperviousness, 1. impénétrabilité; 2. étanchéité, imperméabilité.

impetus, élan, impulsion.

implement, outil, instrument; **bone i.,** outil en os; **flint i.,** outil en silex; **paleolithic i.,** outil préhistorique.

implication (obsolete, see symplectite), texture imbriquée, graphique (*minér.*).

imponded lake, lac de barrage.

imporosity, absence de porosité.

imporous, non poreux.

imposed metamorphic fabric, structure héritée anté-métamorphique.

imposed stream, rivière surimposée, épigénique.

impound, lac de retenue.

impoundment, barrage-réservoir, retenue.

impound (to), barrer une rivière pour construire un lac de retenue, un réservoir.

impoverishment, dégradation, appauvrissement; **i. of the soil,** appauvrissement du sol.

impregnate (to), imprégner, imbiber, saturer.

impregnated rock, roche minéralisée de façon diffuse, épigénétiquement.

impregnation, imprégnation, imbibition; **i. ore,** minerai d'imprégnation.

impressed pebbles, galets impressionnés.

impressed stream, rivière surimposée, épigénique.

impression, 1. figure sédimentaire; 2. empreinte végétale (de feuille).

imprint, 1. figure sédimentaire; 2. empreinte (fossile).

impsonite, impsonite (= pyrobitume).

impulse, impulsion, choc, secousse; **i. meter,** compteur d'impulsion; **i. period,** période d'impulsion; **i. recorder,** enregistreur d'impulsion; **seismic i.,** secousse sismique.

impure, impur.

impurity, impureté.

inalterability, inaltérabilité.

inalterable, inaltérable.

Inarticulata, Inarticulés (Brachiopodes, *pal.*).

inbreak, arrivée soudaine et brutale (d'eau).

incavation, 1. excavation, creusement; 2. creux, dépression.

inceptisol, sol jeune à horizons rapidement formés (ex. : sols humiques à gley, sol brun acide, etc.).

inch, unité de longueur = 25,4 mm; 12 ins. = 1 foot; 36 ins = 1 yard (0,914 m); **i. of mercury,** pouce de mercure = 0,03453 kgf/cm²; **i. of water,** pouce d'eau = 0,00254 kgp/cm²; **i. pound,** pouce-livre = 0,011298 mcsn (mètre centisthène); **i. ton,** pouce-tonne = 0,253086 msn (mètre sthène); **cubic i.** = 16,387 cm³; **square i.** = 6,452 cm².

incidence, incidence; **angle of i.,** angle d'incidence.

incident, incident (rayon lumineux).

incise (to), s'enfoncer, s'encaisser.

incised meander, méandre encaissé.

inclination, 1. inclinaison; 2. plongement, pente; **i. of the needle,** inclinaison magnétique; **magnetic i.,** inclinaison magnétique.

incline, 1. pente, inclinaison, plan incliné, rampe; 2. descenderie, puits incliné (*mine*).

incline (to), 1. incliner, pencher; 2. s'incliner, se pencher.

inclined, incliné; **i. contact,** contact pétrole/eau incliné; **i. extinction,** extinction oblique (optique); **i. fold,** pli oblique, pli déjeté; **i. level,** galerie inclinée; **i. shaft,** puits incliné; **i. well,** forage oblique.

inclinometer, 1. clinomètre, inclinomètre; 2. boussole d'inclinaison.

inclosed meander, méandre encaissé.

inclosing rock, roche encaissante.

inclusion, inclusion; **gaseous i.,** i. gazeuse; **fluid i.,** i. fluide; **liquid i.,** i. liquide; **solid i.,** i. solide.

incoherent, 1. sans consistance; 2. non consolidé, meuble.

incoming, 1. arrivée, venue; **2.** incident (a.); **i. of water,** venue d'eau; **i. tide,** marée montante.

incompatible element, élément incompatible (avec la composition d'un magma).

incompetent, incompétent, tendre; **i. rock,** roche incompétente.

incompressibility, incompressibilité; **i. modulus,** module d'élasticité.

incompressible, incompressible.

incongruent melting, fusion incongruente.

inconsequent drainage, réseau fluviatile non adapté à la structure régionale (réseau surimposé, réseau antécédent).

incretion, concrétion creuse cylindrique.

incrust (to), 1. incruster, encroûter, entartrer; **2.** s'incruster, s'encroûter.

incrustating spring, source pétrifiante.

incrustation, incrustation, encroûtement, entartrage.

incrusting water, eau pétrifiante.

indentation, entaille, échancrure, indentation (d'une côte).

indeterminable, indéterminable.

index, 1. aiguille; **2.** indice, exposant; **3.** repère, signe (indicateur); **4.** répertoire, liste alphabétique; **i. contour,** courbe maîtresse principale; **i. ellipsoid,** ellipsoïde des indices; **i. fossil,** fossile caractéristique; **i. map,** tableau d'assemblage; **i. mineral,** minéral caractéristique; **i. of flatness,** indice d'aplatissement; **i. of refraction,** indice de réfraction; **i. plane,** plan de référence (*tect.*); **i. zone,** niveau stratigraphique, repère, guide; **crystallographic i.,** indice cristallographique; **refractive i.,** indice de réfraction; **structure i.,** coefficient de stabilité (de la structure).

indian pipestone, catlinite; **i. summer,** été indien, fin de l'été (U.S., Canada).

indicative boulder, bloc erratique (dont on peut retrouver l'origine pétrographique).

indicator, 1. indicateur (de pression, etc.); **2.** détecteur de grisou; **3.** indicateur coloré; **4.** bloc erratique dont l'origine géologique est connue; **i. plant,** plante poussant sur un sol donné; **i. vein,** filon guide; **depth i.,** indicateur de profondeur; **trap i.,** indice de piège (*pétrole*).

indicated ore, minerai à tonnage et teneur estimés, calculés.

indicated reserves, réserves minérales théoriques, nominales calculées.

indicatrix, ellipsoïde des indices, indicatrice (optique).

indicolite, indicolite (var. de tourmaline).

indigenous, autochtone, indigène, in situ.

indigo copper, covellite, covelline.

indigolite, indigolite, indicolite (var. de tourmaline).

indissolubility, indissolubilité.

indissoluble, indissoluble.

indraught, 1. entrée d'air; **2.** courant remontant (d'un estuaire).

induced, induit; **i. flow,** débit induit (par des méthodes de récupération du pétrole); **i. magnetization,** aimantation induite; **i. polarization,** polarisation induite, provoquée; **i. radioactivity,** radioactivité induite.

induction, induction; **i. coil,** bobine d'i.; **i. current,** courant d'i.; **i. log,** diagramme de conductivité par i.; **i. logging method,** diagraphie par i.

inductolog, inductolog (diagramme de conductivité par induction).

indurate (to), durcir, se durcir, s'endurcir.

indurated clay, argile indurée, durcie; **i. red earth,** latérite vraie; **i. talc,** talcschiste.

induration, 1. induration, durcissement; **2.** lithification (en général); **3.** induration illuviale.

industrial waste, déchet industriel.

inelastic, non élastique.

inequigranular, à cristaux de tailles différentes; **i. texture,** texture porphyrique.

inequivalve, inéquivalve (*pal.*).

inert, inerte; **i. gas,** gaz inerte.

inertinite, inertinite (cf. charbon).

inexhaustible, inépuisable, intarissable.

inexploitable, inexploitable.

inexplorable, inexplorable, impénétrable.

inface, infacing slope, front de cuesta.

infancy, enfance (sous-stade de cycle, géomorphol.); **i. stage,** stade infantile.

infant stream, rivière au stade infantile.

inferred reserves, réserves présumées (par déduction de caractéristiques géologiques).

infilling, 1. remplissage (d'un filon); **2.** colmatage.

infiltrate (to), s'infiltrer, imprégner, pénétrer.

infiltration, 1. infiltration (d'eau); **2.** dépôt minéral dans les pores d'une roche par percolation; **3.** remplissage filonien hydrothermal (déposé à partir d'une solution dans l'eau); **i. capacity,** taux d'absorption de l'eau de pluie; **i. gallery,** galerie captante; **i. rate,** taux d'infiltration de l'eau de pluie, dans un sol; **i. vein,** filon d'origine hydrothermale; **i. velocity,** vitesse d'infiltration de l'eau.

inflow, arrivée, venue; **i. of water,** venue d'eau.

inflowing stream, affluent.

influent, affluent; **i. tide,** marée montante.

influx, 1. entrée; **2.** affluence (d'un cours d'eau).

infra, infra-sous (préfixe).

infrabasal plates, plaques infrabasales (Crinoïdes, pal.).

Infracambrian, Infracambrien (= Éocambrien) voir Riphéen.

infraglacial, sous-glaciaire.

Infralias, Infralias (∼ Rhétien).

infralittoral, infralittoral.

inframundane, souterrain.

infraneritic, infranéritique (entre − 40 et − 195 m).

infrared, infrarouge; **i. photography,** photographie infrarouge; **i. radiation,** rayonnement infrarouge; **i. spectroscopy,** spectroscopie infrarouge.

infrastructure, infrastructure, structure géologique des niveaux profonds de la croûte terrestre.

infusibility, infusibilité.

infusible, infusible.

infusorial earth, terre à infusoires, diatomite.

ingenite (obsolete), roches éruptives (et métamorphiques).
ingot, lingot.
ingress, pénétration, entrée.
ingression (German), transgression.
ingrown meander, méandre creusé lors de rajeunissement du relief.
inhalent, inhalant (*pal.*, Spongiaires).
inherited, hérité; **i. features**, « héritage »; **i. meander**, méandre h.; **i. river course**, tracé fluviatile surimposé.
inhibitor, inhibiteur.
inhomogeneity, hétérogénéité.
initial, initial; **i. dip**, pendage originel, i. (avant déformation ultérieure du lit ou des couches); **i. motion**, mouvement i. d'un séisme; **i. open flow**, débit i. (d'un puits); **i. production**, production initiale (d'un puits); **i. rating of well**, débit i. d'un puits.
injected igneous body, massif intrusif complètement entouré de roches encaissantes.
injection, injection; **i. complex**, complexe intrusif; **i. gneiss**, migmatite; **i. metamorphism**, métamorphisme d'i.; **i. structure**, figure d'i. (en flamme); **i. well**, puits d'i.; **ribbon i.**, intrusion rubanée.
injection breccia, brèche d'intrusion.
injectivity, injectivité; **i. profile**, profil d'injectivité.
ink stone, mélantérite, vitriol vert (*minér.*).
inland, intérieur des terres; **i. basin**, dépression fermée, playa; **i. ice**, inlandsis, calotte glaciaire; **i. sea**, mer intérieure.
inlet, 1. entrée, arrivée, admission; 2. petit bras de mer, crique.
inlet delta (tidal), delta de marée montante.
inlier, fenêtre (dans une nappe de charriage).
inner, interne, intérieur; **i. core**, noyau interne (de la terre), graine; **i. lowland**, revers de cuesta; **i. moraines**, moraines internes.
Inocerami, Inocérames.
inorganic, inorganique, minéral.
inosilicate, inosilicate, silicate en chaînes.
inoxidizable, inoxydable.
input minerals, minéraux originels.
inrush, irruption, arrivée soudaine; **i. of water**, venue d'eau.
Insect, Insecte.
Insectivora, Insectivores (*pal.*).
inselberg, inselberg, relief résiduel.
insequent, inséquent, inadapté (*géomorph.*); **i. stream**, rivière inadaptée à la structure.
inset, 1. phénocristal; 2. flux de la marée; **i. terrace**, terrasse emboîtée; **i. valley**, vallée emboîtée.
insolubility, insolubilité.
insolubilize (to), insolubiliser.
insoluble, insoluble; **i. residue**, résidu insoluble après attaque acide.
inspissation, asphaltisation.
insubmersibility, insubmersibilité.
insular, insulaire; **i. shelf**, plateau continental i.
insulated stream, tronçon fluviatile isolé de la zone de saturation (par une couche imperméable).

insulation, 1. détachement (d'une île par rapport au continent); 2. isolement (électrique, ou thermique).
intake, entrée, arrivée, recharge d'une nappe; **i. area**, région d'alimentation d'une nappe; **i. place**, région d'alimentation (*hydrol.*); **i. shaft**, puits d'entrée d'air; **i. well**, sondage d'injection; **air i.**, prise d'air.
integripalliate, intégripallié (*pal.*).
intensity, intensité; **i. scale**, échelle d'i. (des tremblements de terre).
interambulacral, interambulacral area, aire interambulacraire (*pal.*).
interarc basin, bassin inter-arc.
interbanded, zoné.
interbed, lit repère mince, interstratifié.
interbedded, interstratifié.
interbedding, interstratification.
interbrachial plates, plaques interbrachiales (Crinoïdes) (*pal.*).
intercalate texture, texture intercalaire, intersertale.
intercepts, paramètres cristallographiques.
intercept time, ordonnée à l'origine (sismique réfraction).
intercrystalline, intercristallin.
interdigitation, interdigitation (ex. relation stratigraphique).
interdistributary bay, baie interbras (dans un delta).
interestuarine, situé entre deux estuaires, interestuarien.
interface, interface.
interfacial, interfacial; **i. angle**, angle dièdre, angle des faces; **i. tension**, tension interfaciale, tension superficielle.
interfere (to), interférer.
interference, interférence; **i. figure**, figure d'i. ; **i. ripple-mark**, ride d'i.
interferometer, interféromètre.
interferometry, interférométrie.
interfinger (to), s'interstratifier.
interfingering, interdigitation.
interfluve, interfluve.
interfolding, plis simultanés d'orientation différente.
interfoliated, intercalé entre les feuillets d'une roche métamorphique.
interformational, intraformationel; **i. conglomerate**, conglomérat i.; **i. sheet**, filon couche.
interfraction, interférence.
interglacial, interglaciaire, entre 2 glaciations.
intergranular, intergranulaire; **i. texture**, var. de structure doléritique (*pétrol.*) ou intersertale.
intergrowth, enchevêtrement.
interior, 1. adj : intérieur, interne; 2. n : intérieur, dedans, région intérieure d'un continent; **i. basin**, dépression endoréïque; **i. salt domes**, diapirs éloignés du golfe du Mexique; **i. sea**, mer intérieure; **i. valley**, vallée karstique abrupte, poljé.
interlacing channel, chenal anastomosé.
interlayer, couche interstratifiée, intercalation.
interlensing, existence de lentilles interstratifiées.

interlimb angle, angle formé par les deux flancs d'un pli.

interlobate deposit, moraine située entre des lobes glaciaires, interlobaire.

interlocked crystals, minéraux engrenés (structure entrecroisée).

interlocked texture, structure entrecroisée.

interlocking seismic recording intermediate, mixage intermédiaire; **i. contour,** courbe de niveau intercalaire; **i. focus earthquake,** séisme de profondeur intermédiaire; **i. rock,** roche neutre, équilibrée (= 10 % de silice libre, syénite, diorite).

intermittence, intermittence (d'une source).

intermittent, intermittent; **i. spring,** source i.; **i. stream,** rivière intermittente, à écoulement intermittent.

intermitter, puits à débit intermittent.

intermont, intermontane, 1. adj : intramontagnard; **2.** n : entremont.

intermount area, région intramontagnarde; **i. basin,** bassin structural, cuvette structurale; **i. glacier,** glacier de confluence; **i. trough,** fossé de subsidence, dans un arc insulaire.

internal, interne, intérieur; **i. cast,** moule interne; **i. drainage,** endoréisme; **i. magnetic field,** champ magnétique terrestre; **i. mold,** moule interne; **i. moraine,** moraine interne; **i. water,** eau de profondeur; **i. wave,** vague de fond.

interpenetrant twin, mâcle d'interpénétration.

interpenetration twin, mâcle d'interpénétration.

interplanal distance, distance réticulaire.

interplanar spacing, distance réticulaire.

interplane bonding, liaison atomique interfeuillets.

interplanetary, interplanétaire.

interpluvial, interpluvial.

interposed, intercalé, interstratifié.

interradial, interradial (*pal.*).

interradials plates, plaques interradiales (Crinoïdes).

interrupted stream, rivière karstique, à pertes.

intersect (to), 1. recouper, entrecouper, intersecter, entrecroiser; **2.** s'entrecroiser, etc.

intersecting, entrecroisé, recoupé; **i. peneplains,** surface polygénique; **i. vein,** filon croiseur.

intersection, intersection, recoupement (de filons).

interseptal, interseptal.

intersertal, intersertal; **i. texture,** texture intersertale (*pétro.*).

interspace, espacement; **i. of time,** intervalle de temps.

interstade, interstade (recul glaciaire dans une glaciation).

interstadial, interstadiaire (cf. supra, adj.).

interstellar space, espace interstellaire.

interstice, 1. interstice, intervalle, vide; **2.** espace disponible entre les phénocristaux dans les structures intersertales; **capillary i.,** interstice capillaire, pore capillaire; **supercapillary i.,** interstice, pore supracapillaire.

interstitial, interstitiel; **i. deposits,** gisements minéraux d'imprégnation; **i. matrix,** matrice interstitielle; **i. solid solution,** solution solide interstitielle (*cristallo.*).

interstratal, interstrate, entre couches.

interstratification, interstratification, intercalation.

interstratified, interstratifié.

interstratify (to), s'interstratifier.

intertidal, intertidal.

intertonguing zone, interdigitation.

intertwinning channels, chenaux anastomosés.

interval, 1. intervalle; **2.** distance verticale entre deux couches de référence; **i. transit time,** temps de traversée d'une couche; **c. interval,** équidistance entre deux courbes.

interveined, traversé par des filons.

interzonal soil, sol interzonal.

intraarc basin, bassin intraarc.

intrabasinal, déposé dans un bassin fermé.

intraclast, intraclaste (débris calcaires remaniés in situ).

intraclastic, intraclastique.

intracontinental sea, mer fermée.

intracratonic, intracratonique; **i. basin,** bassin i.

intracratonic geosyncline, autogéosynclinal.

intracyclothem, sous-cyclothème, cyclothème secondaire (d'un cyclothème principal).

intrafacies, sous-faciès secondaire (d'un faciès plus important).

intraformational, intraformationel; **i. conglomerate,** conglomérat i.; **i. sheet,** filon-couche.

intrageosyncline, géosynclinal intracontinental.

intraglacial, intraglaciaire.

intramagmatic, intramagmatique.

intramontane basin, intramount basin, bassin d'entremont.

intraplate, intraplaque.

intrastal solution, dissolution chimique postérieure à la sédimentation (diagénétique).

intratellural, intratelluric crystallization, cristallisation précoce et profonde de phénocristaux.

intrathecal, endothécal (*pal.*).

intrazonal soil, sol intrazonal.

intrenched meander, méandre encaissé.

intrenched stream, rivière encaissée (par surimposition).

intrude (to), faire intrusion, s'introduire dans.

intruded, intrusif; **i. rock,** roche intrusive; **to be i. by,** être pénétré par.

intrusion, intrusion, injection; **i. breccia,** brèche d'intrusion; **i. of the sea,** transgression; **i. displacement,** faille coïncidant avec une intrusion.

intrusive, intrusif; **i. contact,** contact i.; **i. rock,** roche intrusive; **i. sheet,** sill, filon-couche; **i. vein,** filon i.

intrusives, roches éruptives (cf. granite), roches intrusives, roches magmatiques.

intumescence, 1. intumescence, boursouflure, dôme; **2.** bourgeonnement.

inundate (to), inonder.

inundation, inondation (fleuve).

invade (to), envahir, pénétrer, faire intrusion.

invaded rock, roche hôte, roche encaissante.

inverse zonation, zonation inverse (des feldspaths).

inversion, 1. inversion; 2.déversement; 3. changement de phase, transformation cristalline; **i. of a stratum,** renversement d'une couche; **i. point,** température de transformation d'une phase cristalline en une autre (α en β) (cristaux polymorphes).

invert (to), renverser, retourner; **i. fold,** pli déversé, pli renversé.

Invertebrate, Invertébré (*pal.*).

inverted, renversé, inversé, **i. fold (overturned fold),** pli renversé, **i. limb,** flanc inverse; **i. order,** position inversée; **i. relief,** inversion du relief.

involute, involute (*pal.*).

involution, 1. involution (périglaciaire); 2. plissement de nappes de charriage postérieurement à leur mise en place.

iodargyrite, iodargyrite, iodyrite (*minér.*).

iodide, iodure.

iodine, iode.

iodobromite, iodibromyrite, iodobromite (*minér.*).

iodyrite, iodargyrite, iodyrite (*minér.*).

iolite, iolite, cordiérite (*minér.*).

ion, ion; **i. activity,** activité ionique; **i. concentration,** concentration ionique; **i. exchange,** échange d'ions; **i. probe,** microsonde ionique.

ionic, ionique; **i. bond,** liaison ionique; **i. potential,** potential i.; **i. radius,** rayon ionique; **i. substitution,** substitution i.

ionization, ionisation; **i. chamber,** chambre d'i.; **i. potential,** potentiel d'i.

ionize (to), ionizer.

ionosphère, ionosphère.

ionospheric, ionosphérique.

Iowan (N. Am.), Iowan (stadiaire inférieur du Wiscounsinien).

Ipswichian (U.K.), Ipswichien (interglaciaire Riss-Wurm) = Eemien (G.B.).

iridium, iridium.

iridosmine, idirosmium, iridosmine, iridosmium.

iron, fer; **i. age,** âge du f.; **i. alum,** alun de f.; **i. bacteria,** bactérie ferrugineuse; **i. bearing,** ferrifère; **i. carbonate,** sidérose; **i. deposit,** gisement de f.; **i. disufide,** pyrite; **i. dolomite,** ankérite; **i. glance,** oligiste, hématite; **i. gossan,** chapeau ferrugineux; **i. hat,** chapeau ferrugineux; **i. magnesia mica,** biotite; **i. meteorite,** météorite en f.-nickel; **i. mine,** mine de f.; **i. pan,** alios (ferrugineux); **i. phosphate,** vivianite; **i. pyrite,** pyrite; **i. quarry,** mine de f.; **i. red ocher,** ochre de f.; **i. sand,** sable ferrugineux; **i. spar,** sidérose; **i. spherule,** sphérule (oolite), ferrugineuse; **i. spinel,** magnétite; **i. stone,** 1. minerai de f. 2. roche sédimentaire ferrugineuse; **i. sulfide,** sulfure de f.-pyrite; **i. tourmaline,** tourmaline noire (schorl); **i. yellow ocher,** limonite.

irradiate (to), irradier, rayonner.

irradiation, irradiation.

irrecoverable, irrécupérable (pétrole, etc.).

irrigable, irrigable.

irrigate (to), irriguer, arroser.

irrigation, irrigation.

irrotationnel wave, onde de compression, onde P.

irruption, 1. venue, irruption (d'eau, etc.); 2. intrusion magmatique.

irrupt (to), faire irruption, être injecté (pétrole).

irruptive rock, roche éruptive, intrusive.

isanomalous line, isanomaly, isonomaly, isanomale (météo.).

iserin, iserine, iserite, isérite (fer titané) (= ilménite détritique dans les minéraux lourds des sables).

isinglass, mica (en minces feuillets transparents); **i. stone,** m.

island, île; **i. arc,** arc insulaire; **i. chain,** arc insulaire; **i. shelf,** plate-forme insulaire; **i. slope,** talus insulaire; **caldera i.,** île caldeira; **continental i.,** île continentale; **crater-i.,** île cratère.

islet, îlot.

isobar, isobare (courbe d'égale pression).

isobaric line, ligne isobare.

isobase, isobase (ligne d'égal relèvement ou enfoncement).

isobath, isobathe (courbe d'égale profondeur).

isocal, isocal (ligne, joignant les points d'égale valeur calorifique d'un gisement de charbon).

isocarb, isocarb (ligne joignant les points de même teneur en carbone fixe d'un gisement de charbon).

isochemical metamorphism, métamorphisme topochimique, isochimique, sans changement chimique.

isochore, isochore (ligne d'égale épaisseur établie d'après les données brutes non corrigées d'un forage à la verticale des unités).

isochore map, carte d'isochores.

isochron, isochronal, isochrone, isochronic, isochronous, isochrone (ligne de même âge).

isochron map, carte d'isochrones.

isochrononeity, isochronism, isochronisme.

isoclinal, isoclinal; **i. folding,** plis droits serrés, plis isoclinaux; **i. line,** pli à flancs parallèles.

isocline, isocline, isoclinal.

isoclinic line, isocline.

isocon, courbe d'égale concentration en sels.

isodont dentition, dents isodontes (*pal.* Lamellibranches).

isodynamic line, isodynamique (ligne joignant les points d'égale intensité magnétique).

isofacies, courbe d'isofaciès (ligne joignant les points de faciès identiques).

isogal, isogal (lignes d'égales valeurs de gravité).

isogam, isogamme.

isogeotherm, isogéotherme, courbe d'égale géothermie.

isogeothermal, isogéothermal.

isogon, isogonic line, 1. isogone (ligne d'égale déclinaison magnétique); 2. ligne joignant les points de direction constante du vent.

isograd, 1. isométamorphe ou ligne joignant des points de même faciès de métamorphisme; 2. ligne joignant des valeurs égales de température et de pression; **3.** ligne joignant les points de même faciès pétrographiques.

isogram, courbe de niveau.

isohaline, ligne d'égale salinité (des eaux).

isohel, ligne d'égal ensoleillement.

isohyet, isohyète (ligne d'égale pluviosité).

isohyetal, isohyetal line, isohyétal, isohyète (courbe d'égale précipitation).

isohypse, isohypse, courbe de niveau (d'égale altitude).

isolate (to), isoler, dégager, extraire.

isolation, isolement (*pal.*).

isoline map, cartes à variables figurées en courbes.

isolithic lines, lignes d'égal faciès pétrographique.

isolith maps, cartes de variations d'épaisseur de faciès pétrographiques.

isomagnetic line, ligne isomagnétique.

isomer, isomère.

isomeric, isomerical, isomère.

isomerism, isomérie.

isomerization, isomérisation.

isometric, system, système cubique.

isomorph, isomorphic, isomorphous, isomorphe.

isomorphism, isomorphisme.

isomyarian, isomyaire (*pal.*).

isopach, 1. adj : isopaque, d'égale épaisseur; **2.** n : isopaque.

isopachous line, isopachyte, isopaque (ligne d'égale épaisseur d'une couche).

isopach map, carte isopaque; **i. strike,** direction indiquée par les isopaques.

isopached interval, équidistance de deux isopaques voisines.

isopical deposit, couche isopique (de même âge et de même faciès).

isopiestic level, profondeur de compensation isostatique; **i. line,** ligne d'égale pression hydrostatique.

isopleth map, carte isoplète, carte en isothermes.

isoporic line, ligne d'égale variation de déclinaison magnétique.

isopolls, ligne d'égale fréquence en pollen fossile.

isopycnic, isopycnal, ligne d'égale densité (des eaux marines).

isorad, ligne d'égale radioactivité.

isoresistivity map, carte d'isorésistivité.

isoseism, isoseismal line, isoséiste (courbe d'égale intensité des séismes).

isoseist, isoséiste.

isostacy, isostasy, isostasie.

isostatic, isostatical, isostatique; **i. adjustment,** compensation; **i. anomaly,** anomalie i.; **i. compensation,** compensation i.

isostructural crystals, cristaux de structures cristallines semblables.

isotherm, isotherme.

isothermal, isothermic, isothermique.

isothermal line, isotherme.

isotime, isochrone (ligne de même âge).

isotope, isotope; **i. dating,** datation isotopique; **i. effect,** effet i.; **i. geology,** géochimie, i.; **i. ratio,** rapport i.; **i. fractionation,** fractionnement i.

isotopic, isotopique; **i. fractionation,** séparation isotopique.

isotropic, isotrope, isotropous, isotropique, isotrope.

isotropy, isotropie.

isthmus, isthme.

itabirite, itabirite, spécularite, hématite lamellaire (*minér.*).

itacolumite, itacolumite, grès micacé flexible.

ivory, ivoire (*pal.*).

J

jacinth, hyacinthe, zircon (*minér.*).

jack, 1. cric, vérin; 2. engin (très variable); **3.** blende, sphalérite.

jack-knife derrick, derrick repliable.

jack-well, puits en pompage par dispositif à balancier; **pumping jack,** dispositif de pompage à balancier.

Jacksonian (N. Am.), Jacksonien (étage Éocène sup.).

jacupirangite, jacupirangite, (variété de pyroxénite) (*pétrole*).

jad, havage, sous-cavage (*mine*).

jad (to), haver, sous-caver (*mine*).

jade, jade (*minér.*).

jadeite, jadéite (= pyroxène) (*minér.*).

jagged, déchiqueté, découpé, dentelé.

jalpaite, jalpaïte (*minér.*).

jam, 1. embâcle; **2.** coincement; **ice jam,** embâcle glaciel.

jamesonite, jamesonite (*minér.*).

janosite, copiapite (*minér.*).

japanite, pennine (espèce de chlorite).

jar, 1. choc, secousse; **2.** coulisse (*mine*).

jarosite, jarosite, utahite (*minér.*).

jaspagate, jaspachate, agate jaspée.

jasper, jasperite, jaspe.

jasper opal, jaspopal, opale jaune ressemblant au jaspe.

jasperated, jaspé.

jasperisation, transformation en roches ferrugineuses et siliceuses zonées (formation rubanée, taconite, etc.).

jasperize (to), transformer en jaspe ou en agate.

jasperoid, 1. roche siliceuse, soit en calcédoine cryptocristalline, soit à cristaux de quartz résultant de la substitution d'autres cristaux; **2.** calcaire siliceux.

jaspidian, contenant du jaspe.

jaspilite, formation rubanée à hématite, formation fer, taconite, chert ferrugineux, jaspe (U.S.), itabirite (*minér.*) (Brésil).

jaspoid , ressemblant à du jaspe.

jasponyx, onyx à couches de jaspe.

jaspure, marbre zoné comme du jaspe.

Jatulian, Jatulien (division du Précambrien Balte).

jaw, mâchoire, mors.

jebel (jabal : Arabic), djebel (colline, montagne).

jellyfish, méduse.

jet, 1. jais, jaïet (var. de charbon); 2. jet, injection; **3.** veine fluide (physique); **j bit,** trépan à jet hydraulique; **j. bit drilling,** forage au trépan à jet; **j. coal,** « cannel coal », charbon contenant du jais; **j. flow,** écoulement concentré très rapide; **j. stone,** tourmaline noire, schorl.

jetty, 1. épi littoral (U.S.), jetée; **2.** quai (G.B.).

jewel, gemme, pierre précieuse, bijou; **j. stone,** pierre précieuse.

jeweler, jeweller, joailler.

jewelry, bijouterie.

jewstone, 1. marcasite; **2.** roche dure à cassure inégale.

jig, crible hydraulique, hydrotamis, calibre, gabarit; **j. crabe,** grue à flèche; **j. table,** crible vibrant.

jig (to), cribler, laver.

jigger, 1. mach : crible hydraulique, crible oscillant; **2.** pers : cribleur.

jigger work, lavage au crible.

jigging, criblage, lavage.

jigging machine, crible hydraulique.

jog, 1. coup, secousse, ébranlement; **2.** ressaut topographique.

jogging-table, table à secousses (*mine*).

joint, 1. diaclase, fissure; **2.** fragment de tige de Crinoïde, entroque; **3.** raccord, articulation, joint, élément de tubage; **j. pattern,** réseau de diaclases; **j. plane,** plan de séparation; **j. set,** ensemble de joints parallèles; **bedding j.,** joint (plan) de stratification; **cross tectonic j.,** diaclase transversale (à l'axe du pli); **longitudinal tectonic j.,** diaclase longitudinale; **shear j.,** fissure de cisaillement; **sheet j.,** joint (plan) de décompression des roches en surface; **shrinkage j.,** fissure de retrait, de contraction; **tension, tensional j.,** joint de tension, joint tectonique.

joint up (to), unir des sections de tube (pétrole).

jointed, diaclasé, fissuré.

jointing, 1. fissuration, diaclasage; **2.** assemblage; **3.** plan de diaclase.

jökuhlaup (icel.), libération soudaine d'eau glaciaire.

jökull (icel.), petite calotte glaciaire.

jointy, fissuré, diaclasé.

jolt, secousse, soubresaut.

jolting machine, crible laveur à secousses.

jordanite, jordanite (*minér.*).

Jotnian, Jotnien (division du Précambrien balte).

joule, joule (électricité).

jovian geology, géologie de la planète Jupiter.

jug, détecteur sismique, géophone.

jump, discontinuité, anomalie, rejet, ressaut, saut.

jump drilling, forage au câble, par battage.

jump (to), 1. forcer la pierre au fleuret (mine); **2.** sauter.

jump (to) a claim, s'emparer d'une concession (appartenant à autrui).

jumper, barre à mine, fleuret; **j. drill,** barre de mineur.

junction, confluent, raccordement, jonction, bifurcation; **deferred j.,** confluent déplacé vers l'aval;

hanging j., confluent discordant; **shifted j.,** confluent déplacé vers l'aval.
junkerite, sidérose, sidérite (*minér.*).
junk retriever, tube à sédiments.
junk sub, panier à sédiments (forage).
junked hole, forage abandonné.

Jurassic, Jurassique (période Mésozoïque).
jut out (to), surplomber, faire saillie.
jutting out wall, paroi surplombante.
juvenile, juvénile; **j. gas,** gaz j.; **j. water,** eau j. (eau magmatique souterraine).

K

kainite, kaïnite (*minér.*).

kainotype rock, roche récente (Tertiaire ou Quaternaire).

Kainozoic (Cenozoic) era, Cénozoïque, ère Tertiaire.

Kalevian, Kalévien (division du Précambrien balte).

kali, potasse.

kalinite, kalinite (alun de potasse).

kaliophilite, kaliophylite, kaliophilite (feldspathoïde).

kalirhyolite, rhyolite alcaline (à feldspaths essentiellement potassiques).

kalisyenite, syénite alcaline (à feldspaths potassiques).

kalium, potassium.

kallait, turquoise.

kalomel, calomel.

kamacite, kamacite (fer météorique).

kame, kame, monticule de sédiments fluvio-glaciaires de contact.

kame and kettle topography, topographie de kame et kettle, à monticules et dépressions fermées.

kame terrace, terrasse de kame, terrasse construite entre roc et glace.

kand, fluorine (Cornouailles).

Kansan (N. Am.), glaciation de Kansan.

kaolin, kaolin (argile).

kaolinic, kaolinique.

kaolinite, kaolinite (minéral argileux).

kaolinization, kaolinisation (des feldspaths).

kaolinized, kaolinisé.

K/Ar age, datation radiométrique potassium argon.

karat, carat (quantité d'or : x/24).

Karelian, Karélien (division du Précambrien balte).

Karnian (Eur.), Carnien, Karnien (étage, Trias sup.).

karren (syn. lapiès), lapiès.

karrenfeld (Germ.), paysage des lapiès.

karst, karst; **k. erosion,** érosion karstique; **k. filling,** remplissage karstique; **k. hydrology,** hydrologue karstique; **k. pool,** lavogne; **k. scenery,** paysage karstique; **k. spring,** source karstique; **k. tower,** tour karstique, hum, mogote; **k. valley,** vallée karstique; **confined k.,** k. barré; **covered k.,** k. couvert; **deep k.,** k. profond; **naked k.,** k. nu; **shallow k.,** k. superficiel.

karstenite, anhydrite (*minér.*).

karstic, karstique.

karstification, karstification.

katabatic, catabatique; **k. wind,** vent c.

kataclastic, cataclastique, bréchique, mylonitique; **k. texture,** structure cataclastique, structure de mylonite.

katagenese, katagénèse.

katagenic, katagénique.

katagneiss, gneiss catazonal, gneiss formé en catazone.

katalysis, catalyse.

katamorphic zone, zone catamorphique (fracturation, décomposition et altération des roches).

katamorphism, catamorphisme (altération et cimentation des roches).

Katangian, Katangien; **k. orogeny,** orogénèse Katangien(ne) (ou assyntique).

katatectic layer, couche formée de résidus gypseux de dissolution dans la couverture d'un gisement de pétrole.

katathermal, catathermal.

katazone, catazone, zone la plus profonde du métamorphisme régional.

kathode, cathode; **k. rays,** rayons cathodiques.

kathodic, cathodique.

kation, cation.

katogene, formé par décomposition et altération de roches.

kawk, fluorine (Angl.).

Kazanian (U.S.S.R.), Kazanien (étage, Permien moy. à sup.).

K bentonite, argile interstratifiée à montmorillonite-illite et riche en potassium.

keel, 1. carène (*pal.*, Mollusques); **2.** crête, carène, quille (de banquise).

keeps, taquets (de cage) (*mine*).

Keewatin, kewatinian, Keewatin (série, Archéen) du Bouclier canadien.

kelly bar, tige carrée d'entraînement (*forage*).

kelp, 1. algue, varech; **2.** soude des varechs.

kelve, fluorine (Angl.).

Kelvin scale, échelle Kelvin (température : oC + 273 oK).

kelyphite, kélyphite (*pétro.*); **k. rim, keliphitic border,** auréole réactionnelle d'amphibole ou pyroxène autour de cristaux d'olivine ou de grenat.

kennel coal, cannel coal.

Kenoran orogeny (Can.), orogénèse Kénorienne : — 2 390 à — 2 600 MA (Algomien).

kerabitumen, kérabitume.

kerargyrite, cérargyrite (*minér.*).

keratophyre, kératophyre (variété de laves trachytiques).

keratophyric, kératophyrique.

kerf, havage, saignée (*mine*).

kermesite, kermésite, antimoine rouge (*minér.*).

kerogen, kerosen, kérogène (dans schistes bitumineux); **k. shale,** schiste bitumineux.

kerosene sand, sable bitumineux contenant encore des produits volatils (Australie).

kersantite, kersantite (*pétro.*).

kerve (to), haver, sous-caver.

kettle, kettle, dépression fermée; **k. drift,** moraine terminale à nombreuses dépressions en « marmites »; **k. hole,** marmite de géant, dépression fermée glaciaire; **glacial k., giant's kettle,** marmite de géant.

kettled, ayant des dépressions fermées (kettles).

Keuper (Eur.), Keuper (série ou faciès germanique = Trias sup.).

Kewatinian (Can.) (see Keewatin), Keewatin.

Keweenawan (N. Am.), Keweenawien (série, Précambrien sup., ou Protérozoïque).

key, 1. clé; 2. clavette touche; 3. îlot de corail (Floride, Caraïbes); **k. bed,** couche repère; **k. horizon,** horizon repère; **k. rock,** roche guide; **k. stone,** clef de voûte (d'un horst).

Keyserian (N. Am.), Keysérien (étage, Silurien sup.).

K feldspar, feldspath potassique.

Kibara Orogeny, orogenèse Kibarienne.

kibble, benne.

kick, arrivée d'une secousse sismique.

kidney ore, hématite rouge en rognons.

kidney stone, 1. concrétion ferrugineuse dans argiles (G.B.); 2. néphrite (*minér.*); 3. nodule en forme de rognon.

kies, minerai sulfuré.

kieselguhr (Germ.), kieselghur, diatomite (roche).

kieserite, kiesérite (*minér.*).

kilkenny coal, anthracite.

killed lime, chaux éteinte.

kiln, four, étuve.

kiln (to), cuire; **k. bricks,** cuire des briques.

kimberlite (blue ground), kimberlite, variété de péridotite.

Kimmeridgian, Kimméridgien; 1. (étage, Jurassique sup.); 2. (s.s., sous-étage du précédent).

kimolite, cimolite (*minér.*).

kind, espèce, genre, sorte.

Kinderhookian (N. Am.), Kinderhookien (étage, Mississipien inf.).

kindly bed, couche présumée riche en minerai (*mine*).

kindly ground, terrain riche en minerai (G.B.).

kindred, cortège, clan, série, pétrographique.

kinematic fabrics, structures de dynamométamorphisme.

kinetic, kinetical, cinétique (adj); **k. energy,** énergie cinétique; **k. metamorphism,** dynamométamorphisme.

kinetics, cinétique.

kink band, zone des plis en chevrons; **k. fold,** pli en chevrons; **k. plane,** plan de déformation de la schistosité, de gneissosité.

kinked crystal, cristal zoné en chevrons.

kinks, flexures répétées (tectonique).

kip, kip américain (unité de masse) = 453,59 kg.

kirrolite, cirrolite (*minér.*).

kirve (to), haver, sous-caver.

klastic, clastique, détritique.

kliachite, cliachite, bauxite alumine colloïdale.

klinkstone, phonolite.

klinoklas, clinoclase, feldspath monoclinique.

klinozoisite, clinozoïsite (*minér.*).

klint (pl: klintar), récif calcaire, bioherme (en relief après érosion des roches environnantes).

klippe, klippen, klippe, lambeau de charriage.

kloof, ravin, gorge (Afr. du S.).

knap, colline, éminence.

knapping hammer, marteau-pic (marteau de géologue).

knead (to), malaxer, travailler (de l'argile).

kneading, malaxage (de l'argile).

knee fold, pli en genou.

knee-shaped twin, mâcle en genou.

knick-crack, sol marécageux à sous-sol dur.

knick-marsh soil, sol marécageux à sous-sol dur.

knickpoint, rupture de pente (géomorphologie fluviale).

knifing, sous-solage.

knits, petites particules de minerai.

knob, 1. bosse de terrain; 2. morceau (de charbon); 3. tubercule (*pal.*); **k. and basin topography,** topographie irrégulière, topographie en creux et en bosses; **glaciated k.,** roche moutonnée.

knobly limestone, calcaire noduleux.

knock, détonation, choc.

knoll, monticule, butte; **reef k.,** pinacle corallien.

knolite, zéophyllite (*minér.*).

knot, 1. nœud (unité de vitesse) = 1 mille marin (1 852 m) par heure; 2. nodule, concrétion (de minéraux).

knotted schist, schiste tacheté (= knotenschiefer), schistes noduleux.

knotty rock, roche « tachetée » (par minéraux formés par métamorphisme de contact).

kolm, kolm (cannel-coal radioactif).

kopje, inselberg (Afr. du S.).

kraton (obsolete, see craton), craton, zone stable de l'écorce.

kreep volcanism, activité volcanique lunaire "antemare".

krohnkite, kroehnkite (*minér.*).

kryokonite, cryoconite = poussière nivéoéolienne (*glaciol.*).

kryptogranitish, cryptocristallin (Allemagne).

krystallinohyalin, hyalocristallin.

K section, section transversale circulaire de l'ellipsoïde de la déformation.

kugeldiorit, corsite (Allemagne).

kum (Turk.). koum, désert sableux.

Kummerian (U.S., Wash st.), Kummérien (étage floral, Oligocène inf.).

Kungurian (U.S.S.R.), Kungurien, Kougourien (étage Permien inf.).

kunkar, croûte calcaire.

kunzite, kunzite (var. de spodumène).

kurum (rock stream), nappe de fragments rocheux (cf. glacier rocheux).

kyanite, cyanite, disthène.

kyle, détroit, chenal (Écosse).

L

laboratory, laboratoire; **l. bench,** paillasse de l.; **l. glass-ware,** verrerie de l.; **l. test,** essai en l.

labrador feldspar, labradorite, labradorite; **l. hornblende,** hypersthène.

labrum, labre (*pal.*, Échinodermes).

Labyrinthodontia, Labyrinthodontidés (*pal.*).

La Casitan (N. Am.), La Casitien (étage, Jurassique sup. = Portlandien + Kimméridgien).

laccolite, laccolith, laccolithe.

laccolithic, laccolitic, laccolithique.

Lacertilia, Lacertiliens (*pal.*).

lacuna, 1. lacune stratigraphique (hiatus = érosion); 2. lacune (*pal.*).

lacus (Lune), petite étendue, basaltique isolée.

lacustrine, lacustral, lacustre; **l. chalk,** craie l.; **l. limestone,** calcaire l.; **l. marl,** marne l.

ladder veins, dépôt minéral « en échelle » (perpendiculaire au filon intrusif).

Ladinian, Ladinien (étage, Trias sup.).

Ladogian, Ladogien (Précambrien balte).

lag, retard, décalage; **l. deposit,** résidu de déflation; **l. fault,** faille de charriage; **l. gravel,** pavement désertique; **phase l.,** retard de phase (sismique).

lag (to), 1. retarder, être déphasé; 2. calorifuger.

lagging, garnissage (*mine*).

lagoon, 1. lagune; 2. lagon; 3. dépression intérieure à végétation (U.S.); étang; **l. island,** atoll; **l. moat,** lagon annulaire; **l. reef,** atoll; **cliff l.,** abrupt de lagon.

lagoonal environment, milieu lagunaire.

lagoonar, lagunaire.

laguna, 1. mare éphémère, lac temporaire; 2. lac, mare; 3. lac de doline (sur argile de décalcification); 4. bassin alimenté par source chaude.

lagunal deposits, dépôts lagunaires, dépôts d'atolls, sédiments formés entre les récifs-barrières et le continent.

lahar, coulée de boue (dans débris volcaniques).

laid down, déposé.

lake, lac; **l. basin,** bassin lacustre; **l. bed placers,** « placers » lacustres, gisements lacustres; **l. deposit,** sédiment lacustre; **l. dwelling,** station lacustre préhistorique; **l. iron ore,** minerai de fer des lacs; **l. ore,** minerai de fer des lacs; **l. pitch,** asphalte lacustre; **l. rampart,** levée à galets formée en hiver par la glace; **l. terrace,** terrasse lacustre; **alkali l.,** l. salé; **barrier l.,** l. de barrage; **bitter l.,** l. sulfaté; **borax l.,** l. boraté; **cirque l.,** l. de cirque glaciaire; **crater l.,** l. de cratère; **dammed l.,** l. de barrage; **deflation l.,** l. de cuvette éolienne; **erosion l.,** l. d'érosion; **glacial l.,** l. glaciaire; **ice-dammed l.,** l. de barrage glaciaire; **ice-marginal l.,** l. de front glaciaire; **ice-ponded l.,** l. de barrage glaciaire; **imponded l.,** l. de retenue; **karst l.,** l. karstique; **morainal, morainic l.,** l. de barrage

morainique; **overflow l.,** l. de trop-plein; **oxbow l.,** l. en croissant; **playa l.,** playa, l. temporaire; **ponded l.,** l. de retenue; **proglacial l.,** l. proglaciaire, de front glaciaire; **residual l.,** l. résiduel; **salt, saline l.,** l. salé; **underground l.,** l. souterrain.

lakelet, petit lac.

Lamarckism, Lamarckisme.

Lambert conformal, conic map projection, projection conique conforme de Lambert.

Lambert equal area map projection, projection équivalente de Lambert.

lamellae, lamines.

lamellate, lamellated, lamelleux, lamellé, feuilleté en lamelles.

Lamellibranchiata, Lamellibranchs, Lamellibranches.

lamina (pl laminae), 1. lame, lamelle; 2. feuillet de dépôt, straticule.

laminar, 1. laminaire; 2. lamellaire; **l. flow,** écoulement laminaire; **l. structure,** structure fluidale.

laminary flow, écoulement laminaire.

laminated, en lamines.

lamination (adj), 1. feuilleté; 2. laminé; **l. quartz,** filons de quartz dans structure rubanée; **l. shale,** schiste argileux feuilleté; **l. structure,** structure feuilletée, rubanée.

lamination (n.), 1. lamination, feuillet de dépôt; 2. structure lamellaire; 3. stratification fine, laminage, feuilletage.

laminite, laminite.

lamprophyre, lamprophyre.

lamprophyric, lamprophyrique.

lamp shell (syn. Brachiopod), Brachiopode.

Lanarkian (N. Am.), Lanarkien (étage, Pennsylvanien).

Lancastrian (N. Am.), Lancastrien (étage, Pennsylvanien).

lance head, pointe de lance (Préhist.).

land, terre, pays, contrée; **l. ablation,** érosion; **l. breeze,** brise de terre; **l. bridge,** pont continental (*pal.*); **l. chain,** chaîne d'arpenteur; **l. fall,** éboulement; **l. fill,** remblaiement; **l. forms,** topographie, forme du paysage; **l. ice,** glace d'eau douce; **l. levelling,** nivellement; **l. locked sea,** mer intérieure; **l. mark,** 1. borne, limite; 2. point côté (*topogr.*); **l. measuring,** arpentage; **l. measuring chain,** chaîne d'arpentage; **l. plant,** végétal terrestre; **l. plaster,** roche gypseuse utilisée comme engrais; **l. register,** cadastre; **l. scape,** paysage; **l. sculpture,** façonnement du relief; **l. sediments,** sédiments continentaux; **l. shooting,** sismique terrestre; **l. slide,** glissement de terrain; **l. slide surge,** raz de marée provoqué par glissement de terrain; **l. slip,** glissement de terrain; **l. strip,** piste d'atterrissage; **l. subsidence,** affaissement du sol; **l. survey,** étude de terrain, levé; **l. surveying,** géodésie, arpentage; **l. tied island,** île rattachée; **l. use,** utilisation des sols; **l. ward,** vers le continent, vers la terre; **l. waste,** matériel détritique.

land (to), 1. mettre à terre, descendre, débarquer, décharger; **2.** atterrir.

landing, 1. débarquement; **2.** atterrissage; **3.** recette *(mine);* **l. chart,** carte d'atterrissage; **l. site,** site d'atterrissage (sur une planète).

Landenian, Landénien (étage, Paléocène sup. = Thanétien + Sparnacien).

landsat thematic mapper, cartographie thématique landsat.

langbeinite, langbeinite *(minér.).*

lanthanides, lanthanides, terres rares.

lanthanum, lanthane.

lap, 1. chevauchement, recouvrement; **2.** polissoir (lapidaire).

lap-out map, carte de répartition de formations discordantes.

lapiaz, lapiez, lapié.

lapidary, 1. adj : lapidaire; **2.** n : lapidaire, diamantaire.

lapidification, lapidification (diagénèse).

lapidify (to), lapidifier, se lapidifier.

lapies, lapiez.

lapilli, lapilli (1 à 64 mm).

lapilli tuff, conglomérat volcanique à lapilli dans une matrice fine.

lapis-lazuli, lapis-lazuli, outremer, lazurite *(minér.).*

lapis ollaris, talc.

Laramian orogeny, orogénèse Laramienne.

Laramide orogeny, (Crétacé terminal-fin Paléocène).

lardite, lardite *(minér.),* agalmatolite, talc massif.

lard stone, var. d'agalmatolite, talc massif.

large-scale, à grande échelle.

large solution sink, doline.

larkspur gypsum, gypse pied d'alouette.

larnite, larnite *(minér.).*

larva, larve *(pal.).*

larval, larvaire *(pal.).*

larvickite, laurvickite (var. de syénite) *(pétro.).*

laser, laser.

laser ranging, distancemètre à laser.

lasionite, wavellite *(minér.).*

latching, levé à la boussole *(mine).*

lateglacial, tardiglaciaire.

late magmatic minerals, minéraux tardimagmatiques, minéraux formés en dernier.

latent, latent; **l. heat of crystallization,** chaleur latente de cristallisation; **l. heat of fusion,** chaleur latente de fusion.

late Pleistocene, Pleistocène tardif.

lateral, latéral; **l. cone,** cone adventif; **l. corrasion,** corrasion latérale (fluviatile); **l. crater,** cratère adventif; **l. cutting,** érosion latérale; **l. erosion,** érosion latérale; **l. fault,** décrochement; **l. frost thrust,** poussée de gel horizontale; **l. moraine,** moraine latérale; **l. migration,** migration latérale (du pétrole); **l. planation,** aplanissement des interfluves; **l. separation,** distance des faces d'une

faille; **l. shift,** décrochement; **l. spread,** glissement de terrain; **l. teeth,** dents latérales *(pal.,* Lamellibranches); **l. variation,** variation latérale (stratigraphie).

later arrival (seism.), arrivée ultérieure (d'une onde).

laterite, latérite; **l. material,** matériau latérisé; **l. soil,** sol latéritique.

lateritic, latéritique; **l. red loam,** limon rouge l.; **l. soil,** sol l.

lateritization, latéritisation.

lateritized, latérisé.

laterolog, latérolog (mesure de résistivité dans les sondages).

lath, 1. cristal prismatique, cristal allongé et mince; **2.** palplanche (construction).

latite, latite (var. de trachy-andésite).

latitude, latitude; **l. correction,** correction de latitude.

latitudinal, transversal.

latosol, sol ferralitique, sol latéritique.

lattice, réseau cristallin; **l. cell,** maille (du réseau); **l. drainage,** réseau fluviatile orthogonal « en treillis »; **l. orientation,** orientation cristalline; **l. plane,** plan réticulaire.

Lattorfian (or Latdorfian), Lattorfien (étage, Oligocène inf.).

laumonite, laumontite, laumontite (= zéolithe) *(minér.).*

Laurasia, Laurasie (protocontinent de l'hémisphère nord).

Laurentian, Laurentien (peu usité, cf. Précambrien canadien : granitisation vers — 1 000 MA).

lava, lave; **l. ball, bomb,** bombe volcanique en fuseau; **l. blister,** boursouflure de l.; **l. cascade,** cascade de l.; **l. cave,** caverne de l.; **l. channel,** chenal de l.; **l. cone,** cone de l.; **l. dam lake,** lac de barrage de coulée volcanique; **l. discharge,** débit de l.; **l. dome,** dôme de l.; **l. field,** champ de l.; **l. flood,** coulée de l.; **l. flow,** coulée de l.; **l. fountain,** fontaine de l.; **l. lake,** lac de l.; **l. mesa,** mésa; **l. pit,** fond de cratère rempli de lave active ou figée; **l. plateau,** plateau de l.; **l. pool,** lac de l.; **l. plug,** culot de l.; **l. sheet,** nappe de l.; **l. shield,** bouclier de l.; **l. streak,** filon intrusif de l.; **l. stream,** coulée de l.; **l. tube or tunnel,** tunnel dans coulée de l.; **l. volcano,** volcan de l.; **aa l. or block l.,** l. chaotique, blocailleuse, cheire; **congealed l.,** l. figée (par refroidissement); **glassy l.,** l. vitreuse; **holohyaline l.,** l. vitreuse; **mud l.,** coulée boueuse volcanique; **pillow l.,** l. en coussins; **ropy l.,** l. cordée.

lavic, lavique.

law, loi; **l. of constancy of interfacial angles,** loi de constance des angles des faces cristallines; **l. of correlation of facies (Walther),** loi de corrélation des faciès; **l. of crosscutting relationship,** datation relative des roches ou sédiments (ex. des intrusions); **l. of original continuity,** loi de continuité originelle des couches; **l. of original horizontality (Steno),** loi d'horizontalité originelle des couches (Sténo); **l. of priority,** loi de priorité (taxonomie); **l. of refraction,** loi de réfraction; **l. of stream**

numbers (Horton), loi de rapport inverse entre ordre et nombre des cours d'eau; **l. of superposition**, loi de superposition des couches.

lawsonite, lawsonite (*minér.*).

Laxfordian, Laxfordien (cf. Précambrien d'Écosse).

lay (of the land), configuration du terrain.

lay out a mine (to), aménager une mine.

lay out a map, a curve (to), tracer, dessiner un plan, établir une courbe (sur un graphique).

layer, couche, lit, niveau géologique; **l. corrosion**, corrosion en strates; **l. lattice structure**, structure réticulaire feuilletée; **active l.**, mollisol; **boundary l.**, couche limite; **iron l.**, niveau à concrétions ferrugineuses.

layered igneous rocks, roches intrusives stratifiées (cf. gabbros).

layered intrusions, roches intrusives stratifiées.

layered series, séries éruptives litées.

layered silicate, phyllosilicate.

layering, stratification, litage; **phase l.**, litage d'un niveau caractérisé par une espèce minérale; **rhythmic l.**, stratification rythmique.

layout, tracé (d'une courbe).

lazuli, lapis-lazuli.

lazulite, lazulite (*minér.*).

lazurfeldspar, orthose bleutée (Sibérie).

lazurite, lasurite, lazurite, lasurite (feldspathoïde).

lea land, terre en jachère.

leachate, solution obtenue par lessivage pédologique.

leach hole, doline, fissure de dissolution karstique.

leach (to), filtrer, lessiver, lixivier.

leachy soil, sol perméable, sol lessivé.

lead, 1. plomb; 2. indice de minéralisation, filon; **l. alloy**, alliage de plomb; **l. alpha age method**, datation au plomb alpha; **l. bearing**, plombifère; **l. carbonate**, cérusite; **l. chromate**, crocoïte; **l. glance**, galène; **l. luster**, oxyde de plomb; **l. marcasite**, blende, sphalérite; **l. ore**, minerai de plomb; **l. spar**, anglésite, cérusite; **l. sulphide**, galène; **l. uranium ratio**, rapport isotopique, plomb-uranium; **l. vitriol**, anglésite; **black l.**, graphite; **blind l.**, filon n'affleurant pas; **deep l.**, gravier aurifère (Austr.).

lead pipe, conduite d'amenée.

leaf, 1. feuille; 2. feuillet.

leaflike structure, structure feuilletée.

league, lieue; **land l.**, lieue terrestre = 3 « statute miles » = 4 828 m; **nautical l.**, lieue marine = 3 « geographical miles » = 5 559 m.

leak, fuite, écoulement, perte; **l. proof**, étanche.

leak (to), fuir, couler, suinter.

leakage, 1. fuite, perte d'eau; **l. water**, eau d'infiltration.

lean, maigre, pauvre; **l. clay**, argile peu plastique; **l. coal**, houille maigre; **l. ore**, minerai pauvre, à faible teneur.

leap, 1. bond (d'un grain de sable); 2. rejet (d'une couche); **l. ore**, minerai d'étain de mauvaise qualité (G.B.); **down-l.**, compartiment affaissé; **upl.**, compartiment soulevé.

learned society, société savante.

lease, 1. concession; 2. périmètre d'exploitation; 3. fermage; **oil l.**, concession pétrolière.

least squares method, méthode des moindres carrés.

least-time path, trajet le plus rapide (pour une onde).

lea stone, grès schisteux, grès feuilleté (Angl.).

leat, 1. canal de dérivation, bief; 2. cours d'eau (Cornouailles).

leather bed, brèche de faille d'argile compacte (G.B.).

leavings, stériles (*mine*).

lechatelierite, lechateliérite (*minér.*).

lectotype, lectotype (*pal.*).

Ledian, Lédien = Auversien (sous-étage, Éocène sup.).

ledge, 1. rebord, saillie, corniche; 2. couche affleurant dans une carrière; 3. filon minéralisé; **l. rock**, véritable soubassement; **l. wall**, mur, sol (*mine*).

ledger, partie inférieure d'un filon.

ledger wall, mur d'une couche.

lee, 1. côté sous le vent; 2. côté abrité du courant (eau, glace); **l. eddy**, tourbillon sous le vent; **l. side**, côté sous le vent; **l. ward**, sous le vent.

left-handed crystal, cristal lévogyre.

left-lateral fault, décrochement senestre.

left strike-slip fault, décrochement senestre.

leftwards, vers la gauche.

leg, 1. montant (d'un appareil); 2. appui, support (de boisage); 3. jambage, flanc (d'un anticlinal).

lehm, lehm (parfois syn. de loess).

lenad, feldspathoïdes.

length, longueur; **wave l.**, l. d'onde.

lengthwise section, coupe longitudinale.

lens, 1. lentille, verre (optique); 2. objectif (photo); 3. masse lenticulaire, lentille (*mine*).

lens-like, lenticulaire; **l. bedding**, litage hétérogène..

lens shaped, lenticulé.

lensing, stratification lenticulaire.

lenticle, masse lenticulaire (de terrain).

lenticular, lenticulaire.

lenticule, amas lenticulaire.

lenticuline, lenticulaire (*pal.*).

lentil, 1. lentille rocheuse; 2. subdivision lenticulaire d'une strate.

lentoïd, lenticulé, lenticulaire.

Leonardian (N. Am.), Léonardien (série, Permien inf.).

leonhardite, leonhardite (= zéolithe) (*minér.*).

lepidoblastic, lépidoblastique (= texture foliée de roche métamorphique).

lepidocrocite, lepidocrosite, lépidocrocite (var. de goethite).

Lepidodendron, Lépidodendron (*paléobot.*).

lepidolite, lépidolite (micas litinifères).

lepidomelane, lépidomélane (var. de biotite).

Lepidosauria, Lépidosauriens.

Lepospondyles, Lépospondyles.

leptinite, granulite.

leptochlorite, leptochlorite (chlorites riches en fer).

leptothermal, leptothermal.

leptynite, leptinite, leptite *(Sweden, Finland)*, leptynite (*pétro.*).

leptynolite, leptinolite, leptynolite (*pétro.*).

leucite, leucite (feldspathoïde); **l. basalt,** basalte à leucite; **l. phonolite,** phonolite à leucite (sans néphéline); **l. tephrite,** téphrite à leucite et néphéline; **l. trachyte,** trachyte à leucite.

leucitic, leucitique.

leucitite, leucitite (= phonolite très riche en leucite).

leucitohedron, trapézoèdre (*cristallo.*).

leucitophyre, leucitophyre, phonolite à leucite et néphéline.

leuco *(prefix)*, 1. blanc, sans couleur; 2. leuco (préfixe).

leucocratic, leucocrate, riche en minéraux clairs (minéraux blancs ou acides).

leucogranite, leucogranite (granite clair à muscovite); **l. aplite,** aplite leucogranitique; **l. pegmatite,** aplite pegmatitique.

leucogranodiorite, leucogranodiorite.

leucopyrite *(cf. loellingite)*, leucopyrite, lollingite (var. de marcasite).

leucorhyolite, leucorhyolite, rhyolite leucocrate (*pétro.*).

leucosome, leucosome.

leucotephrite, téphrite à leucite.

leucoxene, leucoxène, titanomorphique (*minér.*).

Levallois flake, éclat Levallois (*préhist.*).

Levalloisian, levalloisien (Paléolothique moy.).

levee, 1. levée (naturelle); 2. digue.

leveed channel, chenal sous-marin, endigué, dans le prolongement de canyons sous-marins.

level, 1. niveau (appareil); 2. teneur; 3. niveau, étage d'une mine, galerie de mine; 4. partie plate (*géogr.*); **l. country,** terrain plat; **l. course,** dans la direction d'une couche; **l. gage,** indicateur de niveau; **l. ground,** terrain plat; **l. of zero amplitude,** seuil du gel permanent (dans le sol), niveau supérieur du pergélisol; **l. seam,** couche horizontale; **l. surface,** surface plane; **l. vial,** niveau à huile; **l. with the ground,** à raz de terre; **air l.,** niveau à bulle d'air; **change of l.,** variation de niveau; **hand l.,** niveau à main; **high energy l.,** milieu sédimentaire à haute énergie; **hydrostatic l.,** niveau hydrostatique; **low energy l.,** milieu sédimentaire à faible énergie; **overflow l.,** niveau de débordement; **sea l.,** niveau marin; **shifting of l.,** variation de niveau; **terrace l.,** niveau de terrasse; **water l.,** niveau de l'eau.

level (to), niveler, aplanir, égaliser.

leveler, niveleuse.

levelling, nivellement, aplanissement; **l. book,** carnet de nivellement; **l. instrument,** niveau à lunette; **l. point,** point de mire; **l. pole,** mire de nivellement, balise; **l. rule,** mire graduée; **l. screw,** vis de réglage; **l. survey,** nivellement topographique; **barometric l.,** nivellement barométrique.

levigate (to), léviger, provoquer la lévigation de minéraux.

levigation, lévigation.

Lewisian, Lewisien (subdivision, Précambrien d'Écosse).

lherzite, lherzolite (roche ultrabasique).

lherzolite, lherzolite (= péridotite à ortho et clinopyroxène).

Lias, Lias (série, Jurassique inf.).

Liassic, liasique, du Lias.

libethenite, libéthénite (*minér.*).

Lichenes, Lichens.

licks, prés salés (U.S.).

lichenometry, lichénométrie.

lidstone *(G.B.)*, toit (d'une mine de fer).

lie, 1. disposition du terrain, gisement, 2. tracé (d'une route).

liesegang ring banding, précipitation rubanée de fer en roche poreuse.

lie (to), reposer sur, être sus-jacent.

lievrite, liébvrite, ilvaïte, (*minér.*).

life, vie, durée; **half l.,** période de décomposition radioactive.

lift, 1. élévation, hauteur d'élévation; 2. étage, niveau (*mine*).

lift (to), 1. lever, soulever, 2. remonter le minerai (*mine*); 3. se lever, s'élever.

lifting, 1. soulèvement, levage; 2. remontée; 3. élévation; **l. way,** puits d'extraction; **gas l.,** procédé d'extraction de pétrole par injection de gaz.

ligament, ligament; **l. area,** région ligamentaire (*pal.*).

Ligerian *(Eur.)*, Ligérien (sous-étage, Turonien inf.).

light, 1. adj : léger, faible; clair, pâle; 2. n : lumière, éclairage; **l. colored mineral,** minéral clair; **l. fraction,** fraction légère (densimétrie); **l. isotope,** isotope léger; **l. mineral,** 1. minéral léger; 2. minéral pâle; **l. oil,** pétrole léger (paraffinique); **l. rays,** rayons lumineux; **l. red silver ore,** proustite; **l. ruby silver ore,** proustite; **l. silt,** limon fin; **l. soil,** sol léger; **l. wave,** onde lumineuse; **polarized l.,** lumière polarisée.

lighting, 1. éclairage; 2. allumage (d'une charge explosive); **l. gas,** gaz d'éclairage.

lightning, éclair, foudre; **l. tube,** fulgurite tubulaire.

ligneous, ligneux.

lignified mor, humus brut ligneux.

lignin, lignine.

lignite, lignite; **l. bearing,** lignitifère; **l. tar oil,** huile de goudron de lignite.

lignitiferous, lignitifère.

liman, liman (estuaire envasé ou fond marin vaseux); **l. coast,** côte alluviale lagunaire.

limb, 1. lèvre (d'une faille); 2. flanc (d'un pli); 3. membre; 4. limbe d'un astre; **normal l.,** flanc normal; **roof l.,** flanc supérieur; **reversed l.,** flanc inverse; **stretched l.,** flanc étiré.

limburgite, limburgite (roche volcanique ultrabasique partiellement vitreuse).

lime, chaux; **l. burning,** cuisson de la c.; **l. clast,** roche calcaire détritique; **l. concretion,** concrétion calcaire; **l. craig,** front de taille d'une carrière de calcaire (Écosse); **l. crust,** croûte calcaire; **l. feldspar,** anorthite; **l. harmstone,** christianite, phillipsite; **l. kiln,** four à c.; **l. milk,** lait de c.; **l. mortar,** mortier de c.; **l. mud mounds,** îlots de boue carbonatée (Floride); **l. mudrock,** calcilulite; **l. nodule,** concrétion calcaire, poupée; **l. pan,** horizon d'accumulation calcaire; **l. pit,** carrière de pierre calcaire; **l. rock,** calcaire; **l. sandrock,** calcarénite; **l. secreting alga,** algue calcaire; **l. sink,** doline; **l. soda feldspar,** feldspath calcosodique (plagioclase); **l. uranite,** autunite; **l. water,** eau de c.; **caustic l.,** c. vive; **fat l.,** c. grasse; **quiet l.,** c. maigre; **slacked l.,** c. éteinte; **soda l.,** c. sodée.

limestone, calcaire; **l. cave,** aven, bétoire; **l. cavern,** bétoire, ponor; **l. pavement,** lapiés; **l. quarry,** carrière de pierre c.; **l. red loam,** limon rouge c.; **l. rock,** roche c.; **l. sink,** doline, poljé; **l. solution,** dissolution du c.; **l. wash,** lait de chaux; **algal l.,** c. à algues; **argillaceous l.,** c. argileux; **banded l.,** c. rubané; **bioclastic l.,** c. bioclastique; **biostromal l.,** c. à faciès récifal; **bituminous l.,** c. bitumineux; **chemically precipitated l.,** c. d'origine chimique; **cherty l.,** c. à silex; **clastic l.,** c. détritique; **coquinoid l.,** c. lumachellique; **coral l.,** c. corallien; **crinoidal l.,** c. à Crinoïdes; **crystalline l.,** c. cristallin, marbre; **dolomitic l.,** c. dolomitique; **encrinitic l.,** c. à entroques; **foraminiferal l.,** c. à Foraminifères; **glauconitic l.,** c. glauconieux; **gryphite l.,** c. à Gryphées; **knobby l.,** c. noduleux; **marly l.,** c. marneux; **nummulitic l.,** c. à Nummulites; **oolitic l.,** c. oolithique; **pellet l.,** c. à petites concrétions; **phosphatic l.,** c. phosphaté; **reef l.,** c. corallien; **sandy l.,** c. sableux; **siliceous l.,** c. siliceux; **shelly l.,** c. coquillier, lumachelle; **skeletal l.,** c. à débris coquillier (bioclastique).

limey, calcaire, calcique.

liming, chaulage.

limit angle, angle limite.

limit (elastic), limite d'élasticité.

limnetic, limnic, limnique, d'eau douce.

limnic, limnique, lacustre.

limnic peat, tourbe limnique, lacustre.

limnology, limnologie.

limonite, limonite (*minér.*); un mélange de divers minéraux ferrugineux.

limonitic, à limonite, contenant de la limonite.

limy, calcaire, calcique.

linarite, linarite (*minér.*).

line, **1.** ligne, alignement, trait; **2.** conduite, canalisation; **3.** raie du spectre; **l. of bearing,** direction d'affleurement; **l. of dip,** direction de pendage; **l. of force,** ligne de force; **l. of fracture,** ligne de fracture; **l. of growth,** strie d'accroissement; **l. of latitude,** parallèle; **l. level,** ligne de niveau; **l. of lode,** direction d'un filon; **l. of magnetic force,** ligne de force; **l. of section,** trait de coupe (d'une carte); **l. of sight,** ligne de visée; **l. of strike,** ligne de direction; **l. scale,** échelle graphique d'une carte; **l. up, 1.** mise en phase (*sism.*); **2.** mise en ligne (pétrole); **coast l.,** ligne de rivage; **crest l.,** ligne de crête; **dashed l.,** ligne en tirets; **dotted l.,**

ligne en pointillés; **drilling l.,** câble de forage; **edge water l.,** limite eau-pétrole; **fault l.,** ligne de faille; **flow l.,** conduite d'écoulement; **gas l.,** gazoduc, conduite de gaz; **geodesic l.,** ligne géodésique; **oil l.,** oléoduc, conduite de pétrole; **seismic l.,** profil sismique; **shore l.,** ligne de rivage; **snow l.,** limite des neiges; **sounding l.,** ligne de sondage; **strand l.,** ligne de rivage; **timberline,** ligne de temps (*sism.*); **wire line coring,** carottage au câble.

line defect, dislocation cristalline.

line up (to), rayer, marquer de lignes, faire le carroyage (d'un plan).

lineage, lignée évolutive; **l. zone,** phylozone.

lineament, linéament, alignement structural (décelé par photographie aérienne).

linear, linéaire; **l. dunes,** dunes longitudinales; **l. vent,** fissure éruptive.

lineated anomaly, anomalie linéaire.

lineated fabric, structure à linéations.

lineation, linéation.

linguiform, en forme de langue.

Lingulid, Lingulidés (*pal.*).

lining, enduit calcaire (karst).

link, **1.** chaînon, lien; **2.** unité de mesure = 20 cm (environ).

linkage, liaison (chimique); **l. analysis,** analyse des groupes (*statistique*).

linked veins, filons réticulés, en gradins, anastomosés.

linking, liaison (chimique).

linn, chute d'eau, petite cataracte.

linnaeite, linnéite (*minér.*).

linophyric structure, structure à phénocristaux alignés.

lip, lèvre, bord, rebord; **inner l.,** bord interne ou columellaire (*pal.*); **outer l.,** bord externe, labre, péristome (*pal.*).

liparite, **1.** liparite (*pétro.,* var. de rhyolite); **2.** liparite (*minér.,* var. de talc).

liparitic, liparitique.

lipid , lipide.

liquefaction, liquéfaction.

liquefaction potential, risque de coulée boueuse (par fluidification).

liquefiable, liquéfiable.

liquefied natural gas, gaz naturel liquéfié.

liquefy (to), **1.** liquéfier, fluidifier; **2.** se liquéfier.

liquid, **1.** adj : liquide; **2.** n : liquide; **l. head,** pression hydrostatique; **l. hydrocarbon,** hydrocarbure liquide; **l. inclusion,** inclusion liquide; **l. limit,** limite de liquidité; **l. paraffin,** huile de paraffine; **l. phase,** phase liquide; **l. seal,** joint hydraulique; **l. waste,** déchet liquide.

liquidity limit, limite de liquidité.

liquidus, liquidus.

liquor, **1.** liqueur; **2.** solution; **heavy l.,** liqueur dense.

listric fault, faille courbe (en forme de cuiller), listrique.

litharge, litharge artificiel, oxyde de plomb.

lithia emerald, hiddénite (var. de spodumène, *minér.*).

lithia mica, mica lithinifère, lépidolite.

lithic, dépôt à nombreux fragments de roches plus anciennes (pyroclastiques); **l. arenites,** grès à débris schisteux divers; ex : grauwacke; **l. tuff, 1.** conglomérat volcanique, tuf volcanique; **2.** brèche de roches éruptives.

lithiclast (syn. lithoclast), débris carbonaté (remanié).

lithification, lithification (diagénèse).

lithify (to), se transformer en roche, se cristalliser, se consolider.

lithionite, lithionite, lépidolite.

lithium, lithium; **l. mica,** lépidolite; **l. tourmaline,** tourmaline lithinifère, elbaïte.

lithoclast (syn. lithiclast), lithoclast, débris carbonaté (remanié).

lithofacies, lithofaciès; **l. map,** carte de l.

lithofraction, fragmentation des roches (par action des vagues ou des eaux courantes).

lithogenesis, lithogénèse.

lithogenic, lithogenetic, lithogénétique; **l. sequence,** lithoséquence.

lithologic, lithological, lithologique.

lithologic log, colonne stratigraphique.

lithologist, pétrographe.

lithographic, lithographique; **l. limestone,** calcaire l.; **l. stone,** calcaire l.; **l. texture,** à cristallinité très fine (micritique).

lithology, 1. pétrographie; **2.** étude microscopique, étude pétrographique.

lithomarge, lithomarge : var. de kaolin (avec halloysite).

lithophile, lithophilic, lithophylic, lithophile (élément géochimique à forte affinité pour l'oxygène).

lithophosphor mineral, minéral thermoluminescent.

lithophyl, feuille pétrifiée ou son moulage.

lithophysa, lithophyse, sphérolite creuse.

lithophysal fabric, texture à sphérolites.

lithophyte, litophyte.

lithosol, lithosol, sol squelettique.

lithosolic soil, lithosol.

lithosphere, 1. lithosphère (opposé à atmosphère) **2.** croûte terrestre.

lithospheric, lithosphérique; **l. plate,** plaque l.

lithostatic pressure, pression géostatique exercée par les terrains sus-jacents.

lithostratigraphic unit, unité lithostratigraphique.

lithostratigraphy, lithostratigraphie.

lithostrome, lithostrome.

Lithothamnion, Lithothamnion (algue rouge).

lithotope, lithotope (sédiment ou roche d'un biotope).

lithotype, lithotype, catégorie ou type de charbon.

lithozone, lithozone (unité non formelle de chronostratigraphie).

lit-par-lit injection, migmatisation, injection lit-par-lit.

litter, litière (*pédol.*).

little fold, plissotement.

little ice age(s), petit âge(s) glaciaire(s).

littoral, littoral; **l. current,** courant de dérive littorale; **l. drift.,** dérive littorale; **l. environment,** milieu littoral; **l. sedimentation,** sédimentation littorale; **l. zone,** zone littorale (estran + zone de déferlement).

live lode, filon richement minéralisé.

liver opal, ménilite (var. d'opale grise).

liver ore, 1. cuprite (variété); **2.** cinabre (parfois).

liver peat, sol lacustre acide.

liverstone, var. de barytine.

living, vivant; **l. chamber,** chambre, loge d'habitation (*pal.*); **l. fossil,** fossile vivant; **l. rock,** roche non exploitée.

lixiviate (to), lessiver (*pédol.*).

lixiviation, lixiviation, lessivage.

L layer, litière pédologique.

Llandeilian, Llandeilien (étage, Ordovicien moy.).

Llandoverian, Llandovérien (étage Silurien inf.).

Llanvirnian, Llanvirnien (étage, Ordovicien moy.).

load, charge; **l. capacity,** c. utile; **l. carrying ability,** capacité de c.; **l. cast,** figure, marque de c; **l. curve,** courbe de c.; **l. test,** essai de compression; **l. metamorphism,** metamorphisme régional; **bed l.,** c. de fond (fluv.); **bottom l.,** c. de fond (fluv.); **dead l.,** c. statique; **solid l.,** c. solide (fluv.); **suspended l.,** c. en suspension; **traction(al) l.,** c. de fond (fluv.); **useful l.,** c. utile.

loaded stream, fleuve ayant atteint sa charge limite.

loading, chargement; **l. dig,** pelle excavatrice; **l. point,** limite de charge; **l. terminal,** station de chargement.

loadstone, 1. magnétite; **2.** aimant naturel.

loam, 1. limon; **2.** terre glaise; **3.** torchis, pisé; **l. clay,** glaise; **l. rim,** dune argileuse.

loamy, limoneux; **l. fine soil,** sol fin l.; **l. sand,** sable l.; **l. soil,** limon.

lob (to), scheider.

lobate, lobé.

lobation, formation de lobes glaciaires à partir d'un inlandsis.

lobe, lobe (*pal.*; glaciol.).

local metamorphism, métamorphisme de contact.

location survey, tracé topographique.

loch (an), loch, petit lac (Écosse); **sea l.,** fjord.

lock, écluse.

Lockportian (N.Y. state), Lockportien (étage, Silurien moy.).

loculus, loge initiale (*pal.*).

lode, filon; **l. claim,** concession minière; **l. deposit,** gisement filonien; **l. filling,** remplissage filonien; **l. mining,** exploitation de filons; **l. ore,** minerai filonien; **l. plot,** filon horizontal; **l. stone, 1.** magnétite; **2.** aimant naturel; **l. stuff,** matière filonienne.

lodgement till, till (moraine) de fond.

lodranite, lodranite (météorite).

loellingite, löllingite, leucopyrite (*minér.*).

loess, loess; **l. doll**, poupée du l.; **l. kindchen**, concrétion calcaire, poupée du l.; **l. soil**, sol loessique; **fluvial l.**, limon fluvio-glaciaire.

log, 1. diagraphie, diagramme, enregistrement continu; 2. grume, tronc d'arbre; **l. book**, journal de sondage; **l. normal distribution**, répartition lognormale; **l. strip**, bande d'enregistrement; **acoustic l.**, diagramme acoustique; **acoustic velocity l.**, diagramme de vitesse acoustique; **calcilog**, diagramme de calcimétrie; **continuous velocity l.**, diagramme continu de vitesse; **electric l.**, diagramme électrique; **gamma ray l.**, diagramme de rayonnement gamma; **geothermal l.**, diagramme géothermique; **inductolog**, inductolog, diagraphie par induction; **laterolog**, latérolog; **neutron l.**, diagramme neutronique; **neutron-gamma l.**, diagramme neutron-gamma; **neutron-neutron l.**, diagramme neutron-neutron; **nuclear l.**, diagramme nucléaire; **permeability l.**, diagramme de perméabilité; **photoelectric l.**, diagramme photoélectrique; **resistivity l.**, diagramme de résistivité; **self-potential l.**, diagramme de polarisation spontanée; **spontaneous potential l.**, diagramme de polarisation spontanée; **temperature l.**, diagramme de variation thermique; **well l.**, diagramme de forage.

logan stone, loggan stone, logging stone, roche branlante.

logging, 1. diagraphie de sondages; 2. exploitation des forêts; **chlorine l.**, diagraphie de teneur en chlorure; **induction l.**, diagraphie par induction; **electric l.**, diagraphie électrique; **mud analysis l.**, détection des indices dans les boues (forage); **radioactive l.**, diagraphie nucléaire.

long, long; **l. clay**, argile très plastique; **l. flame coal**, houille flambante; **l. limb**, flanc long (d'un pli dissymétrique); **l. profile**, profil longitudinal; **l. range order**, état cristallin idéal; **l. ton**, tonne forte = 1 016 kg.

longitude, longitude.

longitudinal, longitudinal; **l. dune**, dune longitudinale; **l. fault**, faille longitudinale; **l. moraine**, moraine longitudinale; **l. section**, coupe longitudinale; **l. stream**, cours d'eau subséquent; **l. wave**, onde longitudinale.

Longmyndian, Longmyndien (subdivision, Précambrien d'Angleterre).

longshore, parallèle au littoral; **l. bar**, cordon littoral sableux dans la zone intertidale; **l. current**, courant de dérive littorale; **l. drift**, dérive littorale; **l. through**, dépression entre deux cordons littoraux.

loop, boucle, maille; **closed l.**, maille fermée (*sism.*); **morainic l.**, rempart morainique; **oscillation l.**, ventre de vibration.

loose, 1. meuble, inconsistant; 2. détendu, mal fixé, libre; 3. flou, vague; **l. ground**, terrain meuble, ébouleux; **l. sédiment**, sédiment meuble, non consolidé; **l. sand**, sable meuble, boulant.

loosen (to), 1. ameublir; 2. dégager, desserrer, libérer.

lophophore, lophophore (*pal.*).

lopolith, lopolite.

lopsided, déjeté, déversé.

losing stream, rivière dont l'eau s'échappe (vers la nappe phréatique).

loss, perte; **l. of circulation**, p. de circulation; **l. of head**, p. de charge; **l. of pressure**, p. de charge; **l. on ignition**, p. au feu; **fluid l.**, p. de fluide; **gross l.**, p. totale; **power l.**, p. d'énergie.

lost, perdu; **l. head**, perte de charge; **l. oil**, pétrole perdu, irrécupérable; **l. pressure**, chute de pression; **l. record**, lacune stratigraphique; **l. river**, perte karstique; **l. volcano**, volcan éteint.

Lotharingian, Lotharingien (sous-étage, Jurassique inf.).

lough, 1. lac, bras de mer (Irlande); 2. cavité irrégulière de mine de fer (Lancashire).

Love waves, ondes sismiques de Love.

low, 1. bas, de faible altitude; 2. à faible teneur; **l. angle fault**, faille subhorizontale; **l. density**, faible densité; **l. dip**, pendage faible; **l. energy environment**, milieu à faible énergie; **l. flow channel**, chenal d'étiage; **l. grade ore**, minerai à faible teneur; **l. gradient**, pente faible; **l. ground**, bas fond; **l. humic gley soil**, sol à gley peu humifère; **l. lands**, bas pays, plaines littorales; **l. lime much**, sol organo-minéral pauvre en calcaire; **l. moor**, tourbière basse; **l. moor soil**, sol tourbeux neutre; **l. pressure**, basse pression; **l. quartz**, quartz de basse température (quartz α); **l. rank graywacke**, grauwacke sans feldspath; **l. rank metamorphism**, faible degré de métamorphisme; **l. tide**, marée basse; **l. velocity zone**, asthénosphère; **l. viscosity**, faible viscosité; **l. water**, basses eaux, marée basse.

lower, 1. inférieur (en position); 2. inférieur, plus ancien (chronologiquement); **l. limb**, flanc inférieur; **l. subsoil**, sous-sol profond; **l. track of a river**, tronçon inférieur (d'une rivière).

lows, 1. dépression barométrique; 2. bassin, synclinal (terme général).

lubrifiant, lubrifiant.

Ludhamian, Ludhamien (interglaciaire Biber-Danube).

Ludian, Ludien (étage, Éocène sup.).

Ludlovian, Ludlovien (étage, Silurien sup.).

lugarite, lugarite (var. de théralite).

Luisian, Luisien (étage, miocène).

lumachelle, lumachelle, calcaire très riche en coquilles de mollusques (ex. huîtres).

lumber, bois de charpente.

luminance, réflectance (*télédétection*).

luminescence, luminescence.

lump, masse, morceau, bloc, motte, aggrégat; **l. aggregate structure**, structure en mottes (*pédol.*); **l. coke**, coke en morceaux; **l. limestone**, calcaire graveleux; **l. ore**, minerai en morceaux; **l. structure**, structure en aggrégats en mottes.

lumpy, formé de mottes, grumeleux.

lunar, lunaire; **l. basalt**, basalte l.; **l. crater**, cratère l.; **l. crust**, croûte l.; **l. disk**, disque l.; **l. geology**, géologie l.; **l. lithosphere,** lithosphère l.; **l. module,** véhicule l.; **l. nearside,** face visible de la lune; **l. ocean,** océan l.; **l. regolith,** régolithe l.; **l. sedimentation,** sédimentation l.; **l. station,** station l.; **l. terra,** massif l.; **l. tide,** marée produite par la force gravitationnelle de la Lune.

lunate, en forme de croissant; **l. bar,** banc de sable ou flèche en croissant; **l. fracture,** cassure glaciaire en croissant; **l. ripple-mark,** ride de plage en forme d'une petite barkhane, ride en croissant.

lunule, lunule (*pal.*).

luscladite, luscladite (var. de théralite).

Lusitanian, Lusitanien (étage, Jurassique sup.).

lusitanite, lusitanite (var. de syénite).

lustrous, lustré, luisant; **l. shale,** schiste lustré.

lutaceous, argileux.

lutecin, lutecite, lutécite (var. calcédoine).

lutecium, lutécium (lanthanide ou terre rare).

Lutetian, Lutétien (étage, Éocène moy.).

lutite, lutyte, lutite (particule $\leqslant$ 0,063 mm).

luvisol, luvisol, sol lessivé tempéré, sol lessivé fersiallitique.

luxullianite, luxullianite ou luxulyanite (granite à tourmaline sphérolithique).

Lycopodiales, Lycopodiales.

lydian stone, lydienne.

lydite, lydite.

lying wall, mur (d'une couche) (mine).

lyophilic, hydrophile.

lyophobic, hydrophobe.

lysimeter, lysimètre.

Lytoceratides, Lytocératidés (*pal.*).

M

maar (pl. maars), maar ou maare (cratère volcanique).

Maastrichtian, Maestrichtian, Maestrichtien (étage, Crétacé sup.).

macaluba (syn. mud volcano), volcan de boue (Sicile).

maceral, macéral.

machine, machine; **m. coal mining,** abattage mécanique du charbon; **m. drilling,** forage mécanique; **m. mining,** abattage mécanique.

macigno, macigno (flysch de l'Éocène supérieur des Alpes italiennes).

macle, mâcle.

macled, mâclé, hémitrope, mâclifère; **m. shale,** schiste mâclifère.

macro, macro- (préfixe signifiant grand).

macroaggregate, macroagrégat.

macroaxis, axe b des systèmes orthorhombique et triclinique.

macroclimate, macroclimat, climat d'ensemble.

macrocosm, macrocosme.

macrocrystal, phénocristal.

macrocrystalline, macrocristallin.

macrodome, macrodôme (*cristallo.*).

macrofacies, facies tract, macrofaciès, faciès différents interdépendants.

macrofossil, macrofossile.

macrography, macrographie.

macrolepidolite, macrolépidolite, lépidolite en grands feuillets (pegmatitiques).

macromeritic (obsolete), macrocristallin.

macromolecular, macromoléculaire.

macropinacoid, macropinacoïde (*cristallo.*).

macropolyschematic rock, roche à diverses structures, roche à texture hétérogène.

macropore, macropore (*pal.*).

macroprism, macroprisme.

macropinacoid, macropinacoïde (*cristallo.*).

macroscopic, macroscopique.

macroseism (syn. earth-quake), macroséisme, tremblement de terre.

macroseismic, macrosismique.

macrospore (syn. megaspore), mégaspore.

macrostructure, macrostructure.

macroturbulence, macroturbulence.

maculose, tacheté (roche, etc.).

made ground, sédiment récent.

Madrepore, Madrépore; **m. marble,** marbre à Madrépores (Dévonien).

Madreporia, Madréporaires.

madreporic, madréporique.

madreporite, madréporite, plaque madréporique (*pal.*).

Maentwrogian, Maentwrogien (étage, Cambrien sup.).

Maestrichtian, Maestrichtien (étage, Crétacé sup.).

mafic, mafique (ferro-magnésien).

mafite, mafite, minéral ferromagnésien.

Magarian, Silurien (E.U.).

Magdalenian (Eur.), Magdalénien (subdivision, Paléolithique récent).

maghemite, oxyde ferrique magnétique, maghémite.

magma, magma; **m. blister,** intumescence, boursouflure magmatique; **m. chamber,** réservoir magmatique; **m. column,** colonne magmatique; **m. differenciation,** différenciation magmatique; **m. granite,** granite d'origine magmatique; **low-pressure m.,** magma à faible pression.

magmatic, magmatique; **m. association,** association m.; **m. differenciation,** différenciation m.; **m. emanations,** substances volatiles m.; **m. fluids,** fluides m.; **m. hearth,** foyer m.; **m. ore-deposit,** gîte intra-magmatique; **m. segregation,** différenciation m.; **m. stoping,** digestion des terrains encaissants; **m. water,** eau juvénile.

magmatogene, magmatogène.

magnesia, magnésie (oxyde de magnésium); **m. alum,** pickéringite; **m. cordierite,** cordiérite sans fer; **m. mica,** mica magnésien, phlogopite.

magnesian, magnésien; **m. limestone, 1.** calcaire m.; **2.** nom d'une formation stratigraphique du Permien anglais; **m. lower,** Ordovicien inférieur (Mississipi, U.S.); **m. marble,** marbre à dolomite, calcaire magnésien marmorisé à dolomite.

magnesioanthophyllite, anthophyllite magnésienne (var. d'amphibole).

magnesiochromite, picotite, chromite (var. de spinelle).

magnesiodolomite, var. de dolomite magnésienne.

magnesioferrite, magnésioferrite (*minér.*).

magnesite, magnésite, giobertite.

magnesium, magnésium; **m. aluminium garnet,** grenat aluminomagnésien; **m. diopside,** diopside magnésien; **m. iron mica,** mica ferro-magnésien; **m. mica,** mica magnésien; **m. rendzine,** rendzine magnésienne.

magnet, 1. aimant; **2.** masse ferro-magnétique aimantée; **m. separator,** séparateur magnétique.

magnetic, magnétique, aimanté; **m. anomaly,** anomalie m.; **m. anomaly pattern,** réseau d'anomalies m.; **m. attraction,** attraction m.; **m. azimuth,** azimut m.; **m. bearing,** direction m.; **m. compass,** boussole m.; **m. concentrator,** concentrateur m.; **m. declination,** déclinaison m.; **m. dial,** boussole; **m. dip,** inclinaison m.; **m. disturbance,** perturbation m.; **m. division,** strate définie par son paléomagnétisme; **m. element,** élément m.; **m. equator,** équateur m.; **m. event,** événement paléomagnétique; **m. field,** champ m.; **m. flux,** flux m.; **m. force,** force m.; **m. inclination,** inclinaison m.; **m. intensity,** intensité m.; **m. interval,** période entre deux inversions m.; **m. iron,** magnétite (*minér*.); **m. iron pyrite,** pyrrhotine (*minér*.); **m. lineation,** alignement m.; **m. meridian,** méridien; **m. needle,** aiguille aimantée; **m. north,** nord m.; **m. oxide of iron,** magnétite (*minér*.); **m. pattern,** réseau m.; **m. perturbation,** perturbation m.; **m. pole,** pôle m.; **m. pyrite,** pyrrhotine, pyrite m.; **m. reversal,** inversion m.; **m. separator,** séparateur m.; **m. spherule,** sphérule m.; **m. storm,** orage m.; **m. stratigraphy,** magnétostratigraphie; **m. stripe,** bande de fond océanique à paléomagnétisme défini; **m. survey,** prospection m.; **m. susceptibility,** susceptibilité m.; **m. tape,** bande m.

magnetism, magnétisme; **diamagnetism,** diamagnétisme; **ferrimagnetism,** ferrimagnétisme, ferromagnétisme; **fossil m.,** paléomagnétisme; **paleomagnetism,** paléomagnétisme; **terrestrial m.,** magnétisme terrestre.

magnetite, magnétite (*minér*.).

magnetization, aimantation; **m. graph,** courbe de magnétisme; **remanent m.,** aimantation rémanente; **reversed m.,** aimantation inversée; **thermoremanent m.,** aimantation thermo-rémanente.

magnetize (to), **1.** aimanter, magnétiser; **2.** s'aimanter.

magnetochronology, magnétochronologie.

magnetograph, magnétographe.

magnetohydrodynamics, magnétohydrodynamique.

magnetoilmenite, magnétoilménite, ilménomagnétite, titanomagnétite (*minér*.).

magnetometer, magnétomètre; **air-borne m.,** m. aéroporté; **astatic m.,** m. astatique; **horizontal m.,** m. horizontal; **proton m.,** m. à protons; **vertical m.,** m. vertical.

magnetometric survey, prospection magnétométrique.

magnetometry, magnétométrie (mesure du champ magnétique terrestre).

magnetomotive, magnétomoteur, magnétomotrice; **m. force,** force m.; **m. gradient,** intensité de champ magnétique.

magnetopause, magnétopause.

magnetosphere, magnétosphère.

magnetostratigraphy, magnétostratigraphie.

magnetostriction, magnétostriction.

magnetotail, queue magnétique de la magnétosphère.

magnetotelluric, magnétotellurique.

magnification, grandissement, grossissement, amplification.

magnifier, loupe grossissante.

magnifying glass, loupe grossissante; **m. lens,** œilleton de visée (photographie).

magnitude, **1.** grandeur; **2.** magnitude (d'un tremblement de terre); **m. scale,** échelle de magnitude.

magnochromite, magnésiochromite (*minér*.).

magnophyric, à gros phénocristaux.

main, **1.** adj : principal; **2.** n : conduite principale; **m. airway,** galerie p. d'aérage; **m. chute,** couloir p.; **m. drive,** galerie p.; **m. gangway,** galerie p.;

m. lode, filon-mère, filon p.; **m. road,** galerie p.; **m. rope,** cable de tête (*mine*); **m. shaft,** puits central, p.; **m. way,** galerie p.

mainland, **1.** continent, terre ferme; **2.** île principale.

main sea, large, grand large.

major-element analysis, méthode d'analyse des éléments majeurs.

make, gisement de matériel filonien utilisable; **m. of quartz,** dépôt de quartz; **m. up gas,** gaz d'appoint (pétrole); **m. up water,** eau d'appoint (pétrole).

malachite, malachite (*minér*.).

malacolite, malacolite (= variété de diopside translucide).

malacology, malacologie (étude des Mollusques).

malacon, zircon altéré brun rouge, malacon.

Malacostraca, Malacostracés.

Malaspina type glacier, glacier de piedmont, glacier de type alaskien.

malchite, malchite (microdiorite à grain très fin) (*pétrogr*.).

maldonite, maldonite (var. de minerai aurifère bismuthé d'Australie).

malleability, malléabilité.

Malm, Malm (série, ou époque Jurassique sup.).

malpais, région volcanique désolée (Espagne, Mexique, Sud-Ouest des E.U.).

maltha, malthe (var. de bitume).

malthenes, malthènes.

mamelon, mamelon (*pal*., Échinodermes).

mamelonated, mamelonné (*géogr*.).

Mammal, Mammalia (pl.), Mammifère(s).

mammillary, mamelonné (pour agrégats minéraux).

mammilated rock, roche moutonnée.

mammilated surface, surface mamelonnée (par action des glaciers).

Mammoth, Mammouth.

mandibule, mandibule.

mandrel, mandril, **1.** pic à deux pointes (*mine*); **2.** mandrin.

manganese, manganèse; **m. aluminium garnet, manganese garnet,** grenat spessartine (*minér*.); **m. hydrate,** psilomélane (*minér*.); **m. nodule,** nodule de manganèse; **m. spar,** dialogite, rhodonite (*minér*.); **m. steel,** acier au manganèse.

manganesian, manganésien.

manganiferous, manganésifère.

manganite, manganite, acerdèse (*minér*.).

manganocalcite, manganocalcite, dialogite calcique (*minér*.).

manganous, manganeux.

mangrove, mangrove; **m. tree,** palétuvier; **m. soil,** m., sol de m.; **m. swamp,** étang des palétuviers.

manhole, **1.** passage pour un homme (*mine*); **2.** regard.

manjak, manjak (var. de bitume).

man machine, échelle mécanique (*mine*).

mantle, **1.** manteau terrestre; **2.** manteau (du corps des mollusques); **m. derived magma,** magma d'origine mantélique; **m. line,** ligne palléale (cf. 2); **m. plumes,** points chauds du manteau; **m.**

rock, sol formé de débris, d'altération, régolithe; **inner m.,** manteau interne; **outer m.,** manteau externe; **waste m.,** couverture détritique.

mantled by, recouvert par.

mantled gneiss dome, dôme de gneiss avec couverture de terrains métamorphiques.

many-celled, pluricellulaire.

map, carte; **m. drawing,** minute cartographique; **m. grid,** quadrillage; **m. scale,** échelle de la c.; **m. series,** ensemble cartographique; **aero-radioactivity m.,** c. de radioactivité levée par avion; **base m.,** fond de c.; **contour m.,** c. en courbes de niveau; **dissected m.,** c. montée sur toile; **geologic m.,** c. géologique; **hypsographic m.,** c. hypsométrique; **isogon m.,** c. d'isogrammes; **outline m.,** fond de c.; **paleogeographic m.,** c. paléogéographique; **paleotectonic m.,** c. paléotectonique; **raised m.,** c. en relief; **relief m.,** c. topographique; **sketch m.,** c. schématique; **structural m.,** c. structurale; **topographic m.,** c. topographique; **m. projections,** projections cartographiques; **conformal m. projection,** projection conforme; **conical m. projection,** projection conique; **cylindrical m. projection,** projection cylindrique; **equal area m. projection,** projection équivalente; **equidistant m. projection,** projection équidistante; **perspective m.,** projection perspective; **stereographic m.,** projection stéréographique.

map (to), cartographier.

mappable, cartographiable.

mapper, cartographe.

mapping, cartographie; **m. camera,** chambre métrique; **m. photography,** photographie (aérienne) pour levé de carte; **aerial m.,** c. aérienne.

marble, marbre; **m. pavement,** dallage de marbre; **m. quarry,** carrière de marbre.

marbled, marbré, jaspé; **m. gley-like soil,** sol à pseudo-gley; **m. soil,** sol bigarré, pseudo-gley.

marcasite, marcassite, marcasite, marcassite (*minér.*).

march, marche (vers l'avant); **m. of a glacier,** avancée d'un glacier.

marching dune, dune mobile.

mare, maria (pl.), mer lunaire; **m. basin,** dépression des mers lunaires; **m. material,** matériel (ferromagnésien des mers lunaires).

margarite, margarite (var. de mica dioctahédrique).

margaritiferous, perlier.

margin, bord, limite, marge, rive; **m. of a glacier,** rive d'un glacier; **continental m.,** marge continentale; **sea m.,** zone littorale, littoral; **active m.,** marge active.

marginal, marginal; **m. crevasse,** crevasse marginale, rimaye; **m. facies,** faciès marginal, côtier; **m. fissure,** fracture péri-intrusive; **m. moraine,** moraine « marginale » = frontale; **m. sea,** mer bordière, limitrophe; **m. trench,** fosse océanique de la marge continentale.

margino-littoral, margino-littoral.

marialite, marialite (terme sodique de la série des wernérites) (*minér.*).

marigram, marégramme.

marikarst, karst supra-tidal.

marine, marin; **m. abrasion,** érosion m.; **m. band,** intercalation m.; **m. bed,** couche m., sédiment m.; **m. cave,** grotte littorale; **m. cutterrace,** plaine d'érosion m., plate-forme littorale; **m. denudation,** érision, abrasion m.; **m. deposit,** sédiment m.; **m. ecosystem,** écosystème m.; **m. environment,** milieu m.; **m. erosion,** érosion m.; **m. facies,** faciès m.; **m. formations,** formations m.; **m. geology,** géologie m.; **m. gyttja,** boue m.; **m. marsh,** marais littoral; **m. methods,** études en mer; **m. platform,** plate-forme en mer; **m. sediment,** sédiment m.; **m. sill,** seuil sous-marin; **m. terrace,** terrasse m.

Marinesian, Marinésien = Bartonien (Éocène sup.).

maritime, maritime.

mariupolite, mariupolite (var. de syénite alcaline à néphéline et albite).

mark, marque, signe empreinte (figure sédimentaire); **backwash m.,** marque de retour de vague; **brenching m.,** marque détritique de courant; **brush m.,** figure de frottement; **chatter m.,** rabotage glaciaire, marques en croissants; **flute m.,** cavité allongée formée par érosion; **flute rill m.,** cavité allongée formée par ruissellement; **glacial m.,** « coup de gouge »; **prod m.,** prod m. (marque de coup d'objet tombant sur le fond avec un angle fort); **rill m.,** rigole de plage; **ripple m.,** ride; **scour m.,** figure d'affouillement; **swash m.,** marque de vague déferlante; **tool m.,** figure sédimentaire formée par un objet dans le courant; **wave m.,** marque de vagues.

mark (to) on a map, indiquer sur une carte.

mark (to) out a route, tracer un itinéraire.

marker, marqueur, horizon repère, horizon sismique; **m. band,** niveau repère; **m. bed, marker horizon,** lit repère, horizon repère, marqueur stratigraphique.

markfieldite, markfieldite (var. de diorite).

markstone, pierre de bornage.

marl, marne; **m. lake,** lac à dépôts marneux; **m. pit,** marnière; **m. shale,** schiste argileux calcaire; **m. soil,** sol marneux; **cherty m.,** marne à silex; **dolomitic m.,** marne dolomitique.

marlaceous, marneux.

marling, marnage.

marlstone, marlite, marne indurée.

marly, marneux.

marmorize (to), transformer en marbre (un calcaire par métamorphisme).

marmorosis, marmorisation.

marsh, marais; **m. deposit,** sédiment marécageux polustre; **m. gas,** gaz des m., méthane; **m. land,** terrain marécageux; **m. ore,** limonite; **salt m.,** m. salant; **tidal m.,** m. maritime.

marshy, marécageux; **m. waste land,** fagne, terrain marécageux.

marsquake, séisme martien.

Marsupalia, Marsupidés.

martian geology, géologie de la planète Mars; **m. ice caps,** calottes glaciaires martiennes; **m. volcanism,** activité volcanique m.; **m. water-erosion** (canyons), érosion de l'eau m.; **m. winds (dust)**, vents de poussière m.

martite, martite (pseudomorphose de magnétite en hématite).

mascon, site lunaire à forte attraction gravimétrique.

mass, masse, bloc amas, massif; **m. flow**, écoulement en masse; **m. mouvement**, mouvement de masse, glissement de terrain; **m. number**, nombre de masse; **m. spectrography**, spectrographe de masse; **m. susceptibility**, susceptibilité magnétique par unité de masse; **m. wasting**, mouvement de masse, solifluxion; **m. copper**, cuivre natif (E.U.); **m. of ore**, amas de minerai.

massicot, massicot, litharge (*minér.* PbO).

massif, massif montagneux.

massive, 1. massif, ive; 2. homogène; 3. épais; 4. sans structure cristalline définie, homogène; 5. en masse; 6. non stratifié; **m. deposit**, amas minéralisés.

mast, mât (de forage).

master, principal; **m. joint**, diaclase p.; **m. lode**, filon-mère, p.; **m. river**, rivière p.; **m. wind**, vent dominant.

mat (to), clayonner (la berge d'une rivière).

matched terrace (syn. paired terrace), terrasses couplées, appariées.

material, 1. matière, matériau; 2. matériel; **building m.**, matériaux de construction.

matlockite, matlockite (*minér.*).

matrix, matrice, gangue, ciment, pâte.

matrix-gem, pierre précieuse liée à sa gangue.

matrix-rock, nodules de phosphate.

matterhorn, pic montagneux pyramidal, en aiguille (nom du Cervin).

mattock, pioche, pic.

mattress, clayonnage.

mature, mûr, évolué (sédiment), parvenu au terme d'une évolution géomorphologique; **m. landscape**, paysage au stade de maturité; **m. sandstone**, grès évolué, mature; **m. sediment**, sédiment mature (ne contenant que des minéraux stables); **m. soil**, sol évolué, à horizons distincts.

matureland, région ayant atteint le stade de maturité.

maturity, maturité; **late m.**, maturité avancée.

maundril, pic à deux pointes.

maxilla, maxillaire (*pal.*).

maximum, 1. adj : maximum; 2. n : maximum; **m. capacity**, charge limite; **m. load**, limite de charge; **m. moisture capacity**, capacité en eau maximum; **m. thermometer**, thermomètre à maxima.

maxwell, maxwell (unité électromagnétique de flux magnétique = 1 gauss/cm²).

Mayanian (N. Am.), Mayanien (subdivision de l'Albertien, Cambrien moy.).

Maysvillian (N. Am.), Maysvillien (étage, Ordovicien sup.).

M-boundary, *M-discontinuity*, discontinuité de Mohorovicic.

M-crust, partie de la croûte sous-jacente à la discontinuité de Mohorovicic, épaisse de 8,3 km et basaltique.

meadow, prairie; **m. bog soil**, sol de p. marécageuse; **m. land soil**, sol marécageux noir; **m. ore**, minerai de fer des marais, limonite; **m. podzolic soil**, sol lessivé de p.; **m. soil**, sol de p. à gley (sol intrazonal formé sur plaine d'inondation).

mealy sand, sable limoneux.

mealy zeolite, natrolite, mésolite (*minér.*).

mean, 1. adj : moyen; 2. n : moyenne; **m. dessication**, écart moyen; **m. latitude**, latitude moyenne; **m. sea level (MSL)**, niveau moyen de la mer; **m. stress**, moyenne des tensions.

meander, méandre; **m. aperture**, ouverture du m.; **m. apex**, sommet du m.; **m. arc**, arc du m.; **m. belt**, zone à m., lit des m.; **m. core**, lobe de m.; **m. cross-over**, point d'inflexion du m.; **m. curvature**, courbure du m.; **m. cusp**, zone érodée par un m.; **m. lobe**, lobe de m.; **m. neck**, racine, pédoncule d'un m.; **m. scar**, échancrure de m., concavité de m.; **m. scroll**, lac d'ancien m.; **m. terrace**, terrasse de concavité de m.; **m. train**, train de m.; **m. valley**, vallée à m; **abandoned m.**, m. abandonné; **compound m.**, m. composé; **cutoff m.**, m. recoupé, réséqué; **enclosed m.**, m. encaissé; **entrenched m.**, m. encaissé, enfoncé sur place; **free m.**, m. libre; **incised m.**, m. encaissé; **inherited m.**, m. hérité.

meander (to), serpenter, décrire des méandres.

meandering stream, fleuve à méandres, fleuve dans sa maturité.

measure, 1. mesure; 2. lit, couche; **coal m.**, couches houillères (Carbonifère supérieur); **m. head**, galerie creusée dans diverses strates.

measure (to), 1. mesurer, arpenter; 2. doser; **m. a field**, arpenter un champ; **m. by the meter**, métrer; **m. solids**, cuber.

measured reserves, réserves estimées.

measurement, 1. mesure, mesurage; 2. dosage (chimie); **m. of land**, arpentage; **m. in meters**, métrage; **m. of solids**, cubage.

mechanical, mécanique; **m. analysis**, analyse granulométrique; **m. efficiency**, rendement m.; **m. origin**, origine m. (par opposition à chimique); **m. twinning**, mâcle d'origine tectonique, m.; **m. weathering**, désagrégation physique des roches.

medial, médian; **m. moraine**, moraine médiane.

median, médian; **m. mass (Zwischengebirge, Germ.)**, zone axiale d'un orogène (moins déformée), horst; **m. particle diameter**, diamètre moyen des particules (d'un sédiment); **m. ridge**, crête médio-océanique; **m. valley**, vallée centrale de la crête médio-océanique.

medicinal water, eau thermominérale.

Medinian (obsolete) (N. Am.), Médinien = Alexandrien (série ou époque, Silurien inf.).

mediophyric, roches porphyriques à phénocristaux (de taille maximum entre 1 et 5 mm).

mediosilicic, moyennement siliceux.

mediterranean red soil, sol rouge méditerranéen.

medium, 1. adj : moyen; 2. milieu; **m. grade**, teneur moyenne; **m. grained**, à grain moyen; **m. sized grain**, grain de taille moyenne; **m. sand**, sable moyen **a.** 0,250 à 0,500 mm (*sédimentol.*); **b.** 0,420 à 2,000 mm (E.U., *géotechnique*).

Medusa, Méduse (*pal.*).

meerschaum, 1. magnésite ou giobertite (*minér.*); **2.** sépiolite ou écume de mer (*minér.*).

meeting, 1. réunion, assemblée; **2.** rencontre (de deux rivières).

megacontinent, continent ancien de grandes dimensions (ex. : Pangaea, Laurasia, Gondwanaland).

megacryst, phénocristal.

megacylothem, mégacyclothème.

megafacies, mégafaciès.

megafauna, macrofaune.

megaflora, macrofossile.

megafossil, macroflore.

megalineament, alignement, structure linéaire de grande dimension.

megalith, mégalithe.

megalitic, mégalithique.

megalospheric, megaspheric, macrosphérique.

megaphenocrysts, mégaphénocristaux.

megaripple, mégaride, macroride grande ride ($\geqslant 0,6$ ou 1 m selon auteurs); **rhomboid m.,** mégaride rhomboïdale.

megascopic, mégascopique (à structure visible à l'œil nu).

megashear, faille à très grand déplacement horizontal (plus de 100 km).

megaspore, macrospore.

Megatherium, Megatherium (*pal.*).

Meiocene (obsolete), Miocène (série ou époque, Néogène inf.).

meionite, méionite (var. wernérite) (*minér.*).

meizoseismal, meizoséismique.

mela, (préfixe) noir, foncé; **m. basalt,** basalte mélanocrate; **m. diorite,** diorite mélanocrate; **m. gabbro,** gabbro mélanocrate (à plagioclases basiques, labradorite et bytownite).

melaconite, mélaconite, ténorite (*minér.*).

melanic, sombre, de couleur foncée.

melanite, mélanite (var. de grenat andradite).

melanization, noircissement par imprégnation d'humus (*pédol.*).

melanized, humifère; **m. gley loam,** limon h. à gley; **m. lateritic soil,** sol latéritique h.; **m. soil,** sol h.

melanocratic, mélanocrate.

melanterite, mélantérite (*minér.*, sulfate de fer hydraté).

melaphyre, mélaphyre (roche éruptive felsitique foncée).

melilite, mélilite (sorosilicate); **m. basalt,** basalte alcalin (à mélilite, au lieu de feldspath).

melioration, amendement (*pédol.*).

mellow, meuble, poreux.

mellowing, ameublissement.

melt, magma.

melt (to), fondre, se fondre, se dissoudre.

melteigite, melteïgite (var. syénite néphélinique).

melting, melt, fusion, fonte; **m. cup,** cuvette de fusion nivale; **m. ice,** glace fondante; **m. point,**

température de fusion; **snow m.,** fusion de la neige.

melts, produits fondus, laves.

meltwater, eau de fonte de neige ou de glace.

meltwater channel, chenal d'eau de fonte.

member, membre (unité lithostratigraphique).

mendip, 1. colline littorale, autrefois île; **2.** colline enfouie dégagée par l'érosion.

mendozite, mendozite, alun sodique (*minér.*).

menhir, menhir.

menilite, ménilite (var. d'opale).

Meotian, Méotien (étage, Miocène inf., région Mer Noire).

mephitic, méphitique, toxique.

mephitic gas, gaz méphitique.

Meramecian (N. Am.), Méramécien (étage, Mississipien sup.).

Mercator's projection, projection de Mercator.

mercurial, mercuriel; **m. barometer,** baromètre à mercure; **m. horn ore,** calomel; **m. thermometer,** thermomètre à mercure.

mercuric, mercurique; **m. sulphide,** sulfure de mercure, cinabre.

mercuriferous, mercurifère.

mercurify (to), extraire le mercure d'un minerai.

mercurous, mercureux; **m. chloride,** calomel, chlorure mercureux.

mercury, mercure; **m. ore,** minerai de m., cinabre; **m. sulphide,** cinabre.

mere (U.K.), petit lac ou marais.

merestone, pierre de bornage.

meridian, méridien; **first m.,** m.-origine; **magnetic m.,** m. magnétique; **prime m.,** m.-origine; **principal m.,** m. principal (E.U.); **standard m.,** m.-origine; **zero m.,** m.-origine.

mero, préfixe : partie, fraction de.

merocrystalline (syn. rypocrystalline), semi-cristallin (état d'une roche contenant à la fois des cristaux et une pâte amorphe).

merohedral, merohedric, mériédrique (*cristallo.*).

merohedrism, mérihédrie.

meromictic lake, lac méromictique.

Merostomata, classe des Mérostomes (*pal.*).

mesa, mésa, plateau.

mesenteries, cloisons radiales (des Coralliaires).

meseta, meseta, plateau.

mesh, maille; **m. analysis,** analyse granulométrique; **m. sieve,** tamis à mailles; **m. texture,** structure réticulée.

meshed, réticulé.

meso, préfixe : milieu de.

mesocratic, mésocrate, mésotype.

mesocrystalline, mésocristallin (diamètre des cristaux entre 0,20 et 0,75 mm).

mesoderm, mésoderme.

Mesodevonian, Mésodévonien.

mesogene, mésogène, issu de profondeur moyenne.

mesohaline, mésohalin, saumâtre.

mesolite, mésolite (var. de zéolite).

Mesolithic, Mésolithique (subdivision de l'Âge de la pierre).

101

mesolithic rock, roche neutre.

mesolittoral zone, estran.

mesonorm, norme minéralogique des roches métamorphiques de mésozone.

mesophyle, mésophyle (*pal.*).

mesosiderite, météorite semi-ferrifère (fer, nickel + hypersthène, anorthite, etc.).

mesosilexite, var. de silexite à minéraux foncés.

mesosphere, mésosphère.

mesostasis, mésostase (= ciment mylonitique entre les fragments d'une brèche).

Mesosauria, Mésosauriens (*pal.*).

mesotheca, mésothèque (*pal.*, Bryozoaires).

mesothermal, mésothermal.

mesotill, sol glaciaire intermédiaire entre gumbotil et silttil.

mesotype, 1. mésocrate, de constitution minéralogique équilibrée; 2. mésotype (var. zéolite).

Mesozoic, Mésozoïque (ère).

mesozonal, mésozonal.

mesozone, mésozone (dans le métamorphisme).

meta-, préfixe désignant le métamorphisme d'une roche.

metabasalt, métabasalte.

metabasite, métabasite (roche basique métamorphique).

metachemical metamorphism, métamorphisme chimique.

metaclay, ancienne argile métamorphisée.

metacrystal, metacryst, porphyroblaste, phénoblaste (dans roches métamorphiques).

metacrystic fabric, structure porphyroblastique.

metadiabase, métadiabase.

metadiorite, 1. diorite métamorphisée; 2. gabbro métamorphisé; 3. roche sédimentaire métamorphisée à structure de diorite.

metadolomite, dolomie métamorphique.

metagabbro, métagabbro.

metakaolin, kaolin artificiellement déshydraté.

metal, 1. métal; 2. minerai, roche (désuet); **m. bearing**, métallifère; **m. drift**, galerie au rocher (mine); **m. mine**, mine de métaux; **m. mining**, exploitation minière de métaux; **road m.**, matériaux d'empierrement.

metalimestone, calcaire métamorphique.

metalling, empierrement (génie civil).

metallic, métallique; **m. iron**, fer métal; **m. luster**, éclat m.; **m. vein**, filon métallifère.

metalliferous, métallifère.

metallization, métallisation.

metallize (to), métalliser (MEB).

metallogenetic, métallogénétique, métallogénique; **m. epoch**, époque favorable à la métallogénèse; **m. province**, province métallogénique.

metallogeny, métallogénie.

metallographer, métallographe.

metallographic, métallographique.

metallography, métallographie.

metallurgic, métallurgique.

metallurgy, métallurgie.

metamerism, métamérie (*pal.*).

metamict, métamicte (minéral radioactif).

metamorphic, métamorphique; **m. aureole**, auréole de métamorphisme; **m. differentiation**, différenciation m.; **m. facies**, faciès de métamorphisme; **m. grade, rank**, degré de métamorphisme; **m. gradient**, gradient, degré de métamorphisme; **m. rock**, roche m.; **m. schist**, schiste m.; **m. water**, eau associée au métamorphisme; **m. zone**, zone m.

metamorphism, métamorphisme; **m. aureole**, auréole de m.; **autometamorphism**, autométamorphisme; **burial m.**, métamorphisme d'enfouissement; **contact m.**, métamorphisme de contact; **dynamometamorphism**, dynamometamorphisme; **dynamothermal m.**, m. général; **load m.**, m. général; **regional m.**, m. général; **retrograde m.**, rétrométamorphisme; **shear m.**, m. de tension; **sub-sea m.**, m. de dorsale sous-marine.

metamorphous, métamorphique.

metapepsis (obsolete), métamorphisme régional, général (désuet).

metaquartzite, quartzite métamorphique.

metargilite, pélite faiblement métamorphisée (pas de schistosité).

metarhyolite, métarhyolite.

metasediment, roche sédimentaire métamorphisée.

metashale, schiste argileux, pélite faiblement métamorphisée (pas de schistosité).

metasilicate, métasilicate (désuet), sel d'acide métasilicique.

metasilicic, métasilicique.

metasomatic, métasomatique.

metasomatism, metasomatosis, métasomatose, remplacement, substitution.

metasomatite, roche formée par métasomatose.

metasome, métasome.

metastable, métastable, instable (équilibre chimique).

metastase, metastasis, metastasy, 1. déplacement latéral de la croûte terrestre (tectonique des plaques); 2. transformation par recristallisation ou dévitrification.

metatropy, métatropie (changement des caractères physiques d'une roche).

metatype, métatype (*pal.*).

metavolcanics, roches volcaniques métamorphisées.

Metazoa, Métazoaires.

meteor, météorite; **m. shower**, averse météoritique.

meteoric, météorique; **m. iron**, fer m.; **m. stone**, météorite; **m. water**, eau atmosphérique.

meteorite, météorite, aérolithe.

meteorite (or meteor) crater, cratère météorique.

meteoritic, météoritique.

meteorologic, meteorological, météorologique; **m. station**, station m.

meteorology, météorologie.

meter, 1. compteur, jaugeur; 2. mètre; **m. rod**, baguette d'arpentage; **flow m.**, débitmètre.

meterage, mesurage.

methane, méthane; **m. series**, paraffènes.

methanometer, grisoumètre.

method, méthode; **m. of working,** m. d'exploitation d'un gisement.
metric, metrical, métrique; **m. carat,** carat m. = 200 milligrammes; **m. quintal,** quintal m. = 100 kilogrammes; **m. system,** système m; **m. ton,** tonne m. = 1 000 kilogrammes.
miargyrite, miargyrite (*minér.*, sulfure).
miarolithic, miarolithique; **m. cavity,** cavité miarolithique.
mica, mica; **m. bearing,** contenant du m.; **m. book,** m. clivable; **m. diorite,** diorite micacée; **m. flake,** paillette de m.; **m. sheet,** lamelle de m.; **m. syenite,** syénite à m.; **m. trap,** lamprophyre; **rhombic m.,** m. phlogopite.
micaceo-calcareous, micacé et calcaire.
micaceous, micacé; **m. chalk,** tuffeau; **m. flagstone,** grès m. en plaquettes; **m. iron-ore,** hématite m., spécularite; **m. sandstone,** grès m.; **m. schist,** micaschiste; **m. structure,** structure feuilletée.
micaphyre, roche porphyrique à phénocristaux de mica.
micaschist, micaschiste.
micaschistose, micaschistous, micaschisteux.
micaslate, micaschiste.
micatization, micatisation, altération en mica.
micrite, micrite (boue calcaire).
micro (prefix), micro- (préfixe), petit.
microanalysis, microanalyse.
microbreccia, microbrèche.
microchemical, microchimique.
microclastic, microdétritique.
microclimate, microclimat.
microcline, microcline (feldspath potassique).
microconglomerate, microconglomérat.
microcosm, microcosme d'un point de vue détaillé.
microcross lamination (cross-bedding), stratifications entrecroisées de petites dimensions.
microcrystal, microcristal.
microcrystalline, microgrenu; **m. limestone,** calcaire microcristallin.
microcryptocrystalline, très finement cryptocristallin.
microdiabase, diabase aphanitique.
microdiorite, microdiorite.
microearthquake, microtremblement de terre.
microfabric, microstructure.
microfacies, microfaciès.
microfault, microfaille, microdécrochement.
microfauna, microfaune.
microfelsitic (see microcryptocrystalline), microfelsitique.
microfissuration, microfissuration.
microflora, microflore.
microfluidal, microfluxion, microfluidal.
microfold, micropli.
microfoliation, microschistosité.
microforms, micromodèles.
microfossil, microfossile.
microgabbro, microgabbro.
microgeology, microgéologie.
micrograined texture, texture microgrenue.
microgranite, microgranite.

microgranitic, microgranitique.
microgranitoid, microgranitoïde.
microgranodiorite, microgranodiorite.
microgranular, microgrenu.
microgranulitic, microgranulitique; **m. porphyry,** granulophyre.
micrographic, micrographique.
microlite, microlith, microlite (*minér.*).
microlitic, microlithic, microlithique.
microlog(ging), microdiagraphie de la résistivité.
micrometric (obsolete), microméritique (désuet = microcristallin).
micrometeorite, micrométéorite.
micrometer, micromètre; **m. eye-piece,** oculaire m.; **m. screw,** vis micrométrique.
micrometric, micrometrical, micrométrique; **m. screw,** vis micrométrique.
micrometry, micrométrie.
micromineralogy, minéralogie microscopique.
micromorphology, micromorphologie (*pédol.*).
micron, micron = 0,001 mm.
micronutrient, oligoélément.
microorganism, microorganisme.
micropaleontology, micropaléontologie.
micropegmatite, micropegmatite.
micropegmatitic, micropegmatitique.
microperthite, microperthite.
microphagous, microphage (*pal.*).
microphenocryst, phénocristal visible seulement au microscope.
microphone, microphone.
microphotograph, microphotography, microphotographie.
microphyric, microporphyric, microphyrique, microporphyrique.
microplate, microplaque.
micropore, micropore.
microporosity, microporosité.
microprobe, microsonde; **electron m.,** microsonde électronique.
microscope, microscope; **m. slide,** lame de verre pour examen microscopique; **binocular m.,** m. binoculaire; **crushing m. stage,** platine à écrasement; **electron m.,** m. électronique; **glass m. slide,** lame de verre pour examen microscopique; **light m.,** m. optique; **metallurgical m.,** m. métallographique; **mineragraphic m.,** m. à lumière réfléchie; **photonic m.,** m. optique; **petrologic m.,** m. polarisant; **polarization m.,** m. polarisant; **reflected light m.,** m. à lumière réfléchie; **scanning m.,** m. électronique à balayage.
microscopic, microscopical, microscopique; **m. examination,** examen m.
microscopically, microscopiquement.
microscopy, microscopie; **phase contrast m.,** m. à contraste de phase.
microsection, plaque mince.
microseism, microséisme.
microseismic, microseismical, microsismique.
microskeleton, squelette du sol.
microsphere, microsphère (*pal.*).

microspheric, microsphérique.

microspherulithic, microsphérolithique.

microsplitter, microséparateur d'échantillons.

microstructure, microstructure.

microsyenite, microsyénite.

microtectonics, microtectonique.

microtexture, microtexture, microstructure.

microwave, microonde.

mid, moyen; **m. Atlantic ridge,** crête médio-atlantique; **m. ocean ridge or oceanic ridge,** crête médio-océanique, chaîne médio-océanique; **m. oceanic rift,** fossé médio-océanique, rift médian; **m. workings,** travaux à mi-pente (*mine*).

midden (*archeo*), midden, accumulation ou colline de débris domestiques; **shell m.,** accumulation de coquilles anthropogènes.

middle, 1. ad. : moyen, intermédiaire, central, médian; **2.** n : milieu; **m. latitude,** latitude moyenne; **m. sized,** de dimension moyenne.

middlings, minerai de seconde qualité obtenu par lavage.

mid-gley soil, sol à gley moyennement profond.

midwall, cloison moyenne (*mine*).

migma, 1. mélangé (préfixe); **2.** matériel provenant de la différenciation granitique.

migmatite, migmatite.

migmatization, migmatisation.

migration, migration; **m. of divides,** m. des lignes de partage (*géogr.*); **primary m.,** m. primaire; **secondary m.** m. secondaire.

migratory dune, dune mobile.

mil, millième de pouce anglais = 0,025 mm.

mild, doux, tempéré (*météo.*).

mildness, douceur (de la température).

mile, mille; **statute m.,** mille anglais = 1609,31 m; **nautical m.,** mille marin = 1 852 m.

mileage, distance en milles, millage.

milestone (to), jalonner, borner.

milk of lime, lait de chaux.

milky quartz, quartz laiteux.

mill, 1. usine; **2.** moulin; **3.** broyeur; **4.** cheminée à minerai; **5.** concentrateur (minier); **m. hole,** cheminée à minerai; **m. result,** rendement du bocard; **m. stone,** pierre à aiguiser, pierre de meule; **sea m.,** moulin de mer (*karstol.*).

mill (to), 1. moudre, broyer; **2.** fraiser.

mill (to) ore, bocarder du minerai.

milled ore, minerai bocardé.

millerite, millérite (*minér.*, sulfure de nickel).

millibar, millibar (*météo.*).

millidarcy, millidarcy.

milling, broyage, fraisage; **m. ore,** minerai de broyage.

Miliolacea, Miliolidés.

Miller indices, indices de Miller (*minér.*).

millman, bocardeur, pique-mine.

millstone grit, 1. grès meulier; **2.** étage stratigraphique anglais correspondant au Namurien (Carbonifère moy.).

mima pound, petite colline circulaire, 3 à 30 m de diamètre, construite par les rongeurs.

mimetite, mimétite, mimétésite, mimétèse (*minér.*).

Mindel (*Eur.*), Mindel (glaciation de) : Pléistocène.

Mindel-Riss, Mindel-Riss : interglaciaire, cf Tyrrhénien.

mine, mine, exploitation; **m. adit gallery,** galerie au jour; **m. captain,** chef-mineur; **m. can,** wagonnet de m.; **m. chamber,** chambre de m., trou de m.; **m. dial,** boussole de mineur; **m. digger,** mineur; **m. engineer,** ingénieur civil des mines (G.B.); **m. face,** front de taille; **m. field,** district minier; **m. fires,** feux de mines; **m. foreman,** contremaître de mines; **m. head,** front de taille; **m. hoist,** treuil d'extraction; **m. inspector,** ingénieur des m.; **m. inspection,** services des m.; **m. level,** galerie de niveau de m.; **m. levelling,** nivellement dans les m.; **m. master,** chef mineur; **m. resistance,** résistance de la m. (au passage d'un courant d'air); **m. run,** tout venant; **m. salt,** sel gemme; **m. shaft,** puits de m.; **m. stone,** minerai; **m. surveying,** levé minier; **m. timber,** bois des mines; **m. transit,** théodolite à boussole.

mine (to), creuser, fouiller, exploiter.

mineable, minable, exploitable.

miner, mineur, ouvrier du fond.

miner's bar, barre à mine; **m. pick,** pic de mineur; **m. shovel,** pelle de mineur.

mineragraphy, 1. minéralogie; **2.** étude microscopique des minerais métalliques.

mineral, 1. minéral; **2.** minerai (*mine*); **m. amber,** ambre; **m. assemblage,** cortège minéralogique, paragenèse; **m. bearing,** minéralisé; **m. belt,** zone minéralisée; **m. blossom (drusy quartz),** quartz en géodes; **m. caoutchouc,** élatérite, bitume; **m. charcoal,** charbon fossile; **m. claim,** concession minière; **m. cleavage,** clivage type d'un système minéralogique; **m. crop,** échantillon minéralogique; **m. deposit,** gisement minéral, gîte minéral; **m. exploration,** prospection minière; **m. facies,** faciès minéral, faciès métamorphique; **m. inclusion,** inclusion minérale; **m. naphta,** pétrole; **m. oil,** pétrole; **m. parent rock,** roche mère minérale; **m. pitch,** asphalte minéral; **m. resource,** ressources minérales; **m. rights,** droits miniers; **m. spring,** source minérale; **m. tar,** goudron minéral; **m. vein,** filon minéralisé; **m. water,** eau minérale; **m. wax,** cire minérale, ozokérite; **m. white,** blanc de Meudon, gypse; **accessory m.,** minéral accessoire; **authigenic m.,** minéral authigène; **facies m.,** minéral indicatif de faciès; **guest m.,** métasome (*minér.*); **heavy m.,** minéral lourd ou dense; **host m.,** palasome; **light m., 1.** minéral léger; **2.** minéral clair; **mafic m.,** minéral ferro-magnésien.

mineralizable, minéralisable.

mineralization, minéralisation.

mineralize (to), minéraliser.

mineralized zone, zone minéralisée.

mineralizer, agent minéralisateur.

mineralizing fluid, fluide minéralisateur.

mineralogic, mineralogical, minéralogique.

mineralogically, minéralogiquement.
mineralogist, minéralogiste.
mineralography, minéralographie.
mineraloid, minéraloïde.
minerogenic, d'origine minérale.
minerogenetic, minerogenetical, minérogénétique : métallogénique.
minery, région minière.
minestuff, pierre de mine, minerai.
minette, minette (syénite micacée : var. de lamprophyre).
minimum, 1. adj : minimum; 2. n : minimum (thermique, etc.), faible valeur de gravité (anomalie); **m. thermometer,** thermomètre à minima; **m. time path,** tracé le plus rapide de propagation (sismique).
mining, exploitation minière, travaux miniers; **m. act,** loi minière; **m. appliances,** matériel de mines; **m. bucket,** benne d'extraction; **m. claim,** concession minière; **m. code,** code minier; **m. concession,** concession minière; **m. contractor,** entrepreneur de travaux; **m. crew,** équipe de mineurs; **m. debris,** rebuts d'exploitation; **m. district,** région minière; **m. engineering,** technique minière; **m. geology,** géologie minière; **m. hole,** trou de mine, chambre de mine; **m. field,** gisement minier; **m. lease,** bail minier; **m. licence,** concession de mines; **m. outfit,** matériel de mines; **m. region,** région minière; **m. regulation,** règlement minier; **m. retreating,** exploitation en retraite, exploitation en retour, dépilage en retraite; **m. shaft,** puits de mine; **m. timber,** bois de mine; **m. village,** coron; **m. work,** travaux miniers.
miniphyric, microporphyrique.
minium, minium (*minér.*).
minor-element, élément trace.
minor feature, forme mineure (*géogr.*).
minute folding, plissotement, plication.
minverite, minvérite (*pétro.*).
Miocene, Miocène (époque ou série, Néogène inf.).
miogeosyncline, miogéosynclinal.
miohaline, saumâtre.
mirabilite, mirabillite, sel de Glauber.
mire, 1. bourbier, fondrière; 2. boue, vase.
mirror, réflecteur (*géoph.*).
mirror-stone, muscovite, mica muscovite.
miscibility, miscibilité; **m. gap,** immiscibilité.
miscible, miscible, mélangeable.
misfit stream, rivière inadaptée (ou sur-adaptée).
mispickel, mispickel, arsénopyrite.
Mississippian (N. Am.), Mississipien (période du Carbonifère).
Missourian (N. Am.), Missourien (étage, Pennsylvanien sup.).
missourite, missourite (syénite alcaline).
mist, 1. brume, embrun; 2. buée.
mix, mélange.
mix crystal, cristal mixte (composé de divers constituants isomorphes), solution solide.
mix (to), mélanger, mêler, malaxer.
mixability, miscibilité.

mixable, mixible, miscible.
mixed, mixte; **m. base crude oil,** pétrole brut à base m.; **m. layer clay mineral,** minéral interstratifié; **m. rendzina,** sol rendziniforme m.; **m. soil,** sol mélangé; **m. volcanoe,** volcan m. (à projections et coulées).
mixing, 1. mélange; 2. malaxage.
mixture, mélange.
moat, dépression, fossé interne dans un récif corallien.
mobile, mobile; **m. belt,** zone orogénique; **m. dune,** dune m.; **m. sand,** sable m.
mobility, mobilité (d'un élément).
mobilisation, mobilisation.
mock lead, more ore, sphalérite, blende.
mock quartz, quartz zoné (utilisé comme gemme).
modal, modal; **m. class,** classe modale; **m. composition,** composition m.; **uni-m.,** uni-modal; **multi-m.,** multi-modal; **m. analysis,** analyse modale.
mode, 1. mode (statistique); 2. méthode; **m. of a rock,** composition minéralogique réelle exprimée quantitativement.
model, modèle (géophysique, informatique, etc.).
moder gley soil, sol à gley à humus de moder.
modulate (to), moduler (une onde).
modulus, module, coefficient; **m. of compression,** module de compression; **m. of rupture,** module de rupture; **m. of volume elasticity,** module d'élasticité; **bulk m.,** module de compression; **elasticity m.,** module d'élasticité; **rigidity m.,** module de rigidité; **shear m.,** module de cisaillement.
mofette, mofette, gaz nocifs (volcanisme).
mogote, relief résiduel calcaire.
Mohawkian (N. Am.), Mohawkien (étage, Ordovicien moy.).
Mohnian (N. Am.), Mohnien (étage, Miocène).
Moho, discontinuité de Mohorovičić.
Mohorovičić discontinuity, discontinuité de Mohorovičić.
Moh's scale, échelle de dureté de Mohs.
Moinian, Moinien (subdivision du Précambrien d'Écosse).
moist soil, sol humide, sol hydromorphe.
moisten (to), humidifier, mouiller, humecter.
moistness, humidité.
moisture, humidité; **m. capacity,** capacité en eau; **m. content,** état hygrométrique; **m. equivalent,** équivalent d'humidité; **m. proof,** à l'épreuve de l'humidité; **m. tension,** force de rétention de l'eau.
mol, mole, molécule-gramme.
molar, molaire; **m. volume,** volume moléculaire.
molasse, molass, molasse, sédiments clastiques postorogéniques, « Nagelfluh ».
mold (US), moule, moulage; **external m.,** moule externe; **internal m.,** moule interne; **natural m.,** moulage laissé après dissolution du fossile; **open m.,** moulage externe.
molding, moulage; **m. sand,** sable de fonderie.
mole, môle, saillie, promontoire du littoral.

molecular, moléculaire; **m. attraction,** attraction m.; **m. bond,** liaison m.; **m. replacement,** minéralisation molécule par molécule d'une substance organique; **m. spectroscopy,** spectroscopie m.; **m. weight,** poids m.

molecule, molécule.

molehill, taupinière.

mollassic, molassique.

mollisol, mollisol (périglaciaire).

mollusc, mollusca, mollusk (U.S.), mollusque; **m. amphineura,** mollusques amphineures; **m. scaphopoda,** mollusques scaphopodes; **m. gastropoda,** mollusques gastéropodes; **m. lamellibranchiata,** mollusques lamellibranches; **m. pelecypoda,** mollusques lamellibranches; **m. cephalopoda,** mollusques céphalopodes; **boring m.,** mollusque lithophage.

Mollweide projection, projection cartographique de Mollweide.

molten, fondu, en fusion; **m. magma,** magma fondu; **m. iron,** fonte en fusion.

molybdenite, molybdénite.

molybdenum, molybdène.

molybdite, molybdite.

molybdic ocher, molybdénocre.

moment of a force, moment d'une force.

monadnock, butte résiduelle, monadnock.

monazite, monazite (*minér.*).

monchiquite, monchiquite (*pétro.*).

monitor, appareil de surveillance.

monitoring, surveillance, prévention, contrôle (*mine*).

monkey drift, galerie de prospection (*mine*).

mono, 1. mono- (préfixe); 2. isolé, seul.

monoaxial, uniaxe; **m. crystal,** cristal uniaxe.

monochromatic, monochromatique.

monochromator, monochromateur (*minér.*).

monoclinal, monoclinal; **m. flexure,** flexure m.; **m. fold,** pli m.; **m. scarp,** escarpement tectonique; **m. stream,** rivière subséquente.

monocline, 1. monoclinal; 2. flexure de couches subhorizontales.

monoclinic, monoclinique; **m. system,** système m.

monoclinous, monoclinal.

Monocots, Monocotyledoneae, Monocotylédones (*paléobot.*).

monocular, monoculaire.

monocyclic, monocyclique (*pal.*).

monogenetic, monogenic, monogénique; **m. breccia,** brèche m.; **m. conglomerate,** conglomérat m.; **m. soil,** sol m.; **m. volcano,** volcan m.

monogeosyncline, monogéosynclinal.

monoglaciation, glaciation à une seule avancée glaciaire.

monograph, monographie.

monolith, monolithe (roche mise en relief par érosion).

monometric system, système cubique.

monomict rock, roche détritique monominérale.

monomictic lake, lac monomictique.

monomineral rock, monomineralic rock, roche monominérale.

monomorphic, monomorphe.

monomyarian, monomyaire, à un seul muscle (*pal.*).

monophase, monophasé.

monophyletic, monophylétique.

Monoplacophore, Monoplacophore (*pal.*).

monorefringence, monoréfringence.

monorefringent, monoréfringent; **m. crystal,** cristal monoréfringent.

monoschematic rock or deposit, roche ou gisement minéral de structure identique dans toute sa masse.

monosymmetric system, système monoclinique.

monothem, unité stratigraphique (peu usité : ~ sous-étage).

Monotremata, Monotrèmes (*pal.*).

monotype, monotype (= holotype décrit à partir d'un seul taxon).

monotypical, monotypique (relatif à un genre comprenant une seule espèce ou à une espèce décrite sur un seul spécimen).

monovalent, monovalent.

mons (pl. montis, montes), mont (sur les planètes).

monsoon, mousson.

monsoonal, saisonnier.

Montana Group (E.U.), groupe du Montana (Crétacé sup.).

montebrasite, montébrasite (*minér.*).

Montian, Montien (étage, Paléocène inf.).

monticellite, monticellite (*minér.*).

monticle, monticule.

montmorillonite, montmorillonite (groupe ou variété d'argile).

monzodiorite, monzodiorite.

monzogabbro, monzogabbro.

monzonite, monzonite (*pétro.*).

monzonitic, monzonitique.

monzosyenite, monzosyénite.

moon, lune; **m. circus,** cirque lunaire; **moonrock,** roche d'origine lunaire; **moonquake,** séisme l.; **moonscape,** relief lunaire; **moonstone,** adulaire, pierre de lune.

moor, 1. lande; 2. terrain marécageux; **m. band pan,** croûte pédologique ferrugineuse; **m. coal,** variété de lignite tendre; **m. land,** 1. adj : couvert de landes; 2. lande; **m. pan,** niveau concrétionné humique; **m. peat,** tourbe formée de mousses; **m. stone,** granite de Cornouailles.

moory, marécageux, terrain couvert de landes.

morainal, morainique; **m. lake,** lac m.; **m. rampart,** arc morainique.

moraine, moraine; **border m.,** m. marginale; **deposited m.,** m. déposée; **dump m.,** m. terminale; **end m.,** m. frontale; **englacial m.,** m. interne; **flank m.,** m. riveraine, latérale; **frontal m.,** m. frontale; **glacial m.,** m. glaciaire (peu usité); **ground m.,** m. de fond, till de fond; **intermediate or interlobate m.,** m. interlobaire; **lateral m.,** m. latérale; **medial m.,** m. médiane; **push m.,** m. de poussée; **reces-**

sional m., m. de retrait; **retreatal m.**, m. de retrait; **stadial m.**, stade morainique; **subglacial m.**, m. de fond, till de fond; **superficial or superglacial m.**, m. superficielle; **surface m.**, m. superficielle; **transverse m.**, m. transversale; **terminal m.**, m. frontale; **terrace m.**, m. fluvioglaciaire; **weathered m.**, m. altérée.

morainic, morainal, morainique; **m. lake,** lac morainique; **m. loop,** vallum arqué.

morass, marais, fondrière; **m. ore,** minerai de fer des marais.

morganite, morganite (var. de béryl).

morion, quartz morion (fumé).

morphochronology, morphochronologie, histoire géomorphologique.

morphogenesis, morphogénèse.

morphogeny, morphogénie.

morphographic map, carte géomorphologique.

morphologic, morphological, morphologique.

morphologically, morphologiquement.

morphologic unit, 1. strate définie par des caractères morphologiques; 2. surface de sédimentation, surface d'érosion.

morphology, morphologie.

morphometry, morphométrie.

morphoscopy, morphoscopie.

morphostructure, morphostructure.

morphotectonic, morphotectonique.

Morrowan (N. Am.), Morrowien (étage, Pennsylvanien inf.).

mosaic, mosaïque (de photographies aériennes); **m. gold,** bisulfure d'étain; **m. texture,** structure en mosaïque.

Moscovian (U.S.S.R.), Moscovien (étage, Carbonifère moy.).

mosor, relief résiduel, monadnock.

moss, 1. marais, tourbière; 2. mousse; **m. agate,** agate dendritique; **m. mor,** humus brut de mousse; **m. peat,** tourbe constituée de mousses; **raised m.,** tourbière haute.

mother, mère; **m. gate,** galerie principale (*mine*); **m. liquor,** solution résiduelle (après cristallisation des minéraux); **m. lode,** filon m.; **m. of coal,** charbon fossile; **m. of emerald,** m. d'émeraude; **m. of pearl,** nacre de perle; **m. of rock,** roche m.; **m. water,** eaux mères.

mottle, 1. tache, marbrure; 2. nodule sédimentaire.

mottled, tacheté, bigarré, bariolé; **m. clay,** argile bigarrée; **m. limestone,** calcaire à tubulures de dolomite; **m. structure,** structure mouchetée.

mottling, marbrure (*pédol.*).

mould (U.K.), 1. moule; 2. moule (externe); **external m.,** moule externe; **internal m.,** moule interne.

mouldered rock, roche décomposée, altérée.

moulder's sand, sable de fonderie, sable de moulage.

moulding, moulage; **m. sand,** sable de moulage, sable de fonderie.

moulin, moulin kame, moulin, accumulation fluvioglaciaire conique, colline fluvioglaciaire.

mound, monticule, butte, tertre.

mount, 1. mont, montagne; 2. montage microscopique; 3. monture (d'un stéréoscope); **sea mount,** guyot, volcan sous-marin.

mountain, montagne; **m. blue,** azurite, minerai bleu de cuivre; **m. brown ore,** limonite; **m. building,** orogénèse; **m. chain,** chaîne de m.; **m. cork,** amiante; **m. crystal,** cristal de roche; **m. flax,** amiante; **m. glacier,** glacier de type alpin; **m. green,** malachite; **m. leather,** amiante; **m. limestone,** calcaire d'âge carbonifère inférieur (G.B.); **m. mahogany,** obsidienne brune (G.B.); **m. meal,** diatomite; **m. milk (lublinite),** var. spongieuse de calcite; **m. pediment,** pédiment d'une m.; **m. side,** flanc de m.; **m. slope,** pente d'un versant montagneux; **m. soap,** var. d'alloysite hydratée; **m. soil,** sol forestier de m.; **m. stream,** torrent; **m. system,** chaîne majeure de m., en Andes, Hymalaya; **m. tallow (hatchettine),** hatchettite; **m. track (of a stream),** tronçon supérieur (d'un cours d'eau); **m. wood,** amiante compacte; **m. waste,** éboulis.

mountainous, montagneux.

mounting, monture, montage; **m. of a lens,** monture d'un objectif.

mouse-eaten quartz, quartz corrodé à cavités.

Moustierian (Eur.), Moustiérien (subdivision du Paléolithique moy.).

mouth, 1. orifice, embouchure, entrée; 2. bouche, gueule (*pal.*); **m. of a drift,** entrée d'une galerie; **bay m.,** embouchure d'une baie.

move-out (or stepout time), augmentation du temps dû à la distance du point de tir à la trace considérée (*géoph.*).

moving drift, moraine mouvante.

muck, 1. gadoue, déblais, fumier; 2. sol organique (avec 40 % de matières organiques), sol humifère; 3. morts-terrains de recouvrement.

mucky peat, sol organique très décomposé.

mucky soil, sol humique à gley.

mucro (pl. mucrones or mucros), mucron, pointe (*pal.*).

mucronate, pointu, à mucron.

mud, boue, vase; **m. avalanche,** coulée de boue; **m. bank,** banc de vase; **m. bit,** tarière à glaise; **m. cone,** volcan de boue, salse; **m. crack,** fente de dessication, de retrait; **m. cracked,** fissuré par dessication; **m. crack polygon,** polygone boueux de dessication; **m. flat,** slikke; **m. flood,** coulée de boue; **m. flow,** lave boueuse; **m. geyser,** volcan de boue; **m. lava,** lave boueuse; **m. line,** ligne de partage entre l'eau claire et boueuse au-dessus de la plate-forme continentale; **m. lumps (of the Mississippi),** îlots de boue (du Mississipi); **m. pellet,** débris d'argile remaniés; **m. polygons,** sol polygonal (dans matériel fin); **m. pot,** solfatare, source volcanique chaude et boueuse; **m. rock,** pélite; **m. slide,** glissement boueux; **m. spring,** source boueuse, volcan de boue; **m. stone,** 1. terme général pour argile, limon, « shale », pélite, schiste argileux (peu précis); 2. pélite (indurée); 3. micrite, calcilutite; **m. stream,** avalanche de boue; **m. volcano, 1.** volcan de boue, salse; **2.** soufflard.

mudding, envasement.

muddy, boueux.

mudflow cone, cône d'avalanches.

mudrock, pélite.

mugearite, mugéarite (*pétro.*).

mull, mull, humus forestier; **m. like mor,** humus intermédiaire; **m. like peat,** humus intermédiaire; **m. soil,** sol brun fortement lessivé.

mullion structure, **1.** structures linéaires (sillons et crêtes) de roches plissées ou métamorphiques; **2.** striations sur plan de faille (parallèles à la direction du déplacement).

mullite (syn. porcelainite), mullite (*minér.*).

mullock, **1.** roche non aurifère (Austr.); **2.** déblais, stériles.

mullocker, mineur au rocher.

mullocking, travail au rocher (*mine*).

mullocky, stérile.

multichannel, multicanal, multitrace; **m. filtering,** filtrage multicanal; **m. processing,** traitement multicanal; **m. reflection seismic profiles,** profils sismiques multicanaux.

multigelation, cycle répété gel-dégel.

multigranular particle, particule polycristalline.

multilamina, formé de plusieurs feuillets.

multilayer system, système multicouches.

multilocular, pluriloculaire (*pal.*).

multipartite map, carte de variation verticale d'un lithofaciès.

multiphase, polyphasé.

multiple, multiple; **m. dike,** filon formé par plusieurs phases intrusives (de la même roche); **m. faults,** failles m.; **m. geophones,** géophones m.; **m. metamorphism,** métamorphisme m.; **m. reflection,** réflection m.; **m. series connection,** montages en séries parallèles.

multiple coast, littoral polycyclique.

multispectral analysis, télédétection multispectrale.

multithrow fault, faille à rejets multiples.

multituberculate, multituberculés (*pal.*).

Murderian (N.Y. State), Murdérien (étage, Silurien sup.).

Muschelkalk, Muschelkalk (étage, Trias moyen).

muscle, muscle (*pal.*); **m. scar,** empreinte musculaire; **adductor m.,** muscle adducteur.

muscovite, muscovite (mica).

museum, musée.

mushroom rock, roche champignon.

mushy, spongieux, détrempé, bourbeux.

musked, **1.** zone marécageuse mal drainée et boisée; **2.** fondrière, marécage (Can.); **m. soil,** sol marécageux de toundra.

mussel, moule, (Mollusque, Lamellibranches).

mylonite, mylonite, structure bréchique.

mylonitic, mylonitique.

mylonitization, mylonitisation, formation de mylonite.

Myoida, Myoïdés.

Mytiloida, Mytiloïdés.

Myriapoda, Myriapodes.

myrmékite, myrmékite (interpénétration quartz feldspath); **m. texture,** structure myrmékitique.

N

nablock, nodule, rognon (de silex, etc).

nacre, nacre.

nacreous, nacré.

nacrite, nacrite (minéral argileux).

nadir, nadir.

nagyagite, nagyagite (*minér.*).

nail-head spar, calcite « en tête de clou », calcite « à pointes de diamant ».

naked, dénudé; **n. eye,** à l'œil nu.

namma hole, puits naturel (Austr.).

Namurian (Eur.), Namurien (étage, Carbonifère).

nannofossil, nannofossile.

nannoplankton, nannoplancton.

Nansen bottle, bouteille de Nansen (Océanogr.).

napalite, napalite (bitume fossile).

naphta, 1. naphta, essence lourde; **2.** naphte; **heavy n.,** essence lourde; **shale n.,** naphte de schiste.

naphtabitumen, naphtabitume.

naphtene, naphtène; **n. base crude,** pétrole brut naphténique; **n. serie,** série naphténique.

naphtenic, naphténique.

naphtenicity, teneur en naphtène.

naphtine, naphtein, naphtéïne (var. de hatchettite).

naphtology, étude du pétrole brut.

Napoleonville (N. Am.), Napoléonville (étage, Miocène).

napoleonite, napoléonite (*pétro.*), corsite, diorite orbiculaire.

nappe, nappe de charriage; **n. inlier,** fenêtre; **n. outlier,** lambeau de recouvrement; **n. root,** racine de la nappe; **overlapping or overthrust n.,** nappe de chevauchement.

Narizian (N. Am.), Narizien (étage, Éocène sup.).

nascent state, état naissant (d'un gaz, etc).

native, natif, pur; **n. element,** élément natif; **n. gold,** or natif; **n. paraffin,** ozokérite; **n. prussian blue,** vivianite; **n. silver,** argent natif.

natroborocalcite, ulexite (*minér.*).

natrolite, natrolite (var. de zéolite).

natron, natron (*minér.*).

natural, naturel; **n. arch, bridge,** arche, voûte n.; **n. coke,** charbon cuit par intrusion de roches éruptives; **n. current,** courant tellurique; **n. earth current,** courant tellurique; **n. gas,** gaz naturel; **n. glass,** verre naturel (dans une obsidienne); **n. levee,** levée n.; **n. magnet,** magnétite; **n. oil,** pétrole brut; **n. remanent magnetization,** aimantation rémanente naturelle; **n. selection,** sélection n.; **n. tilth,** bon état structural (du sol).

naumanite, naumannite (*minér.*).

nautiloid, a) ressemblant au Nautile, relatif au Nautile; **b)** Nautiloïdé (*pal.*).

Nautilus, Nautile.

Navarroan (N. Am.), Navarroien (étage, Crétacé sup.).

navite, navite (*pétro.*).

Neanderthal man, homme de Néanderthal (Paléolithique moy.).

neap tide, marée de morte-eau.

near shore, littoral; **n. circulation,** circulation océanique littorale; **n. current,** courant littoral.

neat, pur, non mélangé; **n. line,** encadrement d'une carte.

Nebraskan (N. Am.), Nébraskien (1re glaciation, Quaternaire).

nebula, nébuleuse.

nebulite, nébulite (roche métamorphique, à texture nébulitique).

neck, 1. « neck », remplissage de cheminée volcanique; **2.** isthme, tombolo; **3.** pédoncule de méandre; **n. cutoff,** recoupement des racines d'un méandre; **n. furrow,** sillon occipital.

necking, rétrécissement.

necrocoenosis, nécrocœnose (assemblage de fossiles avec indices du mode de vie originel).

necrolysis, histoire des fossiles postmortem.

Needian (obsolete), Needien (interglaciaire européen Mindel-Riss).

needle, aiguille, piton, éperon; **n. ice,** glace d'exsudation; **n. ironstone,** goethite filamenteuse; **n. ore,** aciculite, aikinite (*minér.*); **n. shaped,** aciculaire, en aiguille; **n. spar,** aragonite; **n. stone,** natrolite; **n. tin,** cassitérite en fins cristaux; **n. zeolite,** zéolite aciculaire; **compass n.,** aiguille de boussole; **ice n.,** aiguille de glace.

neep tide, marée de morte-eau.

negative, négatif; **n. area,** région subsidente; **n. crystal,** cristal négatif; **n. gravity anomaly,** anomalie négative de gravité; **n. movement,** subsidence; **n. movement of sea level,** mouvement négatif du niveau marin.

nekton, necton (animal nageur).

nektonic, nectonique.

nektoplanctonic, pélagique.

nelsonite, nelsonite (*pétro.*).

nemaline, fibreux, filamenteux (*minér.*).

nematoblastic, nématoblastique (métamorphisme).

Neocene (obsolete), Néogène (période, Cénozoïque).

Neocomian, Néocomien (étage, Crétacé inf.).

Neodarwinism, Néodarwinisme.

Neodevonic, Dévonien supérieur.

neodymium, néodyme.

Neogene, Néogène (période, Cénozoïque).

neogenic, récemment formé (*pétro.*).

Neoglaciation (see Little Ice Age), Néoglaciaire (voir Petit Age Glaciaire).

Neojurassic, Jurassique supérieur.
Neolithic, Néolithique (fin de l'âge de la pierre).
neomagma, magma nouvellement formé.
neomineralization, remplacement, substitutions chimiques et minéralogiques dans une roche.
neomorphic, néomorphique.
neomorphism, néogenèse.
Neoornithes, Néoornithes (*pal.*).
neon, néon (gaz rare).
Neopaleozoic, Paléozoïque supérieur (Silurien, Dévonien et Carbonifère).
neotectonics, néotectonique.
neoteny, neotony (rare), néoténie.
neotype, néotype (remplaçant l'holotype : *pal.*).
neovolcanic, néovolcanique (volcanisme tertiaire ou quaternaire).
Neozoic (obsolete), Cénozoïque, Tertiaire.
nephanalysis, analyse des types de nuages par photographies de satellites météorologiques.
nepheline, nephelite, néphéline (variété de feldspathoïde); **n. basalt,** basalte à néphéline; **n. syenite,** syénite à néphéline.
nephelinite, néphélinite (*pétro.*).
nephelite, néphélite (*minér.*).
nepheloid (layer), niveau marin profond avec boue en suspension.
nephrite, néphrite, jade.
nepionic, postembryonnaire, larvaire.
neptunian dyke (dike, U.S.), remplissage sédimentaire de fissures verticales d'une roche préexistante; **N. geology,** géologie de la planète Neptune; **N. theory,** théorie neptunienne.
Neptunism, Neptunisme (théorie basée sur l'hypothèse de l'origine marine des roches de la Terre).
neptunic, neptunique.
nereite, piste de ver fossile.
neritic, néritique; **n. zone,** zone n. (0 à − 200 m).
neritopelagic, néritopélagique.
nesh (G.B.), friable, pulvérulent, poudreux.
nesosilicate, nésosilicate, silicates à tétraèdres isolés.
ness (G.B.), cap, promontoire.
nest of ore, poche de minerai.
nested calderas, nested craters, caldeiras emboîtées; **n. cone,** cône volcanique emboîté.
net-like stone soil, sol polygonal (à parois formées de pierres).
net plane, plan réticulaire.
net slip, rejet net (= plus petite distance mesurée entre 2 points préalablement adjacents).
netted, réticulé; **n. texture,** structure réticulée (du sol).
nettle cell, cellule urticante.
network, réseau, lacis; **n. of seismological stations,** r. sismique; **distributive n.,** r. fluviatile diffluent.
neutral, neutre; **n. rock,** roche n.; **n. soil,** sol n.
neutron, neutron; **n. activation,** activation neutronique; **n. logging,** diagraphie par neutrons; **n. neutron log,** diagraphie neutrons-neutrons; **n. soil moisture meter,** sonde à neutrons.
Nevadan orogeny (or: nevadian, nevadic), orogénèse névadienne (Jurassique et début Crétacé).

nevadite, névadite (variété de rhyolite).
Neve firn (Germ.), névé; neige quasi-perpétuelle.
new field wildcat, sondage d'exploration.
new land soil, marais littoral récent, polder.
New Red Sandstone, « nouveaux grès rouges » (Permien et Trias).
new stone age (syn. Neolithic), Néolithique.
Newton's scale, échelle de Newton.
Niagaran (N. Am.), Niagarien (série, Silurien moy.).
niccolite, niccolite, nickéline (*minér.*).
niche, niche (écol.).
nick, entaille, rupture de pente.
nickpoint, rupture de pente.
nickel, nickel; **n. bloom,** annabergite (arséniate de nickel); **n. glance,** gersdorffite (NiAsS); **n. iron,** ferronickel; **n. linnaeite,** polydymite (*minér.*); **n. ocher,** annabergite (*minér.*); **n. silver,** alliage de zinc, nickel, cuivre; **n. steel,** acier au nickel; **n. vitriol,** morénosite (*minér.*).
nickeliferous, nickelifère.
nicking, forte flexure avec début de cassure.
nickings, charbons en petits morceaux.
nicol, nicol; **n. prism,** polariseur.
nicopyrite, nicopyrite, pentlandite, pyrite nickelifère.
nife, nife (noyau en fer-nickel de la Terre).
Niggli's classification, classification de Niggli (*pétro.*).
nigrine, nigrine (var. ferreuse de rutile).
nigritine, nigrite, nigrite (var. d'asphalte).
niobate, niobate.
niobium, nobium, niobium, columbium.
niobpyrochlore, pyrochlore (*minér.*).
niobtantalpyrochlore, néotantalite (*minér.*).
nip, 1. resserrement, étranglement; **2.** encoche de sapement (falaise littorale); **n. out,** disparition d'une couche, amincissement.
nipped bed, couche amincie.
niter, salpêtre, nitrate de potasse; **n. cake,** sulfate de sodium brut.
nitral, couche de nitrate (Espagne).
nitrate, nitrate.
nitratite, nitraline, nitrate de soude.
nitre, salpêtre, nitrate de potasse.
nitric, nitrique.
nitrification, nitrification.
nitrobarite, nitrobarite, nitrate de barium.
nitrogen, azote; **n. cycle,** cycle de l'a.; **n. fixation,** fixation de l'a. (par les bactéries des nodosités).
nitrogenous, azoté.
nitrous, nitreux.
nival, nival.
nivation, nivation; **n. hollow,** creux de nivation.
niveo-eolian, nivéo-éolien.
noble metal, métal précieux.
nodal, nodal; **n. line,** ligne n.; **n. point,** point n.; **n. zone,** zone n.
nodular, noduleux, nodulaire; **n. limestone,** calcaire noduleux.
nodule, nodule, rognon, concrétion.

nodulous, noduleux; **n. limestone**, calcaire noduleux.
noise, 1. bruit, son; 2. perturbation magnétique superficielle; **ground n.**, bruit de fond.
nomenclature, nomenclature (*pal.*).
nomen dubium, nom douteux (*pal.*).
nominal, nominal; **n. diameter**, diamètre n.; **n. output**, débit n., production nominale.
nomogenesis, nomogenèse.
nomograph, nomogram, abaque.
nonartesian ground water, eau souterraine libre.
nonbaking coal, charbon maigre.
nonbedded, non stratifié.
non-calcic, non calcaire; **n. brown soil**, sol brun non calcique.
non-capillary porosity, porosité non capillaire, macroporosité.
non-coherent soil, sol meuble divisé.
noncombustible, incombustible.
noncondensable, incondensable.
nonconducting rock, roche non conductrice.
nonconformable, discordant.
nonconformity, discordance.
nonconsolute, immiscible.
noncrystalline, non cristallin, amorphe.
nonferrous metals, métaux non-ferreux.
nongaseous coal, charbon maigre.
non-graded sediment, sédiment mal trié, mal calibré.
non-hydrostatic stresses, processus tectonique (dynamo métamorphique).
non-indurated pan, couche dure, friable du sous-sol, fragipan.
nonmagnetic, non magnétique.
nonmarine, continental.
non-metals, élément non métallique.
non-metallic deposit map, carte des substances utiles.
nonoxidizable, inoxydable.
nonpiercement salt dome, dôme de sel non intrusif.
nonplunging fold, pli à axe horizontal.
nonpolarizable, non polarisable, impolarisable.
nonpolarizing, impolarisable.
nonprocessed gas, gaz non traité.
nonproducing well, puits improductif.
nonrecoverable oil, pétrole non récupérable.
non-saturated zone, zone non saturée.
non-sequence (G.B.), brève lacune stratigraphique, lacune simple.
nonsorted circle, sol polygonal non trié.
nonsorted sets, sols polygonaux non triés.
nonsorted stripes, sols striés mal triés.
non-steady flow, écoulement irrégulier.
nontronite, nontronite (= variété d'argile smectite).
non-uniform pressure, pression orientée.
nonwaxy crude oil, pétrole brut non paraffinique.
nonwetted, non mouillable.
nordmarkite, nordmarkite (= syénite quartzitique hololeucocrate).
Norian, Norien (étage, Trias sup.).
norite, norite (= gabbro à hypersthène).
norm, norme (exprimée sous forme de minéraux virtuels).

norm system, classification des roches éruptives d'après la norme (C.I.P.W.).
normal, normal; **n. anticlinorium**, anticlinorium n.; **n. atmospheric pressure**, pression atmosphérique n. (au niveau de la mer); **n. class**, holohédrie, holosymétrie; **n. dip**, pendage général, pendage n.; **n. displacement**, déplacement n. (d'une faille); **n. erosion**, érosion n., érosion géologique, érosion du sol; **n. fault**, faille n.; **n. fold**, pli n., symétrique; **n. granite**, granite à biotite; **n. gravity**, pesanteur n. (au niveau de la mer); **n. horizontal separation**, composante horizontale du rejet; **n. hydrostatic pressure**, pression hydrostatique n.; **n. limb**, flanc n. d'un pli; **n. metamorphism**, métamorphisme régional; **n. move cut**, augmentation du temps dû à la distance du point de tir à la trace considérée; **n. position**, position n. (d'une couche); **n. sequence**, séquence stratigraphique n.; **n. shift**, composante horizontale n. du rejet; **n. superposition**, superposition n.; **n. throw**, projection verticale du rejet; **n. zoning**, zonation n. (dans un feldspath).
normative composition, norme (d'une roche éruptive).
normative mineral, minéral normatif, minéral virtuel.
north, nord; **n. atlantic drift**, branche n.-atlantique du gulf-stream; **n. east**, n.-est; **n. west**, n.-ouest; **magnetic n.**, n. magnétique; **true n.**, n. géographique.
northeaster, tempête par vents soufflants du N.E. (U.S.).
northerly, vers le nord.
northern dwarf podzol, micropodzol, nanopodzol.
nose, 1. pointe, promontoire, « nez », surplomb; 2. nez, saillant d'un anticlinal; 3. anticlinal à moitié formé (une extrémité ouverte); 4. courbure maximum d'une couche d'un pli sur une carte.
nosean, noséane (= feldspathoïde); **n. phonolite**, phonolite à n.
nose in (to), s'incliner, plonger (pour un terrain).
notch, 1. encoche, entaille, cannelure; 2. défilé, gorge (U.S.).
noumeite, garniérite, nouméite (*minér.*).
nourishment area, région d'alimentation glaciaire.
novaculite, novaculite.
nuclear, nucléaire; **n. energy**, énergie n.; **n. fuel**, combustible n.; **n. log**, diagraphie radiométrique; **n. magnetic resonance**, résonance magnétique n.; **n. power plant**, centrale n.
nucleate (to), former des germes cristallins.
nucleation, 1. nucléation, formation de germes cristallins; 2. accroissement des continents (par adjonction d'orogènes marginaux).
nucleus, germe, noyau; **n. crystal**, g. de cristallisation.
Nuevoleonian (N. Am.), Nuevoléonien (étage; Crétacé inf.).
nugget, pépite d'or.
nullah, ravin, cours d'eau, lit, oued de rivière (Inde).
numbering, comptage, dénombrement.
Nummulite, Nummulite (*pal.*).

Nummulitic, Nummulitique; syn. Paléogène (Eur.);
n. limestone, calcaire à Nummulites.
Nummulitids, Nummulitidés.
nunatak, nunatak (*glaciol.*).
nutation, nutation (de l'axe de la Terre).

nutrient, élément nutritif (*pédol.*); **n. content,** teneur
en substances nutritives.
nutriment, substance nutritive (*pédol.*), nutriment.
nutrition chain, chaîne alimentaire.
nutritive humus, humus nourricier.

O

oasis, oasis.
obduction, obduction.
obducting, faire obduction.
object, objet; **o. glass,** objectif; **o. lens,** objectif; **o. slide,** porte-objet (*microscopie*).
objective, objectif.
oblique, oblique; **o. bedding,** stratification o.; **o. extinction,** extinction o.; **o. fault,** faille o.; **o. lamination,** stratification o.; **o. air photograph,** photographie o.; **o. projection,** projection o.; **o. system,** système monoclinique.
obsequent, obséquent (*géomorph.*); **o. fault-line scarp,** escarpement obséquent de ligne de faille, **o. valley,** vallée obséquente.
observatory, observatoire.
obsidian, obsidienne; **o. lava,** lave d'o.
obsidianite, obsidianite : terme peu usuel (*pétro.*).
obstruct (to), obstruer, barrer.
obstruction dune, dune d'obstacle.
occidental diamond, cristal de roche, quartz limpide (G.B.).
occipital, occipital (*pal.*); **o. furrow,** sillon o. (Trilobites).
occlude (to), fermer, obstruer, retenir dans les pores.
occlusion, occlusion (des gaz dans un solide).
occurence, **1.** gisement, venue, présence de; **2.** événement; **mode of o.,** mode de gisement.
océan, océan; **o. basement,** socle océanique; **o. basin,** bassin o.; **o. bottom,** fond sous-marin; **o. circulation,** circulation o.; **o. current,** courant o.; **o. drilling program, (O.D.P.),** programme de forages o.; **o. floor,** fond o.; **o. floor spreading,** expansion des fonds o.; **o. pan,** cuvette o.; **o. tide,** marée o.; **o. trench,** fosse o.
oceanic, océanique; **o. bank,** guyot, mont sousmarin; **o. crust,** croûte o.; **o. island,** île o.; **o. sill,** seuil o.
oceanicity, influence océanique, caractère océanique.
oceanite, océanite (roche basaltique à phénocristaux d'olivine).
oceanization, océanisation (écartement de plaques).
oceanographic, océanographique.
oceanography, océanographie.
oceanology, océanologie.
ocellar structure, structure ocellaire.
ocher (U.S.), ochre (G.B.), ochre.
ocherous deposit, dépôt ferrugineux.
Ochoan (N. Am), Ochoéen (étage; Permien sup.
ochrept, sol brun tempéré (var. d'inceptisol).
Ocoee (N. Am. Georgia, Virginia), série d'Ocoee (Précambrien).
octahedric layer, couche octaédrique.
octahedrite, octahédrite.
Octocorallia, Octocoralliaires.

odontolite, odontolite (var. de turquoise).
odontology, odontologie.
oersted, oersted (unité électromagnétique).
offing, large, pleine mer.
offlap, régression, en régressivité; séquence sédimentaire négative; **o. deposit,** dépôt régressif.
offset, **1.** déplacement de masses contiguës; **2.** composante horizontale du déplacement d'une faille; **3.** distance entre le point de tir et le géophone (géograph.); **4.** imprimé photographique.
offset shot, tir déporté.
offsetting, décalage de couches.
offshoot, **1.** ramification, apophyse (d'un filon); **2.** bifurcation d'un pli.
offshore, au large du rivage; **o. bar,** cordon littoral émergé; **o. current,** courant dirigé vers le large; **o. drilling,** forage sous-marin; **o. ramp,** talus continental; **o. slope,** talus continental; **o. through,** sillon d'avant-côte; **o. well,** puits d'exploitation au large (offshore); **o. zone,** plate-forme continentale, zone néritique.
offtake, galerie d'écoulement (*mine*).
ogive, ogive glaciaire.
ohm meter, ohm/mètre (unité de résistivité).
oid, suffixe : en forme de, ayant la structure de.
oihocryst, cristal orbiculaire.
oil, pétrole; **o. accumulation,** gisement de p.; **o. basin,** bassin pétrolifère; **o. bearing,** pétrolifère; **o. column,** hauteur imprégnée de la p. (*forage*); **o. deposit,** gisement de p.; **o. field,** champ pétrolifère; **o. finding,** recherche pétrolière; **o. gas,** gaz d'éclairage produit par distillation; **o. gas interface,** interface huile-gaz; **o. geologist,** géologue pétrolier; **o. horizon,** niveau pétrolifère; **o. indication,** indice de p.; **o. layer,** couche pétrolifère; **o. lease,** concession de recherches pétrolières; **o. lens,** lentille de sable pétrolifère; **o. man,** pétrolier; **o. measures,** couches pétrolifères; **o. mining,** exploitation des gisements de p.; **o. producing deck,** pont producteur sur plate-forme flottante; **o. prospecting,** prospection pétrolifère; **o. refinery,** raffinerie de p.; **o. region,** région pétrolifère; **o. reserves,** réserves pétrolifères; **o. reservoir rock,** roche réservoir; **o. rock,** roche pétrolifère; **o. sand,** sable pétrolifère; **o. seepage,** suintement de p.; **o. shale,** schiste bitumineux; **o. show,** indice de p.; **o. showings,** indices de p.; **o. spill,** fuite, décharge de p.; **o. spring,** source de p.; **o. string,** colonne de production; **o. structure,** structure pétrolifère; **o. tar,** goudron de p.; **o. trap,** piège de p.; **o. water,** eau de gisement; **o. water contact,** contact eau-p.; **o. water interface,** surface limite eau-p.; **o. water ratio,** proportion huile-eau; **o. well,** puits de p.; **o. well blowing,** jaillissement

d'un puits de p.; **o. well derrick**, tour de forage de puits de p.; **o. wet**, imprégné de p.; **o. yielding**, pétrolifère; **crude oil**, pétrole brut.

oily, graisseux, huileux.

okenite, okénite (*minér.*).

old, vieux; **o. alluvium**, alluvions anciennes; **o. land**, craton; **o. red sandstone**, vieux grès rouge (Dévonien : Eur.); **o. stage**, stade de vieillesse.

Older Dryas (Eur.), Dryas ancien (intervalle entre Bölling et Allerod).

Oldest Dryas (Eur.), Dryas le plus vieux (précédent l'intervalle Bölling).

olefiant gas, éthylène.

Olenekian East (Eur.), Olénékien (Trias).

oligist, oligiste, hématite (*minér.*).

oligocene, oligocène (époque ou série; cénozoïque).

oligoclase, oligoclase (*minér.*).

oligoclasite, oligoclasite (diorite à oligoclase).

oligohaline, oligohalin.

oligomict, roche détritique monominérale.

oligomictic, monogénique (conglomérat); **o. lake**, lac thermiquement stable.

oligosiderite, oligosidérite (var. de météorite).

oligotrophic, oligotrophique; **o. brown soil**, sol brun oligotrophe; **o. moor soil**, tourbière haute, marais oligotrophe.

olistolith, olistolithe (bloc exotique dans un olitostrome).

olitostrome, olitostrome (séd); chaos sédimentaire formé par glissement.

olivine, olivine; **o. basalt**, basalte à o.; **o. bomb**, bombe à o.; **o. diabase**, diabase à o.; **o. gabbro**, gabbro à o.; **o. leucitite**, basalte à leucite; **o. nephelinite**, basalte à néphéline; **o. nodule**, nodule d'o.; **o. norite**, norite à o.; **o. rock**, péridotite.

olivinite, olivinite (péridotite riche en olivine).

olivinophyre, porphyre à phénocristaux d'olivine.

omission of beds, suppression de couches (par faille).

omission solid solution, réseau cristallin incomplet.

omphacite, omphacite (variété de diopside riche en soude).

oncolith, oncolite, oncolithe.

on (opp. off), en exploitation, exploité.

Onesquethawan (North Am.), Onesquethawan (étage; Dénovien inf.).

onion weathering (onion-skin weathering), altération en écailles d'oignon, en sphéroïdes.

onlap, 1. chevauchement, débordement, transgression; 2. en transgressivité; 3. séquence sédimentaire positive.

onset, départ (d'une réaction chimique).

onsetter, ouvrier d'accrochage (*mine*).

onsetting, accrochage, encagement (*mine*).

onshore (opp. offshore), vers le rivage, à terre; **o. reef**, récif côtier; **o. shelf**, ancien plateau continental émergé; **o. wind**, vent de mer.

Ontarian (U.S.A.; N.Y. St.), Ontarien (étage, Silurien moy.).

ontogenesis, ontogénèse.

ontogenetic stage, stade ontogénétique.

ontogeny, ontogénie, développement individuel.

Onychophora, Onychophores (*Pal.*).

onyx, onyx 1. quartz cryptocristallin; 2. calcite en couches (Mexique).

onyx marble, marbre ou travertin rubané (souvent en stalagmite).

ooid, oolithe.

oolite, calcaire oolithique.

oolith, oolithe (particule subsphérique).

oolithic, oolitic, oolithique; **o. iron ore**, minette, fer oolithique.

oomicrite, oomicrite.

oosparite, oosparite.

ooze, vase, boue pélagique; **calcareous o.**, boue calcaire (+ de 30 % de débris d'organismes).

ooze (to), suinter, s'infiltrer, filtrer.

oozing, suintement.

oozy, 1. vaseux, bourbeux; 2. suintant, humide.

opacite, opacite (substance opaque).

opacity, opacité.

opal, opale; o. oil, distillat lourd de pétrole; **jasper o.**, opale jaspe.

opalescence, opalescence.

opalized wood, bois silicifié.

opaqueness, opacité.

open ouvert; **o. bay**, baie exposée au large; **o. cast**, exploitation à ciel ouvert; **o. country**, pays découvert; **o. cut**, exploitation à ciel ouvert; **o. fault**, faille ouverte; **o. flow capacity**, débit en écoulement naturel; **o. flow potential**, potentiel maximal d'un puits; **o. flow pressure**, pression de gisement en écoulement; **o. fold**, pli ouvert; **o. graded aggregate**, agrégat grossier poreux; **o. hole**, forage à découvert; **o. mining**, exploitation à ciel ouvert; **o. pack**, banquise fragmentée; **o. pit**, 1. mine à ciel ouvert; 2. aven; **o. quarry**, carrière à ciel ouvert; **o. rock**, roche poreuse; **o. sand**, sable poreux; **o. sea**, océan exposé; **o. stope**, taille sans remblayage (*mine*); **o. work**, exploitation à ciel ouvert; **o, working**, exploitation à ciel ouvert.

opening, ouverture; **o. operations**, travaux de traçage (*mine*); **o. out of a ravine**, débouché d'un ravin.

operating, ingénieur d'exploitation.

operation, fonctionnement, exploitation.

operator, exploitant (d'une mine).

operculum, opercule.

ophicalcite, marbre à serpentine (cf. « vert antique »).

Ophidia, Ophidiens (*pal.*).

ophiolite, ophiolite (roche ignée ultrabasique plutonique et volcanique très riche en serpentine); **o. complex**, complexe ophiolitique.

ophite, ophite (diabase à texture ophitique).

ophitic, ophitique; **o. texture**, structure doléritique.

Ophiuridea, Ophiurides (*pal.*).

Opistobranchia, Opistobranches.

Opistocoelous, Opistocèle (*pal.*).

optical character, signe optique d'un minéral.

optical constants, constantes optiques (d'un minéral); **o. extinction**, extinction optique.

optical pyrometer, pyromètre optique, **o. spectroscopy**, spectrométrie optique; **o. spectrum**, spectre optique.

optimum, optimum.

orals, **1.** plaques interradiales (Crinoïdes); **2.** examen universitaire.

orbicular, orbiculaire; **o. diorite**, diorite o.; **o. structure**, structure o.

orbit, orbite.

orbital, orbital.

orbite, orbite (variété d'amphibolite).

Orbitoid, **1.** Orbitoïnidae (*pal.*); **2.** appartenant aux Orbitoïnidae.

ordanchite, ordanchite (= téphrite à haüyne) (*Pétro.*).

order, ordre; **o. of crystallization**, ordre de cristallisation; **normal o. of superposition**, succession normale de couches; **original o. of succession**, disposition originelle des couches.

ordinary ray, *o. ray*, rayon ordinaire.

ordinate, ordonnée.

Ordovician, Ordovicien (période; Paléozoïque).

ore, minerai; **o. apex**, sommet du gîte; **o. bands**, zones minéralisées; **o. bearing**, métallifère, minéralisé; **o. bed**, couche minéralisée; **o. belt**, zone métallifère; **o. benefication**, enrichissement de m.; **o. bin**, trémie à m.; **o. body**, masse minéralisée, massif de m.; **o. breaking**, abattage, concassage de m.; **o. bringer**, venue minéralisante; **o. briquetting**, aggloméral de m.; **o. bunch**, poche de m.; **o. carrying**, minéralisateur; **o. carrying spring**, source minéralisante; **o. channel**, fissure profonde minéralisée; **o. chimney**, cheminée de m.; **o. chute**, cheminée de m.; **o. concentrate**, concentré de m.; **o. content**, teneur en m.; **o. course**, filon minéralisé; **o. current**, courant de minéralisation; **o. deposits**, gisement métallifère; **o. dike**, dyke de m.; **o. dressing**, traitement du m.; **o. dump**, déblais de m., halde; **o. enrichment**, enrichissement du m.; **o. flotation**, flottation du m.; **o. grade**, teneur d'un m.; **o. guide**, indice métallogénique; **o. horizon**, niveau minéralisé; **o. leaching**, lixiviation du m.; **o. leave**, droits miniers; **o. lode**, filon; **o. magma**, magma minéralisé; **o. microscope**, microscope à m.; **o. mine**, mine de m.; **o. mineral**, minéral métallique; **o. mining**, exploitation de m.; **o. pipe**, cheminée de m.; **o. placer**, gîte de m.; **o. pocket**, poche de m.; **o. separator**, séparateur de m.; **o. sheet**, filon-couche de m.; **o. shoot**, passe minéralisée; **o. sill**, passe minéralisée; **o. sintering**, agglomération de m.; **o. slag**, scorie de m.; **o. streak**, bande minéralisée; **o. vein**, filon minéralisé; **o. washing**, lavage des m.; **bog-iron o.**, fer des marais; **cockade o.**, m. en cocarde; **crude o.**, m. brut; **high-grade o.**, m. à forte teneur; **lean o.**, m. pauvre; **low grade o.**, m. pauvre.

organic, organique; **o. compound**, composé o.; **o. matter [muck (N. Am.)]**, matière o. [décomposée]; **o. origin**, origine o.; **o. rock**, roche d'origine o.; **o. sediment**, sédiment o.; **o. slime**, sapropel;

o. soil, sol tourbeux; **o. weathering**, altération biologique.

organism, organisme; **euryhaline o.**, o. euryhalin; **stenohaline o.**, o. sténohalin.

organogenic, *orgagenous*, organogène.

orient (to), orienter.

oriental, oriental; **o. agate**, agate o.; **o. emerald**, émeraude o.; **o. ruby**, rubis véritable; **o. topaz**, topaze jaune.

orientation, orientation.

oriented specimen, **1.** échantillon orienté; **2.** fossile à position orientée.

original, d'origine, primaire; **o. dip**, pendage p.; **o. stratification**, stratification p.; **o. horizontality**, horizontalité p. (de couches).

origofacies, faciès primaire.

Oriskanian (N. Am.), Oriskanien (groupe Dévonien).

ornamentation, ornementation (d'un fossile).

ornemental stone, matériau d'ornementation.

Ornithischia, Ornithischiens (*pal.*).

ornoite, ornoïte (var. de diorite).

orocline, zone oroclinale.

orocratic, orocratique.

orogen, orogène.

orogenesis, orogénèse.

orogenic, *orogenetic*, orogénique; **o. belt**, zone o.; **o. cycle**, cycle o.; **o. facies**, faciès o.; **o. phase**, phase o.; **o. vulcanism**, volcanisme o.

orogeny, orogénie, orogénèse.

orographic, *orographical*, orographique.

orography, orologie.

orohydrographic, orohydrographique.

orohydrography, orohydrographie.

orology, orologie, étude de la formation des reliefs (montagnes).

orometer, oromètre (var. de baromètre).

oromorphic soil, sol de montagne.

orpiment, orpiment (*minér.*, sulfure d'arsenic).

orthamphibole, amphibole orthorhombique.

orthaugite (G.B.), pyroxène orthorhombique.

orthent, régosol, lithosol (*pédol.*).

orthite, orthite, allanite (= variété d'épidote) (*minér.*).

ortho, préfixe : droit, régulier.

orthoamphibolite, orthoamphibolite.

Orthoceratidea, Orthocératidés (*pal.*).

orthochem, orthochème.

orthochromatic, orthochromatique.

orthoclase, orthose; **sanidized o.**, orthose transformée en sanidine.

orthoclasite, orthoclasite (*pétro.*).

orthoclastic, orthoclastique.

orthodome, orthodôme.

orthodromy, orthodromie.

orthofelsite, orthofelsite, orthophyre (*pétro.*).

orthoferrosilite, orthoferrosilite (var. de pyroxène).

orthogenesis, orthogénèse.

orthogeosyncline, orthogéosynclinal.

orthogneiss, orthogneiss.

orthogonal, orthogonal, perpendiculaire.

orthogonals, orthogonales (à un front de houle).
orthographic projection, projection orthographique.
orthomagmatic stage, stade orthomagmatique (de la cristallisation).
orthomorphic projection, projection conforme (*cartogr.*).
orthophyre, orthophyre (porphyre à orthose).
orthophyric, orthophyrique.
orthopinacoid, orthopinacoïde (*cristallo.*).
orthoprism, orthoprisme.
orthopyroxene, pyroxène orthorhombique.
orthoquartzite, grès quartzite uniquement siliceux (sédimentaire).
ortho rocks, roches métamorphiques dérivant de roches ignées.
orthorhombic, orthorhombique.
orthose, orthose.
orthosilicate, orthosilicate.
orthosite, orthosite (= syénite exclusivement orthosique).
orthosymmetric system, système orthorhombique.
orthotectic, orthotectique.
orthotectonic, tectonique de type alpin.
orthox, sol ferralitique de climat humide (*pédol.*).
ortlerite, ortlérite (désuet : diorite altérée en roche verte).
oryctognostic (obsolete), minéralogique.
oryctognosy (obsolete), minéralogie.
Osagean, Osagian, Osagéen (série; Mississipien inf.).
osannite, osannite (var. d'amphibole).
os (pl. osar), os, esker (etym. äs : Suède).
oscillation, oscillation; **o. cross-ripple marks,** rides de plages entrecroisées; **o. ripple,** ride de plage à versant symétrique, ride d'o.; **o. of beaches,** o. des rivages; **o. theory,** théorie oscillatoire; **climatic o.,** o. climatique.
oscillatory current, courant oscillatoire (marin); **o. twinning,** mâcle polysynthétique (*cristallo.*); **o. zoning,** zonation périodique (d'un cristal).
oscillograph, oscillographe.
oscilloscope, oscilloscope.
osculum, oscule (*pal.*).
ose (see os), os, esker.
osmium, osmium.
osseous, osseux, à ossements; **o. breccia,** brèche à o.
ossicle, plaque (du squelette des Échinodermes), entroque.
ossipite, or ossypite, ossypite (var. de troctolite).
Osteichthyes, Ostéichthiens (Poissons osseux).
osteology, ostéologie.
Osteostraci, Ostéostracés.
Ostracod, Ostracoda, Ostracode.
Ostracodermi, Ostracodermes.
Ostreacea, Ostracés.
otolith, otolithe (*pal.*).
ounce, once; **o. avoir-du-poids,** = 28,35 g; **o. troy,** = 31,10 g.
outbreak, **1.** éruption; **2.** affleurement d'un filon (mine); **o. coal,** affleurement de charbon.

outburst, **1.** explosion, dégagement, éruption; **2.** affleurement; **o. of gas,** dégagement rapide de gaz; **o. of mud,** éruption de boue; **glacial o.,** débâcle glaciaire.
outcrop, affleurement; **o. bending,** fauchage des couches; **o. curvature,** fauchage des couches; **o. line,** ligne d'a.; **o. spring,** source d'a.
outcrop (to), affleurer.
outcropping, affleurement.
outer, extérieur; **o. bank,** rive concave e. (d'un méandre); **o. core,** noyau externe du globe terrestre; **o. lip,** lèvre externe (Gastéropodes); **o. moraine,** moraine externe.
outfall, **1.** embouchure, issue; **2.** déversoir, décharge; **3.** couche affleurant à un niveau inférieur (G.B.).
outflow, **1.** écoulement, coulée (de laves); **2.** débit, écoulement externe (pétrole); **o. valley,** vallée de débâcle périglaciaire (sur Terre et aussi sur Mars); **flank outflow,** effusion de flanc (d'un volcan), effusion latérale.
outflowing, épanchement de laves.
outgas (to), dégager.
outgassing, dégazage (produits volcaniques).
outgush (to), répandre, épancher à l'extérieur.
outlet, orifice, sortie, embouchure; **glacier o.,** langue émissaire de glacier.
outlier, **1.** avant-butte; **2.** lambeau de recouvrement.
outline, contour, dessin, tracé (d'une côte); **o. minerals,** minéraux formés par altération (résiduels).
outtake, puits de sortie d'air (mine).
outwash, épandage fluvio-glaciaire; **o. apron, plain,** apports fluvio-glaciaires; **o. drift,** plaine d'épandage fluvio-glaciaire.
ouvala, ouvala (karst).
ouvarovite, ouwarovite (variété de grenat chromifère).
overbank deposit, dépôt alluvial d'inondation, du lit majeur.
overburden, morts-terrains, couverture de dépôts meubles.
overburdened stream, fleuve surchargé.
overcast, **1.** adj : chevauché, couvert, nuageux; **2.** n : croisement de manches à air (*mine*).
overconsolidated material, matériau surconsolidé.
overdeep (to), surcreuser.
overdeepening, surcreusement.
overfault (obsolete), pli-faille inverse.
overfeed (to), suralimenter (en neige, eau, etc.).
overfloat (to), surnager.
overflow, débordement, inondation; **o. spring,** source déversante, source de trop plein; **o. summit overflow,** effusion terminale (volcan).
overflowing well, puits artésien.
overflow (to), déborder, inonder.
overfold, pli déversé.
overfrozen (frozen over), couvert de glaces.
overgrowth, accroissement secondaire d'un cristal, liseré d'accroissement.
overhand, gradins renversés (exploitation en); **o. stopes,** en gradins renversés (mines).

overhang, surplomb.

overhang (to), surplomber.

overland flow, ruissellement pluvial en nappe.

overlap, chevauchement, recouvrement; **o. fault,** faille inverse; **fold o.,** flèche de recouvrement; **regressive o.,** régression; **transgressive o.,** transgression.

overlap (to), 1. chevaucher, recouvrir; 2. dépasser.

overlapping, chevauchement, recouvrement; **o. folds,** plis en échelons.

overlay, graphique, calque transparent.

overlay (to), recouvrir, couvrir, être superposé sur.

overlie (to), couvrir, recouvrir, reposer sur, être sus-jacent sur.

overlying, sus-jacent.

overload, surcharge.

overloaded stream, fleuve surchargé.

overloading, surcharge (d'un édifice).

overplacement, 1. superposition; 2. morts-terrains (*mine*).

overpressure, surpression.

overriding, chevauchement.

oversaturated rock, roche hypersiliceuse, hypersaturée en silice.

oversaturation, sursaturation.

oversize, refus d'un tamis.

overspilling, déversement.

overstep (G.B.), discordance stratigraphique due à une transgression.

oversteepening, surraidissement (d'une vallée glaciaire).

overstress, surcharge.

overthrow fold, pli déversé.

overthrust, chevauchement, charriage; **o. block,** nappe de charriage; **o. fault,** plan de charriage; **o. fold,** pli-faille couché; **o. nappe,** nappe de charriage; **o. sheet,** nappe de charriage; **o. slice,** lambeau de poussée.

overtop, overtopping, débordement d'un fleuve (crue).

overturn, déversement, renversement.

overturn (to), renverser.

overturned fold, pli déversé.

overturned limb, flanc inverse.

overwash, débordement; **o. drift,** épandage fluvioglaciaire; **o. fan,** cône de d. (dans une lagune littorale U.S).

overwater, eau de toit.

overweight, surcharge.

oviparous, ovipare (*pal.*).

oxbow lake, lac en croissant (dans méandre abandonné).

Oxfordian, Oxfordien (étage; Jurassique sup.).

oxic horizon, horizon B (*pédol.*).

oxidability, oxydabilité.

oxidates, sédiments précipités par oxydation.

oxidate (to), oxyder.

oxidation, oxydation.

oxide, oxyde.

oxidizable, oxydable.

oxidization, oxydation.

oxidize (to), oxyder.

oxidized zone, zone oxydée.

oxidizing, oxydant.

oxisol, oxysol (sol tropical altéré).

oxybitumen, oxybitume.

oxygen, oxygène; **o. ratio,** rapport isotopique $^{18}O/ ^{16}O$.

oxygenate (to), oxygéner.

oxymagnite, magnétite oxydée.

oxysphere, lithosphère.

oyster bank, bar, banc ou récif d'huîtres; **o. shell,** coquille d'huître.

Ozarkian (obsolete), Ozarkien (entre Cambrien et Ordovicien).

ozocerite, ozokerite, ozocérite, ozokérite, cire fossile.

ozone, ozone.

P

pacific-type coast, côte concordante.

pacific suite, province pacifique (*pétro.*).

pack, remblai (*mine*); **ice p.**, embâcle glaciaire; **p. ice,** glace de dérive.

packer, 1. remblayeur (*mine*); **2.** garniture d'étanchéité (forage).

packing, 1. tassement, compaction; **2.** remblai, remblayage.

packs, remblai (*mine*); **supporting p.,** piliers de support du toit (*mine*).

packstone, « packstone ».

packwall, mur de remblai (*mine*).

paedomorphose, néoténie.

Pahrump (U.S.A., Cal.), série de Pahrump (Précambrien).

paint pot, source chaude à limons et argiles bariolées.

palaeo, (préfixe U.K., voir paleo).

Palaeozoic, Paleozoic era, Paléozoïque, ère primaire.

palagonite, palagonite.

palagonite tuffs, tufs à palagonite.

palagonitization, polagonitisation.

palagonitized, altéré en palagonite.

palasome (syn. host), palasome.

Palatinian orogeny (see Pfalzian orogeny), orogénèse palatine (Permien terminal).

paleoatmosphere, paléoatmosphère.

paleobotany, palaeobotany, paléobotanique.

Paleocene, Palaeocene, Paléocène (série; base du Cénozoïque).

paleochannel, palaeochannel, paléochenal.

paleoclimatology, palaeoclimatology, paléoclimatologie.

paleocurrent, paléocourant.

paleoecology, palaeocology, paléoécologie.

Paleogene, Palaeogene, Paléogène (intervalle ou période = Paléocène + Éocène + Oligocène).

paleogeographic map, carte paléogéographique.

paleogeography, palaeogeography, paléogéographie.

paleogeomorphology, paléogéomorphologie.

paleoichnology, palaeoichnology, paléoichnologie.

Paleolithic, Palaeolithic, Paléolithique (division de l'âge de la pierre).

paleolithologic, palaeolithologic map, carte de paléofaciès lithologiques.

paleomagnetic, palaeomagnetic north pole, pôle nord paléomagnétique.

paleomagnetic pattern, structure paléomagnétique des zones parallèles à la crête médioatlantique.

paleomagnetism, palaeomagnetism, paléomagnétisme.

paleo-oceanography, paléo-océanographie.

paleontologic, palaeontologic, paléontologique; **p. facies,** faciès p.; **p. province,** province p.; **p. species,** espèce déterminée sur des échantillons fossiles.

paleontology, palaeontology, paléontologie.

paleopedology, étude des sols anciens et des processus originels.

Paleophytic, Palaeophytic, âge des Ptéridophytes (division de paléobotanique).

paleoplain, palaeoplain, paléoplaine.

paleopole, ancien pôle.

paleoreconstruction, reconstitution paléontologique.

paleorelief, paléorelief.

paleosalinity, ancienne salinité.

paleoslope, palaeoslope, pente d'une ancienne surface continentale.

paleosoil, paleosol, palaeosoil, paléosol.

paleostructure, surface antédiscordante d'un terrain ancien.

paleotectonic, palaeotectonic, paléotectonique.

paleotectonic map, carte paléotectonique.

paleotemperature, ancienne température, paléotempérature.

paleovolcanic, palaeovolcanic, paléovolcanique.

paleozoology, palaeozoology, paléontologie animale.

palimpsest structure, structure résiduelle (roches métamorphiques).

palin (prefix), à nouveau, renouvelé.

palingenesis, palingénèse (cf. anatexie).

palinspastic map, carte paléogéographique et paléotectonique.

palisade(s), falaise abrupte formée de roches à débit prismatique.

palladium, palladium.

pallasite, pallasite.

pallial sinus, sinus palléal (*pal.*).

palsa (pl. palsen), palse : monticule de tourbe gelée; hydrolaccolithe de tourbe.

paludal, palustre, marécageux.

palustral, palustrine, palustre.

palygorskite, palygorskite (minéral argileux).

palynology, palynologie.

palynomorph, palynomorphe.

pampa, pampa.

pan, 1. horizon induré, concrétionné (dans le soussol); **2.** dépression boueuse à sec lors des périodes séches (Afrique du Sud); **3.** battée, cuvette; **hardpan,** niveau concrétionné; **ironpan,** alios ferrugineux; **limepan,** horizon d'accumulation calcaire; **settling pan,** bac à décantation.

pancake-ice, glace en « crêpe » (formation d'une banquise).

panel, panneau (mine).

panfan (syn. pediplain), pédiment généralisé (sous climat désertique).

Pangaea, Pangea, Pangea primitif, hypothétique : (supercontinent Laurasie + Gondwanaland).

panidiomorphic, pandiomorphic, pandiomorphique, panidiomorphique (à cristaux très bien formés).

panhumus, horizon d'accumulation humifère.

panning, lavage à la battée (prospection de l'or).

Pannonian (Eur.), Pannonien (étage; Miocène sup. à Pliocène inf.).

panplain, plaine alluviale composite (coalescence de plaines) d'inondation.

pansoil, sol à horizon d'accumulation.

pantellerite, pantellerite (var. de rhyolite hyperalcaline).

pantograph, pantographe.

Pantotheria, Pantothériens.

paper, 1. article, communication; **2.** papier; **p. coal**, houille carton; **p. schist**, schiste carton induré; **p. shale**, schiste carton; **filter p.**, p. filtre; **logarithmical p.**, p. logarithmique; **scale p.**, p. millimétré.

parabolic, parabolique; **p. dune**, dune p.

paraclase (obsolete), paraclase (cf. faille).

paraconformity, discordance simple, « paraconcordance » (à relief enterré).

paracrystalline deformation, changement structural contemporain de la recristallisation (pétro.).

paraffin, paraffine.

paraffin-base crude, pétrole brut à base paraffinique.

paraffin dirt, dépôt paraffinique; **p. hydrocarbon**, hydrocarbure saturé p. (à chaîne ouverte linéaire); **p. series**, séries p.

paragenesis, paragenèse, séquence minérale.

paragenetic, paragénétique (cf. ordre de cristallisation des minéraux).

parageosyncline, paragéosynclinal (géosynclinal intracratonique).

paraglacial, périglaciaire.

paragneiss, paragneiss.

paragonite, paragonite, muscovite sodique.

paraliageosyncline, paraliagéosynclinal (géosynclinal profond installé près des marges continentales).

paralic, paralique (milieu de transition); **p. environment**, milieu littoral p.

parallel, parallèle; **p. bedding**, litage p.; **p. drainage pattern**, réseau hydrographique composé de cours d'eau p.; **p. evolution**, évolution de phylums p.; **p. extinction**, extinction droite; **p. faults**, failles p. de même pendage; **p. fold**, pli formé de couches ayant gardé la même épaisseur; **p. growth**, croissance de cristaux suivant la même orientation; **p. retreat**, érosion des versants suivant une pente p. à la pente originale; **p. shot**, tir p. (géophys.); **p. transgression**, transgression concordante.

paramagnetic, paramagnétique.

paramagnetism, paramagnétisme.

paramarginal resource, ressources minérales presque (mais pas tout à fait) exploitables avec rentabilité.

parameter, paramètre (cristallo.).

paramorph, paramorphic, paramorphique.

paramorphism, paramorphose.

para rock (obsolete), roche métamorphique d'origine sédimentaire.

paraschist, schiste d'origine sédimentaire.

parasitic cone, cône volcanique adventif; **p. crater**, cratère adventif; **p. fold**, pli d'entraînement.

parataxitic, parataxitique (relatif au litage d'une roche pyroclastique).

paratectonic recrystallization, recristallisation concomitante de la déformation tectonique.

paratectonics, tectonique de régions stables (de cratons).

paratype, paratype (autre que l'holotype).

paraunconformity, discordance stratigraphique (consécutive à une lacune stratigraphique; pas de changement de pendage).

parautochtonous rocks, roches subautochtones (déplacées sur une courte distance par des nappes de charriage).

parental, originel, roche-mère.

parental magma, magma originel.

parent, parent, d'origine; **p. element**, élément radioactif; **p. material**, matériau originel; **p. rock**, roche-mère.

pargasite, pargasite (var. d'amphibole sodique).

parietal art, art pariétal (préhistoire).

park, 1. parc herbu entouré de bois (Montagnes Rocheuses); **2.** doline évasée et peu profonde, poljé (Arizona).

paroptesis, cuisson (des roches).

paroxysm, paroxysme.

partial, partiel; **p. crystallization**, cristallisation p.; **p. fusion**, fusion p.; **p. melting**, fusion p.

partial pressure, pression partielle (d'un gaz dans un mélange).

particle, particule; **p. shape**, forme des p.; **p. size**, granulométrie; **p. size analysis**, analyse granulométrique; **p. size histogram**, histogramme des classes granulométriques; **p. velocity**, vitesse des p. (d'eau de mer dans les vagues).

parting, 1. petite diaclase; **2.** mince couche de schiste intercalée dans du charbon; **3.** plan de séparation d'un cristal (autre qu'un plan de clivage); **4.** décollement (géogr.); **5.** plan de fracture préférentiel.

parting plane, plan de stratification.

Pasadenan orogeny, orogénèse pasadénienne (Pléistocène, Californie).

pass, 1. col, défilé; **2.** chenal navigable; **3.** passe, goulet; **4.** passe corallienne.

passage beds, couches de transition.

passband, bande passante (géoph.).

past mature soil, sol sénile, dégénéré.

patch, morceau, lopin de terre.

patch reef, petites constructions coralliennes.

patera (Mars), ancienne caldeira martienne.

paternoster lakes, lacs en chapelet.

path, trajectoire (d'une onde).

patine, patine (d'un silex).

pattern, 1. modèle; **2.** réseau, structure.

patterned ground, sol polygonal, sol structuré (périglaciaire).

paved land, paved soil, pavage désertique, régolithe.

pavement, dallage; **desert p.**, pavage désertique.

paving stone, pierre pour pavés.

pay, concentration, gisement (U.S.A.); **p. ore**, g.

minéral rentable; **p. sand,** sable productif; **p. streak,** filon exploitable; **p. zone,** zone productive.

payable, exploitable (U.S.A.).

pea-coal, houille fine.

pea-iron, **1.** limonite; **2.** pisolithe ferrugineux.

pea-like, pisiforme.

pea-like iron, fer pisolithique.

pea-ore, minerai pisiforme; **p. stone,** pisolite.

peach stone (G.B.), schiste chloritique.

peacock ore, bornite (*minér.*).

peak, **1.** sommet, pic; **2.** pointe, pic (enregistrement).

peak to background ration, rapport signal-bruit de fond.

peak flood, maximum de crue (inondation du lit majeur); **p. flow,** débit de pointe; **p. value,** valeur maximum (d'un phénomène).

pearl, perle; **p. ash,** carbonate de potassium; **p. diabase,** variolite; **p. spar,** dolomite, ankérite; **p. stone,** perlite.

pearly, nacré.

peat, tourbe; **p. bog,** tourbière; **p. bog soil,** sol de marais tourbeux; **p. bog clay,** vase de marais; **p. clod,** motte de t.; **p. coal,** charbon de t.; **p. gas,** gaz obtenu à partir de la t.; **p. hillock,** bombement dans une tourbière; **p. humus,** humus tourbeux; **p. moor,** tourbière; **p. podsol,** podzol tourbeux; **p. soil,** sol tourbeux; **limnic p.,** t. lacustre; **moss p.,** t. formée de mousse; **upland p.,** t. d'altitude (en montagne).

peatification, tourbification.

peaty, tourbeux; **p. forest humus,** humus forestier t.; **p. gley podzol,** podzol gleyifié à horizon concrétionné; **p. loam,** limon t.; **p. meadow soil,** sol tourbiforme de prairie; **p. mor,** humus brut t.; **p. muck,** sol t. neutre, sol organique très décomposé; **p. podzolic soil,** podzol t.; **p. soil,** sol t.

pebble, galet (20 à 64 mm); **p. armor,** pavement désertique; **p. beach,** plage de g.; **p. culture,** industrie humaine (début paléolithique); **p. gravel,** cailloutis; **p. jack,** blende; **p. phosphate,** phosphorites; **p. stone,** g.; **facered p.,** caillou à facettes; **rounded p.,** g. émoussé; **striated p.,** g. strié.

pebbly, caillouteux.

pechblende, pechblende (*minér.*) var. uraninite.

pechstein, pechstein.

Pecten, Pecten (Lamellibranche).

pectolite, pectolite (*minér.*).

ped, agrégat de particules de sol.

pedal gape, sinus (correspondant au passage de la sole pédieuse, *pal.*).

pedalfer (cf. pedocal), pédalfer (terme ancien).

pedestal, **1.** pente douce dans une roche; **2.** roche champignon; **p. boulder, rock,** bloc perché, roche champignon.

pedicellaria, pédicellaire (*pal.*).

pedicle, pédicule (*pal.*).

pedicle valve, valve ventrale (Brachiopodes).

pediment, pédiment, glacis rocheux; **p. embayment,** golfe de p.; **p. mantle,** débris en surface du p.;

inset p., p. emboîté; **rock p.,** p.; **desert p.,** p. désertique.

pedimentation, processus de formation d'un pédiment.

pediplain, pediplane, pédiplaine, plaine de corrosion.

pedocal (cf. pedalfer), pédocal (ancien terme), corrosion.

pedogenesis, pédogenèse.

pedology, pédologie, science des sols.

pedon, plus petite unité pédologique avec tous les niveaux du profil.

peel thrust, écaille tectonique.

peeling, desquamation, écaillage.

pegmatite, pegmatite.

pegmatitic, pegmatitique; **p. stage,** phase p. de cristallisation.

pegmatization, pegmatisation.

pegmatoid, pegmatoïde, à faciès de pegmatite.

pegmatophyre, pegmatophyre.

pelagic, pélagique; **p. ooze,** vase p. à débris d'organismes (Globigérines, Radiolaires, Diatomées, etc.).

Pelecypoda, Lamellibranches (*pal.*).

Pelee type eruption, éruption de type péléen.

Pele's hair, cheveux de Pelée (produits pyroclastiques).

Pele's tears, larmes de Pelée.

pelit, pelite, pelyte, pélite (équiv. lutite).

pelitic, pélitique, argileux.

pelitic-metamorphite sequence, métapélite.

pellet, granule, petite concrétion, boulette; **p. limestone,** gravelle, calcaire graveleux; **p. structure,** structure en agrégats, structure micro-noduleuse (des argiles); **fecal p., faecal p. (U.K.),** coprolithe.

pelletizing, formation de nodules, de boulettes.

pellicular water, eau pelliculaire.

Pelmatozoa, Pelmatozoaires (*pal.*).

pelmicrite, pelmicrite.

pelsparite, pelsparite.

pen, stylet (d'un séismographe).

pencil stone, pyrophyllite (*minér.*).

pendant, apophyse de roches encaissante dans roche intrusive, enclave.

penecontemporaneous, pénécontemporain (après sédimentation, avant lithification synsédimentaire).

peneplain, pénéplaine; **incipient p.,** p. embryonnaire; **rejuvenation of p.,** rajeunissement d'une p.; **stripped p.,** p. dégagée.

peneplanation, pénéplanation.

penetration, pénétration, intrusion; **p. test,** pénétrométrie.

penetration twin, mâcle d'interpénétration (*cristallo.*).

penetrative rock, roche intrusive.

penetrometer, pénétromètre.

peninsula, péninsule.

pennite, penninite, pennite, penninite (minéraux du groupe des chlorites).

Pennsylvanian (N. Am.), Pennsylvanien (période, Carbonifère sup.).

pennystone, sphérosidérite.

pentagonal dodecahedron, dodécaèdre pentagonal (*cristallo.*).

pentamerous symmetry, symétrie de type cinq, pentagonale.

pentlandite, pentlandite, pyrite nickelifère ou nicopyrite.

pepino hill (syn. mogote, hum), butte témoin karstique (Puerto-Rico).

pepperite, pépérite.

pepita, pépite.

peptize (to), maintenir en suspension une solution colloïdale.

peptizer, défloculant.

peralkaline, hyperalcalin.

peraluminous rock, roche hyperalumineuse (classif. de Shand).

percentage log, diagraphie de forage basée sur des pourcentages pétrographiques des déblais (« cuttings »).

perched, perché; **p. block,** bloc p.; **p. boulder,** bloc p.; **p. ground water,** nappe phréatique perchée; **p. valley,** vallée perchée; **p. water,** eau perchée; **p. water table,** nappe phréatique perchée.

percolation, percolation, infiltration (d'eau).

percrystalline (obsolete), percristallin (porphyrique).

percussion, percussion; **p. boring,** sondage par p.; **p. drilling,** forage à p.; **p. mark,** marque de p. (d'un galet contre un autre).

perdigon (obsolete), concrétion ferrugineuse.

perennial stream, cours d'eau permanent; **perennial snowline,** limite des neiges persistantes.

perennially frozen ground, pergélisol.

perfect, parfait; **p. elasticity,** élasticité parfaite; **p. gas,** gaz parfait; **p. plasticity,** plasticité parfaite.

pergelisol, pergélisol.

peribac, (lorsque la Lune et la Terre sont rapprochées au maximum et alignées avec le Soleil).

periclase, périclase (*minér.*).

periclinal, périclinal (*tecto.*).

pericline, pericline.

peridot, péridot, olivine (*minér.*).

peridotite, péridotite (*pétro.*); **p. shell,** manteau terrestre.

perigee, périgée; **p. syzygy tide,** marée de vives eaux.

periglacial, périglaciaire; **p. geomorphology,** géomorphologie périglaciaire.

perihelion, périhélion.

perimagmatic, périmagmatique.

period (i.e. Jurassic period), période (ex. période Jurassique).

periodicity, périodicité (climatique).

periostracum, périostracum (*pal.*).

periproct, périprocte, anus (*pal.*).

Perissodactyla, Périssodactyles (*pal.*).

peristerite, péristérite (variété translucide d'albite).

peristome, péristome (*pal.*).

peritectic point, péritectique.

peritidal, relatif à l'estran (zone de balancement des marées).

perlite, perlite (*pétr.*).

perlitic, perlitique.

permafrost, pergélisol; **p. aggradation,** accroissement du p.; **p. subsoil,** p.; **p. table,** niveau supérieur du p.

permanent frozen ground, frozen soil, pergélisol.

permeability, 1. perméabilité (à un fluide); 2. perméabilité magnétique; **p. coefficient,** coefficient de perméabilité; **p. trap,** piège stratigraphique (à toit imperméable); **lateral p.,** perméabilité latérale; **relative p.,** perméabilité relative; **vertical p.,** perméabilité verticale.

permeable, perméable.

permeate (to), filtrer à travers, imprégner.

permeation, imprégnation, pénétration (d'un fluide).

Permian, Permien (période; Paléozoïque).

permineralization, minéralisation secondaire (des fossiles, etc.) par infiltration et précipitation dans les pores.

permineralized plant, végétal minéralisé (en silice, etc.).

permit , permis (d'exploitation de forage, etc.).

Permo-Triassic, 1. adj : permo-triasique; 2. n : Permo-Trias.

perovskite, pérovskite (*minér.*).

peroxide, peroxyde.

perpendicular, perpendiculaire, vertical; **p. separation,** rejet perpendiculaire; **p. slip,** composante perpendiculaire du rejet net; **p. throw,** rejet vertical mesuré perpendiculairement aux couches.

perpetually frozen soil, pergélisol.

persilic rock, roche hypersiliceuse (% de Si compris entre 69 et 80).

persistent water-table soil, sol perpétuellement imbibé d'eau, sol à gley.

perthite, perthite, association de feldspath potassique et sodique par interpénétration (cf. microcline).

perthitic texture, texture perthitique.

pervade (to), s'infiltrer dans, pénétrer.

pervasive, pénétrant.

pervious, perméable.

perviousness, perméabilité.

pesticide, pesticide.

pestle (to), broyer au mortier.

petalite, pétalite (*minér.*).

peter (saltpeter, petre), pierre.

peter out (to), s'épuiser, diaparaître en biseau.

petrification, petrifaction, pétrification, lapidification.

petrified (desert) rose, rose des sables (en gypse ou en barytine).

petrified wood, bois silicifié.

petrify (to), pétrifier, fossiliser.

petrocalcic horizon, horizon carbonaté induré.

petrochemical, pétrochimique.

petrochemistry, pétrochimie.

petrofabric, pétrographie structurale, structurologie; **p. analysis,** pétrographie structurale; **p. diagram,** diagramme structural.

petrofacies, lithofaciès.

petrogenesis, petrogenèse.
petrogenetic, pétrogénétique, lithogénétique.
petrogenic grid, diagramme pression-température d'équilibre de phases minérales.
petrogeny, petrogenèse.
petrogeny's residual system, liquide magmatique résiduel.
petrograph, *petrographer*, pétrographe.
petrographic(al), pétrographique; **p. microscope,** microscope polarisant; **p. province,** province pétrographique.
petrography, pétrographie.
petrogypsic horizon, horizon gypseux.
petroleum, pétrole; **p. basin,** bassin pétrolifère; **p. crude,** p. brut; **p. deposit,** gisement de p.; **p. engineering,** génie pétrolier; **p. geologist,** géologue pétrolier; **p. geology,** géologie du p.; **p. seep,** suintement de p.; **p. well,** puits de p.
petroliferous, pétrolifère; **p. bed,** couche p.; **p. province,** région p.; **p. shale,** schiste bitumineux; **p. structure,** structure p.
petrologic, pétrographique, pétrologique; **p. microscope,** microscope polarisant.
petrologist, pétrographe.
petrology, pétrographie, pétrologie.
petromict, *petromictic (obsolete)*, hétérogène, polygénique.
petrophysics, pétrophysique.
petrous (obsolete), pierreux.
petzite, petzite (*minér.*).
pewter, étain.
pewtery, stannifère.
phacoid, *phacoides (inusited)*, phacoïde (massif intrusif lenticulaire).
phacoidal, lenticulaire.
phacolite, phacolite (var. de chabasite) (*minér.*).
phacolith, phacolite (massif intrusif).
Phaeophyta, Phéophycées (algues brunes).
phaneric, *phaneritic*, phanéritique, phanérocristallin.
phanero (prefix), visible.
phanerocrystalline, phanérocristallin (à cristaux visible à l'œil nu).
Phanerogams, Phanérogames (*paléobot.*).
Phanerozoic, Phanérozoïque (la vie est visible).
phantom, fantôme; **p. horizon,** horizon f., **p. reflection,** réflection interne (*optique; métallo.*).
pharmacolite, pharmacolite (*minér.*).
phase, phase; **p. boundary,** limite de p.; **p. change,** déphasage; **p. diagram,** diagramme d'équilibre de p.; **p. equilibria,** équilibre de p.; **p. lag,** retard de phase, déphasage; **p. rule,** règle des p.; **p. shift,** changement de p.; **p. system,** système de p.; **p. velocity,** vitesse de p.
phasing, interférence.
phenacite, phénacite (*minér.*).
phenoclast, grand fragment détritique dans une matrice sédimentaire plus fine.
phenocryst, phénocristal.
phenotype, phénotype.
phial, fiole, flacon.

phi grade scale, échelle granulométrique logarithmique des unités « phi ».
phillipsite, phillipsite (var. de zéolite).
phlogopite, phlogopite, mica magnésien.
phonolite, phonolite (*pétrogr.*).
phorogenesis, phorogenèse.
phosgenite, phosgénite (*minér.*).
phosphate, phosphate; **p. phosphate rock,** roche phosphatée.
phosphatic, phosphaté; **p. conglomerate,** conglomérat phosphaté; **p. deposits,** dépôts phosphatés, phosphorites; **p. foecal pellet,** coprolite; **p. nodule,** nodule p.; **p. oolite,** oolithe phosphatée; **p. sand,** sable phosphaté; **p. sandstone,** grès phosphaté.
phosphatization, phosphatisation.
phosphor, phosphor.
phosphorescence, phosphorescence.
phosphorescent decay, radiophosphorescence.
phosphoric, phosphoré.
phosphorite, 1. phosphorite; 2. phosphates.
phosphorous, phosphoreux.
phosphorus, phosphore.
photic zone, zone photique (0 à − 200 m).
photocontour, carte topographique obtenue par photorestitution.
photogeological map, carte photogéologique.
photogeology, photogéologie.
photogeomorphology, photogéomorphologie.
photogrammetric, photogrammétrique; **p. mapping,** cartographie photogrammétrique.
photogrammetry, photogrammétrie.
photointerpreter, photointerprétateur.
photomap, photoplan.
photometer, photomètre.
photometry, photométrie.
photomicrograph, *photomicrography*, microphotographie.
photomicroprob analysis, analyse à la microsonde photographique électronique.
photomorphology, photogéomorphologie.
photosynthesis, photosynthèse.
phragmocone, phragmocône (*pal.*).
phreatic, phréatique; **p. activity,** manifestation p. volcanique; **p. eruption,** éruption volcanique p.; **p. water,** eau souterraine (sous la nappe p.); **p. zone,** zone de saturation.
phreatomagmatic, phréatomagmatique.
phreatophyte, phréatophyte (*pal.*).
phtanite, phtanite (*pétro.*).
phyla, pluriel de phylum.
phyletic, phylétique; **p. evolution,** évolution p.
phylliform, feuilleté; **p. structure,** structure f.
phyllite, phyllite.
phyllitic metatuff, tuf volcaniclastique métamorphisé (avec des linéations de micas microscopiques).
phyllitization, formation de roches feuilletées.
Phylloceratids, Phyllocératidés (*pal.*).
phylloid, foliacé.
phyllonite, mylonite recristallisée.
phyllose, foliacé.

phyllosilicates, phyllosilicates.
phylogenic species, espèce phylogénique.
phylogeny, phylogénie.
phylogerontism, dégénérescence d'un phylum.
phylum, phylum.
phyric, phyrique, porphyrique, porphyroïde.
physical, physique; **p. geography,** géographie p.; **p. weathering,** désagrégation mécanique.
physical process maps, carte des risques naturels et artificiels.
physiographic, géomorphologique; **p. province,** région géographique à climat et structure homogènes.
physiography, géographie physique, géomorphologie.
phytocoenose, phytocénose (*pal.*).
phytogenic, d'origine végétale; **p. soil,** sol phytogène.
phytokarst, karst des algues, phytokarst.
phytolith, phytolite (structure minérale formée par une plante).
phytomorphic soil, sol d'origine végétale.
phytophagous, phytophage (*pal.*).
phytoplankton, phytoplancton.
Piacentian (see Plaisancian), Plaisancien (étage; Pliocène inf.).
Pick, 1. pic, pioche; 2. pic (d'un diagramme); **stone dressing p.,** pic de tailleur de pierre.
picked ore, concentré de triage.
picker, 1. trieur (mine); 2. pic, pioche.
pickhammer, marteau piqueur.
picking, piochage, triage à main.
pickle (to), décaper.
picotite, picotite, spinelle chromifère (*minér.*).
picrite, picrite (roche ultrabasique).
picritic, picritique.
picromerite, picromérite (*minér.*).
pictograph, diagramme de variabilité, diagramme de dispersion.
piecemeal stoping, assimilation magmatique (par digestion des enclaves).
piedmont, piédmont; **p. alluvial plain,** plaine alluviale de p.; **p. glacier,** glacier de p.; **p. slope,** glacis de p.; **p. steps,** gradins de p.
piedmontite, piémontite, épidote manganésifère.
pier, épi littoral.
pierce (to), transpercer, pénétrer.
piercement fold, pli diapir.
piercing fold, diapir intrusif.
piezoclase, fracture de pression, piézoclase.
piezocrescence, piézocristallisation (accroissement de minéraux suivant une direction cristallographique déterminée sous l'influence de pressions).
piezocrystallization, piézocristallisation.
piezoelectric, piézoélectrique; **p. detector,** détecteur piézoélectrique, sismographe piézoélectrique.
piezoelectricity, piézoélectricité.
piezometer, piézomètre.
piezometric, piézométrique; **p. level,** niveau p.; **p. surface,** niveau p.

piezometry, piézométrie.
pigeonite, pigeonite (variété de diopside).
pike, pic, pioche.
pikeman, abatteur, piqueur (mine).
pile up (to), amonceler, entasser.
pillar, pilier; **p. drawing,** dépilage (mine); **p. mining,** exploitation par p.; **p. rock,** roche champignon.
pillar (to), soutenir, consolider par des piliers.
pillow, coussinet, oreiller; **p. lava,** lave en oreiller; **p. structure,** désagrégation en boules; **salt p.,** coussinet de sel.
pillowed cone, évent volcanique sous-marin, recouvert de pillow-lavas.
pillowed unit, formation en coussinet.
pilotaxitic, pilotaxitique.
pilot balloon, ballon sonde.
pimple mound, petit monticule.
pimple plain, plaine mamelonnée (Texas et Louisiane).
pinacoid, pinacoïde.
pinacoidal, pinacoïde.
pinch, pinch out, 1. amincissement en coin, rétrécissement; 2. biseau (*stratigr.*).
pinch and swell, boudinage.
pinching, rétrécissement, amincissement.
pinch out (to), disparaître progressivement en biseau (*stratigr.*).
pingo, pingok (pl: pingos), pingo(s), monticule de glace en forme de petit volcan couvert de sédiment; hydrolaccolithe; **p. like mound,** monticule pingoïde; **p. remnant,** mardelle (dépression ovale laissée après fusion de la lentille de glace); **p. ridge,** rebord d'un pingo annulaire.
pink soil, sol rosé calcaréo-argileux, terra-rossa.
pinnacle, pic, cime, pinacle, pyramide; **p. reef,** pinacle récifal; **p. rock,** «pénitent».
pinite, pinite.
Pinnipedia, Pinnipèdes (*pal.*).
pinnule, pinnule (*pal.*).
pipage, transport par conduites.
pipe, 1. conduite, canalisation, tuyau; 2. cheminée volcanique; **p. clay,** nodule d'argile réfractaire remaniée; **p. ore,** limonite en remplissage vertical dans matrice argileuse, dans fissures karstiques; **p. stone,** catlinite; **ore p.,** colonne minéralisée, «cheminée» de minerai.
pipelaying barge, bateau poseur de conduites sous-marines.
pipeline, oléoduc, pile-line; **p. flow efficiency,** débit de l'oléoduc; **p. run,** quantité transportée; **gas p.,** gazoduc; **oil p.,** oléoduc.
pipeline (to), transporter par pipeline.
piper, soufflard.
piping, 1. canalisation; 2. abatage hydraulique (mine).
pipkrake (swedish), pipkrake, aiguille de glace.
piracy, capture (du cours supérieur d'une rivière).
pirate stream, rivière qui capte le cours supérieur d'une autre rivière.
Pisces, Poissons (*pal.*).
pisoids, pisolithes.

pisolite, pisolith, pisolite.

pisolitic, pisolithic, pisolithique; **p. gravel,** gravier p.; **p. iron,** minerai de fer en grains; **p. limestone,** calcaire p.; **p. tuff,** tuf volcanique pyroclastique composé de lapillis agglomérés.

pistazite, pistacite, épidote finement cristallisée.

pit, 1. trou, cavité, alvéole; 2. puits de mine, mine; 3. carrière; 4. fosse de coulée (métallurgie); **p. and mound soil,** sol à microrelief accidenté; **p. crater,** cratère puits; **p. fire,** feu de mine; **p. gas,** grisou; **p. headframe,** chevalement; **p. man,** mineur; **p. timber,** bois de mine; **clay p.,** glaisière; **coal p.,** mine de charbon; **marl p.,** marnière; **mud p.,** bac à boue; **sand p.,** sablière; **setting p.,** bassin de décantation.

pitch, 1. poix, brai, asphalte; 2. plongement de l'axe d'un pli; 3. angle formé entre un filon principal et une ramification; 4. puits vertical *(spéléo.)*; **p. coke,** coke de brai; **p. length,** longueur d'une apophyse filonienne; **earth p.,** asphalte; **minerai p.,** asphalte; **tar p.,** brai de goudron.

pitchblende, pechblende, uraninite *(minér.).*

pitching, 1. descente, inclinaison, pente; 2. incliné (adj.); **p. of slope,** perré.

pitchstone, pechstein, rétinite *(pétrogr.).*

pitted, corrodé, dépoli; **p. plain,** 1. plaine fluvioglaciaire à dépressions dues à la fusion de la glace; 2. plaine à nombreuses petites dolines rapprochées; **p. surface (of a grain),** surface picotée (d'un grain de sable).

pivot fault, faille « à charnière » (faille normale dont le rejet diminue progressivement).

Placentals, Placentalia, mammifères placentaires *(pal.).*

placer, gisement alluvial, placer; **p. claim,** concession minière dans les alluvions ou autre dépôt meuble; **p. deposit,** gîte alluvionnaire; **p. mining,** exploitation de gisements minéraux dans les terrains superficiels par lavage.

Placoderm, Placodermes *(pal.).*

plagioclase, plagioclase; **p. feldspars,** feldspaths plagioclases.

plagioclasite, plagioclasite *(pétro.).*

plagiogranite, granodiorite.

plagionite, plagionite *(minér.).*

plagiophyre, plagiophyre *(pétro.).*

plain, 1. adj: plat, plan; 2. n: plaine; **p. of denudation,** niveau d'aplanissement; **p. tract,** lit majeur, large, du cours inférieur d'un fleuve; **abyssal p.,** plaine abyssale; **coastal p.,** plaine côtière; **outwash p.,** plaine d'épandage proglaciaire.

Plaisancian (Eur.), Plaisancien (étage Pliocène).

planar, plan, plat; **p. cross-stratification,** stratifications entrecroisées planes; **p. flow structure,** structures planes, litées ou feuilletées de roches magmatiques; **p. preferred orientation,** foliation plane (roche métamorphique); **p. water,** eau laminaire.

planation, 1. aplanissement; 2. élargissement des vallées par érosion latérale; 3. pénéplanation; **p. surface,** surface d'aplanissement.

plane, plan; **p. of cleavage,** p. de clivage; **p. of polarization,** p. de polarisation; **p. of stratification,** p. de stratification; **p. of symmetry,** p. de symétrie; **p. of unconformity,** p. de discontinuité; **p. polarized light,** lumière polarisée dans un p., lumière polarisée non analysée (L.N.P.A.); **p. survey,** levé à la planchette; **p. table,** planchette; **p. table survey, p. tabling,** levé à la planchette; **axial p.,** p. axial; **bedding p.,** de stratification; **boundary p.,** surface limite; **slip p.,** surface de glissement.

planet, planète.

planetary, planétaire; **p. geology,** géologie p., astrogéologie; **p. satellite,** satellite d'une planète; **p. science,** science p.

planetoid, astéroïde.

planetology, planétologie.

planimetric map, carte planimétrique.

planimetry, planimétrie.

planispiraled shell, coquille enroulée dans un plan.

planitia (Mars), « plaine » martienne.

planktivorous, planctivore *(pal.).*

plankton, plancton.

planktonic, planctonique.

planosol, planosol.

plant cover, couverture végétale.

plantae, plantes *(pal.).*

plasma, 1. plasma (calcédoine vert-foncé); 2. plasma *(astro.).*

plaster, plâtre.

plastic, plastique; **p. clay,** argile p.; **p. deformation or flow,** déformation p.; **p. index,** indice de plasticité; **p. limit,** limite de plasticité; **p. relief map,** carte en relief sur p.; **p. strain,** déformation p.; **lower p. limit (G.B.),** limite inférieure de plasticité; **upper p. limit (G.B.),** limite supérieure de plasticité.

plastically strained crystals, cristaux déformés plastiquement.

plasticity, plasticité; **p. index,** indice de p.; **p. limit,** limite de p.

plate, 1. plaque lithosphérique *(tectonique)*; 2. plateau *(industrie)*; **p. boundary,** limite d'une plaque; **p. collision,** collision de plaques; **p. like structure,** structure en plaquettes; **p. margin,** marge d'une plaque lithosphérique; **p. rotation,** rotation de plaques; **p. shale,** schiste en plaquettes; **p. tectonics,** tectonique des plaques; **thin p.,** plaque mince.

plateau, plateau; **p. basalt,** basalte des plateaux; **p. glacier,** inlandsis tabulaire.

platform, plateforme; **p. reef,** récif corallien en p.; **abrasion p.,** p. d'abrasion; **solution p.,** p. de dissolution; **wave-built p.,** p. d'accumulation; **wave-cut p.,** p. d'érosion marine.

platinum, platine.

platy, aplati, en plaquettes; **p. flow structure,** structure pétrographique ou minéralogique en feuillets; **p. fracture,** débit en plaquettes; **p. parting,** débit en plaquettes; **p. structure,** structure en plaquettes.

Platyrrhin, Platyrrhiniens (*pal.*).

playa, playa; **p. lake,** lac temporaire (se transformant en zone boueuse ou playa par évaporation).

Pleistocene, Pléistocène (série ou époque du Cénozoïque) (période du Quaternaire).

pleochroic, pléochroïque; **p. haloe,** halo p.

pleochroism, pléochroïsme.

pleomorphous, pleomorphic, polymorphe.

Plesiosauria, Plésiosaures (*pal.*).

pleural, pleural (*pal.*).

pleura (pl: pleurae), (*pal.*).

plicate, plicated, 1. plissé, plissoté; 2. présentant des côtes (*pal.*).

plication, plication, involution.

Pliensbachian, Pliensbachien (étage; Lias).

plinian (see vulcanian), plinien (*vulcano.*).

plinthite, plinthite (*pédol.*).

Pliocene, Pliocène (série ou époque; Cénozoïque).

plot, 1. tracé, graphique; 2. restitution (photogr. aérienne).

plot (to), 1. reporter des données sur une carte; faire un levé de terrain; dresser un plan; tracer une courbe; 2. restituer (photogr. aérienne).

plotted section, coupe restituée (par photogrammétrie).

plotter, appareil de restitution, restituteur (photogr. aérienne); **photographic p.,** stéréographe; **stereoplotter,** stéréorestituteur.

plotting, levé, tracé, restitution; **p. scale,** échelle de restitution.

pluck (to), 1. écailler, détacher des fragments; 2. déloger (glacier).

plucking, 1. éclatement, débitage par le gel ou par la glace en mouvement; 2. arrachement de blocs de substratum par les glaciers ou les fleuves.

plug, bouchon, obturateur; **p. dome,** cumulo-volcan, dôme volcanique.

plug (to), boucher, colmater.

plugging, obturation, colmatage.

plumasite, plumasite (var. d'anorthosite à corindon) (*pétrogr.*).

plumbagina, graphite.

plumbaginous, graphiteux.

plumbago, graphite.

plumbiferous, plombifère.

plumbogummite, plumbogummite (*minér.*).

plumes, panaches.

plume agate, agate dendritique.

plumose antimonial ore, jamesonite, stibnite (*minér.*).

plunge, plongement, inclinaison (de l'axe d'un pli); **p. point,** rupture de pente sur l'estran; **p. pool,** grande marmite de géant (au pied de chutes d'eau).

plunging, plongeant; **p. fold,** pli p.

plush copper, chalcotrichite, (= cuprite à faciès aciculaire) (*minér.*).

pluton, pluton, intrusion ignée.

plutonic, plutonique; **p. emanations,** substances magmatiques volatiles; **p. metamorphism,** métamorphisme p.; **p. rock,** roche p.; **p. series,** séries

de roches formées à partir d'un magma originel; **p. water,** eau juvénile, eau endogène.

plutonism, plutonisme.

plutonite, plutonite.

plutonium, plutonium.

pluvial, pluvial (*géomorph., météo.*); **p. period, pluvial stage,** période pluviale, époque pluviale.

pluviometer, pluviomètre.

pluviometric, pluviométrique.

pluviometry, pluviométrie.

pneumatogenic, pneumatogène, formé par agent gazeux.

pneumatolysis, pneumatolysm, pneumatolyse.

pneumatolytic, pneumatolytique; **p. metamorphism,** métamorphisme de contact modifié par pneumatolyse; **p. mineral,** minéral pneumatolytique.

pneumotectique, cristallisation du magma modifiée par pneumatolyse.

pocket, poche, amas; **p. beach,** petite plage abritée; **p. stereoscope,** stéréoscope portatif.

pockety, à poches richement minéralisées.

pod, masse minéralisée allongée («en cigare») (mines).

podsol, podzol (*pédol.*); **p. loam,** limon podzolisé.

podsolic, podzolique; **p. horizon,** horizon cendreux ou horizon décoloré; **p. peat,** sol tourbeux podzolique; **p. soil,** sol podzolique; **gray brown p. soil,** sol lessivé.

podsolization, podzolisation.

podsolized, podzolisé; **p. lateric soil,** sol latéritique; **p. red earth,** sol rouge lessivé; **p. rendzina,** rendzine podzolisée; **p. soil,** sol podzolisé.

podsoluvisol, podzoluvisol, sol lessivé, glossique, sol podzolique glossique (*pédol.*).

poecilitic (obsolete), poikilitic, poecilitique (*pétr.*).

poeciloblastic, poikiloblastic, poeciloblastique, texture à gros cristaux dans les roches métamorphiques.

poikiloblasts, porphyroblastes spongieux.

point, 1. point; 2. pointe (*préhist.*); **p. bar,** banc de sable, de lobe convexe de méandre; **p. counting,** comptage par points; **p. diagram,** diagramme structural en points; **p. of the horse,** point de ramification d'un filon; **dew p.,** point de rosée; **melting p.,** point de fusion; **yield p.,** point de rupture.

point (to), être orienté vers.

poised stream, cours d'eau en équilibre de charge (pas d'érosion, pas de sédimentation).

polar, polaire; **p. circle,** cercle p.; **p. ice-cap,** calotte glaciaire; **p. projection,** projection p.; **p. wandering,** migration des pôles; **p. wandering curve,** ligne de déplacement des pôles paléomagnétiques; **p. zenithal gnomonic projection,** projection polaire.

polarity, polarité; **p. period,** période de polarité géomagnétique.

polarizability, polarité.

polarization, polarisation; **p. microscope,** microscope polarisant; **self p., spontaneous p.,** p. spontanée; **p. colours,** couleurs de p.

polarize (to), polariser.
polarized, polarisé; **p. light,** lumière p.
polarizer, polariseur.
polarizing, polarisant; **p. angle,** angle de polarisation; **p. microscope,** microscope p.
polder, polder (Hollande; Belgique).
pole, 1. pôle; 2. poteau; **magnetic p.,** pôle magnétique.
polianite, polianite, pyrolusite (*minér.*).
polished section, lame pétrographique polie (pour examen au microscope métallographique); **p. pebble,** galet poli (ex. éolisé).
polje, polye, poljé (serbo-croate).
pollen, pollen; **p. analysis,** analyse pollinique; **p. diagram,** diagramme pollinique; **p. spectrum,** spectre pollinique; **nonarboreal p., (N.A.P.),** p.-de végétaux herbacés.
pollutant, substance polluante.
pollute (to), polluer.
polluted water, eau polluée.
pollution, pollution.
polonium, polonium.
polybasite, polybasite (*minér.*).
Polychaeta, Polychètes (*pal.*).
polychromatic, polychromatique.
polyconic map projection, projection polyconique ordinaire.
polycrystal, assemblage de cristaux, nodule cristallin.
polycrystalline, polycristallin.
polycyclic, polycyclique; **p. landscape,** relief paysage polycyclique.
polygenic, polygenous, polygenetic, polygénique; **p. breccia,** brèche p.; **p. conglomerate,** conglomérat p.; **p. soil,** sol complexe p.
polygeosyncline, polygéosynclinal (le long d'une bordure continentale).
polygon (of stones), polygone (de pierres).
polygonal fissure soil, polygonal ground, sol polygonal (périglaciaire).
polygonal soil, sol polygonal.
polygonal structured soil, sol polygonal.
polyhalite, 1. adj : polyhalin; 2. n : polyhalite (*minér.*).
polyhedra, polyèdre.
polyhedral structure, structure polyédrique (*pédol.*).
polyhedrous fabric, structure polyédrique.
polymer, polymera, polymère.
polymerization, polymérisation.
polymerize (to), polymériser.
polymetallic deposit, gisement polymétallique.
polymetamorphism, polymétamorphisme.
polymict, hétérogène; **p. breccia,** brèche hétérogène, polygénique.
polymictic, polygénique; **p. conglomerate,** conglomérat p.; **p. lake,** lac à brassage continuel des eaux; **p. rocks,** roches p. (arkoses, etc.).
polymineralic, polyminéral.
polymodal sediment, sédiment à courbe granulométrique plurimodale.
polymorh, polymorphic, polymorphous, poly-

morphe; **p. transitions,** transitions (atomiques), polymorphes (deux formes d'un minéral).
polymorphism, polymorphisme (existence de plusieurs formes cristallines pour une espèce minérale).
polyp, polype (*pal.*).
polyparium, colonie de polypiers.
polyphase deformation, orogénèse polyphasée.
polyphyletic, polyphylétique.
polysynthetic, polysynthétique; **p. twinning,** mâcle polysynthétique, mâcles multiples (*cristallo.*).
polytypic species, espèce formée d'un groupe de sous-espèces.
polytypism, polytypisme.
Polyzoan (see Bryozoan), Bryozoaires.
pond, mare, étang.
ponding, formation de lac par barrage naturel (soulèvement tectonique, glacier, glissement de terrain, etc.).
pondlet, petit marécage, petit étang.
ponor, aven.
Pontian, Pontien (étage; Miocène).
pontic (cf. euxinic), euxinique.
pool, gisement (de fluide liquide).
pool and riffles, topographie de lit fluviatile en bancs et en bassins.
poor coal, charbon pauvre.
poorly drained soil, sol mal drainé.
porcelain clay, kaolinite, kaolin.
porcellanite, porcellanite (*pétrol.*).
pore, 1. pore, interstice, vide; 2. pore (*pal.*), foramen; **p. pressure,** pression interstitielle; **p. size,** diamètre des pores; **p. space,** volume des pores, porosité; **p. water,** eau interstitielle.
Poriferan (cf. sponge), Spongiaires.
poriferous, poreux.
porosimeter, porosimètre.
porosity, porosité; **p. log,** diagramme de porosité; **fracture p.,** porosité de fracture (U.S.A.); **effective p.,** porosité effective; **intergranular p.,** porosité intergranulaire; **primary p.,** porosité primaire; **secondary p.,** porosité secondaire.
porous, poreux, perméable; **p, soil,** sol poreux.
porphyraceous, porphyroïde (= structure de roche ignée).
porphyre, porphyre.
porphyrite, porphyrite, roche ignée à phénocristaux.
porphyritic, porphyrique, porphyroïde.
porphyroblast, porphyroblaste.
porphyroblastic, porphyroblastique; **p. texture,** structure porphyroblastique.
porphyroid, porphyroïde.
porphyry, porphyre (roche éruptive à phénocristaux); **p. copper,** minerai de cuivre porphyrique, porphyre cuprifère.
portal, 1. détroit; 2. entrée d'une mine.
Porterfield (N. Am.), groupe de l'Ordovicien.
Portlandian, Portlandien (étage; Jurassique sup. ou Malm).

positive, **1.** adj : positif; **2.** n : voûte d'un craton; **p. area**, craton, zone terrestre longtemps émergée; **p. crystal**, cristal à allongement positif; **p. element**, craton, région ayant tendance au soulèvement; **p. elongation**, allongement positif (*cristallo.*); **p. gravity anomaly**, anomalie gravimétrique positive; **p. movement**, soulèvement; **p. movement of sea level**, soulèvement du niveau marin; **p. ore**, réserves prouvées de minerai; **p. segment**, craton; **p. shoreline**, rivage de submersion (par transgression marine, ou enfoncement littoral).

possibilities, possibilités, éventualités d'exploitation minière.

possible ore, gisement éventuel de minerai, minerai probable.

possibles reserves, réserves minérales probables.

post, poteau, pilier (*mine*); **substructure p.**, poteau de soutènement (*mine*).

postcumulus minerals, minéraux formés après coup qui ont été engendrés par cristallisation fractionnée.

post-deuteric alteration, altération post-deutérique.

post-drill, perforatrice à colonne.

postglacial, postglaciaire.

posthumous fold, pli posthume; **p. structure**, structure affectant des roches jeunes, apparaissant sur une structure ancienne.

post-kinematic mineral, minéral formé par métamorphisme après déformation.

post magmatic, tardimagmatique, hydrothermal.

post orogenic, post tectonique.

post tectonic, post tectonique.

pot clay, argile réfractaire.

pot hole, marmite de géant.

potamic, fluvial, fluviatile.

potamology, potamologie (étude des rivières; étym. grecque).

potash, potasse; **p. feldspar**, feldspath potassique; **p. fixation**, adsorption de potasse par les argiles; **p. mica**, mica potassique (muscovite).

potassic, potassique.

potassium, potassium; **p.-argon dating**, datation potassium-argon; **p. bentonite**, métabentonite.

potato stone, géode.

potential, **1.** adj : potentiel, susceptible de; **2.** n : potentiel électrique, potentiel énergétique; **p. barrier**, barrière de potentiel; **p. difference**, différence de potentiel; **p. drop**, chute de potentiel; **p. electrode**, électrode de potentiel; **p. energy**, énergie potentielle; **p. gradient**, gradient de potentiel; **p. ore**, gisement p.; **p. ratio method**, méthode des rapports de chute de potentiel; **p. sonde**, sonde de potentiel; **p. test**, essai d'écoulement (d'un puits de pétrole); **magnetic p.**, force magnétomotrice; **membrane p.**, potentiel de membrane; **open flow p.**, débit maximum d'un puits; **oxido-reduction p.**, potentiel d'oxydoréduction.

potentiometer, potentiomètre.

potentiometric map, carte de résistivité, ou d'équipotentiel.

potentiometric surface, surface piézométrique.

potentiometry, potentiométrie.

pothole, **1.** marmite de géant (petit kettle); **2.** trou rempli de sels ou de saumure (Vallée de la Mort); **3.** doline, dépression karstique; **4.** fondis, éboulement du toit d'une mine; **5.** glaisière abandonnée.

Potsdamian (N. Am.), Potsdamien (ancien étage; Cambrien sup.).

potter's clay, argile plastique.

potter's ore, galène.

pottery clay, argile plastique.

pounce, ponce.

pound, livre; **p. avoirdupois**, = 0,45359 kg; **p. troy**, = 0,37324 kg.

pounding, broyage, concassage.

pour, averse, grosse pluie.

powder, poudre; **p. diffraction method**, méthode des poudres Debye-Scherrer (diffractométrie rayons X); **p. diffractometer**, diffractomètre à poudre; **p. ore**, minerai disséminé pulvérulent; **p. snow**, neige poudreuse.

powdered, à l'état de poudre, en poudre.

powdery, pulvérulent, poudreux.

power, **1.** énergie, puissance; **2.** pouvoir grossissant; **p. plant**, centrale nucléaire; **dispersive p.**, pouvoir dispersif; **low p.**, faible pouvoir grossissant; **magnifying p.**, pouvoir grossissant; **nuclear p.**, puissance nucléaire; **resolving p.**, pouvoir de résolution.

pozzuolana, pozzolan, pozzolana, pozzuolane (Italy), pouzzolane, tuf.

prairie grey soil, sol gris de prairie; **p. soil**, sol brun-gris; **p.-steppe brown soil**, sol brun steppique.

prase, prase (variété de quartz vert).

prasinite, prasinite (variété de schistes verts).

Pratt isostasy, théorie isostatique de Pratt.

Preboreal (Eur.), Préboréal (intervalle; entre Dryas récent et Boréal).

Précambrian, Précambrien (ensemble des temps antépaléozoïques).

precession, précession; **p. camera**, chambre photographique de diffraction aux rayons X.

precious, précieux; **p. garnet**, grenat pyrope; **p. metal**, métal précieux; **p. stone**, pierre précieuse.

precipitate, précipité.

precipitates, calcaires d'origine chimique, calcaires de précipitation.

precipitate (to), précipiter.

precipitation, précipitation; **p. variability**, variation de la moyenne pluviométrique annuelle.

precipitous, escarpé, abrupt, à pic.

preconsolidation pressure, pression des sédiments sus-jacents (tassement diagénétique).

predation, prédation (*pal.*).

predator, prédateur (*pal.*).

predazzite, prédazzite (marbre magnésien).

preferred orientation, orientation préférentielle (de minéraux).

pregeologic, antérieur à l'histoire géologique de la Terre (imprécis).

preglacial, préglaciaire, antéglaciaire (spécifiquement anté-pléistocène).

prehistoric, préhistorique (avant les données historiques); **p. tool assemblage,** industrie p.

prehnite, prehnite (*minér.*).

Pre-Imbrian, Pré-Imbrien (plus vieille division des temps lunaires).

prekinematic mineral, minéral formé avant déformation tectonique.

preliminary survey, étude préliminaire; **p. waves,** ondes sismiques préliminaires.

preorogenic, 1. antérieur à l'orogenèse; 2. datant du début d'une phase orogénique.

preservation, conservation (*pal.*).

press, 1. pression; 2. presse; **hydraulic p.,** presse hydraulique.

press (to), comprimer, serrer.

pressure, pression; **p. arches,** ogives glaciaires; **p. decline,** baisse de p.; **p. drilling,** forage sous p.; **p. gauge,** manomètre; **p. gradient,** gradient de p.; **p. head,** p. hydrostatique; **p. loss,** perte de p.; **p. metamorphism,** métamorphisme de p.; **p. release,** diminution de pression, décompression tectonique; **p. release jointing,** fissuration par décompression; **p. ridge,** ride longitudinale de progression d'une coulée de lave; **p. shadow,** ombre de pression tectonique; **p. solution,** dissolution par pression (en : formation de stylolite); **p. texture,** structure cataclastique; **p. wave,** onde P, onde de compression; **p. well,** puits d'injection; **geostatic p.,** pression géostatique; **hydrostatic p.,** pression hydrostatique; **input p.,** pression d'injection.

pressured, comprimé.

pretectonic fabric, structure anté-orogénique; **p. pluton,** intrusion anté-orogénique; **p. recrystallization,** recristallisation anté-orogénique.

prevailing wind, vent prédominant.

Priabonian, Priabonien (étage; Éocène sup.).

primary, 1. primaire, originel; 2. Paléozoïque, Primaire (*stratigr.*); **p. basalt,** magma basaltique originel; **p. dip,** pendage originel; **p. dolomite,** dolomie primaire (formée par précipitation en milieu marin); **p. Era (obsolete),** Primaire, ère Paléozoïque; **p. flowage,** déformation magmatique des roches éruptives à l'état plastique; **p. magma,** magma primitif; **p. melt,** magma primitif; **p. mineral,** minéral primaire; **p. openings,** pores, cavités primaires; **p. recovery,** récupération primaire (pétrole); **p. sedimentary structure,** structure sédimentaire primitive; **p. soil,** sol autochtone; **p. stratification,** stratification primaire; **p. wave,** onde P (*sismol.*).

Primate, Primate.

prime meridian, méridien d'origine (Greenwich).

prime white oil, variété de kérosène.

primitive, primitif, primaire; **p. soil,** sol embryonnaire, lithosol; **p. water,** eau juvénile.

principal, principal; **p. axis,** axe de symétrie principale (système hexagonal); axe optique (cristaux uniaxes); **p. stresses,** composantes de la tension suivant trois axes perpendiculaires; **p. shock,** secousse principale d'un séisme.

print, empreinte, marque.

prism, prisme, solide cristallin à faces parallèles à l'axe vertical.

prism-like fabric, microstructure prismatique (*pédol.*); **p.-like structure,** structure prismatique (*pédol.*).

prismatic, prismatique; **p. iron pyrite,** marcassite; **p. jointing,** fissuration prismatique (des basaltes); **p. manganese ore,** pyrolusite; **p. structure,** structure prismatique (*pédol.*).

probable ore, gisement minéral probable.

probe, sonde, capteur.

Proboscidea, Proboscidiens.

process, processing, traitement (d'une matière, d'un minerai ou d'une donnée mathématique).

prodelta, prodelta; **p. clays,** argiles, limons, vases déposées en avant d'un delta; **p. slope,** talus prodeltaïque.

prod mark, marque de frottement.

producer, 1. puits de pétrole productif; 2. gazogène; 3. générateur.

producing, productif; **p. expenses,** frais d'exploitation; **p. horizon,** couche productrice de pétrole, roche réservoir.

production, production; **p, control,** gestion de la p.; **p. figures,** chiffres de p.; **p. horizon,** horizon productif; **p. rate,** taux de p.; **p. sand,** sable pétrolifère; **p. well,** puits productif.

productive pool, gisement productif; **p. well,** puits productif.

productivity, productivité; **p. index (P.I.),** indice de p.; **p. test,** essai de p.

profile, profil, coupe; **p. of equilibrium,** profil d'équilibre; **p. section,** coupe transversale géologique; **p. paper,** papier quadrillé; **p. recorder,** enregistreur de profils; **bore p.,** profil d'un sondage; **cross p.,** profil transversal, coupe; **longitudinal p.,** profil longitudinal; **soil p.,** profil pédologique.

profiler, profileur, enregistreur de profil.

profiling, enregistrement d'un profil sismique.

proglacial, proglaciaire; **p. lake,** lac proglaciaire.

progradation, avancée, progression du rivage vers la mer.

prograde, prograde (adj.).

prograde (to), s'avancer, progresser (vers la mer).

prograding, progradation.

prograding shore-line, rivage en progression vers la mer.

progressive, progressif; **p. development,** anagenèse; **p. evolution,** évolution progressive; **p. overlap,** transgression.

projection, projection (*cartogr.*).

promontory, promontoire, cap.

proof stress, limite élastique.

prop, étai, support, poteau; **p. drawing,** déboisage (*mine*); **p. stay,** étai (*mine*).
prop (to), boiser, étayer (*mine*).
propagation, propagation; **p. of waves,** p. des ondes.
propane, propane.
propylite, propylite (= variété d'andésite altérée).
propylitic alteration, propylitisation.
propylitization, propylitisation (altération hydrothermale de roches éruptives à grain fin).
Prosimii, Prosimiens (*pal.*).
prospect, 1. prospection; **2.** région en cours de prospection; **p. hole,** puits d'exploration; **p. well,** puits d'exploration.
prospect (to), prospecter.
prospecting, prospection; **p. hammer,** marteau de géologue-prospecteur; **p. licence,** permis de prospection, permis de recherche.
prospection, prospection, exploration.
prospective, prometteur, susceptible de contenir des réserves.
prospector, prospecteur.
protalus rampart, éboulis de blocs glissés sur congères de neige dans zones montagneuses.
Proterophytic, (or Archeophytic), Pteridophytic (or Paleophytic), temps des Algues (division de Paléobotanique).
Proterozoic, Protérozoïque (division du Précambrien); **p. mobile belts,** zones orogéniques mobiles du Précambrien.
Proterozoides, Protérozoïdes (orogénèse précambrienne).
Protist, Protista, Protistans, Protistes.
protoclase, protoclase.
protoclastic, protoclastique.
protoconch, protoconque (*pal.*).
protodolomite, dolomite primaire (*minér.*).
protogenic, protogenous, protogénique.
protogine, protogine (*pétro.*).
protolith, roche antérieure au métamorphisme.
protomylonite, protomylonite (mylonite de contact métamorphique).
proton, proton.
protopetroleum, protopétrole.
protore, gisement minéral à trop faible teneur pour être exploité (sauf s'il y a enrichissement secondaire).
prototype, archétype (*pal.*).
Protozoa, Protozoan, Protozoaire (unicellulaire).
protract (to), faire le relevé, établir un plan, une carte.
protrude (to), sortir, faire sortir, faire saillie (relief).
protrusion, protrusion, protubérance (relief).
protrusive dome, dôme d'extrusion.
proustite, proustite (*minér.*).
proved, proven, prouvé; **p. oil land,** terrain pétrolifère reconnu; **p. ore,** gisement minéral p.; **p. reserves,** réserves p.
provincial stage, unité stratigraphique locale.
proximate analysis, analyse quantitative approximative.

psamment, régosol sableux.
psammite, psammite, grès micacé.
psammitic texture, texture psammitique.
psammoblastic, psammoblastique.
psephitic, pséphitique; **p. rock,** rudite métamorphisée.
psepho (prefix), faux.
psephyte, psephite, pséphite (*pétro.*).
pseudoanticline, pseudoanticlinal.
pseudo-bedding, pseudo-stratification.
pseudobreccia, pseudobrèche, fausse brèche.
pseudo-clivage, pseudo-clivage.
pseudocrossbedding, structure ressemblant à la stratification entrecroisée.
pseudocrystalline, pseudocristallin.
pseudofossil, pseudofossile.
pseudo galena, sphalérite, blende (*minér.*).
pseudogley, pseudogley.
pseudogleyed, à pseudogley, pseudogleyifié.
pseudokarren, pseudo-karst.
pseudokarst, pseudo-karst.
pseudolamination, pseudoschistosité.
pseudomanganite, pyrolusite (*minér.*).
pseudomorph, pseudomorphe.
pseudomorphism, pseudomorphisme.
pseudonodule, structure concrétionnée.
pseudopodium, pseudopodia (pl.), pseudopode.
pseudosolution, pseudosolution, fausse solution.
pseudostratification, structure litée de roche ignée.
pseudotachylite, pseudotachylite (variété de mylonite).
pseudovolcano, cratère d'origine douteuse (astroblème, cratère d'explosion phréatique).
psilomelane, psilomélane (*minér.*).
Psilopsida, Psilopsidés (*paléobot.*).
psycrometer, psychromètre.
psycrophyte, psychrophyte.
Psylophyte, Psylophytale.
Pteridophyta, Ptéridophytes (*paléobot.*).
Pteridospermae, Ptéridospermées (*paléobot.*).
Pterobranchia, Ptérobranches (*pal.*).
Pterocerian, Ptérocérien (faciès du Kimméridgien inférieur).
Pterodactyl, Ptérodactyle (*pal.*).
Pteropod, Ptéropode (*pal.*); **P-ooze,** vase calcaire abyssale à P.
Pterosaur, Pterosauria, Ptérosauriens (*pal.*).
ptygmatic fold, pli ptygmatique.
puckering, plissotement.
pudding ball, concrétion argileuse durcie.
pudding stone, poudingue.
puddle, flaque d'eau, petite mare (sur glace ou glacier).
puddle (to), glaiser (le cœur d'une digue).
puddled, compacifié, damé (sol).
puf, 1. grisou; **2.** explosif, explosion (U.S.A.).
puffed soil, sol à microrelief accidenté.
puffing hole, trou souffleur, soufflard.
pug, 1. argile, glaise; **2.** zone broyée; **p. mill,** broyeur, malaxeur.
pug (to), 1. malaxer, pétrir; **2.** glaiser.

pulaskite, pulaskite (= syénite à néphéline hololeucocrate).

pull (to), 1. traîner, tirer, extraire; 2. détuber ou remonter des tiges (forage).

pull rod, tige d'entraînement.

pulling, remontée, extraction, enlèvement; **p. casing,** décuvelage; **p. test,** essai de traction (forage).

pulsar, pulsar (*astro*).

pulsate (to), entrer en vibrations, avoir des pulsations.

pulsation, pulsation, vibration.

pulse, vibration, signal, impulsion; **seismic p.,** onde sismique.

pulverulite, pulverite, calcaire pulvérulent, roche sédimentaire formée d'aggrégats construits (silteuse ou argileuse).

pumice, pierre ponce; **p. fall,** chute de fragments de pierre ponce; **p. flow,** avalanche de ponce en suspension; **p. tuff,** tuff pyroclastique consolidé formé de fragments de pierre ponce.

pumiceous, ponceux.

pumicite, pumilith, pumicite, cendres volcaniques lithifiées.

pump, pompe (*mine*); **p. out,** assèchement, épuisement; **p. station,** station de pompage; **drainage p.,** station d'exhaure; **slush p.,** pompe à boue.

pump a well dry (to), assécher un puits.

pumpellyite, pumpellyite (*minér.*).

pumping, pumpage, pompage, assèchement, exhaure.

pumping test, essai de pompage.

puncher, haveuse à pic.

punky crystal, cristal pourri.

puppet, poupée calcaire (dans un sol loessique).

Purbeckian, Purbeckien (faciès; Jurassique sup.).

pure quartz sandstone, grès pur (très riche en quartz).

purple copper ore, bornite (*minér.*).

push, poussée, impulsion; **p. moraine,** moraine d'avancée glaciaire, de poussée; **p. wave,** onde P.; **ice p.,** poussée de gel.

put (to) a well on, mettre un puits en production (forage).

put (to) down a shaft, creuser un puits (*mine*).

puzzolan, pouzolane.

P. wave delay, retard d'arrivée des ondes P.

pycnometer, pycnomètre.

pycnometry, pycnométrie.

pygidium, pygidium (*pal.*).

pyralspite, pyralspite (variété de grenat).

pyramid, pyramide, prisme pyramidal (*cristallo.*).

pyramidal garnet, idocrase (*minér.*); **p. manganese ore,** hausmannite (*minér.*); **p. peak,** pic abrupt, massif escarpé; **p. zeolite,** apophyllite (*minér.*).

pyrargyrite, pyrargyrite (*minér.*).

Pyrenean orogeny, orogénèse pyrénéenne (Éocène terminal).

pyribole, pyribole (ensemble pyroxène + amphibole indifférencié).

pyritaceous, pyriteux.

pyrite (U.K.), pyrite (*minér.*, FeS_2); **p. cockscomb,** marcassite (*minér.*); **p. copper,** chalcopyrite (*minér.*); **magnetic p.,** pyrrhotite (*minér.*); **spear p.,** marcassite; **tin p.,** stannite (*minér.*); **white iron p.,** marcassite (*minér.*).

pyritic, pyriteux; **p. smelting,** fusion p.

pyritiferous, pyriteux.

pyritization, pyritisation.

pyritize (to), pyritiser.

pyritobituminous, pyrobitumineux.

pyritohedron, pyritoèdre, dodécaèdre pentagonal.

pyritous copper, chalcopyrite (*minér.*).

pyrobitumen, pyrobitume.

pyrochlore, pyrochlore (*minér.*).

pyroclast, fragment pyroclastique.

pyroclastic, pyroclastique; **p. breccia,** brèche volcanique; **p. deposits,** dépôts de nuées ardentes; **p. flow,** ignimbrite, nuée ardente; **p. rock,** roche p.; **p. surge,** épanchement p., nuée ardente descendante (à la base du volcan); **p. tuff,** tuf volcanique.

pyrocrystalline, pyrocristallin (cristallisé à partir d'un magma fondu).

pyrogenese, pyrogénèse.

pyrogenic, pyrogenetic, pyrogénétique, pyrogène; **p. mineral,** minéral pyrogénétique (formé à haute température).

pyrogenous, pyrogène, igné.

pyrognostic, pyrognostique.

pyrolite, pyrolite.

pyrolusite, pyrolusite (*minér.*).

pyrolysis, pyrolyse.

pyromagnetic, pyromagnétique.

pyrometamorphism, pyrométamorphisme.

pyrometasomatic, pyrométasomatique.

pyrometric cone equivalent, essai au cône pyrométrique (pour les argiles).

pyromorphite, pyromorphite (*minér.*).

pyronaphta, pyronaphte.

pyrope, pyrope (= variété de grenat).

pyrophyllite, pyrophyllite (*minér.*).

pyroschist, schiste bitumineux.

pyroshale, schiste bitumineux.

pyrosphere, pyrosphère.

pyroxene, pyroxène; **p. hornfels facies,** faciès de cornéennes à pyroxène.

pyroxenite, pyroxénite (*pétro.*).

pyrrhotite, pyrrhotine, pyrrhotite (*minér.*).

Q

Q, symbole en séismologie indiquant l'atténuation des ondes.

Q period, Quaternaire.

Q wave, onde sismique de Love.

quadrangle, quart de carte (d'une échelle donnée).

quadrant, **1.** quart de cercle, un arc de 90 degrés; **2.** un instrument d'arpentage ancien (cf., sextant).

quadrature, **1.** une des deux positions autour de l'orbite lunaire (le premier ou le troisième quart); **2.** quand l'alignement Terre-Lune est perpendiculaire à l'alignement Terre-Soleil; **3.** position de 2 ou plusieurs planètes à 90°.

quaggy, marécageux.

quagmire, fondrière, marécage.

quake, tremblement; **earth q.**, tremblement de terre.

quake (to), trembler.

quanat (Arab.), conduit ancien souterrain.

quantitative, quantitatif; **q. analysis**, analyse quantitative.

quaquaversal (old term), dirigé dans tous les sens; **q. dip**, pendage rayonnant dans tous les sens; **q. fold**, dôme; **q. structure**, structure périclinale, dôme, coupole.

quarfeloid (in feldspars), roche contenant quartz, feldspaths, et feldspathoïdes.

quarrier, ouvrier carrier.

quarry, carrière; **q. face**, front de c.; **q. spall**, débris de c.; **q. stone**, moellon, pierre de taille; **q. wastage**, déblais de c.,; **q. water**, eau de c; **clay q.**, c. d'argile; **open q.**, c. à ciel ouvert; **rock q.**, c. de roches.

quarry (to), extraire la pierre, exploiter une carrière, creuser.

quarryman, carrier.

quartation, quartation (séparation de l'or et de l'argent dans un minerai).

quarter, **1.** quart; **2.** unité de mesure U.S.A. = 11,34 kg; système livre-avoir-du-poids = 12,7 kg.

quartering, méthode d'échantillonnage du minerai; **q. down**, division, fragmentation.

quarter-wave plate, lame mica quart d'onde.

quartile, quartile (*séd.*).

quartz, quartz; **q. anorthosite**, anorthosite avec un peu de q.; **q. arenite**, quartzite sédimentaire;

q. basalt, basalte à q; **q. claim**, concession minière dans roches éruptives à filons; **q. conglomerate**, poudingue siliceux; **q. diabase**, diabase quartzique; **q. diorite**, diorite quartzique; **q. drift**, dépôt détritique quartzeux; **q. felsite**, rhyolite ou porphyre quartzeux; **q. gabbro**, gabbro avec un peu de quartz; **q. mine**, mine d'or filonien; **q. monzonite**, monzonite quartzifère; **q. porphyry**, porphyre quartzifère, microgranite; **q. reef**, filon de q.; **q. rock**, quartzite; **q. trachyte**, rhyolite; **q. vein**, filon de q., souvent aurifère; **q. wedge**, coin lame de q.; **free-milling q.**, q. à or libre; **rose q.**, pierre précieuse de q. titanifère rougeâtre; rubis de Bohême; **rutilated q.**, q. à aiguille de rutile; **smoky q.**, q. enfumé.

quartziferous, quartzifère.

quartzite, **1.** quartzite (métamorphisme des grès, etc.); **2.** grès quartzitique (cimenté par silice secondaire) ou quartzite.

quartzitic sandstone, grès quartzite.

quartzoid, cristal de quartz bipyramidé (cristal à 6 faces).

quartzose, quartzeux; **q. sand**, sable q.; **q. sandstone**, grès q.

quartzous, (rare) quartzeux.

quartzy, (rare) quartzeux.

quasar, quasar (*astro.*).

quasi-cratonic (tect.), région ou processus de tectonisme limité (« germanotype »).

quasi-equilibrium, condition temporaire d'équilibre d'un cours d'eau.

quaternarist, spécialiste du Quaternaire.

Quaternary, Quaternaire (période; Cénozoïque);

quay, banc de sable côtier (U.S.).

quernstone (U.K.), grès meulier.

quest (to), prospecter.

quick, rapide; **q. clay**, argile saline thixotropique; **q. ground**, sol mouvant; **q. hardening cement**, ciment à prise r.; **q. lime**, chaux vive; **q. mud**, vase mouvante; **q. sand**, sable boulant, mouvant; **q. silver**, mercure; **q. vein**, filon productif.

quiescence, repos (*sismique*).

quiet reach, secteur calme d'un cours d'eau.

quinary system, système chimique à cinq composants.

R

rabban (rare), chapeau ferrugineux.
race, **1.** raz, ras (de courant); **2.** course (du soleil).
racemization age, méthode géochronologique basée sur la racémisation d'acides aminés.
rachis, rachis, axe (*pal.*).
rack, râtelier; **pipe r.**, parc à tiges (forage).
rack (to), ranger les tiges dans le derrick.
radar, radar, mécanisme et image de télédétection à distance; **r. astronomy**, radarastronomie; **r. imagery**, cartographie radar.
raddle, hématite (terreuse) (terme peu fréquent).
radial, radial; **r. drainage**, réseau hydrographique r.; **r. dyke**, dyke irradiant à partir de l'orifice d'un volcan punctiforme à cratère; **r. fault, 1.** faille r; **2.** faille à mouvement vertical prédominant; **r. symmetry**, symétrie r.
radials, plaques radiales (*pal.*).
radian, radian.
radiant energy, énergie radiante.
radiate (to), irradier, rayonner.
radiated pyrite, marcasite (*minér.*).
radiation, irradiation, radiation, rayonnement; **r. balance**, équilibre entre rayonnement transmis et réfléchi; **r. logging**, diagraphie par radiation; **r. meter**, détecteur de radiations; **r. rate**, intensité de rayonnement.
radii, côtes (des coquilles de Lamellibranches).
radioactive, radioactif; **r. age determination**, datation radiométrique; **r. dating**, datation radiométrique; **r. decay**, désintégration r.; **r. element**, élément r.; **r. natural background**, niveau r. naturel; **r. pollutant**, polluant r.; **r. series**, séries d'éléments r.; **r. tracer**, traceur r.; **r. waste**, déchet r.
radioactivity, radioactivité; **r. log**, diagraphie par r. (naturelle).
radioastronomy, radioastronomie.
radiocarbon dating, datation au carbone radioactif.
radioelement, élément radioactif.
radiogenic, radiogénique, formé par décomposition radioactive; **r. isotope**, isotope formé par décomposition radioactive; **r. lead**, plomb provenant de la décomposition radioactive.
radiography, radiographie (rayons X).
radio halo, halo pléochroïque.
radiointerferometer, radiointerféromètre (*astro.*).
radioisotope, radioisotope.
Radiolaria, Radiolaires.
Radiolarian, à Radiolaires; **r. chert**, radiolarite; **r. ooze**, boue océanique à radiolaires.
radiolarite, radiolarite (roches à Radiolaires).
radiole, radiole (*pal.*).
radiolitic structure, structure radiée.
radiometallography, radiométallographie.

radiometer, radiomètre; **scanning r.**, r. à balayage.
radiometric datation, datation radiométrique.
radionuclides, radionucléides.
radiophotography, radiophotographie.
radiosonde, radiosonde.
radiotelescope, radiotélescope (*astro.*).
radium, radium.
radius, rayon; **r. ratio**, rapport de deux r. ioniques.
radon, radon (gaz).
Radstockian (Eur.), Radstockien (étage; Carbonifère moy.).
radula, radula (*pal.*).
rafted ice block, gros bloc de pierre transporté par radeaux de glace.
rafting, **1.** transport de matériaux par la glace, les plantes ou autre matériel flottant; **2.** chevauchement de plaques de glace.
rag (U.S.), roche dure (siliceuse).
ragstone, **1.** roche à grain grossier; **2.** parfois, calcaire oolithique fossilifère (G.B.); **3.** pierre débitée en minces dalles.
ragged, déchiqueté (relief).
rain, pluie; **r. beat**, battage par la p.; **r. bow**, arc en ciel; **r. bow chalcedony**, calcédoine zonée irisée; **r. chart**, carte pluviométrique; **r. drop**, goutte de p.; **r. drop imprint**, empreinte de gouttes de p.; **r. fall**, chute de p.; **r. gauge**, pluviomètre; **r. glass**, baromètre; **r. pits**, empreintes d'impact de gouttes de p.; **r. print**, empreinte d'une goutte de p.; **r. rill**, rigole de ruissellement; **r. shadow**, zone abritée des p.; **r. shower**, averse; **r. splash**, impact des gouttes de p.; **r. wash**, ruissellement.
raised, soulevé, élevé; **r. beach**, plage soulevée; **r. bog**, sol de tourbière; **r. reef**, récif situé au-dessus du niveau de la mer; **r. shoreline**, ligne de rivage soulevée, côte soulevée.
rake, inclinaison d'une couche.
ram (to), battre, damer, tasser (le sol).
ram in (to), enfoncer (un pieu).
ramification, ramification (d'un filon).
ramifying, ramification.
rammer, damoir, pilon.
ramp, **1.** rampe, pente, talus; **2.** plateau continental; **3.** compartiment soulevé; **4.** faille normale en surface, mais à pendage inverse en profondeur; **r. valley**, vallée limitée par 2 failles de chevauchement.
rampart, rempart, levée (de terre, de galets), crête de plage; **r. craters**, cratères d'impact martiens à rempart.
Rancholabrean (U.S.A., Cal.), Rancholabrien (étage; Pléistocène sup.).
rand, **1.** suite de collines; **2.** le Rand, district minier aurifère d'Afrique du Sud.

randanite, variété sombre de diatomite.

random, hasard; **r. noise**, bruit de fond; **r. orientation**, orientation au h.; **r. stone**, blocs de toutes dimensions.

range, 1. série, rangée, file; 2. gamme, étendue; 3. distance, portée; 4. intervalle; **r. resolution**, précision dans l'appréciation des distances (échosondage); **r. zone**, unité biostratigraphique définie par la répartition d'un taxon; **mountain r.**, chaîne de montagnes; **stratigraphic r.**, répartition stratigraphique; **tidal r.**, amplitude de la marée (entre haute et basse mer).

rank, classe (d'un charbon), pourcentage de carbone d'un charbon à l'état sec; **r. variety**, gamme de catégories de charbon.

ranker, ranker (*pédol.*).

rapakivi, granite, ou monzonite quartzique contenant des phénocristaux d'orthose revêtus de plagioclase.

rapid, rapide (d'un fleuve).

rare earths, terres rares.

rashing, schiste houiller (tendre, contenant de la matière carbonnée).

rasp (to), racler, frotter (glacier).

rat hole, trou pour tige carrée (forage).

rate of production, taux de production.

rate of firing, fréquence de mise à feu (*sism.*).

ratemeter, dosimètre.

Ratites, Ratites (*pal.*).

Rattle stone, concrétion stratifiée creuse.

Rauracian, Rauracien (sous-étage; Oxfordien).

Ravenian U.S.A., Wash., Ravénien (étage floral; Éocène sup.).

ravine, ravin.

ravine (to), raviner.

raw, brut; **r. humus**, matte; **r. materials, stocks**, matières premières; **r. soil**, sol jeune (profil AC).

ray, 1. rayon lumineux, etc.; 2. bras (d'un Crinoïde); **r.-finned fish**, poisson actinoptérygien; **cosmic r.**, r. cosmique; **r. path**, trajectoire d'un rayon lumineux.

Rayleigh waves, ondes de Rayleigh (var. d'ondes superficielles).

razorback, crête aiguë (à sommet tranchant).

reach, 1. bief (d'un canal); 2. partie rectiligne d'un fleuve.

reactant, réactif (chimie).

reaction, réaction; **r. boundary, curve, line**, ligne cotectique; **r. point**, point de réaction péritectique; **r. rim**, auréole kélyphitique, bordure réactionnelle; **r. velocity**, vitesse de r.; **balanced r.**, r. équilibrée; **reversible r.**, r. réversible; **side r.**, r. parasite.

readvance moraine, moraine de réavancée.

reagent, réactif.

realgar, réalgar (*minér.*).

reamer, trépan aléseur (forage).

reaming, reforage, alésage.

rebore (to), réaléser.

recase (to), recuveler.

reccurence, récurrence (réapparition d'un ancien phénomène).

receiver, récepteur (*géophys.*).

recent, récent; syn. de Holocène.

receptor, géophone.

recess, rentrant (d'un pli, etc.).

recession, 1. régression (marine); 2. recul du rivage; 3. recul du front glaciaire; **ice r.**, recul glaciaire, retrait glaciaire.

recessional moraine, moraine de retrait.

recessive character, caractère récessif.

recharge, apport, alimentation d'eau (à la nappe phréatique).

reclamation (of coal), 1. récupération; 2. réaménagement.

reconnaissance, exploration, reconnaissance; **r. map**, carte de reconnaissance; **r. survey**, étude de reconnaissance.

reconstruction, reconstitution paléogéographique.

record, archives, documents, enregistrement, film (*géophys.*).

record (to), enregistrer, relever.

recorder, appareil enregistreur.

recording, enregistrement; **r. cylinder**, cylindre enregistreur; **r. truck**, camion laboratoire.

recover (to), récupérer (du pétrole).

recoverable, récupérable.

recovery, récupération; **r. factor**, facteur de récupération (du pétrole); **r. ratio**, taux de récupération (de charbon, de minerai).

recrystallization, recristallisation.

recrystallize (to), recristalliser.

rectangular drainage, réseau hydrographique orthogonal.

rectification, redressement (de photographies aériennes obliques).

rectify (to), 1. redresser (un courant électrique); 2. redresser (une photographie); 3. éliminer les parasites (d'un sismogramme, etc.).

recumbency, renversement, position couchée.

recumbent, couché, renversé; **r. anticline**, anticlinal renversé; **r. fold**, pli couché.

recurrence interval, intervalle de récurrence (temps séparant l'apparition de séismes majeurs).

recurrent fauna, faune récurrente (réapparaissant); **r. faulting**, rejeu de la faille.

recurved spit, flèche littorale recourbée.

recycled, polycyclique.

red, rouge; **r. antimony**, kermésite (*minér.*); **r. arsenic**, réalgar (*minér.*); **r. beds**, formations rouges (Permo-Trias); **r. chalk**, hématite (*minér.*); **r. clay**, argile des grands fonds; **r. cobalt**, érythrite (*minér.*); **r. copper ore**, cuprite (*minér.*); **r. earth**, terre r. latéritique; **r. hematite**, hématite r.; **r. iron ore**, hématite r.; **r. iron vitriol**, botryogène; **r. lead**, minium de plomb; **r. manganese**, rhodonite, rhodocrosite (*minér.*); **r. measures**, couches permo-triasiques; **r. mud**, boue r.; **r. ochre**, hématite (*minér.*); **r. oxide of iron**, hématite (*minér.*); **r. oxide of zinc**, zincite (*minér.*); **r. shift**, décalage vers le r. (*astro.*); **r. silver ore**, pyrargyrite, proustite (*minér.*); **r. sulphure of arsenic,** realgar; **r. tide,** eaux r. (de microorganismes); **r. vitriol,** biéberite (*minér.*); **r. vitriol,** zincite (*minér.*).

reddening, rubéfaction.

reddle, ochre rouge, hématite.

redeposit (to), redéposer (*séd.*).

redeposition, remaniement.

redrill (to), reforer.

reducible, réductible (*chimie*).

reducing, réduction; **r. agent,** agent réducteur.

reducibility, réductibilité.

reduction, réduction (enlèvement d'oxygène); **r. index,** taux d'abrasion d'un matériel; **r. of rocks,** émiettement des roches; **free-air r.,** correction à l'air libre; **isostatic r.,** correction isostatique.

reduzates, roches sédimentaires formées par réduction ou dans un milieu réducteur (charbon, pétrole).

reef, **1.** récif, formation récifale; **2.** filon, minéralisé; **r. atoll,** atoll récifal; **r. belt,** ceinture récifale; **r. breccia,** brèche récifale; **r. builder,** animal constructeur de récif; **r. complex,** complexe récifal; **r. drive,** galerie de prospection (Austr.); **r. face,** front récifal; **r. facies,** faciès récifal; **r. flat,** platier; **r. front,** front récifal; **r. knoll,** pinacle corallien; **r. limestone,** calcaire construit; **r. platform,** plateforme récifale; **r. wash,** exploitation de moraines aurifères; **apron r.,** biostrome; **back r.,** arrière récif; **bank r.,** banc corallien; **barrier r.,** récif barrière; **fore r.,** avant récif; **fringing r.,** récif frangeant; **inner r.,** récif interne; **platform r.,** banc corallien; **sand r.,** banc de sable (haut fond); **shore r.,** récif littoral; **table r.,** banc corallien.

reefal, récifal.

reefing, **1.** exploitation des filons aurifères; **2.** formation de récifs.

reefoid, récifal, semblable.

reentrant, rentrant, indentation, cavité (du littoral).

reference axes, axes principaux de l'ellipsoïde de déformation.

refine (to), **1.** raffiner (pétrole); **2.** affiner (les métaux).

refined, raffiné; **r. copper,** cuivre fin; **r. steel,** acier fin.

refinement, affinage (d'un minerai).

refinery, raffinerie.

refining, raffinage.

refining plant, raffinerie.

reflect (to), réfléchir (la lumière), renvoyer, refléter.

reflectance, réflectance (taux de réflexion); **r. spectra,** spectre de réflectance (*astro.*).

reflected light, lumière réfléchie; **r. wave,** onde réfléchie.

reflecting, réfléchissant; **r. horizon,** miroir; **r. point,** point miroir (*géoph.*); **r. surface,** surface réfléchissante.

reflection, réflexion; **r. profile,** profil de sismique r.; **r. seismics,** sismique r.; **r. shooting,** sismique r.; **r. survey,** campagne de sismique r.; **r. wave,** onde réfléchie; **seismic r. method,** méthode de sismique r.

reflectivity, réflectivité.

reflector, réflecteur.

reflux, **1.** reflux (distillation); **2.** reflux (mer); **tide r.,** jusant.

refolded, replissé.

refolding, replissement.

refract (to), réfracter.

refracted wave, onde réfractée.

refraction, **1.** réfraction (*optique*); **2.** sismique réfraction; **r. shooting,** sismique réfraction; **r. survey,** campagne de sismique réfraction; **broadside r.,** réfraction en arc.

refractive index, indice de réfraction.

refractivity, réfringence.

refractometer, réfractomètre.

refractometry, réfractométrie.

refractor, **1.** lunette astronomique; **2.** marqueur (*géoph.*).

refractories, refractory materials, produits réfractaires.

refractoriness, qualité réfractaire.

refractory, réfractaire; **r. clay,** argile réfractaire; **r. ore,** **1.** minerai difficile à traiter; **2.** minerai réfractaire; **r. sand,** sable réfractaire.

refreeze (to), regeler.

refringence, réfringence.

refringent, réfringent.

Refugian (N. Am.), Réfugien (étage; Éocène-Oligocène).

refuse, **1.** rebuts, déchets (de carrière); **2.** refus d'un tamis.

refuse dump, terril, halde.

refusion, fusion.

reg, reg, surface désertique à pavage de cailloux.

regelation, regel.

regenerated crystal, grand cristal recristallisé (dans une brèche); **r. glacier,** glacier régénéré; **r. rock,** **1.** roche régénérée; **2.** roche détritique.

regeneration, régénération (*géomorph.*).

regime, regimen, **1.** régime (fluviatile); **2.** régime (alimentation, écoulement, déclin) d'un glacier.

regolite, regolith, régolithe (roches ou sédiments altérés sur place).

regosol (U.S.A.), régosol (sans horizon défini).

regress, recul du rivage.

regression, régression; **depositional r. (marine),** r. stratigraphique.

regressive, régressif.

regressive overlap (better: offlap), en régressivité.

regular system, système cubique (*cristallogr.*).

regulus antimony, antimoine métallique.

regulus metal, plomb antimoine.

reinjection, réinjection (forage).

reintrusion, intrusion répétée.

rejuvenate (to), rajeunir (par reprise d'érosion).

rejuvenated water, eau remise en circulation.

rejuvenation, rajeunissement, rejeu (d'une faille); **r. head,** point de rupture du profil; **r. of crystals,** recristallisation.

related rocks, roches en rapport mutuel, en relation.

relative, relatif; **r. age**, âge relatif, datation relative; **r. chronology**, chronologie relative; **r. dating**, datation relative; **r. humidity**, degré hygrométrique; **r. permeability**, perméabilité relative.

relaxation of energy, libération d'énergie.

release, dégagement, détente; **r. joints**, fissures de décompression; **r. of strain**, détente des contraintes.

release (to), libérer (de l'énergie).

relic, relict, relicit, résidu d'érosion, structure résiduelle; **r. clast**, fragment minéral résiduel antémétamorphique; **r. permafrost**, pergélisol résiduel; **r. texture**, structure résiduelle (*pétro*.); **r. sediments**, sédiments résiduels (formés dans un autre milieu que le milieu actuel); **r. structure**, structure résiduelle.

reliction, émersion des terres (régression), découvrement, retrait des eaux (non saisonnier).

relief, 1. relief (terrestre); 2. relief (des minéraux sous le microscope); **r. inversion**, inversion de r.; **r. map**, carte en r.; **r. well**, puits de secours, puits d'intervention; **optical r.**, r. d'un minéral observé en lame mince; **relative r.**, rapport du r. d'un bassin hydrographique à son périmètre.

reliquiae, organismes fossiles, reliques paléontologiques.

Relizian (N. Am.), Relizien (étage, Miocène, équiv. Burdigalien).

remanent magnetization, aimantation rémanente.

remanié (french), sédiment ou fossile remanié ou mélangé.

remelting, nouvelle phase de fusion, refonte.

remnant magnetism, magnétisme rémanent.

remote, éloigné, lointain; **r. control**, télécommande aéroportée; **r. processing**, télétraitement; **r. sensing**, télédétection (par satellite); **r. sensor**, télédétecteur.

remove (to), enlever, extraire.

rendoll, sol calcaire, rendzine (*pédol*.).

rendzina, rendzine (*pédol*.).

rent, 1. fente, fissure, fracture; 2. loyer, rente.

reopened vein, filon de remplissage secondaire.

repeated reflections, réflexions multiples; **r. twinning**, mâcle polysynthétique.

Repettian (N. Am.), Répettien (étage, Pliocène inf.).

replacement, remplacement, substitution; **r. deposit**, gîte de s.; **r. vein**, filon de s.

replenishment of ground-water, réalimentation de l'eau souterraine.

repository, dépôt souterrain de matériaux radioactifs.

representative fraction (R.F.), échelle d'une carte.

reprint (offprint, U.K.), tiré à part.

reprocessing, retraitement.

Reptile, Reptilia, Reptiles.

reptilian, reptilien.

rerun (to), redistiller.

resampling, échantillonnage répété.

resection, 1. relèvement (topographie); 2. recoupement.

resequent, reséquent (*géomorph*.); **r. stream**, rivière reséquente, enfoncée dans le lit antérieur.

reserves, réserves; **primary r.**, r. primaires; **proved r.**, r. prouvées; **secondary r.**, r. secondaires.

reservoir, roche réservoir (de pétrole); **r. engineering**, étude de réservoir; **r. pressure**, pression de réservoir; **r. rock**, roche réservoir; **r. water**, eau de formation; **multilayer r.**, réservoir multicouches.

residual, résiduel; **r. clay**, argile r. d'altération; **r. deposit**, gîte r.; **r. hill**, hum; **r. liquid**, magma r.; **r. magma**, magma r.; **r. magnetism**, magnétisme rémanent; **r. melt**, magma r.; **r. soil**, sol r. d'altération.

residuary, résiduel, qui reste.

residue, residuum, résidu, reste; **weathering r.**, éluvions.

resin, résine; **true r.**, résine naturelle.

resinous, résineux; **r. shale**, schiste bitumineux.

resistance, résistance (des matériaux).

resistates, roches sédimentaires «résistantes à l'altération» (telles que roches détritiques arénacées).

resistivity, résistivité; **r. method**, méthode de résistivité des roches (*géophys*.); **actual r.**, r. vraie.

resolution, 1. résolution, dissolution, décomposition; 2. pouvoir de résolution (*opt*.).

resolving limit, limite de résolution, pouvoir séparateur.

resorption, résorption, dissolution (d'un minéral dans un magma).

resource, ressource; **r. concepts**, notion de ressource minérale (épuisable, inépuisable, etc.).

rest magma, magma résiduel.

resurgence, résurgence.

resurgent, résurgent (sens varié, cf. contexte).

resurgent spring, river (karst), résurgence (de rivière).

retaining wall, mur de retenue.

retardation, 1. retard, ralentissement; 2. évolution régressive (*pal*.); **optical r.**, retard de longueur d'onde.

reticle, réticule.

reticulate structure, structure réticulaire (*pétro*.).

reticulated veins, filons anastomosés en réseau.

reticulation, réticulation (*fluviatile*).

reticulite, ponce basaltique.

retimber, reboiser (*mine*).

retinite, rétinite, pechstein (roche volcanique vitreuse riche en eau).

retouch, retouche (sur outil préhistorique).

retreat of a cliff, recul d'une falaise.

retreatal moraine, moraine de retrait.

retrievable, récupérable (minerai).

retroarc basin, bassin arrière-arc.

retrograde, métamorphisme régressif.

retrograding shoreline, rivage en recul, d'abrasion (par suite de l'érosion).

retrogression, recul de la ligne de rivage.

retrogressive erosion, érosion régressive.

retrogressive metamorphism, rétromorphose, diaphtorèse.

retrometamorphism, rétrométamorphisme.

retrosiphonate, rétrosiphoné (*pal.*).

reversal, changement, inversion; **r. of dip**, changement de pendage (local); **magnetic r.**, inversion magnétique.

reverse, inverse; **r. bearing**, visée arrière; **r. dip**, pendage i.; **r. fault**, faille i.; **r. limb**, flanc i.; **r. stream**, fleuve obséquent.

reversed polarity, polarité inverse.

reversible process, processus réversible.

reversing thermometer, thermomètre à renversement.

revived stream, fleuve réactivé, surimposé (par soulèvement).

revolution, révolution (orogénique).

revolving nosepiece (with the objectives), monture tournante d'un microscope.

rework (to), remanier (des sédiments).

rework, reworking, 1. remaniement; 2. rejeu.

rhabdosome, rhabdosome (colonie de Graptolithes).

Rhaetian (syn. Rhaetic), Rhétien (étage : Trias sup. ou Lias inf.).

rheid, substance déformée par écoulement visqueux.

rheidity, comportement plastique, déformation plastique.

rheology, rhéologie (étude des déformations de la matière).

rheomorphism, rhéomorphisme, déformation à l'état visqueux ou plastique.

Rhodanian Orogeny, orogénèse Rhodanienne, (fin du Miocène).

rhodium, rhodium.

rhodocrosite, rhodocrosite, dialogite (*minér.*).

rhodolite, rhodolite (var. rose de grenat).

rhodonite, rhodonite (*minér.*).

Rhodophyta, Rhodophycées (Algues rouges).

rhomb, abréviation de rhomboèdre; **r. porphyry**, porphyres à phénocristaux (Norvège); **r. spar**, dolomite.

rhombenporphyric, rhomboporphyre.

rhombic dodecahedron, dodécaèdre rhomboïdal; **r. mica**, phlogopite; **r. symmetry, r. system.**, système orthorhombique.

rhombochasms, fossé tectonique dans la croûte continentale rempli par la croûte océanique.

rhombohedral, rhomboédrique.

rhombohedron, rhomboèdre.

rhomboidal, rhomboédrique.

rhumb line, loxodromie.

rhums, schistes bitumineux (Écosse).

rhyacolite, sanidine (feldspath potassique).

Rhynchonelloid Rynchonellida, Rhynchonellacés (Brachiopodes).

Rhynchocephalia, Rhynchocéphales.

rhyodacitic, rhyodacitique.

rhyodacite, rhyodacite (*pétro.*).

rhyolite, rhyolite (roche effusive).

rythmic sedimentation, série sédimentaire rythmique.

rhythmite, 1. roche litée (dans une séquence sédimentaire); 2. séquence sédimentaire à répétition multiple.

ria, ria, vallée submergée: **r. coast**, côte à rias.

rib, 1. pilier de support de galerie (*mine*), charbon solide le long d'une paroi; 2. minerai solide d'un filon; pilier pour supporter le toit d'une exploitation de filon; 3. mince lit de pierre dans une couche de charbon (Écosse); 4. veinule de minerai dans un filon; 5. résidu de dissolution dans un passage (*spéolo.*); 6. côte (*pal.*).

ribbed, strié, cannelé, nervuré.

ribbon, ruban; **r. diagram**, coupes géologiques transversales perspectives et sériées; **r. injection**, apophyse de roche éruptive injectée suivant plans de schistosité; **r. jasper**, jaspe zoné; **r. rock**, roche rubanée, litée; **r. structure**, structure rubanée.

rice coal, poussier d'anthracite.

rich coal, houille grasse; **r. lime**, chaux grasse.

Richmondian (N Am.), Richmondien (étage: Ordovicien sup.).

richterite, richtérite (*minér.*).

Richter magnitude scale, échelle de magnitude de Richter.

ricolite, serpentine (Nouveau Mexique).

riddle, tamis, crible à main.

riddle (to), tamiser, cribler.

riddling, criblage, tamisage.

rider, 1. couche mince de houille sus-jacente à une couche plus épaisse; 2. gîte minéral sus-jacent au filon principal.

ridge, chaîne, crête, dorsale (sous-marine); **r. and swale topography**, topographie en creux et en bosses; **r. fold**, dôme allongé, anticlinal symétrique; **medio-atlantic ridge**, ride médio-atlantique.

riebeckite, riébeckite (amphibole sodique).

Riedel fracture (shear), faille de Riedel (en échelon, précoce).

riegel, verrou glaciaire.

riffle, 1. chenal, sillon; 2. ride sous-aquatique; 3. haut-fond; 4. cavité au fond d'un sluice (pour arrêter l'or).

rift, 1. grande faille à décrochement horizontal; 2. intersection d'un plan de faille avec la surface; 3. plan de fissilité (granites); 4. crevasse, fissure; 5. haut-fond (N.E. des U.S.A.); **r. trough**, fossé tectonique, graben; **r. valley**, fossé central de la crête; **central r.**, fossé central d'une dorsale médio-atlantique; **wall r.**, paroi d'un fossé tectonique.

rifting, distension, séparation des continents.

rig, tour de forage; **r. floor**, plancher de forage du derrick; **r. time**, temps de forage; **semi-submersible r.**, plate-forme de forage semi-submersible.

rigging up, aménagement du derrick.

right, droit, dextre; **r. angle**, à angle droit; **r. handed separation**, faille à déplacement dextre; **r. lateral fault**, décrochement dextre; **r. lateral separation**, faille à déplacement dextre; **r. strike-slip fault**, décrochement dextre.

rigid crust, niveau crustal rigide (granitique).

rigidity, rigidité (d'une couche).

rill, 1. sillon, rigole; **2.** petit ruisseau; **3.** filet de courant de retour de vagues; **4.** sillon tectonique lunaire.

rille, petite vallée lunaire.

rillenstein (germ.), roches à très petites cannelures de dissolution.

rill erosion, formation d'un réseau de ruissellement.

rillmark, ravinement de courant.

rim, frange, liseré, bord, rebord; **r. cement,** ciment recristallisé avec la même orientation cristallographique que les grains (*pétrogr.*); **r. rock, 1.** limite naturelle d'une région (ex : falaise escarpée); **2.** remontée du substratum; **r. stone,** travertin calcaire (dans alluvions); **r. syncline,** synclinal bordier (autour de diapirs), dépression périclinale.

rime, givre, gelée blanche.

rimmed pool, mare à rebords.

ring, anneau, cercle; **r. coal,** charbon bitumineux; **r. complex,** complexe annulaire; **r. dikes,** filons annulaires en cônes; **r. fracture,** fracture annulaire; **r. hydrocarbon,** hydrocarbure cyclique; **r. of fire,** ceinture de feu péripacifique; **r. ore,** gisement à couches concentriques; **r. silicate,** cyclo-silicate; **r. structure,** structure annulaire; **r. structures,** cyclosilicates, silicates en chaînes fermées.

rip, clapotis, bouillonnement; **r. current,** courant d'arrachement, de déchirure; **r. tide,** courant de marée.

Riphean (U.S.S.R.), Riphéen (ère, Précambrien récent, équiv. Sinien, Beltien, Éocambrien).

ripple, ride, ondulation; **r. bedding,** stratification entrecroisée à petite échelle; **r. lenght,** longueur d'onde des r.; **r. mark,** r. de plage; **r. train,** réseau de r.; **current r.,** r. de courant; **giant current r.,** r. géante de courant; **lunate r.,** r. en croissant; **undulatory r.,** r. sinueuse; **wave r.,** r. de vague; **wind r.,** r. éolienne.

rippled, ridé, ondulé.

rippling stream, ruisseau murmurant, gazouillant.

ripply, couvert de rides.

rip-rap, enrochement.

rise, 1. pente océanique (reliant les grands fonds au talus continental); **2.** crête océanique (sans fossé central); **3.** source (dans terrains calcaires); **4.** résurgence (d'eau souterraine); **5.** crue; **6.** galerie ascendante (*mine*), cheminée (*mine*); **fast-spreading r.,** crête océanique à expansion rapide (Pacifique).

rise (to), 1. se lever, se soulever; **2.** creuser ou exploiter vers le haut (*mine*).

riser, 1. talus (de terrasse fluviatile, etc.), élément vertical; **2.** galerie montante (*mine*); **3.** faille inverse.

riser pipe, colonne montante (pétrole), tube de forage à double circulation de fluides.

rising spring, exsurgence, émergence.

Riss (Eur.), glaciation de Riss (avant dernière g. pléistocène).

rissian, rissien.

river, rivière, fleuve; **r. bank,** berge, rive; **r. bar,** banc d'alluvions fluviatiles; **r. bar placer,** gîte alluvionnaire; **r. basin,** bassin fluvial; **r. bed,** lit du f.; **r. bottom,** plaine alluvionnaire; **r. capture,** capture fluviatile; **r. cut plain,** plaine d'érosion fluviale; **r. drift,** alluvions fluviatiles; **r. flat,** lit majeur, plaine alluvionnaire; **r. load,** charge fluviale; **r. mouth,** embouchure; **r. system,** réseau hydrographique; **r. terrace,** terrasse fluviatile; **r. valley,** vallée fluviatile.

riverine, fluvial.

riving seams, fissures ouvertes entre les couches affleurant dans une carrière.

rivulet, ruisseau.

roadbed, soubassement, assiette d'une route.

road metal, rock, matériaux de revêtement routier.

roaring forties, « quarantièmes rugissants ».

roast (to) ore, griller, calciner le minerai.

roasting, calcination, souvent avec oxydation.

rob (to), déboiser, extraire les piliers laissés antérieurement (dépiler).

robbing pilars, dépilage.

robble (G.B.), faille.

roche moutonnée (french), roche moutonnée (*géomorph. glaciaire*).

rock, roche; **r. analysis,** analyse géochimique; **r. bar,** verrou; **r. bench,** plate-forme d'abrasion; **r. bend,** pli; **r. bit,** trépan tricône; **r. breaker,** broyeur à pierres; **r. breaking,** désagrégation de r.; **r. boring organism,** organisme perforant; **r. burst,** coup de toit; **r. clay,** argilite; **r. color chart,** code des couleurs des r.; **r. cork,** amiante; **r. crystal,** cristal de r.; **r. failure,** fissuration de la r.; **r. fall,** glissement d'éboulis rocheux; **r. fall chute,** couloir d'avalanches; **r. fan,** cône d'ablation; **r. fill dam,** barrage d'enrochement; **r. filling,** remblayage avec des déblais; **r. floor,** fond rocheux; **r. flour,** poussière de r. effritée par les glaciers; **r. gangway,** galerie au rocher; **r. gas,** gaz naturel; **r. glacier,** glacier rocheux, champ de pierres; **r. hound,** minéralogiste amateur; **r. in place,** substratum rocheux; **r. knob,** éperon rocheux de dénudation; **r. leather,** amiante; **r. metal,** poussière de r.; **r. mechanics,** mécanique des r.; **r. milk,** poussière de r.; **r. oil,** pétrole; **r. pediment,** glacis rocheux désertique; **r. plane,** glacis rocheux désertique; **r. phosphate,** phosphorite; **r. quartz,** cristal de r.; **r. rubble,** brèche de faille; **r. salt,** sel gemme; **r. sample,** carotte d'échantillon; **r. series,** séries de r. éruptives; **r. shaft,** puits à remblais; **r. shelter,** abri sous r.; **r. silk,** amiante soyeuse; **r. slide,** glissement de terrain; **r. soap,** montmorillonite, saponite; **r. step,** seuil rocheux; **r. stratigraphic unit,** unité lithostratigraphique; **r. stream,** glacier rocheux; **r. suite,** série magmatique; **r. tar,** pétrole brut; **r. terrace,** niveau d'érosion; **r. tower,** kopje; **r. type, 1.** catégorie majeure de r. (ex : éruptive, sédimentaire, etc.); **2.** catégorie spécifique; **r. unit,** unité lithostratigraphique; **r. waste,** débris rocheux; **cap r.,** r. couverture; **roof r.,** r. couverture; **sealing r.,** r. couverture; **source r.,** r. mère.

rockallite, rockallite (granite sursaturé à aegyrine).

rocker, crible-laveur.

rocking, criblage, lavage par balancement.

rocky, rocailleux, rocheux.

Rocky Mountain Orogeny (Can. U.S.), orogénèse des Montagnes Rocheuses (Crétacé terminal et Paléocène, équiv. de l'orogénèse Laramide).

rod, 1. tige; 2. baguette (sédiment.); 3. unité de longueur ou perche = 5,0292 m; 4. rouleau (tectonique).

roddon (U.K.), levée fluviatile produite à marée montante.

Rodenta, Rodent, Rongeurs (pal.).

Roentgen, Roëntgen (unité de radiation). ·

roestone, 1. oolithe; 2. étym. : forme d'œuf de poisson.

roily oil, pétrole brut émulsionné (avec de l'eau).

roke, filon de minerai.

roll mill, laminoir.

roller, tourbillon à axe horizontal.

roller bit, trépan à molettes.

rollers, tempêtes.

rolling ground, terrain ondulé.

rolling transport, transport fluviatile par roulement des matériaux.

roll-over (U.S.A.), flexure.

rollover (syn. dip reversal) (U.S.A.), retournement des couches contre faille.

roof, toit, plafond (*mine*); **r. arch,** voûte; **r. collapse,** effondrement du toit; **r. foundering,** effondrement du toit dans un réservoir magmatique; **r. limb,** flanc supérieur (d'un pli couché); **r. pendant,** enclave géante, apophyse de roche encaissante dans roche éruptive sous-jacente; **r. rock,** roche de couverture; **r. thrust,** flanc supérieur d'un chevauchement.

roofing slate, ardoise de couverture; **r. tile,** tuile de couverture.

room, salle souterraine (*spéléo.*).

room and pillar system, exploitation par chambres et piliers (*mine*).

room timbering, boisage de chambres (*mine*).

root, racine; **r. of a Crinoid,** partie basale de la tige d'un Crinoïde; **r. of a fold,** r. d'un pli, d'une nappe; **r. zone,** zone d'enracinement (d'un pli couché).

rope, corde, câble; **r. drilling,** sondage, forage au câble; **r. haulage hoist,** treuil de traction au câble; **r. roll,** tambour d'extraction; **r. winch,** treuil à câble.

ropy lava, ropey lava, lave cordée, pahoehoe.

rose diagram, diagramme circulaire, ou semi-circulaire.

roselite, rosélite (*minér.*).

rosette, 1. rose des sables (gypse); 2. autre forme minérale ressemblant à une rose (barite, etc.).

rosin, resin, colophane, résine; **r. jack,** sphalérite, blende; **r. tin,** cassitérite (var. rougeâtre).

rostrum, rostre (*pal.*).

rot (to), pourrir, se décomposer.

Rotaliina, Rotaliidés.

rotary, rotatif; **r. boring,** forage rotary; **r. fault,** faille rotationnelle; **r. drilling,** faille rotationnelle; **r. table,** table de rotation (forage).

rotating-crystal photograph, diagramme de cristal tournant.

rotational, de rotation, rotationel; **r. fault,** faille rotationnelle; **r. flow,** écoulement turbulent; **r. movement,** mouvement de rotation (d'un compartiment d'une faille); **r. wave,** onde transversale.

rotatory dispersion, dispersion rotatoire (de la lumière).

Rotliegende (Eur.), Rotliegende (série, Permien inf. et moy.).

rotten rock, roche pourrie, décomposée; **r. stone,** résidu d'altération siliceux des roches calcaires (mais cohérent).

rough, 1. brut; 2. rugueux, accidenté; 3. grossier, approximatif; **r. coal,** charbon tout-venant; **r. diamond,** diamant brut.

roughness, rugosité.

rounding, degré d'émoussé (d'un galet), arrondi.

roundness, émoussé; **r. ratio,** degré d'é.

roundstone, galet émoussé, usé.

royal agate, variété d'obsidienne (tachetée).

rubbish, décombres, déblais, déchets, remblais.

rubble, blocaille, moellons; **r. drift, 1.** dépôt de cryoturbation soliflué; **2.** dépôt grossier et anguleux d'origine glaciaire, à matrice terreuse; **r. ice,** glace en fragments; **volcanic r.,** produits pyroclastiques non consolidés.

rubbly, blocailleux; **r. reef,** filon très fragmenté (Austr.).

rubefaction, rubéfaction (*pédol.*).

rubellite, rubellite (var. rouge de tourmaline).

rubicelle, rubicelle (variété de spinelle magnésien).

rubidium, rubidium.

rubidium-strontium dating, datation rubidium-strontium.

rubification, rubéfaction.

ruby, rubis, corindon; **r. blende,** sphalérite rouge; **r. copper,** cuprite; **r. silver ore,** pyrargyrite, proustite; **r. spinel,** rubis, spinelle; **r. sulphur,** réalgar; **r. zinc,** sphalérite transparente et rouge.

rudaceous deposits, rudite.

rudaceous rocks, roches sédimentaires détritiques à éléments grossiers, rudites.

ruddle, ochre rouge.

Rudistae, Rudistids, Rudistes (*pal.*).

rudite, rudyte, rudite (particule plus grande que 2 mm).

rugose, rugueux, ridé; **r. coral,** Tétracoralliaire.

Rugosa, Tétracoralliaires (*pal.*).

rugged, anfractueux, accidenté.

run, 1. petit ruisseau; 2. filon subhorizontal interstratifié; 3. direction d'un filon; 4. terrain meuble mouvant; 5. éboulement de terrain; 6. tout venant (charbon).

run-off, 1. débit; **2.** eau de ruissellement; **ground water r.,** écoulement souterrain.

run (to), 1. parcourir; **2.** s'écouler (liquide).

run derrick (to), monter un derrick.

run in (to), redescendre le train de tiges.

run up, jet de rive.

runnel, 1. ruisseau (petit); **2.** petite dépression littorale; **3.** sillon d'estran.

running ground, terrain boulant.

running sand, 1. sable mouvant; **2.** sable en suspension (dans eau ou pétrole).

running water, eau courante.

runs, pourcentage de métal dans un minerai.

rupes, escarpement planétaire (Lune).

Rupelian, Rupélien (étage, Oligocène moy., syn. Stampien).

rupture front, front de rupture (entre une partie non déplacée et une autre décalée par un séisme); **r. strength,** résistance à la rupture.

rush, 1. mouvement brusque; **2.** affaissement du toit (*mine*).

rush gold, or revêtu d'oxyde de fer ou de manganèse.

rush spring, source jaillissante.

rutile, rutile (*minér.*).

ruttles (rare), brèche de faille.

R wave, onde de Rayleigh (de surface) (*sismol.*).

rythmites, dépôts sédimentaires rythmés.

S

saalband, salbande (mine).

Saalian, saale (Eur.), glaciation Saalienne, équiv. du Riss.

Saalic Orogeny, orogénèse Saalienne (début du Permien).

Saamian, Saamien (partie du Précambrien).

sabana, mine dans alluvions fluviatiles émergées (U.S.).

Sabinas (N. Am.), série de Sabinas (Jurassique sup.).

Sabinian (N. Am.), Sabinien (étage, Éocène inf.).

sabkha (var. of sebkha), sebkha.

sabuline, sableux.

sabulous, sablonneux.

saccharoid, saccharoidal, saccharoïde; s. fracture, cassure s.; s. marble, marbre s.; s. texture, structure s.

saddle, 1. selle (courbure vers l'avant de la ligne de suture des Ammonites); 2. voûte, pli anticlinal; 3. col; s. back, courbure en dos d'âne; s. bend, charnière anticlinale; s. fold, flexure anticlinale transverse; s. reef, voûte anticlinale (Austr.); s. shaped, en forme de pli anticlinal; inverted s., courbure en forme de synclinal; s.-backed, en dos d'âne.

safety board, passerelle d'accrochage (forage); s. lamp, lampe de sûreté (mine).

safe yield, taux (ou vitesse) tolérable de prélèvement d'eau d'une nappe phréatique (sans introduire de nuisance).

safranite, safranite, quartz jaune.

sag, abaissement, niveau bas (d'une couche), affaissement, fléchissement; down-sagging, tassement.

sag-and-swell topography, topographie en creux et en bosses.

sag (to), s'affaisser, fléchir, ployer.

sagenite, sagénite (variété de rutile).

sagenetic quartz, quartz à inclusions aciculaires de rutile.

sagging, affaissement.

sagittal, sagittal (pal.).

sagpond, affaissement tectonique.

sagvandite, sagvandite (carbonatite à giobertite et bronzite).

sahlite, sahlite (var. de pyroxène).

Sakmarian (Eur.), Sakmarien (étage, Permien inf.).

salamander's hair, amiante.

sal ammoniac, chlorure d'ammonium.

salamstone, saphir de ceylan.

salband, salbande (mine).

salcrete, croûte de sel (cf. calcrète).

sal gemmae, sel gemme, halite.

salic (minerals), aluminosilicates (classif. C.I.P.W.).

salic horizon, horizon pédologique avec plus de 2 % de sel (U.S.).

salient, en saillie.

saliferous, salifère; s. system, Trias (désuet).

salimeter, salimètre, doseur de sels.

salina, marais salant, saline.

saline, salin, salé; s. clay soil, sol argileux, salé; s. dome, diapir; s. lake, lac salé, chott; s. peaty soil, sol organique salin; s. soil, sol salin; s. spring, source salée, saumâtre; s. turfy soil, sol organique salin; s. water, eau salée, saumâtre.

salines, 1. terrain salin; 2. dôme de sel.

saliniferous, salinifère, salifère.

salinity, salinité.

salinization, salinisation (d'un sol).

salinometer, salinomètre.

salite, salite (minér.).

saliter, nitrate de soude.

salmiak, salmiac, chlorure d'ammonium.

Salmian, Salmien (étage = Trémadoc; Ordovicien inf.).

sal mirabile, mirabilite, sel de Glauber.

sal natro, sal natron, carbonate de sodium brut anhydre.

Salopian (Eur.), Salopien (étage, Silurien moy. à sup.).

salpeter, salpetre, salpêtre (nitrate de potassium).

salse, salse, volcan de boue.

salt, 1. adj : salé; 2. n : sel; s. bearing, salifère; s. bed, couche de s.; s. block, usine à s.; s. bottom, bassin salifère, chott; s. brine, saumure; s. content, salinité; s. crust, croûte de s.; s. dome, dôme de s.; s. flat, étendue salifère, plaine de s.; s. flour, nitrate de potassium; s. intrusion, intrusion (diapir) de s.; s. lake, lac s.; s. marsh, marais salant; s. mine, mine de s.; s. pan, 1. lac saumâtre peu profond; 2. grande battée à évaporer; 3. couche dure salée; s. pasture, pré s.; s. pit, saline, mine de s.; s. plug, culot de s.; s. refinery, saunerie; s. rock, halite, s. gemme; s. salt, s. marin; s. spring, source salée; s. swamp, marais salant; s. tectonism, tectonique salifère; s. upwelling, diapir de s.; s. wall, filon vertical de s.; s. water, eau salée; s. weathering, haloclastie; s. well, puits de saumure (donnant de l'eau salée); s. works, saline.

saltation, saltation (séd.).

saltative evolution, évolution par sauts, par bonds.

saltatory evolution, évolution par bonds.

salted, salé; s. soil, sol s.

saltern, saline, marais salant.

saltness, salinité, salure.

salpeter, salpêtre, nitre; s. bed, salpêtrière, nitrière; s. works, salpêtrerie.

salpetrous, salpêtreux.

salty rock, roche saline.

samarium, samarium.

sample, échantillon; **core s.,** carotte; **s.-splitting device,** séparateur de matériaux meubles.

sampler, échantillonneur (personne), échantillonneuse (appareil), carottier, dispositif de prélèvement.

sampling, échantillonnage, prise d'échantillons; **s. machine,** échantillonneuse; **s. works,** laboratoire d'essais.

sand, sable; **s. bank,** banc de s.; **s. bar,** cordon littoral sableux; **s. bath,** bain de s. (laboratoire); **s. blasted,** décapé par le s.; **s. bodies,** dunes oolithiques sous-marines; **s. boil,** masse de s. et d'eau éjectée par liquéfaction (suite à un séisme); **s. dredge,** drague à s.; **s. drift,** vent de s.; **s. dune,** dune de s.; **s. field,** erg; **s. flag,** grès à débit en dalles; **s. flood,** masse de s. mouvant; **s. gall,** tubulure de s.; **s. hill,** dune; **s. lens,** lentille sableuse; **s. pack,** remblai d'ensablage; **s. pipe,** cavité tubulaire sableuse; **s. pit,** sablière; **s. plain,** plaine de s. (podsolisé); **s. ribbon,** mégaride sous-marine; **s. ripple,** ride éolienne sur s.; **s. rock,** grès; **s. scratches,** stries éoliennes; **s. spit,** flèche sableuse littorale; **s. storm,** tempête de s.; **s. streak,** filonnet de muscovite et quartz; **s. stream,** dépôt sableux alluvial marin; **s. volcano,** injection de s.; **s. waves,** dune sous-marine; **s. well,** puits creusé dans couche sableuse; **air borne s.,** s. éolien; **barren s.,** s. stérile; **coarse grained s.,** s. grossier; **cross bedded s.,** s. à stratifications entrecroisées; **dirty s.,** s. argileux; **drifting s.,** s. mouvant; **flying s.,** s. éolien; **impregnated s.,** s. imprégné; **oil s.,** s. pétrolifère; **open s.,** s. poreux, perméable; **quick s.,** s. fluent; **running s.,** s. mouvant; **shaly s.,** s. argileux; **shelly s.,** s. coquillier; **tar s.,** s. bitumineux; **tight s.,** s. compact.

sandblasted pebble, galet façonné par le vent, galet à facettes.

sanding up, ensablement.

sandr (sandur), plaine d'épandage proglaciaire (Islande).

sandstone, grès; **s. dike,** filon de g.; **s. grit,** g. grossier; **s. lens,** lentille sableuse; **s. quarry,** grésière; **argillaceous s.,** g. argileux; **arkosic s.,** g. arkosique; **bituminous s.,** g. bitumineux; **calcareous s.,** g. calcaire; **coquina s.,** g. lumachellique; **dolomitic s.,** g. dolomitique; **ferruginous s.,** g. ferrugineux; **gypsiferous s.,** g. gypseux; **micaceous s.,** g. micacé; **phosphatic s.,** g. phosphaté; **quartzitic s.,** g. quartzite; **quartzose s.,** g. quartzeux; **shelly s.,** g. coquillier; **siliceous s.,** g. siliceux.

sandstorm, tempête de sable.

sand up (to), s'ensabler.

sandwich layer, couche intermédiaire.

sandy, arénacé, sableux, sablonneux; **s. clay,** argile sableuse; **s. limestone,** calcaire sableux; **s. loam,** limon sableux; **s. marl,** marne sableuse; **s. muck,** sol organo-minéral sableux; **s. mud,** vase sableuse; **s. regolith,** arène; **s. rock,** roche arénacée; **s. soil,** sol sableux.

Sangamonian (N. Am.), Sangamonien (interglaciaire).

sanguinaria, hématite (Espagne).

sanidine, sanidine (feldspath).

Sannoisian, Sannoisien (étage, Oligocène, équiv. Tongrien).

Santonian, Santonien (étage, Crétacé sup.).

sap (to), saper, miner.

saponite, saponite (= variété d'argile smectite).

sapper, disthène.

sapphire, saphir; **s. quartz,** quartz saphirin bleuté, pseudo-saphir.

sapphirine, **1.** adj : saphirin; **2.** n : sapphirine (*minér.*).

sapping, sapement.

saprogenous endohumus, humus doux.

saprolite, roche résiduelle, décomposée, altérée saprolite.

saprolith, régolithe (riche en argile).

sapropel, sapropel, sapropèle.

sapropelic coal, charbon sapropélique (charbon d'algues).

sapropelite, sapropélite.

sard, sardoine (silice calcédonieuse).

sardachate, agate zonée de calcédoine (rouge).

sardic orogeny, orogénèse sarde (fin du Cambrien).

sardonyx, sardoine, sardonyx (= variété de calcédoine translucide brun-rouge).

Sarmatian, Sarmatien (étage, Miocène sup.).

sarsen stone, pierre monolithique, menhir en grès (Angleterre).

satellite, **1.** satellite (artificiel); **2.** filon-satellite d'un massif éruptif.

satin spar, **1.** spath satiné; **2.** var. fibreuse d'aragonite; **3.** var. fibreuse de gypse.

saturated, saturé; **s. core,** carotte imprégnée; **s. hydrocarbon,** hydrocarbure s.; **s. steam,** vapeur d'eau s.; **s. vapour pressure,** pression de vapeur saturante; **s. zone,** zone souterraine s. d'eau.

saturation, saturation; **s. degree,** degré de s.; **s. factor,** facteur de s.; **s. indice,** taux de s.; **s. level,** nappe aquifère, nappe phréatique; **s. line,** **1.** sommet de la nappe phréatique; **2.** ligne de s. (en silice); **s. pressure,** pression de s.; **s. zone,** zone de s.; **oil s.,** s. en pétrole; **water s.,** s. en eau.

Saucesian (N. Am.), Saucésien (étage, Oligocène à Miocène).

sauconite, sauconite (minéral argileux du groupe des montmorillonites).

Saurischia, Dinosauriens saurischiens.

Sauropterygia, Sauroptérygiens (*pal.*).

saussurite, saussurite, zoïsite (minéral d'altération des feldspaths); **s. gabbro,** gabbro à feldspaths altérés, saussuritisés.

saussuritization, saussuritisation.

savane red soil, sol rouge de savane.

savanna(h), savane (savanna = forêt, Savannah, Georgia, U.S.A. = ville).

Savic orogeny, orogénèse save (Oligocène terminal).

sawback, chaîne montagneuse accidentée, déchiquetée.

saw-toothed, déchiqueté, en dents de scie (relief).

saxicavous shell, mollusque lithophage.

Saxonian, Saxonien (étage, Permien moy.).

scabble (to), tailler, dégrossir (la pierre de carrière).
scabbler, tailleur de pierres.
scabbling hammer, marteau de carrier.
scabland, **1.** terre érodée par le vent et le ruissellement; **2.** terrain recouvert de fragments rocheux anguleux, sans sol; **s. topography,** relief irrégulier.
scad, pépite.
scale, **1.** écaille (zoologie), fragment rocheux; **2.** tartre, incrustation, dépôt calcaire; **3.** échelle; **4.** paraffine brute; **s. coated,** entartré; **s. copper,** cuivre en paillettes; **s. correction,** correction d'échelle; **s. crust,** tartre; **s. deposit,** tartre; **s. drawing,** dessin à l'échelle; **s. model,** maquette, modèle réduit; **s. of hardness,** échelle de dureté; **s. of height,** échelle des hauteurs; **s. stone,** wollastonite; **s. wax,** paraffine brute (en écailles); **s. unit,** unité de l'échelle; **according to s.,** à l'échelle.
scale (to), **1.** détartrer, nettoyer; **2.** écailler (mine); **3.** s'écailler, se desquamer.
scaled, écailleux.
scalenohedron, scalénoèdre.
scaling, **1.** écaillage, exfoliation, desquamation; **2.** détartrage; **3.** comptage électronique de pulsation; **4.** relatif à l'échelle d'une carte.
scallop, **1.** Pecten (*pal.*); **2.** échancrure, cavité, cupule.
scallop (to), festonner, découper, denteler.
scalped anticline, anticlinal érodé avant dépôt de couches discordantes.
scaly, écailleux, en écailles; **s. fracture,** cassure écailleuse, cassure esquilleuse.
scamy post, grès micacé lité (psammite) (G.B.).
scan (to), balayer, explorer, parcourir (opt.).
scandium, scandium.
scan lines, lignes de balayage (télédection).
scanner, capteur, détecteur à balayage (télédection); **multispectral s.,** détecteur à balayage multispectral.
scanning microscope, microscope électronique à balayage; **s. speed,** vitesse de balayage; **s. system,** dispositif d'étude à balayage (microscope, sonde).
Scaphopoda, Scaphopodes.
scapolite, scapolite, wernérite (*minér.*).
scapolitization, scapolitisation.
scar, **1.** cicatrice; **2.** rocher escarpé, piton rocheux isolé (Écosse); **3.** scorie de cuisson de pyrite (G.B);. **4.** niche de décollement (d'un glissement de terrain).
scarcement, ressaut, saillie.
scares, feuillets de pyrite dans le charbon (G.B).
scarn, skarn, tactite (roche métamorphique carbonatée).
scarp, escarpement, gradin (parfois d'origine tectonique, faille); **fault-line s.,** e. de ligne de faille; **fault s.,** e. de faille; **inward-facing s.,** e. internes en vis-à-vis (de part et d'autre d'un rift sous-marin); **resurrected fault-s.,** e. de faille dégagé par l'érosion.
scarped, escarpé, abrupt.
scarped face (of cuesta), front (de cuesta).
scarplet, ressaut de faille; **reverse s.,** r. inverse.
scarring, formation de scories par brûlage de pyrite (G.B.).

scatter, **1.** dispersion, éparpillement; **2.** diffraction; **s. light,** lumière diffuse.
scatter (to), disperser, éparpiller, disséminer.
scattering, **1.** dispersant; **2.** diffusion, dispersion.
scavenger, nécrophage (pal.).
scheelite, scheelite (*minér.*).
schiller spar, schillerspath (enstatite ou bronzite altérée).
schist, schiste (métamorphique); **s. rock,** roche schisteuse; **amphibolitic s.,** s. à amphibole; **bituminous s.,** s. bitumineux; **chloritic s.,** s. chloritique; **clay s.,** s. argileux; **crystalline s.,** s. cristallin; **garnitiferous s.,** s. grenatifère; **graphitic s.,** s. graphitique; **mica s.,** micaschiste; **sericitoschist,** séricitoschiste; **spotted s.,** schiste tacheté.
schistic rock, roche schisteuse.
shistoid, schistoïde; **s. fracture,** cassure s.
schistose, schisteux; **s. crystalline rocks,** roches cristallophylliennes.
schistosity, **1.** foliation (des schistes cristallins); **2.** schistosité.
schistous, schisteux.
Schmidt net, projection de Schmidt.
Schizodont, Schizodonte (pal.).
schorl, schorl, tourmaline; **s. rock,** roche à t. et à quartz.
schorlaceous, schorlous, schorlacé.
schorliferous, schorlifère.
schorlite, tourmaline.
schorlomite, schorlomite (var. de grenat).
schorl rock (G.B.), schorlite.
schorre (Neth.), schorre, niveau littoral recouvert (uniquement lors des marées maximum).
schuppe, écaille (*tect.*).
schuppen structure, structure imbriquée.
science of engineering, science de l'ingénieur.
scintillation, scintillation; **s. counter,** compteur à s.; **s. layer,** couche luminescente.
scintillometer, scintillomètre.
scissors fault, faille en ciseaux.
Scleractinian coral, Madréporaire (Trias moy. à actuel).
sclerenchyma, sclérenchyme (cf. coraux).
scleroclase, scléroclase, sartorite.
scolecite, scolécite.
scolecodont, scolécodonte (mâchoire d'annélide).
scolite, scolite (trace de vers).
scoop, godet, cuiller (*mine*); **s. dredger,** drague à g.
scoop (to), excaver, évider, puiser, épuiser.
scooping, dragage.
scoria, scoriae (pl.), **1.** scorie (volcanique ou industrielle); **2.** laitier; **s. cone,** cône de scories.
scoriaceous, scoriated, scoriacé.
scoring (glacial), marque glaciaire.
scorious, scoriacé.
scorodite, scorodite (*minér.*).
scour, affouillement, érosion, creusement, dégradation; **s. and fill,** creusement et remblayage; **s. mark,** trace d'affouillement; **s. side,** face érodée

d'un verrou; **crescent s.**, marque en croissant; **flute s.**, marque en creux allongé; **tidal s.**, a. dû aux courants de marée.

scour (to), décaper, affouiller, nettoyer.

scouring, décapage, affouillement.

scout, ingénieur prospecteur, informateur (pétrole).

scouting, reconnaissance, prospection.

scrap, déchet, débris, rebut.

scrap (to), mettre au rebut, se débarrasser de.

scrape (to), gratter, racler.

scraper, grattoir (préhist.), racloir, décapeur, engin de travaux publics.

scratch, 1. strie, rayure, raie; 2. dépôt d'ébullition d'eau de mer.

scratched, strié; **s. boulder**, bloc s.; **s. pebble**, galet s.

scratching, rayure, striage.

scree, éboulis, talus; **s. breccia**, brèche d'é.; **s. of frost-shattered debris**, é. de matériel gélivé.

screen, 1. écran; 2. tamis, crible, filtre; 3. abri météorologique; **s. analysis**, analyse granulométrique; **s. pipe**, tube filtre; **s. sizing**, calibrage au tamis; **jigging s.**, crible à secousses; **mud s.**, filtre à boue; **vibrating s.**, tamis vibrant.

screen (to), cribler, passer au crible.

screened, criblé, tamisé; **s. coal**, charbon c.; **s. ore**, minerai classé; **s. sand**, sable nettoyé, classé par l'action des vagues.

screening, criblage, tamisage, filtrage; **s. drum**, trommel classeur; **s. machine**, crible mécanique; **s. plant**, installation de c.

screenings, déchets de criblage, refus de tamisage.

screenman, cribleur.

screw auger, tarière rubanée, à vis, tarière torse.

screw (to) the rods, visser les tiges de forage.

scroll, croissant d'alluvionnement, méandre à croissant.

scrub, brousse, savane.

scrubbed gas, gaz purifié.

scrubber, épurateur.

scrubber plant, usine d'épuration de gaz naturel.

scrubbing, épuration; **gas s.**, épuration du gaz naturel.

scrubstone (G.B.), grès calcaire.

scuba, soucoupe plongeante (*océano.*).

scud, 1. minces couches d'argile ou de charbon (*mine*); 2. pyrite interstratifiée dans les couches de charbon; 3. fractostratus (*météorol.*).

scum, écume, mousse.

scum (to), se couvrir d'écume.

Scyphozoa, Scyphozoaire (*pal.*).

Scythic stage, Scythian, Scythien (étage, Trias inf.).

sea, mer; **s. arm**, bras de m.; **s. basin**, bassin océanique; **s. beach**, plage, grève; **s. beam**, sondeur multifaisceaux; **s. board**, littoral, côte; **s. bottom**, fond marin; **s. breeze**, vent de m.; **s. built levee**, cordon littoral; **s. cave**, grotte littorale; **s. cliff**, falaise littorale; **s. coast**, côte, littoral; **s. flood**, inondation, raz de marée; **s. floor**, fond marin; **s. floor spreading**, expansion océanique; **s. foam**, sépiolite; **s. gate**, détroit; **s. level**, niveau marin; **s. level change**, variation du niveau

marin; **s. line**, conduite sous-marine; **s. mount**, guyot, mont sous-marin; **s. mud**, vase marine; **s. quake**, séisme sous-marin; **s. salt**, sel marin; **s. sand**, sable marin; **s. scape**, paysage marin; **s. slide**, glissement sous-marin; **s. stack**, éperon d'érosion marine; **s. valley**, canyon sous-marin; **s. wall**, rempart de cordon littoral, digue; **s. way**, détroit; **s. weed**, algue plante marine; **choppy s.**, m. dure; **enclosed s.**, m. fermée; **eperic shelf s.**, m. épicontinentale; **epicontinental s.**, m. épicontinentale; **heavy s.**, grosse m.; **high s.**, haute m., grand large; **hollow s.**, m. creuse, m. à grands creux; **inland s.**, m. intérieure; **land locked s.**, m. intérieure; **main s.**, grand large (marin); **marginal s.**, m. bordière; **minor s.**, m. secondaire; **shelf s.**, m. épicontinentale; **sub s. drilling**, forage sous-marin; **wild s.**, m. démontée.

seal, 1. scellement, obturation, dispositif d'étanchéité; 2. barrage étanche; **s. fluid**, fluide obturateur; **water s.**, fermeture hydraulique.

seal off (to) a water bearing formation, obturer un aquifère.

sealed, étanche, scellé.

sealing, scellement, procédé d'étanchéité; **s. formation**, roche couverture.

seam, filon, veine, couche; **coal s.**, couche de charbon; **s. of high dip**, dressant (*mine*); **s. work**, travaux en couche (*mine*).

seamy rock, roche fissurée.

seashore, bord de mer, littoral.

seasonal run-off, écoulement périodique, saisonnier.

seat rock, mur (*mine*), semelle (d'une mine).

sebkha (sabkha, Arabie), sebkha.

secondary, secondaire; **s. clay**, argile remaniée; **s. enlargement**, accroissement s. (*cristallo*); **s. enrichment**, enrichissement s.; **s. gley like soil**, sol à simili gley s.; **s. limestone**, calcaire redéposé secondairement; **s. migration**, migration s.; **s. mineral**, minéral d'altération; **s. podzol**, sol podzolique s.; **s. recovery**, récupération s.; **s. reflections**, réflexions secondaires multiples; **s. soil**, sol allochtone; **s. structures**, structures secondaires; **s. waves**, ondes secondaires transversales.

Secondary Era (obsolete), better: Mesozoic, Secondaire: mieux : Mésozoïque (Ere).

second bottom, basse terrasse; **s. mining**, abatage, dépilage; **s. working**, abatage, enlèvement des piliers, dépilage.

secretion, 1. sécrétion; 2. concrétion (*minér.*).

section, 1. coupe, profil (*cartogr.*); 2. coupe (d'une carrière); 3. plaque mince (*minér.*); **columnar s.**, profil stratigraphique; **cross s.**, coupe transversale; **geological s.**, coupe géologique; **polished s.**, section polie.

secular variation, variation séculaire du champ magnétique.

secunda oil, pétrole provenant de schistes bitumineux.

sedentary detrital soil, sol détritique autochtone.

sediment, sédiment; **s. trap,** piège à s., piège sédimentaire; **s. vein,** filon rempli de s. sous-jacents; **clastic. s.,** s. détritique; **neritic s.,** s. néritique.

sedimental, sédimentaire.

sedimentary, sédimentaire; **s. basin,** bassin s.; **s. break,** lacune de sédimentation; **s. cover structure,** tectonique de couverture; **s. cycle,** cycle s.; **s. deposit,** sédiment; **s. mantle,** manteau s. (structure); **s. rock,** roche s.; **s. structure,** structure s.; **s. trap,** piège s.

sedimentation, sédimentation; **s. analysis,** analyse par s.; **s. balance,** balance à s.; **s. curve,** courbe de s.; **s. rate,** vitesse de s.; **s. tank,** bassin de s.; **s. test,** essai de s.

sedimentogenesis, genèse sédimentaire.

sedimentology, sédimentologie.

seed-bearing plants, plantes à graines; **s. fern,** Ptéridospermée.

Seelandian (Eur.), Seelandien (étage, Paléocène inf.).

seep, suintement, infiltration.

seep (to), suinter, filtrer, s'infiltrer.

seepage, 1. suintement, infiltration; 2. fuite, déperdition; 3. indice de pétrole ou de bitume.

seeping, suintement, filtration, infiltration.

seepy, suintant.

segment, segment (*pal.*).

segregate (to), se déposer, se cristalliser.

segregated vein, filon de ségrégation.

segregation, 1. agrégat minéral authigène; 2. ségrégation; 3. séparation; **s. magmatic,** ségrégation magmatique; **s. layers,** feuillets minéraux formés par ségrégation; **s. vein,** filon de ségrégation.

seiche, seiche (*géogr.*).

seif, seif, dune longitudinale.

seislog, seislog.

seism, séisme.

seismic, seismical, seismal, sismique; **s. activity,** séismicité; **s. array,** rangée de séismomètres; **s. basement,** socle acoustique, socle s.; **s. detector,** détecteur s. (séismographe); **s. discontinuity,** discontinuité s. (entre 2 couches); **s. event,** tremblement de terre, secousse s.; **s. exploration,** prospection s.; **s. gap,** lacune s.; **s. log,** diagraphie s.; **s. map,** carte s.; **s. method,** méthode s.; **s. noise,** bruit s.; **s. profil,** profil s.; **s. prospection,** prospection s.; **s. record,** enregistrement s.; **s. reflection,** réflection s.; **s. reflection method,** sismique réflection; **s. refraction,** s. réfraction; **s. refraction method,** sismique réfraction; **s. risk,** risque s.; **s. shadow,** zone de silence; **s. source,** source s.; **s. spectral-analysis,** analyse d'un spectre s.; **s. station,** station s.; **s. stratigraphy,** stratigraphie s.; **s. surface, s. waves,** ondes superficielles; **s. transition, s. zone,** zone de transition s.; **s. travel time,** temps de propagation des ondes s.; **s. velocity,** vitesse d'une onde s.; **s. wave,** onde élastique (crée par un tremblement de terre); **s. zone,** zone s.

seismicity, sismicité; **s. chart,** carte sismique.

seismogram, sismogramme.

seismograph, séismographe; **lunar s.,** s. lunaire; **pendulum-mounted s.,** s. à mouvement pendulaire; **spring-mounted s.,** s. à ressort; **strain s.,** s. détecteur des allongements et compressions terrestres.

seismographic, sismographique, séismographique.

seismography, sismographie.

seismologic, seismological, séismologique.

seismologist, sismologue, séismologue.

seismology, seismologie.

seismometer, seismomètre.

seismotectonics, sismotectonique.

Selbornian, Selbornien (étage, Crétacé inf., Albien).

Seldovian (U.S.A., Al.), Seldovien (étage floral, Oligocène à Miocène).

selenide, séléniure (composé minéral).

seleniferous, sélénifère.

selenious, sélénieux.

selenite, sélénite (var. de gypse).

selenitic, séléniteux.

selenium, sélénium.

selenodesy, sélénodésie.

selenogeomorphology, géomorphologie de la lune.

selenography, sélénographie.

selenology, sélénologie.

self-capture, autocapture.

selfedge, salbande.

self-potential method, méthode électrique de polarisation spontanée.

self rain-gauge, pluviomètre enregistreur.

self-registering apparatus, appareil enregistreur.

selvage, salbande.

sem data, données M.E.B.

semianthracite, houille anthraciteuse.

semi-arid, semi-aride; **s. soil,** sol semi-aride.

semibituminous coal, houille demi-grasse.

semi-crystalline, semi-cristallin.

semi-mature soil, sol à demi évolué, sol jeune.

semi-opacity, demi-opacité.

semi-opal, opale commune.

semi-planosol, semi-planosol.

semiprecious stone, pierre fine.

semistratified, demi-stratifié.

semi-submersible rig, plate-forme semi-submersible.

semi-swamp soil, sol semi-marécageux à gley.

semitransparent, demi-transparent.

Senecan series (North Am.), série de Senecan (Dévonien sup.).

senile, sénile (*géogr. physique*).

senility, sénilité (*géogr. physique*).

Senonian, Sénonien (étage; Crétacé sup).

sense (to), détecter.

sensitive, 1. sensitif; 2. sensible; **s. clay,** argile fluable; **s. plaque,** plaque sensible (photo); **s. tint (gypsum plate),** gypse teinte sensible.

sensor, détecteur (télédétection).

separate (to), séparer, extraire.

separation, 1. séparation; 2. rejet vertical (*tecto.*); **s. funnel,** entonnoir à séparation.

separator, séparateur; **s. funnel,** entonnoir à séparation.

sepiolite, sépiolite, magnésite (argile).

septal, septal.

septarium, septaria (pl.), concrétion, nodule, septaria.

septate, cloisonné, à septum.

septum, septa (pl.), cloison, septum, septe (*pal.*).

Sequanian (G.B. & Fr.), Séquanien (sous-étage; Kimméridgien).

sequence, séquence, suite, série en succession.

sequential, successif.

serac, sérac (*glaciol.*).

serial samples, échantillons sériés.

sericite, séricite; **s. gneiss,** gneiss à s.; **s. schist,** séricito-s.

sericitic, séricitique, sériciteux.

sericitization, séricitisation.

series of strata, ensemble de couches (de même âge).

serozem, sierozem, sol grisâtre des steppes.

serpentine, serpentine (*minér.*); **s. marble,** marbre serpentin (marbre vert antique).

serpentinic, serpentineux.

serpentinite, serpentinite.

serpentinization, serpentinisation.

serpentinous, serpentinisé.

Serpula, Serpule (Annélide, *pal.*).

serrate, serrated, accidenté, déchiqueté (relief), strié, cannelé, dentelé.

set, série, ensemble; **s. of faults,** ensemble de failles.

set (frame), cadre, châssis de mine.

setting, 1. compaction (du sol); 2. coucher d'un astre.

settle (to), 1. se déposer, sédimenter, précipiter; 2. se tasser, tasser.

settled production, production stable de pétrole d'un puits quelque temps après le début de production.

settlement, 1. précipitation, dépôt; 2. tassement du sol; **s. rate,** vitesse de précipitation.

settler, cuve de lavage (traitement des minerais).

settling, sédimentation, décantation, dépôt, séparation, tassement, affaissement; **s. basin,** bassin de décantation; **s. tube,** tube à sédimentation; **s. rate,** vitesse de sédimentation.

sextant, sextant.

sexual dimorphism, dimorphisme sexuel.

shadow zone, zone d'ombre sismique.

shady side, ubac.

shaft, puits (*mine*), excavation verticale, cheminée; **s. bottom,** fond du puits; **s. boring,** fonçage d'un puits; **s. frame,** cadre du puits; **s. head,** chevalement du puits; **s. hoist,** treuil d'extraction; **s. set,** cadre de puits; **s. sinking,** fonçage de puits; **s. tackle,** chevalement; **s. timbering,** boisage du puits; **s. tower,** chevalement.

shake, 1. secousse; 2. caverne (dans calcaire); 3. fissure.

shake (to), 1. secouer, ébranler; 2. trembler, chanceler, branler.

shaking, secousse; **s. screen,** crible oscillant; **s. sieve,** tamis à secousses; **s. table,** table à secousses.

shale, schiste argileux, argile feuilletée, argile à feuillets; **s. industry,** exploitation de schistes; **s. naphta,** napthe de schistes; **s. oil,** pétrole de schistes bitumineux; **s. rock,** roche schisteuse; **bituminous s.,** s. bitumineux; **carbonaceous s.,** s. charbonneux; **oil s.,** s. bitumineux; **sandy s.,** argilite sableuse.

shallow, peu profond; **s.-focus earthquake,** tremblement de terre peu profond (inférieur à 65 km); **s. ground,** terrain superficiel aurifère (Austr.); **s. water environment,** milieu d'eau peu profonde.

shallows, haut-fond, banc de sable.

shaly, schisteux; **s. bedded,** sédiment s.; **s. lamination,** schistosité; **s. parting,** débit s.

shaping, façonnement, modelé (du relief).

shard, 1. fragment courbe et aciculaire de roche volcanique; 2. fragment, éclat.

sharpstone, roche sédimentaire formée de fins fragments anguleux.

Shasta (U.S.A. Cal.), série de Shasta (Crétacé inf.).

shatter, fragment, morceau; **s. breccia,** brèche de dislocation; **s. mark,** marque de choc.

shatter (to), fracasser, briser, fragmenter.

shattering, éclatement, fragmentation, broyage; **frost s.,** gélivation.

sheaf of fold, faisceau de plis.

sheaf-like structure, structure en gerbes.

shear, cisaillement; **s. cleavage,** clivage par plifracture; **s. fold,** plis de c.; **s. fracture,** fracture de c.; **s. joint,** fracture de c.; **s. modulus,** module de c.; **s. surface,** plan de c.; **s. structure,** structure de c.; **s. thrust,** charriage de c.; **s. waves,** ondes transversales (onde S); **s. zone,** zone cisaillée (*géophys.*).

shear (to), cisailler.

shearing, cisaillement; **s. displacement,** c.; **s. force,** force de c.; **s. strain,** contrainte due au c., déformation tangentielle; **s. strength,** résistance au c.; **s. stress,** effort de c., contrainte de c.

shed line, ligne de partage des eaux.

sheepback rocks, sheepbacks, roches moutonnées.

sheep-tracks, « pieds de vaches » (*géomorph.*).

sheer, 1. adj. abrupt; 2. adv. abruptement, perpendiculairement.

sheet, 1. feuille, feuillet; 2. couche, nappe; 3. amas mince de galène (U.S.); **s. deposit,** gîte, minéral stratiforme (U.S.); **s. erosion,** érosion en nappes; **s. flood,** inondation en nappes; **s. flow,** écoulement laminaire; **s. ground,** filon-couche; **s. jointing,** séparation en bancs; **s. joints,** diaclases horizontales; **s. like-intrusion,** intrusion stratiforme; **s. mineral,** minéral en feuillets stratifiés; **s. sand,** couche de grès marin; **s. structure,** structure lamellaire; **s. vein,** filon-couche; **s. wash,** ruissellement en nappes; **s. water,** eaux des nappes; **intrusive s.,** filon-couche intrusif; **thrust s.,** nappe de charriage.

sheeted, stratifié; **s. vein,** groupe de filons séparés de stériles; **s. zone,** zone fracturée.

sheeting, 1. découpage en strates, découpage par diaclases et joints de stratification, stratification;

2. débit, débitage superficiel des roches; **s. pile,** palplanche; **s. plane,** plan de stratification.

shelf, **1.** plateau continental; **2.** hauts-fonds, banc de sable; **s. break,** limite entre plate-forme continentale et talus continental; **s. ice,** plate-forme flottante de glace d'origine continentale; **s. edge,** limite supérieure du talus continental; **s. facies,** faciès continental; **s. sea,** mer épicontinentale; **continental s.,** plate-forme continentale.

shell, **1.** coquille, coquillage; **2.** enveloppe; **s. band,** argile ferrugineuse fossilifère; **s. breccia,** lumachelle; **s. hash,** lumachelle; **s.-like,** conchoïdal; **s. limestone,** calcaire coquillier; **s. marl,** vase calcaire lacustre fossilifère; **s. rock,** lumachelle; **s. sand,** sable à coquilles.

shelled animal, mollusque.

shellfish, invertébré aquatique à coquille ou carapace.

shelly, coquillier; **s. limestone,** calcaire c.; **s. sand,** falun.

shelter, lieu de refuge, abri.

shelter-cave, abri sous roche.

shelving, en pente, incliné.

shield, bouclier (*tectonique*); **s. volcanoe,** volcan surbaissé, en bouclier; **canadian s.,** bouclier canadien.

shift, composante horizontale du rejet d'une faille; **red s.,** décalage spectral vers le rouge (*astro.*).

shift (to), déplacer, changer, remuer.

shifting, déplacement (des côtes), migration; **s. river bed,** lit fluviatile mobile, changeant; **s. dune,** dune mobile; **s. sand,** sable mouvant; **s. soil,** sol mouvant; **lateral s.,** décrochement; **monoclinal s.** glissement monoclinal.

shingle, galet, caillou; **s. bar,** cordon de galets; **s. beach,** plage de galets; **s. spit,** levée de galets; **s. structure,** structure filonienne imbriquée, en échelons.

shingling, structure imbriquée.

shingly, à galets, caillouteux.

shipboard magnetometer, magnétomètre marin.

shipping ton, tonneau (= 1,132.674 m^3).

shiver, **1.** éclat, fragment; **2.** pierre schisteuse.

shoal, haut-fond, banc, gué; **s. reef,** banc récifal; **s. water,** eau peu profonde.

shoaliness, caractère peu profond (d'une eau).

shoaly, plein de hauts-fonds (cours d'eau).

shock, secousse, choc sismique; **s. breccia,** brèche sismique, tectonique; **s. metamorphism,** dynamométamorphisme; **s. wave,** onde de choc.

shode, **1.** guidon (*mine*); **2.** fragment meuble de filon (Cornouailles).

shod soil, sol à sous-sol rocheux.

shoestring, cordon littoral; **s. gully erosion,** érosion en rigoles; **s. sand,** grès en bandes allongées.

shonkinite, shonkinite (var. de syénite).

shoot, **1.** mise à feu (d'un explosif); **2.** couloir (d'avalanches); **3.** cheminée (*mine*); **4.** veine, filon exploitable; **s. of ore,** filon oblique minéralisé.

shoot (to), mettre à feu, abattre à l'explosif, faire un tir sismique.

shooter, boutefeu.

shooting, tir (*géoph.*); **s. boat,** bateau boutefeu; **s. flow,** écoulement torrentiel; **s. off the solide,** abatage par des explosifs (mine); **s. truck,** camion boutefeu; **s. a well,** explosion de dynamite à l'entrée d'un puits, enregistrement d'un carottage sismique; **reflection s.,** réflexion sismique; **refraction s.,** réfraction sismique; **seismic s.,** tir sismique; **well s.,** carottage sismique.

shore, rivage, littoral, côte; **s. cliff,** falaise littorale; **s. deposit,** dépôt littoral; **s. drift,** dérive littorale; **s. dune,** dune littorale; **s. face,** zone infratidale; **s. feature,** morphologie littorale; **s. platform,** plate-forme littorale, terrasse d'abrasion; **s. profil,** profil littoral; **s. protection,** protection du littoral; **s. terrace,** terrasse littorale; **back s.,** arrière plage; **fore s.,** avant plage; **ashore,** à terre; **offshore,** au large; **on shore,** à terre.

shoreline, ligne de rivage; **s. management,** aménagement du littoral; **s. of depression,** côte de submersion; **s. of elevation,** côte d'émersion; **s. of emergence,** côte d'émersion; **s. of progradation,** côte d'accumulation; **s. of retrogradation,** côte d'abrasion; **s. of submergence,** côte de submersion; **cliffy s.,** côte à falaise; **depressed s.,** côte affaissée; **fault s.,** côte de faille; **graded s.,** côte régularisée; **raised s.,** côte soulevée; **smooth graded s.,** rivage à tracé régularisé; **submerged s.,** côte submergée.

short, **1.** court; **2.** cassant; **s. coal,** charbon friable; **s. cut,** court-circuit d'un méandre; **s. limb,** flanc inverse; **s. range order,** désordre atomique (état amorphe).

shortening, rétrécissement, raccourcissement (de la lithosphère, etc.).

shot, coup de feu, mine, tir, explosion; **s. break,** signal enregistré de l'explosion, temps zéro; **s. boring,** sondage à la grenaille; **s. depth,** profondeur de tir; **s. drill,** explosion à la grenaille; **s. exploder,** exploseur; **s. firer, 1.** ouvrier boutefeu, tireur de coups de mines; **2.** allumeur; **s. firing,** tirage des coups de mine; **s. hole,** trou de tir; **s. instant,** moment de l'explosion; **s. hole drilling,** forage sismique; **s. metal,** alliage de plomb et d'arsenic; **s. point,** point de tir.

shotty gold, or en granules.

shoulder, épaulement.

shove, poussée.

shoved moraine, moraine de poussée.

shovel, excavateur, pelle mécanique.

shovel (to), pelleter, ramasser à la pelle.

show, indice, traces; **oil s.,** indice de pétrole.

show case, vitrine à échantillons.

shower, pluie, averse; **ash s.,** pluie de cendres.

showings, traces; **oil s.,** indices de pétrole.

shrave, concrétion ferrugineuse.

shred, fragment, filament.

shreddy, filamenteux.

shrink, rétrécissement, retrait.

shrink (to), (se) rétrécir, (se) contracter, (se) resserrer.

shinkage, contraction, retrait, tassement; **s. crack**, fissure de retrait; **s. hole**, cavité de retrait.

shrink-swell, sols ou argiles alternativement rétractés ou gonflés.

shut in, défilé, gorge fluviatile.

shut in pressure, pression statique (dans un puits).

shut-in well, puits fermé.

shut-off, vanne, fermeture; **water s.**, fermeture des eaux (forage).

shut off (to), fermer, obturer.

shutter, vanne.

sial, sial.

sialic, silico-alumineux, sialique.

siallitic soil, sol à fort rapport silice alumine.

siallitization, siallitisation.

sialma, couche terrestre comprise entre sial et sima.

siberian ruby, rubellite (var. rouge).

siberite, rubellite (var. rouge-violet).

Sicilian, Sicilien (formation de la Méditerranée; Pléistocène).

side, 1. côté; 2. flanc, paroi; 3. versant; 4. lèvre (d'une faille); **s. dump car**, wagonnet basculant de côté; **s. entry**, galerie latérale; **s. looking sonar**, sonar latéral; **s. moraine**, moraine latérale; **s. of shaft**, paroi d'un puits; **s. scan sonar**, sonar latéral; **s. scanning method**, télédétection; **s. stoping**, abatage latéral; **s. tracked hole**, forage dévié (pétrole); **s. wall**, paroi latérale; **s. wall coring**, carottage latéral; **s. wall sampler**, carottier latéral; **fault s.** lèvre d'une faille; **fold s.**, flanc d'un pli.

siderite, sidérite; 1. minéral; 2. météorite.

siderolite, sidérolite.

sideromagnetic, sidéromagnétique, paramagnétique.

sideromelane, verre basaltique (dans tufs palagonitiques)

siderophyllite, sidérophyllite (biotite noire ferrique).

siderophyre, sidérophyre (météorite).

sideroscope, détecteur magnétique de fer.

siderosphere, noyau ferreux central de la terre.

siderurgy, sidérurgie.

sidewall core, échantillon (ou carotte) prélevé latéralement dans un forage.

sidewall sampling, prélèvement latéral (dans un forage).

Siegenian, Siégénien (étage, Dévonien inf.).

Sienna, terre de Sienne (argile jaune orangée).

sierozem, serozen, sol gris désertique.

sierra, sierra (*géomorph.*).

sieve, tamis, crible; **s. analysis**, analyse granulométrique; **s. filter**, filtre à t.; **s. flange**, rebord de t.; **s. plate**, plaque criblée; **s. test**, essai de criblage; **s. texture**, structure diablastique; **bolting cloth s.**, t. à trous filtrants; **mesh s.**, t. à mailles; **woven wire s.**, t. à fil métallique.

sieve (to), tamiser, cribler.

sieving, tamisage, criblage; **wet s.**, tamisage à l'eau.

sif, seif, dune longitudinale.

sifflet bed, couche en biseau.

sift (to), passer au tamis, tamiser, cribler.

sifter, 1. tamis, crible; 2. ouvrier cribleur.

sifting, 1. tamisage, criblage; 2. produit de tamisage.

sight-bar, 1. alidade; 2. aiguille de mire du viseur (*photogr.*).

sight-compass, boussole à pinnule.

sight-distance, distance de visée.

sighter, aiguille de mire d'un viseur.

sighting, vue, visée, pointage; **s. apparatus**, appareil de visée; **s. aperture**, voyant d'un instrument; **s. board**, alidade, voyant; **s. tube**, viseur; **s. line**, ligne de visée.

sigmoidal fold, pli à axe longitudinal en S.

signal, signal (*géophys.*).

silcrete, silcrète (croûte siliceuse).

silcrust, croûte siliceuse.

Silesian, Silésien (Carbonifère sup.).

silex, silex.

silexite, silexite.

silica, 1. silice; 2. roche siliceuse pulvérulente (U.S.); **s. alumina ratio**, quotient silice-alumine; **s. pan**, horizon dur silicifié (*pédol.*); **s. sand**, sable siliceux; **s. saturation**, saturation en silice.

silicarenite, silicarénite, grès quartzeux.

silicate, silicate; **s. rock**, roche silicatée.

silicated, silicaté.

silicatization, silicatisation, transformation en silicates.

siliceous, siliceux; **s. concretion**, concrétion siliceuse; **s. earth**, diatomite; **s. laterite**, latérite siliceuse; **s. limestone**, calcaire s.; **s. matrix**, ciment s.; **s. ooze**, boue siliceuse; **s. sinter**, geysérite, opale; **s. shale**, schiste s.; **s. soil**, sol s.; **s. rock**, roche siliceuse.

silicic, 1. siliceux (*pétro.*); 2. silicique (*chimie*); **s. acid**, acide silicique; **s. rock**, roche siliceuse.

silicicalcareous, silicocalcaire.

siliciclastic, siliciclastique.

siliciferous, siliceux.

silicification, silicification.

silicified, silicifié; **s. wood**, bois s.

silicify (to), silicifier, imprégner de silice.

silicious, siliceux.

Silicispongiae, Éponges siliceuses.

silicium, silicium.

silicocalcareous, silicalcaire.

Silicoflagellates, Silicoflagellés (*pal.*).

silicon, silicium.

silky luster, éclat soyeux.

sill, 1. filon-couche; 2. seuil; **s. drift**, galerie de fond (*mine*); **s. floor**, niveau de fond (*mine*).

sillimanite, sillimanite (*minér.*).

silt, 1. G.B. : limon (0,02-0,002 mm) U.S. : limon (0,05-0,002 mm), normal : (0,0625-0,004 mm); 2. sol à 80 % ou plus de limon; 3. envasement; **s. agglomeration**, formation de grumeaux de limon; **s. loam**, lehm (limono-argileux); **s. pan**, horizon pédologique concrétionné souvent siliceux.

siltil, siltill, limon morainique, sol glaciaire (avec sous-sol perméable).

silting up, envasement, colmatage, remblayage par embouage (*mine*).

siltstone, aleuronite, microgrès.

silt up (to), envaser.

silty, silteux, vaseux, boueux; **s. bog soil,** sol marécageux noir; **s. clay soil,** sol limono-argileux; **s. clay loam,** limon argileux fin; **s. clayed much,** sol organo-minéral limono-argileux; **s. gravel soil,** sol limono-caillouteux; **s. loam,** lehm.

Silurian, Silurien (période, Paléozoïque).

silver, argent; **s. bearing,** argentifère; **s. chloride,** chlorure d'a.; **s. glance,** argentite, argyrite; **s. lead,** plomb argentifère; **s. mine,** mine d'a.; **s. ruby,** pyrargyrite; **s. state,** Névada.

silvery, argenté.

sima, sima.

similar folding, pli isoclinal.

simoon, simoun.

simple vein, filon monominéral.

sine wave, onde sinusoïdale.

Sinemurian, Sinémurien (étage, Lias).

single, unique, simple.

single-crystal analysis, diagramme monominéral.

single channel filtering, filtrage monocanal.

single grain structure, structure monoparticulaire à grains indépendants; **s. vein,** filon simple.

sinistral, sénestre; **s. fault,** faille à déplacement latéral gauche; **s. fold,** pli s. (déversé vers la gauche).

sink, **1.** doline, creux, poljé; **2.** courant océanique descendant; **s. hole,** effondrement, puisard, doline; **s. hole pond,** lac de doline; **s. hole lake,** lac de doline.

sink (to) a hole, faire un sondage; **s. a shaft,** foncer un puits.

sinker, ouvrier fonceur (*mine*); **s. bar,** barre de surcharge (*mine*).

sinking, **1.** fonçage, foncement, creusement; **2.** affaissement, tassement; **3.** agent agglutinant; **s. a shaft,** creusement d'un puits; **s. a well,** creusement du puits; **s. by boring,** fonçage au trépan; **s. creek,** canyon; **s. current,** courant descendant (*océano.*); **s. of bore hole,** perforation du trou de mine; **s. of the roof,** affaissement du toit de la mine; **s. of water,** infiltration des eaux; **s. pump,** pompe de fonçage (*mine*).

sinople, sinople (argile ferrugineuse).

sinter, tuf, travertin, geysérite; **s. deposit,** concrétion; **calcareous s.,** tuf calcaire; **siliceous s.,** tuf siliceux.

sintered, aggloméré.

sintering, agglomération (de particules).

sinuosity, sinuosité (d'un cours d'eau).

sinuous, sinueux (cours d'eau).

sinupalliate, sinupallié (*pal.*).

siphon, siphon (*pal.*); **excurrent s.,** s. exaltant; **incurrent s.,** s. inhalant.

siphonal funnel, canal siphonal (*pal.*).

siphuncle, canal siphonal.

Sipunculoïda, Siponculé (*pal.*).

sit, affaissement (mine).

size, grandeur, taille, dimension; **s. frequency curve,** courbe de fréquence granulométrique; **s. grade,** classe granulométrique; **natural s.,** g. nature.

sized, calibré, dimensionné.

sizing, classement granulométrique, criblage; **s. screen,** crible classeur; **s. trommel,** trommel classeur.

skapolith, scapolite (*minér.*).

skarn, **1.** skarn (calcaire argileux de métamorphisme de contact); **2.** gangue silicatée (Suède).

skeletal soil, sol squelettique.

skeleton, squelette (*pal.*).

skerry, récif, rocher isolé, îlot rocheux.

sketch map, croquis de reconnaissance.

skid the derrick (to), déplacer une tour de forage sans la démonter.

skim, écume (*métallurgie*).

skimming, séparation (de deux types de solutions).

skimming plant, installation de distillation non poussée (jusqu'aux fractions légères).

skimpings, refus du crible (traitement des minerais).

skin, pellicule, croûte, film d'eau; **s. dust,** mince croûte de poussière.

skip, monte-charge à godets, benne, skip (*mine*); **s. hoist,** monte-charge à g.

skull, crâne (*pal.*).

Skythian (see Scythian), Scythien (étage du Trias).

Skytic stage (obsolete sp.), Scythien (étage du Trias).

slab, plaquette, dalle.

slab (to), trancher (une roche).

slabbing, tranchage.

slabstone, pierre à débit en dalles, en plaquettes (3 à 12 cm d'épaisseur).

slack, mou, lâche; **s. lime,** chaux vive; **s. sea,** mer étale; **s. tide,** marée étale, étal; **s. water,** eau étale.

slack coal, charbon fin, fines.

slade, vallon, clairière.

slag, scorie, laitier; **s. shingle,** scorie d'empierrement.

slaggy, scoriacé.

slaked lime, chaux.

slaking, émiettement et décomposition de matériaux par altération atmosphérique (U.S).

slant, **1.** adj : oblique, en biais, incliné; **2.** n : inclinaison, pente, pendage, dénivellation; **s. drilling,** forage oblique; **s. time,** temps de trajet oblique (*sismol.*).

slant (to), **1.** être en pente, être incliné; **2.** s'incliner.

slanted drill hole, forage oblique.

slanting, oblique.

slash, slash : ancien sillon de plage, sillon d'estran (marécageux).

slat, lame, lamelle, ardoise mince.

slate, ardoise, schiste; **s. clay,** argile schisteuse; **s. coal,** charbon schisteux; **s. oil,** pétrole de schistes bitumineux; **s. pit,** ardoisière, carrière d'a.; **s. quarry,** carrière d'a., ardoisière; **s. quarryman,** ardoisier.

slaty, **1.** ardoisier; **2.** schisteux; **s. clay,** argile schisteuse; **s. cleavage,** schistosité, clivage ardoisier; **s. coal,** charbon schisteux; **s. structure,** structure schisteuse.

sledge hammer, masse, marteau à frapper devant.

sleet, **1.** neige fondue; **2.** nodules de glace (U.S.A.).

sleeve exploder, dispositif à tri sismique tracté.
slice, 1. tranche, lame, écaille; 2. coup en biais; **thrust s.,** lambeau de charriage.
sliced structure, structure finement particulaire.
slicing, exploitation par tranches (*mine*).
slickenside, surface de friction, miroir de faille.
slide, 1. glissement, éboulement; 2. matériel glissé; **s. debris,** matériel éboulé; **s. fault,** faille subhorizontale; **s. furrow,** couloir d'avalanche; **s. gauge,** pied à coulisse; **s. movement,** mouvement du glissement; **s. projector,** projecteur à diapositives; **s. rock,** éboulis, cône de déjection.
slide (to), glisser, faire glisser, coulisser.
sliding, glissant, coulissant; **s. caliper,** pied à coulisse; **s. erosion,** érosion par glissement; **s. object stage,** platine porte-lames (microscopie); **s. tectonics,** tectonique de glissement.
slikke, slikke (partie vaseuse inférieure de l'estran).
slime, limon, vase, boue, poussière de minerai; **s. concentration table,** table à secousse (*mine*); **s. pit,** bassin de dépôt de boues (forage); **s. separator,** séparateur de schlamms; **s. water,** eau trouble.
slim hole, forage à diamètre réduit.
slimming, broyage fin à l'état humide.
slimy, limoneux, vaseux, boueux.
slip, 1. glissement, éboulement; 2. faille (U.S.); 3. rejet d'une faille; **s. bedding,** plissotement de couches, formé par glissement; **s. cleavage,** clivage de crénulation; **s. erosion,** érosion par glissement; **s. face,** 1. face sous le vent; 2. face de glissement intracristalline; **s. fault,** faille de cisaillement; **s. fold,** pli de cisaillement; **s. plane,** plan de glissement; **s. sheet,** couche formée par affaissement; **s. vein,** 1. filon de faille; 2. filon fissure; **s. tectonite,** glissement majeur selon les plans S dominants; **dip s.,** projection du rejet net d'une faille; **land s.,** glissement de terrain; **net s.,** rejet net; **perpendicular s.,** projection du rejet net sur une perpendiculaire aux bancs dans le plan de faille; **trace s.,** projection du rejet net sur une parallèle aux bancs dans le plan de faille.
slip (to), glisser, s'ébouler.
slippage, glissement.
slipping, glissement.
slipp-off, décollement; **s. correction,** correction de pente; **s. decline,** érosion des versants; **s. erosion,** érosion par glissement; **s. slope,** d.
slippy (G.B.), fissuré.
slips, surface de glissement, miroir de glissement.
slit, fente, fissure, rainure.
slit (to), fendre.
slitter, pic.
sliver, tranche, éclat, fragment.
sloam, couche d'argile intercalée (dans le charbon).
slob, vase, limon.
slop, fondrière, boue, fange.
slope, 1. pente, inclinaison; 2. talus; 3. versant; **s. concavity,** partie inférieure concave d'un versant; **s. convexity,** partie supérieure convexe d'un versant; **s. deposit,** dépôt de pente; **s. failure,** glissement de terrain; **s. of equilibrium,** pente d'équilibre; **s. line,** ligne de plus grande pente; **s. loam,** limon de pente; **s. movement,** glissement de terrain; **s. stability,** stabilité des pentes; **s. wash,** 1. matériel soliflué; 2. ruissellement sur versant; **alluvial s.,** glacis alluvial; **back s.,** revers, contre-pente; **break of s.,** rupture de pente; **continental s.,** talus continental; **counter s.,** contre-pente; **reversed s.,** contre-pente.
sloped, en pente, incliné.
sloping, en pente.
sloppy, détrempé, bourbeux, fangeux.
slot, 1. encoche, entaille, fente, rainure, saignée; 2. émission sismique; **s. gorge,** défilé.
slot (to), entailler, fendre, haver (mine).
slotted, à fente, à encoche, à rainure.
slotting, entaillage, rainurage.
slough, 1. bourbier, fondrière; 2. mue (d'un Reptile).
sloughy, bourbeux, marécageux.
sludge, boue, bouillie, vase, limon, déchets de sondage.
sludger, pompe à boue, pompe à sable, désensableur (*mine*).
sludgy, vaseux, fangeux, bourbeux.
sluggish, lent, paresseux; **s. stream,** cours d'eau paresseux.
sluice, écluse, bonde, canal.
sluice (to), laver au sluice (*mine*).
sluicing, concentration de minéraux lourds et minerai par lavage au sluice.
slump, 1. glissement, éboulement, effondrement; 2. matériel glissé; **s. bedding,** stratification déformée (terme général); **s. fault,** faille normale; **s. fold,** plissement intraformationnel (par glissement sur talus continental); **s. scar,** niche de décollement.
slump (to), s'enfoncer, s'effondrer, glisser.
slumping, glissement de terrain.
slurry, bouillie, boue.
slush, 1. neige à demi-fondue, fange, bourbe, boue glacée; 2. remblai d'embouage; 3. glace de banquise en formation; **s. pit,** bac à boue (forage).
slush flow, coulée de neige et d'eau.
slushing, remblayage hydraulique par embouage.
slushy, détrempé par la neige, boueux, bourbeux.
small coal, charbon pulvérulent.
small ore, minerai pulvérulent; **s. scale,** à petite échelle.
smalls, fines (de minerais).
smaltite, smaltite (*minér.*).
smash (to), 1. cogner, heurter; 2. briser en morceaux, fracasser.
smasher, broyeur.
smectite, smectite (variété d'argile).
smelt (to), fondre (le minerai).
smeltable, fusible.
smelter, 1. fondeur; 2. fonderie.
smeltery, fonderie.
smelting, fusion, fonte; **s. plant,** fonderie; **s. works,** fonderie.
smithsonite, smithsonite (*minér.*).
smoker chimney, cheminée minéralisante avec émanations gazeuses (fond du pacifique).

smokestone, quartz fumé.

smoking coal, houille grasse, charbon bitumineux.

smoky, fuligineux; **s. quartz,** quartz enfumé; **s. topaz,** topaze enfumée.

smoothing (of landscape), aplanissement (du paysage).

smut, 1. suie; 2. charbon terreux.

Snail, Gastéropode, Escargot.

snout, nez; **glacial s.,** front glaciaire.

snow, neige; **s. avalanche,** avalanche de n.; **s. bank,** tache de n.; **s. drift,** amas de n., congère; **s. fall,** chute de n., enneigement; **s. free,** sans n.; **s. field,** névé; **s. glacier,** glacier de névé; **s. line,** limite des n. persistantes; **s. melt,** fonte nivale; **s. patch,** congère; **s. patch erosion,** nivation; **s. settling,** tassement de la n.; **s. slide,** avalanche; **s. slip,** avalanche; **dry s.,** n. sèche; **powdery s.,** n. poudreuse.

snowy, enneigé, neigeux.

soak, 1. imbibition; 2. bassin de réception pluvial, lac, dépression (Austr.).

soak (to), tremper, imbiber, détremper, imprégner; **s. away,** disparaître par infiltration; **s. in water,** absorber de l'eau.

soakage water, eau d'infiltration, eau d'imbibition.

soaker, pluie battante, déluge.

soaking, trempage, imbibition.

soapstone, 1. stéatite, talc; 2. toute roche tendre au toucher gras.

soapy luster, éclat gras.

socket, fossette (d'une dent de la valve opposée, *pal.* Lamellib.).

soda, 1. soude; 2. carbonate de soude; 3. préfixe indiquant une teneur en pyroxènes ou amphiboles sodiques; **s. alum,** alun sodique; **s. ash,** carbonate de sodium anhydre; **s. deposit,** gisement de soude; **s. feldspar,** feldspath sodique, albite; **s. lake,** lac salé; **s. lime,** chaux sodée; **s. lime feldspar,** feldspath calco-sodique; **s. microcline,** anorthose; **s. niter,** salpêtre (du Chili), nitrate de sodium; **s. orthoclase,** anorthose; **s. rhyolite,** rhyolite à pyroxènes sodiques; **s. salt,** sel sodique; **s. syenite,** syénite sodique à albite.

sodalite, sodalite (= feldspathoïde) (*minér.*).

sodium, sodium; **s. alkalic rock,** roche alcalinosodique; **s. alum,** alun de s.; **s. chloride,** chlorure de s.; **s. feldspar,** feldspath sodique; **s. hydroxide,** hydroxyde de s.; **s. salt,** sel sodique.

soft, mou, malléable, tendre; **s. bodied animal,** animal à corps mou; **s. clay,** argile plastique; **s. coal,** charbon gras; **s. ground,** 1. terrain mauvais à consolider; 2. minerai meuble; **s. ore,** hématite meuble (U.S.); **s. rock,** roche tendre; **s. tertiary sandstone,** molasse tendre; **s. water,** eau dépourvue de sels de calcium et de magnésium.

softening, adoucissement, amollissement; **s. of lead,** enlèvement de l'antimoine du minerai de plomb; **water s.,** adoucissement des eaux.

sohlband, cycle sédimentaire à banc majeur à la base (cf. dachband).

soil, 1. sol, terrain; 2. morts-terrains de recouvrement; **s. analysis,** analyse de sol; **s. auger,** tarière pédologique; **s. auger sample,** échantillon pris à la tarière; **s. cap,** morts-terrains de recouvrement; **s. cementation,** stabilisation du sol; **s. compaction,** compactage du sol; **s. creep,** lent glissement du sol; **s. densification,** compactage du sol; **s. discharge,** taux d'évaporation; **s. engineering,** technique des sols; **s. erosion,** érosion du sol; **s. flow, soil flowage,** solifluxion; **s. horizon,** horizon pédologique; **s. improvement,** amélioration des sols; **s. leaching,** lessivage des sols; **s. map,** carte pédologique; **s. mechanics,** mécanique des sols; **s. profile,** profil pédologique; **s. sampler,** échantillonnage des sols; **s. science,** pédologie; **s. scientist,** pédologue; **s. separate,** constituant granulométrique; **s. series,** série de sols; **s. skeleton,** squelette du sol; **s. slip,** glissement de terrain; **s. solution,** solution du sol; **s. stabilization,** stabilisation des sols; **s. stripes,** sol strié; **s. type,** type de sol; **s. veins,** fentes de gel à remplissage terreux; **s. with impeded drainage,** sol mal drainé; **s. with pattern design,** sol structuré, sol polygonal; **azonal s.,** sol azonal; **brown s.,** sol brun; **brown forest s.,** sol brun forestier; **fen s.,** tourbière basse; **fossil s.,** paléosol; **gleyed forest s.,** sol forestier à gley; **gray brown podzolic s.,** sol lessivé; **gray wood s.,** sol gris forestier; **intrazonal s.,** sol intrazonal; **laterite s.,** sol latéritique; **polygonal s.,** sol polygonal; **prismatic s.,** sol polyédrique; **secondary s.,** sol allochtone; **solonetz s.,** sol désertique; **stationary s.,** sol superficiel; **top s.,** sol superficiel; **zonal s.,** sol zonal.

solar, solaire; **s. cell,** cellule s.; **s. corona,** couronne s.; **s. collector,** collecteur s.; **s. constant,** constante s.; **s. disk,** disque s.; **s. energy,** énergie s.; **s. flare,** éruption s.; **s. heat,** chaleur s.; **s. radiation,** radiation s.; **s. salt,** sel obtenu par évaporation au soleil; **s. spot,** tâche s.; **s. wind,** vent s.

sole, 1. semelle, sole (d'une mine); 2. plan de chevauchement inférieur dans une série de nappes; **s. fault,** faille chevauchante subhorizontale; **s. injection,** intrusion plutonique subhorizontale; **s. mark,** figure basale; **s. thrust,** surface inférieure de chevauchement.

soled cobble, galet glaciaire à facettes.

soled pebble, galet à facettes.

solemarked bed, lit à partie inférieure présentant des figures de charge.

Solenhofen stone, calcaire lithographique de Solenhofen (Jurassique sup., Allem.).

solfatara, solfatare, fumerolle.

solfataric, solfatarique.

solid-solid reaction, réaction à l'état solide.

solid, 1. adj : solide, dur, consistant; 2. n : corps solide, solide; **s. beds,** couches dures; **s. fuel,** combustible solide; **s. geology,** géologie du substratum (U.K.); **s. ground,** terrain ferme; **s. inclusion,** inclusion solide (*minér.*); **s. load,** charge solide; **s. phase,** phase solide; **s. soil,** sol dur, ferme; **s.**

solution, solution monocristalline, solution solide; **s. stage,** état solide; **s. waste,** débris, déchets solides.

solid-earth geophysics, physique du globe.

solidification, solidification.

solidify (to), solidifier, se solidifier.

solidifying point, point de solidification.

solidus, solidus.

solidus line, courbe solidus.

solifluction, solifluxion; **s. bench,** banquette de s.; **s. deposit,** dépôt de s.; **s. festoon,** guirlande de s.; **s. flow,** coulée de s.; **s. lobe,** lobe de s.; **s. loess,** loess soliflué; **s. pocket,** poche de s.; **s. sheet,** manteau de s. head (U.K.); **s. slope,** pente formée par s.; **s. stream,** coulée de s.; **s. terrace,** replat de s.; **s. wrinkle,** bourrelet de s.; **checked s.,** s. entravée; **impeded s.,** s. entravée.

solitary coral, coralliaire isolé.

solitary wave, vague solitaire, lame de fond.

solod, podzol de terrain salé.

solodic soil, sol salin lessivé.

solodization, solodisation.

solodized solonetz, solonetz solodisé.

solonchak, solonchak (sol salin blanc).

solonetz, solonetz (sol alcalin noir).

solubility, solubilité; **s. curve,** courbe de solubilité.

solubilize (to), solubiliser.

soluble, soluble.

solum, partie supérieure du profil pédologique.

solute, 1. adj : dissous; 2. n : soluté.

solute (to), dissoudre (chimie).

solution, 1. dissolution, solution; 2. altération chimique pédologique; **s. basin,** cuvette de d.; **s. cavity,** cavité de d.; **s. facet,** facette de d.; **s. mining,** exploitation par d.; **s. pan,** alvéole de d.; **s. pit,** cupule de d.; **s. residue,** résidu de d.; **s. sink,** dolines; **s. tubes,** galeries karstiques; **s. valley,** vallée karstique; **aqueous s.,** s. aqueuse.

Solutrean, Solutréen (industrie humaine, Paléolithique récent).

Solvan (Eur.), Solvien (étage, Cambrien moy.).

solvent, solvant, dissolvant; **s. extraction,** extraction au s.; **s. power,** pouvoir s.; **s. treating,** raffinage par s.

solvus line, courbe de solvus (*cristallo.*).

sonar, sonar; **s. target,** écho sismique; **long range s.,** s. à longue portée.

sonic, acoustique (adj.).

sonic altimeter, écho sonde.

sonic drill, appareil de forage par vibrations.

sonic log, enregistrement acoustique dans un forage; **s. profile,** échogramme; **s. wave,** onde acoustique.

sonobuoy, bouée sonore.

sonograph, sonographe (enregistreur de sismique réflexion).

sonometer, sonomètre.

sonoprobe, sonde acoustique, échosondeur enregistreur.

sooty coal, charbon gras.

sorb (to), adsorber.

sorosilicate, sorosilicate, silicates en chaînes.

sorption, adsorption.

sorptive, adsorbant.

sort (to), trier, classer, cribler, séparer.

sorted circles, sols polygonaux (à bords de pierres triées).

sorted polygons, polygones de pierres.

sorting, triage, classement; **s. coefficient,** coefficient de t.; **s. hammer,** massette de sheidage; **s. index,** indice de t. granulométrique.

sound, 1. adj : solide, sain; 2. n : son; 3. n : détroit, bras de mer; **s. velocity,** vitesse du son; **s. velocity log,** diagraphie de vitesse; **s. wave,** onde sonore.

sounding, sondage; **s. balloon,** ballon sonde; **s. borer,** tarière pédologique; **s. line,** ligne, fil de sondage (hydrographie).

soundstone, phonolite (*pétrogr.*).

sour, 1. corrosif, contenant du soufre; 2. acide, aigre; **s. humus,** humus acide; **s. natural gas,** gaz naturel corrosif non désulfuré; **s. oil,** pétrole sulfuré.

source, 1. source (d'un fleuve); 2. foyer (d'un tremblement de terre); 3. origine d'un tir (sismique); **s. beds,** roches formatrices de pétrole; **s. rock,** roche mère.

sourdough (U.S.), chercheur d'or vétéran.

southerly exposure, adret, versant exposé au sud.

spa, eau minérale.

space, espace; **s. lattice,** réseau cristallin.

spaciation, espacement du réseau hydrographique.

spacing, espacement, écartement; **interplanar s.,** écart réticulaire.

spade, bêche, pelle.

spall, fragment, éclat.

spall (to), 1. faire éclater, briser; 2. broyer; 3. s'écailler, s'effriter, se desquamer.

spalling, 1. effrittement, desquamation; 2. broyage; 3. scheidage.

spalls, minerais de sheidage.

span, 1. court espace de temps; 2. petite étendue de terrain, intervalle; **life s.,** durée de vie.

spangle, paillette.

spar, spath; **adamantine s.,** s. adamantin, corindon; **diamond s.,** s. adamantin; **fluor s.,** fluorine; **greenland s.,** cryolithe; **heavy s.,** barytine; **satin s.,** gypse fibreux; **slate s.,** argentite; **tabular s.,** wollastonite.

sparagmite, sparagmite (pétro.).

sparite, sparite (pétro.).

spark array profile, profil de sismique réflexion.

sparker, étinceleur, sparker (appareil de sismique marine).

sparkle (to), étinceler, produire des étincelles.

Sparnacian, Sparnacien (étage, Paléocène sup.).

sparry, à cristaux de spath, spathique; **s. coal,** charbon à diaclases remplies de calcite; **s. iron,** sidérite, sidérose; **s. lode,** filon de fluorine, calcite ou barytine; **s. limestone,** 1. marbre à gros grains; 2. calcaire cristallin.

spate, crue, avalaison.

spathic, spathique; **s. iron,** sidérose.

spathose, spathique.

spatter, éclaboussure; **s. cone,** cône volcanique secondaire surbaissé formé de laves; **s. lava,** lave projetée à l'état liquide; **s. works,** abatage hydraulique (*mine*).

spear pyrite, marcasite crêtée, pyrite crêtée.

speciation, spéciation (*pal.*).

species, espèce (*pal.*).

specific, spécifique; **s. capacity (of a well),** débit d'un puits; **s. conductance,** conductivité s.; **s. gravity,** densité; **s. heat,** chaleur s.; **s. name,** nom d'espèce, nom s.; **s. retention,** rétention d'eau (d'un sol).

specimen, spécimen, échantillon.

speck, 1. tache, point, goutte, moucheture; 2. grain, particule.

speckled, tacheté, moucheté, marbré.

spectral, spectral; **s. analysis,** analyse s.; **s. gamma-ray log,** diagraphie s. par rayons gamma (dans un forage).

spectrogram, spectrogramme.

spectrograph, spectrographe.

spectrographic analysis, analyse spectrographique.

spectroheliograph, spectrohéliographe.

spectrometer, spectromètre; **grating s.,** s. à réseau.

spectrometric, spectrométrique.

spectrometry, spectrométrie.

spectrophotography, spectrophotographie.

spectrophotometer, spectrophotomètre.

spectrophotometry, spectrophotométrie.

spectroscope, spectroscope; **prism s.,** spectroscope à prisme.

spectroscopic, spectroscopique.

spectroscopy, spectroscopie.

spectrum, spectre; **absorption s.,** spectre d'absorption; **diffraction s.,** spectre de diffraction; **magnetic s.,** spectre magnétique.

specular, spéculaire, miroitant; **s. coal,** houille piciforme, pechkohle; **s. hematite,** oligiste; **s. iron,** hématite; **s. iron ore,** fer spéculaire, hématite; **s. schist,** itabirite; **s. slate ore,** itabirite; **s. stone,** mica.

specularite, fer spéculaire, oligiste.

speiss cobalt, cobalt arsenical.

speleochronology, datation des dépôts des cavernes.

speleological, spéléologique.

speleologist, spéléologue.

speleology, spéléologie.

speleothem, dépôts de caverne par précipitation.

sperrylite, sperrylite (*minér.*).

spessartite, spessartite (var. de grenat).

spew ice, aiguille de glace, pipkrake.

Sphagnum peat, tourbe de Sphaignes.

sphalerite, sphalérite, blende.

sphene, sphène, titanite.

sphenoid, sphénoèdre (*cristallo.*).

sphenolith, sphénolite; 1. coccolite; 2. lame intrusive.

Sphenopsida, Sphénopsidés (*pal.*).

spheric, spherical, sphérique; **s. wave,** onde s.; **s. weathering,** désagrégation en boules.

spheroidal, sphéroïdal; **s. jointing,** 1. prismation basaltique; 2. fissuration s.; **s. structure,** structure orbiculaire; **s. weathering,** désagrégation en boules.

spherolite, sphérolite.

spherosiderite, sphérosidérite.

spherule, sphérule.

spherule texture, texture sphérolitique.

spherulite, sphérolite.

spherulitic, sphérolitique; **s. ore,** particule de minerai à structure radiale; **s. texture,** structure de dévitrification (de verres volcaniques).

spicular chert, gaize siliceuse.

spicularite, roche sédimentaire composée de spicules d'éponges.

spicule, spicule (*pal.*).

spiculite, gaize.

spilite, spilite (basalte altéré).

spilitic, spilitique.

spill (to), 1. répandre, renverser (un liquide), déborder; 2. s'écouler, se répandre.

spill point, point de débordement.

spillage, 1. débordement; 2. déchet.

spilling, 1. débordement; 2. palplanches (construction); **s. flow lines,** directions d'échappement de gaz ou de pétrole à partir d'un gisement; **s. point,** point de d.

spillway, canal de trop plein.

spin, rotation, tournoiement.

spinal column, colonne vertébrale (*pal.*).

spindle bomb, bombe en fuseau.

spindle stage, platine à aiguille.

spindrift, embruns.

spine (volcanic), aiguille (volcanique).

spinel, spinelle; **s. twin,** mâcle des s.

spiniform, en forme d'épine.

spinning, 1. mouvement de tournoiement; 2. affolement de l'aiguille aimantée d'une boussole.

spinning fiber, asbeste (variété).

S.P. interval, intervalle de temps entre les arrivées des ondes P et S.

spiracle, spiracle (*pal.*).

spiralium, lophophore (*pal.*).

spire, spire (*pal.*).

Spiriferida, Spiriféridés.

spirit, alcool, essence; **s. of alum,** solution aqueuse de SO_2; **s. salt,** acide chlorhydrique; **s. of tin,** chlorure stannique.

spirotheca, muraille (*pal.*).

spit, flèche littorale.

splash (to), éclabousser, jaillir en éclaboussures.

splashy, bourbeux.

splay deposits, dépôts de crue.

splendent luster, éclat resplendissant.

spliced veins, filons anastomosés.

split (to), se fendre, se fissurer.

splitting, fissuration.

spodic horizon, horizon spodique.

spodosol, podzol, sol à B spodique.

spoil, déblais, halde.

spongolite, spongolite (roche composée de spicules d'éponges).
sporangia, sporange.
sporomorph, sporomorphe.
spot, tache solaire.
spot height, point côté.
spotted schist, spotted slate, schistes tachetés.
spouting horn, trou souffleur.
spreading of plates, dérive des plaques (*tecto.*).
spring, source; **s. line,** ligne de s.; **s. tide,** grande marée; **thermal s.,** source thermale.
spudding, forage par battage.
squall, coup de vent, bourrasque.
squat of ore (G.B), nid de minerai.
squeeze, compression, tassement.
squeezer, presse.
squeeze out (to), comprimer, extraire.
squeezing, compression.
squib, canette, raclette (tir de mine).
squibbing, agrandissement par explosion (d'un trou de mine).
S surface, surface de foliation, surface plane.
stab pipe (to), ajouter une nouvelle longueur de tige.
stability, stabilité, solidité; **s. field,** domaine de stabilité (d'un minéral); **s. series,** classement des minéraux en fonction de leur résistance à l'altération.
stabilized dune, dune fixée.
stable isotope, isotope stable.
stack, 1. pilier, pinacle, éperon rocheux; 2. cheminée; 3. données sismiques devant subir une sommation; **s. gases,** gaz brûlés.
stake, électrode.
staked trace, trace somme (*géoph.*).
stacking, 1. entassement, empilage; 2. sommation (*sism.*).
staddle (to), boiser, étayer.
stade (syn. stadial), sous-stade (glaciaire), stade secondaire, stadiaire.
stadial, stadiaire; **s. moraine,** moraine de recul.
staff, 1. équipe de recherches; 2. mire; **s. holder,** porte mire; **cross s.,** équerre d'arpenteur.
Staffordian (eur.), Staffordien (étage, Carbonifère sup.).
stage, 1. phase, stade, degré; 2. étage géologique; 3. palier (mine); 4. platine d'un microscope; **s. flotation,** flottation étagée; **s. of maturity,** stade de maturité (du relief); **s. working,** exploitation en gradins, exploitation à ciel ouvert; **late s. crystallization,** dernier stade de cristallisation.
staggered wells, puits disposés en quinconce.
stagnant, stagnant, dormant.
stagnation, stagnation (d'un glacier); **s. degree,** degré de stagnation (de l'eau dans un sol).
stain (to), 1. tacher, souiller; 2. colorer, teinter.
staining test, test, essai de coloration (de minéraux).
stake, piquet, pieu, poteau.
stake (to), jalonner, piqueter, borner (une concession).
stalagmite, stalagmite (spéléo.).

stalactite, stalactite (spéléo.).
stalagtitic, stalagtitique.
stalked, pédonculé (*pal.*).
stall, 1. chambre de grillage du minerai; 2. taille (mine).
stamp (to), 1. broyer, briser, concasser, bocarder (le minerai); 2. estamper (le métal), emboutir.
Stampian, Stampien (étage, oligocène, équiv. Rupélien).
stamping, bocardage.
stamp mill, bocard, pilon.
stamp milling, bocardage, broyage au bocard.
stanchion, étai, appui, béquille.
stanchness, étanchéité.
stand, 1. longueur de tiges de forage vissées; 2. support, socle; 3. statif d'un microscope; 4. association végétale.
stand of tide, étale de la marée, marée étale.
stand pipe, colonne montante, circuit de refoulement d'une pompe à boue.
standage, puisard (*mine*), collecteur d'eau.
standard, 1. adj : de série, ordinaire; 2. n : norme, étalon; **s. barometer,** baromètre-étalon; **s. curve,** abaque; **s. deviation,** écart type; **s. error,** écart-type; **s. mineral,** minéral normatif, minéral virtuel; **s. parallel,** parallèle de latitude; **s. solution,** solution titrée; **s. stratum,** couche de référence.
standing water, eau stagnante.
standing water table, nappe phréatique.
standing waves, ondes stationnaires.
stannate, stannate.
stannic, stannique.
stanniferous, stannifère.
stannite, stannite.
stannous, stanneux.
stannum, étain.
star, 1. adj : étoilé; 2. n : étoile, astre; **s. antimony,** antimoine métallique; **s. connection,** montage en étoile; **s. metal,** antimoine métallique; **s. quartz,** quartz étoilé; **s. ruby,** rubis étoilé; **s. sapphire,** saphir astérique; **shooting s.,** étoile filante (*astro.*).
Starfish, Astérie, Étoile de mer.
stark landscape, paysage nu.
start (to) drilling, commencer un forage.
starting place, point de départ (d'une avalanche).
starved basin, bassin sédimentaire à faible remplissage sédimentaire.
static, statique; **s. gravimeter,** gravimètre non astatisé; **s. head,** hauteur d'élévation (hydraulique); **s. metamorphism,** métaporphisme régional; **s. pressure,** pression statique; **s. zone,** zone inférieure à la source d'une nappe d'eau.
statics, 1. perturbations atmosphériques; 2. mécanique statique.
stationary, stationnaire; **s. rig,** installation de forage fixe; **s. wave,** onde stationnaire.
statistics, statistique; **s. analysis,** méthode s.; **s. distribution,** distribution s.; **s. function,** loi s.; **s. population,** population s.; **s. trend,** tendance s.
statute mile, mille terrestre = 1 609,31 m.
staurolite, staurotide (minéral).

stave in (to), défoncer, enfoncer, effondrer.

stay, support, étai, hauban.

stay (to), haubanner, étayer; **2.** s'arrêter, demeurer.

stayer, puits à production stable.

steady, régulier, constant (débit).

steady state, état d'équilibre.

steady-state stream, rivière régularisée.

steam, vapeur; **s. coal**, charbon pour production de v.; **s. gauge**, manomètre; **s. pressure**; pression de v.; **s. power station**, station thermique.

steatite, stéatite, talc.

steatitic, stéatiteux.

steatitization, altération avec formation de talc.

S tectonite, roche présentant des structures planes.

steel, acier; **s. band**, lit de pyrite dans charbon (illinois, U.S.A.); **s. foundery**, fonderie d'a.; **s. mill**, aciérie; **s. plant**, aciérie; **s. works**, fonderie d'a.

steep, escarpé, raide, à forte pente; **s. seam**, couche fortement inclinée.

steephead, niche (de source), surplomb, reculée.

steeply, en pente forte; **s. dippind lode**, « dressant », filon fortement redressé.

steepness, abrupt, raideur, escarpement.

steer, à fort pendage, fortement incliné (G.B.).

Stegocephalia, Stégocéphales (*pal.*).

Stegosauria, Stégosauriens (*pal.*).

stellerite, stéllerite (*minéral.*).

stem, tige (de trépan); **s. stream**, axe fluvial; **drill s.**, train de tiges; **drilling s.**, maîtresse tige.

stemmer, bourroir (mine).

stenohaline, sténohalin (*pal.*).

step, **1.** gradin, marche; **2.** graduation; **3.** phase, stade; **s. bit**, trépan à redans; **s. fault**, faille en gradins, en escalier; **s. fold**, flexure, pli monoclinal; **s. line**, ligne de brisants; **s. out well**, puits d'extension; **s. scanning**, exploration point par point; **s. vein**, filon en gradins; **confluence s.**, gradin de confluence glaciaire; **rock s.**, gradin structural.

Stephanian, Stéphanian (étage; Carbonifère sup.).

stephanite, stéphanite (*minér.*).

steppe, steppe; **s. black earth**, sol noir steppique, tchernozem; **s. bleached earth**, sol salin lessivé acide; **s. soil**, sol steppique; **s. solonchack**, sol salin blanc de s.

stepped, à gradins, en gradin, échelonné.

steptoe, relief résiduel, îlot (entouré de laves volcaniques).

stereoautograph, appareil de restitution photogrammétrique.

stereocomparator, stéréocomparateur.

stereocompilation, stéréorestitution.

stereographic projection, projection stéréographique.

stereo net, stéréogramme (de Wulff).

stereophotogrammetry, stéréophotogrammétrie.

stereoscope, plaquette stéréoscopique, stéréoscope; **scanning s.**, stéréoscope à balayage.

stereoscopic, stéréoscopique; **s. pair**, couple de photographies pour s.; **s. vision**, vision en relief, stéréoscopique.

stereoscopy, stéréoscopie.

stereotriangulation, triangulation aérienne.

sterile, stérile, terre stérile; **s. coal**, schiste noir et argile sus-jacent au charbon (G.B.).

Stevenson screen, abri météorologique de Stevenson.

stibial, antimonieux.

stibic, stibique, antimonique.

stibium, antimoine.

stibnite, stibine (antimoine sulfuré).

stickiness, viscosité.

sticky, collant, visqueux, gluant; **s. point**, point d'adhésivité.

stiff clay, argile peu plastique.

stiffness, **1.** rigidité, dureté; **2.** raideur (d'une pente).

stilbite, stilbite (zéolite).

still coke, résidu de distillation de pétrole de schiste bitumineux.

still gas, gaz de distillation.

still residue, résidu de distillation.

still stand, position stationnaire de la mer.

still water, eau stagnante.

still water level, niveau marin en eaux calmes.

stilpnomelane, stilpnomélane (*minér.*).

stilpnosidérite, limonite.

stink coal, résine fossile soufrée (dans lignite).

stink damp, **1.** gaz d'explosion; **2.** hydrogène sulfuré.

stinking schist, schiste fétide.

stinkstone, **1.** calcaire bitumineux; **2.** roche fétide; **3.** anthraconite.

stock, **1.** stock, matière première; **2.** petit massif intrusif; **s. pile**, empilement de minerai; **s. tank**, réservoir de stockage.

stockwork, ensemble, réseau de fissures minéralisées dans la roche hôte.

Stokes' law, loi de Stockes (de chute des particules).

stolon, stolon (*pal.*).

stomach stone, gastrolite.

stone, **1.** pierre, caillou; **2.** unité de poids = 6 348 kg; **s. age**, âge de la pierre; **s. beach**, galet; **s. brash**, amas de pierres; **s. bubbles**, sphérolithes; **s. circle**, polygone de pierres; **s. coal**, anthracite; **s. crusher**, concasseur de pierres; **s. dresser**, tailleur de pierres; **s. drift**, galerie au rocher (*mine*); **s. dust**, pulvérin rocheux; **s. flax**, amiante; **s. field**, perrier, chaos de blocs; **s. fragment**, débris pierreux; **s. gall**, concrétion d'argile dans des grès; **s. garland**, guirlande de pierres (périglaciaire); **s. iron**, sidérolite; **s. lattice**, chaos de blocs; **s. man**, mineur au rocher; **s. mill**, concasseur de pierres; **s. net**, polygone de pierres; **s. oil**, pétrole; **s. packing**, perré; **s. pavement**, dallage de pierres (périglaciaire); **s. pick**, pic au rocher; **s. pit**, **1.** carrière de roche; **2.** trou de pierres (périgl.); **s. polygon**, polygone de pierres; **s. polygon soil**, sol polygonal; **s. quarry**, polygone de pierres; **s. ring**, cercle de pierres; **s. roofing slab**, lauze; **s. saw**, scie à pierres; **s. sawyer**, scieur de pierres; **s. slide**, éboulement, ava-

lanche de pierres; **s. stripes,** sol strié; **s. work,**
1. travail au rocher; **2.** maçonnerie; **s. wreath,**
polygone de pierres; **s. yellow,** ochre jaune;
drip s., stalactite; **meteoric s.,** aérolithe.

stoniness, abondance en fragments pierreux.

stoneware, grès, poterie de grès.

stony, pierreux, rocailleux; **s. desert,** désert pierreux; **s. land,** pays à sous-sol rocheux; **s. loam,**
limon pierreux; **s. marl,** marne caillouteuse;
s. meteorite, météorite silicatée; **s. soil,** sol
pierreux.

stoop, pilier (mine); **s. and room system,** méthode
des piliers et galeries.

stopcoking, fermeture périodique d'un puits de
pétrole.

stope, **1.** gradin; **2.** exploitation en gradin;
s. face, front d'abatage, front d'attaque; **un-**
derhand s., gradin droit; **overhand s.,** gradin
renversé.

stope (to), **1.** exploiter une mine en gradins;
2. abattre le minerai.

stoped out workings, chantiers épuisés.

stoping, stopping, **1.** abatage, exploitation;
2. barrage; **s. ground,** minerai prêt à être
extrait; **cliff s.,** sapement de falaise; **magmatic**
s., 1. assimilation magmatique; **2.** intrusion
magmatique (acide).

storm, orage, dépression, tempête; **s. beach,** levée
de plage; **s. cloud,** nuée d'orage; **s. roller (sed.),**
structure intraformationnelle diapirique en
forme d'oreiller; **s. surge,** raz de marée; **s. tide,**
raz de marée; **s. zone,** zone des tempêtes;
s. wave, vague de tempête; **s. wave platform,**
banquette de tempête; **magnetic s.,** orage
magnétique.

stoss and lee topography, topographie glaciaire
dissymétrique (roches moutonnées).

stoss side, côté en pente douce, face exposée au
courant.

stove, étuve; **s. coal,** anthracite, charbon pour
chauffage domestique; **s. sand,** sable étuvé.

stow (to), remblayer.

stowage, remblayage.

stower, remblayeur.

stowing, remblayage.

straight, droit, rectiligne; **s. chain hydrocarbon,**
hydrocarbure à chaîne droite; **s. down,** à
la verticale; **s. extinction,** extinction droite
(*minér.*); **s. hole,** forage vertical; **s. run product,**
produit de distillation directe; **s. through flow,**
écoulement direct (de pétrole); **s. well,** puits
vertical.

straighten a drill hole (to), redresser un forage
dévié.

strain, **1.** contrainte effort; **2.** déformation; **s.**
breaks, fractures de compression; **s. ellipsoid,**
ellipsoïde des déformations; **s. gauge,** tensiomètre; **s. hardening,** résistance progressive
des roches à la déformation; **s. limit,** limite
d'allongement; **s. release,** relâchement des

contraintes; **s. shadows,** extinction ondulante
(du quartz mylonitisé); **s. slip cleavage,** clivage
par pli-fracture; **s. waves,** ondes de déformation.

strain (to), **1.** exercer un effort, tendre, sur-
tendre; **2.** filtrer, passer (un liquide).

strainer screen, tamis, filtre, crépine, tamis.

straining, **1.** tension, filtration; **2.** filtrage.

strainmeter, jauge de déformation.

strait, **1.** adj. : étroit; **2.** n. : détroit.

strand, **1.** rive, grève; **2.** estran, plage; **3.** fil, fibre,
toron; **s. dune,** dune d'estran; **s. flat,** plate-
forme côtière; **s. line,** ligne de rivage; **s. plain,**
plaine littorale.

stranded moraine, moraine latérale perchée.

strap (to), **1.** déterminer le volume d'un réservoir
(pétrole); **2.** mesurer la profondeur d'un puits.

strap in (to), mesurer les longueurs de tubes
d'un sondage.

stratabound deposit, gîte stratiforme.

strath, **1.** vallée large régularisée (U.S.A.); **2.**
vallée remplie de dépôts fluvio-glaciaires
(Écosse); **s. terrace,** terrasse rocheuse recou-
verte d'un mince placage alluvial (U.S.A.).

stratic, stratigraphique.

straticulate, finement stratifié, zoné.

stratification, stratification; **s. foliation,** disposition en feuillets de minéraux parallèlement
au litage; **s. plane,** plan de stratification.

stratified, stratifié; **s. cone,** strato-volcan; **s. debris**
slope, éboulis s.; **s. drift,** moraine fluvio-
glaciaire; **s. rock,** roche sédimentaire; **s. screes,**
grèzes litées (périgl.); **s. water,** eau composée
de couches de densité variable.

stratiform, stratiforme.

stratify (to), être stratifié.

stratigrapher, stratigraphe.

stratigraphic, stratigraphical, stratigraphique; **s.**
boundary, limite s.; **s. break,** lacune s.; **s.**
classification, classification s.; **s. column,** colonne s.; **s. control,** contrôle s.; **s. correlation,**
corrélation s.; **s. gap,** lacune s.; **s. hiatus,**
lacune s.; **s. isochore,** isochore; **s. leak,** infiltration de fossiles dans strates inférieures; **s. map,**
carte structurale; **s. paleontology,** paléontologie s.; **s. pile,** colonne s.; **s. range,** répartition
s.; **s. section,** coupe structurale; **s. separation,**
rejet s.; **s. sequence,** séquence s.; **s. throw,** rejet
s.; **s. trap,** piège s.; **s. unit,** unité s.

stratigraphy, stratigraphie.

stratocone, stratocône, volcan stratifié.

stratocumulus cloud, strato-cumulus (*météo.*).

stratosphere, stratosphère.

stratotype, stratotype.

stratovolcano, stratovolcan.

strat-trap, piège stratigraphique.

stratum, strata (pl.), couche géologique, strate,
lit; **s. plain,** pénéplaine structurale; **overlying**
s., couche sus-jacente; **underlying s.,** couche
sous-jacente.

stratus, stratus (*météo.*).

stray, 1. adj. : diffus, dispersé; **2.** n. : dispersion (électricité); **s. block,** block erratique; **s. lines,** lignes de dispersion; **s. current,** courant vagabond.

streak, 1. raie, strie; **2.** trait, couleur de la poudre de minerai; **3.** bande de minerai, lentille de sable; **4.** panachure (*pédol.*); **s. plate,** plaque de porcelaine pour essai à la touche.

streak (to), rayer, strier.

streaked, 1. rayé, rayuré; **2.** zoné.

streakiness, 1. linéation; **2.** état rayé.

streaky, 1. rayé, rayuré; **2.** zoné, bariolé; **s. structure,** structure zonée, bariolée.

stream, 1. fleuve, rivière; **2.** courant, cours d'eau, filet d'eau; **s. bank erosion,** érosion des berges; **s. bed,** lit d'un fleuve; **s. bottom soil,** sol fluviatile; **s. capacity,** capacité fluviatile; **s. frequency,** densité de drainage fluviatile; **s. gradient,** pente d'un fleuve; **s. flow,** écoulement fluviatile; **s. gage,** jauge de hauteur d'eau; **s. gold,** or alluvionnaire; **s. lava,** coulée de lave; **s. load,** charge fluviatile; **s. number,** rapport inverse entre ordre et nombre (classif. réseau fluviatile); **s. order,** ordre, catégorie d'un cours d'eau (ex. : ruisselet = 1er ordre); **s. pattern,** réseau hydrographique; **s. piracy,** capture d'un cours d'eau; **s. sediment,** sédiment fluviatile; **s. terrace,** terrasse fluviatile; **s. tin,** cassitérite alluvionnaire; **s. transportation,** transport fluviatile; **s. tributary,** affluent; **s. works,** exploitations alluvionnaires; **boulder s.,** coulée de blocs; **branch s.,** fleuve anastomosé; **collecting s.,** collecteur; **lava s.,** coulée de lave; **main s.,** rivière principale; **master s.,** rivière principale; **side s.,** cours d'eau secondaire; **trunk s.,** axe fluvial.

stream (to), couler, s'écouler, ruisseler.

streamer, flûte sismique, traîne sismique (cable enregistreur sismique tracté par bateau).

streaming, 1. séparation du minerai du gravier par eau courante; **2.** exploitation d'étain alluvionnaire; **s. flow,** écoulement fluviatile.

streamlet, petit ruisseau, ruisselet.

streamline flow, écoulement laminaire.

streamy, riche en cours d'eau.

strength, 1. résistance, rigidité, force; **2.** richesse en minerai; **3.** intensité; **s. of a lode,** richesse d'un filon; **s. of magnetic field,** intensité du champ magnétique; **breaking s.,** résistance à la rupture; **compressive s.,** résistance à la compression; **impact s.,** résistance au choc; **shearing s.,** résistance au cisaillement; **yield s.,** limite d'élasticité; **wind s.,** force du vent.

stress, effort, tension, contrainte; **s. diagram,** diagramme des forces; **s. ellipsoid,** ellipsoïde des pressions; **s. mineral,** minéral de pression; **s. pressure,** pression orientée (tectonique, non hydrostatique); **bending s.,** tension; **breaking s.,** résistance à la rupture, effort de rupture; **permissible s.,** contrainte admissible; **tensile s.,** effort de traction; **yield s.,** effort de torsion.

stretch, 1. allongement extension; **2.** étendue de pays, bande de terrain; **3.** direction d'un filon (mine).

stretch thrust, pli faille.

streched, étiré; **s. limb,** flanc étiré; **s. modulus,** coefficient d'élasticité; **s. pebbles,** galets étirés; **s. texture,** linéation; **s. thrust,** flanc inverse étiré.

strewing sand, sablage.

stria, striae (pl.), strie, rayure.

striate (to), strier.

striated, strié; **s. boulder,** bloc strié; **s. pebble,** galet strié; **s. soil,** sol strié terreux.

striation, striation.

strike, 1. direction, orientation (d'une couche de terrain ou d'une faille); **2.** découverte d'un nouveau gisement; **s. fault,** faille directionnelle longitudinale; **s. joint,** diaclase longitudinale; **s. line,** isohypse, ligne d'égale altitude; **s. separation,** rejet horizontal d'une faille; **s. shift,** composante horizontale du rejet; **s. slip,** rejet horizontal d'une faille; **s. slip fault,** décrochement; **s. stream,** fleuve monoclinal; **s. valley,** vallée longitudinale; **line of s.,** direction.

strike (to), 1. frapper, faire jaillir des étincelles; **2.** suivre la direction de; **3.** découvrir (un filon).

strike (to) bed rock, rencontrer le fond solide.

string, 1. filon, petite veine (de minerai), veinule; **2.** tige, train de tiges (forage); **s. galvanometer,** galvanomètre à fil; **s. of casing,** colonne de tubage; **s. of rods,** train de tiges; **oil s.,** colonne de production; **water s.,** colonne de tubage.

stringer, 1. petit filon, veinule; **2.** chef de chantier de pose; **s. zone,** zone d'écrasement à nombreux filonnets minéralisés.

string up (to), appareiller (forage).

stringing, 1. alignement des tiges de forage placées bout à bout; **2.** bardage (des tubes d'un pipeline).

strip, bande; **s. mine,** mine à ciel ouvert; **s. mining, 1.** exploitation par excavateurs; **2.** déblaiement, décapage des morts-terrains; **s. of land,** bande de terre.

strip (to), dépouiller, extraire, enlever, décaper, découvrir, exploiter.

strip off (to), décaper.

stripped, décapé; **s. ground,** sol strié; **s. plain,** plaine décapée des roches tendres, ayant pour soubassement une roche plus dure; **s. structural terrace,** terrasse structurale décapée de ses terrains superficiels.

stripper, excavateur (mine), déboiseur (mine).

stripping, 1. extraction; **2.** décapage des morts terrains, dépilage; **3.** dénudation (du relief); **glacial s.,** ablation glaciaire.

stripping plant, installation de distillation primaire, de rectification (pétrole).

stromatotactis, stromatotactis.

strombolian eruption, éruption de type strombolien.

Stromatolite, Stromatolite.

Stromatoporoids, Stromatoporoïdés (*pal.*).

strontian, strontianite (*minér.*).

strontianiferous, contenant du strontium.

strontianite, strontianite.

strontium, strontium.

structural, structural; **s. basin**, cuvette synclinale; **s. bench**, atténuation de pendage, replat; **s. bulge**, gonflement anticlinal; **s. closure**, fermeture s.; **s. contour**, isohypse (d'un niveau repère); **s. contour line**, courbe de niveau; **s. contour map**, carte s.; **s. control**, **1.** contrôle s.; **2.** gîtologie s. (des minerais); **s. discordance**, discordance tectonique; **s. geology**, géologie s.; **s. high**, dôme anticlinal; **s. low**, synclinal; **s. nose**, saillant anticlinal; **s. plain**, pénéplaine s., faiblement inclinée; **s. saddle**, abaissement axial; **s. style**, style tectonique; **s. terrace**, terrasse tectonique; **s. trap**, piège s.; **s. valley**, vallée tectonique.

structure, structure; **s. contour**, isohypse; **s. contour map**, carte structurale à courbes de niveau; **s. hole**, sondage géologique; **s. section**, diagramme structural; **banded s.**, s. rubanée; **columnar s.**, s. prismée (des basaltes); **flow s.**, s. fluidale; **folded s.**, s. plissée; **homoclinal s.**, s. homoclinale; **imbricate s.**, s. en écailles; **laminated s.**, s. en feuillets; **layer lattice s.**, s. réticulaire feuilletée; **monoclinal s.**, s. monoclinale; **on s.**, s. favorable (au gisement de pétrole); **prismatic s.**, s. prismée; **salt s.**, dôme de sel; **superimposed s.**, s. superposée; **table-like s.**, s. tabulaire; **thrust s.**, s. charriée; **unconformable s.**, s. discordante.

structured salt soil, solonetz (sol salin).

structureless, amorphe; **s. saline soil**, sol salin blanc non structuré, solontchak.

Strunian (Eur.), Strunien (étage; Dévonien sup. terminal).

strut, étai, traverse, arc-boutant.

strutting, étaiement (mine).

stuff, **1.** minerai mélangé à la gangue; **2.** produits d'extraction.

stuffed mineral, minéral à structure spongieuse.

stulled stope, taille étayée (mine).

stump, pilier (de charbon).

stylolite, stylolite (joint irrégulier de pression-dissolution).

stylolitic, stylolitique.

subaerial, subaérien; **s. erosion**, érosion subaérienne.

subalkali, subalcalin.

suballuvial bench, pédiment sous-jacent au remblaiement de bajada.

subaluminous composition, composition subalumineuse.

subangular, subanguleux, légèrement anguleux.

subaqueous, subaquatique; **s. gliding**, glissement sous-aquatique; **s. slide**, glissement sous-aquatique.

subarctic, subarctique.

subarid, subaride.

subarkose, grès arkosique (ayant de 10 à 25 % de feldspath).

Subatlantic, Subatlantique (intervalle; − 2 500 ans).

sub-bituminous coal, lignite.

Subboreal, Subboréal (intervalle; entre − 5 000 et − 2 500 ans).

subclass, sous-classe.

subcoastal plain, plate-forme continentale submergée.

subcontinental, sous-continental.

subcrop, contact de couches sous jacentes discordantes avec d'autres plus récentes, sous-affleurement.

subcropping stratum, couche sous-jacente à une discordance.

subcrustal, subcrustal.

subcrystalline, subcristallin.

subdelta, sous-delta.

subdrift, galerie costresse (mine).

subduct (to), passer dessous.

subducted plate, plaque enfoncée par subduction.

subduction, subduction (de plaques); **s. trench**, fosse de s.; **s. zone**, zone de s.

subdued landscape, relief adouci.

subfacies, sous-faciès, faciès secondaire.

subfamily, sous-famille.

subfluvial, sous-fluvial.

subgelisol, subpergélisol (niveau non gelé sous pergélisol).

sub-genus, sub-genera (pl.), sous-genre.

subglacial, sous-glaciaire; **s. planing**, rabotage glaciaire; **s. polishing**, rabotage glaciaire; **s. sapping**, rabotage glaciaire; **s. tunnel**, tunnel sous-glaciaire; **s. wash**, moraine.

subgrade, **1.** soubassement; **2.** sous-sol.

subgraywacke, grauwacke appauvri en feldspath.

subgroup, sous-groupe.

subhedral, hypidiomorphe.

subhercyanian orogeny, orogénèse sub-hercynienne.

subjacent, sous-jacent (terrain).

subirrigation, irrigation souterraine.

subkingdom, sous-règne.

sublacustrine, sous-lacustre.

sublayer, couche sous-jacente.

sublevel, galerie intermédiaire (mine).

sublimate, sublimé.

sublimate (to), sublimer.

sublimation, sublimation; **s. vein**, gîte d'émanation.

sublittoral, sublittoral, néritique.

submarine, sous-marin; **s. bar**, crête s.-m.; **s. bulge**, delta s.-m.; **s. canyon**, canyon s.-m.; **s. coast**, côte de submersion; **s. fan**, delta de canyon s.-m.; **s. morphology**, morphologie s.-m.; **s. oil**, formation pétrolifère s.-m.; **s. ridge**, dorsale s.-m.; **s. rise**, haut-fond s.-m.; **s. valley**, canyon s.-m.; **s. volcano**, volcan s.-m.

submerge (to), submerger, noyer.

submerged shore line, côte de submersion.

submergence, submersion, affaissement.

submersion, submersion.

submetallic luster, éclat sub-métallique.

submorainic deposits, moraine stratifiée.

suboceanic, sous-océanique.

suborder, sous-ordre.

subphylum, sous-embranchement.

subpolar, subpolaire.
subrosion, dissolution de roches salines.
subrounded, subarrondi, presque arrondi.
subsequent, subséquent; **s. fault,** faille s.; **s. stream,** fleuve s.; **s. valley,** vallée s.
subset, sous-ensemble (de valeurs).
subside (to), 1. tomber, s'affaisser, se tasser, s'enfoncer; 2. baisser, s'abaisser.
subsidence, 1. affaissement, subsidence, effondrement, fondis; 2. décrue d'une rivière.
subsidiary anticline, anticlinal secondaire.
subsidiary stream, rivière tributaire.
subsiding area, zone d'affaissement.
subsilicate, silicate basique.
subsilicic rock, roche basique (teneur en silice inférieure à 52 %).
subsoil, sous-sol.
subsoiling, soussolage.
subspecies, sous-espèce.
substage, 1. **sous-étage** (*stratigr.*); 2. sous-palatine (*minér.*).
substitution, substitution, remplacement; **s. vein,** filon minéralisé de substitution.
substratum, substratum, couche inférieure.
substructure, infrastructure.
subsurface, subsurface; **s. contours,** isohypses; **s. eluviation,** infiltration et écoulement souterrain; **s. geology,** géologie de subsurface; **s. irrigation,** irrigation souterraine; **s. storage,** stockage souterrain; **s. water,** eau souterraine.
subterposition, sous-jacence.
subterrane, substratum.
subterranean, subterraneous, souterrain; **s. water,** eau souterraine.
subtidal, subtidal.
subtilling, soussolage.
subtropical, subtropical; **s. high,** anticyclone.
subtrusion, subtrusion.
sub-type, sous-type (*pal.*).
subvitreous, subvitreux.
succession of strata, succession de couches.
succinite, succinite, ambre.
sucked stone, pierre poreuse.
sucker rod, tige de pompage.
suck up (to), sucer, aspirer, pomper (un liquide).
suction, aspiration, succion; **s. dredge,** drague aspirante; **s. pump,** pompe aspirante.
Sudetan (Sudetic) orogeny, orogénèse sudète (Carbonifère inf. à moy,).
suffione, soufflard.
sugar loaf, relief en pain de sucre.
sugary grain, saccharoïde.
sugary quartz, quartz granuleux et friable (G.B.).
suite, 1. ensemble, cortège (*pétrogr.*, *minér.*); 2. série (*stratigr.*).
sulcus, sinus, sillon (*pal.*).
sulfate, sulfate.
sulfide, sulfure; **s. zone,** zone sulfurée.
sulfur, soufre; **s. bacteria,** thiobactérie; **s. balls,** concrétion de pyrite dans le charbon; **s. dioxyde,** anhydride sulfureux.

sullage, 1. limon fluviatile; 2. eau d'égout.
sulphate, sulfate.
sulphated, sulfaté.
sulphation, sulfatation.
sulphatize (to), transformer en sulfate.
sulphide, sulfure; **s. mineral,** minéral sulfuré; **s. ore,** minéral sulfuré; **s. zone,** zone sulfurée (d'un minéral); **antimony s.,** stibine.
sulphidic, sulfuré.
sulphohalite, sulfohalite (*minér.*).
sulphur, soufre; **s. ball,** rognon de pyrite; **s. bearing,** sulfurifère; **s. dioxide,** anhydride sulfureux; **s. fume,** vapeur sulfureuse; **s. mine,** soufrière; **s. mud,** boue sulfureuse; **s. ore,** pyrite; **s. pit,** soufrière; **s. ratio,** rapport isotopique S_{34}/S_{32}; **s. water,** eau sulfureuse; **s. works,** raffinerie de soufre; **ruby s.,** réalgar.
sulphureted, sulfuré, soufré.
sulphuric, sulfurique; **s. spring,** source sulfureuse.
sulphurite, soufre natif.
sulphurous, sulfureux.
sulphury, sulfureux.
summation curve, courbe granulométrique cumulative.
summit, sommet, crête.
sump, 1. puisard (mine); 2. bassin à boue (forage).
sun, soleil; **s. crack,** fissure de dessication; **s. opal,** opale de feu; **s. spot,** tache solaire.
sunken block, graben.
sunny side, adret, versant ensoleillé.
sunstone, var. d'oligoclase.
supercool (to), surfondre.
supercooled, surfondu.
supercooling, surfusion.
superficial, superficiel; **s. crust,** encroûtement s.; **s. deposits,** sédiments quaternaires ou actuels; **s. erosion,** érosion s., érosion en nappe; **s. mull,** humus doux s.
superfluent lava, lave débordante (du cratère).
supergene, supergène (ex. dissolution, hydratation, oxydation); **s. enrichment,** enrichissement secondaire.
superglacial, supraglaciaire (ex. transport).
superheating, surchauffe (magmatique).
superimposed, surimposé; **s. river,** rivière s.; **s. valley,** vallée s.
superimposition, surimposition, épigénie.
superincumbent, superposé.
superjacent, superposé.
supermorainic deposits, moraine stratifiée.
superplate, superplaque.
superpose (to), superposer.
superposed fold, pli superposé.
superposed stream, fleuve surimposé.
superposition, superposition (principe de ...).
supersaturate, sursaturer.
supersaturated solution, solution sursaturée.
supersaturation, sursaturation.
superstage, surplatine (de microscope).
superstratum, couche supérieure, couche susjacente.

superstructure, plis superficiels.

superterranean, subaérien.

supplementary contour, courbe de niveau inter-calaire.

support, soutènement, support (mine).

support (to), soutenir, étayer (mine).

supporting, point d'appui, soutènement, pilier.

supracapillary space, macroporosité.

supracretaceous, supracrétacé.

supracrustal sequence, série supracrustale (dans terrains archéens).

supragelisol, suprapergélisol.

supralittoral zone, zone supralittorale.

supratenuous folding, plissottement par tassement différentiel.

supratidal, supratidal.

surf, ressac, brisants; **s. zone,** zone des brisants.

surface, surface; **s. anomalies,** anomalies géophysiques de s.; **s. bed,** couche superficielle; **s. break,** affaissement de s., fondis; **s. correction,** correction des mesures géophysiques relatives aux couches superficielles; **s. cover,** couverture végétale; **s. geology,** géologie des formations superficielles et des horizons supérieurs des couches sous-jacentes; **s. mine,** mine à ciel ouvert; **s. mining,** exploitation à ciel ouvert; **s. moraine,** moraine superficielle; **s. of no strain,** s. de contrainte nulle; **s. of unconformity,** s. de discordance; **s. plant,** installation de s., carreau; **s. pressure,** pression superficielle; **s. runoff,** écoulement superficiel; **SH. wave,** onde de Love; **s. shooting,** tir en s.; **s. soil,** sol superficiel; **s. staff,** personnel du jour (mine); **s. tension,** tension superficielle; **s. termination,** affleurement; **s. velocity,** vitesse superficielle des ondes; **s. water,** eau de ruissellement; **S. wave,** onde superficielle; **s. work,** travail au jour; **s. working,** exploitation à ciel ouvert.

surficial deposit, dépôt superficiel.

surfuse (to), surfondre.

surfusible, surfusible.

surge (to), s'élever brusquement (mer).

surge, **1.** houle, lame de fond; **2.** avancée rapide (d'un glacier); **3.** éruption phréatomagmatique de vapeurs et de cendres fines, à direction horizontale; **s. channel,** chenal, entaille (de récif) par déferlement; **s. zone,** zone de balayage sous-marin (entre 0 et 18 m).

surging sea, mer houleuse.

surimposition, surimposition.

survey, **1.** étude, examen attentif; **2.** levé, lever, relevé (topographique, cartographique); **3.** bureau d'études; **s. company,** section topographique; **s. net,** canavas topographique; **s. vessel,** navire hydrographe; **aerial s.,** levé photogrammétrique; **geological s.,** bureau d'études géologiques.

survey (to), **1.** examiner, mettre une question à l'étude; **2.** faire le levé de, arpenter; **3.** inspecter, visiter.

surveying, étude (topographique, géologique, géodésique, hydrographique), prospection, arpen-tage; **s. camera,** appareil de prises de vues; **s. compass,** boussole; **s. instrument,** instrument topographique; **s. rod,** jalon (d'arpenteur); **s. ship,** navire hydrographique; **geophysical s.,** prospection géophysique; **naval s.,** hydrographie; **photographic s.,** photogrammétrie.

surveyor, **1.** ingénieur (topographe, géographe, hydrographe); **2.** inspecteur; **3.** arpenteur, géomètre; **s. chain,** chaîne d'arpenteur; **s. level,** niveau à lunette; **s. of mines,** inspecteur des mines; **s. transit,** théodolite à boussole.

susceptibility, susceptibilité, sensibilité; **magnetic s.,** susceptibilité magnétique.

suspended, **1.** en suspension, flottant; **2.** suspendu; **s. load,** matériaux transportés en suspension, charge en suspension; **s. sediment,** sédiment en suspension; **s. solids (SS),** polluants solides restant en suspension; **s. water,** eau vadose.

suspension, suspension; **s. current,** courant de turbidité; **s. load,** charge en suspension.

suture joint, stylolite.

suture line, **1.** ligne de suture (*pal.*); **2.** ligne tectonique, structurale majeure.

suturing, soudure (de plaques lithosphériques).

Svecofennian, Svécofennien (division du Protérozoïque, bouclier baltique).

swale, **1.** dépression marécageuse; **2.** bas-fond; **3.** dépression morainique.

swallet, **1.** nappe d'eau souterraine (*mine*), irruption d'eau; **2.** fissure aquifère karstique, aven.

swallow hole, **1.** aven. abîme, avaloir, bétoire, gouffre, puits naturel; **2.** perte d'un fleuve; **s. tail twin,** mâcle en fer de lance.

swamp, marais, marécage, bas-fond; **s. much,** tourbe peu consolidée; **s. ore,** limonite; **s. theory,** théorie autochtone de formation du charbon.

swampy, marécageux; **s. ground,** fondrière; **s. toundra soil,** sol marécageux de toundra.

swarm, groupe, essaim; **fissure s.,** groupe de fissures.

swash, **1.** jet de rive; **2.** onde de translation; **3.** chenal navigable; **s. bar,** cordon littoral de haut de plage; **s. mark,** marques de plage (par déferlement).

swash (to), faire jaillir de l'eau, clapoter.

S wave, onde transversale.

sweep, balayage (électronique); **s. of meanders,** glissement des méandres vers l'aval; **s. zone,** zone d'estran.

sweeping row, ligne de balayage; **s. line,** ligne de balayage.

sweet, doux; **s. crude,** pétrole brut non sulfuré; **s. gas,** gaz non corrosif; **s. soil,** sol neutre; **s. water,** eau douce.

swell, **1.** anticlinal à grand rayon de courbure; **2.** bombement, renflement; **3.** mont sous-marin; **4.** houle; **s. length,** longueur d'onde de la houle; **ground s.,** houle de fond.

swell and swale topography, topographie en creux et bosse, topographie morainique bosselée.

swelling, 1. adj. : gonflant; 2. n. : gonflement, renflement; s. clay, argile gonflante.

swimming stone, variété d'opale de densité inférieure à 1.

swinestone, calcaire fétide, bitumineux.

swing, oscillation, va-et-vient; s. moor, marais tremblant; s. sieve, tamis oscillant.

swinging, oscillation, balancement; s. of meander, déplacement du méandre.

swirl, remous, tourbillon, tourbillonnement, brassage.

swirl (to), tourbillonner, faire tournoyer, brasser.

swirling, 1. adj. : tourbillonnant; 2. n. : tourbillon.

swivel, 1. touret; 2. tête de sonde.

syenite, syénite (*pétro.*); s. aplite, aplite syénitique; s. pegmatite, pegmatite syénitique.

syenitic, syénitique.

syenodiorite, monzonite (*pétro.*).

syenogabbro, gabbro alcalin (à orthose).

syenoid, syénite à felspathoïdes.

sylvanite, sylvanite (*minér.*).

sylvinite, sylvinite (halite et sylvite).

sylvite, sylvite, sylvine (*minér.*).

sylvogenic soil, sol forestier.

symmetrical, symétrique; s. fold, pli symétrique.

symmetry, symétrie; s. axis, axe de s.; s. plane, plan de s.; bilateral s., s. bilatérale; pentamerous s., s. de type 5; plane of s., plan de s.; radial s., s. radiale.

symplektic intergrowth, association symplectique (*pétr.*).

synchronal, synchronous, synchrone; s. deposits, dépôts synchrones.

synchroneity, synchronisme (de 2 couches).

synclase, synclase, fissure de retrait.

synclinal, synclinal; s. axis, axe du s.; s. bend, charnière s.; s. closure, cuvette s; s. limb, flanc s.; s. trough, dépression s.; s. truncation, terminaison périsynclinale tronquée; upstanding s., s. perché.

syncline, synclinal, pli synclinal; closed s., synclinal fermé.

synclinore, synclinorium, synclinorium.

syncrude, pétrole brut (de schistes bitumineux).

syndiagenesis, syndiagenèse, altération synsédimentaire.

syneresis, synérèse (départ d'eau au cours de la diagénèse d'une roche).

syneresis crack, fente de synérèse.

synform, structure synforme (sans connaissance de la polarité stratigraphique).

synformal anticline, faux synclinal.

syngenesis, syngénèse (diagénèse précoce).

syngenetic, syngénétique.

syngenite, syngénite (*minér.*).

synkinematic, syntectonique.

synorogenic, syntectonique.

synplutonic dike, filon plus ou moins contemporain de la mise en place du pluton.

synsedimental, intraformationnel.

synsedimentary, synsédimentaire.

syntaxis, convergence de chaînes montagneuses.

syntaxy, syntaxie (*minér.*).

syntectic magma, magma syntectique.

syntectonic, syntectonique; s. fabrics, structures syntectoniques, synorogéniques.

synthem, unité stratigraphique limitée par des discordances.

synthexis, syntexis, assimilation magmatique.

synthetic, synthétique; s. crude, pétrole synthétique; s. fault, 1. faille conforme; 2. faille secondaire parallèle à la faille principale (U.S.A.).

syntype, syntype (*pal.*).

system, 1. système (*stratigr.*); 2. système en cours de cristallisation (*minér.*).

systematics, classification et taxonomie (*pal.*).

T

tabetisol (cf talik), tabétisol, zone profonde du sol non gelée.

table, table, plateau; t. land, plateau, relief tabulaire; t. like-structure, structure tabulaire; t. mount, guyot, volcan sous-marin; t. mountain, montagne tabulaire; t. reef, plature corallienne; t. slate, schiste ardoisier; t. spar, wollastonite; glacier t., table de glacier; inclined oil-water t., surface de contact pétrole-eau; tide t., annuaire des marées; turn t., table de rotation (forage).

tabular, tabulaire; t. crystal, cristal t.; t. jointing, division en bancs; t. spar, wollastonite; t. structure, structure t.

Tabulata, Tabulés (pal.).

tachygenesis, tachygénèse.

tachylyte, tachylite, tachylite (verre volcanique).

tachymeter, tachymètre.

taconic, taconian, taconique; T. orogeny, orogénèse Taconique (Ordovicien moy. à sup.).

taconite, taconite (minerai de fer).

tactite, tactite (pétr.).

tafoni, taffoni, forme d'érosion alvéolaire.

tag (to), marquer (radioactivité).

tagged atom, atome traceur.

Taghanican (N. Am.), Taghanicien (étage; Dévonien moy.).

taiga (U.S.S.R.), taïga.

tail, extrémité, queue; t. of a bar, extrémité aval d'un banc de sable.

tailing out, biseautage (d'une couche).

tailings, tails, 1. résidus de distillation; 2. refuts, rebuts de tamisage ou de triage de minerais; 3. résidu, rejet; t. heap, tas de résidu, halde.

take (to), prendre; t. a core, prélever une carotte; t. down, démonter; t. out the pillars, dépiler, enlever les piliers de charbon, déhouiller.

take-off post, bras de balancier (pétrole).

talc, talc; t. shist, talcschiste; t. slate, talcschiste.

talcite, 1. talcite (minéral.); 2. talcschiste.

talcochloritic, talcochloritique.

talcomicaceous, talco-micacé.

talcose, talcous, talqueux; t. granite, protogine; t. schist, schiste talqueux.

talcum, talc.

talcy, talqueux.

talik (Russ.), niveau de sol non gelé au-dessous ou au-dessus du pergélisol (cf. tabétisol).

talus (scree, U.K.), talus; t. accumulation, cône d'éboulis; t. cone, cône d'éboulis; t. creep, solifluxion; t. fan, cône de déjection; t. glacier, talus d'éboulis mouvant, glacier rocheux; t. slope, glaces d'éboulis.

taluvium, colluvions de bas de pente.

tame landscape, relief pénéplané.

Tamiskamian (see Timiskanian), Timiskanien (Précambrien).

tamp (to), damer, pilonner, tasser, bourrer, compacter.

tamper, bourroir.

tamping, bourrage (mine); t. plug, cartouche de bourrage; t. rod, bourroir.

tangential, tangentiel; t. fault, faille subhorizontale, chevauchement; t. stress, effort t.; t. thrust, poussée, charriage t.; t. wave, onde de cisaillement.

tangled channels, réseau fluviatile entrelacé, en tresse.

tangue (french), tangue (fines particules calcaires).

tank, 1. réservoir, citerne; 2. dépression aquifère.

tantalite, tantalite (minér.).

tantalum, tantal, tantale.

tape, ruban; t. line, décamètre magnétique; t. recorder, enregistreur magnétique.

taper out (to), s'effiler, se terminer en biseau.

taphonomy, taphonomie (spécialité de la paléoécologie).

taphrogenesis, taphrogénèse (formation des rifts).

taphrogeny, taphrogénie (cf. taphrogénèse).

taphrogeosyncline, taphrogéosynclinal (géosynclinal entre failles).

tapiolite, tapiolite (minér.).

tapping, soutirage (fluviatile).

tar, goudron, bitume; t. pit, lac, trou de bitume; t. pitch, goudron de houille; t. pool, lac de bitume; t. sands, sables bitumineux; bituminous t., goudron bitumineux; coal t., goudron de houille; wood t., goudron végétal.

Tarannon (Eur.), Tarranien (étage; silurien inf.).

Tardiglacial, Tardiglaciaire (mal défini : entre dernière glaciation et post-glaciaire).

tarn (icelandic), lac de cirque.

tarry, goudronneux.

Tartarian (U.S.S.R.) (or Tatarian), Tartarien (étage; Permien sup.).

tasmanite, tasmanite (minéral résineux).

taurite, taurite (var. de rhyolite).

taxite, taxite (pétro.).

taxitic structure, structure taxitique (roche volcanique).

taxodont dentition, taxondontie (pal.).

taxon (pl.: taxa, taxons), taxon (pal.).

taxonomic, taxonomique.

taxonomy, taxonomie, classification systématique.

Tayloran (North Am.), Taylorien (étage; Crétacé sup.).

T.D. (total depth), profondeur totale (d'un forage).

T. direction, direction du mouvement d'un plan tectonique.

tear, déchirement; **t. fault**, faille de déchirement, décrochement, faille transversale.

tectiform, tectiforme.

tectofacies, tectofaciès : faciès lié à l'évolution tectonique.

tectogene, tectogène : zone faillée profonde.

tectogenesis, tectogénèse (formation des montagnes).

tectomorphic, tectomorphique, deutéromorphique.

tectonic, tectonique; **t. basin**, dépression t.; **t. breccia**, brèche t.; **t. control**, contrôle t.; **t. cycle**, cycle orogénique; **t. fabric**, faciès t. (ex : mylonite); **t. facies**, disposition structurale; **t. framework**, cadre t.; **t. map**, carte t.; **t. style**, style t.; **t. window**, fenêtre d'une nappe (*tecto.*).

tectonicist, tectonicien.

tectonics, tectonique; **folding t.**, t. de plissement.

tectonite, mylonite, roche mylonitisée.

tectonophysics, tectonophysique.

tectonosphere, tectonosphère (équiv. : croûte).

tectonostratigraphic divide, limite tectonostratigraphique.

tectorogonenic event, événement tectorogénique.

tectosilicate, tectosilicate (silicates, à tétraèdres en structure à 3 dimensions).

tectotope, tectotope : milieu de même condition tectonique (terme déconseillé).

teeth, dents, dentition (*pal.*).

tektite, tectite (*pétro.*).

telemagmatic rocks, roches magmatiques éloignées du centre instrusif.

telemeter, télémètre.

telemetering, télémesure.

telemetry, télémétrie.

Teleosts, Teleostei, Téléostéens (poissons).

telescope, téléscope.

telescoping deposits, sédiments chevauchant, recouvrant d'autres terrains.

teleseism, téléséisme.

teleseismic signal, signal télésismique.

telethermal ore deposits, gisements minéraux superficiels produits par solutions hydrothermales ascendantes.

telethermal water, eau téléthermale, eau endogène d'origine lointaine.

tellurian, tellurien (des profondeurs de la Terre).

telluric, tellurique; **t. acid**, acide t.; **t. current**, courant t.; **t. method**, méthode t.

telluride, tellurure (composé).

telluriferous, tellurifère.

tellurite, tellurite (*minér.*).

tellurium, tellurium; **t. dioxide**, tellurite.

tellurobismuthite, tellurobismuthite.

temperate, tempéré; **t. glacier**, glacier de type tempéré.

temperature, température; **t. correction**, correction de t.; **t. drop**, chute de t.; **t. fluctuation**, fluctuation t.; **t. gradient**, gradient thermique; **t. inversion**, inversion de t. (météo.); **t. range**, gamme

de t.; **t. recorder**, enregistreur de t.; **average t.**, t. moyenne; **low t.**, basse t.; **mean t. difference**, écart moyen de t.

tempering, 1. trempe, durcissement; 2. gachâge (du plâtre).

template method, méthode de triangulation radiale.

temporary hardness of water, dureté d'une eau carbonatée.

tennantite, tennantite (*minér.*)

tenor, teneur d'un minerai.

tenorite, ténorite (*minér.*).

tensibility, extensibilité.

tensile, extensible, extensile; **t. bending test**, essai de rupture par flexion; **t. force**, force de traction; **t. strain**, déformation due à la traction; **t. strength**, résistance à la traction; **t. stress**, effort de traction; **t. test**, essai à la traction.

tensility, extensibilité.

tension, 1. extension, traction, tension; 2. pression; **t. bar**, éprouvette de traction; **t. cracks**, fissures d'extension; **t. fault**, faille normale d'extension; **t. fracture**, fracture d'extension; **t. gash**, fissure d'extension tectonique; **t. test**, essai de rupture; **t. of steam**, pression de vapeur.

tensional, d'extension.

tentacle, tentacule (Invertébrés).

tephra, projections volcaniques.

tephrite, téphrite (roche effusive à caractère basaltique).

tephrochronology, téphrochronologie (datation par les cendres volcaniques).

tephroite, téphroïte (*minér.*).

teratology, tératologie.

Terebratula, Térébratule (*pal.*).

Teredo attack, corrosion par les Tarets.

terminal basin, amphithéâtre morainique.

terminal moraine, moraine frontale.

terminal velocity, vitesse limite de chute.

ternary diagram, diagramme triangulaire.

ternary system, système à trois composants (ex : $CaO-Al_2O_3-SiO_2$).

terra, terre; **t. cotta clay**, argile réfractaire; **t. empelitis**, schiste ampéliteux; **t. fusca**, sous-sol calcaire sombre, riche en fer; **t. ponderosa**, barytine; **t. rossa**, terra rossa.

terrace, terrasse; **t. height**, altitude relative d'une t.; **t. level**, niveau de t.; **t. scarp**, talus de t.; **aggradational t.**, t. alluviale construite; **alluvial t.**, t. alluviale; **climatic t.**, t. climatique; **eustatic t.**, t. eustatique; **fill t.**, t. d'accumulation; **fill and fill t.**, t. emboîtées; **fluvial t.**, t. fluviatile; **fluvioglacial t.**, t. fluvio-glaciaire; **inner-valley t.**, t. emboîtée; **inset t.**, t. emboîtée; **matched t.**, couplées; **river t.**, t. fluviatile; **rock t.**, t. rocheuse; **slipp-off t.**, t. polygénique; **stopped, stopping t.**, t. étagée; **strath t.**, t. rocheuse; **stream t.**, t. fluviatile; **structural t.**, replat structural; **tectonic t.**, t. tectonique; **valley-plain terrace**, t. rocheuse.

terraced, terraciform, en terrasse.

terracette, terrassette (gradin de solifluxion).

terrain, 1. terrain; **2.** ensemble de strates; **t. analysis,** analyse de terrain; **t. correction,** correction topographique; **t. factors,** facteurs de terrain.

terrane (U.S.), 1. terrain, région; **2.** formation géologique.

terraqueous water, eau souterraine.

terrene (rare), 1. terreux; **2.** terrestre.

terrestrial, terrestre, continental; **t. environment,** milieu continental; **t. magnetism,** 1. champ magnétique terrestre; **2.** magnétisme terrestre; **t. planet,** planète ayant des caractéristiques communes avec la terre.

terrigenous, terrigene, terrigène.

Tertiary, Tertiaire (période Cénozoïque).

tervalent, trivalent.

teschenite, teschénite (pétro.).

tesseral, cubique; **t. system.** système c., régulier.

test, essai, épreuve, expérience, test; **t. bar,** éprouvette (pour essai); **t. boring,** sondage d'exploration; **t. glass,** éprouvette (de laboratoire); **t. paper,** papier réactif; **t. pit,** puits de recherche, trou d'exploration; **t. tube,** éprouvette, tube à essais; **t. well,** puits d'exploration; **bending t.,** essai de flexion; **breaking t.,** essai de rupture; **laboratory t.,** essai en laboratoire; **production t.,** essai de production (pétrole); **tensile t.,** essai de traction.

testing, essai d'épreuve; **t. drill,** sonde de prospection.

Tethyan ocean, Téthys (géosynclinal : Permien à début Tertiaire).

Tethys, Téthys.

Tetracoral, Tetracorallia, Tétracoralliaires (pal.).

tetradymite, tetradymite (minér.).

tetragonal system, système quadratique.

tetrahedral, tétraédrique; **t. site,** emplacement d'un atome dans le tétraèdre silice oxygène.

tetrahedrite, tétraédrite (minér.).

tetrahedron, tétraèdre.

tetrahexahedron, tétrahexaèdre, cube pyramidé.

Tetrapoda, Tétrapodes (pal.).

tetravalent, tétravalent.

Textularina, Textularinés (pal.).

texture, 1. structure, agencement des minéraux; **2.** texture; **banded t.,** s. zonée; **brecciated t.,** s. bréchoïde; **dendritic t.,** s. dendritique; **foliate t.,** s. foliacée; **granule t.,** s. granulaire; **serrate t.,** s. engrenée.

textural, textural.

thalassic, thalassique; **t. rock,** roche abyssale.

thallasocratic, thallasocratique.

thallasocraton, lithosphère océanique bien stabilisée.

thallasoid, bassin lunaire.

Thallophyta, Thallophytes (Algues et Champignons).

thallus, thalle (paléobot.).

thalweg, thalweg, fond de vallée.

thanatocoenosis, thanatocoenose (accumulation fossilifère post mortem, cf. biocenosis, necrocoenosis).

Thanecian (obsolete), Thanétien (étage; Paléocène).

Thanetian, Thanétien (étage; Paléocène).

Thanet sands, sables thanétiens.

thaw, dégel; **t. depression,** dépression thermo-karstique; **t. front,** front de dégel; **t. lake,** lac thermokarstique.

thaw (to), dégeler, fondre.

thawing, dégel, fonte.

theca (pl. thecae), thèque (pal.).

Thecamoeba, Thécamébiens (pal.).

Thecodontia, Thécondontes (pal.).

thenardite, thénardite (minér.).

theodolite, théodolite (topog.).

theralite, théralite (var. de gabbro).

Therapsida, Thérapsidés (pal.).

therm, therm = 25 200 calories.

thermae, sources thermales.

thermal, 1. thermal; **2.** thermique; **t. analysis,** analyse thermique; **t. capacity,** chaleur spécifique; **t. demagnetization,** désaimantation; **t. depression,** dépression atmosphérique thermique; **t. effect,** effet thermique (dilatation, etc.); **t. efficiency,** rendement thermique; **t. equator,** équateur thermique **t. expansion,** dilatation thermique; **t. gaze,** émanations gazeuses thermales; **t. gradient,** gradient thermique; **t. logging,** diagraphie thermique; **t. metamorphism,** thermométamohphisme; **t. pollution,** pollution thermique; **t. shrinkage,** retrait par refroidissement; **t. spring,** source thermale; **t. stability limit,** limite de stabilité thermique; **t. stratification,** stratification thermique; **t. unit,** unité thermique (calorie); **t. water,** eau thermale; **t. weathering,** thermoclastic.

thermality, thermalité.

thermic, 1. thermique; **2.** thermal.

thermochemical, thermochimique.

thermochemistry, thermochimie.

thermocline, thermocline (limnologie).

thermodynamic, thermodynamical, thermodynamique; **t. metamorphism,** thermométamorphisme.

thermodynamics, thermodynamique.

thermogram, thermogramme.

thermograph, thermomètre enregistreur.

thermography, thermographie; **aerial t.,** thermographie aérienne.

thermogravitational diffusion, diffusion thermogravitationnelle.

thermokarst, thermokarst; **t. pit,** dépression thermokarstique.

thermohaline circulation, circulation thermique des eaux marines.

thermoluminescence, thermoluminescence.

thermolysis, thermolyse.

thermomagnetic analysis, analyse thermomagnétique.

thermometamorphism, thermométamorphisme.

thermometer, thermomètre; **t. calibration,** étalonnage du t.; **t. screen,** abri météorologique; **t. well,**

puits thermométrique; **bimetallic t.**, t. à bilame; **recording t.**, t. enregistreur.

thermometric, thermometrical, thermométrique; **t. hydrometer,** densimètre à corrections thermométriques; **t. scale,** échelle thermométrique.

thermomineral, thermomineral.

thermonatrite, thermonatrite (*minér.*).

thermoremanent magnetization, magnétisme thermorémanent.

thermosphere, thermosphère.

thermostable, thermostable.

Thetis' hair-stone, quartz à inclusions aciculaires d'actinolite.

thick, épais, puissant; **t. bands,** lits de vitrain (cf. charbon); **t. beds** couche épaisse; **t. seam,** gîte important.

thick-bedded, à lits épais (de 65 cm à 120 cm).

thicken (to), épaissir.

thichening, épaississement.

thickness, épaisseur, puissance; **actual t.,** épaisseur réelle; **working t.,** ouverture de la taille (*mine*).

thief formation, formation fissurée, poreuse.

thill, 1. mur (d'une couche), Écosse; 2. mince couche d'argile réfractaire.

thin, mince, fin, ténu; **t. bedded,** finement stratifié (lits de 0,5 à quelques cm d'épaisseur); **t. plate,** lame mince; **t. section,** lame lince; **t. sectioning,** confection de lames minces; **t. walled,** à parois minces.

thin (to), 1. amincir, diminuer, amenuiser; 2. s'amincir, s'amenuiser; **t. down,** amincir, diluer; **t. out,** se terminer en biseau.

thinly bedded (thin bedded), finement stratifié.

thinness, faible épaisseur, minceur.

thinning, amincissement.

thixotropic, thixotropique.

thixotropy, thixotropie (forte cohésion des sédiments fin non dérangés).

tholeiite, tholéiite, tholéyite.

tholeiitic, tholéiitique; **t. basalt,** basalte t.; **t. magma;** magma t.; **t. suite,** lignée t.

tholoide, cumulo-dôme.

thomsonite, thomsonite (zéolite).

thorianite, thorianite (*minér.*).

thorite, thorite (*minér.*).

thorium, thorium.

thorium-lead dating, datation thorium-plomb.

thread of ore, veinule de minerai.

thread of water, filet d'eau.

three dimensional map, carte en relief.

three-dimension dip, pendage réel d'une couche souterraine (sismique).

three-faceted stone, galet à 3 facettes, galet éolisé, dreikanter.

three-layer structure, phyllite à trois feuillets.

threshold, seuil, haut-fond; **t. pressure,** seuil de déformation; **t. velocity,** vitesse limite, vitesse minimale de transport (éolien, etc.).

thrible board, plate-forme d'accrochage (forage).

throat, cheminée magmatique.

through coal, charbon tout venant; **t. glacier,** glacier bifurqué; **t. valley,** vallon bidirectionnel.

throughput, débit, consommation.

throw, 1. rejet vertical d'une faille; 2. composante verticale du rejet.

throwing clay, argile plastique.

thrown side, compartiment affaissé, lèvre affaissée (d'une faille).

thrown wall, lèvre abaissée.

thrust flaw, copeau de charriage.

thrusting, charriage, poussée.

thulite, thulite (var. de zoïsite).

thunder egg (popular: U.S.A.; Or.), concrétion siliceuse, géode siliceuse dans laves.

Thuringian (Eur.), Thuringien (étage Permien sup.).

tickle (Can.), passe (dans un cordon littoral).

tidal, de marée, à marée; **t. bore,** mascaret; **t. channel,** chenal de marée, passe; **t. compartment,** tronçon fluviatile à marée; **t. current,** courant de marée; **t. creek,** chenal de schorre; **t. delta,** delta de marée; **t. estuary,** estuaire de marée; **t. flat,** estran; **t. inlet,** passe; **t. marsh,** marais littoral; **t. pool,** sillon de plage, sillon prélittoral; **t. prism,** amplitude des marées, marnage; **t. range,** amplitude des marées; **t. river,** rivière à marée; **t. scour,** affouillement, érosion de marée; **t. stream,** courant de marée; **t. wave,** raz de marée, tsunami; **t. zone,** zone intertidale, zone de balancement des marées.

tidalite, tidalite (sédiment formé dans la zone de battement des marées).

tide, marée; **t. gage (U.S.),** gauge (U.K.), marémètre, marégraphe; **t. island,** presqu'île; **t. lines,** laisses de marée; **t. marks,** laisses de marées; **t. pole,** marémètre; **t. power,** énergie marémotrice; **ebb t.,** jusant; **falling t.,** marée descendante; **high t.,** m. haute; **low t.,** m. basse; **neep t.,** m. de mortes-eaux; **rising t.,** m. montante; **spring t.,** m. de vives-eaux.

tideland area, terrain inondable.

tideless, sans marée.

tiemannite, tiemannite (*minér.*).

tiff (U.S.A., S.W., Missouri), calcite cristalline; **t. S.E. Missouri,** barytine cristalline.

tiger eye, tiger's eye, œil de tigre, crocidolite (ou sa pseudomorphose en quartz).

tight, étanche, serré, bien ajusté, imperméable; **t. fold,** pli fermé; **t. sand,** sable compact, peu perméable; **t. sandstone,** grès compact, colmaté.

tightly compressed fold, pli fermé, serré.

tightness, étanchéité.

Tiglian, Tiglien.

tile, tuile, carreau, brique; **t. clay,** argile à briqueterie; **t. earth,** terre à brique; **t. ore,** cuprite terreuse, cuivre rouge; **t. stone,** tuile, brique.

till, till, moraine glaciaire, argile à blocaux (ancien); **t. fabric,** structure du till, structure tridimensionnelle (des éléments grossiers); **subglacial t.,** moraine till de fond, sous-glaciaire; **subaqueous t.,** moraine till glacio-marine; **upper t.,** moraine till superficiel.

tillite, tillite, till consolidé.

tilt, 1. inclinaison, pente, basculement; 2. distorsion photogrammétrique due à l'inclinaison de prise de vue; t. **meter,** clinomètre; t. **slide,** éboulis de pente, éboulis de gravité.

tilt (to), 1. incliner, pencher; 2) être incliné, se pencher; t. **up (to),** basculer.

tilted block, bloc basculé.

tilting of strata, inclinaison des couches.

timber, bois, bois de charpente; t. **drawer,** déboiseur (mine); t. **drawing,** déboisage; t. **line,** limite de la zone forestière; t. **preservation,** traitement préventif des bois de mine; t. **set,** cadre de boisage (mine); **mine t.,** bois de mine.

timber (to), boiser (un puits de mine).

timbering, boisage.

time, temps; t. **belt,** fuseau horaire; t. **break,** instant de tir (*géophys.*)*;* t. **correlation,** chronostratigraphie, chronologie; t. **depth curve,** courbe temps-profondeur; t. **distance curve,** courbe temps-distance; t. **gradient,** variation du t. de propagation avec la profondeur; t. **firing,** instant d'explosion; t. **interval S-P,** retard entre les arrivées des ondes S et P; t. **lag.** retard; t. **span, lapse,** laps de t.; t. **of transit,** t. de propagation; t. **to depth conversion,** conversion temps-profondeur (*géophys.*).

Timiskamian (N. Am.), Timiskamien (division du Précambrien : Archéen sup.).

tincal, tinkal, tincal (borax brut natif).

tinguaite, tinguaïte (var. de phonolite).

tin, étain; t. **bearing,** stannifère; t. **deposit,** gîte stannifère; t. **dredge,** drague à é.; t. **dredging.,** draguage stannifère; t. **dressing,** préparation du minerai d'é.; t. **ground,** terrain stannifère; t. **lode,** filon d'é.; t. **mine,** mine d'é.; t. **ore, placer,** gîte alluvionnaire de cassitérite; t. **pyrite,** stannite; t. **smeltery.** fonderie d'é.; t. **spar,** cassitérite; t. **stone,** cassitérite; t. **stuff,** cassitérite mélangée à sa gangue; t. **white cobalt,** smaltite; t. **works,** smaltite.

Tintinnidae, Tintinnoïdés.

tintometer, colorimètre.

Tioughiogan (N. Am.), Tioughiogien (étage Dévonien moy.).

tip, 1. extrémité, pointe; 2. basculeur, verseur; 3. dépôt de déblais; 4. accumulation de neige d'avalanche.

tipped block, bloc basculé.

tipping car, wagonnet basculant (mine).

tipple car, culbuteur (mine).

titanate, titanate.

titanaugite, augite titanifère.

titangarnet, grenat titanifère.

titanhornblende, hornblende titanifère.

titanic, titané; t. **anhydrite,** dioxyde de t. (rutile, brookite); t. **iron ore,** ilménite, fer titané; t. **oxide,** t. oxydé, rutile; t. **schore,** t. oxydé, rutile.

titaniferous, titanifère; t. **iron ore,** fer titané, ilménite.

titanite, titanite, sphène.

titanitic, titanitique.

titanium, titanium, titane.

titanmica, mica titanifère.

titanoferrite, ilménite, fer titané.

titanomagnetite, titanomagnétite.

titanomorphite, titanomorphite.

titanous, titaneux.

titantourmaline, tourmaline titanifère.

titer test, essai de titrage.

Tithonian (Eur.), Tithonique (faciès du Portlandien).

title of gold, titre de l'or.

titrate (to), titrer.

titration, titrage.

tjäle, (swedish), tjaele, tjäle, gélisol.

toad's eye tin, cassitérite en nodules zonés.

toadstone, 1. crapaudine, pierre de crapaud; 2. roches ignées ou pyroclastiques (Derbyshire).

Toarcian, Toarcien (étage; Lias sup.).

toe, front d'éboulement, rebord d'éboulis; **glacier t.,** front d'une langue glaciaire (glacier de vallée).

tombolo, tombolo, isthme.

ton, tonne.

ton avoir-du-poids (G.B.); long t., gross t., = 1 016,05 kg (tonne forte); **short t. (U.S., Can.),** = 907,185 kg (tonne courte); **metric t., tonne,** = 1 000 kg (tonne métrique).

tonalite, tonalite, diorite quartzique.

Tongrian, Tongrien (étage; Oligocène inf.).

tongue of land, 1. langue de terre, péninsule; 2. biseau, lentille (de sédiments); **mud t.,** langue de boue (périglaciaire).

Tonolowayan (N. Am.), Tonolowayien (étage; Silurien sup.).

Tonowandan (U.S.A., N.Y. St.), Tonowandien (étage; Silurien moy.).

tool, outil; t. **mark,** marque de courant, figure sédimentaire en cavité parallèle au courant.

tooth, dent (*pal.*).

toothing, dentition (*pal.*).

top, cîme, sommet, point le plus haut, limite supérieure d'une couche; t. **hole,** trou de toit (mine); t. **pressure,** pression du toit; t. **rod,** tête de sonde; t. **set beds,** lits deltaïques sommitaux; t. **soil,** couche arable; t. **wall,** toit (mine); **bed t.,** toit de la couche; **formation t.,** toit de la formation.

topaz, topaze; t. **safranite,** citrine, quartz jaune; t. **quartz,** citrine; **false t.,** citrine, quartz jaune.

topazite, 1. andradite (var. de grenat); 2. roche filonienne à quarte et topaze.

topazolite, topazolite (var. de grenat).

topographer, topographe.

topographic, topographical, topographique; t. **adjustment,** adaptation du réseau; t. **adolescence,** stade d'adolescence; t. **correction,** correction de terrain; t. **high,** hauteur; t. **infancy,** stade infantile; t. **low,** dépression; t. **map,** carte t.; t. **maturity,** stade de maturité; t. **old age,** stade de sénilité, de pénéplanation; t. **sheet,** planchette t.; t. **survey,** levé t.; t. **unconformity,** discordance t.; t. **youth,** stade de jeunesse.

topography, topographie.

topotype, topotype (*pal.*).

topping, distillation fractionnée; **t. plant,** unité de fractionnement; **t. tower,** colonne de fractionnement.

topsoil, sol superficiel horizon A.

tor (G.B.), **1.** pic, sommet, pinacle; **2.** blocs, roches isolées par érosion.

tor (U.S., obsolete), roche moutonnée.

torbanite, torbanite (var. de schiste bitumineux).

torbernite, torbernite (*minér.*).

tornado (tornadoes, pl.), tornade.

torpedo (to) a well, torpiller un puits.

torrent, torrent, gave; **t. track,** trajet, tracé torrentiel.

torrential, torrentiel.

torrentially, torrentiellement.

torrid (obsolete), torride (climat, région).

torsion, torsion; **t. balance,** balance de t.; **t. coefficient,** coefficient de t.; **t. period,** période d'oscillation; **t. shear test,** essai de cisaillement par t.; **t. wire,** fil de t.

torsional, de torsion; **t. strain,** déformation due à la t., **t. strength,** résistance à la t.; **t. test,** essai de t.

Torridonian (Scotland), Torridonien (division du Précambrien sup.).

Tortonian, Tortonien (sous-étage; Miocène moy.).

toscanite, toscanite (*pétro.*).

total, total; **t. depth,** profondeur limite d'un puits; **t. displacement,** rejet net; **t. recovery,** quantité t. de pétrole extrait d'un puits; **t. reflection,** réflexion totale; **t. solids,** résidus; **t. throw,** rejet t.; **t. time correction,** correction globale du temps de propagation.

touchstone, pierre de touche, lydienne.

toughness, ténacité, résistance (d'une roche).

toundra polygon, macropolygone (*périgl.*).

tourmaline, tourmaline; **t. granite,** granite à t.; **t. rock,** roche à quartz et t.

tourmalinization, tourmalinisation.

Tournaisian, Tournaisien (sous-étage ou étage; Carbonifère inf.).

tower (U.S.A.: Wyo.), pic, piton; **t. karst,** tour karstique.

towhead, haut fond, banc émergé (U.S. Mississipi).

T. plane, plan de réarrangement cristallophyllien (des gneiss).

trabeculae, trabécule (*pal.*).

trace, **1.** enregistrement graphique, trace, spot; **2.** trace, tracé (d'une faille); **3.** trace, petite quantité infinitésimale; **t. element,** élément trace; **t. fossil,** empreinte fossile, structure fossile; **t. intensity,** intensité minimum (d'enregistrement sismique).

trace (to) a lode, suivre un filon.

tracer, traceur; **radioactive t.,** t. radioactif.

trachiae, trachée (*pal.*).

trachyandesite, trachyandésite (*pétro.*).

trachybasalt, trachybasalte.

trachyte, trachyte (*pétro.*).

trachytic, trachytique; **t. lava,** lave t.

trachytoid, trachytoïde; **t. phonolite,** phonolite à néphéline; **t. structure,** structure t.

track, tracé, piste, voie; **avalanche t.,** couloir d'avalanche; **to lay the t.,** poser la voie (*mine*).

trackless system, système sans rails (*mine*).

traction, transport de fond (ex : fluviatile), transport par traction.

tractional load, charge de fond, charge fluviatile transportée au fond.

tractive current, courant de traction, de transport fluviatile de fond.

trade-wind, trades, alizé.

trail, trace, piste, sentier; **t. of fault,** brèche de faille indiquant la direction de la faille.

train, **1.** alignement de blocs glaciaires; **2.** série (d'ondes); **3.** série de réflections (sur séismogramme).

train (to), tracer, suivre (*mine*); **t. a lode,** suivre un filon.

training of a river, contrôle d'une rivière.

tram for ore, berline, benne à minerai.

transceiver, transducteur (*géophys.*).

transcrystalline, intracristallin, transcristallin.

transcurrent fault, décrochement, faille à déplacement horizontal, faille d'arrachement, faille transversale.

transect, **1.** profil transversal (sismique); **2.** méthode de carroyage (botanique).

transection glacier, glacier réticulé, transfluent.

transfer, transport (de sédiments).

transformation, changement de phases, transformation (ex : quartz α-quartz β).

transform fault, décrochement, faille transformante, faille transversale.

transformism, transformisme (théorie du).

transcient methods, méthodes électriques.

transgression, **1.** G.B. discordance; **2.** U.S. transgression.

transgressive, **1.** G.B. discordant; **2.** U.S. transgressif.

transit compass, théodolite à boussole.

transit instrument, lunette méridienne.

transition bed, couche de transition.

translated rock-sheet, nappe de charriage.

translation gliding, glissement intracristallin (de couches atomiques les unes sur les autres); **t. plane,** plan de glissement, translation, plan de translation.

translatory wave, vague de translation.

transmutation, transmutation, décomposition radioactive.

transponder, transpondeur (*géoph.*).

transport, transport (fluviatile, etc.).

transportation by water, transport par les eaux.

transported soils, sols alluviaux peu évolués.

transporting power, puissance de transport.

transposition (of fabric), structure héritée de roches antémétamorphiques.

Transvaal jade, grenat grossulaire.

transverse, transversal; **t. crevasse,** crevasse transversale; **t. dune,** dune transversale; **t. fault,** faille transversale; **t. fold,** pli croisé; **t. lami-**

nation, stratification entrecroisée; **t. section,** coupe transversale; **t. valley,** cluse, vallée transversale; **t. wave,** onde transversale.

trap, **1.** piège (de pétrole), réservoir fermé, étanche; **2.** trapp, roche éruptive sombre; **t. door,** porte d'aération (*mine*); **t. fault (U.S.),** faille circulaire; **t. rock,** roche effusive, ou intrusive (diabase); **fault t.,** piège de faille; **oil t.,** piège à pétrole; **réservoir t.,** piège à pétrole; **stratigraphic t.,** piège stratigraphique; **structural t.,** piège structural.

trapezohedron, trapézoèdre.

trapped oil, pétrole piégé.

trash, détritus, déchets, décombres; **t. ice,** débris de banquise.

travel time, temps de propagation.

travel-time curve, courbe temps-distance, hodochrone.

travelled boulder, bloc erratique.

traverse, ligne transversale (levé).

travertine, travertin, tuf calcaire.

tray, plateau de colonne (pétrole).

treating plant, usine de purification.

tree agate, agate arborescente.

tree-like river system, réseau hydrographique dendritique.

tree-ring, anneau de croissance (dendrochronologie).

trellis drainage pattern, réseau orthogonal, type appalachien.

trellised drainage, réseau fluviatile rectangulaire, en grillage.

Tremadocian (Eur.), Trémadocien, Trémadoc (étage; Ordovicien inf.).

tremolite, trémolite; **t. actinolite series,** série d'amphiboles (trémolite-actinote).

tremometer, sismographe.

tremor, tremblement, secousse, trépidation; **t. earth,** secousse sismique.

Trempealeauian (N. Am.), Trempéaléauien (étage : Cambrien terminal).

trench, **1.** fosse sous-marine, fosse de subduction; **2.** fossé, tranchée; **t. excavator,** excavatrice.

trenching, excavation de tranchée, terrassement.

trend, **1.** direction (d'un pli); **2.** tendance, direction générale; **fault t.,** direction d'une faille; **change in t.,** changement de direction.

Trentonian (N. Am.), Trentonien (étage; Ordovicien moy.).

trepan, trépan.

trestle, chevalet (*mine*).

triactial diagram, diagramme triangulaire.

trial, essai, expérience; **t. boring,** sondage de reconnaissance; **t. hole,** sondage d'exploration; **t. pit,** puits d'exploration.

triangular diagram, diagramme triangulaire.

triangulate (to), trianguler.

triangulation, triangulation; **t. point,** point géodésique.

Trias, Trias (période; Mésozoïque).

triassic, triasique.

tributary river, affluent.

trichite, trichite (*pétro.*).

trickle (to), couler, suinter, s'infiltrer.

triclinic, triclinique; **t. system,** système t.

tricone rock bit, trépan tricône.

tridymite, tridymite (*minér.*).

trigonal system, système rhomboédrique.

Trilobita, Trilobite, Trilobite (*pal.*).

Trilobitomorpha, Trilobitomorphes (*pal.*).

trimetric, orthorhombique.

trimorphism, trimorphisme (*cristallogr.*).

Trinitian (N. Am.), Trinitien (étage; Crétacé inf.).

trioctahedral, trioctaédrique.

triphane, spodumène, triphane.

triphyline, triphylite, triphylite (*minér.*).

triple junction, rencontre de trois plaques lithosphériques; **t. point,** point triple (*thermodynamique*).

triploidite, triploïdite (*minér.*).

tripoli, tripoli, diatomite.

tripolite, tripoli, diatomite.

trisoctahedron, octaèdre pyramidé.

tristetrahedron, tétraèdre pyramidé.

tritiated water, eau enrichie en tritium.

tritium, tritium.

triturate (to), triturer, réduire en poudre.

trivalent, trivalent.

trivial name, nom d'espèce (*pal.*).

trochiform, conique (*pal.*).

troctolite, troctolyte, troctolite (var. de gabbro).

troglodytic, troglodytical, troglodytique.

troilite, troïlite (minéral des météorites).

trommel, trommel, trieur, crible rotatif.

trommel (to), passer au trommel.

troostite, troostite (*minér.*).

tropic, tropical, **1.** adj : tropical; **2.** n : tropique; **t. cyclone,** cyclone tropical; **t. easterlies,** alizés tropicaux.

tropopause, tropopause.

troposphere, troposphère.

trottoir (french), petite terrasse ou corniche organique, formée le long de la zone intercotidale.

trouble, **1.** faille, dislocation; **2.** difficulté, panne.

trough, **1.** dépression, axe synclinal, gouttière synclinale; **2.** auge glaciaire; **3.** creux de vague; **t. axis,** axe synclinal; **t. bend,** charnière synclinale; **t. limb,** flanc inverse d'un pli; **t. of a syncline,** charnière synclinale; **t. point,** creux maximum d'une ride (séd.); **t. valley,** vallée en auge, vallée synclinale; **fault t.,** fossé tectonique; **glacial t.,** auge glaciaire; **oceanic t.,** fosse océanique.

true, vrai, véritable, réel; **t. dip,** pendage vrai; **t. folding,** flexure; **t. lode,** fissure minéralisée; **t. north,** nord astronomique; **t. ruby,** rubis oriental.

truncation, truncature.

trunck glacier, glacier principal; **t. stream,** fleuve principal (central par rapport au réseau fluviatile).

tschernosem, tschernozem (cf. chernozem), chernozem (*pédol.*).

tsunami, raz de marée, tsunami.
tub, benne, berline, wagonnet.
tube extractor, extracteur de tubes; **decantation t.**, tube à décantation; **settling t.**, tube à sédimentation.
tubercule, tubercule (*pal.*).
tubing, colonne d'exploitation, tube de pompage; **t. a well**, scellement du train de tiges; **t. board**, plate-forme de garage des tiges; **t. head**, tête de colonne de production.
tubular spring, source karstique.
tubule, concrétion calcaire, poupée du loess.
Tubulidentata, Tubulidentés (*pal.*).
tufa, tuf sédimentaire calcaire ou siliceux, travertin.
tufaceous, tuffaceous, tufacé.
tuff, tuf (volcanique), cendre volcanique cimentée; **t. ball**, cinérite; **t. breccia**, brèche volcanique pyroclastique; **t. cone**, cône de cendres, cinérite; **t. lava**, ignimbrite, tufs soudés; **t. palagonite**, brèche éruptive, palagonite; **t. ring**, couronne pyroclastique; **welded t.**, tufs soudés, ignimbrite.
tuffite, tuffite, roche volcanodétritique.
tugger, treuil; **air t.**, t. pneumatique.
tumble, chute, éboulis.
tumble (to), tomber, culbuter, s'ébouler.
tumbler, cylindre tournant.
tumescence, gonflement, bombement, intumescence (volcan).
tumulose, accidenté, bosselé, à monticules.
tumulus (tumuli, pl.), **1.** petit dôme de lave; **2.** tumulus (tombeau).
tundra, toundra; **t. placer**, gisement de moraines ou d'alluvions; **t. polygon**, grand polygone de t.
tungstate, tungstate.
tungsten, tungstène.
tungstenium, tungstène.
tungstite, tungstite; **t. ocher**, tungstite.
tunnel, galerie à flanc de coteau (mine), tunnel, travers-banc, souterrain; **t. disease**, maladie des mineurs; **t. heading**, extrémité de fendue; **t. valley**, ravin sous-glaciaire.
tunnel (to), percer un tunnel à travers une colline.
turbary (G.B.), tourbière.
turbid, trouble (eau).
turbidimeter, opacimètre, turbidimètre.
turbidite, dépôt de courant de turbidité.
turbidity, turbidité, opacité; **t. current**, courant de turbidité; **t. size analysis**, analyse densimétrique.
turbodrill, trépan à turbine hydraulique.
turbodrilling, turboforage.
turbulence, turbulence (*météo.*).
turbulent flow, écoulement turbulent.
turf, tourbe séchée; **t. mound**, butte gazonnée; **t. pit**, tourbière.
turfary (G.B.), tourbière.

turgite, limonite (variété de).
turkey-fat ore (U.S.A.: Arkansas and Missouri), smithsonite.
turkey stone, **1.** schiste (variété); **2.** turquoise.
turm karst, pitons karstiques en relief résiduel (cf. hum, mogote).
turmalin, tourmaline (*minér.*).
turn table, table de rotation.
turnover, brassage (de couches d'eaux).
Turonian, Turonien (étage; crétacé sup.).
turpentite, thérébentine.
turquoise, turquoise.
turrelite (U.S.A.: Texas), schiste bitumineux.
Turitella, Turitelle (*pal.*).
turtle stone, nodule, concrétion, septaria (grande taille).
tusculite, tusculite (*pétro.*).
twin, mâcle, cristal mâclé; **t. axis**, axe de mâcle; **t. crystals**, cristaux mâclés; **t. crystallization**, hémitropie; **t. dolines**, dolines jumelées; **t. gliding**, dislocation cristalline; **t. plane**, plan de mâcle; **albite t.**, mâcle de l'albite; **Carlsbad t.**, mâcle de Carlsbad; **contact t.**, mâcle par accolement; **interpenetrant t.**, mâcle d'interpénétration; **juxtaposition t.**, mâcle par accolement; **penetration t.**, mâcle d'interpénétration; **polysynthetic t.**, mâcle polysynthétique; **repeated t.**, mâcle polysynthétique; **rotation t.**, mâcle par rotation autour d'un axe; **swallow-tail t.**, mâcle en fer de lance; **x-shaped t.**, mâcle en croix.
twin (to), mâcler, se mâcler.
twinned, mâclé, hémitrope.
twinning, formation de mâcles; **t. axis**, axe d'hémitropie; **t. lamella**, lamelle de mâcle; **t. plane**, plan de mâcle; **mechanical t.**, mâcle mécanique.
twist, torsion.
twisting, torsion, rotation.
twitch, étranglement (d'un filon).
two-cycle valley, vallée façonnée par deux cycles d'érosion.
two-layer structure, phyllite à deux couches.
two-phase flow, écoulement diphasé.
Tyler standard grade, échelle granulométrique de Tyler.
type, type (*pal.*); **t. fossil**, fossile stratigraphique; **t. genus**, genre t.; **t. locality**, localité t.; **t. section**, stratotype; **t. species**, espèce t.; **t. specimen**, échantillon t., holotype.
typhoon, typhon.
typic, typical, typique.
typology, typologie.
typomorphic, typomorphique (*minér.*).
Tyrrel sea, mer holocène (− 6 800 B.C.) dans la région de la baie d'Hudson (Canada).
Tyrrhenian, Tyrrhénien (I et II : formations méditerranéennes, Pléistocène).

U

udalf, sol lessivé de climat humide (*pédol.*).

udoll, brunizem (sols de prairie) (*pédol.*).

udometer, pluviomètre, pluviographe.

udometric, pluviométrique.

udometry, pluviométrie.

udult, ultisol de climat humide (*pédol.*).

Ulatisian (N. Am.), Ulatisien (étage, Éocène moy.).

ulexite, ulexite (*minér.*).

Ulsterian (N. Am.), Ulstérien (série, Dévonien inf.).

ultimate base level, niveau de base final; **u. load,** charge limite; **u. recovery,** production totale (du début à la fin d'un puits); **u. strength,** résistance limite (à la rupture); **u. bending resistance,** limite à la rupture par flexion.

ultisol, sol rouge d'altération (ferrisol), sol à B argillique (*pédol.*).

ultrabasic, ultrabasique (*pétro.*).

ultrabasite, ultrabasite.

ultrahaline, hypersalin.

ultramafic, ultrabasique, ultramafique.

ultramafites, roches ultrabasiques.

ultramafitolites, roches ultrabasiques.

ultrametamorphic rock, roche ultramétamorphique.

ultrametamorphism, anatexie.

ultramylonite, ultramylonite.

ultrared, infrarouge.

ultrasonics, méthode des ultrasons.

ultrasound, ultrason.

ultrastructure, microstructure.

umber, ombre, terre d'ombre, terre de Sienne.

umbilicate, ombiliqué.

umbilicus, umbilic (*pal.*).

umbo, crochet.

umbonal region, région du crochet (*pal.*).

umbrept, ranker, sol brun à horizon umbrique (*pédol.*).

umlaufberg (U.S., germ.), colline isolée par l'érosion fluviatile.

unaffected, inaltéré.

unalterable, inaltérable.

unaltered, inaltéré, non altéré.

unary system, système à un seul composant.

unassisted eye, à l'œil nu.

unattackable, inattaquable (à l'acide).

unbalanced, non équilibré (chimie).

unbedded, non stratifié.

unbroken ore, minerai non abattu.

uncap (to), découvrir, décaper (les morts-terrains).

uncased, non tubé (forage).

unchambered, sans loge, non cloisonné.

unchanged, inaltéré, non modifié.

uncoiling, déroulement.

unconfined aquifer, nappe libre.

unconformable, discordant; **u. bed,** couche d.; **u. stratification,** d. statigraphique.

unconformity, discordance; **u. by erosion,** d. d'érosion; **u. by dip,** d. angulaire; **u. of overlap,** transgression; **u. of transgression,** transgression; **u.-bounded units,** unités stratigraphiques limitées par des d.; **angular u.,** d. angulaire; **erosional u.,** d. d'érosion; **mechanical u.,** d. mécanique; **non-angular u.,** d. simple; **non-depositional u.,** lacune stratigraphique; **parallel u.,** d. parallèle, d. simple : strates parallèles; **stratigraphic u.,** d. stratigraphique; **structural u.,** d. angulaire; **topographic u.,** d. topographique.

unconsolidated, meuble, non consolidé, non cimenté.

uncover (to), découvrir, mettre à découvert (un filon, etc.).

uncovered area, région découverte, sans formations superficielles.

uncrystalline, amorphe.

unctuous clay, argile « grasse ».

uncut diamond, diamant non taillé.

unda, zone infratidale.

undaform zone, zone infratidale, deltaïque distale.

undation, ondulation terrestre à grand rayon de courbure; **u. theory,** théorie des ondulations de la croûte terrestre.

undecomposed, non décomposé.

underbed, couche sous-jacente, « mur ».

underclay, « mur » argilo-schisteux d'une couche de charbon (ancien sol).

undercliff (G.B.), falaise éboulée en masse.

undercool (to), surfondre.

undercooling, surfusion.

undercut, 1. adj : havé, sous-cavé; **2.** n : havage, sous-cavage; **u. etching,** affouillement inférieur; **u. slope,** versant de lobe concave de méandre.

undercut (to), sous-caver, haver.

undercutter, 1. machine : haveuse, déhouilleuse; **2.** personne : haveur.

undercutting, sapement, excavation, havage.

underearth, 1. couche d'argile formant le mur d'une couche de charbon; **2.** sol sous la surface; **3.** profondeur de la terre.

underfit river, rivière inadaptée.

underflow, sous-écoulement.

underfold, pli secondaire.

underground, 1. adj : souterrain; **2.** adv : souterrainement; **u. drainage,** drainage s.; **u. hands,** ouvriers du fond; **u. haulage,** roulage s.; **u. staff,** personnel au fond; **u. storage,** stockage s.; **u. stream,** cours d'eau s.; **u. tapping,** dérivation souterraine (karstique); **u. water,** eaux souterraines, nappe souterraine; **u. workings,** travaux de fond; **u. workman,** ouvrier de fond.

underhand stope, gradin droit (mine).

underhand stoping, abatage, dépilage par gradins droits, abatage descendant.

underhole (to), haver, sous-caver.
underlay, inclinaison (par rapport à la verticale).
underlay (to), être incliné par rapport à la verticale.
underlay shaft, puits incliné.
underlie (to), être sous-jacent à.
underlier (G.B.), puits incliné.
under limb, flanc intérieur (d'un pli déversé).
underlying, sous-jacent; **u. bed**, couche sous-jacente; **u. rock**, soubassement.
undermine (to), caver, sous-caver, miner, saper.
underplate (to), être sous-jacent à une plaque et remonter.
underplating, remontée de magma sous les plaques.
underream (to), élargir un trou tubé.
underreamer, trépan élargisseur.
undersaturated rock, roche non saturée (sous-saturée).
undersea, sous-marin.
under seam, filon inférieur, profond.
underset, **1.** adj : déversé (*mine*); **2.** n : déversement.
underside, lèvre inférieure, face inférieure.
undersoil, sous-sol.
understratum, couche inférieure.
underthrust, sous-charriage.
undertow, courant de retour, flot de fond.
underwater, sous l'eau, marin.
undetected layer, couche non détectée (*géoph.*).
undeveloped, non développé, non exploité.
undiscovered resources, gisements présumés, non identifiés.
undissolved, non dissous.
undrained, non drainé.
undressed stone, pierre non taillée.
undulating country, région accidentée.
undulation, **1.** ondulation, accident de terrain; **2.** ondulation (*physique*).
undulatory, ondulatoire (*physique*); **u. extinction**, extinction onduleuse, roulante.
unearth (to), déterrer, découvrir.
unequigranular, à grain inégal.
uneven, inégal, irrégulier, accidenté, anfractueux.
unevenness, inégalité, anfractuosité.
unexhausted, inépuisé.
unexploited, inexploité.
unexplored, inexploré.
unfailing spring, source intarissable.
unfilled, vide, non remblayé.
unfossiliferous, non fossilifère.
unfossilized, non fossilisé.
ungaite, ungaïte (var. de dacite).
Ungulate, Ongulé (*pal.*).
Unguligrade, Onguligrade.
unhewn stone, pierre brute, non taillée.
uniaxial, uniaxe.
unicellular, unicellulaire (*pal.*).
uniclinal, monoclinal; **u. fold**, flexure, pli m.
uniformitarianism, uniformitarisme, actualisme (théorie de).
unilocular, uniloculaire, à une seule loge.
unindurated, non consolidé.
uniserial, unisérié (*pal.*).

unit, **1.** unité de mesure; **2.** installation, unité de fabrication; **u. cell**, maille d'un réseau cristallographique.
univalent, monovalent.
univalve, univalve (*pal.*).
universal stage, platine universelle.
unkindly lode (Austr.), filon peu prometteur.
unlined, sans revêtement, non boisé.
unlithified, meuble.
unmelted, non fondu.
unmined, non exploité.
unmix (to), ne pas se mélanger, faire exsolution.
unmixing, exsolution.
unoriented texture, texture non orientée.
unoxydized, inoxydé, inaltéré.
unpenetrated, non traversé, non perforé.
unpolished stone, pierre non polie.
unproductive, improductif, stérile.
unproved territory, région non étudiée par forages.
unram (to), débourrer.
unrecoverable, irrécupérable (pétrole).
unrefined, non raffiné.
unrest caldera, caldeira en activité.
unroasted ore, minerai non grillé.
unroofed arche, anticlinal érodé.
unsaturated, non saturé; **u. hydrocarbon**, hydrocarbure non saturé; **u. zone**, zone non saturée (au-dessus de la nappe phréatique).
unscreened, non criblé, non tamisé.
unsealed, non fermé, non étanche.
unset (to), démonter, dessertir (un diamant).
unsettled, trouble, non sédimenté.
unsifted, non tamisé.
unslaked lime, chaux vive.
unsoiling, découverte (*mine*), découverture, enlèvement du terrain de couverture.
unsolidified, non solidifié.
unsorted, non trié, tout venant.
unstable equilibrium, équilibre instable.
unsteady, irrégulier (écoulement).
unstem (to), débourrer.
unstrained, non comprimé.
unstratified, non stratifié.
unsymmetric, unsymetrical, asymétrique, dissymétrique (pli).
untimber (to), déboiser (mine).
untop (to), enlever les terrains superficiels, découvrir.
unwashed, non lavé.
unwater (to), assécher (mine).
unweathered, non altéré.
unworkable, inexploitable.
unworked, **1.** inexploité; **2.** non étudié.
unwrought, non façonné, inexploité.
uparching, courbure vers le haut, voussure, bombement.
up block, compartiment soulevé (tectonique).
upburst (rare), éruption.
upcast, soulevé; **u. side**, lèvre de faille soulevée; **u. shaft**, puits de retour d'air; **u. ventilation**, sortie d'air.

up-current, vers l'amont, amont.

up-dip, amont-pendage.

updoming, bombement, renflement.

up-down criteria, critères de polarité verticale (*séd.*).

up-fault, faille inverse.

up-fold, pli anticlinal, voûte.

upgrade (to), améliorer (la teneur).

upgrading, accroissement (artificiel) de teneur.

upheaval, soulèvement.

upland, hauteur, terrain élevé, moyenne montagne.

uplead, lèvre soulevée.

uplift, soulèvement, redressement, dôme, bombement; **u. block**, compartiment soulevé.

uplifted, soulevé; **u. side**, compartiment, bloc soulevé.

upper, supérieur; **u. bend**, charnière anticlinale; **u. layer**, couche s.; **u. side**, lèvre s.

upper Carboniferous (Eur.), Carbonifère sup. (~ Pennsylvanien).

upraise, remontée.

upraising, remontage, remontée (mine).

upright fold, pli droit.

uprising, soulèvement.

uprush, jet de rive.

up-shaft, puits de retour d'air (mine).

up-shoaling, haut-fonds.

up-side, lèvre soulevée.

upstanding, en relief, dégagé par l'érosion.

upstream, en amont.

upswelling, gonflement, bombement.

uptake shaft, puits de sortie d'air.

upthrown, soulevé (compartiment); **u. fault**, faille inverse; **u. lip**, lèvre soulevée; **u. side**, compartiment s.; **u. wall**, compartiment s.

upthrust, soulèvement.

uptrusion, intrusion vers le haut.

upturned, retourné, rebroussé.

upturning, rebroussement.

upward, **1.** adj : ascendant.

upward, **2.** n : bombement (anticlinal).

upwarping, bombement.

upwelling, remontée d'eau marine; **u. current**, courant marin ascendant; **u. spring**, source jaillissante.

up-wind side, face au vent (d'une dune).

Uralien (U.S.S.R.), Ouralien (étage, Carbonifère terminal).

uralite, ouralite (*minér.*); **u. diorite**, diorite ouralitisée; **u. gabbro**, gabbro ouralitisé.

uralitization, ouralitisation.

uranate, uranate.

uraninite, uraninite (pechblende).

uraniferous, uranifère.

uranite, uranite (*minér.*).

uranium, uranium; **u. galena**, galène uranifère; **u. minerals**, minéraux d'u.

uran mica, uranite.

uranophane, uranophane.

Urgonian (Fr.), Urgonien (faciès, Crétacé inf.).

Uriconian (G.B.), Uriconien (division du Précambrien).

urinestone, anthraconite.

urtite, urtite (var. d'ijolite).

use up (to), épuiser.

U-shaped profile, profil transversal en U; **U. valley**, vallée en U, d'origine glaciaire.

ustalf, sol fersiallitique ferrugineux.

ustert, vertisol de climat chaud, à saisons contrastées.

ustoll, chernozem méridional.

ustox, sol ferrallitique de climat chaud et sec.

ustult, ultisol de climat chaud.

uvala (Serbo-Croate), uvala : grande dépression fermée karstique.

uvarovite, ouvarovite (*minér.*).

V

vacancy, lacune (*cristallo.*).

vacant site, lacune (*cristallo.*).

vadose canyon, canyon, rivière souterraine (karst).

vadose water, eau vadose.

vadose zone, zone non saturée en eau (au-dessus de la nappe phréatique).

Valanginian, Valanginien (étage, Crétacé inf.).

vale, 1. vallon, vallée; **2.** gouttière, canal; **3.** dépression tectonique.

valence, valence.

Valentian (Llandoverian), Valentien (étage, Silurien inf.).

valentinite, valentinite (*minér.*).

valley, vallée; **v. bottom,** fond de v.; **v. drift,** matériaux fluvio-glaciaires; **v. fill,** remblayage d'une v. par des matériaux meubles; **v. flat,** fond de v.; **v. floor,** fond de v.; **v. glacier,** glacier de v. (alpin); **v. head,** partie amont d'une v.; **v. profile,** profil (longitudinal) d'une v.; **v. side,** versant de v.; **v. sink,** dépression allongée; **v. system,** réseau fluviatile; **v. tract,** cours moyen d'un fleuve; **v. rain,** moraine glaciaire; **antecedent v.,** v. antécédente; **anticlinal v.,** v. anticlinale, combe; **collapse v.,** v. d'effondrement; **consequent v.,** v. conséquente; **construction v.,** v. structurale; **destructional v.,** v. d'érosion; **dry v.,** v. sèche (sans cours d'eau), (périglaciaire, karstique); **entrenched v.,** v. encaissée; **epigenetic v.,** v. épigénétique; **fault v.,** v. de faille; **flat-floored v.,** v. en fond de bateau; **glacial v.,** v. glaciaire; **glaciated v.,** v. englacée; **hanging v.,** v. suspendue; **incised v.,** v. encaissée; **inner v.,** v. emboîtée; **inset v.,** v. emboîtée; **obsequent v.,** v. obséquente; **perched v.,** v. suspendue; **polycyclic v.,** v. polycyclique; **rift v.,** v. (ou fossé) d'effondrement; **subsequent v.,** v. subséquente; **superposed v.,** v. surimposée; **surimposed v.,** v. surimposée; **transverse v.,** v. transversale; **trough v.,** v. en auge; **underfit v.,** v. inadaptée; **U-shaped v.,** v. en U; **wide v.,** v. évasée.

Valmeyeran (U.S.A. illinois), série de Valmeyeran (Mississipien).

value of an ore, richesse d'un minerai.

valvate, valvé, possédant une valve (*pal.*).

valve, valve (*pal.*).

van, 1. pelle à vanner, van; **2.** essai de valeur d'un minerai (G.B. Cornouailles).

van (to), vanner (le minerai).

vanadate, vanadate.

vanadic, vanadique; **v. ocher,** vanadinite.

vanadiferous, vanadifère.

vanadinite, vanadinite (groupe des apatites).

vanadium, vanadium.

vanthoffite, vanthofite (*minér.*).

vapor, vapour, vapeur; **v. phase,** phase v.; **v. pressure,** pression de v.; **v. tension,** pression de v.; **v. tight,** étanche à la v.

vaporizable, vaporisable.

vaporization, vaporisation; **v. specific temperature,** chaleur spécifique de v.

vaporize (to), vaporiser, évaporer.

variation, variation, écart; **v. of the compass,** déclinaison magnétique; **v. recording,** enregistrement des variations; **abnormal v.,** anomalie magnétique.

variegated, bariolé, bigarré; **v. copper ore,** cuivre panaché, bornite; **v. pyrite,** bornite, bornine.

variolated structure, structure variolitique (dans roche ignée basique).

variolite, variolite.

variolite structure, structure variolitique.

variometer, variomètre, magnétomètre; **gravity v.,** variomètre de gravité, gravimètre; **magnetic v.,** variomètre magnétique, magnétomètre.

Variscan, Variscian orogeny, orogénèse Varisque : Carbonifère et Permien (équiv. : orogénèses hercynienne, armoricaine, des Altaïdes).

vari-size grained, de granulométrie variable.

varve (swedish), varve : dépôt glacio-lacustre; **v. chronology,** chronologie suivant le comptage des varves.

varved, varvé, à varves; **v. clay,** argile à v.; **v. slate,** schiste v. (argile à v. métamorphisée).

varvity, disposition en varves.

varying in grade, à teneur variable.

vascular plant, plante vasculaire (*pal.*).

vat, cuve, bac.

vauclusian spring, source vauclusienne.

vault, voûte.

V bar, cordon littoral en V (en pointe de flèche).

Vectian (Aptian), Vectien (étage, Crétacé inf.).

vegetable soil, terre végétale.

vein, veine, filon; **v. breccia,** brèche filonienne; **v. dyke,** dyke instrusif; **v. filling,** remplissage de filon; **v. gold,** or filonien; **v. matter,** remplissage de filon; **v. mineral,** minéral filonien; **v. mining,** exploitation filonienne; **v. ore,** minerai en filon; **v. wall,** salbande.

veined, veiné.

veinlet, petit filon, veinule.

veinstone, gangue.

veinstuff, gangue.

veinule, veinule, filet.

veiny, veiné.

velocity, vitesse; **v. change,** variation de v.; **v. determination,** calcul de la v.; **v. discontinuity,** discontinuité de v. sismique; **v. function,** loi de v.; **v. hole,** forage pour sismo-sondage; **v. log,** diagraphie de v.; **v. of flow,** débit; **v. of propaga-**

tion, v. de propagation; **v. survey,** sismo-sondage; **propagation v.,** v. de propagation; **terminal v.,** v. limite.

veneer, plaquage, pellicule.

Veneris crinis, cheveux de Vénus, sagénite.

vent, **1.** orifice volcanique, cheminée volcanique; **2.** ouverture, évent; **v. breccia,** brèche volcanique; **v. hole,** évent; **v. hotter-smoker,** source chaude avec émanations gazeuses (fond du Pacifique), « fumeur » noir; **eruption v.,** orifice éruptif.

ventifact, caillou façonné par le vent.

ventilate (to), ventiler, aérer (mine).

ventilating shaft, puits d'aérage.

ventilation, aérage, aération (mine); **v. bore hole,** trou d'aération; **v. funnel,** cheminée d'aération; **v. pipe,** conduit d'aération; **v. shaft,** puits d'aération.

Venturian (N. Am.), Venturien (étage, Pliocène moy.).

Venus' hair crystals, cristaux filamenteux de rutile (dans cristaux de quartz).

venusian geology, géologie de la planète Vénus.

vergence, vergence (*tecto.*).

vermicular quartz, quartz de corrosion.

vermiculite, vermiculite (argile).

vernier calipper, pied à coulisse.

versant, versant (géomorph.).

Vertebrate, Vertebrata, Vertébré.

vertical, vertical; **v. closure,** fermeture structurale; **v. deformation,** amplitude de déplacement; **v. dip,** pendage v.; **v. displacement,** rejet v.; **v. exaggeration,** rapport de l'échelle v. d'une coupe à l'échelle horizontale; **v. fault,** faille v.; **v. photograph,** photographie aérienne v.; **v. scale,** échelle des hauteurs; **v. section,** coupe v.; **v. separation,** rejet v.; **v. shift,** composante v. du rejet; **v. throw,** rejet v.; **v. time propagation,** temps de propagation v.; **v. well,** forage v.

verticals, photographies aériennes verticales.

vertisol, vertisol (*pédol.*).

vesicle, vésicule, vacuole, cavité, vide (des roches magmatiques).

vesicular, vésiculaire, vacuolaire.

vesiculation, **1.** structure vacuolaire (des laves); **2.** exsolution.

Vesulian (G.B.), Vésulien (étage, Jurassique moy.).

vesuvian garnet (old term), leucite.

vesuvianite, vésuvianite, idocrase (*minéral.*).

vibrating screen, tamis vibrant.

vibrational, vibratoire (mouvement).

vibro-classifier, vibro-classeur.

vibroseis, vibrosismic, méthode vibrosismique.

Vicksburgian (N. Am.), Vicksburgien (étage, Oligocène moy.).

view, vue; **front v.,** vue de face; **sectional v.,** vue en coupe; **top v.,** vue d'en haut.

viewing, visée.

Villafranchian (Eur.), Villafranchien (étage continental, équiv. de Calabrien, Pléistocène inf.).

Vindobonian, Vindobonien (étage, Miocène moy.).

Vinean, Vinéen (sous-étage, Carbonifère sup.).

virgation, virgation (*géomorph.; tectonique*).

Virgilian (N. Am.), série de Virgilian (Pennsylvanien terminal).

virgin, natif, pur, vierge; **v. field,** gisement vierge; **v. gold,** or natif; **v. land,** terre vierge.

Virglorian, Virglorien = Anisien (étage, Trias moy.).

Virgulian, Virgulien (sous-étage, Kimméridgien).

viscoelastic, viscoélastique.

viscoelasticity, viscoélasticité.

viscoplasticity, viscoplasticité.

viscosimeter, viscosimètre.

viscosity, viscosité; **v. chart,** abaque de v.; **v. coefficient,** coefficient de v.; **v. gage,** jauge de v.; **v. index,** indice de v.

viscous, visqueux; **v. flow,** écoulement v.; **v. remanent magnetization,** aimantation rémanente visqueuse.

Visean (Eur.), Viséen (étage, Carbonifère inf.).

Vishnu (N. Am.), schistes de Vishnu (groupe du Précambrien).

visor, encorbellement.

visual examination, étude à l'œil nu.

vitrain, vitrain (cf. charbon).

vitreasity, vitrosité.

vitreous, vitreux; **v. copper,** chalcosine; **v. luster,** éclat v.; **v. silver,** argentinite.

vitric, vitreux; **v. tuff,** cinérite (de fragments de verre).

vitrification, vitrification.

vitrinite, vitrinite (cf. charbon).

vitriol, vitriol; **blue v.,** sulfate de cuivre; **green v.,** sulfate de fer; **white v.,** sulfate de zinc.

vitrite, vitrite (cf. charbon).

vitroclastic, structure bréchique (à fragments vitreux).

vitrophyre, vitrophyre (roche porphyrique à pâte vitreuse).

vitrophyric, vitrophyrique.

vivianite, vivianite (*minér.*).

vlei, vley, vliy (etym. Dutch), petit marécage (U.S.A., Afrique du Sud).

void, vide, pore; **v. ratio,** taux de porosité.

voidage, porosité.

volatile, volatil; **v. combustible,** gaz combustible (de charbon).

volatiles, composants volatifs résiduels du magma.

volatilization, volatilisation.

volatilize (to), volatiliser.

volcanello, petit volcan (souvent secondaire).

volcanic, volcanique; **v. activity,** activité v.; **v. ash,** cendre v.; **v. belt,** zone v.; **v. bomb,** bombe v.; **v. breccia,** brèche v.; **v. chimney,** cheminée v.; **v. conduit,** cheminée v.; **v. cone,** cône v.; **v. conglomerate,** tuf v.; **v. crater,** cratère v.; **v. debris flow,** lahar; **v. dome,** dôme v.; **v. ejecta,** projections v.; **v. ejecta blanket,** couverture pyroclastique; **v. eruption,** éruption v.; **v. flow,** coulée de laves; **v. foam,** ponce v.; **v. front,** arc v.; **v. funnel,** cheminée v.; **v. gases,** gaz v.; **v. glass,** verre v., obsidienne; **v. intrusion,** intrusion v.; **v. lake,** lac de barrage v.; **v. mudflow,** lahar; **v. neck,** culot v.; **v. plug,** culot v.; **v. scoria,** scories v.; **v. sink,** caldeira; **v. slag,** scories v.; **v. spine,** aiguille v.; **v. tuff,** tuf v.; **v. vent,** orifice v.

volcanicite, volcanicité, volcanisme.

volcaniclastic, volcanoclastique.

volcaniclastics, dépôts volcanoclastiques.

volcanist, vulcanologue.

volcanize (to), volcaniser.

volcano, volcanoe, volcan; **abortive v.,** volcan avorté; **actif v.,** volcan actif; **central v.,** volcan puncti-forme; **dormant v.,** volcan assoupi; **embryonic v.,** volcan embryonnaire; **explosive v.,** volcan de type explosif; **extinct v.,** volcan éteint; **mixed v.,** volcan mixte; **monogenic v.,** volcan monogénique; **mud v.,** volcan de boue; **stratified v.,** volcan stratifié, stratovolcan.

volcanogenic, d'origine volcanique.

volcano-karst, formes karstiques dans roches vol-canoclastiques, en tufs.

volcanologist, vulcanologist, vulcanologue.

volcanology, vulcanologie.

volcano-tectonic depression, caldeira, dépression volcano-tectonique.

volume, volume; **v. control,** réglage de gain (sismi-que); **v. flow rate,** débit volumique; **v. susceptibility,** susceptibilité magnétique, taux de magnétisation; **river v.,** débit; **tidal v.,** débit du flot ou du jusant.

voog, géode, druse, cavité cristallisée.

vortical, tourbillonnaire.

vortice, tourbillon.

vosegite, voségite.

vough (obsolete, see vug), druse, géode.

Vraconian, Vraconien (étage, Crétacé sup. basal).

V-shaped valley, vallée en V.

vug, vugh, druse, géode, poche à cristaux.

vuggy fabric, structure vésiculaire.

vugular, vacuolaire, caverneux.

vulcanian, vulcanien, plutonien.

vulcanism, volcanisme.

vulcanorium, vulcanorium.

W

Waalian, Waalien (interglaciaire Donau-Günz).

wacke, grès avec 15 à 75 % de matrice boueuse.

wad, wad **1.** (hydroxyde de manganèse); **2.** pl. wadden (hollandais) : estran.

wadi (arabic) (pl. wadis, etc.), oued (Afrique du nord).

wake, remous.

walk a bed (to), suivre un banc.

walking beam, balancier (pétrole), levier de battage.

wall, 1. paroi, versant, éponte; **2.** salbande; **3.** mur (d'une couche); **4.** lèvre de faille; **w. face,** front de taille; **w. moraine,** rempart morainique; **w. pillar,** pilier du mur (mine); **w. rock,** roche encaissante; **w. salpetre,** nitrate de calcium; **w. sample,** carotte latérale; **back w.,** mur de rimaye; **boulder w.,** levée de blocs glaciaires; **crater w.,** paroi du cratère; **foot w.,** mur, lèvre inférieure de faille; **hanging w., 1.** compartiment soulevé; **2.** toit d'une faille oblique; **head w.,** mur de rimaye; **morainic w.,** rempart morainique; **ring w.,** enceinte, vallum volcanique.

Wallachian orogeny, orogénèse Wallache (Pliocène terminal).

walling, revêtement de puits.

wandering, migration, déplacement, méandrisation; **w. coal,** petite couche de charbon (Écosse); **w. dune,** dune mouvante; **polar w.,** migration des pôles.

waning flood, fin de crue.

want, wants (G.B.), lacune sédimentaire ou d'érosion par un chenal (avec colmatage par d'autres matériaux).

warner, détecteur de grisou, indicateur de grisou, grisoumètre.

warning, signe avant-coureur d'un séisme.

warp, 1. gauchissement, courbure, flexure; **2.** dépôt vaseux alluvionnaire, limon pédologique, limon périglaciaire.

warp (to), 1. déformer, déjeter, déverser; **2.** se déformer.

warpage, gauchissement, déformation.

warped, déformé, gauchi, ondulé, plissoté.

warping, gauchissement, déformation, ploiement, flexure, plissement; **w. up,** bombement.

Wasatchian (U.S.), Wasatchien (= Yprésien).

wash, 1. produits de lavage, produits lavés; **2.** U.S. : dépôt meuble détritique superficiel; **3.** U.S. : gravier aurifère; **4.** U.S. alluvions grossières, cône alluvial; **5.** limon de ruissellement; **6.** dégradation, affouillement; **7.** cours d'eau éphémère, oued, vallée désséchée (U.S.A.); **w. boring,** forage à injection; **w. dirt,** terrain, gravier aurifère; **w. fault (G.B.),** filon de charbon remplacé par des schistes; **w. gravel,** graviers d'alluvions à laver; **w. load,** limon fin; **w. pan,** batée; **w. plain,** plaine fluvio-glaciaire; **rain w.,** ruissellement diffus; **rill w.,** ruissellement en filets; **sheet w.,** ruissellement en nappe.

wash (to), 1. laver, débourber; **2.** affouiller (les berges); **w. away (to),** emporter; **w. out (to),** épuiser.

washboard moraine, moraine bosselée (étym. : en planche à laver).

washed moraine, moraine délavée.

washerman, laveur, débourbeur.

washery, atelier de lavage, installation de lavage.

washing, 1. lavage, débourbage; **2.** chantier de lavage; **w. dish,** batée de lavage; **w. drum,** tambour laveur (trommel débourbeur); **w. of ore,** lavée de minerai; **w. stuff,** terrain aurifère à laver; **w. table,** table de lavage.; **w. trommel,** trommel débourbeur; **gold-w.,** chantier de lavage de minerai aurifère.

Washitan (N. Am.), Washitien (étage, limite Crétacé inf. à sup.).

washlands (wetlands), terres inondables.

washout, chenal, ou paléochenal sédimentaire, poche de dissolution, partie érodée d'une couche de charbon.

washover, 1. surforage; **2.** delta intérieur dans un lagon; **w. fan,** delta marin intralagunaire (sédiments marins déposés par des tempêtes à l'intérieur des terres).

washover (to), surforer.

wastage, 1. ablation; **2.** érosion, ou ablation, glaciaire; **3.** décrue; **4.** perte; **w. area,** zone d'ablation nivale.

waste, 1. débris, déchets, rebuts stériles; **2.** remblai, vieux travaux; **3.** perte; **w. chute,** cheminée à remblais; **w. coal,** charbon récupéré sur les déblais; **w. dump,** terril, halde; **w. fill,** remblayage; **w. gas,** gaz brûlés; **w. heap,** halde; **w. material,** déchets; **w. ore,** minerai de rebut; **w. plain,** nappe alluviale de piedmont, talus d'éboulis, glacis montagneux; **w. rocks,** déchets stériles; **w. shaft,** puits de remblai.

wasting, gaspillage; **w. process,** processus destructeur.

water, eau; **w. adit,** galerie d'écoulement; **w. authority,** agence de bassin; **w. balance,** équilibre hydrique; **w. bearing,** aquifère; **w. bearing formation,** formation aquifère; **w. biscuit,** concrétion lacustre algaire; **w. body,** nappe d'e.; **w. budget,** bilan de l'e; **w. cement,** ciment hydraulique; **w. concentration,** concentration de l'e. de minerai; **w. content,** teneur en e.; **w. course,** cours d'e., drain, chenal; **w. cycle,** cycle hydrologique; **w. divide,** bassin hydrographique; **w. eye,** petite dépression d'altération dans une roche éruptive; **w. feader,** poche d'e.; **w. flooding,** injection d'e.

dans un forage; **w. flush drilling**, sondage à injection d'e.; **w. free**, sec, anhydre; **w. gauge**, indicateur de niveau, échelle d'étiage; **w. gap**, cluse, vallée transversale; **w. grade**, pente d'un drain, niveau d'e.; **w. hardness**, dureté de l'e.; **w. hole**, 1. taffoni; 2. trou d'e. sur glace; **w. humus**, matière organique aquatique; **w. inflow**, venue d'e.; **w. injection**, injection d'e.; **w. insoluble**, insoluble dans l'e.; **w. jump**, chute d'e.; **w. layer weathering**, altération par l'e. de mer; **w. level**, 1. niveau hydrostatique; 2. niveau de l'e.; 3. niveau d'e. (instrument); 4. galerie d'écoulement (mine); **w. lime**, chaux hydraulique; **w. load**, pression hydrostatique; **w. logged bed**, couche imbibée d'e.; **w. management**, gestion des ressources en e.; **w. mark**, laisse de marée (sur plage); **w. notch**, encoche de sapement; **w. wave**, vague; **w. of constitution**, e. de constitution; **w. of dehydratation**, e. de constitution; **w. of external origin**, e. d'origine externe; **w. of hydration**, e. d'hydratation; **w. of imbibition**, 1. e. de saturation; 2. e. de carrière; **w. of retention**, e. connée; **w. of saturation**, e. de saturation; **w. oil contact**, contact e.-pétrole; **w. opal**, hyalite; **w. pack**, remblai hydraulique; **w. packing**, remblayage; **w. parting**, ligne de partage des e.; **w. plane**, sommet d'une nappe phréatique; **w. pocket**, taffoni, trou dans la roche (le long de cours d'e.); **w. pollution**, pollution des e.; **w. power**, force hydraulique; **w. proof**, imperméable, hydrofuge; **w. recession**, tarissement de l'e.; **w. rolled**, roulé par les e.; **w. sand**, sable aquifère; **w. sapphire**, cordiérite; **w. saturation**, saturation en e.; **w. shaft**, puits d'exhaure; **w. sink**, marmite de géant; **w. slip**, fissure aquifère; **w. soluble**, soluble dans l'e.; **w. source**, source d'e.; **w. spout**, trombe marine; **w. supply**, apport en e., ressources en e.; **w. surface**, contact e.-pétrole (forage); **w. table**, surface de nappe d'e. (phréatique); **w. table rock**, roche formée au niveau de la nappe phréatique; **w. tight**, étanche; **w. tightness**, étanchéité; **w. vapour**, vapeur d'e.; **w. vein**, petit cours d'e. souterrain, fissure aquifère; **w. wash**, lavage à l'e. d'un minerai; **w. well**, puits produisant de l'e.; **w. witch**, sourcier; **w. worn**, usé par l'e.; **atmospheric w.**, e. atmosphérique; **artesian w.**, e. artésienne; **brackish w.**, saumâtre; **capillary w.**, e. capillaire; **confined ground w.**, e. captive; **connate w.**, e. connée, e. de constitution; **film w.**, e. pelliculaire; **fresh w.**, e. douce; **ground w.**, e. souterraine, de fond; **hygroscopic w.**, e. hygroscopique; **infiltration w.**, e. d'infiltration; **interstitial w.**, e. interstitielle; **juvenile w.**, e. juvénile; **karst w.**, e. karstique; **magmatic w.**, e. magmatique; **mean w.**, e. moyenne; **melt w.**, e. de fusion; **meteoric w.**, e. atmosphérique, météorique; **mineral w.**, e. minérale; **native w.**, e. d'origine; **percolation w.**, e. d'infiltration; **phreatic w.**, e. phréatique; **quarry w.**, e. de roche, de carrière; **rain w.**, e. de pluie; **resurgent w.**, e. résurgente; **run off w.**, e. de ruissellement; **running w.**, e. courante; **salt w.**, e. salée; **slack w.**, étale

(de marée); **spring w.**, e. de source; **standing w.**, e. stagnante; **subsurface w.**, e. souterraine, e. lithosphérique (U.S.); **sulfurous w.**, e. sulfureuse; **surface w.**, e. superficielle, e. hydrosphérique (U.S.); **suspended w.**, e. vadose; **resurgent w.**, e. de résurgence; **vadose w.**, e. vadose, e. de percolation.

water (to), 1. arroser; 2. diluer avec de l'eau; 3. apporter de l'eau.

watered area, région arrosée.

water-bearing, aquifère; **w. bed**, terrain a.; **w. strata**, terrain a.

waterfall, chute d'eau.

waterless, sans eau, dépourvu d'eau.

waterlogging, engorgement du sol par l'eau.

waterproof (to), imperméabiliser.

watershed, 1. ligne de partage des eaux; 2. bassin versant (U.S.), bassin hydrographique.

waterway, 1. cours d'eau; 2. voie d'eau; 3. voie d'eau navigable.

watery, humide, aqueux, aquifère.

Waucoban (N. Am.), série de Waucoban (Cambrien inf.).

wave, 1. onde; 2. vague; **w. amplitude**, amplitude d'une vague; **w. attack**, érosion par les vagues; **w. attenuation**, amortissement (d'une onde); **w. base**, niveau de base des vagues; **w. built terrace**, terrasse construite, d'accumulation; **w. camber**, cambrure de la vague; **w. crest**, crête de vague; **w. current**, courant de vagues; **w. current ripple mark**, ride de courant; **w. cut bench**, banquette littorale; **w. cut notch**, entaille de sapement; **w. cut shelf**, plate-forme d'abrasion; **w. cut terrace**, plate-forme littorale d'érosion; **w. delta**, delta de tempête, delta intérieur; **w. depht**, profondeur d'action des vagues; **w. front**, front d'onde, surface d'onde; **w. height**, hauteur de la vague; **w. hollow**, creux de vague; **w. impact**, impact, attaque des vagues; **w. incidence**, incidence des vagues; **w. length**, longueur d'onde, des vagues; **w. load**, charge des vagues; **w. mark**, 1. laisse de marée; 2. trace des vagues; **w. path**, trajet des ondes; **w. period**, période des vagues; **w. platform**, terrasse d'abrasion; **w. propagation**, propagation des ondes; **w. ramp**, talus de déferlement; **w. ray**, orthogonale de houle; **w. refraction**, réfraction d'ondes; **w. ripple-marks**, rides symétriques; **w. set**, train de vagues; **w. spilling**, déversement de la vague; **w. steepness**, rapport entre la hauteur des vagues et leur longueur d'onde; **w. train**, 1. train de vagues; 2. train d'ondes; **w. trough**, creux de vague; **w. velocity**, vitesse d'une onde; **airborne w.**, onde sonore; **breaking w.**, vague déferlante; **collapsing w.**, vague déferlante; **compressional w.**, onde de compression; **current w.**, vague de courant; **direct w.**, onde directe; **distortional w.**, onde de distorsion; **earthquake w.**, onde sismique; **longitudinal w.**, onde longitudinale; **plunging w.**, vague déferlante; **primary w.**, onde primaire; **reflected w.**, onde réfléchie; **seaquakes w.**, ondes sismiques; **standing w.**, onde stationnaire; **storm w.**, vague de tempête; **translation w.**, onde de translation; **wind-driven w.**, vague de vent; **refracted w.**, onde réfractée; **secondary w.**, onde secondaire; **seismic w.**, onde sismique; **sound w.**, onde sonore; **transverse w.**, onde transversale.

wavellite, wavellite (*minéral.*).

wavy, ondulé, onduleux; **w. extinction**, extinction ondulante, roulante; **w. vein**, filon d'épaisseur irrégulière.

wax, paraffine, cire minérale; **w. coal**, lignite; **w. opal**, opale jaune à éclat cireux; **w. shale**, schiste kérobitumineux; **w. tailings**, résidu de distillation du pétrole; **mineral w.**, cire minérale; **paraffin w.**, paraffine; **shale w.**, paraffine de schiste.

way, 1. galerie, chemin; 2. méthode, procédé; **w. shaft**, descenderie (*mine*).

way up criteria, critères de polarité verticale (de stratification).

weak solution, solution diluée.

weakness planes, plans, zones de faiblesse (mécanique).

Wealdian, Wealdien.

Wealden (G.B.), Crétacé inférieur fluviatile.

wear away (to), s'user, se ronger, se creuser, s'éroder.

wear resistance, résistance à l'usure.

weather, temps, intempérie; **w. conditions**, conditions météorologiques; **w. chart**, carte météorologique; **w. glass**, baromètre; **w. map**, carte météorologique; **w. pit**, cupule de dissolution; **w. side**, face exposée à l'altération; **w. stain**, coloration due à l'altération superficielle; **w. station**, station météorologique.

weather (to), 1. altérer; 2. s'altérer, se désagréger.

weatherable, altérable.

weathered granite, granite décomposé, désagrégé.

weathering, altération par les agents atmosphériques, altération climatique, désagrégation et altération; **w. agent**, agent d'altération; **w. correction**, correction d'altération (sismologie); **w. deposits**, dépôts d'altération; **w. index**, indice d'altération; **w. layer**, couche altérée superficielle à faible vitesse sismique; **w. profile**, profil d'altération; **w. velocity**, vitesse sismique dans la zone superficielle d'altération; **w. zone**, zone d'altération; **cavernous w.**, altération sous-cutanée, taffoni; **honeycomb w.**, désagrégation alvéolaire, taffoni; **mechanical w.**, désagrégation mécanique; **salt w.**, haloclastie, désagrégation saline; **spheroidal w.**, décomposition en boules, exfoliation.

webbed eyepiece, oculaire à réticule.

websterite, webstérite, aluminite.

wedge, coin, biseau; **w. work of roots**, travail de fissuration des racines; **ice w.**, fente de gel « en coin » à remplissage de glace; **loess w.**, ancienne fente de gel à remplissage de limon.

wedge down the face (to), faire tomber le front de taille au moyen de coins (*mine*).

wedge out (to), se terminer en biseau.

wedging out, amincissement en biseau.

wedgework of ice, fissuration par la glace.

weeping rock, roche suintante.

Weichselian (Eur.), Weichsélien (équiv. Wurmien ou Wisconsinien).

weight, poids, charge; **atomic w.**, poids atomique; **breaking w.**, charge de rupture; **molecular w.**, poids moléculaire; **specific w.**, poids spécifique, densité.

weir, barrage, déversoir, réservoir.

welded pumice, welded tuff, ignimbrite, dépôts pyroclastiques lithifiés.

welding, lithification de dépôts de nuées ardentes.

well, puits, forage; **w. bore**, puits de forage; **w. boring**, forage de puits; **w. casing**, tubage de puits; **w. core**, carotte; **w. drilling**, forage de puits; **w. head, well pressure**, pression en hauts de puits; **w. log**, diagraphie de forage; **w. logging**, diagraphie de forage; **w. sample**, échantillon de forage; **w. screen**, crépine; **w. shooting**, carottage sismique; **w. sinking**, fonçage de puits; **w. ties**, corrélations des données sismiques et géologiques de forage; **w. tubing**, tubage de puits; **w. velocity survey**, carottage sismique; **artesian w.**, puits artésien; **depleted w.**, puits épuisé; **exploration w.**, puits d'exploration; **flowing w.**, puits jaillissant; **gas w.**, puits de gaz; **gushing w.**, puits jaillissant; **pumping w.**, puits semi-artésien; **input w.**, puits d'injection; **natural w.**, puisard; **oil w.**, puits de pétrole; **water w.**, puits d'eau; **wild w.**, puits à jaillissement non maîtrisé.

well-graded soil, sol bien calibré.

well-rounded pebble, galet bien arrondi.

well-sorted, bien trié (séd.).

well up (to), sourdre (eau).

welt, bombement, géanticlinal.

Wemmelian (U.S.), Wemmelien (= Éocène supérieur).

Wenlockian, Wenlockien (étage, Silurien moy.).

Wentworth scale, échelle granulométrique logarithmique de Wentworth.

Werfenian (= Scythian), Werfénien (étage, Trias inf.).

wernerite, wernérite, scapolite (*minér.*).

westerly, ouest, de l'ouest, vers l'ouest; **w. wind, westerlies**, vent d'ouest (opp. alizé).

Westphalian (Eur.), Westphalien (étage, Carbonifère sup.).

wet, humide, mouillé; **w. analysis**, analyse par voie humide; **w. crushing**, broyage à l'eau; **w. essay**, essai par voie humide; **w. gas**, gaz humide; **w. grinding**, broyage à l'eau; **w. sorting**, triage à l'eau; **w. stamping**, bocardage à l'eau; **w. treatment**, traitement par voie humide.

wet (to), mouiller, humidifier.

wetlands, terres inondables, schorre.

wetness, humidité.

wettability, mouillabilité.

wetted, mouillé; **w. nappe**, nappe noyée; **w. perimeter**, périmètre m.

wetting agent, agent de mouillage.

whaleback, 1. roche moutonnée (géomorph. glaciaire); 2. relief allongé et arrondi; w. **dune,** mégadune longitudinale.

Wheelerian (N. Am.), Wheelérien (étage, Pliocène sup.) (équiv. Calabrien).

wheelerite, wheelérite (var. de résine).

whetstone, pierre à aiguiser.

whim-shaft, puits à cabestan.

whinstone, roche foncée (basalte, dolérite).

whirl, tourbillon, remous (*océano.*).

whirling pillars of dust, colonnes de poussières.

whirlpool, tourbillon.

whirlwind, tourbillon, trombe.

Whitbian (G.B.), Whitbien (étage, Jurassique inf. terminal).

white, blanc; w. **agate,** calcédoine; w. **antimony,** valentinite; w. **clay,** kaolin; w. **coal,** houille blanche; w. **cobalt,** smaltite; w. **damp,** oxyde de carbone; w. **frost,** gelée blanche; w. **garnet,** 1. var. de grenat grossulaire; 2. leucite; w. **iron ore,** sidérite; w. **iron pyrite,** marcasite, pyrite blanche; w. **jura,** jurassique supérieur; w. **lead ore,** cérusite; w. **lime,** lait de chaux; w. **mica,** mica b.; w. **olivine,** forstérite; w. **pyrite,** marcasite; w. **scale,** paraffine en écaille; w. **shorl,** albite; w. **tellurium,** 1. sylvanite; 2. krennerite.

whitecap, crête d'écume d'une vague, mouton.

Whiterock (U.S.A.), série de Whiterock (Ordovicien moy. basal).

Whitneyan (U.S.), Whitneyien (= Chattien).

whorl, spire (*pal.*).

Wichita Orogeny, orogénèse de Wichita (Mississipien à Pennsylvanien).

wide meshed, à grosse maille.

widen (to), 1. élargir, étendre; 2. s'élargir, s'étendre, s'évaser.

widening, élargissement.

Wilcoxian (N. Am.) (= Sabinian), Wilcoxien (étage Éocène).

wild, sauvage, non contrôlé, désordonné; w. **coal,** schiste interstratifié dans une couche de charbon; w. **lead,** blende; w. **sea,** mer démontée; w. **well,** puits à éruption non maîtrisée.

wildcat well, forage de reconnaissance.

wildfire, grisou, incendie, feu de mine.

wildflysch, wildflysch (dépôt d'avalanche sous-marine, au front des nappes de charriage).

willemite, willémite (minér.).

wimble, tarière à glaise.

win (to) ore, extraire du minerai.

winch, treuil; w. **for raising ore,** t. d'extraction.

wind, vent; w. **action,** éolisation; w. **blown sand,** sable éolien; w. **borne deposit,** dépôt éolien; w. **break,** brise-vent; w. **carved pebble,** galet éolisé à facettes; w. **carving,** corrosion éolienne; **corrasion,** érosion éolienne; w. **cut pebble,** galet à facettes; w. **deposit,** dépôt éolien; w. **dial,** anémomètre; w. **drift,** dépôt éolien; w. **erosion,** érosion éolienne; w. **faceted pebble,** galet éolisé à facettes; w. **gage, gauge,** anémomètre; w. **gap,** cluse sèche, abandonnée; w. **grinding,** usure éolienne; w. **polish,** poli éolien; w. **ripple,** ride éolienne; w. **scale,** échelle de vitesse des vents; w. **scouring,** décapage éolien; w. **shadow,** partie abritée des vents; w. **shaped,** façonné par le vent; w. **ward side,** face au vent; w. **wear,** usure éolienne; w. **worn pebble,** caillou éolisé.

w. **rose,** rose des vents; w. **scoured,** déblayé par le vent; w. **tide,** élévation du niveau d'eau par le vent.

wind (to), 1. extraire, enlever; 2. enrouler, dévider; 3. s'enrouler, se dérider; 4. serpenter (rivière).

winding, 1. adj : sinueux, en lacet; 2. n : extraction, remontée; 3. n : enroulement, bobinage; w. **engine,** machine d'extraction; w. **ore,** remontée de minerai; w. **rope,** câble d'extraction; w. **shaft,** galerie de descente, puits d'extraction.

windings, sinuosités, méandres, détours.

windkanter, caillou éolisé à facettes, caillou à poli éolien.

windlass, treuil.

windmill, éolienne.

window, fenêtre tectonique (dans une nappe de charriage).

windway, galerie d'aérage (mine).

wings, flancs (d'un anticlinal).

winning, 1. extraction, abatage; 2. champ d'exploitation; w. **gold,** extraction de l'or; w. **headway, level,** galerie de traçage.

winnowing, vannage, séparation (de particules de tailles différentes); w. **gold,** vannage de l'or, poussière d'or (par soufflerie).

winter dumps (U.S.), tas de minerai aurifère stocké en hiver pour traitement en été (Alaska).

winter moraine, moraine mineure de poussée hivernale.

winze, descenderie, descente.

wire rope, câble métallique.

Wisconsin (N. Am.), Wisconsin (dernière glaciation).

Wisconsinan, Wisconsinian, Wisconsinien (étage; Weichselien, Würmien, Pléistocène).

withdraw (to), extraire, enlever.

withdrawal, extraction, enlèvement; w. **of the ore,** enlèvement du minerai; w. **of the sea,** recul de la mer.

witherite, withérite.

within-plate basalt, basalte intra-plaque.

withstand a pressure (to), supporter une pression.

withstand (to), résister (à un séisme).

wold, 1. plateau, plaine ondulée; 2. U.S. : plateau terminé par une cuesta.

Wolfcampian (N. Am.), Wolcampien (série; Permien inf.).

wolfram, wolfram, tungstène; w. **ocher,** wolframocre, tungstite; w. **steel,** acier au tungstène.

wolframite, wolframite (*minéral.*).

wolframium, tungstène.

wollastonite, wollastonite (*minéral.*).

wood, 1. bois; 2. bois silicifié; w. **agate,** bois silicifié; w. **charcoal,** charbon de bois; w. **coal,** 1. charbon de bois; 2. lignite (U.S.A.); w. **copper,** olivénite fibreuse; **w. hematite,** hématite zonée; **w. iron,** sidérite fibreuse; **w. opal,** bois opalisé, bois silicifié; **w. resin,** résine naturelle; **w. rock,** amiante fibreuse; **w. stone,** bois pétrifié; **w. tin.** cassitérite finement cristallisée en nodules.

Woodbinian (N. Am.), Woodbinien (étage crétacé sup.).

work in depth, travaux en profondeur.

work (to), travailler, exploiter; **w. a coal seam (to),** exploiter une couche de charbon; **w. a quarry,** exploiter une carrière; **w. in gassy places (to),** travailler dans le grisou; **w. in the broken (to),** déhouiller, déplier les piliers; **w. open cast (to),** travailler à ciel ouvert; **w. out a mine (to),** épuiser une mine; **w. underground (to),** travailler au fond.

workability, exploitabilité.

workable, exploitable.

workableness, exploitabilité.

worked out, épuisé.

working, chantier, exploitation, abatage; **w. beam,** balancier (pétrole), levier de battage (mine); **w. face,** front de taille, front d'abatage; **w. a gold mine,** exploitation d'une mine d'or; **w. in fiery seams,** travail dans couches grisouteuses; **w. in** the whole, travaux de traçage; **w. order (in),** en exploitation normale; **w. outwards,** exploitation par la méthode directe; **w. pit,** puits d'extraction; **w. shaft,** puits d'extraction.

worm auger, tarrière rubannée.

worm's eye map, carte des structures sédimentaires et des transgressions.

worn, usé; **water w.,** usé, façonné par l'eau; **wind w.,** usé, façonné par le vent.

worthless, sans valeur (minerai).

wrench fault, faille de décrochement.

wrinkle, 1. ondulation, ride, sillon; 2. plissement.

wrinkle (to), rider, plisser.

wrinkling, plissement, gauchissement.

wulfenite, wulfénite (*minér.*).

Wulff net, cannevas (stéréographique) de Wülff.

Würm (Eur.), Würm (glaciation de).

Wurmian (Eur.), Wurmien (étage; Pléistocène) (équiv. Weichsel; Wisconsin).

Würtherian, Scythien.

wurtzilite, wurtzilite (var. de pyrobitume).

wurtzite, wurtzite (*minér.*).

wyomingite, wyomingite (*pétro.*).

X

xanthiosite, xanthiosite (*minér.*).
xanthophyllite, xanthophyllite (mica).
xanthosiderite, xanthosidérite, goethite.
xenoblast, xénoblaste (*minér.*).
xenoblastic, xénoblastique.
xenocryst, xenocrystal, xénocristal.
xenolith, enclave, xénolite.
xenomorph, xénomorphe (*minér.*).
xenomorphic, xénomorphe.
xenon, xénon.
xenothermal, xénothermal (hydrothermal de haute température et faible profondeur).
xenotime, xénotime (*minér.*).
xeralph, sous-ordre d'alfisol en régions arides, (sol fersiallitique de climat sec).
xerochimenic, climat sec, associé à des journées courtes.
xeroll, sol marron (à horizon A_1 mollique).

xerophile plant, plante xérophile (adaptée à un climat sec).
xerorendzine, sol de rendzine, grisâtre des régions steppiques.
xerosene, sol poreux de zones arides.
xerothermic period, période xérothermique (de sécheresse).
xerult, ultisol de climat sec.
X-rays, rayons X.
X-ray diffraction, analyse diffractométrique aux rayons X.
X-ray fluorescence, fluorescence aux rayons X.
X-ray tracing, diagramme diffractométrique de rayon X.
xylanthite, xylanthite (var. de résine fossile).
xylanthrax, charbon de bois.
xylopal, bois silicifié.

Y

yank, secousse.

yard, yard = 0,914 m; **cubic y.**, yard cube = 0,764555 m³; **square y.**, yard carré = 0,836127 m².

yardage, métrage, cubage.

yardang, yarding, yardang (*géomorph.*).

Yarmouth (N. Am.), Yarmouth (interglaciaire).

Yarmouthian, Yarmouthien (entre glaciations du Kansas et de l'Illinois).

yazoo stream (U.S.A. Missis.), affluent parallèle (au fleuve principal).

year-round mining, exploitation continue.

yellow, jaune; **y. arsenic**, orpiment; **y. brass**, laiton; **y. copper**, 1. laiton; 2. chalcopyrite; **y. copper ore**, chalcopyrite; **y. earth**, ocre jaune; **y. ground**, kimberlite oxydée (Afrique du Sud); **y. lead ore**, wulfénite; **y. metal**, or; **y. ochre**, ocre jaune, limonite; **y. ore**, chalcopyrite; **y. pyrite**, chalcopyrite; **y. ratebane**, orpiment; **y. spinel**, rubicelle.

Yeovilian (G.B.), Yévilien (étage; sommet du Lias supérieur).

yield, 1. limite élastique, fléchissement; 2. débit, production, rendement; **y. limit**, limite élastique; **y. point**; limite d'élasticité; **y. strength**, limite élastique; **y. stress**, limite élastique.

yield (to), 1. céder, fléchir; 2. débiter, produire.

yielding, 1. élasticité, déformation; 2. production, rendement; **y. point**, limite de fluage.

Ynezian (N. Am.), Ynézien (étage; Paléocène inf.).

yoked basin, zeugogéosynclinal.

Yoldia sea (Eur.), mer à Yoldia (post-glaciaire).

young, jeune, récent; **y. plain (U.S.A.)**, plaine marécageuse à topographie irrégulière; **y. valley**, vallée à son stade de jeunesse.

Younger Dryas (Eur.), Dryas récent (intervalle tardiglaciaire; entre Allerod et Préboréal).

Young's modulus, module de Young (*mécanique*).

youth, stade de jeunesse (fluviatile).

youthful stage, stade de jeunesse (*géomorphol.*); **y. topography**, topographie au stade de jeunesse.

Ypresian, Yprésien (étage; Éocène sup., syn : Landénien).

Y-tombolo, tombolo double, enfermé un petit lagon ou étang.

yttrialite, yttrialite (*minér.*).

yttrium, yttrium; **y. apatite**, apatite yttrifère; **y. garnet**, grenat yttrifère.

yttrocalcite, yttrocalcite (*minér.*).

yttrocerite, yttrocérite (*minér.*).

yttrofluorite, yttrofluorite (var. de fluorite).

yttrotantalite, yttrotantalite (*minér.*).

yttrotitanite, yttrotitanite (*minér.*).

Z

zaffer blue, bleu de cobalt.

zastruga, zastrugi (glacial.), sastrugi, forme d'érosion éolienne de névé.

zebra dolomite, dolomite rubanée; **z. rock,** roche rubanée.

Zechstein (eur.), Zechstein (série; Permien sup.).

zellen dolomit (germ.), dolomite bréchique (cf. rauchwacke).

Zemorrian (N. Am.), Zémorrien (étage; Oligocène à Miocène).

zenith telescope, lunette zénithale.

zenithal projection, projection azimuthale.

zeolite, zéolite (minér.).

zeolite facies, faciès de métamorphisme zéolitique.

zeolitic, zéolithique.

zeolitization, zéolitisation.

zero, zéro; **z. bias,** polarisation nulle; **z.-energy coast,** côte d'énergie z. protégée complètement contre l'action des vagues; **z. isobase,** ligne de mouvement z. dans une zone de réajustement isostatique; **z. level,** niveau z.; **z. line,** trait z.; **z. meridian,** méridien z.; **z.-order stream,** cours d'eau minimum (hypothétique); **z. setting,** remise à z. d'un instrument.

zeuge (german), roche champignon, pyramide de fée.

zigzag fold, pli en zigzag, en accordéon.

zinc, zinc; **z. bearing,** zincifère; **z. blend,** blende; **z. bloom,** hydrozincite; **z. deposit,** gisement de z.; **z. spar,** smithsonite; **z. spinel,** spinelle zincifère, gahnite; **z. vitriol,** sulfate de z.; **z. white,** oxyde de z.

zinciferous, zincifère.

zincite, zincite (minér.).

zinckiferous, zincifère.

zincky, zincifère.

Zingg's classification (séd.), système de classification des galets (sphériques, aplatis, etc.).

zinkite, zincite.

zinky, zincifère.

zinnwaldite, zinnwaldite (mica lithifère).

zircon, zircon (minéral lourd); **z. pyroxenes,** silicates zirconifères; **z. syenite,** syénite à néphéline et à zircon.

zirconium, zirconium.

zirconiferous, zirconifère.

Zoantharia, Zoanthaires (pal.).

zoarcium, colonie de bryozoaires.

zoecium, zoecia (pl.) (var. of zooecium), zoécie (pal.).

zoic, fossilifère.

zoisite, zoïsite (minér.).

zonal, zonal; **z. circulation,** circulation zonale (des vents d'ouest); **z. guide fossil,** fossile stratigraphique; **z. soil,** sol z.; **z. structure,** structure zonée; **z. wind,** alizé.

zonation, 1. disposition en zones stratigraphiques, zonation (pal.); **2.** disposition en zones écologiques ou climatologiques.

zone, 1. zone; **2.** biozone; **fossil z.,** fossile stratigraphique; **low-velocity z.** zone à faible vitesse de propagation d'onde sismique; **metamorphic z.,** zone de métamorphisme; **stratigraphic z.,** zone stratigraphique (ex. biozone).

zone melting, fusion zonale (au sommet d'une colonne magmatique).

zone of accumulation, 1. zone d'accumulation (nivale); **2.** horizon b (pédol.); **z. of aeration,** zone d'aération; **z. of eluviation,** horizon a (pédol.); **z. of flow, 1.** zone à écoulement plastique (géologie dynamique); **2.** zone d'écoulement (d'un glacier); **z. of fracture,** croûte terrestre cassante; **z. of leaching,** horizon de lessivage des sols; **z. of mobility,** asthénosphère; **z. of saturation,** zone saturée, nappe phréatique; **z. of weathering,** couches superficielles soumises à l'altération.

zoned, zoné.

zoning of crystals, zonation des cristaux.

zooecium (see zoecium), zoécie.

zooecology, écologie animale.

zoogene, zoogène, d'origine animale.

zoogenic rock, roche d'origine biologique.

zooid, zooïde (pal.).

zoolite, zoolith, animal fossile.

zooplankton, zooplancton, plancton animal.

zooturbation, bioturbation.

Zuloagan (North Am.), Zuloagien (étage; Jurassique sup.; équiv. Oxfordien).

zweikanter (german), galet éolisé à deux facettes.

zwischengebirge (germ.), unité géotechnique entre deux zones orogéniques.

Zyrjanian, étage glaciaire russe ($\simeq$ Würmien).

Part 2
FRENCH-ENGLISH

2e partie
FRANÇAIS-ANGLAIS

A

Aalénien, Aalenian.

abaissement axial d'un pli, downdipping of the axis of a fold.

abaissement de la température, drop in temperature, fall of temperature, cooling.

abandonnée (carrière), disused (quarry).

abandonné (méandre), abandoned meander.

abaque, diagram, chart, monograph.

abatre, abattre, to break down, to hew; **a. du minerai,** to stope the ore, to mine.

abattage, blasting, rock blasting.

abdomen (pal.), abdomen.

aber (Wales), drowned river valley.

aberration (opt.), aberration.

abîme, abyss.

abiotique, abiotic.

ablation, ablation, denudation; **a. glaciaire,** glacial ablation.

aboral (pal.), aboral.

aborder (un continent), to land.

abrasif, abrasive.

abrasion, abrasion; mechanical erosion.

abri, shelter, rock-shelter; **a. sous-roche,** s.-cave.

abrité, sheltered, protected.

abrupt, precipitous, scarped, steepy; **a. d'éboulement,** scar.

absarokite, absarokite.

absence (d'une couche), lack, stratigraphic gap, hiatus, diastem.

absolu, absolute; **chronologie a.,** a. age; **datation a.,** a. dating.

absorbable, absorbable.

absorbant, **1.** adj: absorbant; **2.** n: absorber; **complexe a.,** absorption complex.

absorber, to absorb, to soak up.

absorption, absorption; **a. atmosphérique,** atmospheric a.; **capacité d'a.,** a. capacity; **coefficient d'a.,** a. coefficient; **raies d'a.,** a. lines; **spectre d'a.,** a. spectrum.

absorptivité, absorptivity, absorptiviness.

abyssal, abyssal, deep-sea; **profondeurs a.,** abyssal depths.

abyssobenthique, abyssobentic.

Acadien (Cambrien moyen), Acadian (m. Cambrian).

acanthite, acanthite.

Acanthodiens (pal.), Acanthodii.

accélérations de la pesanteur, constant of gravitation, intensity of gravity, acceleration of gravity, gravitational acceleration.

accélérogramme, accelerogram.

accès (d'une mine), access, entrance, entry.

accessible (région), accessible.

accessoire (minéral), accessory (mineral); **a. (lame),** a. (plate).

accident de terrain, ground feature, undulation, unevenness; faulted-terrane (U.S.); **a. (rejet de faille),** throw.

accidenté (relief), uneven, rough, undulating, hilly, hillocky, faulted (terrain).

accordéon (plissement en), accordian folding.

accore (côte), steep coast.

accrétion (des continents), accretion, accreted terrain (a. terrane, U.S.).

accroissement (pal.), growth, increase.

accumulation, accumulation, deposition; **a. de blocs, en bas d'une falaise,** rock fan; **a. d'éboulis,** rock fan, talus; **a. d'éboulis de gravité,** talus, debris fall; **a. de débris par solifluxion,** mass wasting; **a. de débris gélivés,** congelifract, deposit accumulation; **a. de pétrole,** petroleum accumulation; **a. détritique en bas de pente,** colluvium; **a. littorale,** shore deposit.

accumuler, s'accumuler, to pile up, to accumulate.

acerdèse, acerdese, manganite.

Acheuléen, Acheulian.

achondrite (météorite), achondrite.

achroïte (var. de tourmaline), achroite.

achromatique, achromatic.

aciculaire (minéral), acicular, needle-shaped.

aciculé, aciculate.

aciculite, aikinite, aciculite.

acide, **1.** adj: acid; **2.** n: acid; **a. carbonique (grisou),** carbonic a. (choke damp); **a. chlorhydrique,** hydrochloric a.; **a. faible,** weak a.; **a. fort,** strong a.; **a. fulvique,** fulvic a.; **a. humique,** humic a.; **roche a.,** a. rock, acidic rock; **sol a.,** a. soil; **traitement a.,** a. treatment.

acidification, acidification.

acidifier, s'acidifier, to acidify.

acidité, acidity.

acier, steel.

aciérie, steel works.

aclinal, table-like, flat.

aclinique, aclinic, aclinal.

acmite, acmite.

acoustilog, acoustilog.

acoustique, acoustic(al); **diagraphie a.,** a. well-logging; **écho-sondage,** a. echo-sounding; **fréquence a.,** a. frequency; **horizon a.,** a. horizon; **levé a.,** a. survey; **méthode a.,** a. method.

acquis (caractère), acquired (character).

acrisol, acrisol.

Acritarches (pal.), Acritarcha.

actif, active; **cluse a.,** a. gap; **faille a.,** a. fault; **marge a.,** a. margin.

Actinaires (pal.), Actinaria.

actinolite (minér.), actinote, actinolite.

actinoschiste, actinolite-schist.

actinote, actinote, actinolite.

actinotique, actinolitic.
Actinoptérygiens (pal.), Actinopterygii.
action, action; **a. chimique**, chemical a.; **a. climatique du gel**, frost weathering; **a. du gel**, frost work; **a. éolienne**, wind a.; **a. glaciaire**, glacial a.; **a. tampon**, buffering, buffer action.
activité, activity; **a. chimique**, chemical a.; **a. solaire**, solar a.; **a. volcanique**, volcanic a.
actualisme (pal.), actualism.
actuel, recent, present (jamais "actual").
acyclique, acyclic.
adamantin, adamantine.
adamellite (var. de granite), adamellite.
adamite (minér.), adamine, adamite.
adaptation, **1.** adaptation *(pal.)*; **2.** adjustment *(geogr.)*; **a. au milieu** *(pal.)*, adaptation; **a. structural**, structural adjustment.
adapter, s'adapter (rivière), to fit, to adjust.
addition (géoph.), stack.
adducteurs (muscles), adductor (muscles).
adduction (d'eau), water supply.
adhésivité (d'un sol), stickiness, adhesive capacity.
adiabatique, adiabatic.
adinole (pétro.), adinole.
adipite (minér.), chabazite.
adoucissement de l'eau, water softening.
adret, sunny southern slope (in northern hemisphere; inversed in south hemisphere).
adsorbant, adsorbant.
adsorber, to adsorb.
adsorption, adsorption; **a chimique**, chemical a.; **a. d'eau**, water a.
adulaire (minér.), adularia, adular.
adventif (cône), parasitic cone, adventive (cone).
aegyrine (minér.), aegyrite, aegirite, aegyrine.
aérage (mine), ventilation.
aération (mine), ventilation, aeration.
aérer (mine), to aerate, to ventilate.
aérien, aerial; **levé a.**, a. mapping; **photogrammétrie a.**, a. survey; **photographie a.**, a. photography; air photography.
aérogéologie, photogeology.
aérolite, aérolithe, aerolite, meteoric stone, stony meteorite.
aérolithique, aerolitic, meteoritic.
aéronomie, aeronomy.
aérosite, aerosite, pyrargyrite.
aérosol, aerosol.
aérotriangulation, stereotriangulation.
aetite, eaglestone.
afeldspathique, feldspar free.
affaiblissement (d'ondes), decay.
affaissé (bloc), trough-fault block, graben.
affaissement, collapse, downthrow, downwarp, sinking, subsidence; **a. isostatique**, isostatic subsidence.
affaisser, s'affaiser, to collapse, to subside, to sink.
affinement (d'un métal précieux), refinement.
affiner (du fer, de l'acier), to refine, to fine.
affinité (chimique), chemical affinity.

affleurement, exposure, outcrop, outcropping; **a. altéré**, weathered outcrop, blossom; **a. caché**, buried outcrop, concealed outcrop; **a. oxydé**, blossom.
affleurer, to expose, to outcrop, to crop out.
affluent, tributary river, affluent, inflowing stream; **sous-a.**, subtributary, minor tributary.
affluer (fleuve), to flow into.
affouillement, undermining, erosion, scouring.
affouiller, to undermine, to undercut, to scour, to erode, to wash.
agalmatolite (pyrophyllite), agalmatolite, figure stone.
agate, agate; **a. arborisée**, dendritic a., tree a.; **a. d'Islande**, obsidian; **a. jaspée**, jasper a.; **a. mousseuse**, moss a.; **a. noire**, obsidian; **a. rubanée**, banded a.
agatifère, agatiferous.
âge, age; **a. absolu**, absolute a.; **a. de la pierre polie**, neolithic a.; **a. de la pierre taillée**, old stone a.; **a. isotopique**, isotopic a.; **a. primaire**, paleozoic era; **a. radiométrique**, radiometric a.; **a. relatif**, relative a.; **a. secondaire**, mesozoic era; **a. tertiaire**, tertiary era.
agent, agent; **a. atmosphérique**, atmospheric a.; **a. d'érosion**, erosion a.; **a. de dispersion**, dispersion a.; **a. de flottation**, flotation a.; **a. minéralisateur**, mineralizing a.
agglomérat, agglomerate.
agglomération, agglomeration; **a. de minerai**, ore sintering.
aggloméré, agglomerated, conglomerated, sintered.
agglomérer, s'agglomérer, to agglomerate, to conglomerate.
agglutinant, agglutinant, binding.
agglutiner, s'agglutiner, to agglutinate, to stick together.
aggradation, aggradation, accumulation; progradation (coastal).
agitateur (labo.), stirring rod, agitator.
agitation sismique, microseism.
agmatite, agmatite.
Agnathes (pal.), Agnatha.
Agoniatidés (pal.), Agoniatides.
agpaïte (pétro.), agpaite.
agrégat, aggregate, cluster.
agrégation, aggregation.
agrégé (sédiment), aggregated.
agrogéologie, agrogeology.
agrologie, agronomy.
aigue-marine, aquamarine.
aiguille (de cadran), needle, pointer, index; **a. (géogr.)**, needle; aiguille; **a. aimantée**, magnetic needle; **a. de glace**, icicle, pipkrake; **a. volcanique**, volcanic spine.
aimant, magnet, lodestone; **a. naturel**, lodestone, magnetite.
aimantation, magnetization; **a. inverse**, reversed m.; **a. rémanente**, remanent m.; **a. spontanée**, spontaneous m.
aimanter, to magnetize.

aimanté, magnetized, magnetic; **barreau a.,** bar magnet.

air, air; **arrivée d'a.,** a. inlet; **extraction par l'a. comprimé,** a. hoisting; **forage à l'a. comprimé,** a. drilling; **injection d'a.,** a. lift.

aire, area, region, belt; **a. continentale,** continental area; **a. d'alimentation,** drainage area, catchment area; **a. de déflation éolienne,** blownout land; **a. d'inondation,** flood plain.

akérite (var. de syénite), akerite.

aklé, akle.

alabandine, alabandite.

alabastrite, gypseous alabaster.

alaskite (var. de granite), alaskite.

alass (cryokarst), alas.

albâtre, alabaster; **d'a.,** alabasterine; **a. calcaire,** travertine; **a. gypseux,** gypseous a.

albédo, albedo.

albertite, albertite.

Albien (Crétacé), Albian (l. Cretaceous).

albite (minér.), albite, sodafeldspar.

albite-oligoclase, albiclase; **mâcle de l'albite,** albite twin.

albitisation, albitization.

albitisé, albitized.

albitite (var. de syénite), albitite.

albitophyre, albitophyre.

alboll (pédol.), alboll.

alcalin, alkaline, alkalic, alkalinous, alkalous; **feldspath a.,** a. feldspar; **granite a.,** alkali granite.

alcalinisation, alkalization.

alcaliniser, s'alcaliniser, alkalify.

alcalinité, alcalinity.

alcalino-terreux, earth alkali.

Alcyonaires (pal.), Alcyonaria.

aleurite, silt.

aleurolite, silstone.

alfisol (pédol.), alfisol.

alguaire, algal.

Algue, Alga, algae (pl.); **a. bleue,** a. cyanophycea; **a. calcaire,** lime-secreting alga; **a. fossile,** fossil alga; **a. incrustante,** incrusting alga; **charbon d'a.,** algal coal; **récif d'a.,** algal bank.

Algonkien, Algonkian (Precambrian, Canada).

alidade, alidade, sighting board.

alignement de blocs (périgl.), blocktrain.

alignement de failles, fault alignment.

alios, ironpan, hardpan; **a. ferrugineux,** ironpan; **a. humique,** humic ironpan.

aliphatique, aliphatic, open chain compound.

alizé, trade wind.

alkane, alkane.

allanite (var. d'épidote), allanite.

Alleröd, Allerod (stage, latest Pleistocene).

allite, allitic soil.

allitisation (altération latéritique), allitization.

allitique, allitic.

allivalite (var. de gabbro.), allivalite.

allochem, allochème, allochem.

allochimique, allochemical.

allochroïte, allochroite.

allochtone, allochton, allochtonous, allogenic, allothigenous.

allogène, allogeneous, allogenic, allogenous.

allométrie (pal.), allometry.

allomorphe, allomorphic.

allongement (d'un minéral), elongation; **a. positif,** positive e.; **a. négatif,** length-slow e., negative lenght fast.

allothigène, allothigeneous.

allotriomorphe, allotriomorphic.

allotrope, allotropique, allotrope, allotropic.

allotropie, allotropy.

alluvial, alluvial; **cône a.,** cone, a. fan; **glacis a.,** slope; **nappe a.,** apron; **piémont a.,** piedmont; **plaine a.,** plain; **terrasse a.,** terrace.

alluvionnement, alluviation, aggradation.

alluvions, river deposits, alluvial deposits, alluvium; **a. anciennes,** old alluvium; **a. aurifères,** gold bearing alluvium; **a. côtières,** coastal deposits; **a. fluviatiles,** river deposits; **a. fluvio-glaciaires,** fluvio-glacial deposits; **a. glaciaires,** glacial till, drift; **a. marines,** marine deposits; **a. modernes,** holocene deposits; **a. post-glaciaires,** post-glacial alluvium; **a. quaternaires,** pleistocene deposits; **a. récentes,** holocene deposits; **a. stannifères,** tin-bearing alluvium.

alluvionnaire, alluvial.

alluvionnement, alluviation.

almandin (grenat), almandine, almandite.

alnoïte (pétrol.), alnoite.

alouette (gypse pied d'), twinned gypsum crystal, larkspur gypsum.

alpin, alpine; **glacier a.,** a. glacier.

alstonite [$CaBa(CO_3)_2$], alstonite.

altérable, alterable.

altération, alteration, weathering; **a. atmosphérique,** weathering; **a. chimique,** chemical weathering; **a. hydrothermale,** hydrothermal weathering; **a. kaolinique,** kaolinization; **a. profonde,** deuteric alteration; **a. subaérienne,** weathering; **a. superficielle,** weathering.

altérer, s'altérer (roche), to alter, to weather, to be weathered, to be decayed.

altéré, altered, weathered.

altérite, weathered and decayed rock.

alternance (de faciès), alternation.

alternance gel-dégel, freeze and thaw cycle.

alterné, alternate, alternating; **stratification a.,** a. bedding.

altimètre, altimeter, height gauge.

altimétrie, altimetry.

altimétrique, altimetric(al).

altiplanation, altiplanation.

altitude, elevation, altitude, heigth; **a. absolue,** absolute elevation, altitude; **a. relative,** relative elevation, altitude.

altocumulus, altocumulus cloud.

altostratus, altostratus cloud.

alumine, alumina, alumine oxide; **gel d'a.,** alumogel; **silicates d'a.,** aluminum silicates.

alumineux, aluminous; **épidote a.,** aluminum epidote; **grenat a.,** aluminum garnet; **shiste a.,** alum schist.
aluminière, alum mine.
aluminifère, aluminiferous, aluniferous.
aluminite, aluminite.
aluminium, aluminium (U.K.), aluminum (U.S.).
aluminosilicate, aluminosilicate.
alvéolaire, alveolar, cellular, honeycomb.
alvéole, alveolus, alveole, cell, tafoni (taffoni).
Alvéolinidés (pal.), Alveolinidae.
amalgame natif, native amalgam.
amas, accumulation, cluster, heap, pile, pocket; **a. cristallin,** crystalline nodule; **a. de galaxies,** cluster of galaxies; **a. de gélifraction,** cryogenic deposit; **a. de minerai,** ore heap, ore pocket, mass of ore; **a. de neige (congère),** snow drift; **a. de quartz,** mass of quartz.
amasser, s'amasser, to heap up, to pile up, to gather, to collect.
amazonite, amazonite, amazonstone.
amblygonite, amblygonite.
ambre, amber, succinite; **a. jaune,** yellow amber; **a. noir,** jet, black onyx.
ambrite, ambrite.
ambulacraire (pal.), ambulacrar; **aire a.,** a. area, a. field; **plaque a.,** a. plate; **zone a.,** a. area.
ambulacre, ambulacrum.
aménagement du littoral, coastal development.
aménagement (d'une pente), grading.
amendement, manure.
amenée (d'eaux), conducting, bringing.
amenuisement (d'une roche), comminution.
améthyste, amethyst; **quartz a.,** amethystine quartz.
ameublissement (d'un sol), mellowing, tillage.
amiante, amianth, amianthus, asbestos, moutain cock.
amiantoïde, amiantoid.
amincie (couche), thinned out bed.
amincissement, thinning out, pinching.
aminé (acide), amino-acid.
ammoniac (gaz), ammonia.
ammoniaque, ammonium hydroxide.
ammonisation, ammonification.
Ammonite, Ammonite; **a. enroulée,** involute a.
Ammonitidés, Ammonoidea, Ammonoids, Ammonitida.
ammonium, ammonium.
Ammonoïdés, Ammonoidea.
amodiation (de droits miniers), leasing, subleasing.
amonceler, s'amonceler, to heap up, to pile up, to bank up, to drift.
amoncellement (de sédiments), accumulation, heap, pile.
amont, upstream, upper part; **a. pendage,** back, backs, updip.
amorphe, amorphous, amorphic, amorphose, uncrystalline, structureless.
amortissement (d'un pendule, d'oscillations), damping, reduction.
amortissement d'une faille, dampening of a fault.

amortisseur d'oscillations, pulsations dampener.
ampélite, ampelite, bituminous black shale, carbonaceous shale.
ampélitique, ampelitic.
ampère, ampere; **a. mètre,** amperemeter.
Amphibiens, Amphibia.
amphibole, amphibole; **a. calcique,** calcic a.; **a. ferro-magnésienne,** mafic a.; **a. sodique,** soda a.; **shiste à a.,** hornblende shist.
amphibolifère, amphiboliferous.
amphibolique, amphibolic.
amphibolite, amphibolite.
amphibolitisation, amphibolitization.
amphiboloschiste, hornblende schist.
amphidromique, amphidromic.
amphigène, leucite.
amphigénite, leucitite.
Amphineure, Amphineura.
amphithéâtre morainique, terminal basin.
amphotère, amphoteric.
amplification, 1. amplification (of a wave); 2. enlarging, enlargement, magnification (of a photograph).
amplitude, amplitude, extent, height, size; **a. des marées,** tidal range.
amygdalaire, amygdaloïde, amygdaloid(al).
anabatique, anabatic.
anaclinal, anaclinal.
anadiagénèse, anadiagenesis; deep-seated alteration (diagenesis).
anaérobie, anaerobic.
anaglaciaire, anaglacial stage.
anal (pal.), anal.
analcime, analcite; **basalte à a.,** a. basalt.
analcitite (var. de basalte), analcitite.
analogique, analog.
analyse, analysis; **a. au chalumeau,** blow-pipe assay, blow-pipe a.; **a. de la trace (géoph.),** trace a.; **a. densimétrique,** float and sink a.; **a. diffractométrique,** X-ray a.; **a. granulométrique,** granulometric, particle size or sieve a.; **a. macroscopique,** macroscopic a.; **a. mécanique,** particle size a.; **a. microscopique,** microscopic a.; **a. quantitative,** quantitative a.; **a. pollinique,** pollen a.; **a. spectral,** spectral a.; **a. structurale,** tectonic a.; **a. thermique,** thermal a.
analyser, to analyse, to assay.
analyseur (micro. pol.), analyser.
anamorphique, anamorphic.
anamorphisme, anamorphism.
anastomosée (rivière), anastomosed river, river with anabranches.
anastomosés (chenaux), braided channels.
anatase (minér.), anatase, octahedrite.
anatectique, anatectic.
anatexie, ultrametamorphism; **granite d'a.,** ultrametamorphic granite.
anatexite, anatexite.
anchimétamorphisme, very low grade metamorphism, anchimetamorphism.
ancrage (d'un forage), anchorage, anchor.

andalousite, andalusite, chiastolite; **cornéenne à a.,** a. hornstone.

andept (pédol.), andept.

andésine (minér.), andesine.

andésite, andesite, andesyte; **a. à augite,** a. with augite.

andésitique, andesitic.

andosol (pédol.), endosol.

andradite (grenat), andradite.

anémographe, anemograph; **a. enregistreur,** recording a.

anémographique, anemographic.

anémographie, anemography.

anémomètre, anemometer, wind gauge; **a. enregistreur,** anemograph.

anémométrie, anemometry.

anémométrique, anemometric.

anéroïde (baromètre), aneroid (barometer).

anfractueux, craggy.

anfractuosité, cragginess.

Angiospermes, Angiospermae.

angle, angle; **a. critique,** critical a.; **a. de direction,** a. of strike, strike direction; **a. de discordance,** a. of unconformity; **a. de pendage,** a. of dip, dip a.; **a. de pente,** a. of slope, slope a.; **a. d'équilibre,** a. of rest; **a. de réflection,** a. of reflection; **a. de réfraction,** a. of refraction; **a. de repos,** a. of repose; **a. de stratification,** a. of bedding; **a. de talus,** a. of repose; **a. d'équilibre,** a. of rest, of repose; **a. d'extinction,** extinction a.; **a. d'incidence,** a. of incidence; **a. d'inclinaison,** a. of hade; **a. du plan de pendage avec la verticale,** a. of hade; **a. entre faces,** interfacial a.; **a. limite,** limit a.

anglésite, lead, spar, anglesite.

angrite, angrite.

angström, angstrom = 10^{-8} cm.

angulaire, angular; **discordance a.,** unconformity, angular unconformity.

angularité (séd.), angularity, angularness (rare).

anguleux, angular.

anhydre, anhydrous, anhydric.

anhydride carbonique, carbonic dioxyde.

anhydrite, anhydrite, anhydrit.

anion, anion.

Anisien, Anisian, anisic (m. Triassic).

anisométrique, anisometric.

anisomyaire (pal.), anisomyarian.

anisopaque, anisopachyte.

anisotrope, anisotropic, anisotropal, anisotropous.

anisotropie, anisotropy.

anisotropiquement, anisotropically.

ankaramite (var. de basalte), ankaramite.

ankérite (minér.) ankerite.

annabergite (minér.), annabergite.

Annélides (pal.), Annelida.

anneau (astro.), ring.

anneau occipital (pal.), occipital ring.

annuaire des marées, tide tables.

annuaire hydrologique, water year-book.

annuel, annual.

annulaire, annular, ring-shaped, ring dyke; **récif a.,** a. reef.

anode, anode.

anodonte (pal.), anodont; **charnière a.,** a. dentition.

anomalie, anomaly, anormal variation; **a. de Bouguer,** Bouguer a.; **a. de gravité,** gravity a.; **a. de résistivité,** resistivity a.; **a. électrique,** electric a.; **a. électromagnétique,** electromagnetic a.; **a. gravimétrique,** gravity a.; **a. isostatique,** isostatic a.; **a. locale,** local a.; **a. magnétique,** magnetic a.

anormal (contact), **1.** fault; **2.** thrust.

anorthite (plagioclase), anorthite.

anorthose (feldspath.), anorthoclase.

anorthosite (pétro.), anorthosite.

anse, cove, creek.

Antarctique, **1.** Antarctica (name); **2.** antartic (adj.).

Antécambrien, Precambrian.

antécédence, antecedence.

antécédent, antecedent; **rivière a.,** a. stream.

antéclise, anteclise.

antédiluvien, antediluvian.

antédistension, pre-rifting.

antéglaciaire, preglacial.

Antennates (pal.), Antennata.

antenne (pal.), antenne.

antérieur (pal.), anterior.

anthophyllite (amphibole), anthophyllite.

Anthozoaire (pal.), Anthozoa.

anthracifère, anthraciferous.

anthracite, glance coal, hard coal, anthracite.

anthraciteux, anthracitic, anthracitous.

Anthracolitique (désuet), Carboniferous and Permian.

anthraconite, anthraconite, stinkstone.

anthracophore, coal bearing.

Anthracosauriens (pal.), Anthracosauria.

anthropique, anthropogenic.

anthropologie, anthropology.

anthropologiste, anthropologist.

Anthropogène, Quaternary (russian).

anthropologue, anthropologist.

anthropozoïque, **1.** adj: anthropozoic; **2.** n: Pleistocene.

anticathode, anthicathode, target.

anticlinal, **1.** adj: anticlinal; **2.** n: anticline.; **a. asymétrique,** asymmetric anticline; **a. déversé,** overturned anticline; **a. dissymétrique,** assymetric anticline; **a. droit,** upright anticline; **a. en forme de selle,** saddle anticline; **a. faillé,** faulted anticline; **a. fermé,** closed anticline; **a. plongeant,** plunging anticline; **a. renversé,** recumbent anticline; **a. symétrique,** symmetric anticline.

anticlinorium, anticlinorium, composite anticline.

anticyclone, anticyclone.

anticyclonique, anticyclonic.

antidune, antidune.

anti-épicentre, anticentre.

antiforme, antiform.

antigorite, antigorite.

antimoine, antimony, stibium; **a. natif,** a. ore; **a. oxyde,** valentinite; **a. sulfure,** antimonite, antimony glance, stibnite.
antimonial, antimonial.
antimoniate, antimonate.
antimonié, antimoniated.
antimonieux, antimonious, stibial.
antimonifère, antimoniferous.
antimonique, antimonic.
antimoniure, antimonide, stibnite.
antiperthite, antiperthite.
antipodal, antipodal.
antipode, antipode.
antithétique (faille), antithetic (fault).
antlérite, antlerite.
apatite, apatite.
aphanitique (texture), aphanitic, aphyric.
aphotique (océano.), aphotic, without light.
aphyrique, aphyric.
apical, apical.
aplanir (géomorpho.), to level.
aplanissement, aplanation, planation, flattening, levelling.
aplati, flattened, platy, tabular.
aplatissement (séd.), flattening, flatness; **indice d'a.,** flatness index, flatness ratio.
aplite, aplite.
aplitique, aplitic.
aplome, aplome (var. of melanite).
apobac (astro.), apobac (position of sun at its maximum distance from the solar system center of mass).
apogée (astro.), apogée (maximum distance between the sun and the earth).
apogranite, apogranite.
apomagmatique (gîte), apomagmatic deposit.
apophyllite, apophyllite.
apophyse, apophysis, offshoot, outgrowth.
appalachien (relief), appalachian structure.
appareil, apparatus, device; **a. branchial (pal.),** branchial skeleton; **a. de forage,** drilling rig; **a. de prises de vues aériennes,** aerial camera; **a. de prises de vues en bandes continues,** continuous strip camera; **a. de prises de vues cartographiques,** mapping camera; **a. de sondage,** drilling rig; **a. enregistreur,** recording instrument.
apparent, apparent; **pendage a.,** a. dip.
appauvrissement (du sol, etc.), impoverishment.
appliquée (géologie), applied geology.
appliquée (hydrogéologie), applied hydrogeology.
apport (fluviatile), river drift, river deposits.
apporter (des sédiments), to drift, to lay down.
approfondir (le lit d'une rivière), to deepen, to excavate.
Aptien, Aptian (l. Cretaceous).
aptitude au gel, freezability.
aptychus (pal.), aptychus.
aqualf (pédol.), aqualf.
aqueduc, aqueduct.
aqueux, aqueous, hydrous; **solution aqueuse,** aqueous solution.

aquiclude, aquiclude.
aquifère, **1.** adj: aquiferous, water-bearing; **2.** n: aquifer, nappe.
aquifère multicouche, multilayer aquifer.
aquod (pédol.), aquod.
aquo-igné, aqueo-igneous, hydrothermal.
aquult (pédol.), aquult.
aquifuge, aquifuge.
Aquitanien, Aquitanian (l. Miocene).
aragonite, aragonite.
araser (géomorpho.), to level down, to wear flat, to plane.
arasé, levelled, truncated.
arasement (géomorpho.), levelling, erosion.
arborescence (minér.), dendrite.
arborescent, dendritic, arborescent, tree-like.
arborisation (minér.), dendrite.
arborisé, arborized, dendritic.
arc (insulaire), island arc.
arc-boutement, arching.
arc morainique, moraine rampart.
arc résiduel, remnant arc.
Archéocyathidé (pal.), Archaeocyathid.
Archéosauriens (pal.), Archaeosauria.
arche, arch; **a. naturelle,** rock a.
Archéides, Archeides.
Archéen, Archean (early Precambrian).
archéologie, archeology.
archéologique, archeological.
Archéopterygiens, Archaeopteryges.
archipel, archipelago.
architecture (d'une roche), texture.
archives, records, files.
arctique, arctic; **banquise a.,** a. pack, ice-pack.
ardente (nuée), glowing cloud.
ardoise, slate, roofing slate; **a. tachetée,** spotted slate.
ardoisier (adj.), slaty.
ardoisière, slate quarry.
aréique (sans écoulement), arid, without any river system, arheic.
aréisme, arheism, arheic regime.
arénacé, arenaceous, sandy.
arène, granitic sand, quartz sand, grit, sandy regolith.
arène « ménagée » (géomorpho.), half, developed weathered mantle.
arénisation, granular disintegration.
arénite, arenite, sandstone.
arénolutite, arenolutite.
arénorudite, arenorudite.
arénosol, arenosoil.
aréographie (sur Mars), areography.
aréolaire (altération), superficial weathering.
aréole, areola (pal.).
aréomètre, araeometer.
aréométrie, araeometry.
Arenig, Arenigian (l. Ordovician).
arête (de montagne), crest, edge, ridge; **a. (de poisson),** fishbone; **a. anticlinale,** anticlinal crest;

a. d'un polyèdre, edge of a polyhedron; **a. vive (tranchante),** sharp edge.
argent (métal), silver; **a. arsenical,** proustite; **a. corné,** hornsilver, cerargyrite; **a. natif,** native s.; **a. rouge,** red s., ruby s.; **a. rouge antimonial,** pyrargyrite; **a. rouge arsenical,** proustite; **a. sulfuré,** argentite.
argentée (couleur), silvery.
argentifère, argentiferous; *pyrite a.,* argentopyrite.
argentite, argentine, argentite, argyrite.
argentopyrite, argentopyrite.
argid (pédol.), argid.
argile, clay; **a. à blocaux,** boulder c.; **a. à briques,** brick c.; **a. à chailles,** Oxford c. (Jura); **a. à poterie,** potter's c.; **a. à silex,** flint c.; **a. à varves,** varved c.; **a. de frottement tectonique,** gouge, fault-c., shale; **a. ferrugineuse,** iron c.; **a. feuilletée,** shale; **a. figuline,** pottery c.; **a. glaciaire,** glacial till; **a. gonflante,** swelling c.; **a. grasse,** unctuous c.; **a. gypseuse,** gypsiferous c., gypseous c.; **a. latéritique,** lateritic c.; **a. limoneuse,** silty c.; **a. litée,** bedded c.; **a. lourde,** heavy c.; **a. marbrée,** mottled c.; **a. marneuse,** marly c.; **a. plastique,** plastic c., potter's c.; **a. réfractaire,** fire c., refractory c.; **a. rouge des grands fonds,** red ooze c. (deep sea); **a. rouge des plateaux,** decalcification c.; **a. sableuse,** sandy c.; **a. schisteuse,** shale c., smectitic c.; **a. smectique,** fuller's earth; **a. stratifiée,** laminated shale; **a. téguline,** tile c.; **a. teneur (en),** c. content.
argileux, clayey argillous, argillaceous.
argilière, clay pit.
argilisation, argillization.
argilique (horizon), argillic (horizon).
argilite, argillite, rock clay, mudrock, mudstone.
argilo-calcaire, argillocalcareous.
argiloferrugineux, argilloferruginous.
argilomagnésien, argillomagnesian.
argilo-sableux, argilloarenaceous.
argon, argon.
Argovien, Argovian.
argyrite, argyrite, argyrose.
argyropyrite, argyropyrite.
argyrose, argentite.
argyrythrose, argyrythrose, pyrargyrite, dark-red silver ore.
aride, aride, dry.
aridification, aridification.
aridité, aridity, aridness, dryness; **indice d'.,** arid index.
aridosol, aridosol.
ariégite (pétro.), ariegite (var. of pyroxenite).
arkose, arkose, granite wash.
arkosique, arkosic.
arménite, armenite.
armoricaine (orogénie), Armorican orogeny.
aromatique, aromatic; **hydrocarbure a.,** aromatic hydrocarbon, aromatics.
arpenter, to survey, to measure.
arpenteur, surveyor, land surveyor.
arqué, arched, bent, curved.

arrache-carottes (for.), core extractor, core lifter.
arrachement (courant d'), rip-current.
arrachement (par les eaux superficielles), washing out; **niche d'arrachement,** scar.
arrache-tube (for.), tubing spear.
arrangement (d'appareils), array, pattern.
arrière, back; **a. pays,** b. land; **a. plage,** b. shore; **a. plan,** b. ground; **a. récif,** b. barrier.
arrivée (sism.), arrival; **a. d'un fluide,** inlet, inflow, intake; **a. d'air,** air inlet; **a. d'eau,** water inlet.
arrondi (séd.), rounded; **degré d'a.,** roundness ratio.
arrondissement, roundness; **indice d'a.,** r. ratio.
arrosé (à fortes pluies), watered.
arséniate, arseniate.
arsenic, arsenic; **a. blanc,** white a.; **a. sulfuré rouge,** red a., ruby a.
arsenical, arsenical; **argent a.,** a. silver blend.
arsénieux, arsenious.
arsénifère, arseniferous.
arséniosidérite, arseniosiderite.
arséniosulfure de cobalt, cobaltine.
arsénite, arsenite, arsenolite.
arséniure, arsenide.
arsénopyrite (minér.), arsenopyrite.
artésien, artesian; **aquifère a.,** a. aquifer; **bassin sédimentaire a.,** a. basin; **nappe a.,** a. water, a. water aquifer; **puits a.,** a. well; **source a.,** a. spring; **structure a.,** a. structure.
Arthrodires, Arthrodiran.
Arthropodes, Arthropoda.
article (pal.), article, ossicle.
articulation, hinge, link, joint.
articulé, articulate, jointed, hinged.
Articulées (pal.), Articulata.
Artinskien, Artinskian (l. Permian).
Artiodactyles, Artiodactyla.
art pariétal (préhist.), parietal art.
as (swed.), asar (pl.), esker, asar.
asbeste (minér.), asbestos.
asbestoïde, asbestoid.
ascendante (source), ascending (spring).
ascension capillaire, capillary ascent.
ascensionniste, climber.
aséismique, asismique, aseismic.
Ashgillien, Ashgillian (u. Ordovician).
aspect extérieur (d'un cristal), habitus.
aspérité, roughness, unevenness.
asphalte, asphalt, glance pitch, mineral p., petroleum p.; **a. naturel,** native, naturel asphalt; **gisement d'a.,** asphalt deposit; **goudron d'a.,** asphalt tar; **suintement d'a.,** asphalt seepage.
asphaltène, asphaltene.
asphaltique, asphaltic; **bitume a.,** asphalt bitumen; **calcaire a.,** a. limestone; **charbon a.,** a. coal, albertite; **pétrole brut a.,** asphalt base crude; **roche a.,** a. rock; **sable a.,** a. sand; **schiste pyrobitumineux a.,** pyrobituminous a. shale.
asphaltite, asphaltite.
asphaltoïde, asphaltoid.
assèchement, drainage, dewatering, drying.

assécher, to drain, to dry, to dewater, to pump out.

assiette (soubassement), basis, support, foundation, bottom, bed.

Assilines (calcaire à), Assiline (limestone).

assimilation, assimilation; **a. magmatique**, magmatic a.

assise, 1. seating, laying (of foundation); **2.** bed, stratum (geology); **3.** course, layer (masonry).

association, association; **a. de minéraux**, mineral a.; **a. symplectique**, symplecktic intergrowth.

assolement, rotation of crops.

assoupi (volcan), quiescent (volcano).

assynthienne (orogénèse), Assynthian (orogeny u. Precambrian).

Astartien, Astartian (u. Jurassic).

astatique, astatic.

asténolite, asthenolith.

Astéridés, Asteroidea.

Astérie (pal.), Star-fish.

astérie (à), asteriated.

astérisme, asterism.

asthénosphère, asthenosphere.

Astien, Astian (pliocene).

astre, star.

astroblème, astrobleme.

astrogéologie, astrogeology.

astrographe, astrograph.

astronautique, astronautics.

astronomique, astronomical.

astronomie, astronomy.

astrophysicien, astrophysician.

astrophysique, astrophysics.

asymétrique, asymmetric(al); **pli a.**, a. fold.

atacamite (minerai de cuivre), atacamite.

ataxique, ataxic.

ataxite (pétro.), ataxite.

atectonique, anorogenic.

atmophile (élément), atmophile (element).

atmosphère, atmosphere.

atmosphérique, atmospheric; **agent a.**, a. agent; **perturbations a.**, atmospherics; **pollution a.**, a. pollution; **pression a.**, a. pressure; **radiation a.**, a. radiation.

atoll, atoll.

atome, atom; **a. marqué (par radioactivité)**, tagged a.

atomicité, atomicity.

atomique, atomic; **liaison a.**, a. bond; **masse a.**, a. mass; **nombre a.**, a. number; **poids a.**, a. weight; **rayon a.**, a. radius; **structure a.**, a. structure.

attache musculaire, muscle attachment.

attapulgite, attapulgite.

attaque (chimique), etching.

attaquable, attackable.

attaquer, to corrode, to etch; **a. à l'acide**, to etch.

atterrissage, landfall.

atterrissement, alluviation.

Attique (orogénèse), Attic (orogeny Miocene: post-Sarmatian).

attraction, attraction, pull; **a. de la gravité**, gravitation; **a. magnétique**, magnetic attraction; **a. moléculaire**, cohesive force.

attrition, attrition.

Aturien, Aturian (u. Cretaceous).

aubrite (météorite), aubrite.

auge glaciaire, glacial trough.

augite (minér.), augite; **a. aegyrinique**, aegirite a.; **a. titanifère**, titanaugite; **basalte à phénocristaux d'a.**, augitophyre.

augitique, augitic.

augitite, augitite.

aulacogène (sillon), taphrogeosyncline.

au large, offshore.

auréole, aureole, halo; **a. de contact**, contact aureole; **a. métamorphique**, metamorphic aureole; **a. pléochroïque**, pleochroic halo; **a. réactionnelle**, reaction rim.

aurifère, gold bearing.

auro-argentifère, auri-argentiferous.

auroferrifère, auroferriferous.

auroplombifère, auroplumbiferous.

aurore polaire, polar aurora.

australite (pétro), australite.

Australopithèque, Australopithecus.

authigène, authigénique, authigenic, authigenous, authigenic.

authigenèse, authigenesis.

auto-capture, self capture.

autochtone (terrain), autochtonous.

autolyse, autolyse.

automatique (cartographie), automated cartography.

autométamorphique, autometamorphic.

autométamorphisme, autometamorphism.

automorphe, automorphic, automorphous, euhedral.

autopneumatolyse, autopneumatolysis.

autotrophe, autotrophic.

Autunien, Autunian (Permian).

autunite (minér.), autunite, lime uranite.

Auversien, Auversian (= Ledian: Eocene).

aval (d'un fleuve), down stream; **a. pendage**, down dip, bottom; **face aval (d'une dune)**, downwind side, leeside.

avalaison, freshet.

avalanche, avalanche; **a. boueuse**, lahar, mud stream; **a. de neige**, snow slip; **a. de neige poudreuse**, drift a.; **a. de rochers**, rock slide; **a. nivale de gravité**, snow a.; **a. sèche**, dry a.; **cône d'a.**, a. fan; **couloir d'a.**, a. chute, track.

avalanchite, 1. avalanche deposit; **2.** grain flow.

avaleresse, shaft, sinking.

avaloir, swallow-hole.

avancement d'un forage (exprimé en pieds), footage.

avancement d'une galerie, drift avance.

avant, fore; **a. butte**, outlier; **a. côte**, offshore; **a. dune**, f. dune; **a. fosse**, f. deep; **a. mont**, foot-hills; **a. pays**, f. land; **a. plage**, f. shore; **a. puits**, f. shaft; **a. récif**, f.-reef.

aven (karst), swallow hole, sink, aven, swallet, solution chimney, abyss.

aventurine, aventurine.

averse météorique, meteor shower.

avulsion (perte d'une terre par érosion), avulsion.

axe, axis; **a. anticlinal,** anticlinal a.; **a. cinématique,** structural a.; **a. cristallographique,** crystal a.; **a. d'hémitropie,** twin a.; **a. de rotation,** a. of rotation; **a. de symétrie,** a. of symmetry; **a. d'un pli,** fold a.; **a. hydrographique,** trunk-stream; **a. optique,** optic a.; **a. polaire,** polar a.; **a. synclinal,** synclinal a.

axial, axial; **lobe a.,** a. lobe; **plan a.,** a. plane.

axinite (minér.), axinite.

azimut, **1.** azimuth; **2.** strike.

azimutal, azimuthal.

azoïque, azoic, Precambrian (obsolete).

azonal (sol), azonal (soil).

azote, azote, nitrogen.

azoté, nitrogenous.

azoteux, nitrous.

azotique, nitric.

azurite, azurite.

B

bac, jar; **b. à boue**, mud tank; **b. de lavage**, wash tank.

bacalite (pétro.), bacalite.

bâche (sillon de plage), runnel offshore through.

bactériologique (origine), bacteriogenic (origin).

bactérie, bacterium, bacteria.

baddeleyite, baddeleyite.

baguette d'arpentage, 1. surveyor's rod; meter rod; 2. Jacob's staff.

baguette de sourcier, doodle bug, divining rod, dowsing stick (of rhabdomancy).

bahada, bahada.

bahamite (pétro.), bahamite.

baie, bay, embayment; **entrée de b.**, bay entrance; **fond de b.**, bay head.

bail minier, mining lease.

bain, bath; **b. de fusion**, melting b.; **b. marie**, laboratory b., water b.; **b. thermostaté**, thermostated b.

baisse, fall, falling, decline; **b. des eaux**, fall of flood, of water; **b. de pression**, pressure decline; **b. de production**, output fall.

baissée (du niveau de la mer), fall.

baisser (rivière, température), to fall.

Bajocien, Bajocian (m. Jurassic).

balance (appareil), balance; **b. à fléau**, beam b.; **b. de précision**, analytical b.; **b. de torsion**, torsion b.

balancier (de pompe), pumping beam.

balayage (sur un écran), sweep, sweeping.

balayage (géoph.), scan.

balise de transmission, data collection platform.

balise répondeuse, transponder.

ballast de pierre cassée, broken-stone ballast.

ballastière, gravel-pit, ballast pit.

balme, rock-shelter.

banatite (pétro.), banatite.

banc, bank, bar, bench; **b. à vérins**, Campanile giganteum bed (Lutetian); **b. corallien**, platform reef; **b. de graviers**, gravel bank, gravel bar; **b. de sable**, sand bank, sand bar; **b. fossilifère**, fossiliferous bed; **b. induré**, hardground; **b. marin**, bar, shingle, shoal; **b. royal**, Miliole bed (middle Lutetian bed).

bande (de terrain), belt, stretch, strip; **b. d'absorption**, band; **b. de sédiments**, streak, stripe; **b. de boue (glaciol.)**, dirt-band; **b. de spectre**, band; **b. de terrain**, land strip; **b. magnétique**, magnetic tape; **b. perforée**, punched tape; **spectre de b.**, band spectrum.

banque de données, data bank.

banquette, bench, bank; **b. de glace**, ice foot; **b. d'érosion marine**, wave-cut bench; **b. littorale**, berm.

banquise, ice-pack, pack, sea ice.

banquise côtière, fast ice.

bar, bar (unit of pressure).

bariolé, variegated, mottled.

barkévicite (minér.), barkevicite.

barkhane, barchan, crescentic dune; **aile de b.**, b. arm; **groupe de b.**, b. swarm.

barocline, baroclinic.

baromètre, barometer; **b. altimétrique**, orometer, moutain b.; **b. anéroïde**, aneroid b.; **b. enregistreur**, barograph, self-recording b.

barométrie, barometry.

barométrique, barometric; **pression b.**, b. pressure.

barothermographe, barothermograph.

barrage (construction), dam, weir; **b. d'écrêtement**, flood control reservoir; **b. de retenue**, retention dam; **b. de terre**, earth d.; **b. déversoir**, overflow d.; **b. naturel (d'une rivière par des rochers)**, barrier; **b. volcanique**, volcanic dam; **b. voûte**, arch dam.

barranca, barranco, barranca, dry ravine, gully (on volcanic cone).

barre, bar, barrier beach; **b. à mine**, miner's bar; **b. d'accrétion successive**, beach accretion ridge (of a delta); **b. d'eau (mascaret)**, tidal bore; **b. de méandre**, point bar; **b. de plage**, surf, barrier beach; **b. du relief**, hog back; **b. de roche**, rock bar; **b. de sable**, sand bar; **b. d'estuaire**, bar, tidal bore; **b. frontale**, b. front; **b. frontale deltaïque**, deltaic face of a bar.

Barrémien, Barremian (l. Cretaceous).

barrer, to dam, to bar.

barrière, barrier, fence; **b. de glace**, ice-barrier; **récif-barrière**, barrier reef.

barril (pétrole), barrel (1 bbl = 42 gallons U.S.; = 35 gallons U.K.; 0,15876 m^3).

Bartonien, Bartonian (u. Eocene).

barycentre, barycenter, center (centre, U.K.) of mass. of two or more celestial bodies.

barylite (m. lourds), barylite.

barysphère, barysphere.

barytifère, barytic.

barytine (minér.), barite, heavy baryte, heavy spar, cawk; **b. crêtée**, crested barite.

barytique, barytic.

barytocalcite (minér.), barytocalcite.

barytocélestine, barytocélestite (minér.), barytocelestite.

baryum, barium.

bas, low; **b. de pente**, foot slope; **b. de plage**, fore shore; **b. fond**, shallow, flat, low ground, deep, pit; **b. pays**, low land.

basale (couche), basal (bed).

basalte, basalt; **b. à leucite**, leucite b.; **b. à néphéline**, nepheline b.; **b. à olivine**, olivine b.; **b. de plateau**, plateau b.

basaltique, basaltic; **lave b.,** b. lava; **orgue b.,** columnar basalt; **primation b.,** b. jointing; **tuf b.,** b. tuff; **verre b.,** b. glass.

basaltoïde, basaltiform.

basanite, basanite.

basculement (de couches), tilting, flip.

base (chimie), base; **base (niveau de) de l'érosion,** base level of erosion; **échange de b.,** base exchange.

base de banc, sole cast; **b. (soubassement),** basis, bottom, foot.

basicité, basicity.

basique, basic; **b. roche,** b. rock.

basse (banc sous-marin), bank.

basses eaux, low water.

basse mer, low tide.

basse pression, low pressure.

basse teneur, low grade.

bassin, basin; **b. alluvial,** alluvial b.; **b. aréïque,** arheic b.; **b. artésien,** artesian b.; **b. cryptoréique,** cryptorheic b.; **b. de bordure,** bordering b.; **b. de décantation,** settling b.; **b. d'effondrement,** fault-basin, rift; **b. de surcreusement,** river b.; **b. endoréique,** endorheic b.; **b. exoréique,** exorheic b.; **b. fluvial,** river b.; **b. fluviatile,** river b.; **b. géologique,** geological b.; **b. houiller,** coal field; **b. hydrographique,** watershed, drainage b., catchment b.; **b. intraarc,** intraarc b.; **b. limnique,** limnic b.; **b. marginal,** marginal sea; **b. marginal interne,** back-arc b.; **b. marginal externe,** fore-arc b.; **b. océanique,** ocean b.; **b. paralique,** coastal b.; **b. pétrolifère,** oil b.; **b. rétro-arc,** behind-arc b.; **b. sédimentaire,** b., sedimentary b.; **b. structural,** structural b., synclinal; **b. tectonique,** fault b., tectonic b.; **b. versant,** watershed, river b.

bastite, bastite, schiller-spar.

batée, battée, pan, wash pan, batea, wash trough, washing dish.

batholite, batholith.

batholithique, batholithic.

Bathonien, Bathonian.

bathyal, bathyal.

bathygraphique, bathygraphic.

bathymètre, bathymeter, bathometer.

bathymétrie, bathymetry.

bathymétrique, bathymetric(al), bathometric(al); **carte b.,** b. map.

bathypélagique, bathypelagic.

bathyscaphe, bathyscaphe.

Batracosauriens (pal.), Batracosauria.

battage au câble, spudding.

battage d'or, gold beating.

battance, soil capping.

battement (d'ondes), beat.

bauéritisation, micas weathering (with loss of Fe).

baume du Canada, Canada balsam.

baume (abri), rock shelter.

bauxicroûte, bauxicrust.

bauxite, bauxite.

bauxitique, bauxitic.

bauxitisation, bauxitization.

bayou, bayou.

béant, béante (fissure), gaping, yawning (fissure).

Beaufort (échelle de), Beaufort scale.

« bec de l'étain », twinned cassiterite.

Becke (frange de), Becke line.

bédière, glacial rill.

Bédoulien, Bedoulian (l. Cretaceous).

beerbachite (pétro.), beerbachite.

beidellite (minér.), beidellite.

beine, top-beds (of a delta).

Bélemnite (pal.), Belemnite.

Bélemnoïdés (pal.), Belemnoidea.

Beltienne (orogénèse), Beltian Orogeny.

belvedere, look-out, vantage point, plateau rim, terrace.

Bénioff (plan de), Benioff plane.

« bénitier » (pal.), Tridacna.

benne, bucket; **b. à bascule,** tipping b.; **b. de creusement,** sinking b.; **b. excavatrice,** excavating or hoisting b.

Bennettitales (paléobot.), Bennettitales.

benthique, benthic.

benthonique, benthonic.

benthos (pal.), benthos.

bentonite, bentonite, Denver mud.

benzène, benzene.

berceau (vallée en), dale.

berge, bank.

berline, mine car, colliery wagon.

berme, berm.

Berriasien (Néocomien), Berriasian (l. Cretaceous).

Bertrand (lentille de), Bertrand lens.

béryl, beryl.

béryllium, beryllium.

bêta (rayonnement), beta (radiation).

bétain, concretionary sand.

bétoire (karst), swallow-hole, sink-hole, sink.

béton, concrete; **b. armé,** reinforced c., armoured ferroconcrete; **b. précontraint,** prestressed c.

biaxe, biaxial; **cristal b.,** b. crystal.

bicarbonate, bicarbonate.

bichromate, bichromate.

biéberite, bieberite, cobalt vitriol.

bief (canal), reach, level; **b. de rivière,** stillwater reach; **b. à silex (Picardie),** clay with flints, decalcification residue.

biface (préhist.), handaxe.

bigarré, variegated, mottled; **grès b.,** variegated sandstone, « Buntsandstein » (strati.).

bilan géochimique, evaluation.

bilan glaciaire, glacial balance.

bilan hydraulique, water balance.

bilatéral, bilateral; **symétrie b.,** bilateral symmetry.

biloculaire, bilocular.

bindheimite, bindheimite.

binoculaire, binocular; **loupe b.,** b. lens; **microscope b.,** b. microscope.

binominal (système), binomial (system).

biocénose, biocoenose, biocoenosis, life association.

biochimie, biochemistry.
biochimique, biochemical; **sédiment b.,** biochemical deposit.
biochore, biochore.
bioclaste, bioclast.
bioclastique, bioclastic.
biodégradable, biodegradable.
biodétritique (sédiment), biomechanical (deposit).
biofaciès, biofacies.
biogenèse, biogenesis.
biogéochimie, biogeochemistry.
biogéochimique, biogeochemical.
biogéographie, biogeography.
bioherme, bioherm.
biolithite (calc. récifal), biolithite.
biologique, biologic(al); **d'origine b.,** biogenic; **espèce b.,** b. species.
biologie, biology.
biomasse, biomass.
biome, biome.
biométrie, biometry.
biomicrite, biomicrite.
biominéral, biomineral.
biorhexistasie, biorhexistasy.
biosparite, biosparite.
biosphère, biosphere.
biostratigraphie, biostratigraphy, biostratonomy.
biostratigraphique, biostratigraphic; **zone b.,** b. zone.
biostrome (pétro. séd.), biostrome.
biostructure, biostructure.
biotique, biotic.
biotite (minér.), biotite.
biotope, biotope.
bioturbation, bioturbation.
bioturbé, bioturbate.
bioxyde, dioxide.
biozone, biozone.
bipolaire, bipolar.
bipyramidé (quartz), bipyramidal (quartz).
biréfringence, birefringence.
biréfringent, birefringent, double refracting.
biseau (cristallographique, stratigraphique), bevelment, wedge, pinching out, nip.
biseautage (d'une couche), pinching out, wedging out.
bisérié (pal.), biserial.
bismuth, bismuth.
bismuthifère, bismuthiferous.
bismuthinite, bismuth glance.
bismuthique, bismuthic.
bitume, bitumen, asphalt; **b. de pétrole,** asphalt petroleum; **b. lacustre,** lake asphalt; **b. naturel,** natural bitumen.
bitumineux, bituminous; **calcaire b.,** b. limestone; **charbon b.,** b. coal; **sable b.,** b. sand; **schiste b.,** oil shale, b. shale.
bituminisation, bituminization.
bivalence, bivalence.
bivalent, divalent, bivalent.
Bivalves, Bivalvia, Pelecypods.

blanc, white; **b. de chaux,** whitewash, limewash; **b. de plomb,** w. lead, ceruse; **b. de zinc,** w. zinc.
blanchiment (pédo), bleaching.
blastèse, blastesis, blastogenesis (of porphyroblasts).
Blastoïdés, Blastoidea.
blastique (suffixe), blastic.
blastomylonite, blastomylonite.
blaviérite, sericitized rhyolite.
blende, sphalerite, blend, blende, false galene, jack, black jack, zinc blend.
bleu, blue; **algues b.,** Cyanophyta; **vase b.,** b. mud.
bloc (sédimentologie, tectonique), block, boulder block, fault block, small klippe; **b. charrié,** overthrust block; **b. continental,** craton; **b. démesuré,** ice-rafted block, oversized boulder; **b. diagramme,** block diagram; **b. erratique,** drift boulder, erratic block, drop stone; **b. perché (glaciaire),** perched block, stray block; **b. (cheminée de fée),** chimney rock, earth pillar; **b. résiduel d'altération (matacoe, Brésil),** residual weathered boulder; **b. soulevé,** up-thrown block; **b. transporté par glaces flottantes (bl. démesuré),** ice-rafted block.
blocaille, rubble stone, scree.
blocailleux, rubbly.
bocard, stamp, ore-crusher.
bocard à minerai, ore-stamp.
bocardage, milling, stamping.
bocardage à l'eau, wet stamping.
bocardage à sec, dry stamping.
bocarder, to mill, to stamp.
bocardeur, millman.
boehmite, boehmite.
bogaz, deep cleft, gorge, channel (Turk.).
boghead, boghead.
bois, wood; **b. fossile,** fossil w.; **b. opalisé,** opalized w.; **b. pétrifié,** petrified w., w.-stone; **b. silicifié,** silicified w.
boisage (de galeries), timbering; **b. de chambres,** room t.; **b. de puits,** shaft t.; **b. jointif,** close t.
boisé (puits, galerie), timbered, wooded.
boiser (une galerie), (une région), to timber, to wood.
bojite (gabbro), bojite.
bombé (relief), arched, bulged, cambered.
bombe volcanique, volcanic bomb; **b. en croûte de pain,** bread-crust bomb.
bombement, upwarp, upwarping, upswell, bulge, camber, swelling; **b. anticlinal,** anticline bulge.
bombement volcanique (à la surface d'une coulée), tumulus.
bomber, se bomber, to bulge, to swell, to camber.
bonne qualité (de minerai), high grade.
Bononien, Bononian (u. Jurassic, l. Portlandien).
boracite, boracite.
boralf (pédol.), boralf.
borate, borate.
boraté (lac), borax lake.
borax, borax.
bord, edge, border, rim; **b. d'une rivière,** bank, riverside; **b. de mer,** seabord.
bord abattu (préhist.), blunted side.

bordière (faille), boundary (fault).
bordure (d'un gisement), edge, border, margin, rim; **b. d'un continent**, continental margin; **b. figée**, congealed rim, chilled margin; **b. réactionnelle**, reaction rim.
bore, bore.
boréal, boreal, borealis.
borinage, coal-mining district.
borique (acide), boric (acid).
bornage (d'un terrain), boundary.
bornite (minerai de cuivre), bornite, erubescite, peacock copper, variegated copper ore.
boroll (pédol.), boroll.
borosilicate, borosilicate.
bort (diamant), bort.
bosse (de terrain), mound, hill.
Bothnien, Bothnian (Precambrian).
botryogène, botryogen.
botryoïde, botryoidal.
bouche (de cheminée volcanique), vent.
boucher (un puits de mine), to seal (a shaft).
bouchon (de cheminée volcanique), plug, neck.
boucle cosmique (astro.), cosmic loop.
boucle (de méandre), loop.
bouclier, shield; **b. céphalique (pal.)**, head-s.; **b. continental**, continental s.; **b. de laves**, lava s.
boudinage (tecto.), boudinage.
boue, mud, sludge, slush, slime; **b. à Diatomées**, Diatom ooze; **b. à Globigérines**, Globigerine ooze; **b. à Radiolaires**, Radiolarian ooze; **b. argileuse**, clay mud; **b. calcaire**, lime mud; **b. d'injection**, mud fluid; **b. de forage**, drilling mud; **b. glaciaire**, glacier silt, boulder clay; **b. marine**, ooze, sea mud; **coulée de b.**, mud flow; **cratère de b.**, mud crater; **volcan de b.**, mud volcano.
boueux, muddy, sludgy, silty.
Bouguer (anomalie de), Bouguer anomaly.
bouillant, boiling.
bouillie (périgl.), sludge.
bouillir (magma), to boil.
bouillonner, to boil, to bubble.
boulant (sable), quick (sand).
boulbène (S.O. France), loam.
boule (désagrégation en), spheroidal weathering; (cf. **onion-skin weathering**, exfoliation).
boulette, pellet.
bouleversement (tectonique), convulsion.
bourbeux, splashy, muddy, miry.
bourbier, slough, mire, mud-pit.
bourgeonnement (pal.), gemmation.
bournonite, bournonite.
bourrage (d'un trou de mine), bulling.
bourrage tectonique, tectonic thickening.
bourrelet de plage, beachridge, natural levee, shore rampart; **b. de poussée glacielle**, ice-shoved ridge; **b. de gélifluction**, gelifluction step, solifluction bench, terracette.
bourrer (mine), to stem, to tamp, to ram.
boursouflement, heave, heaving.
boursouflure de gel, du sol, upheaving; **b. de lave (pustule)**, hornito.

boussole, compass, dial; **b. d'inclinaison**, dipping compass; **b. géologique**, geologic compass; **b. de mine**, mine dial.
bout-du-monde (karst), closed valley, cul-de-sac, dead end.
boutefeu (ouvrier), fireman, blaster.
boutonnière (géomorphol.), exhumed and eroded anticlinal fold.
boyau (mine), pipe, trench, breakthrough.
brachidium (pal.), brachidium.
Brachiopode, Brachiopod.
brachyanticlinal, brachy-anticline, dome, quaquaversal fold.
brachycéphale (anthropo.), brachycephalic.
brachyodonte (Mollusque), brachyodont.
brachypli, brachyfold.
brachysynclinal, brachysyncline, centroclinal fold, basin.
bradygenèse, bradygenesis.
bradyséisme, bradyseism.
brai, tar, pitch; **b. de houille**, coal tar; **b. de pétrole**, petroleum tar.
branche (d'un filon), offshoot, ramification branching.
branchial (pal.), branchial.
branchie (pal.), branchiae, gills (fish).
Branchiopodes, Branchiopods.
bras, arm; **b. de mer**, sound; **b. de rivière**, arm, distributary; **b. mort (rivière)**, oxbow-lake, cut-off meander; **faux-b.**, blind arm.
braunite, braunite.
Bravais (réseau de), crystallographic lattice.
brèche, breccia; **b. de faille**, fault b.; **b. de friction**, friction b., crush b., friction gouge; **b. de pente**, avalanche b.; **b. d'effondrement**, collapse b.; **b. d'extrusion**, injection b.; **b. éruptive**, eruptive b.; **b. intraformationnelle**, intraformational b.; **b. monogénique**, monogenic b.; **b. osseuse**, bone bed, osseous b.; **b. polygénique**, polygenic b.; **b. pyroclastique**, pyroclastic b.; **b. récifale**, reef b.; **b. salifère**, saliferous b.; **b. sédimentaire**, sedimentary b.; **b. tectonique**, cataclastic b.; **b. volcanique**, pyroclastic b.; **formation de b.**, brecciation; **fausse b.**, pseudo-b.
bréchification, brecciation.
bréchiforme, bréchoïde, brecciated.
bréchique, brecciated; **conglomérat b.**, breccio-conglomerate.
brillance, brightness, brilliancy.
Briovérien, Brioverian (Precambrian).
brique, brick; **b. cuite**, burnt b.; **b. réfractaire**, fire b., refractory b.; **terre à b.**, b. earth.
briqueterie, brickworks.
briquette, brick, briquette; **b. de charbon**, cake of carbon; **b. de tourbe**, peat brick.
brisant (océano.), breaker, reef.
brise-glace, ice-breaker.
briser (par le gel), to break, to shatter.
brocatelle, brecciated marble.
brochantite, brochantite.
broiement (de minerai), crushing, grinding.

bromargyrite, bromargyrite, bromyrite, bromite.
brome, bromine.
bromite, bromite.
bromoforme (liqueur dense), bromoform.
bromure, bromide.
bromyrite, bromyrite, bromite.
Brontosaure (pal.), Brontosaurus.
bronze, bronze.
bronzite, bronzite.
bronzitite (pétro.), bronzitite.
brookite (minér.), brookite.
brouillard, fog, mist.
brownien, brownian (motion).
broyable, grindable.
broyage, grinding, crushing, milling, shattering; **b. humide,** wet crushing; **b. par cylindres,** crushing by rolls; **b. sec,** dry grinding; **b. secondaire,** regrinding.
broyer, to crush, to grind, to mill.
broyeur, mill, crusher, breaker; **b. à cylindre,** rolling crusher, crushing rolls; **b. à minerais,** ore-crusher.
brucelles (pince), tweezers.
brucite (minér.), brucite.
bruine (météo.), drizzle.
bruit, noise; **b. de fond,** ground n., random n.; **b. parasite,** background n.; **b. superficiel,** random n.
brûler, to burn, to burn down, to calcine.
brûleur, burner.
brume, thick haze, mist.
brumeux, hazzy, foggy.
brun, brown; **sol brun,** brown soil; **s. b. calcaire,** calcareous b. soil; **s. b. forestier,** b. forest soil; **s. b. podzolique,** b. podzolic soil.
brut, **1.** adj : raw; **2.** n : crude (oil).
brut à base mixte, mixed base oil; **b. léger,** light crude; **b. sulfuré,** high sulfur crude.
Bruxellien, Bruxellian (m. Eocene).
Bryophytes (paléobot.), Bryophyta.
Bryozoaires, Bryozoa, bryozoair.
bulle (de gaz), bubble; **niveau à b.,** bubble level; **formant des b.,** bubbly.
bulleux : à petites cavités, vuggular, vuggy.
Bunsen (bec), Bunsen (burner).
Burdigalien, Burdigalian (l. Miocene).
bureau d'études géologiques, geological survey.
burette, burette.
burette de Mohr, Mohr pipet.
burin : moderne, chipper, chisel; **b. préhistorique,** burin.
buriner (géogr.), to chisel.
burmite, burmite.
bustite, bustite.
butane, butane.
butte, conical hill, butte, mound, knoll, hillock; **b. à lentille de glace,** ice mound; **b. de terre,** earth hummock; **b. gazonnée,** earth mound, thufur; **b. résiduelle (karstique),** hum haystock, hum; **b. témoin,** residual hill, small inselberg; **avant-b.,** outlier.
bysmalite, bysmalith.
byssus, byssus.
bytownite, bytownite.

C

câble, cable, rope; **c. d'extraction,** hoisting cable, hoisting rope; **c. de curage (for.),** bailing rope; **c. de forage,** drilling cable, drilling line.

cadastre, **1.** land register, cadastral map, property map; **2.** cadastral survey.

cadmifère, cadmiferous.

Cadomien, Cadomian (Precambrian).

cadoueyre (S.O. France), blow out.

cadre, frame, casing, timber set; **c. de boisage,** frame set; **c. de puits,** shaft frame, shaft set.

caesium, caesium.

cafémique (pétro.), cafemic.

cage, cage; **c. d'extraction,** drawing c., hoisting c.

caillasse, **1.** brackish marl and limestone of upper Lutetian (Paris basin); **2.** hard siliceous bed; **3.** broken stones.

caillou, pebble; **c. à facettes,** wind faceted p.; **c. éolisé,** ventifact; **c. émoussé,** rounded p.; **c. vermiculé,** water-worn p.

cailouteux, pebbly, gravelly; **plage c.,** shingly beach.

caillotis, gravel, broken stone; **c. d'empierrement,** ballast stone; **c. émoussés,** gravel.

cairn, cairn.

caisse de criblage, screening box.

caisson à minerai, ore bin.

Calabrien, Calabrian (l. Pleistocene).

calage (d'une séquence), calibration.

calaïte, turquoise.

calamine, calamine; **gisement de c.,** calamine deposit.

calanque, cala.

calcaire, **1.** adj: calcareous, limy; **2.** n: limestone; **c. à ciment,** cement stone; **c. à Crinoïdes,** Crinoidal l.; **c. à entroques,** encrinitic l.; **c. à polypiers,** coral l.; **c. à silex,** cherty l.; **c. à tubulures,** burrowed l.; **c. argileux,** clayed l., argillaceous l.; **c. asphaltique,** asphaltic l.; **c. bitumineux,** bitumous c.; **c, bréchique,** brecian l., brecciated l.; **c. cavernous,** cavernous l.; **c. compact,** compact l.; **c. concrétionné,** ballstone; **c. construit,** reef l., bioherm; **c. coquillier,** coquina, coquinoid l., shelly l.; **c. corallien,** coral l., coralline l.; **c. crayeux,** chalky l.; **c. cristallin,** crystalline l.; **c. détritique,** clastic l.; **c. dolomitique,** dolomitic l.; **c. fétide,** stinkstone l.; **c. fossilifère,** fossiliferous l.; **c. glauconieux,** glauconitic l.; **c. granuleux,** granular l.; **c. gréseux,** sandy, cornstone l.; **c. grossier,** Lutetian l. (Paris basin); **c. hydraulique,** clayey calcareous formation; **c. lacustre,** lacustrine l.; **c. lithographique,** lithographic l.; **c. lumachellique,** shelly l., coquinoid l.; **c. magnésien,** magnesian l.; **c. marneux,** marly l.; **c. microcristallin,** microscrystalline l., microsparite; **c. noduleux,** knobly l.; **c. oolithique,** oolitic; **c. pétrolifère,** oil-bearing l.; **c. phosphaté,** phosphatic l.; **c.**

pisolithique, pisolitic l.; **c. poreux,** porous l.; **c. récifal,** reef l.; **c. sableux,** arenaceous l., sandy l.; **c. siliceux,** siliceous l.; **c. spathique,** spathic l.

calcarénite (pétro.), calcarenite.

calcaréo-argileux, calcareo-argillaceous.

calcaréo-ferrugineux, calcareo-ferruginous.

calcaréo-magnésien, calcareo-magnesian.

calcaréo-siliceux, calcareo-silicious.

calcareux, calcariferous.

calcédoine, chalcedony.

calcédonieux, calcedonic.

calcédonite, calcedonite.

calcifère, calciferous.

calcification, calcification.

calcifié, calcified.

calcifier, to calcify.

calcilutite, calcilutite.

calcimètre, calcimeter.

calcimorphe, calcimorphic.

calcin, calcin, limestone hardpan; **c. en « chou-fleur »,** cauliflower c.

calcinable, calcinable.

calcination, calcination, calcining, roasting.

calciner, to roast, to calcine (ore).

calciocélestine, calciocelestite.

calcioferrite, calcioferrite.

calcique, calcic.

calcirudite, calcirudite.

calcite, calcite, calc spar.

calcite spathique, sparry calcite.

calcium, calcium.

calco-alcalin, calc-alkaline.

calco-sodique, soda-lime.

calcomalachite, calcomalachite.

calcrète, calcrete.

calcschiste, calcareous schist.

calculateur analogique, analog computer.

caldeira, caldera; **c. d'affaissement,** collapse c.

caldérite, calderite.

Calédonien (cycle), Caledonian (orogeny).

Calédonides, Caledonides.

calédonite, caledonite.

calibrage (de matériaux), sizing.

calibre, caliper, gage, gauge.

calibrer, to calibrate, to gauge, to measure.

calice (pal.), calyx.

caliche, caliche.

californite, californite.

callaïnite, callainite.

Callovien, Callovian (m. Jurassic).

calomel, calomel, mercurous chloride.

calorie, calory; **grande c.,** great calory, large calory; **petite c.,** gram calory, lesser calory, small calory.

calorifique, caloric.

calorimétrie, calorimetry.

calotte glaciaire, ice cap, ice sheet, glacial sheet; **c. continentale,** continental ice sheet.

Calpionnelles, Calpionnellidae.

calque, tracing.

cambisol, cambisol.

Cambrien, Cambrian (l. Paleozoic).

camion laboratoire, recording truck.

camp de terrain, training field camp.

campagne de sismique-réflexion, reflection survey.

campagne de sismique-réfraction, refraction survey.

Campanien, Campanian (u. Cretaceous).

canal, canal; **c. d'amenée,** intake channel, supply channel; **c. d'évacuation,** discharge channel; **c. fluviatile, marin,** channel; **c. de drainage,** drainage drain; **c. d'irrigation,** ditch; **c. endosiphonal (*pal.*),** endosiphonal c.; **c. siphonal (*pal.*),** siphuncle.

canalisation, 1. pipe, pipe-works, distribution system; 2. canalization (of a river); **c. d'eau,** water pipe; **c. pétrolière,** oil line; **c. principale,** main line; **c. terminale,** terminal line.

canaliser, 1. to canalize (a river); 2. to pipe (oil), to pipeline.

cancrinite, cancrinite.

canevas stéréographique, stereographic net; **c. structural,** structural pattern.

cannelé, grooved, fluted.

canneler, to flute, to groove, to corrugate.

cannelure, groove, flute, corrugation, furrow.

canon à air (géoph.), air-gun.

canyon, canyon; **c. sous-marin,** sub-marine c.

cap, cape, headland, foreland.

capacité, capacity; **c. capillaire,** capillary c.; **c. d'adsorption,** adsorbing c.; **c. d'échange,** exchange c.; **c. de production,** productive c.; **c. de rétention d'eau,** water-holding c.; **c. de transport fluviatile,** transport c.; **c. électrique,** capacitance; **c. en air,** air c.; **c. en eau,** moisture c.

capillaire, capillary; **eau c.,** capillary water.

capillarité, capillarity.

captage d'eau, water-catchment.

captage d'une source, tapping of a spring.

captation d'eau, catching of water.

capter, 1. to pipe, to catch (water); 2. to collect, to pick-up (electricity).

capteur, scanner, sensor, detector, pick-up, receiver.

capteur à vanne de flux, fluxgate detector.

capture (géogr.), capture, piracy; **c. d'un cours d'eau,** stream piracy; **c. par déversement,** spontaneous capture, overspill; **coude de c.,** elbow of capture; **point de c.,** point of capture.

capturer, to capture.

caractère, character, characteristic, feature, property.

Caradocien, Caradocian (v. Ordovician).

carapace, 1. zool.: carapace; 2. geol.: hardpan; **c. latéritique,** thin laterite crust.

carat, carat.

carbonado, bort, black diamond.

carbonatation, carbonatation.

carbonate, carbonate; **c. de chaux,** c. of lime, calcium c.; **c. de fer,** iron c.; **c. de sodium anhydre,** soda ash.

carbonaté, carbonated.

carbonatite, carbonatite.

carbone, carbon; **c. fixe,** fixed c.; **c. libre,** free c.

carboné, carbonaceous.

carboneux, carbonous.

carbonifère, 1. adj : carboniferous, coal bearing; 2. n : Carboniferous.

carbonification, carbonization, charring.

carbonique, carbonic; **anhydride c.,** carbon dioxyde; **gaz c.,** carbon dioxyde, carbonic gas.

carbonisable, carbonizable.

carbonisation, carbonization, charring, coking.

carbonisé, carbonized, charred.

carboniser, to carbonize, to char.

carborundum, carborundum.

carburant, fuel.

carbure, carbide; **c. de calcium,** c. of calcium.

carburer, to carburize.

cardinal, cardinal; **dent c. (*pal.*),** cardinal tooth; **point c.,** cardinal point.

carène (pal.), carina, keel.

caréné, carinated.

Carlsbad (mâcle), Carlsbad twin.

carnallite, carnallite.

carnet de levé, survey book.

carnet de sondage, bore-holing journal.

carnet de terrain, field book.

Carnien, Carnian (u. Triassic).

carnieule, cellular dolomite.

carnivore (pal.), carnivora.

carnotite, carnotite.

carottage, core drilling, coring; **c. au câble,** cable tool drilling; **c. continu,** continuous coring; **c. électrique,** electric well logging; **c. sismique,** well shooting.

carotte, core, core sample, drill core, boring sample; **c. de forage,** drilling core; **c. latérale,** side well core.

carotter, to core.

carottier, sampler, core barrel, core drill, core bit.

carottier latéral, side-wall sampler.

Carpoïdés (pal.), Carpoidea.

carré, square.

carreau (d'une carrière), head.

carreau (d'une mine), bank, bank-head, surface plant.

carrier, quarryman, quarrier.

carrière, quarry, pit; **c. à ciel ouvert,** open quarry; **c. d'argile,** clay pit; **c. de gravier,** gravel pit; **c. de pierre,** stone quarry; **c. de sable,** sand pit.

carroyage, squarring.

carte, map, chart; **c. bathymétrique,** bathymetric chart; **c. de cadastre,** cadastral m.; property m.; **c. de formation,** formation m.; **c. de surface,** areal m.; **c. dépliante,** folding m.; **c. des déclinaisons magnétiques,** magnetic c.; **c. d'isobathes,** isobath m.; **c. d'isochores,** isochor m.; **c. d'isochrones,** isochronic m.; **c. d'isopaques,** isopach m.; **c. en**

courbes de niveau, contour m.; **c. en relief,** three-dimensional m.; **c. entoilée,** mounted on cloth m.; **c. géologique,** geological m.; **c. géomorphologique,** physiographic m.; **c. gravimétrique,** gravimetric m.; **c. hydrographique,** hydrographic m.; **c. hypsographique,** hypsographical m.; **c. isanomale,** isanomalic m.; **c. isopaque,** isopach m.; **c. lithologique,** lithologic m.; **c. marine,** nautical chart; **c. météorologique,** weather chart; **c. ombrée,** cartogram; **c. orographique,** orographic m.; **c. paléogéographique,** paleogeographic m.; **c. paléotectonique,** paleotectonic m.; **c. pluviométrique,** rain chart, isohyetal m.; **c. structurale,** structural m., structural contour m.; **c. subgéologique,** earthworm m.; **c. topographique,** topographical m.

cartilage (pal.), cartilage.
cartographe, cartographer, mapper.
cartographie, cartography, mapping.
cartographie radar, radar imagery.
cartographique, cartographic, cartographical.
carton (schistes), paper schist.
cartothèque, map library.
cascade, cascade, falls, waterfall.
case lysimétrique, lysimeter.
cassant, brittle, breakable.
casse (inusité), scree.
casser, to break, to fracture.
cassitérite, cassiterite.
cassure, fracture, crack, break; **c. compacte,** compact fracture; **c. conchoïdale,** conchoidal fracture; **c. fibreuse,** fibrous fracture; **c. inégale,** uneven fracture; **c. lamelleuse,** lamellated fracture; **c. nette,** clean fracture; **c. saccharoïde,** saccharoidal fracture; **c. schisteuse,** slaty fracture.
castorite, castorite.
cataclase, cataclasis.
cataclastique, cataclastic.
cataclinal, cataclinal.
cataclysme, cataclysm.
cataclysmique, cataclysmic, cataclysmal.
catagenèse, catagenesis.
cataglaciaire, cataglacial.
catalyse, catalysis.
catalyser, to catalize.
catalyseur, catalyst.
catalytique, catalytic.
cataracte, cataract, falls.
catastrophisme, catastrophism.
catazone, catozone.
cathode, cathode.
cathodique, cathodic; **rayons c.,** cathodic rays.
cation, cation.
catlinite, catlinite, pipestone.
caudal (pal.), caudal.
causse, barren limestone plateau, karst plateau.
cavage (mine), crowding.
caverne, cave, cavern.
caverneux, cavernous, vuggy.
cavitation, cavitation.

cavité, cavity, hole, pit, vough; **c. cristallisée,** bird-eyes; **c. de dissolution,** solution cavity; **c. tourbillonnaire,** pothole.
caye, sand cay.
Caytoniales (paléobot.), Caytoniales.
cédarite, cedarite.
ceinture orogénique, orogenic belt.
ceinture plissée, folded belt.
céladonite, celadonite.
célestine, celestite, celestine.
cellulaire (pal.), cellular.
cémentation, cementation.
cendre, ash, cinder.
cendreux, ashy, ash-like.
Cénomanien, Cenomanian (v. Cretaceous).
Cénozoïque, Cenozoic, Cainozoic, Kainozoic (U.K.).
cénozone (pal.), cenozone.
centigrade, centigrade; **degré c.,** centigrade degree, Celcius [$1\,°C = 1{,}8\,°F$; temperature conversion: $0\,°C = 5/9\,(°F - 32);\ °F = 9/5\,°C + 32$]; **échelle c.,** centigrade scale, Celsius scale.
centigramme, centigram.
centilitre, centiliter.
centimètre, centimeter.
centipoise, centipoise.
centrale, central station, plant; **c. électrique,** power plant, generating plant; **c. nucléaire,** nuclear plant; **c. thermique,** thermal p.
centre, center, centre, midpoint; **c. de gravité,** centre of gravity; **c. éruptif,** eruption point; **c. d'un séisme,** focus; **c. minier,** mining center; **c. de recherches,** research department.
céphalique (pal.), cephalic.
céphalon (pal.), cephalon.
Céphalopodes (pal.), Cephalopoda, Cephalopods; **c. dibranchial,** dibranchial c.; **c. tétrabranchial,** tetrabranchial c.
céphalothorax, cephalothorax.
céramique, ceramic; **industries c.,** pottery industry.
cérargyrite, cerargyrite, horn silver.
Cératite, Ceratite.
Cératidés, Ceratida.
cercle, circle; **c. avec triage du matériel,** sorted c.; **c. de pierre,** stone ring, stone c.; **c. de tourbe,** peat ring; **c. hydrographique,** sextant; **c. polaire,** polar c.
cérine, cerine, allanite.
Cérithe, Cerithium (*pal.*).
cérium, cerium.
cérusite, cerussite, cerusite, cerussite, lead spar.
césium, cesium, caesium.
Cétacés, Cetacea.
ceylanite, ceylonite, ceylonite.
chabasie, chabacite, chabazite, chabasite.
chaille, chert, siliceous concretion.
chaîne, chain, ridge, range; **c. à godets,** bucket chain, conveyor chain; **c. anticlinale,** anticlinal range; **c. cyclique (chimie),** ring chain; **c. d'arpentage,** measuring chain, surveying chain, land chain; **c.**

de liaison (chimie), binding chain; c. de montagne, mountain range system; c. de sols, catenary soil association, catena.

chaînon (pal.), link.

chalcanthite, chalcanthite, blue vitriol.

chalcocite, chalcosite, chalcosine, copper glance.

chalcolite, chalcolite, torbernite.

chalcophile, chalcophile.

chalcophyllite, chalcophyllite, copper mica.

chalcopyrite, chalcopyrite, copper pyrite, yellow copper ore.

chalcosidérite, chalcosiderite.

chalcosphère (manteau sup.), chalcosphere.

chalcosine, chalcocite, copper glance.

chalcostibine, chalcostibite, chalcostibite.

chalcotrichite, chalcotrichite.

chaleur, heat, warmth; c. latente de cristallisation, latent heat of crystallization; c. latente de fusion, latent heat of fusion; c. latente de vaporisation, latent heat of vaporization; c. spécifique, specific heat.

chalumeau, blow pipe.

chalybite, chalybite, spathic iron ore.

chambre, chamber, cavity, camera; c. de grillage, roasting c.; c. de mine, mine c., stope; c. de prise de vues photogrammétriques, surveying camera; c. d'habitation (pal.), body c., living c.; c. magmatique, magmatic c., c. vide (mine), open stope.

chamoïsite, chamosite, chamoisite.

champ, field; c. aurifère, gold f.; c. de blocs, block f.; c. de dunes, dune f.; c. d'exploitation (mine), winning f.; c. de fractures, cluster of faults; c. de gaz naturel, gas f.; c. de glace, ice f.; c. de laves, lava f.; c. de neige, snow f.; c. de pétrole, oil f.; c. de pierres, stone f.; c. dipolaire, dipole f.; c. magnétique, magnetic f.; c. pétrolifère, oil f.; c. visuel, f. of vision.

changement, change, variation; c. de pendage, dip reversal; c. de polarité, phase reversal.

chantier (mine), working, workings, working place; c. d'abattage, stope; c. à ciel ouvert, open-cast working, openwork; c. de lavage (de minerais); washings; c. en gradins, stope; c. épuisés, exhausted workings.

chaos, chaos; c. de blocs, block field, boulder field, tor.

chaotique, chaotic.

chape, 1. cap, cover, lid; 2. coating.

chapeau, cap, capping, c. rock; c. de fer, oxidized c., gossan, ironstone, iron hat; c. de gaz, gas c.

Characées (pal.), Characea.

charbon, coal; c. à coke, coking c.; c. anthraciteux, hard c.; c. asphaltique, asphaltic c.; c. bitumineux, bituminous c.; c. de bois, charcoal, wood c.; c. demi-gras, semi-bituminous c.; c. de tourbe, peat c.; c. feuilleté, foliated c.; c. flambant, flaming c.; c. gras, soft c., bituminous c., smoking c.; c. maigre, non-gaseous c.; c. non lavé, raw c., unwashed c.; c. pyriteux, brassil, brazil; c. tout venant, run c., unsorted c.

charbonnage, 1. coal mining; 2. colliery, coal mine.

charbonner, to carbonize, to char, to coal.

charge, load, weight; c. de fond, bottom l.; c. de mine (explosif), blasting charge; c. de rupture, breaking l., breaking point, break point; c. en suspension, suspended l.; c. hydraulique, pressure, water p., head, p. head, static head, liquid head; c. hydrostatique, water head; c. solide, solid l.; c. static, static l.; c. transportée, l., capacity (of a stream).

chargement, loading, charging.

charger (mine), to load, to fill, to charge.

chargeur (ouvrier), loader, charging man.

chargeuse, charger, loader; c. mécanique, mechanical l.

Charmouthien, Charmouthian (l. Jurassic).

charnière, hinge; c. anticlinale, arch bend, saddle b., upper b., anticlinal crest, axial line; c. des Lamellibranches, hinge; c. inférieure (d'un pli), trough, synclinal fold; c. supérieure (d'un pli), anticlinal fold, saddle; c. synclinale, synclinal b., synclinal trough, trough b.

charnockite (pétro.), charnockite.

Charophytes (pal.), Characea.

charpente silicatée, silicate pattern.

charriage, thrusting, overthrust; c. de cisaillement, shear thrust; c. fluviatile, bed load transport; c. tangentiel, tangential thrust; copeau de c., thrust wedge, thrust slice; faille de c., overthrust fault; nappe de c., thrust sheet; pli de c., overfold; surface de c., thrust plane.

charrier (glacier, etc), to carry along, to drift.

charriot de mine, mine truck.

chassignite, chassignite.

chatain (sol), castanozem.

Chattien, Chattian (v. Oligocene).

chaulage, liming.

chaussée (haut-fond rocheux), shoal.

Chaussée des Géants (Irlande), Giants causeway.

chaux, liming, chalk; c. carbonatée, carbonate of lime, bitter spar; c. éteinte, slack lime, slaked lime, hydrated lime; c. grasse, fat lime; c. hydratée, slack lime, hydrated lime; c. hydraulique, water lime, hydraulic lime; c. maigre, poor lime; c. vive, quick lime, unslaked lime.

chef foreur, boring master.

cheire, spiny lava, aa lava, cheire.

Chélicerates (pal.), Chelicerata.

Chelléen, Chellean (Pleistocene industry early Paleolithic).

Chéloniens (pal.), Chelonia.

cheminée, chimney, vent, neck, pipe, throat; c. à minerai, chute, ore chute, ore pass, shoot; c. de fées, earth pyramid, chimney rock, earth pillar, erosion column; c. diamantifère, diamond pipe; c. minéralisée, mineralized chimney; c. volcanique, volcanic neck, volcanic pipe, diatreme.

cheminement (reptation), creep, creeping.

chenal, channel; c. d'écoulement, drainage c.; c. fluviatile, stream c., c. way; c. de basses eaux (étiage), low flow c.; c. de marée, tidal c., tidal

creek; **c. sous-marin,** submarine c.; **c. sous-lacustre,** sublacustrine c.
chercheur d'or, digger.
chernozem, chernozem.
chert, chert, hornstone, rock flint.
chessylite, chessylite, azurite, blue copper carbonate.
chevalement, **1.** head frame, pit-head frame, headstock (mines); **2.** derrick superstructure (*drill.*).
chevauchant, **1.** overlapping (slates); **2.** overthrusting (*geol.*).
chevauchement, **1.** overlapping, overlap; **2.** overthrust, thrust (*geol.*); **c. en retour,** back thrusting; **surface de c.,** thrust plane.
chevauchement-fracture, fracture-thrust.
chevaucher, **1.** to overlap; **2.** to overthrust (*geol.*).
chevelu hydrographique, stream system.
cheveux de Pelé, pelean hairs.
cheveux de Vénus, Venus hairstone.
chevilles (structures en), peg structure.
chevron (pli en), chevron fold.
chevrons (structure en), herringbone structure.
chiastolite, chiastolite.
chicot, pinnacle.
chilénite, chilenite.
chimico-minéralogique, chemicomineralogical.
chimico-physique, chemico-physical.
chimie, chemistry; **c. appliquée,** applied c.; **c. minérale,** mineralogical c.; **c. du pétrole,** petroleum c.
chimiosynthèse, chemosynthesis.
chimiotrophisme, chemotrophism.
chimique, chemical.
chimiquement, chemically.
chimiste, chemist.
chiolite, chiolite.
chitine, chitin.
chitineux, chitinous.
Chitinozoaires, Chitinozoarians, Chitinozoa.
chloantite, chloantite.
chlorate, chlorate.
chloration, chlorination.
chlore, chlorine.
chloreux, chlorous.
chlorhydrique, chlorhydric, hydrochloric.
chlorination, chlorination, chlorinating.
chlorite, chlorite.
chloriteux, chloritique, chloritic, chloritous.
chloritisation, chloritization.
chloritoïde, chloritoid.
chloritoschiste, chlorite schist, chlorite slate.
chloromélanite, chloromelanite.
chlorophane, chlorophane.
Chlorophycées, Chlorophycea.
chlorophylle, chlorophyll.
chlorophyllite, chlorophyllite.
chlorospinelle, chlorospinel.
chlorure, chloride; **c. d'argent,** silver c.
chlorurer, to chlorinate, to chlorinize.
choanocytes (pal.), choanocytes.
choc, shock, impact.
Chondricthyens, Chondrichthyes.

chondrite (météorite), chondrite.
chondrodite, chondrodite.
Chordés, Chordata.
chott, salt lake basin, salt bottom, salt pan, shott.
christianite, christianite.
chromatographie, chromatography.
chrome, chromium; **acier au c.,** chrome steel.
chromeux, chromous.
chromifère, chromiferous.
chromique, chromic.
chromite (minerai), chromite.
chromosphère (astro.), chromosphere.
chronologie, chronology.
chronostratigraphique, chronostratigraphic.
chronozone, chronozone.
chrysobéryl, chrysoberyl.
chrysocolle (minerai de cuivre), chrysocolla.
chrysolite (olivine), chrysolite.
chrysoprase (var. verte de calcédoine), chrysoprase.
chrysotile, chrysotile.
chute, **1.** fall; **2.** shoot (mine); **c. d'eau,** waterfall; **c. de neige,** snowfall; **c. de pluie,** rain fall; **c. de pression,** pressure drop, pressure loss; **c. de tension,** voltage drop.
cicatrice de décollement (de glissement sous-marin), slump scar.
Cidaroïdes (pal.), Cidaroideae.
ciel ouvert (carrière à), open cast, open cut, open.
Ciliés (pal.), Ciliata.
cime, peak, summit, top.
ciment, cement; **c. à prise lente,** slow setting c.; **c. à prise rapide,** quick hardening c.; **c. argileux,** water c.; **c. calcaire,** calcareous c.; **c. de grains minéraux,** ground mass; **c. hydraulique,** hydraulic c.; **c. latéritique,** lateritic c.; **c. siliceux,** siliceous c.
cimentation, cementing, cementation.
cimenter, to cement.
cimenté (peu), softly cemented.
Cimmérien, Cimmerian (Orogeny, Mesozoic : **1.** late Triassic; **2.** late Jurassic).
cimolite, cimolite.
cinabre, cinnabar, cinabar.
cinérite, lithified ash, cinerite, cinereous tuff, vitric tuff.
cipolin, marble (patterned), cipolin (rare).
circonscrit (massif), intrusive body.
circulation, circulation, travelling; **c. de la boue,** mud circulation; **c. forcée (d'eau karstique),** pressure flow; **c. par gravité,** gravitational flow.
circuler (eau), to circulate, to flow, to travel.
circumlunaire, circumlunar.
circumpolaire, circumpolar.
circumterrestre, terrestrial.
cire, wax; **c. brute,** crude w.; **c. de lignite,** lignite w., stone w.; **c. de paraffine,** paraffin w.; **c. fossile,** ozocerite, ozokerite; **c. minérale,** earth w., mineral w., fossil w.
cireux, waxy.
cirque (1. glaciaire), cirque, corrie; **c. en chaudron,** caldron cirque; **c. en fauteuil,** armchair cirque;

c. en amphithéâtre, amphitheatre; **lac de c.,** cirque lake.

cirque (2. d'érosion), amphitheatre, amphitheater (U.S.), erosion basin, valley head, coomb.

Cirripèdes (pal.), Cirripedia.

cisaillement, shearing, shear; **ondes de c.,** S waves; **plan de c.,** shear plane; **pli de c.,** faulted anticline.

cisailler (une couche), to shear.

ciseler (géomorph.), to chisel, to carve.

citrine (quartz jaune), citrine, citrine quartz.

Clactonien (préhist.), Clactonian.

clapier (Alpes), alluvial cone.

clapotis, oscillatory wave.

clarain, clarain.

clarite (cf. charbon), clarite.

classe, class, grade; **c. de sol,** soil class; **c. granulométrique,** size grade; **c. paléontologique,** class.

classement, classing, classifying, grading, sizing, sorting; **c. granulométrique,** granulometric sorting.

classer, to classify, to grade, to size, to sort, to separate.

classification, classification, classing, sorting.

classifier, to classify, to class, to sort.

claste, clast.

clastique, clastic.

claya (mine), clay band.

clayonnage (hydraul.), mat, mattress, brush matting.

climacique, relating to a climax.

climat, climate; **c. tempéré,** temperate c.

climatique, climatic.

climatologie, climatology.

climatologique, climatologic, climatological.

clinochlore (minér.), clinochlore.

clinoclase, clinoclasite, clinoclase, clinoclasite.

clinodôme, clinodome.

clinomètre, clinometer, inclinometer.

clinométrique, clinometric, clinometrical.

clinopinacoïde, clinopinacoid.

clinoprism, clinoprisme.

clinopyroxene, clinopyroxene.

clinorhombique, clinorhombic.

clinozoïsite (épidote), clinozoisite.

clintonite (mica), clintonite.

clivable, cleavable.

clivage, cleavage, cleat; **c. ardoisier,** axial-plane foliation, slaty c.; **c. cubique,** cubic c.; **c. de crénulation,** crenulation c.; **c. d'un minéral,** mineral c.; **c. de flux,** flow c.; **c. de fracture,** shear c.; **c. d'une roche,** slaty c.; **c. par pli-fracure,** shear c.; **c. prismatique,** prismatic c; **meneaux de c.,** cleavage mullions.

cliver, se cliver, to cleave.

cloison (pal.), septum, septa (*pal.*).

cloisonnage (mine), bratticing, partitioning.

cloisonnée (structure), boxwork.

cloisonner (mine), to brattice, to partition.

cluse, transverse valley, cross valley; **c. sèche, cluse morte,** dry gap, wind gap; **c. vive, cluse active,** water gap.

Clyménidés (pal.), Clymenida.

Cnidaires (pal.), Cnidaria.

cobalt, cobalt; **c. arséniaté,** erythrite, c. bloom; **c. arsenical,** smaltite, gray c.; **c. gris,** cobaltine, c. glance.

cobaltifère, cobaltiferous.

cobaltite (minerai), cobaltite, cobalt glance.

cobaltique, cobaltic.

Coblencien, Coblentzian (l. Devonian).

cocarde (minerai en), cochade ore.

Coccolite (pal.), Coccolith.

Coccolithophoridés (pal.), Coccolithophoridae.

code minier, mining code.

codéclinaison, codeclination.

codeur (géoph.), coder.

coefficient, coefficient, ratio; **c. d'aimantation,** magnetic susceptibility; **c. d'écoulement,** drainage ratio; **c. d'élasticité,** elastic coefficient, modulus of elasticity; **c. de débit,** discharge coefficient; **c. de lessivage,** leaching ratio; **c. de perméabilité,** permeability coefficient; **c. de rupture,** modulus of rupture; **c. de triage,** sorting index; **c. de viscosité,** viscosity coefficient.

Coelentérés (pal.), Coelenterata.

coelome (pal.), coelome.

coesite, coesite.

cœur d'un pli, axial part.

coffrage, casing, coffering.

coffre à minerai, ore bin, ore bunker.

coffré (pli), box fold.

coffrer (mine), to coffer.

cohésion (d'un sol), cohesion, cohesiveness, coherence.

coiffe (pédol.), capping.

coiffer (une couche), to cap.

coin, corner, edge; **c. d'entraînement,** rotary drilling; **c. de glace,** ice wedge; **c. de sable,** sand wedge; **fente de gel en c.,** ice wedge; **fente de froid en c.,** ice wedge.

coincement (mine), wedging, jam.

coke, coke; **c. brut,** raw c.; **c. de brai,** pitch glance; **c. de charbon,** coal c.; **c. de pétrole,** oil c.; **c. de tourbe,** peat c.; **c. maigre,** lean c.; **c. naturel,** native c., c. coal, cokeite.

cokéfaction, coking.

cokéfier, se cokéfier, to coke.

cokerie, coking plant.

col (géogr.), pass, col.

colatitude, colatitude.

colémanite, colemanite.

Coléïdés (pal.), Coleoidea.

collant (sol), sticky, adhesive.

collecter, to collect, to gather.

collecteur, collector, main; **c. (drain. égoût),** main, main drain; **c. d'eau,** sink hole, sump; **c. d'exhaure (mine),** pump out drum; **c. de poussière (mine),** dust collector, dust catcher; **c. glaciaire,** glacial cirque.

collimater, to collimate.

collimateur, collimator.

collimation, collimation.

colline, hill; **c. dénudée,** fell; **c. glaciaire,** glacial h., drumlin; **c. sous-marine,** abyssal h.

collinite, collinite.

collision, collision, collapsing; **c. de plaques lithosphériques,** collapsing of plates.

colloïdal, colloidal.

colloïde, colloid.

collophane (phosphate), collophane.

colluvial, colluvial.

colluvion, colluvium.

colmatage, 1. clogging, choking, blocking; 2. flood deposit, silting.

colmatant (pour boues de forage), plugging agent.

colmater (for.), to clog up (filter, sieve), to choke up (pipe).

colombite, columbite.

colonial (pal.), colonial.

colonie (d'organismes), colony; **c. de polypiers,** polyp c.

colonnade volcanique, columnar lava.

colonne, column, pillar; **c. coiffée,** earth pillar; **c. de basalte,** basaltic column; **c. de distillation,** distillation column; **c. d'érosion,** erosion column, earth pillar; **c. d'exhaure (mine),** rising main; **c. de fractionnement,** fractionating column; **c. de production,** production string; **c. minéralisée,** ore chute, ore chimney, ore shoot; **c. montante de boue,** rising mud column; **c. stratigraphique,** lithologic log; **c. technique,** protection casing; **c. vertébrale (pal.),** spinal column.

colorimètre, colorimeter.

colorimétrie, colorimetry.

columellaire (pal.), columellar.

columelle (pal.), columella.

combe, anticlinal valley.

combinaison (chimique), combination.

combiner, se combiner (chimie), to combine.

comblement (d'un lac, etc.), filling up.

combler, to fill up.

combustibilité, combustibility.

combustible, fuel, combustible; **c. gazeux,** gaseous fuel; **c. liquide,** liquid fuel; **c. nucléaire,** nuclear fuel; **c. solide,** solid fuel.

combustion, combustion.

comète, comet.

commissure (pal.), commissure.

compacification, packing.

compacité, compactness.

compact (sédiment), compact, close grained, tight.

compactage (de matériaux), compaction.

compaction, compaction, packing.

compartiment, compartment; **c. d'extraction,** hoisting c.; **c. de puits,** shaft c.; **c. d'une faille,** faulted block.

compensation isostatique, isostatic compensation.

compétence (d'un courant), competency.

compétence (d'une couche), competence.

compétent (sédiment), competent.

compétente (strate), competent stratum, controlling stratum.

complexe, 1. adj : complexe; 2. n : complex, group; **c. adsorbant,** adsorbing complex, base exchange complex; **c. argilo-humique,** clay humus complex; **c. de sols,** soil complex; **c. sédimentaire,** sedimentary complex.

complexométrie, complexometry.

composant, 1. adj : component, constituent; 2. n : component, constituent (d'une roche).

composante (d'une force), component; **c. horizontale,** horizontal c.

composé, 1. adj : compound, composite; 2. n : compound; **c. aromatique,** aromatic c.; **c. chimique,** chemical c.; **c. en chaîne,** chain c.; **c. non saturé,** unsaturated c.; **c. organique,** organic c.; **faille c.,** c. fault; **pli c.,** composite fold; **volcan c.,** c. volcano.

composition, composition; **c. acide,** acidic c.; **c. granulométrique,** grading, size grading; **c. minéralogique virtuelle,** norm.

compresser (une roche), to compress, to pack.

compressibilité, compressibility.

compressible, compressible.

compression, compression, crushing; **c. adiabatique,** adiabatic c.; **essai à la c.,** crushing test; **faille de c.,** compressional fault.

comprimable, compressible.

comprimer, to compress, to squeeze.

compteur, counter, meter, recorder; **c. à moulinet,** flow meter; **c. à scintillation,** scintillation counter; **c. d'impulsions,** impulse meter, pulse counter; **c. Geiger,** Geiger counter.

comptonite, thomsonite (var.).

concassage, breaking, crushing.

concasser, to break, to crush, to pound.

concasseur, breaker, crusher, stone-breaker, stone-crusher; **c. à minerai,** ore crusher.

concavité de méandres, meander scars.

concentrateur (appareil), concentrator; **c. à boues;** slimes c.

concentration, concentration, concentrating; **c. de minerai,** ore concentration; **c. par voie humide,** water-concentration; **c. par flottement,** concentration by flotation.

concentré (n), concentrate; **c. de minerai,** ore c.

concentrer (un minerai), to concentrate.

concession, concession, claim, grant, lease; **c. de mines,** mining claim, mineral claim; **c. de placer,** placer claim; **c. filonienne,** lode claim; **c. minière,** mining claim, mineral claim; **c. pétrolière,** oil lease; **c. de pétrole sous-marin,** offshore lease.

conchoïdale (cassure), conchoidal.

conchyoline, conchiolin.

concordance, conformability, conformity, concordance.

concordant, conformable, concordant; **stratification c.,** conformable bedding.

concrétion, concretion, travertine, (karst); **c. calcaire,** calcareous concretion; **c. de manganèse,** manganese nodule; **c. ferrugineuse,** iron concretion; **c. siliceuse,** siliceous concretion; **c. tuffeuse,** tuffaceous concretion.

concrétionné, concretionary.
condensabilité, condensability.
condensable (gaz), condensable.
condensation, condensation; **c. atmosphérique**, atmosphere c.
condensée (série), condensed (sequence).
condenser, se condenser (un gaz), to condense.
condition climatique, climatic condition.
conduit, pipe, duct, conduit; **c. d'aération (mine)**, air pipe; **c. volcanique**, volcanic chimney, volcanic pipe.
conduite, duct, line, pipe, pipe line; **c. d'alimentation en eau**, water supply line; **c. d'amenée (mine)**, head pipe, supply pipe; **c. de gaz**, gas line; **c. de pétrole**, oil pipe.
Condylarthres (pal.), Condylarthra.
cône, cone; **c. adventif**, parasitic c.; **c. alluvial**, alluvial fan; **c. alluvial aggloméré**, fanglomerate; **c. d'ablation**, rock fan; **c. couvert (cône de glace recouvert de débris)**, dirt c.; **c. d'avalanche**, avalanche c.; **c. d'éboulis**, fan, talus fan; **c. d'éruption**, c. of eruption; **c. de cendres**, ash c.; **c. de débris**, cinder c.; **c. de déjection**, fan delta, alluvial fan, alluvial c.; **c. de lave**, lava c.; **c. de rabattement (hydro)**, depression c.; **c. de scories**, cincer c., scoria c.; **c. emboîté**, nested c., ringed c.; **c. mixte**, composite c.; **c. proglaciaire**, sandur and outwash plain; **c. sous-marin**, deep sea fan; **c. torrentiel**, alluvial c.; **c. volcanique**, volcanic c.
conelet de scories, hornito.
configuration, configuration, lay, lie, pattern, geometry.
confluence (glacio), confluence; **gradin de c.**, confluence step.
confluent, confluent.
congelable, freezable.
congélation (périgl.), congelation, freezing.
congeler, to congeal, to freeze.
congère, snowbank, snowdrift, snowpatch.
conglomérat, conglomerate; **c. d'écrasement**, crush c.; **c. intraformationnel**, intraformational c.; **c. monogénique**, monogenic c.; **c. polygénique**, polygenic c.
conglomération, conglomeration.
conglomératique, conglomeratic.
congloméré, conglomerated.
Coniacien, Coniacian (v. Cretaceous).
Conifères (paléobot.), Coniferous.
conjuguée (faille), conjugate (fault).
connée (eau), connate (water).
conodonte, conodont.
consanguinité magmatique, consanguinity.
conséquent (réseau fluv.), consequent.
conséquent (cours d'eau), consequent stream.
conservation (de la nature), conservation, conservancy, preservation.
conserver, se conserver (pal.), to conserve, to preserve.
consistance (d'un sol), consistency, firmness.
consistant, consistent, firm.

consolidation (d'un fossile), consolidation, strengthening.
consolider, to consolidate, to strengthen.
consommation de pétrole, oil consumption.
constante (n.), constant; **c. cosmologique (astro.)**, cosmological c.; **c. d'élasticité**, elastic c.; **c. de gravitation**, gravitational c.; **c. réticulaire (cristallo.)**, lattice c.
constellation (astro), constellation.
constituant, component, constituent; **c. du sol**, soil constituent; **c. granulométrique**, soil separate.
constitution (d'une roche), constitution, composition, structure.
constructeur (organisme), reef builder.
construit (calcaire), bioherm.
contact, contact; **c. anormal**, abnormal c.; **c. intrusif**, intrusive c.; **c. pétrole-eau**, water-oil c; **auréole de c.**, contact zone; **métamorphisme de c.**, contact metamorphism.
contamination (d'un magma), contamination.
contemporain (sédiment), contemporaneous.
contenu en substances nutritives du sol, nutrient content.
continent, continent, mainland.
continental, continental, terrestrial; **glacis c.**, continental rise; **marge c.**, continental margin; **plateau c.**, continental shelf; **plate-forme c.**, continental shelf; **sédiment c.**, terrestrial deposit; **talus c.**, continental slope.
continentalisation, continentalisation.
continuité (des couches), continuity.
contourite (séd.), contourite.
contracter, se contracter, to shrink, to contract, to narrow.
contraction (sédiment), contraction, shrinking, shrinkage.
contraction (fente de), contraction crack, shrinkage crack.
contrainte (méc., phys.), stress, force; **c. à la compression**, compressive stress; **c. de cisaillement**, shear stress; **c. de rupture**, breaking stress.
contraire (faille), antithetic fault.
contralizé, antitrade wind.
contraposé, contraposed.
contre-balancier (pétrole), balance bob, counter balance.
contre-courant, counter current.
contrée, country, land; **c. marécageuse**, marshy land; **c. minière**, mining country; **c. pétrolifère**, oil-bearing area; **c. rocheuse**, rock land.
contrefort, buttress, spur, foothill.
contremaître, foreman; **c. de mine**, mine f.; **c. du fond (mine)**, underground f.; **c. du jour (mine)**, surface f.
contre-pente, reversal of slope.
contrôle, control, checking, inspection, monitoring; **c. climatique**, climatic control; **c. d'avance de forage**, drilling control; **c. de débit**, outflow control; **c. de température**, temperature control; **c. de tête de puits**, casing head; **c. du toit (d'une**

couche), roof control; **c. granulométrique,** sieve acceptance; **c. structural,** structural control.

Conularidés, Conularida.

convection (courant), convection (current).

convergence (de plaques), convergency, convergence.

convergente (plaque), converging, convergent (plate).

conversion (d'énergie), conversion, transformation.

convertir (de l'énergie), to convert, to change, to transform.

convoi à blocs (périgl.), blocks field.

convolution (périgl.), convolution.

convoyeur, conveyor, **c. à godets,** bucket c., pan c.; **c. de taille,** face c.

coordinence, coordination number.

coordonnées (math.), coordinates; **c. astronomiques,** astronomical c.; **c. géographiques,** geographic data; **c. polaires,** polar c.

copeau, chip, **c. de charriage,** thrust slice, thrust wedge.

copeau tectonique, thrust slice.

Copépodes (pal.), Copepoda.

copiapite, copiapite.

coprolite, coprolithe, coprolite, faecal pellet.

coquillage (vide), shell.

coquille, shell, test.

coquillier, shelly.

corail, coral; **récif de c.,** coral reef.

Coralliaire, Corallia; **c. isolé,** solitary coral, horn coral; **c. colonial,** compound coral, colonial coral; **massif de c.,** coral head; **récif c.,** coral reef; **squelette c.,** corallum, corallite; **vase c.,** coral mud.

corallien, coral, coralline; **massif c.,** coral head; **récif c.,** coral reef.

Corallien, Corallian (u. Jurassic).

corallifère, coralliferous.

coralligène, coralligenous.

Corallinacées (algues rouges), Corallinaceae.

corbeille vibratile (pal. spong.), flagellated chamber.

Cordaïtales (paléobot.), Cordaitales.

cordée (lava), ropy lava.

Cordés (pal.), Cordata.

cordiérite (minér.), cordierite, iolite.

cordillère, cordillera.

cordon (mine), string, stringer.

cordon littoral, bar, offshore bar, barrier beach, beach ridge; **c. appuyé,** headland bar; **c. de galets,** shingle bar; **c. de tempête,** high storm ridge; **c. en V; V** bar; **c. libre,** offshore bar.

corindon, corundum, diamond spar.

cornaline, cornelian.

corné, horny.

cornéenne, hornfels, non-foliated metamorphic rock.

corniche, ledge, cornice.

coron, mining village.

coronitisation, coronitization.

corps, body, substance; **c. céleste,** celestial body; **c. composé,** compound body; **c. de minerai,** ore body; **c. de sonde,** drilling shaft; **c. extrusif,** extrusive body; **c. intrusif,** intrusive body; **c. simple,** element.

corrasion, corrasion, erosion; **c. eolienne,** wind abrasion, wind erosion, wind carving.

correction, correction; **c. d'altitude,** elevation c.; **c. de Bouguer,** Bouguer c.; **c. de latitude,** lattitude c.; **c. topographique,** topographic c.

corrélation, correlation; **c. de diagraphie,** c. of well logs; **c. stratigraphique,** stratigraphic c.; **c. temporelle,** time c.

corrode (to), 1. to corrode (metals); 2. to erode, to corrade (geol.).

corrosif, corrosive, corroding, caustic, etching.

corrosion, corrosion, etching, attacking; **c. atmosphérique,** atmospheric corrosion; **c. chimique,** chemical corrosion; **c. souterraine,** underground corrosion.

corsite (diorite), corsite.

cortège (pétro.), suite.

cortège minéralogique, mineralogic assemblage.

cortex, cortex.

cosalite (minér.), cosalite.

cosmique, cosmic.

cosmogonie, cosmogony.

costière (mine), drift, drifting-level, drift way drive.

costresse (mine), subdrift, counter-level.

côte (marine), coast, coastline, seaboard, seacoast, shore, shoreline; **c. (pente),** slope; **c. (cuesta),** cuesta, escarpment; **c. à falaise,** cliffy shoreline; **c. affaissée,** depressed shoreline; **c. d'accumulation,** accretion coast, accumulative coast; **c. d'émersion,** shoreline of emergence; **c. de faille,** fault coast; **c. construite;** constructional coast; **c. decoupée,** embayed shore; **c. deltaïque,** deltaic coast; **c. plate,** low coast, flat coast; **c. rocheuse,** rocky coast; **c. soulevée,** raised coast.

cote (géodésie), reading; **c. de nivellement,** height, elevation; **c. d'un sondage,** elevation of the well.

côte (d'un vertébré fossile), costa, rib.

côté sous le vent, lee side.

coteau, hill, little hill.

coticule, coticule (manganiferous garnet-quartzite).

cotidal, cotidal, tidal.

côtier, coastal, coastwise.

cotunnite, cotunnite.

couche, bed, layer, stratum, deposit, seam; **c. active,** mollisol, active layer; **c. altérée,** weathered layer; **c. aquifère,** water bearing stratum; **c. basale,** bottomset bed; **c. compétente,** competent bed; **c. concordante,** conformable layer; **c. concrétionnée argileuse,** claypan; **c. cultivée,** till layer; **c. d'altération,** zone of weathering; **c. d'argile,** clay layer, clay seam; **c. d'arrêt,** blocking layer; **c. d'eau,** water layer; **c. de charbon,** coal seam, coal bed, coal measures; **c. de couverture,** overburden layer, upper layer; **c. de galets,** pebble bed; **c. d'humidification,** humic layer; **c. de minerai,** ore course, ore bed; **c. de transition,** transition bed; **c. discordante,** unconformable bed; **c. du mur,** underbed, bottom layer; **c, du toit,** superincumbent bed, top of bed; **c. encaissante,** enclosing layer; **c. exploitable,** workable bed, seam; **c. filtrante,** filter bed; **c. fossilifère,** fossiliferous bed; **c. frontale,** fore-set bed; **c. granitique,** granitic layer; **c. granoclassée,** graded bedded; **c. grisouteuse,** gassy seam; **c. inclinée,** inclined bed, dipping stratum; **c. horizontale,** horizontal stratum; **c. imprégnée d'eau,** water logged bed; **c. incompétente,** incompetent bed; **c. inférieure,** lower bed; **c. intercalée,** intercalated bed; **c. interstratifiée,** interstratified bed; **c. lacustre,** lacustrine bed; **c. limite,** boundary layer; **c. mince,** thin seam; **c. oblique,** cross bed; **c. pétrolifère,** petroliferous layer; **c. productive,** oil bearing stratum; **c. rapide (géoph.),** high-speed layer; **c. repère,** marked bed; **c. réservoir,** reservoir bed; **c. saisonnièrement dégelée,** thawing layer; **c. salée et dure,** saltpan; **c. salifère,** salt bed; **c. savon,** sole thrust, slip bed; **c. sommitale,** top-set bed; **c. sous-jacente,** underlying bed, subjacent bed, underbed; **c. superficielle,** surface layer, top layer, topsoil; **c. supérieure,** upper bed, overlying stratum; **c. surplombante,** superincumbent bed; **c. sus-jacente,** overlying stratum; **c. toujours gelée,** permafrost.

couché, recumbent; **pli c.,** recumbent fold.

coude (de rivière), bend, elbow.

coulage d'eau (mine), leakage (of water).

coulée, flow, stream; **c. boueuse,** mud flow, lahar; **c. chaotique,** aa; **c. de blocs,** boulder stream, rock glacier; **c. de cendres,** ash flow; **c. d'éboulis,** land slide; **c. de lave,** lava flow, lava stream; **c. de minerai,** ore shoot, ore chute, chimney of ore; **c. de pierres,** block stream, rock flow, stone river; **c. de solifluxion,** solifluction deposit; **c. de terre,** creep, landslide; **c. d'argiles à blocs,** head, coombrock.

couler (rivière), to flow, to run off.

couleur de la poussière d'un minerai, streak.

coulissage, strike-slip fault.

coulisse, slider, slideway; **c. de battage (for.),** bumper sub; **c. de repêchage (for.),** fisching jar.

couloir, passage, passageway; **c. à charbon,** coal chute; **c. à minerai,** ore chute, ore shoot; **c. d'avalanche,** avalanche passageway, slide furrow; **c. de front deltaïque,** delta front gully; **c. karstique,** valley sink; **c. interdunaire,** dune valley; **c. oscillant,** shaker conveyor, swinging conveyor, jigger conveyor.

coup, blow, shot; **c. d'eau,** inrush of water, water inflow, water inrush, water outbreak; **c. de charge (mine),** rock burst bump; **c. de foudre,** thunderbolt; **c. de grisou,** fire damp explosion (U.K.); gas explosion (U.S.); **c. de mine,** shot blast, rib, blast; **c. de point (préhist.),** hand axe; **c. de poussière,** coal dust explosion; **c. de toit (mine),** rock burst; **c. de vent,** gale, gust of wind.

coupe, 1. de mine, cut, cutting; **2.** de terrain, section, cut; **3.** cartographique, section, profile; **4.** de sondage, drill log.

coupe en travers, cross cut, cross section; **c. géologique,** geological section; **c. lithologique,** lithologic log; **c. de profondeur (sism.),** depth section; **c. schématique,** schematic section; **c. sériées,** serial sections; **c. sismique,** seismic s.; **c. stratigraphique,** stratigraphic section; **c. temps (géoph.),** time s.; **c. transversale,** cross section; **c. verticale,** vertical section.

coupeuse rotative (mine), rotary heading machine.

coupure, cut, cutting; **c. de carte,** map sheet; **c. de méandre,** cut-off, avulsion; **c. stratigraphique,** stratigraphic boundary; **c. transversale,** cross valley.

courant, 1. adj: running; 2. n: current, flow; **c. ascendant,** upwelling; **c. boueux,** mud stream; **c. d'arrachement,** rip current; **c. de compensation,** compensation current; **c. de contour,** contour current; **c. de flot,** flood current; **c. de jusant,** ebb current; **c. de marée,** tidal current; **c. de remontée,** upwelling current; **c. de retour,** back flow, back jet current; **c. de turbidité,** turbidity current; **c. fluviatile,** stream current; **c. glaciaire,** ice stream; **c. laminaire,** laminary current; **c. littoral,** longshore current; **c. océanique,** ocean current; **c. sagittal,** rip current; **c. souterrain,** groundwater flow; **c. tellurique,** telluric current; **c. torrentiel,** unsteady current.

courbe (graphique), curve, diagram, graph; **c. cartographique,** contour line; **c. géométrique,** curve, bend; **c. bathymétrique,** depth curve; **c. cumulative,** cumulative curve; **c. de crue,** rising limb; **c. de décrue,** récession limb; **c. de fréquence,** frequency curve; **c. d'égale profondeur,** isobath; **c. de niveau,** contour line; **c. de niveau fermées,** closed, closing contour lines; **c. de niveau intercalaire,** intermediate contour line; **c. de porosité,** porosity curve; **c. de pression,** pressure curve; **c. de production,** production curve; **c. de résistivité,** resistivity curve; **c. de solubilité,** solubility curve; **c. de vitesse,** velocity curve; **c. dromochromique,** time distance curve; **c. du temps de parcours vertical,** vertical travel time curve; **c. en pointillé,** dotted curve; **c. en trait continu,** solid line curve; **c. granulométrique,** grain-size curve; **c. hypsométrique,** contour line; **c. logarithmic,** logarithmic; **c. maîtresse,** index contour; **c. piézométrique,** pressure curve; **c. profondeur-temps,** depth-time curve; **c. temps,** time curve; **c. temps-distance (sism.),** travel-time curve; **c. structurale,** structural contour; **c. vitesse-profondeur,** velocity curve.

courbure (d'une strate), curvature, bending.

couronne, crown, ring, rim; **c. à diamants,** diamond rock drill crown, diamond boring crown; **c. de carottage,** core bit; **c. de carottier,** core cutter head; **c. de sondage,** boring head; **c. sans diamants,** unset crown; **c. solaire,** corona.

cours (d'un fleuve), course, flow; **c. d'eau,** stream, watercourse; **c. d'eau temporaire,** ephemeral stream; **c. de plateau continental,** shelf current; **c. souterrain,** underground stream; **c. inadapté,** misfit river; **c. inférieur,** lower course; **c. moyen,** middle course; **c. supérieur,** upper course.

course (de la houle), fetch.
coussin (de lave), pillow lava.
couverture, blanket, cover, coverage, covering; **c. de morts terrains**, overburden; **c. de photographies aériennes**, aerial coverage; **c. de sable**, coversand; **c. de terrains sus-jacents**, overlying beds; **c. du sol (litière forestière)** litter; **c. glaciaire**, glacial sheet; **c. morainique**, glacial drift; **c. superficielle**, residual soil, waste mantle; **c. végétale**, vegetal cover.
couvre-objet (micro), cover glass.
couvrir (un terrain), to cover, to overlay.
covalence, covalence.
covellite, covelline (minerai de cuivre), covellite, covelline.
craie, chalk; **c. à silex**, c. with flints; **c. blanche**, white c.; **c. bleue**, blue c.;**c. glauconieuse**, greensand marl; **c. lacustre**, calcareous deposit; **c. magnésienne**, magnesian c.; **c. marneuse**, marly c., **c. phosphatée**, phosphatic c.
crâne (pal.), skull, cranium.
cranial (pal.), cranial.
craquage, cracking; **c. catalytique**, catalytic cracking; **c. en phase vapeur**, vapor phase.
craquelure, cracks; **c. de gel**, frost crack; **c. de dessèchement**, suncrack, shrinkage crack.
crassier (métal), slag heap, slag dump.
cratère, crater; **c. actif**, active c.; **c. adventif**, parasitic c.; **c. central**, central c.; **c. d'explosion**, explosion c.; **c. ébréché**, breached c.; **c. emboîté**, nested c.; **c. lac**, c. lake; **c. lunaire**, lunar c.; **c. puits**, pit c.
cratériforme, crateriform.
craton, craton.
cratonique, cratonic.
crayère, chalk pit.
crayeux, chalky.
crénulation (tecton.), crenulation.
Créodontes (pal.), Creodonta.
Crétacé, 1. adj: cretaceous; 2. n: Cretaceous period, chalk period.
crêt monoclinal, hogback.
crête, crest, ridge, summit; **c. alguaire**, algal ridge; **c. anticlinale**, anticlinal ridge; **c. de plage**, storm ridge, beach ridge; **c. médio-océanique**, mid-oceanic ridge; **c. migrante**, offshore bar; **c. prélittorale**, submarine bar, migrating bar, offshore bar.
creusement (trav. pub.), digging, excavation.
creusement (de puits), sinking.
creuser, to dig, to excavate, to hollow out.
creuset (labo), crucible, pot, melting pot; **c. en platine**, platine crucible; **c. en terre réfractaire**, fire clay crucible.
creux, 1. adj: (vide) hollow; (profond) deep; 2. n: hollow, cavity, hole; **c. de déflation**, déflation hole, blowout; **c. (doline de dissolution)**, stone sink; **c. d'effondrement**, collapse sink.
crevasse, crack, crevasse, crevice, fissure, cleft; **c. de gel**, frost crack.
crevassement, cracking.
crevasser, to crevice, to crevasse.

criblage, screening, sieving, sifting, jigging, sizing.
crible, sieve, screen, jigger; **c. à minerais**, jig, jigger, jigging machine; **c. à secousses**, jigging screen, shaking screen; **c. à classeur**, sizing screen; **c. en tôle perforée**, pinched plate screen; **c. oscillant**, oscillating screen, shaking screen; **c. rotatif**, trommel screen.
cribler (avec un crible), to screen, to sift, to sort, to jig.
Crinoïdes (pal.), Crinoids.
Crinozoaires (pal.), Crinozoa.
crique, creek, cove.
cristal, crystal; **c. aciculaire**, acicular c.; **c. anisotrope**, anisotropic c.; **c. biaxe**, biaxial c.; **c. de roche**, rock c., mountain c.; **c. hémitrope**, twin c.; **c. mâclé**, twin c.; **c. négatif**, negative c.; **c. de quartz bipyramidé**, bipyramidal quartz c.; **c. uniaxe**, uniaxial c.
cristallière, rock crystal mine.
cristallifère, crystalliferous.
cristallin, crystalline.
cristallinité, crystallinity.
cristallisabilité, crystallizability.
cristallisable, crystallizable.
cristallisant, crystallizing.
cristallisation, crystallization, crystallizing; **c. de sel**, salt crystallization; **c. fractionnée**, fractional crystallization; **c. par évaporation**, crystallization by evaporation.
cristalliser, se cristalliser, to crystallize.
cristallisoir, 1. crystallizer; 2. chiller (refining apparatus).
cristallite, crystallite.
cristalloblaste, crystalloblast.
cristalloblastique, crystalloblastic.
cristallochimie, crystal chemistry.
cristallogenèse, crystallogeny.
cristallogénique, crystallogenic, crystallogenical.
cristallographe, crystallographer.
cristallographie, crystallography; **c. chimique**, chemical c.; **c. physique**, physical c.
cristallographique, crystallographic, crystallographical.
cristalloïde, crystalloid.
cristallométrie, crystallometry.
cristallométrique, crystallometric.
cristallophyllien, schistose, metamorphic, phyllocristalline (rare), foliated crystalline.
cristobalite (silice), cristobalite.
crochet, 1. umbo (pal.); 2. hook (of a littoral spit).
crochon (tect.), bend, drag fold.
crocidolite, crocidolite.
crocodilage, shrinkage, or contraction of joints, "aligatoring".
Crocodiliens, Crocodilia.
crocoïte (minér.), crocoite, crocoisite, natural lead chromate.
croiser (filon), to cross, to intersect.
croisette, twinned staurolite.
croiseur (filon), cross lode, cross vein, counterlode.
croissant (dune en), crescentic dune.

croissant de choc (éolien), eolian crescent shaped feature.

croissant de plage, beach cusp.

croix (mâcle en), x-shaped twin.

croix de St. André, staurolite twinned.

Crossoptérygiens, Crossopterygii.

croupe (géogr.), ridge.

croûte, crust, coating; **c. altérée**, weathered c.; **c. concrétionnée**, hardpan, concrete bed; **c. de sel**, salt c., salcrete; **c. désertique**, desert varnish; **c. dure**, pan, duricrust; **c. ferrugineuse**, iron pan, ferruginous c.; **c. gypseuse**, gypsum c.; **c. indurée**, duricrust; **c. terrestre**, earth c.; **c. zonaire**, zoned c.; **bombe en c. de pain**, bread c. bomb.

crue (d'un fleuve), rise, rising, flood; **c. brutale**, flash flood; **c. complexe**, multi-peaked flood; **c. nivale**, snowmelt flood; **c. simple**, single-peaked flood; **c. (d'un glacier)**, advance.

crura (pal.), crura.

Crustacés, Crustacea.

crustal, crustal.

cryergie, cryergy.

cryoclastie, cryoclastism.

cryoclastisme, cryoclastism.

cryoconite, cryoconite.

cryodisjonction, frost splitting.

cryogénie, cryogeny.

cryokarst, cryokarst.

cryolite (minér.), cryolite.

cryolithosphère (Mars), cryolithosphere.

cryologie, cryology.

cryonival, cryonival.

cryonivation, cryonivation.

cryopédologie, cryopedology.

cryoplanation (terrasse de), cryoplanation terrace.

cryoreptation (d'un sol), frost creeping.

cryosol, cryomorphic soil, cryosol.

cryosphère, cryosphere.

cryotectonique, cryotectonics.

cryoturbation, geliturbation.

cryptocristallin, cryptocristalline.

cryptodôme, endogeneous dome.

Cryptodonte (pal.), Cryptodonta.

cryptohalite, cryptohalite.

Cténodontes (pal.), Ctenodonta.

cubage (d'un minerai), cubic measurement, cubage, cubature.

cubanite, cubanite.

cuber (un volume de minerai), to cube, to gage, to gauge.

cuboïte (analcime), analcime, analcite.

cuesta, cuesta.

cuiller (sondage), auger, spoon, gouge bit; **c. à sédiments**, sand bucket; **c. de curage**, clean out bailer.

Cuirassés (Poissons), Placodermii.

cuirasse (pédologique), hardpan, crust, duricrust; **c. ferrugineuse**, ferruginous cuirasse, ironpan; **c. latéritique**, laterite.

Cuisien, Cuisian (l. Éocène).

cuisson de briques, baking of bricks.

cuivre, copper; **c. brut**, raw c.; **c. gris**, grey c.; **c. gris antimonial**, panabase; **c. gris arsenical**, tennantite; **c. jaune**, brass, yellow c.; **c. natif**, native c.; **c. noir**, black c.; **c. panaché**, variegated c. ore, bornite, peacock c.; **c. pyriteux**, chacolpyrite, yellow c. ore; **c. rouge**, c., pure c., cuprite.

cuivreux, coppery, cupreous, cuprous.

cuivrique, cupric.

cul de sac, blind valley.

Culm, Culm (facies of l. Carboniferous).

culminant (relief), culminating.

culmination (d'un pli), culmination, high.

culminer (relief), to culminate.

culot de fonte de glace, kettle.

culot (volcanique), plug (volcanic).

cultivable (sol), arable, tillable.

culture suivant courbes de niveau, contour farming.

cumulative (courbe), cumulative.

cumulats (minér.), cumulates.

cumulo-dôme, cumulo-dome.

cumulo-volcan, cumulo volcano, plug dome.

cuprifère, cupriferous, copper bearing.

cuprigène, copperbearing.

cuprite (minerai), cuprite, red copper ore.

cupule, pit, cusp; **c. de dissolution**, solution cusp, weather pit; **c. de fusion**, melt pit.

curage (d'une rivière), cleaning out.

curium, curium.

cuvelage d'un puits de mine, tubing; **c. d'un puits de pétrole**, casing, string of casing.

cuveler un puits de mine, to tube; **c. un puit de pétrole**, to case.

cuvette (géogr.), basin; **c. endoréïque**, bolson; **c. lacustre**, lake b.; **c. océanique**, oceanic b.; **c. synclinale**, centroclinal, structural b., centrocline.

cyanite, cyanite, kyanite.

Cyanophycées, Cyanophyta.

cyanure, cyanide.

Cycadales (pal.), Cycadales.

cycle (géogr.), cycle; **c. d'érosion**, c. of erosion, c. of denudation; **c. fluvial**, river c.; **c. géomorphologique**, geomorphic c., physiographic c.; **c. orogénique**, orogenic c.; **c. sédimentaire**, deposition c.

cyclique, cyclic.

cyclosilicate, ring silicate, cyclosilicate.

cyclothème, cyclothem.

cylindrique, cylindric, cylindrical.

cymophane, cymophane.

cyprine, cyprine.

Cystoïdés (pal.), Cystoidea.

D

dacite, dacite.

dacitique, dacitic.

dahamite, dahamite.

dahlite, dahlite.

dallage, pavement; **d. de pierres,** boulder p.; **d. nival,** snow p.

dalle, slab, flagstone; **d. à Lingules,** Lingula flags; **d. nacrée,** flaggy bathonian limestone.

damage (du sol), ramming, tamping.

dambo (vallée en U sous climat tropical), dambo.

damer (tasser), to ram, to tamp.

damourite, damourite.

damouritisation, damouritization.

Danien, Danian (l. Palaeocene).

darcy (unité de perméabilité), Darcy.

Darcy (loi de), Darcy's law.

Darwinisme, Darwinism.

Dasycladacées, Dasycladaceae.

datation, dating, age dating (never "datation"); **d. absolue,** absolute dating; **d. au radiocarbone,** C^{14} dating; **d. relative,** relative dating.

datation par traces de fission, fission track dating.

datolite (minér.), datolite.

davisienne (théorie), Davisian (theory).

davyne, davyte, davyne.

débâcle glaciaire, glacial outburst, debacle.

débarassé de (minerai), free of.

débit (de liquide), output, flow rate, discharge, yield, flow; **d. annuel,** annual discharge; **d. de pointe,** peak flow; **d. fluviatile,** river discharge; **d. initial (d'un puits),** initial flow; **d. journalier,** daily flow; **d. périodique,** intermittent flow; **d. régularisé,** regulated flow; **d. solide,** solid discharge.

débitage, débit (façon de se séparer), splitting, jointing; **d. en boule,** spheroidal weathering; **d. en crayons,** pencil cleavage; **d. prismatique,** columnar jointing.

débiter (un liquide), to discharge, to yield.

débiter (se) (roches), to split.

débitmètre, flowmeter, flow recorder.

déblai, cutting, muck, dug earth, refuse, rubbish, waste, spoil; **d. de forage,** drill cuttings.

déblaiement, clearing away, removing, scouring.

déblayer, to clear away, to remove.

déboisage (mine), untimbering.

déboisement (d'une région), deforestation.

déboiser (une galerie), to untimber; **d. (une région),** to deforest.

débordement (fluv.), overflow, overflowing.

déborder (rivière), to overflow.

débouché (d'un lac, d'un fleuve), outlet, debouchure.

débourbage (de minerai, etc.), washing.

débourber (minerai), to wash.

débourrage (mine), unramming.

débourrage d'une cheminée volcanique, outburst.

débourrer (mine), to unram.

débris, remains, debris, stone fragments, rock waste; **d. de forage,** cuttings; **d. végétaux,** plant residues, plant remains.

décalage (de couches par faille), offsetting.

décalcification, decalcification; **résidu de d.,** decalcification residue.

décalcifié, decalcified.

décalcifier, to decalcify, to decalsify.

décaler (une couche), to offset, to shift, to displace.

décantation, decantation, settling, elutriation.

décanter (séd.), to decant, to elutriate.

décapage (par acide), etching; **d. (par engins),** scrapping, stripping, clearing, removing; **d. (par érosion),** scouring.

décaper (à l'acide), to etch; **d. (avec un engin),** to scrap, to strip, to clear away.

décapitation (d'une rivière), decapitation, beheading.

décapitée (rivière), beheaded, decapitated (river).

Décapodes (pal.), Decapoda.

décarbonatation, decarbonatation.

décarbonater, to decarbonate.

décharge (d'eaux), discharge, outlet.

décharger (se) dans un lac, to empty itself into a lake.

déchets de raffinage, refinery bottoms; **d. pétrographiques,** waste rocks; **d. radioactifs,** radioactive waste.

déchiqueté (relief), jagged.

déclinaison (astro.), declination.

déclinaison magnétique, magnetic declination.

déclinomètre, declinometer.

déclivité, slope, declivity, incline, gradient, down grade, tilting.

décohésion, weathering, decaying.

décoiffement (tecto.), collapse structure.

décollement (tecto.), decollement, parting, gravity tectonics, slipp off, down-sliding.

décoloration (par des argiles), clay bleaching, clay decolorizing.

décoloré (horizon du sol), bleached.

décomposable, decomposable, decayable.

décomposer, se décomposer, to decay, to decompose.

décomposition (des roches), decay.

décontamination (radioactive), decontamination.

découverte (géologique), discovery, finding; **exploitation par d.,** open-pit mine.

découverture (mine), stripping, uncapping, baring.

découvrir, to find; **d. un gisement,** to find, to detect; **d. les morts-terrains,** to strip, to uncover.

décrépitation, decrepitation.

décrépiter, se décrépiter, to decrepitate, to break up on heating.

décrochement, strike-slip fault, transverse fault, wrench fault.
décroissance (d'une crue, etc.), decrease, decline.
décrue (rivière), fall, falling; **d. (glacier)**, retreat.
dédolomitisation, dedolomitization.
défaut cristallographique, lattice defect.
défense de Mammouth, tusk.
déferlement, surf; **zone de d.**, surf zone.
déferler, to unfurl, to break into foam.
déferrisé, de-ironized.
déficit hydrique, water deficit.
défilé, defile, pass, gorge.
déflagration (mine), deflagration, blast.
déflation, deflation; **creux de d.**, deflation hole.
défloculant, deflocculating agent.
défloculer, to defloculate.
défluent, effluent.
défonçage (mine), digging up, ripping; **d. périglaciaire**, periglacial deep digging.
défoncer, se défoncer (mine), to stave in, to collapse, to break.
déformation, deformation, strain, warping, set; **d. antérieure à la cristallisation**, precrystalline deformation; **d. continue**, plastic strain; **d. élastique**, elastic strain; **d. par cisaillement**, shearing strain; **d. par compression**, compressive strain; **d. par glissement**, collapse structure; **d. par traction**, tensile strain; **d. permanente**, permanent set; **d. plastique**, plastic strain; **d. tangentielle**, shearing strain; **ellipsoîde de d.**, deformation ellipsoid.
déformer (tecto.), to warp, to buckle, to distort, to deform.
défricher (agro.), to reclaim, to clear, to grub.
dégagement (chimie), escape, release, emission; **d. de chaleur**, heat release; **d. de gaz**, gas escape, disengagement, gas discharge; **d. gazeux instantané**, gas outburst.
dégager (un gaz), to disengage, to emit, to liberate.
dégazage, degassing, gas freeing, outgassing.
dégazer, to outgas, to degasify.
dégel, thaw, thawing; **d. du pergélisol**, depergelation.
dégeler (périgl.), to thaw.
dégélifluxion, solifluction.
déglacement, melting of floating ice.
déglaciation, deglaciation.
dégorgeoir, disgorging spring.
dégradation, decay, degradation; **d. de la structure**, structure degradation; **d. par les eaux d'un fleuve**, scouring.
dégradé (pédol.), degraded.
degré, degree, grade; **d. celsius**, celsius degree (= 1,8 degré fahrenheit; 0 °C = 32 °F); **d. centigrade**, centigrade degree; **d. d'aggrégation (d'un sol)**, crumb capacity; **d. de dureté**, degree of hardness; **d. de latitude**, degree of latitude; **d. de longitude**, degree of longitude; **d. de métamorphisme**, metamorphism gradient; **d. de saturation**, saturation degree; **d. géothermique**, geothermal gradient.
déhouilleuse (mine), coal-cutting machine.
déjection (cône de), alluvial cone, alluvial fan.
déjections (volc.), ejectamenta.

déjeté (pli), inclined (fold), oblique (fold).
délaissé (méandre), cut-off meander, ox-bow lake.
délavage (du sol), washing out, outwash.
délayer, to dilute.
délétère (gaz), deleterious, toxic, noxious.
déliquescence, deliquescence.
déliquescent, deliquescent.
délit (strati.), joint.
délitage, spalling, parting.
délitement, breakdown.
déliter (se), to split, to crumble, to disintegrate.
délogement, quarrying.
delta, delta; **d. de flot**, flow d.; **d. de jusant**, ebb d.; **d. de marée**, tidal d.; **d. de tempête**, storm d.; **d. en patte d'oie**, bird-foot d.; **d. en pointe**, cuspate d.; **d. en progression**, protruding d.; **d. fluvial**, bird-foot d.; **d. fluviatile**, river d.; **d. intérieur**, interior d.; **d. lié aux houles**, cuspate d.; **accroissement du d.**, advancing of d.; **bras de d.**, d. distributory; **front de d.**, d. front.
deltaïque, deltaic; **plaine d.**, deltaic plain; **sédiment d.**, deltaic deposit.
delthyrium (pal.), delthyrium.
deltidiale (plaque), deltidial (plate).
deltidium, deltidium.
déluge (biblique), deluge, noarchian flood; **d. (de pluie)**, downpour; **d. (inondation)**, flood.
démagnétiser, to demagnetise.
démaigrissement (de plage), retreat of the beach.
démantèlement des continents, erosion.
démantèlement (glaciaire), stripping.
démanteler (un relief), to dismantle.
démantoïde (var. de grenat), demantoid.
démembré (réseau fluviatile), dismembered drainage.
démesuré (bloc), ice-rafted block.
demi-deuil (basalte), basalt with phenocrysts of black augite and white feldspar.
demi-profondeur (roche de), hypabyssal rock.
demoiselle coiffée, earth pillar, rock pedestal.
démontée (mer), wild (sea).
démonter (un derrick), to dismantle.
Démosponges (pal.), Demospongea.
dendrite, dendrite.
dendritique, dendritic, arborescent; **réseau hydrographique d.**, dendritic drainage.
dendrochronologie, dendrochronology.
dendrogéomorphologique, dendrogeomorphic.
déneigé (terrain), snow-free.
dénivellation, dénivellement, dislevelment, difference in level, drop.
dénivellé (adj.), 1. adj : delevelled; 2. n : relief.
dénoyage (mine), unwatering, dewatering.
dense, dense, heavy.
densimètre, densimeter, hydrometer.
densimétrie, densimetry.
densité, density, specific gravity, denseness; **d. apparente**, natural density, apparent density; **d. de drainage**, drainage density; **d. spectrale**, spectral density; **courant de d.**, density current.
densitomètre, densitometer.
dents cardinales (Lamellibranche), teeth.

denté, dentelé (pal.), serrated, jagged, dentate.

denticulation (pal.), denticle.

dentition, denture (pal.), teeth, dentition.

dénudation (érosion), denudation.

dénudé (terrain), bare, uncovered.

déparaffinage, dewaxing.

déparaffiner, to dewax.

déphasage (d'ondes), phase change, dephasing.

déphosphoration, dephosphoration.

dépilage (mine), pillar drawing, pillar robbing, pillar extraction, robbing pillars, stoping; **exploitation par d.,** pillar mining.

dépiler, to strip, to rob, to remove the pillars.

déplacement, 1. displacement; **2.** throw (= rejet of geologists); **d. apparent,** apparent displacement; **d. horizontal (faille),** strike-slip; **d. latéral (des méandres),** swinging of meanders; **d. en masse (périgl.),** mass-wasting, slipping; **d. suivant la direction (mine),** strike shift.

dépoli (grain), frosted.

déposé (sédiment), laid down, settled.

déposer, se déposer, to settle, to deposit.

dépôt, deposit; **d. abyssal,** abyssal d.; **d. allochtone,** allochtonous d.; **d. alluvial,** alluvial d.; **d. colluvial,** colluvial d.; **d. continental,** land d., continental d.; **d. criblé,** sieve d.; **d. d'eau douce,** fresh water; **d. d'eau de fonte,** fluvio-glacial d.; **d. d'écume,** foam impression; **d. deltaïque,** deltaic d.; **d. de pente,** slope d.; **d. détritique,** detrital d.; **d. d'inondation,** flood d.; **d. éolien,** eolian d.; **d. fluviatile,** river d.; **d. fluvio-glaciaire,** fluvio-glacial d., glacio-fluvial d.; **d. glaciaire,** glacial till, glacial drift; **d. glaciogénique,** till; **d. houiller,** coal d.; **d. lacustre,** lacustine d.; **d. lagunaire,** lagoon d.; **d. littoral,** littoral d., beach d.; **d. marin,** marine d.; **d. non stratifié,** unbedded d.; **d. pélagique,** pelagic d.; **d. salin,** saline d.; **d. sédimentaire,** sedimentary d.; **d. siliceux,** siliceous d.; **d. superficiels,** drift; **d. terrigène,** terrigeneous d.; **d. torrentiel,** torrent d.

dépression (topo.), hollow, basin; **d. aréïque,** desertic basin; **d. barométrique,** fall of barometric pressure; **d. endoréique,** endoreic basin; **d. exoréique,** exoreic basin; **d. fermée,** closed basin, closed depression; **d. halokarstique,** halokarstic pit; **d. karstique (doline),** cockpit, sink hole, sink; **d. océanique,** trough; **d. périclinale,** rim syncline; **d. structurale,** syncline, structural trough; **d. tectonique,** graben; **d. thermokarstique,** thaw depression; **d. volcano-tectonique,** volcano-tectonic trough.

déprimé (affaissé), depressed.

dérivation (géoph.), by-pass.

dérive (géogr.), drift; **d. des continents,** continental d.; **d. littorale,** beach d., longshore drifting.

dérivé (produit), derivative, by-product, derivate.

dérocher (mine), to separate ore from gangue.

déroulement (d'une ammonite), uncoiling.

desagrégation, disintegration, disaggregation, weathering, crumbling; **d. en boules,** spheroidal weathering; **d. granulaire,** granular disintegration; **d.**

mécanique, physical weathering; **d. par le gel,** frost weathering; **d. physico-chimique,** weathering; **d. thermique,** destruction by insolation.

désagrégeable, disintegrable.

désagrégé, crumbled, weathered, desaggregated.

désagréger, se désagréger, to disaggregate, to weather, to crumble.

désaimantation, demagnetisation.

désargentation (d'un minerai), desilverization, desilverizing.

désargenté (plomb), desilverized (lead).

désaturation, desaturation.

descendant (adj.), downward, descending; **exploitation d.,** underhand mining; **source d.,** descending spring.

descenderie, decline, way shaft, winze.

descendre (température), to fall, to drop; **d. (une pente),** to go down.

descente, descente, slope, incline; **d. du tubage,** lowering of the casing; **d. de l'eau dans le sol,** percolation.

descloïzite, descloizite.

désenvaser, to clean out.

désert, 1. adj: deserted; **2.** n: desert; **d. de déflation,** wind-scoured desert, deflation desert; **d. de gélifraction,** frost desert; **d. de sable,** sand desert; **d. littoral,** coastal desert; **d. rocheux,** hammada, gibber plain (Austr.); **d. salé, salin,** salt desert.

désertification, desertification.

désertique, desert; **croûte d.,** desert crust, desert varnish, patina; **pavage d.,** desert pavement; **poli d.,** desert polish, patina; **sol d.,** desert soil; **zone d.,** desert zone.

déshydratation, dehydratation, dessication, dewatering.

déshydrater, to dessicate, to dehydrate.

déshydraté, dessicated.

desilicification, desilication, desilicification.

désintégration, disintegration, decay; **d. en blocs,** block disintegration; **d. granulaire,** mineral disintegration; **d. nucléaire,** nuclear disintegration.

désintégrée (roche), disintegrated.

désintégrer, se désintégrer, to disintegrate.

desmine (minér.), stilbite.

désolée (région), desolate (country).

désoufrage, desulphuration.

désoxyder, to deoxidize.

desquamation (des roches), scaling, desquamation, exfoliation, peeling; **d. en écailles,** exfoliation.

dessalage, desalination, desalting.

dessalement, dessalure, freshening, desalination, lowering of salinity.

désalinisation, desalinization, desalination.

dessèchement, dessication, drying up, draining.

dessécher, se dessécher, to dry up, to dessicate, to drain.

desséché, dried, dessicated.

desserte (mine), haulage way.

dessication, dessication, drying out (up); **fente de d.,** dessication crack, mud crack.

destructeur (processus), destructive (process).

destruction (d'une roche), destruction, disruption.

désulfuration, desulphuration, desulfurization.

désulfurer, to desulphurize, to desulphurate.

détectabilité (géoph.), readibility.

détecter, to detect.

détecteur, detector; **d. à scintillation,** scintillation d.; **d. de grisou,** fire damp d.

détection, detection; **d. aéroportée,** air borne d.; **télédétection,** remote sensing.

détendre (des gaz), to expand.

détente adiabatique, adiabatic expansion.

détermination, determination; **d. de la teneur d'un minerai,** ore content assay, ore grade assay.

déterminisme (théorie), determinism.

déterrer (un fossile), to unearth, to exhume.

détonateur (mine), blaster, cap, blasting cap; **d. à retardement,** delay blasting cap; **d. de mèche,** cap, fuse; **d. électrique à retard,** delay electric blasting cap; **d. ordinaire,** regular blasting cap.

détoner (mine), to detonate, to blast.

détournement (d'un cours d'eau), diversion, diverting.

détourner (un cours d'eau), to divert.

détrempé (sol), satured (soil), soked.

détritique, detrital, detritic, clastic; **roche d.,** derivate rock.

détroit, sound, strait.

détubage (for.), pulling.

deutérique, deuteric.

deutérium, deuterium.

deutéromorphique, deuteromorphic.

déversé (pli), overfold.

déversement (d'eau), discharge, overflow; **d. (de matériaux),** dumping; **d. (gauchissement),** warping.

déverser (des matériaux), to dump, to pour, to discharge; **d. (se pencher, se déformer),** to incline, to warp.

déversoir, overflow, overfall, weir, spillway.

déviation, deviation, deflection; **d. d'un forage,** deflection of drilling; **d. magnétique,** magnetic deflection.

dévier (un forage), to deflect, to deviate.

dévisser (les tiges de forage), to break the pipes down.

dévitrification, devitrification.

Dévonien, Devonian.

diabase, diabase.

diabasique, diabasic.

diablastique (texture), diablastic (texture).

diachrone, diachronous.

diaclasage, joint pattern.

diaclase, joint; **d. de distension,** extension j.; **d. diagonale,** oblique j.; **d. directionnelle,** strike j.; **d. horizontale,** horizontal j., sheet j.; **d. longitudinale,** strike j.; **d. secondaire,** minor j.; **d. transversale,** transverse j.; **réseau de d.,** joint pattern.

diaclasé, jointed.

diaclinal, diaclinal.

diadochite, diadochite.

diaftorèse, diaftoresis.

diagenèse, diagenesis.

diagenétique, diagenetic.

diagnose, determination.

diagramme, diagram, graph, log; **d. de barres magnétiques,** pattern of magnetic reversals; **d. de calcimétrie,** calcilog; **d. de conductivité par induction,** induction log; **d. de cristal tournant,** rotating-crystal photograph; **d. de perméabilité,** permeability log; **d. de polarisation spontanée,** self potential log; **d. de résistivité,** resistivity log; **d. de rayons gamma,** gamma ray log; **d. de vitesse d'avancement d'un forage,** drilling time log; **d. diffractométrique,** X rays tracing; **d. lithologique,** lithologic log; **d. neutron-neutron,** neutron-neutron log; **d. polaire,** pole diagram rose; **d. stéréographique,** time-depth chart; **d. triangulaire,** triangular diagram, triangular plot.

diagraphie, logging, well logging; **d. acoustique,** acoustic well logging; **d. de densité,** densilog; **d. de fracturation,** fracture log electric log; **d. électrique,** electric log; **d. gamma-gamma,** gamma log; **d. nucléaire,** radioactive logging; **d. sismique,** seismic log.

diallage (augite), diallage.

dialogite (rhodocrosite), dialogite, rhodocrosite.

dialyse, dialyse.

diamagnétique, diamagnetic.

diamagnétisme, diamagnetism.

diamant, diamond; **d. brut (non taillé),** rought d.; **d. noir,** bort, bortz; **d. taillé,** cut d.; **couronne au d.,** diamond bit; **forage au d.,** diamond drilling.

diamantaire, diamond cutter.

diamantifère, diamantiferous, diamondiferous.

diamantin, diamantine.

diamétrage (d'un forage), caliper logging.

diamètre d'un sondage, hole size.

diamorphisme, diamorphism.

diaphtorèse (rétromorphose), diaphtoresis; retrograde metamophism.

diapir, diapir, salt dome.

diapir de boue, mud lump; **pli d.,** diapir fold, diapiric fold.

diapirisme, diapirism.

diaspore, diaspore.

diastème, diastem, gap, hiatus.

diastrophique, diastrophic.

diastrophisme, diastrophism.

diathermique, diathermic.

diatomée, diatom; **vase à d.,** diatom ooze.

diatomée centrique, centric diatom.

diatomée pennée, pennate diatom.

diatomite, diatomite, diatomeous earth, infusorial earth, kieselgurh.

diatrème, volcanic pipe, volcanic chimney, diatreme.

dichotomie, dichotomy.

dichotomique, dichotomous.

dichroïte (cordiérite), dichroite.

Dicotylédones (paléobot.), Dicotyledon, Dicots.

dicyclique (pal.), dicyclic.

différenciation magmatique, magmatic differentiation.

diffluence (glaciaire), (glacier) diffluence.

diffracter (optique), to diffract.

diffraction de rayons X, X ray diffraction.

diffractomètre, diffractometer.

diffractométrie (X), X ray diffraction.

diffuseur (glaciaire), glacial tongue.

digitation (tecto.), fingering; **d. deltaïque (levée autour d'un chenal),** barfingers.

digue, levee, spit; **d. courbe,** curved spit; **d. en épi,** straight spit.

dilatation, dilatation, dilatency, expansion.

dilater, se dilater, to dilate, to expand.

diluer, to dilute.

dilué, dilute.

dilution, dilution.

diluvial, diluvial.

diluvium (peu usité), diluvium.

dimension, size, dimension; **d. granulométrique,** particle size.

diminuer, to decrease.

diminution (de pression), decrease, decline.

dimorphe, dimorphous, dimorphic.

dimorphisme, dimorphism.

dimyaire, dimyarian.

Dinantien, Dinantian (l. Carboniferous: Tournesian + Visean).

Dinoflagellés (pal.), Dinoflagellata.

Dinosaures (pal.), Dinosaurs, Dinosauria.

Dinothériens (pal.), Dinotheria.

dioctaédrique (minér.), two layers (mineral).

diogénite, diogenite.

diopside (minér.), diopside.

dioptase (minér.), dioptase.

diorite, diorite; **d. à augite,** augite d.; **d. à feldspathoïde,** foid d.; **d. orbiculaire,** orbicular d.; **d. quartzique,** quartz d.

dioritique, dioritic.

Dipneustes (pal.), Dipnoi.

dipolaire, dipolar.

dipôle (magnétique), dipole.

Diptères (pal.), Diptera.

dipyre, dipyre.

direction, strike, bearing, direction, course of, trend; **d. d'une couche,** direction of bed; **d. d'une faille,** trend of fault; **d. d'un filon,** course of lode; **d. de pendage,** dip line; **d. magnétique,** magnetic bearing; **en direction,** along the strike; **galerie en d.,** drift.

dirigé vers, striked, oriented.

discontinuité, discontinuity; **d. de Mohorovicic,** Mohorovicic d., Moho; **d. de vitesse,** velocity d.; **d. sismique,** seismic d.; **d. stratigraphique,** stratigraphic gap, hiatus.

discordance, discordance, unconformity; **d. angulaire,** angular discordance; **d. d'érosion,** erosional discordance; **d. parallèle,** parallel discordance; **d. tectonique,** structural discordance.

discordant, unconformable, nonconformable.

disharmonie, disharmonic structure.

disharmonique, disharmonic; **pli d.,** disharmonic fold.

disjonction en bancs, sheet jointing.

dislocation (minér.), dislocation.

dislocation cristalline, twin gliding, mechanical twinning.

dislocation tectonique, dislocation, displacement, fault.

disloqué, dislocated, faulted.

disloquer, to dislocate, to disrupt.

disparaître en biseau, to thin out.

disperser (des matériaux), to scatter; **d. (la lumière),** to disperse.

dispersion, dispersion, dispersal, scattering; **d. colloïdale,** colloidal dispersion; **d. de la chaleur,** thermal dispersion; **d. de la lumière,** dispersion of light.

dispositif, device, apparatus, array; **d. sismique,** spread, array; **d. en éventail,** fan spread; **d. en ligne,** in line spread.

disposition structurale, structural arrangement.

disque solaire, solar disk.

dissection (du relief), dissection.

disséminé (minerai), disseminated (ore).

dissépiment (pal), dissepiment.

dissimulé (gisement), covered up, concealed.

dissipateur (glaciaire), glacial tongue.

dissipation (de chaleur), dissipation, (heat) escape.

dissociation (chimique), dissociation; **t. de dissociation,** dissociation point.

dissolubilité, dissolubility.

dissolution, solution, dissolution; **d. sous-pression,** pressure solution; **cupule de d.,** solution pit; **cuvette de d.,** solution pan.

dissoudre, se dissoudre, to dissolve.

dissous, dissolved.

dissymétrique (pli), asymmetric (fold).

distance entre géophones, geophone spacing.

distance focale (astro.), focal length.

distancemètre à laser, laser ranging.

distensif, distensive.

distension (tecto.), overstretching, extension; **faille de d.,** distensional fault, extensional fault.

disthène (minér.), disthene, kyanite.

distillat, distillate; **d. de goudron,** tar d.; **g. léger,** light d.; **d. paraffinique,** paraffin d.

distillation, distillation; **d. du pétrole,** petroleum d.; **d. fractionnée,** fractional d.

distiller, to distil.

distorsion de phase, phase distorsion.

distribution granulométrique, grain-size distribution.

district houiller, coal district.

district minier, mining district.

distrophe (lac), dystrophic (lake).

ditroïte (pétro), ditroite.

divagation (d'une rivière), shifting.

divalence (chimie), divalence.

divalent, divalent.

divergente (évolution), divergent.

diverticulation, branching.

division *(d'une roche)*, parting; **d. en bancs,** sheet structure; **d. en dalles,** tabular jointing, slab jointing; **d. en plaquettes,** platy p.; **d. prismatique,** columnar jointing, basaltic jointing; **d. stratigraphique,** stratigraphic division.

djebel *(Arabie)*, jebel, jabal, hill, mountain.

djezireh *(Arabie)*, zezireh, island, interfluvial land.

dodécaèdre, dodecahedron.

dodécaédrique, dodecahedral.

Dogger, Dogger (m. Jurassic); large concretion (U.K.).

dolérite, dolerite.

doléritique, doleritic.

dolérophanite, dolerophanite.

doline, sink, sink hole, cockpit, lime sink, limestone sink, swallow hole; **d. composée,** compound sink hole; **lac de d.,** sink-hole pond.

dolomie, dolomite, magnesian limestone, dolostone.

dolomicrite, dolomicrite.

dolomite *(minéral)*, dolomite, pearl spar.

dolomitique, dolomitic; **calcaire d.,** dolomite limestone; **marbre d.,** marble.

dolomitisation, dolomitization.

dôme, dome, uplift quaquaversal fold, cupola; **d. adventice,** parasitic dome; **d. de boue,** mud mound; **d. de glace,** ice cap; **d. de lave,** lava dome; **d. de sel,** salt dome; **d. de sel intrusif,** diapir, piercement salt dome; **d. d'extrusion,** protusive dome; **d. halocinétique,** salt dome; **d. lunaire,** lunar dome; **d. volcanique,** puy, dome.

Domérien, Domerian (l. Jurassic).

dominante *(espèce)*, dominant.

domite *(pétro.)*, domite, mica trachyte.

donnée *(géologique)*, datum, data (pl.).

données satellitaires, satellite data.

dopplérite, dopplerite.

dorsale médio-océanique, mid-oceanic ridge.

dorsale médio-pacifique, *(sans fossé central)*, rise.

dosage *(chimie)*, dosage, dosing, titration, test.

doser *(géochimie)*, to titrate, to proportion, to dose.

dosimètre, dosimeter.

double réfraction, double refraction.

doublet, doublet.

dragage, draguage *(fluv.)*, dredging; **pompe de d.,** dredging pump; **profondeur de d.,** dredging depth.

drague, dredge, dredger; **d. à godets,** bucket chain dredge; **d. à mâchoires,** grab dredge; **d. aspirante,** suction dredge; **d. suceuse,** pump dredge, suction dredge.

draguer, to drag.

drainable *(hydro.)*, drainable.

drainage, drainage; **d. cryptoréïque,** cryptorheic d., underground (karst) d.; **d. endoréique,** endorheic d., interior d.; **d. exoréique,** exorheic d., exterior d.; **d. par expansion d'eau (pétrole),** water drive; **d. par poussée d'eau,** water drive; **d. par poussée de gaz,** gas drive; **d. superficiel,** surface d.; **chenal de d.,** d. channel; **densité de d.,** d. texture; **puits de d.,** d. shaft; **réseau de d.,** d. pattern; **d. en treillis,** treillis d.

drainer, to drain.

dravite *(tourmaline)*, dravite.

dressant *(d'une couche, etc.)*, edge, edge seam, steeply dipping lode.

dresser une carte, to map.

dresser un plan, to plot.

dresser (se) *(pic, piton)*, to rise.

droit, 1. adj: upright, erect, right; 2. n: right; **d. de concession,** leasing power; **d. d'extraction de pétrole,** mineral right; **d. minier,** mineral right; **angle d.,** right angle; **ligne d.,** straight line; **pli d.,** dip right fold.

dromochronique, time-distance curve, travel-time curve.

drumlin, drumlin.

druse, druse, voog, vough, vug.

drusique, drusy.

dumortiérite, dumortierite.

dune, dune; **d. d'estran,** beach d.; **d. d'obstacle,** lee d., obstruction d.; **d. embryonnaire,** embryonic d.; **d. en croissant,** crescentic d., barchan; **d. fixée,** fixed d.; **d. hydraulique,** antidune; **d. intérieure,** inland d.; **d. littorale,** coastal d.; **d. longitudinale,** longitudinal d.; **d. mouvante,** migrating d., shifting d.; **d. parabolique,** parabolic d.; **d. sous-marine,** bank, submarine d., mega-ripple; **d, stabilisée,** fixed d.; **d. subtidale,** subtidal d.; **d. transversale,** cross-bar d., transverse d.; **d. vive,** active d.; **avant d.,** fore d.; **chaîne de d.,** d. range, d. ridge; **champ de d.,** d. field.

dunite *(r. ultrabasique)*, dunite.

durain, durain.

durbachite *(pétro.)*, durbachite.

durée de la vie d'un puits de pétrole, life of an oil well; **d. de trajet (sism.),** travel time.

dureté, hardness; **d. de l'eau,** water h; **degré de d.,** hardness degree; **échelle de d.,** hardness scale.

dynamique, dynamic; **géologie dynamique,** dynamical geology.

dynamométamorphisme, dynamometamorphism.

dyscrasite *(minér.)*, dyscrasite.

dysodonte *(pal.)*, dysodont.

dystrophe, dystrophic.

E

eau, water; **e. artésienne,** artesian w.; **e. atmosphé-rique,** atmospheric w.; **e. buvable,** drinkable w., potable w.; **e. capillaire,** capillary w.; **e. captive,** confined ground w.; **e. connée,** connate w.; **e, côtière,** coastal w.; **e. courante,** running w.; **e. de carrière,** quarry w., imbibition w. (by absorption); **e. de chaux,** lime w.; **e. de constitution,** combined w.; **e. de cristallisation,** crystallization w.; **e. de fontaine,** spring w.; **e. de fond,** bottom w.; **e. de fonte,** melt w.; **e. de fonte des neiges,** snowmelt w.; **e. de formation,** formation w.; **e. de gisement,** oil field w.; **e. de gravité,** gravitational w.; **e. de mer,** sea w.; **e. de mine,** mine w.; **e. de pluie,** rain w.; **e. de puits,** well w.; **e. de rétention,** rétention w.; **e. de ruissellement,** run-off w.; **e. de surface,** surface w.; **e. descendante,** percolating w.; **e. de source,** spring w.; **e. distillée,** distilled w.; **e. dormante,** stagnant w.; **e. douce,** fresh w.; **e. dure,** hard w.; **e. ferrugineuse,** ferruginous w.; **e. filtrée,** filtered w.; **e. fossile,** fossil w.; **e. hygroscopique,** hygroscopic w.; **e. incrustante,** incrusting w.; **e. juvénile,** juvenile w.; **e. libre,** free w., gravitational w.; **e. liée,** bound w.; **e. marine,** sea w.; **e. mère,** bitter w.; **e. météorique,** meteoric w.; **e. minérale,** mineral w.; **e. pelliculaire,** pellicular w.; **e. perchée,** perched w.; **e. phréatique,** phreatic w.; **e. pluviale,** rain w.; **e. potable,** drinking w.; **e. salée,** salt w.; **e. saumâtre,** brackish w.; **e. souterraine,** underground w.; ground w., subterranean w.; **e. stagnante,** stagnant w.; **e. sulfureuse,** sulfurous w.; **e. superficielle,** surface w.; **e. suspendue,** suspended subsurface w., hanging w.; **e. thermale,** thermal w.; **e. usées,** waste w.; **e. vadose,** vadose w.; **e. vive,** running w.; **coup d'e.,** inrush of w.; **forage à l'e.,** wet drilling.

éboulant, cavy.

éboulement, slide, rockslide, landslide, landslip, earth slide, fall of stone.

éboulement de rochers, rockslide, rockfall.

éboulement du toit (d'une couche), rockfall.

ébouler, s'ébouler, to fall down, to cave in.

ébouleux (terrain), loose (ground).

éboulis, scree (U.K.), talus (U.S.); **é. de gélifraction,** frost-shattered scree, talus; **é. de gravité,** gravity accumulation; **é. ordonné,** periglacial breccia, bedded scree, stratified debris, slope.

ébranlement (sismique), shock.

ébréché (cratère), breached (crater).

ébullition, ebullition, boiling; **entrer en é.,** to begin to boil.

Éburnéenne (orogenèse), Eburnean (Orogeny).

écaillage (de roches), rock scaling, scaling, spalling, flaking.

écaille, imbrication, slice, scale, flake; **é. d'huître,** oyster shell; **é. de mica,** mica flake; **é. imbriquée,**

tectonic slice; **desquamation en é.,** exfoliation; **structure en é.,** imbricate structure, thrust slices.

écaillement, spalling, desquamation.

écart (strat.), deviation; **é. absolu,** absolute d.; **é. moyen,** mean d.; **é. réticulaire,** interplanar spacing; **é. type,** standard d.

écartement de géophones, spacing.

écartement de plaques, rifting.

échancrure (fluviatile), notch, indentation.

échange, exchange; **é. de base,** base e.; **é. d'ions,** ion e.

échantillon, sample, specimen; **é. carrotté,** core sample; **é. d'eau,** water sample; **é. de forage,** boring sample; **é. de roche,** rock sample; **é. glycériné,** glycolated sample; **é. en poudre,** powder sample; **é. moyen,** average sample; **é. pris au hasard,** random sample; **é. scié,** slabbed sample; **é. type,** représentative sample.

échantillonnage, sampling; **é. au hazard,** random s.; **é. continu,** continuous s.

échantillonner, to sample.

échantillonneur (appareil), sampler.

échauffement, heating.

échelle, scale; **é. centigrade,** centigrade s.; **é. chronologique,** time s.; **é. chronostratigraphique,** chronostratigraphic chart; **é. de densité,** component densité s.; **é. de dureté,** hardness s.; **é. des hauteurs, des longueurs,** height, vertical or length s.; **é. géologique,** geological column; **é. graduée,** graduate s., **é. graphique,** graphic s.; **é. d'intensité** intensity s.; **é. limnimétrique,** gauge; **é. de magnitude,** magnitude s.; **é. thermométrique,** thermometric s.

échelon (failles en), echelon faults; **é. (plis en),** overlapping folds.

échine (géomorph.), ridge.

Échinoderme (pal.), Echinoderm.

Échinoïde, Echinoid.

échogramme, echogram.

échomètre, echometer, sonic depth finder.

échosondage, echosounding.

écho sondeur, echo sonder, echo-ranging sonar.

éclaboussement (d'écume), splash.

éclabousser (magma), to splash, to spatter.

éclat 1. reflet; 2. fragment, 1. glance, lustre, shine; 2. chip, fragment, splinter; **é. brillant,** shining lustre; **é. d'un astre,** brightness; **é. gras,** greasy lustre; **é. mat,** dull lustre; **é. métallique,** metallic lustre; **é. préhistorique,** flake; **é. résineux,** pitch glance; **é. soyeux,** silky lustre; **é. vitreux,** vitreous lustre.

éclatement, shattering, splitting, riving; **é. par le gel,** frost shattering, frost wedging.

éclater, to burst, to split.

éclimètre, clinometer.

éclipse (astro.), éclipse.
écliptique (astro.), ecliptic.
éclogite (r. métamorphique), eclogite.
écologie, ecology.
écologique, ecologic(al); **niche é.**, ecologic niche.
écorce (terrestre), (earth) crust.
écostratigraphique (unité), ecozone.
écosystème, ecosystem.
écotope, ecotope.
écoulement, flow, flowing, out flow, discharge, run-off; **é. boueux**, mud flow; **é. de crue**, flood flow; **é. de sol**, solifluction, soil flow; **é. en nappe**, sheet flood; **é. éphémère**, ephemeral run-off; **é. glaciaire**, ice flow; **é. laminaire**, laminar flow; **é. libre (par gravité)**, gravitational flow; **é. permanent**, perennial run-off; **é. saisonnier**, seasonal run-off; **é. souterrain**, ground water run-off; **é. superficiel**, surface run-off; **é. torrentiel**, shooting flow; **é. turbulent**, turbulent run-off; **é. visqueux**, viscous run-off; **coefficient d'é.**, run-off coefficient, drainage ratio; **sous-é.**, underflow.
écran, screen; **é. de plomb**, lead s.; **é. granitique**, granitic s.
écrasement, crush, crushing, collapse, collapsing.
écraser, s'écraser, to crush, to squeeze out, to collapse.
écroulement, collapse, fall, block falls.
ectinite (pétro.), ectinite.
ectoderme (pal.), ectoderm.
écueil, stack, reef, snag.
écume de mer, sepiolite, meerschaum.
édaphique, edaphic; local ecologic (effect).
edaphologie, edaphology.
edénite, edenite.
Eemien (interglaciaire Riss-Wurm), Eemien (marine Pleistocene zone 5 e).
effervescence (chimie), effervescence; **faire e.**, to effervesce.
effervescent, effervescent.
effet brisant (mine), rending effet; **e. de gel**, frost effect; **e. de socle**, tightening of isogrades upon an uplift of basement; **e. sismoélectrique**, seismic-electric effect.
efficacité de balayage (pétrole), sweep efficiency.
efflorescence, efflorescence; **faire e.**, to effloresce.
efflorescent, efflorescent.
effluent, effluent.
effluve (gazeux), exhalation, emanation.
effondré (terrain), collapsed, fallen in, broken down.
effondrement, collapse, caving in, breaking down, falling in.
effondrement circulaire, caldeira, cauldron subsidence.
effondrer, s'effondrer, to cave in, to fall in, to collapsus, to break down.
efforation (karst), efforation, corrosion.
effort, stress, load, pull; **e. de cisaillement**, shearing stress; **e. de compression**, compressive stress; **e. de flexion**, bending stress; **e. de traction**, tensile stress.

effritement, crumbling, disintegration, fretting.
effusif (processus), extrusive (process); **roche e.**, extrusive rock.
effusion, extrusion; **e. volcanique**, volcanic extrusion.
égueulé (cratère), breached crater
Eifélien, Eifelian (m. Devonian = Couvinian).
ejecter, to eject.
éjection, ejection.
éjection de laves, ejectamenta.
élargissement (d'un forage), underreaming, widening.
élasticité, elasticity; **coefficient d'é.**, elastic coefficient; **limite d'é.**, elastic limit; **module d'é.**, e. modulus.
élastique, elastic; **déformation é.**, e. deformation; **limite é.**, e. strength; **onde é.**, e. wave.
elatérite (var. de bitume), elastic bitumen, mineral pitch.
elbaïte, elbaite.
electrique, electric(al); **carottage é.**, e. coring; **diagramme é.**, e. log; **diagraphie é.**, e. logging; **potentiel é.**, e. potential; **polarisation é.**, e. polarization; **prospection é.**, e. prospecting; **sondage é.**, e. sounding; **tir é.**, e. blasting.
électroanalyse, electroanalysis.
électrocarottier, electrocorer.
électrochimie, electrochemistry.
électrochimique, electrochemical.
électroforage, electric drilling.
électrolysable, electrolyzable.
électrolyse, electrolysis.
électrolyser, to electrolyze.
électrolyte, electrolyte.
électrolytique, electrolytic(al).
électromagnétique, electromagnetic; **champ é.**, e. field; **diagraphie é.**, e. logging; **prospection é.**, electromagnetic prospecting.
électromagnétisme, electromagnetism.
électron, electron.
électronique, electronic; **microscope é.**, electron microscope; **sonde é.**, electron microprobe.
électrophorèse, electrophoresis.
électrostatique, electrostatic.
electrum, electrum.
élément, element; **é. accessoire**, accessory e.; **é. atmophile**, atmophile e.; **é. blanc**, felsic e.; **é. foncé**, mafic e.; **é. lithophylique**, lithophilic e.; **é. néoformé**, authigeneous e.; **é. radioactif**, radio e.; **é. siderophyle**, siderophylic e.; **é. trace**, trace e.; **é. traceur**, tracer e.; **oligo-element**, minor element.
éléolite, elaeolite (= nepheline).
éléolitique, elaeolitic.
Éléphantidés (pal.), Elephantoidea.
élévateur à godets, skip hoist, bucket elevator.
élévation (de température), rise; **é. (de terrain)**, elevation, rise, uplift.
élevé (relief), high, elevated.
élever, s'élever, to raise, to lift up, to elevate.
ellipsoïde, ellipsoid; **e. de révolution**, e. of revolution; **e. des indices (de déformation)**; strain e.; **e. des**

indices optiques, indicatrix; **e. terrestre,** e. of the earth.

élutriation, elutriation.

éluvial, eluvial, (rare) residual; **horizon é.,** eluvial horizon; **sol é.,** eluvial soil.

éluviation (lessivage du sol), eluviation, depletion.

éluvion (peu employé), eluvial deposit, eluvium.

elvan, elvan.

émail (pal.), enamel.

émanation volcanique, volcanic emanation.

embâcle (glaciaire), ice jam, ice dam; **e. glaciel,** ice jam.

emboîté, channeled, cut-an-filled; **auges glaciaires e.,** trough-in-trough; **cônes e.,** nested cones, ringed crater; **terrasses e.,** inset terraces, fill-in-fill terrace; vallées e., valley-in-valley.

emboîtement, channeling, cut-and-fill structure, ravine-filling.

emboîter, to fit into, to nest.

embolite (minér.), embolite.

embouchure, mouth, outfall.

embouchure fluviatile, river mouth.

embourbement, mudding.

embranchement (pal.), phylum.

embrasement, burning.

embraser, to catch fire, to fire; **s'embraser,** to set on fire.

embréchite, embrechite.

embrun, spray.

embryonnaire, embryonic, embryonary.

embut (karst), sink-hole.

émeraude, emerald.

émergence, emergence, emersion, rising spring.

émerger, to emerge.

émersion, emersion, emergence.

émettre, to emit; **é. de la chaleur,** to emit; **é. des gaz,** to exhale, to release; **é. des laves,** to eject.

émiettement (des roches), crumbling, disintegration, comminution.

émietter, s'émietter, to crumble.

éminence, eminence, height, élévation.

émissaire, emissary.

émissaire d'un lac, outlet.

émission, emission; **é. de cendres volcaniques,** outburst of cinders; **é. de chaleur,** heat e.; **é de gaz,** gas escape, gas release; **é de laves,** outflow of lava; **é. de rayonnement,** radiation e.; **é. radioactive,** radiactive e.

émissivité, emissivity.

emmagasinage (de gaz, de pétrole), accumulation, storage.

emmagasiner, to stock, to store, to accumulate.

émoussé, blunt, dull; **indices d'.,** degrees of roundness; **grain de quartz é.-luisant,** blunt shinning quartz grain; **outil é.,** diedged tool.

émousser, to blunt, to dull; **s'émousser,** to become blunted, dull.

empierrement (d'une route), gravelling, ballasting.

empierrer, to pave.

empiéter (chevaucher), to overlap.

empilement de glace, ice pile-up.

empilement (de terrains), piling, stocking.

emplacement, site, spot, location.

emplectite (minér.), emplectite.

emposieu (Karst Jura, Dauphiné), sink-hole, cockpit.

empreinte, print, imprint, impression, mark; **e. de goutte de pluie,** raindrop imprint; **e. de pas,** foot print; **e. fossile,** fossil print; **e. musculaire,** muscle scar.

emprisonné (gaz, eau), entrapped, confined.

émulsibilité, emulsibility.

émulsifier, to emulsify.

émulsion, emulsion; **é. aqueuse,** aqueous e.; **é. de pétrole,** oil e.; **é. vraie,** true e.

émulsionner, to emulsify.

énantiomorphe, enantiomorphic.

énargite, enargite.

encaissant (terrain), enclosing, surrounding (rocks), country rock; **couches e.,** enclosing beds.

encaissé, enclosed, encased, embanked; **méandre e.,** enclosed meander, incised meander; **vallée e.,** encased valley.

encaissement (d'une rivière), entrenchment, embanking, down-cutting.

enchevêtrement (de cristaux), intergrowth.

enclave, enclave, inclusion, enclosure, xenolith; **e. endogène,** endogeneous xenolith, cognate inclusion.

encoche de sapement, wave notch.

encorbellement (surplomb), overhang, undercut, visor.

encrine, encrinus; **calcaire à e.,** encrinitic limestone.

encroûtants (organismes), encrusters.

encroûté (niveau du sol), cemented layer horizon.

encroûtement (pédol.), incrustation, calcretization, encrusting film.

encroûtement bauxitique, bauxicrust.

encroûtement récifal, bindstone.

endémique, endemic.

endémisme, endemism.

endiguement, embanking, embankment.

endiguer, to dam up, to stem, to embank, to impound.

endocycle (Oursin), endocyclic (sea Urchin).

endogène, endogeneous, endogenous.

endokarst, endokarst.

endomorphe, endomorphous, endomorphic.

endomorphisme, endomorphism, endometamorphism.

endoréique, endorheic, endoreic.

endoréisme, endorheism (interior drainage).

endosquelette, endoskeleton.

endothermique, endothermic.

enduit (minér.), coating, coat.

enduit karstique, calcrust.

endurance (limite d'), endurance limit.

énergie, energy; **e. chimique,** chemical e.; **é. géothermique,** geothermal e.; **é. nucléaire,** nuclear e.; **é. solaire,** solar e.; **é. thermique,** thermal e.

enfoncement (du sol), hollow, depression.

enfoncer (dans la boue), s'enfoncer, to sink.

enfoui (anticlinal), hiden anticlinal.
enfumé (quartz), smoked quartz.
engel (périgl.), freezing up.
englacement, formation of floating ice.
englaciation, glacial transport (of debris, ex. englacial till).
engorgé d'eau, water logged.
engorgement (d'eau), water logging.
engorger, s'engorger, to choke up, to block, to clog.
engouffrement, engulfment.
engraissement (d'une plage), growth.
enlever (des déblais), to remove, to carry away, to strip, to clear.
ennalogène (enclave), xenolith.
enneigement, formation of a snow cover.
ennoyage (de l'axe d'un pli), pitching.
enraciné (pli), deep-seated (fold).
enregistrement, record, recording; **e. de carottage**, core record; **e. magnétique**, magnetic record; **e. sismique**, seismic recording.
enregistrer, to record, to register.
enregistreur, recorder; **e. d'échos**, echo-sounder, echo-sounding, recorder; **e. de profil**, profiler; **e. de température**, temperature recorder; **e. graphique**, graphic recorder; **e. sur bande magnétique**, tape recorder.
enrichissement de minerais, ore benefication.
enrochement, embankment with rock, protected with rip-rap.
enrocher, to enrock.
enroulée (coquille), coiled, involute.
ensablement, sanding up, silting up.
ensabler, to silt up.
ensellement, structural saddle, low.
enstatite, enstatite.
entablement (d'un pli), plateau, capping rock.
entaille (géomorpho.), notch, indentation, cut, groove.
entailler (géomorpho.), to notch, to jag, to groove.
entasser (des sédiments), to pile up, to accumulate, to heap up.
enterré (gisement), buried, sunken, deep, underground.
enthalpie, enthalpy.
entonnoir, funnel; **e. de dissolution**, sink hole; **e. séparateur**, separating f.; **cirque glaciaire en e.**, corrie.
entraînement hydraulique, water drive; **pli d'e.**, dragfold.
entraîner par lessivage, to wash down; **e. par les eaux d'un fleuve**, to carry away, to carry along; **e. par un glacier**, to drift.
entrecouper (filons), to intersect.
entrecroisée (stratification), cross-bedding, current bedding, cross lamination.
entrecroiser, s'entrecroiser, to intersect, to interlace, to criss-cross.
entrée, entrance, inflow, inlet, intake; **e. d'air**, air intake; **e. d'eau**, water inflow; **e. de baie**, bay mouth; **e. de mine**, entry, portal.
entropie, entropy.

entroque, ossicle.
entroques (calcaire à), entrochal limestone, encrinite.
envahi (par l'eau), flooded.
envasement, silting, silting up.
envaser, to silt, to choke up, with mud.
enveloppe (d'un pli), envelope (of a fold).
enveloppe terrestre, earth-mantle.
environnement, environment, surroundings; **étude géologique de l'e.**, environmental geology.
Éocambrien, Eocambrian (uppermost Precambrian), Vendian.
Éocène, Eocene.
éolianite, aeolianite (U.K.), eolianite (U.S.).
éolien, eolian; **érosion é.**, eolian erosion; **sédiment é.**, wind-borne deposit; **sédiment é.. consolidé**, eolianite.
éolisation, eolisation.
éolisé, wind worn; **galet é.**, wind facetted pebble, eolian pebble.
éolithe, eolith.
éon, aeon.
épaisseur (d'une couche), thickness.
épaissir (terrain), to thicken, to become thick.
épaississement, thickening.
épanchement (de laves), outpouring; **roche d'é.**, volcanic rock.
épancher, s'épancher, to pour out.
épandage boueux, à blocs, debris flow deposits.
épandage fluvio-glaciaire, outwash deposits.
épaulement (glaciaire), glacial shoulder.
épeirogénèse, epeirogeny.
éperon (rocheux), spur; **é. de dénudation**, rock knob.
éphémère (écoulement), ephemeral (flowing).
épi (naturel), bar, spit; **é. (artificiel)**, jetty, groyne; **é. avancé**, headland bar; **é. latéral**, bay-side bar.
épibiotisme, epibiotism.
épibole (pal.), epibole.
épibolite, epibolite.
épicentre, epicenter.
épicontinental, epicontinental; **mer é.**, epicontinental sea.
épidiagenèse, epidiagenesis (alteration by descending water).
épidosite (pétro.), epidosite.
épidote, epidote.
épidotite (pétro.), epidotite.
épierrement (agro.), clearing out of stones, removing of stones.
épigène, epigene, epigenetic.
épigenèse, epigenesis.
épigénétique, épigénique, epigenetic.
épigénie, epigenesis, surimposition.
épimagmatique, epimagmatic.
épinéritique, epineretic.
épipélagique, epipelagic.
épirogénétique, épirogénique, epeirogenic, epirogenic, epirogenetic; **mouvement é.**, epeirogenic movement.
épirogénie, épirogenèse, epeirogeny.

épitaxie, epitaxy.

épitaxique, epitaxic; **croissance é.,** epitaxic growth.

épithermal, epithermal; **minerai é.,** epithermal deposit.

épizone, epizona.

éponge (pal.), sponge.

Éponge calcaire, Calcispongea.

éponte (d'un filon), wall, selvage, salband; **é. supérieure,** hanging wall.

époque, epoch; **é. glaciaire,** Glacial e.

épreuve (essai en labo.), trial, assay, test proof.

éprouvette (échantillon d'essai), sample, test bar; **é. de tension,** tension bar; **é. de traction,** tensile test piece; **é. graduée,** graduated buret, graduated test tube; **é. mécanique,** test bar.

epsomite, epsomite.

épuisée (mine), exhausted, depleted, worked out.

épuisement (d'une mine), depletion, exhausting.

épuiser (de l'eau), to drain, to pump out, to unwater; **é. (un minerai),** to exhaust; **s'épuiser (mine),** to become exhausted.

épurer, to treat, to refine, to scrub; **é. de l'eau,** to purify, to filter.

équante (structure), isotropous.

équatorial, equatorial; **projection é.,** equatorial projection.

équiaréale (projection), equiareal.

équidimensionnel, equant, equidimensional; **roche à minéraux é.,** equigranular rock.

équidistance (entre deux courbes de niveau), contour interval.

équilibre, equilibrium; **é. chimique,** chemical e.; **é. isostatique,** isostatic e.; **profil d'é.,** equilibrium profile; **talus d'é.,** slope of equilibrium.

équilibrer (chimie), to balance, to equilibrate.

équipe, team, crew, gang; **é. de forage,** drilling crew, drilling gang; **é. de jour,** day shift; **é. de nuit,** night shift; **é. géologique,** geologic crew; **é. sismique,** seismic crew.

équipement de forage, drilling equipment.

équiplanation, equiplanation.

Équisétales (paléobot.), Equisetales.

équivalent en eau, water equivalent of snow.

équivalve, equivalve.

érathème, erathem.

ère, era; **è. Primaire,** Paleozoic e.; **è. Secondaire,** Mesozoic e.; **è. tertiaire,** Cenozoic e.; **è. Quaternaire,** Pleistocene e.

erg, erg, sand desert.

ergeron, pleistocene loam.

érodabilité (d'un sol), erodibility.

érodable, erodible.

éroder, s'éroder, to erode, to abrade.

érosif, erosive.

érosion, erosion, abrasion; **é. alvéolaire,** honeycomb e., tafoni; **é. aréolaire,** areal erosion; **é. des berges d'un fleuve,** stream bank cutting; **é. chimique,** weathering; **é. différentielle,** selective erosion; **é. éolienne,** wind erosion, deflation; **é. de masse,** mass wasting; **é. en nappe,** sheet erosion; **é. en ravins,** gully erosion; **é. fluviatile,** river erosion; **é. glaciaire,** glacial erosion; **é. karstique,** karst erosion, furrowing; **é. latérale fluviatile,** lateral erosion; **é. linéaire,** linear erosion; **é. marine,** marine erosion; **é. mécanique,** mechanical weathering; **é. normale,** normal erosion; **é. pelliculaire,** sheet erosion; **é. régressive,** headward erosion, backward erosion, retrogressive erosion; **é. souterraine,** tunnel erosion; **é. superficielle,** sheet erosion; **é. torrentielle,** gully erosion; **é. tourbillonnaire,** gully excavation; **cycle d'é.** erosion cycle; **escapement d'é.,** erosion scarp; **niveau d'é.,** erosion level; **surface d'é.,** erosion surface; **témoin de l'é.,** erosion remnant.

erratique, erratic, erratic boulder, morainic boulder; **bloc e.,** erratic block, glacially transported block.

érubescite, erubescite, bornite, phillipsite.

éruptif, eruptive; **centre é.,** eruptive point; **cône éruptif,** eruptive cone; **roche é.,** eruptive rock.

éruption, eruption, blow out; **é. centrale,** central e.; **é. de gaz,** gas flow out; **é. fissurale,** fissure eruption; **é. latérale,** flank eruption; **é. phréatique,** phreatic eruption; **é. phréatomagmatique,** phreatomagmatic eruption; **é. punctiforme,** central eruption; **é. solaire,** solar flare; **é. volcanique,** volcanic eruption; **nuage d'é.,** eruption cloud.

érythrite, erythrite.

érythrosidérite, erythrosiderite.

escalier (faille en), step (fault).

escarboucle, almandine (garnet).

escarpé, steep, sheer, precipitous.

escarpement, scarp; **e. de faille,** fault s., fault cliff; **e. de ligne de faille,** fault-line s.; **e. tectonique,** fault-controlled s.

esker, esker.

espace infracapillaire, infracapillary space.

espace supracapillaire, supracapillary space.

espacement de puits, drilling pattern.

espèce (pal.), species.

esquille, chip, splinter.

essai, test, assay, try; **e. à la flamme,** flame test; **e. à la perle,** bead test; **e. aux acides,** acid test; **e. au chalumeau,** blow-pipe analysis; **e. au marteau,** hammer test; **e. d'écoulement,** flow test; **e. de choc,** impact test; **e. de cisaillement,** shear test; **e. de coloration,** flame test; **e. de compression,** compression test; **e. de dureté,** hardness test; **e. de flexion,** bending test; **e. de laboratoire,** laboratory test; **e. de production (pétrole),** production test; **e. de résistance,** strength test; **e. de rupture,** breaking test; **e. de traction,** tensile test; **e. par voie humide,** wet assaying; **e. par voie sèche,** dry assaying.

essaim de météorites, meteor swarm.

essence, gasoline; **e. lourde,** heavy naptha; **e. minérale,** mineral spirit.

essentiel (minér.), chief (mineral).

essexite, essexite.

essonite (minéral), essonite.

estavelle (karst), gushing spring.

estérellite, esterellite.

estimation (de la valeur d'un minerai), estimate, appraisal.

estompage, shading (of a map).

estran, tidal flat, foreshore, strand zone.

estuaire, estuary.

estuarien, estuarine.

étage, stage; **é. d'une mine,** level; **é. du fond,** bottom level; **é. géologique,** geologic(al) stage; **é. houiller,** coal measure.

étagées (terrasses), stepped (terraces).

étagement (de terrasses), stepping.

étai (mine), pit prop, pit post.

étaiement, shoring, propping.

étain, tin; **é. alluvionnaire,** alluvial t.; **é. de roche,** mine t.; **é. oxydé,** cassiterite; **é. pyriteux,** stannite; **mâcle en bec de l'é.,** twinned cassiterite; **minerai d'é.,** t. ore, cassiterite.

étale (marée), slack (tide).

étalonner (un appareil), to adjust, to calibrate, to gauge, to test.

étanche, tight, impervious; **é. à l'air,** air tight; **é. à l'eau,** water tight; **é. au gaz,** gas tight.

étanchéité, tightness.

étang, pond, pool, lagoon.

état (physique), state; **é. colloïdal,** colloidal s.; **é. gazeux,** gaseous s.; **é. liquide,** liquid s.; **é. solide,** solid s.

étayage, étayement, staying, propping.

étayer, to stay, to prop, to buttress.

éteindre (de la chaux), to slake, to slack; **é. (un incendie),** to extinguish.

éteint (volcan), extinct volcano, quiescent volcano.

étendre (couche, etc.), to lay, to extend, to stretch; **é. (diluer),** to dilute.

étendu (terrain), extensive; **é. (dilué),** diluted.

étendue (n.), extent, area, expanse; **e. d'eau,** stretch.

éthane, ethane.

éther, ether; **é. sel,** ester.

éthylène, ethylene.

étiage, low water level.

étier, tide channel.

étinceleur (océano.), sparker.

étirage (d'une couche), stretching, drawing.

étirement (d'un pli), stretching.

étoile (astro.), star; **é. de mer,** starfish.

étranglement (d'un filon), pinching out, nip, narrow, constriction.

étroit (défilé), narrow, confined.

étude, examination, survey, study; **é. au microscope,** microscope examination; **é. de laboratoire,** laboratory study; **é. de terrain,** field survey; **é. diffractométrique,** X ray diffraction analysis; **é. géologique,** geological survey; **é. géophysique,** geophysic(al) survey.

étuve (labo.), exsicator, drying stove, drying oven.

euchroïte, euchroite.

euclase, euclase.

eucrite (pétro.), eucrite.

eudidymite (minér.), eudidymite.

euédrique, euhedral.

eugéosynclinal, eugeosyncline.

eulysite (pétro.), eulysite.

euphotique, euphotic.

euphotide, euphotide.

eurite (pétro.), eurite.

euritique, euritic.

eurhyalin, euryhaline, halo-tolerant, salt tolerant.

Euryptères (pal.), Eurypterida.

eurytherme, eurythermic, eurythermal.

eustasie, eustasy, eutacy (rare).

eustatique, eustatic.

eustatisme, eustasy, eustatism.

eutectique, eutectic; **mélange e.,** eutectic mixture; **point e.,** eutectic point; **température e.,** eutectic temperature.

eutectoïde, eutectoid.

eutrophe (lac), eutrophic (lake).

eutrophique, eutrophic.

eutrophisation, eutrophication.

euxénite, euxenite.

euxinique, euxinic; **faciès e.,** euxinic facies, black sea facies, sapropel facies.

évacuation (d'eau), exhaust, evacuation; **é. (de déchets),** waste disposal.

évaluation (d'un minerai), evaluation, valuation, estimation.

évaporable, evaporable.

évaporation, evaporation; **dépôt d'é.,** evaporated deposit.

évaporer, to evaporate.

évaporite, evaporite, evaporate.

évasée (vallée), widened.

événement (géologique), (geologic) event.

événement (sismique), event.

éventail (cône de déjection en), fan shaped debris cone; **é. de clivages,** cleavage fan; **é. deltaïque profond,** deep sea fan.

évidement, cavity groove.

évider, to hollow out, to scoop out, to groove.

évolute (pal.), evolute.

évolutif (processus), evolutionary process.

évolution (pal.), evolution.

évorsion, evorsion.

examen, examination; **e. aux rayons X,** X ray e.; **e. microscopique,** microscopic e.

examiner (une coupe), to examine.

excavation, hollow, hole, pit, excavation; **e. au front,** face excavation; **e. descendante,** underhand mining.

excavatrice, digger, excavator; **e. à godets,** scoop dredger; **e. de tranchées,** trench digger.

excursion, field trip.

excursion en boucle, looping excursion (of magnetic pole).

exercer une pression, to exert a pressure.

exfoliation, exfoliation, scaling.

exfolier, s'exfolier, to exfoliate, to split spheroidally.

exhalaison, exhalation.

exhaler (des gaz), to exhale, to release.

exhaure, unwatering, dewatering.

exhaussement (du sol), uplift, elevation.

exhumation (d'une structure), exhumation.

exhumé (relief), exhumed.

exine, exine.
exinite (cf. charbon), exinite.
exocycle (Oursin), exocyclic (sea-Urchin).
exogène, exogeneous, exogenic, exogenetic.
exomorphique, exomorphic.
exomorphisme, exomorphism.
exondation, exundation.
exoreïque, exorheic, exoreic; **réseau hydrographique e.**, external drainage.
exoreisme, exorheism.
exoscopie, exoscopy.
exoscopique, exoscopic.
exosquelette, exoskeleton.
exothermique, exothermal, exothermic.
exotique (bloc), erratic (*glac.*), exotic block (*tecto.*).
expansion des fonds océaniques, sea-floor spreading.
expert-géologue, consulting geologist.
expert minier, mining expert.
explicative (notice d'une carte), explanatory (note).
exploitabilité, workability.
exploitable, workable, mineable, gettable, payable.
exploitation, exploitation, winning, working; **e. à ciel ouvert**, open cut, open mining; **e. alluviale**, alluvial working; **e. avec remblayage**, mining with filling; **e. de minerai**, ore mining, **e. de mines**, mining; **e. de mines de sel**, salt mining; **e. en aval pendage**, dip working; **e. hydraulique**, hydraulicing; **e. minière**, mining; **e. par chambres et piliers**, room and pillars system; **e. par chambres magazins**; shrinkage stoping; **e. par découverte**, strip mining; **e. par foudroyage**, caving; **e. par gradins**, bench stoping; **e. par gradins droits**, underhand mining; **e. par piliers**, pillar mining; **e. par recoupes**, cross-cut system; **e. par tranches**, slicing.
exploité (mine), worked.
exploiter, to work, to mine out, to get; **e. à ciel ouvert**, to work open cast; **e. en gradins**, to stope.
explorateur, explorer.
exploration, prospection, prospecting, exploration; **forage d'e.**, exploratory boring.
explorer, to explore, to prospect.
exploser (mine), to explode, to blow up.
exploseur, exploder; **e. électrique**, electric e.
explosible, explosible.

explosif, explosive; **e. de mines**, mining e.; **e. de roche**, e. for rock work.
explosion, explosion, blow up, shot, detonation; **e. de grisou**, fire explosion; **brèche d'e.**, explosion breccia; **caldeira d'e.**, explosion caldera; **cratère d'e.**, explosion crater.
explosivité, explosiveness.
exposer (affleurer), to expose, to lay.
exposition (d'une pente), exposure.
expulser (des laves), to expulse, to eject.
expulsion (de laves), ejection, extrusion.
exsudation, exsudation, salt-weathering, fretting.
exsudation, salt-weathering, fretting.
exsuder (un liquide), to exude.
exsurgence, point of emergence.
extensibilité (d'une roche), extensibility.
extension (géographique), extent.
externe, external, outer; **manteau e.**, outer mantle.
extinction, extinction; **e. droite**, straight e.; **e. oblique**, inclined e., oblique e.; **e. ondulante (roulante)**, undulose e., undulatory e.; **angle d'e.**, extinction angle.
extraclaste, extraclast.
extractible, extractable.
extraction, extracting, hoisting, extraction, drawing; **e. de charbon**, coal winning; **e. de données (*inform.*)**, data retrieval; **e. de pétrole**, oil winning; **e. de pierre de taille**, quarrying; **e. de minerai**, hoisting ore; **e. hydraulique**, hydraulic hoisting; **e. par puits**, shaft hoisting; **e. par solvant**, solvent hoisting, liquid hoisting; **câble d'e.**, hoisting cable; **cage d'e.**, hoisting cage.
extraire, to extract, to get out, to win, to withdraw, to draw out.
extrait (chimie), extract.
extraordinaire (rayon), extraordinary ray.
extraterrestre (géologie), extraterrestrial geology.
extrémité (d'une presqu'île), end, tip.
extrémité d'une coulée boueuse, toe of earth-flow deposit.
extruder, to extrude.
extrusif, extrusive.
extrusion, extrusion; **faire e.**, to extrude.
exurgence, exurgence, rising spring.
exutoire, exsurgence.

F

fabrique (texture), fabric.

face, front, side; **f. abritée du vent**, lee side; **f. amont**, up stream side; **f. au vent**, wind side; **f. aval**, down stream side; **f. d'un cristal**, crystal f.; **f. exposée au courant**, stoss side; **f. inférieure**, underside.

facette, facet; **f. de dissolution**, solution f.; **galet à facettes**, facetted pebble, windworn pebble, eolian pebble; **galet à 2 facettes**, dreikanter.

facette de faille, fault facet.

facetter (une pierre précieuse), to facet.

faciès, facies; **f. continental**, continental f.; **f. corallien**, coralline f.; **f. d'eau douce**, fresh water f.; **f. lacustre**, lacustrine f.; **f. limnique**, limnic f.; **f. marin**, marine f.; **f. métamorphique**, metamorphic f.; **f. néritique**, neritic f.; **f. paléontologique**, fossil assemblage; **f. récifal**, reef f.; **carte de f.**, f. map.

façonnement (géomorpho.), shaping.

façonné par le vent, windworn.

facteur, factor; **f. d'évaporation**, f. of evaporation; **f. de lessivage**, leaching f.; **f. temps**, time f.; **f. volumétrique**, volume.

facule solaire, solar facule.

fagne (Ardennes), marshy waste land.

Fahrenheit (degré), Fahrenheit (degree) $(1,8\,°F = 1\,°C; 1\,°F = 0,5556\,°C; \text{temperature }°F = 9/5 + 32; \text{ex: } 32\,°F = 0\,°C)$.

faille, fault; **f. de décrochement**, strike-slip f.; **f. à charnière**, hinge f., pivotal f.; **f. à faible pendage**, low angle f.; **f. à rejet horizontal**, strike-slip f.; **f. à répétition**, repetitive f.; **f. active**, active f.; **f. anormale**, reversed f.; **f. antithétique**, antithetic f.; **f. béante**, gaping f.; **f. cachée**, buried f.; **f. chevauchement**, thrust f., lapping f.; **f. circulaire**, circular f.; **f. composée**, compound f., composite f.; **f. conforme**, conformable f., dip f.; **f. contraire**, unconformable f.; **f. de chevauchement**, overlap f., overthrust f.; **f. de cisaillement**, shear f.; **f. de compensation**, adjustement f.; **f. de compression**, compressional f.; **f. d'effondrement**, slip f.; **f. d'extension**, tension f., growth f.; **f. de gravité**, gravity f.; **f. de rotation**, rotary f., rotational f.; **f. dextre**, right lateral f.; **f. directe**, gravity f.; **f. directionnelle**, strike f.; **f. disjonctive**, tension f., normal f.; **f. en ciseaux**, scissors f.; **f. en échelon**, en echelon f.; **f. en escaliers**, step f., distributive f.; **f. en gradins**, step f.; **f. en retour**, antithetic f.; **f. fortement inclinée**, high angle f.; **f. horizontale de décrochement**, strike-slip f.; **f. inclinée**, dipping f.; **f. interstratifiée**, bedding f.; **f. inverse**, reverse f.; **f. limite**, boundary f.; **f. listrique**, listric f.; **f. longitudinale**, longitudinal f., strike f.; **f. normale**, down f., gravity f., dip-slip f., normal f.; **f. oblique**, oblique f., diagonal f.; **f. ouverte**, open f.; **f. perpendiculaire**, transverse f., cross f.; **f. pli**, fold f.; **f. radiale**, radial f.; **f. rajeunie**, activated f.; **f. ramifiée**, branching f., splitting f.; **f. secondaire**, minor f., auxiliary f.; **f. sénestre**, left lateral f.; **f. syngénétique**, growth f.; **f. transformante**, transform f.; **f. transversale**, transverse f., cross f., dip f.; **brèche de f.**, f. breccia; **conglomérat de f.**, f. conglomerate; **ensemble de f.**, f. set; **escarpement de f.**, f. scarp; **de ligne de f.**, f. line scarp; **faisceau de f.**, f. bundle; **formation de f.**, faulting; **miroir de f.**, slickenside, f. polish; **paroi de f.**, f. wall; **piège de f.**, f. trap; **plan de f.**, f. plane; **pli-f.**, faulted-anticline; **rejet de f.**, f. throw; **ressaut de f.**, f. scarp; **stries de f.**, f. striae.

faillé, faulted; **bloc f.**, fault block; **filon f.**, fault vein; **zone f.**, f. area.

faire, to make, to do; **f. des bulles (magma)**, to bubble; **f. couler (hydro.)**, to drain off; **f. effervescence (chimie)**, to effervesce; **f. exploser (mine)**, to fire; **f. précipiter (chimie)**, to precipitate; **f. réagir**, to react; **f. saillie (géomorpho.)**, to stand out; **f. sauter (mine)**, to blast, to shoot.

faisabilité (études de), feasibility (study).

faisceau de plis, tectonic bundle.

faîte, crest, top; **ligne de f.**, crest line.

falaise, cliff; **f. de glace**, ice c.; **f. littorale**, shore c.; **f. marine**, sea c.; **f. morte**, abandoned c., fossil c.; **abrupt de f.**, cliff wall; **microf.**, nip.

falun, shelly sand, falun.

famatinite, famatinite.

Famennien, Famennian (u. Devonian).

famille (pal. pétro.), family; **f. de roches éruptives**, f. of igneous rocks.

fange, mud.

fangeux (chantier), miry, muddy.

fanglomérat, fanglomerate.

farine de roche, rock flour; **f. fossile**, fossil f., diatomite; **f. glaciaire**, glacial meal; **f. minérale**, powdered ore.

fassaïte, fassaite.

fathogramme, fathogram.

fathomètre, fathometer.

fatigue (des matériaux), strain, fatigue.

fauchage (des couches), down-bending, curvature, gravitational sagging.

faune, fauna; **f. appauvrie**, depleted f.; **succession de f.**, faunal succession.

faunistique, faunal; **province f.**, faunal province.

faunizone, faunizone.

fausérite, fauserite.

fausse galène, zinc blende, blende, sphalerite.

fausse stratification, false bedding.

fausse topaze, false topaz.

faux anticlinal, antiformal syncline.

faux pendage, false dipping.

faux synclinal, pseudo syncline, synformal anticline.

fayalite, fayalite.

feldsparénite, arkose.

feldspath, feldspar; **f. alcalin,** alkali f.; **f. calcosodique,** lime-soda f.; **f. plagioclase,** plagioclase f.; **f. potassique,** potash f.; **f. sodique,** soda f.; **f. vert,** amazonite; **f. vitreux,** sanidine; **solution solide de f.,** f. exsolution.

feldspathique, feldspathic, feldspathose.

feldspathisation, feldspathization.

feldspathoïdes, feldspathoids, foids.

felsite, felsite.

fémique, mafic.

fendillement (du sol), cracking, fissuring.

fendiller (se), to fissure, to crack.

fendre, se fendre, to split, to slit, to fissure, to crack, to cleave.

fendue, day drift, day level.

fenestra, "bird eyes", keystone-vug.

fenêtre (d'une nappe), nappe inlier, geologic window, fenster.

fénite (pétro.), fenite.

fénitisation, fenitization.

fente, fissure, crack, split, slit; **f. collimatrice,** divergence slit; **f. de compression,** compression joint; **f. de contraction,** shrinkage crack; **f. de dessication,** shrinkage crack; **f. de froid,** ice wedge, frost crack; **f. de remplissage,** fissure vein; **f. de retrait,** sun crack, shrinkage crack; **f. en coin (périgl.),** ice wedge; **f. filonienne,** fissure vein; **f. tectonique,** fault fissure.

fer, iron; **f. arsenical,** arsenopyrite; **f. blanc,** tin, tin plate; **f. de lance (gypse),** arrow-head twin; **f. des marais,** swamp ore, bog i. ore; **f. météorique,** meteoric i.; **f. natif,** native i.; **f. oligiste,** hematite; **f. oolithique,** oolithic i.; **f. pisiforme,** pealike i.; **f. spathique,** siderite, i. spar; **f. spéculaire,** specular hematite; **f. sulfaté,** melanterite; **f. sulfuré,** pyrite; **f. titané,** titanoferrite; **chapeau de f.,** gossan.

ferme (terrain), solid, firm.

fermé, closed; **dépression f.,** closed basin.

fermer, to close, to seal off; **f. un puits,** to close in a well.

fermeture, closure; **f. d'un anticlinal,** anticlinal c.; **f. structurale,** structural c.

ferralite, laterite, allite.

ferralitique, ferralitic; **sol f.,** ferralitic soil.

ferralitisation, ferralitization.

ferriargilane, ferruginous clay skin.

ferreux, ferrous, ferreous.

ferricyanure, ferricyanide.

ferrifère, ferriferous.

ferrique, ferric.

ferrite (minér.), ferrite.

ferrocalcite, ferrocalcite.

ferrocobaltite, ferrocobaltite.

ferrod (podzol ferrugineux), ferrod.

ferrodolomite, ferroandolomite.

ferrogabbro, ferrogabbro.

ferrohypersthène, ironhypersthene.

ferromagnésien, ferromagnesian.

ferromagnétique, ferromagnetic; **minéral f.,** mafic mineral.

ferromagnétisme, ferromagnetism.

ferromanganèse, ferromanganese.

ferronatrite, ferrinatrite.

ferrosilite, ferrosilite.

ferrugineux, ferruginous, ferrugineous; **cuirasse f.,** ferruginous crust, duricrust; **eau f.,** ferruginous water; **sol f.,** ferrimorphic soil, ferrisol; **source f.,** ferruginous spring; **à ciment f.,** iron-cemented (ferricrete).

ferruginisation, iron cementation, ferruginous cementation.

fersiallitique, fersiallitique.

feston de plage, beach cusp.

feston (de solifluxion), festoon, guirland.

feu, fire; **f. de grisou,** f. damp; **f. de mine,** pit f.; **pierre à feu,** flint.

feuille (de métal), sheet, leaf.

feuillet de mica, mica flake; **f. d'argile,** layer; **f. de schiste,** folium, folia (pl.); **f. sédimentaire,** lamina; **silicate en feuillets,** phyllosilicate.

feuilletage, **1.** layering (séd.); **2.** foliation (metamorphic rocks).

feuilleté, lamellar, laminated, foliated; **roche f.,** foliate rock; **structure f.,** foliation structure.

fibre, fiber, fibre.

fibreux, fibrous; **cassure fibreuse,** fibrous fracture.

fibrolite, fibrolite.

figer (lave), se figer, to solidify, to congeal.

figuline (argile), figuline (clay).

figure, figure, diagram; **f. d'affouillement,** scour mark; **f. d'affouillement et de colmatage,** scour and fill structures; **f. de base de bancs,** sole cast; **f. de charge,** load cast; **f. de choc,** prod cast; **f. de courant,** rill-mark, ripple mark; **f. de corrosion,** etch pits; **f. de glissement,** slump structure; **f. de rebond,** bounce cast; **f. de roulement,** roll cast; **f. de saut,** skip cast; **f. d'interface,** interfacial cast; **f. en brosse,** brush cast; **f. en chevron,** chevron cast; **f. en croissant,** crescent cast; **f. en cuvette,** dish structure; **f. en gouttière,** gutter cast.

filamenteux (minér.), filamentous.

filet d'eau, runnel of water.

filet de minerai, veinlet, stringer, thread.

Filicales (pal.), Filicales.

film (d'eau), pellicule; **f. (géophys.),** record.

filon, ledge, lode, seam, sill, vein; **f. annulaire,** ring-dyke; **f. aveugle,** blind lode; **f. clastique,** clastic dyke, sill; **f. composé,** compound vein; **f. couche,** intraformational vein, bedded vein, sill; **f. couche,** sill; **f. de minerai,** ore sill; **f. croiseur,** cross vein, cross, lode, cross course; **f. de faille,** slip vein; **f. de glace,** ice vein; **f. de ségrégation,** segregated lode; **f. de substitution,** replacement vein; **f. en chapelet,** loaded vein; **f. en échelons,** ladder lode; **f. épithermal,** epithermal lode; **f. houiller,** coal seam; **f. hypothermal,** hypothermal vein; **f. intrusif incliné,** dike, steep vein; **f. lenticulaire,** lenticular lode; **f. mère,** mother lode, main lode; **f. mésothermal,** mesothermal lode; **f. métallifère,** metalliferous lode; **f. métasomatique,** metasomatic vein; **f. minéral,** mineral vein; **f. nourricier,** mother lode, feeder; **f. principal,** mother lode, master; **f. ramifié,** branching lode; **f. secondaire,** dropper; **f. stérile,** barren vein.

filonnet, small, thin lode.

filtrage (sism.), filtering.

filtrage monocanal (géoph.), single-channel filtering.

filtrage multicanal, multi-channel filtering.

filtrant, filtering; **couche f.,** filter bed; **sable f.,** filter sand; **tamis f.,** filter sieve, filter screen; **terre f.,** filter clay.

filtrat (chimie), filtrate.

filtration, filtration, percolation.

filtre, filter, screen; **f. à eau,** water filter; **f. à poussière,** dust filter; **f. à sable,** sand filter; **f. à vide,** vacuum filter; **f. d'onde** *(géophys.),* wave filter; **f. passe-bande,** pass band filter; **f. presse** *(pétro.),* filter pressing.

filtrer, to filter, to filtrate, to seep, to screen.

fin, **1.** adj: (petit) small, small-sized; **2.** n: (précieux) fine; **3.** n: end, ending; **broyage f.,** fine crushing; **grain f.,** fine grained, fine textured; **limon f.,** fine loam; **minerai f.,** fine ore; **sable f.,** fine sand.

finement stratifié, thin bedded.

fines (minerai), fines, smalls, slack.

fissile (roche), cleavable, fissile, fissionable.

fissilité, cleavage, splitting, fissility.

fission (d'un continent), rifting.

fission nucléaire, nuclear fission.

fissurale (éruption), fissural eruption, fissure eruption.

fissuration, splitting, cracking, fissuration, fissuring.

fissure, crack, fissure, cleft; **f. aquifère,** water-bearing fissure; **f. de décollement de coulée boueuse,** crown scar; **f. de dessication,** sun crack, heat crack; **f. de gel,** frost crack; **f. d'extension,** tension fissure; **f. de retrait,** contraction crack, mud crack, sun crack; **f. filonienne,** fissure vein; **f. métallisée,** fissure vein; **f. minéralisée,** fissure vein.

fissuré, cracked, fissured.

fixisme (pal.), fixism.

fjord, fjord, fiord.

Flagellés (pal.), Flagellates.

flambant (charbon), flaming (coal).

flamme, flame; **essai à la f.,** flame test; **injection en f.,** flame structure.

« flammes » (des ignimbrites), eutaxitic fabric.

flanc, flank, limb, side; **f. arrière,** back limb; **f. avant,** forelimb; **f. d'anticlinal,** anticlinal limb; **f. de côteau (galerie à),** adit; **f. d'un pli,** limb; **f. inférieur d'un pli-couché,** under limb; **f. inverse,** reversed limb; **f. normal,** normal limb; **f. renversé,** reversed limb; **f. supérieur,** roof limb.

Flandrien, Flandrian (early Holocene).

flaque (d'eau), puddle, small pool.

flèche, spit, bar; **f. littorale,** spit; **f. d'amour,** rutilated quartz; **f. de jonction,** tombolo.

fléchir, to bend, to yield.

fléchissement (d'une strate), bending, yielding.

fleur de cobalt, cobalt bloom.

fleur de silice (micro. à balayage), silica bloom.

fleur de zinc, zinc bloom, hydrozincite.

fleuret (de mine), drill.

fleuve, river, stream; **f. à marée,** tidal river; **f. côtier,** coastal stream; **f. de glace,** glacier; **f. inadapté,** underfit or misfit stream; **f. temporaire,** ephemeral stream.

flexible (roche), flexible, bendable.

flexion, bending, flexion.

flexoforage, flexodrilling.

flexure, down bending, uniclinal, warping; **f. continentale,** shelf edge; **f. monoclinale,** monoclinal flexure; **f. répétée,** kink flexure.

flocon (de neige), flake.

floculation, flocculation, coagulation; **agent de f.,** flocculating agent; **essai de f.,** flock test; **point de f.,** flock point.

floculer, to flocculate.

flore (paléobot.), flora.

floss-ferri, floss-ferri.

flot, flow, rising tide; **f. de retour,** backwash; **f. liquéfié,** liquefied flow; **f. montant,** uprush; **f. turbide,** turbidity flow; **courant de f.,** flow current; **delta de f.,** flow delta.

flottation, flotation; **f. de minerai,** ore flotation; **f. par moussage,** froth flotation; **concentré de f.,** flotation concentrate; **essai de f.,** flotation test.

flotter (océano.), to float, to buoy up.

fluage, creep, flow.

fluctuation (du niveau marin), fluctuation.

fluctuer, to fluctuate.

fluidal, fluidal; **structure f.,** fluidal structure; **texture f.,** fluidal texture.

fluide, fluid; **f. boueux,** mud f.; **f. de forage,** drilling f.; **f. d'étanchéité,** seal f.; **f. mouillant,** wetting f.; **f. obturateur,** seal f.; **inclusion f.,** f. inclusion; **mécanique des f.,** f. mechanics.

fluidification, fluxing.

fluidifier, to fluidify.

fluidité, fluidity.

fluocérite, fluocerite.

fluor, fluorine; **datation au f.,** fluorine dating.

fluorapatite, fluorapatite.

fluorescéine, fluorescein.

fluorescence, fluorescence, bloom; **f. aux ultra-violets,** fluorographic method.

fluorescent, fluorescent.

fluorine, fluorite, fluor, fluor spar, fluorite, calcium fluoride.

fluorure, fluoride.

flûte (sism.), seismic cable, streamer.

fluvent (sol alluvial), fluvent.

fluvial, fluvial, fluviatile cycle; **cycle f.,** fluvial geomorphic cycle; **erosion f.,** fluvial river erosion.

fluviatile, fluviatile; **f. continental,** fluvio-terrestrial; **dépôts f.,** fluviatile deposits; **processus f.,** fluviation.

fluvio-glaciaire, fluvio-glacial.
fluvio-lacustre, fluvio-lacustrine.
fluvio-marin, fluvio-marine.
fluvio-nival, fluvio-nival.
fluvio-périglaciaire, fluvio-periglacial.
flux, flux, flow; **f. magnétique,** magnetic flow; **f. thermique,** heat flow.
flysch, flysch, sedimentary association (orogenic).
focal, focal; **distance f. (opt.),** focal length; **profondeur f. (sism.),** focal depth.
foïdique, feldspathoidic.
foisonnement (d'une couche), expansion, swelling.
foliacé (minéral), foliaceous, foliate, foliated.
foliation, foliation.
fonçage (de puits), shaft sinking.
foncé (minéral), dark (mineral), mafic (mineral).
fond, bottom, floor, ground; **f. abyssal,** abyssal depth; **f. de bateau (pli en),** syncline; **f. de carte,** base map; **f. de puits,** well bottom; **f. de vallée,** thalweg, valley floor; **f. marin,** sea floor; **f. océanique,** oceanic floor, ocean bottom; **haut-f.,** shoal; **pli de f.,** folding of the basement.
fondant (adj.), melting; **glace f.,** m. ice; **neige f.,** m. snow.
fonderie, smeltery, smelting works.
fondis, swallow-hole.
fondre (des minerais), to melt down, to smelt, to fuse.
fondrière, pit, hollow, slough, bog, quagmire.
fondu (minerai, magma), fused, melted, molten, smelted.
fontaine, spring, pool; **f. ardente,** fire well; **f. de laves,** lava fountain; **f. vauclusienne,** vauclusian spring.
Fontainebleau (sables et grès), Fontainebleau (sands and sandstones, middle Oligocene).
fonte (état de fusion), fusing, thawing, melting, smelting; **f. (métal),** cast iron, pig iron; **f. au coke,** coke pig iron; **f. des neiges,** snowmelt; **f. nivale,** snowmelt.
fontis, swallow hole, roof collapsing, subsidence of surface.
forabilité, drillability.
forable, drillable.
forage, bore hole, boring, drill, drill hole, drilling well; **f. à l'air,** air drilling; **f. à l'eau,** wet drilling; **f. au diamant,** diamond drilling; **f. à grand diamètre,** big hole; **f. à injection,** wash boring; **f. à percussion,** percussion drilling; **f. carotté,** core drilling; **f. d'exploitation,** exploitation drilling; **f. d'exploration,** exploratory drilling; **f. de reconnaissance,** wildcat well; **f. dévié,** deflected well; **f. en éventail,** fan drilling; **f. orienté,** directional drilling; **f. par battage,** cable drilling; **f. par percussion,** percussion drilling; **f. par rotation,** rotary drilling; **f. percutant,** percussion drilling; **f. sismique,** shot hole drilling; **f. sous-marin,** offshore drilling; **f. thermique,** fusion piercing; **boue de f.,** drilling sludge; **navire de f.,** drilling vessel; **plate-forme de f.,** drilling rig; **tige de f.,** drill rod; **trou de f.,** drill hole.

foramen (pal.), foramen.
Foraminifères (pal.), Foraminifera; **F. arénacés,** arenaceous F.; **F. benthiques,** benthonic F.; **F. imperforés,** aporous F.; **F. perforés,** perforated F.; **F. planctoniques,** planktonic F.
Foraminifères (à), foraminiferous; **boue à F.,** Foraminiferal ooze.
force, force, power, stress; **f. d'attraction de la pesanteur,** gravity attraction; **f. de cisaillement,** shearing force; **f. de compression,** compression stress; **f. de Coriolis,** Coriolis force; **f. de gravité,** gravity force; **f. de rupture,** breaking stress; **f. de tension,** tensile stress; **f. géostrophique,** Coriolis force; **f. hydraulique,** hydraulic power; **f. magnétique,** magnetic force.
forer, to drill, to bore; **f. par battage au câble,** to spud.
foret, bit, drill; **f. au diamant,** diamond bit; **f. hélicoïdal,** twist drill.
foreur (technicien), drill man, driller, borer.
foreuse (machine), drill, drilling machine; **f. carottier,** core drill; **f. diamantée,** diamond bit; **f. pneumatique sur chenilles,** air-track drill.
formation (couche) (résultat d'action), formation, bed formation; **f. aquifère,** water bearing bed; **f. caractéristique,** guide formation; **f. de brèches,** brecciation; **f. de failles,** faulting; **f. faillée,** faulted bed; **f. marine,** marine bed; **f. métallifère,** ore-bearing formation; **f. non consolidée,** unconsolidated formation; **f. pétrolière,** oil producing formation; **f. poreuse,** thief formation; **f. récifale,** reef; **f. salifère,** saline formation; **f. sédimentaire,** sedimentary bed; **f. superficielle,** superficial deposits, drift; **carte de f.,** formation map.
forme, form, shape; **f. cristalline,** crystal form; **f. d'érosion,** erosional form; **f. du paysage,** land form.
forstérite (péridot.), forsterite.
fort pendage, steep dipping.
forte pente, steep slope; **f. teneur,** high grade.
fosse, hole, pit; **f, à boue (forage),** mud pit; **f. d'avant arc,** fore-arc trench; **f. d'effondrement,** graben, rift valley, taphrogenic trough; **f. d'effondrement remblayée,** back-filled trough, sedimented graben; **f. géosynclinale,** geosyncline; **f. océanique,** trench, deep, hadal zone; **f. sédimentaire,** basin; **f. tectonique,** graben, fault trough; **f. topographique,** depression, hollow; **avant-fosse,** fore-deep.
fossé, ditch, drain, trench; **f. de drainage,** drain; **f. d'effondrement,** rift; **f. médian d'une dorsale,** median rift; **f. tectonique,** rift, fault trough.
fossette (pal.), socket.
fossile, 1. adj: fossil; 2. n: fossil; **f. caractéristique,** guide f.; index f.; **f. de faciès,** facies f.; **f. de zone,** zone f.; **f. remanié,** reworked f., derived f.; **f. roulé,** water-worn f.; **f. stratigraphique,** guide f., index f.; **bois f.,** f. wood; **bon f.,** index f.; **eau f.,** f. water; **faune f.,** f. fauna; **flore f.,** f. flora.
fossilifère, fossiliferous, fossil bearing.
fossilisateur (processus), fossilizing (process).
fossilisation, fossilization.

fossiliser, se fossiliser, to fossilize.
foudre (météo.), lightning.
foudroiement (mine), block caving.
foudroyage (mine), caving.
foudroyer (mine), to cave in.
Fougère (paléobot.), Fern (Filicales).
fouille, digging, excavation; **f. à ciel ouvert,** open pit; **faire des f.,** to excavate.
fouiller, to excavate, to dig.
foulon (terre à), fuller's earth.
four, furnace, oven; **f. à calciner,** calcining furnace; **f. à chaux,** lime kiln; **f. à pyrite,** pyrite oven; **f. de grillage,** roasting furnace.
fourniture d'eau, water supply.
fowlérite, fowlerite.
foyaïte (syénite), foyaite.
foyer, furnace, hearth; **f. d'une lentille optique, focus; f. d'un séisme,** focus; **f. magmatique,** magmatic hearth.
fraction, fraction; **f. légère,** light f.; **f. lourde,** heavy f.; **f. minérale,** mineral f.
fractionné (sédiment), divided, fractionate; **analyse f.,** fractional analysis; **cristallisation f.,** fractional crystallization.
fractionnement (du pétrole brut), fractionating, fractionation; **colonne de f.,** fractionating column, fractionator.
fractionner (un produit pétrolier), to fractionate.
fracturation (d'une couche), formation fracturing; **f. hydraulique,** hydrofracturing; **f. par le froid,** frost breaking, frost splitting.
fracture, fracture; **f. conchoïdale,** conchoidal f; **plan de f.,** fracture plane; **porosité de f.,** fracture porosity.
fracturer, to fracture, to split, to break.
fragile (minéral), brittle.
fragipan (pédol.), fragipan.
fragment (de roche), chip, fragment.
fragmentation, fragmentation.
fragmenter, to divide into fragments.
frais (d'exploitation d'un gisement), operating costs, running costs.
fraisil, coal cinders, frazil.
frange, rim; **f. capillaire,** capillary r.; **f. de Becke,** Becke line; **f. littorale,** continental shelf; **f. réactionnelle,** reaction r.; **f. salée,** fresh-salt water interface.

frangeant (récif), fringing (reef).
franklinite, franklinite.
freibergite, freibergite.
fréquence, frequency; **f. acoustique,** audio f.; **f. gel-dégel,** freeze-thaw; **f. de mise à feu,** rate of firing; **f. de vibration,** vibrational f.; **bande de f.,** f. band; **courbe de f.,** f. curve.
friabilité, friability, grindability.
friable (roche), friable, crumbly.
friche (pédo.), fallow land.
friedélite, friedelite.
froid, 1. adj: cold; 2. n: cold; **action du f. (en dessous de 0 °C),** frost weathering, frost action; **fentes de f. (périgl.),** ice wedge; **fissuration par le f. (gel),** frost splitting.
froncement (d'une couche), puckering.
fronde (Ptéridophytes), frond.
front, front, face; **f. d'avancement de chantier,** heading face; **f. de charriage,** thrust front; **f. de chevauchement,** thrust front; **f. de cuesta,** scarp face; **f. de dégel,** thaw front; **f. de gel,** freeze front; **f. de nappe (de charriage),** brow; **f. de plage,** shore face; **f. de taille,** face, mine face, working face; **f. d'ondes,** wavefront; **f. glaciaire,** glacial front; **f. salé,** fresh-salt water interface; **f. volcanique,** outer volcanic arc.
frontale (moraine), frontal (moraine).
frontière de plaque, plate boundary.
frottement (d'un glacier), friction.
frottis (microscopique), immersion mount.
frustule (de Diatomée), frustule.
fuel, fuel oil.
fuite (d'eau, de gaz), leak, escape, leakage, loss, spill.
fulgurite, fulgurite, lightning tube.
« fumeur » noir (Pacifique), black smoker.
fumerolle, fumarole.
fumerollien, fumarolic.
funiculaire (eau), funicular (water).
fusain, charcoal, fusain.
fusibilité, fusibility, fusibleness; **échelle de f.,** fusibility scale.
fusinite, fusinite.
fusion, fusion, melting, smelting; **courbe de f.,** fusion curve; **température de f.,** fusing point.
Fusuline (pal.), Fusulina.
Fusulinidés (pal.), Fusulinids (Foraminifera).

G

gabbro, gabbro; **g. basique,** alkali g.
gabbroïde, gabbroid.
gabbroïque, gabbroic.
gadolinite, gadolinite.
gahnite (minér.), gahnite, zinc spinel.
gaillettes, lump coal.
gaine, gangue, matrix.
gaize, gaize (a glauconitic or calcitic sandstone cemented with silica: France and Belgium).
galaxie (astro.), galaxy.
galène, galene, galenite, lead glance, lead sulphide.
galerie (mine), gallery, drift, gangway; **g. à flanc de coteau,** adit; **g. au rocher,** stone drift; **g. captante,** infiltration gallery; **g. costresse,** countergangway; **g. d'accès,** adit; **g. d'avancement,** drift stope; **g. de drainage,** drainway water gallery; **g. d'évacuation,** haulage drift; **g. de fond,** deep level, bottom level; **g. de mine,** drift; **g. de recherche,** exploratory drift; **g. de retour d'air,** airway return; **g. de roulage,** haulage; **g. en direction,** drift; **g. en impasse,** blind drift; **g. principale,** main gangway; **g. transversale,** cross gangway, cross heading.
galet, pebble; **g. à facettes,** facetted p.; **g. aménagé,** p. culture; **g. arrondi,** rounded p.; **g. émoussé,** worn p.; **g. éolisé,** wind-worn p.; **g. façonné par le vent,** wind facetted p.; **g. mou,** clay gall, mud ball; **g. noir (vase),** flat tidal p.; **g. strié,** striated ball.
gallon, gallon (U.S.) = 3,785 liter (litre), gallon (U.K.) = 4,55 l.
gamma, gamma; **diagraphie gamma-gamma,** gamma-gamma log; **méthode de diagraphie par les rayons g.,** gamma ray well logging.
gamme de fréquence, frequency range.
gangue, gangue, matrix, enclosing matrix.
ganister (pétro.), ganister.
ganoïde (pal.), ganoid.
Gargasien, Gargasian (Aptian, l. Cretaceous).
garniérite (var. de serpentine nickelifère), garnierite.
garrigue, xerophytic scrub associated with limestone (mediterranean France; cf. maquis).
gaspillage, wasting; **g. de gisement,** gophering, robbing.
gassi (inter-dunaire), passage-way.
Gastéropode (pal.), Gastropod.
gastrolithe, gastrolith.
gauchir (se), to warp, to bend.
gauchissement (d'une courbe), warping, buckling, wrinkling.
gaufrage, corrugation.
Gault, Gault (cf. Albian, l. Cretaceous).
Gauss (courbe de), Gaussian curve.
gaussienne (répartition), gaussian distribution.

gave (pyrénées), torrent.
gaz, gas; **g. brut,** raw g.; **g. captif,** entrapped g.; **g. carbonique,** carbon dioxide; **g. combustible,** fuel g.; **g. de grisou,** stink (damp); **g. de cokerie,** oven g.; **g. de craquage,** cracked g.; **g. de houille,** coal g.; **g. de pétrole,** oil g.; **g. de raffinerie,** refinery g.; **g. des marais,** fire damp, marsh g.; **g. emprisonné,** entrapped g.; **g. liquéfié,** liquefied g.; **g. naturel,** natural g.; **g. corrosif (acide),** sour g.; **g. naturel non désulfuré,** sour g.; **g. occlus,** entrapped g.; **g. pauvre,** lean g., poor g., dry g.; **g. riche,** rich g., wet g.; **g. sec,** dry g.; **g. volcanique,** volcanic g.; **champ de g. naturel,** g. field; **chapeau de g.,** g.-cap; **coke à g.,** g.-coke; **conduite de g.,** g. line; **drainage par g.,** g. drive; **éruption de g.,** g. blow-out; **gisement de g.,** g. pool; **injection de g.,** g. injection; **proportion g.-huile,** g.-oil ratio (G.O.R.); **puits de g.,** g. well; **stockage de g.,** g. storage.
gazéifère, gas bearing; **horizon g.,** gas horizon; **roche g.,** gas rock.
gazéification, gasification, gasifying.
gazéifier, to gasify.
gazeux, gaseous, gassy; **hydrocarbure g.,** gaseous hydrocarbon; **inclusion gazeuse,** gaseous inclusion.
gazoduc, gas (pipe) line.
géant (marmite de), pot-hole.
géanticlinal, geanticline.
Gédinnien (Dévonien inférieur), Gedinnian (l. Devonian).
gédrite (amphibole), gedrite.
Geiger (compteur), Geiger (counter).
gel (action du froid), frost, freezing gel; **g. discontinu,** freeze-thaw; **g. de silice,** silice gel; **action du g.,** frost action, frost weathering; **action du g. et du dégel,** freeze and thaw action; **fentes de g.,** ice wedge; **fissuration par le g.,** frost splitting, frost breaking; **polygones de g.,** earth rings, tundra polygons; **poussée de g.,** ice thrust, ice push.
gelée, frost.
geler, se geler, to freeze.
gélicontraction, frost shrinkage, gelicontraction (rare).
gélidéflation, ablation of frozen ground, gelideflation (rare).
gélidisjonction, frost-shattering.
gélif, easily cracked by frost.
gélification, gelation, gel-formation.
gélifier (se), to gel, coagulate.
gélifluxion, periglacial solifluction, gelifluction (rare).
gélifract, frost fractured chip, gelifract, congelifract.
gélifracté, frost-fractured, shattered, gelifracted (rare).

gélifraction, frost breaking, cryofracture, gelifraction.

géliplaine, periglacial plain, geliplain (rare).

géliplanation, cryoplanation, geliplanation.

gélisol, frozen ground, permafrost, gelisol.

gélisol temporaire, seasonally frozen ground.

gélisolation, gelisolation.

géliturbation, cryoturbation, geliturbation (rare), congeliturbation.

géliturbé, frost-stirred, contorted.

gélivation, frost-breaking, frost-disruption, frost weathering, frost-thaw action, gelivation.

gélivé, frost-shattered.

gélivité, frost susceptibility.

gemme, gem, gemstone; **sel g.,** rock salt, halite; **taille de g.,** gem cutting.

gemmologie, gemmology.

génal (pal.), genal.

« gendarme », rock pinnacle.

génération (pal.), generation.

genèse, genesis.

génétique, **1.** adj: genetic; **2.** n: genetics.

génie civil, civil engineering.

génitale (plaque), genital (plate).

génotype (pal.), genotype.

genou (mâcle en), geniculating twin.

genre (pal.), genus.

géobios, geobios, terrestrial biotope, organic environment (as opposed to hydrobios).

géobotanique (adj.), geobotanical.

géocentrique, geocentric.

géochimie, geochemistry.

géochimique, geochemical; **carte g.,** g. map; **dépression g.,** g. sink; **indicateur g.,** g. indicator; **profil g.,** g. profile; **prospection g.,** geochemical prospecting.

géochronologie, geochronology; **g. isotopique,** isotopic g.

géochronologique, geochronologic, geochronological; **stratigraphie g.,** geochronologic sequence; **unité g.,** geochronologic unit.

géocratique, geocratic.

géocryologie, geocryology.

géode, druse, geode, vough, voog.

géodépression, graben, rift valley (ex: Rhine, E. Africa).

géodésie, geodesy, geodetics.

géodésique, geodesic, geodetic.

géodésiste, geodesist.

géodique, concretionary; with geodes, geodal, geodic.

géodynamique, **1.** adj: geodynamic; **2.** n: geodynamics.

géofracture, geofracture.

géogenèse, geogenesis (rare).

géognosie, geognosy (rare).

géognostique, geognostic.

géographe, geographer.

géographie, geography.

géographie physique, physical geography, physiography, geomorphology.

géographique, geographic(al); **coordonnées g.,** geographic coordinates; **longitude g.,** geographic longitude.

géohydrologie, geohydrology.

géoïde, geoid.

géologie, geology; **g. appliquée,** applied g.; **g. du pétrole,** petroleum g.; **g. de l'environnement,** environmental g.; **g. de surface,** surface g.; **g. de terrain,** field g.; **g. dynamique,** dynamic g.; **g. générale,** general g.; **g. historique,** historic(al) g.; **g. isotopique,** isotopic g.; **g. lunaire,** lunar g.; **g. minière,** mining g.; **g. sous-marine,** marine g.; **g. stratigraphique,** stratigraphic g.; **g. structurale,** structural g.

géologique, geologic(al); **cadre g.,** g. setting; **carte g.,** g. map; **colonne g.,** g. column; **coupe g.,** g. section; **phénomène g.,** g. event; **thermomètre g.,** g. thermometer.

géologue, geologist; **g. conseil,** consulting g.; **g. pétrolier,** petroleum g.; **boussole de g.,** geologist's compass; **marteau de g.,** geologist's hammer.

géomagnétique, geomagnetic; **inversion g.,** geomagnetic reversal.

géomagnétisme, geomagnetism.

géomètre, geometer, surveyor.

géomorphologie, geomorphology, physiography (rare), physical geography; **g. périglaciaire,** periglacial geomorphology.

géomorphologique, geomorphologic(al); **cycle g.,** geomorphologic cycle.

géomorphométrie, geomorphometry.

géopète, geopetal, geopetality (orientation with respect to center of earth).

géophone, geophone.

géophysicien, geophysicist.

géophysique, **1.** adj: geophysic(al); **2.** n: geophysics; **carte g.,** g. map; **diagraphie g.,** geophysic log; **prospection g.,** geophysic survey, geophysic prospecting; **relèvement g.,** geophysic surveying.

géophysique appliquée, exploration geophysics.

géopression, geopressure.

Géorgien, Georgian (l. Cambrian equiv. Waucoban).

géostatique (adj.), geostatic.

géostatistique, geostatistics.

géosynclinal, **1.** adj: geosynclinal; **2.** n: geosyncline, sedimentary trough, oceanic trench, geotectocline (rare); **autogéosynclinal,** sedimentary trough, autogeosyncline basin of large dimensions; **eugéosynclinal,** eugeosyncline, eugeocline; **leptogéosynclinal,** leptogeosyncline; **miogéosynclinal,** miogeosyncline, miogeocline; **monogéosynclinal,** monogeosyncline; **paragéosynclinal,** marginal, para-geosyncline; **paraliagéosynclinal,** marginal basin, paraliageosyncline; **polygéosynclinal,** polygeosyncline; **ride g.,** geosynclinal ridge; **taphrogéosynclinal,** taphrogeosyncline, rift basin, trough; **zeugogéosynclinal,** zeugogeosyncline, yoked, basin, trough of large dimensions.

géotechnique (n.), soil engineering, geotechnics.

géotechnique (adj.), geotechnical; **carte g.,** g. map; **levé g.,** g. survey; **propriété g.,** g. property.

géotectonique, geotectonic.

géothermie, geothermy.

géothermique, geothermal, geothermic; **degré g.,** geothermal degree; **diagramme g.,** geothermal log; **énergie g.,** geothermal energy log; **gradient g.,** geothermal gradient.

géothermomètre, geothermometer.

géothermométrie, geothermometry.

géotrope, geopetal.

géotumeur (tecto.), geotumor.

gerbe (structure en), sheef-like structure.

germanite, germanite.

germanium, germanium.

germe cristallin, crystal nucleus.

gersdorfitte, gersdorfitte.

geyser, geyser; **conduit de g.,** geyser pipe.

geysérien, geyseric.

geysérite, geyserite.

gibbsite, gibbsite.

gicler (liquide), to spout, to spatter.

gieseckite, gieseckite.

gigantolite (minér.), gigantolite.

Gigantostracés (pal.), Gigantostraca.

Ginkgoales (pal.), Ginkgoales.

giobertite (minér.), giobertite.

Girondien, Girondian (rare: l. Miocene, including both Aquitanian and Burdigalian).

gisement, deposit, field, outcrop, pool; **g. alluvial,** placer, alluvial deposit; **g. alluvionnaire d'or,** gold diggings; **g. d'imprégnation,** impregnation deposit; **g. disséminé,** disseminated deposit; **g. de charbon,** coal field, coal measure; **g. de fer,** iron ore deposit; **g. de gaz,** gas pool; **g. de pétrole,** oil accumulation; **g. en filons, filonien,** lode deposit; **g. métallifère,** ore deposit; **g. minéralisé,** ore deposit; **g. fissural,** fissure deposit; **g. minier,** mining deposit; **g. productif,** productive deposit; **g. stratifié,** bedded deposit, stratified deposit; **g. stratiforme,** stratabound deposit; **g. synclinal,** trough vein.

gismondite, gismondite.

gîte, deposit; **g. de contact,** contact d.; **g. d'émanation,** sublimation vein; **g. d'exsudation,** exsudation deposit (due to capillarity); **g. de substitution,** replacement d.; **g. de surface,** surface d.; **g. filonien,** lode d.; **g. leptothermal,** leptothermal d. (just below epithermal); **g. métallifère,** ore d.; **g. périmagmatique,** perimagmatic d.; **g. stratifié,** stratified d.

gitologie, metallogeny.

Givétien, Givetian (m. Devonian).

givre (météo.), hoar frost.

glabellaire (sillon), glabellar (furrow).

glabelle (pal.), glabelle.

glace, ice; **g. bulleuse,** bully i.; **g. dans le sol,** ground i.; **g. de banquise,** pack i., **g. de glacier,** glacier or glacial i.; **g. d'exsudation,** needle i.; **g. enfouie,** buried i.; **g. flottante,** floe or drift i.; **g. fondante,** melting i.; **g. fossile,** fossil i.; **g. littorale,** shore i.; **g. morte,** dead i.; **aiguille de g.,** i. needle; **butte de g.,** pingo; **champ de g.,** i. field; **coin de g.,** i. wedge; **couche de g.,** i. layer; **filon de g.,** i. vein, i. sill; **lentille de g.,** i. lens; **pied de g.,** i. foot.

glacé (par le gel), iced, icy, cold, frozen.

glacer, to freeze, to ice.

glaciaire, glacial; **abrasion g.,** g. scouring; **aiguille g.,** g. horn; **auge g.,** g. trough; **avance g.,** g. advance; **calotte g.,** ice-sheet (extensive), ice-cap (medium or small); **cannelure g.,** g. groove; **cirque g.,** g. cirque; **crevasse g.,** glacier crevasse; **crue g.,** glacier surge; **débâcle g.,** glacier outburst; **dépôt g.,** g. deposit; **écoulement g.,** g. flow; **érosion g.,** g. abrasion, g. scouring; **exutoire g.,** glacier outlet; **fusion g.,** downwasting, g. wastage; **fluvio-g.,** glacio-fluvial; **lac g.,** g. lake; **langue g.,** glacier tongue; **lobe g.,** g. lobe; **marmite g.,** g. pothole; **milieu g.,** g. environment; **moulin g.,** glacier moulin; **moraine g.,** g. drift; **périglaciaire,** periglacial; **période g.,** ice age, g. period, glacials; **phase g.,** g. stage; **poli g.,** g. polish; **proplaciaire,** proglacial; **plaine proglaciaire,** outwash plain; **rabotage g.,** g. planing, g. plucking; **recul g.,** g. retreat; **régression g.,** g. recession; **striage g.,** g. scratching; **strie g.,** g. scratch; **table g.,** glacier table (boulder with ice pedestal); **terrasse g.,** g. terrace; **transport g.,** g. transport; **vallée g.,** g. carved valley, g. canyon; **vallée g. en gradins,** g. stairway.

glacialisme, glacialism, glacial theory.

glaciation, glaciation, ice flood, glacierization (G.B.); **limite de g.,** glaciation limit.

glaciel, owing to floating ice, ice-foot deposit.

glacier, glacier; **g. alpin,** alpine g.; **g. composé,** composite g.; **g. de cirque,** cirque g.; **g. d'entremont,** intermont g.; **g. de névé,** neve g., firn g.; **g. de piémont,** piedmont g.; **g. de plateau,** plateau g.; **g. de vallée,** valley g., moutain g.; **g. polaire,** polar g.; **g. régénéré,** recemented g.; **g. rocheux,** rock g.; **g. suspendu,** hanging g., glacieret; **g. tempéré,** temperate g.; **g. transfluent,** transsection g.; **front de g.,** g. face; **lait de g.,** g. milk; **moulin de g.,** g. shaft, g. mill.

glacière (naturelle), ice pit.

glacieret, small glacier.

glacio-eustatique (oscillation), glacio-eustatic fluctuation.

glacioeustasie, glacioeustasy.

glacioisostasie, glacioisostasy.

glaciolacustre, glaciolacustrine.

glaciologie, glaciology.

glaciologue, glaciologist.

glacionival, glacionival.

glaciotectonique, ice thrust.

glaciovolcanique, glaciovolcanic.

glacis, glacis; **g. continental,** continental rise; **g. d'ablation,** eroded piedmont slope; **g. d'accumulation,** alluvial apron; **g. de piémont,** piedmont slope; **g. désertique,** pediment (cf. bajada); **g. emboîté,** fill and piedmont slope; **g. étagé,** stepped piedmont slope.

glacitectonique, glacier ice thrust.

glaçon, drift ice, icicle, small floe, ice cake.

glaise, clay, loam, loam clay; **g. vertes,** Sannoisian marls (Paris basin); **fausses g.,** upper part of Landenian clay (Paris basin).

glaiser (trav. publics), to puddle.

glaiseux, clayey, loamy, clayish.

glaisière, clay pit.

glasérite, glaserite.

glauber (sel de), mirabilite.

glaubérite, glauberite.

glaucodot, glaucodot.

glauconie (pétro.), glauconite, green earth.

glauconieux, glauconitic; **sable g.,** glauconitic sand.

glauconifère, glauconiferous.

glauconite (minér.), glauconite.

glauconitisation, glauconitization.

glaucophane (amphibole), glaucophane; **schiste à g.,** glaucophane schiste.

glaucophanite, glaucophane schist.

gley, gley.

gley argileux, clayed gley; **pseudogley,** gley like soil; **sol à g.,** gley soil; **sol forestier à g.,** gleyed forest soil.

gleyification, gleying process.

gleyifié, gleyed.

gleyiforme, gley-like.

glissement, slip, slipping, gliding; **g. boueux,** mud slide; **g. coulée,** flow slide; **g. de roches,** rock slide; **g. de solifluction,** solifluxion; **g. de terrain,** landslide, landslip, earth flow; **g. en masse,** slumping; **g. gravitaire,** slumping; **g. intracristallin,** translation gliding; **g. sous-aquatique,** subaquaeous slide; **g. superficiel de sol,** creeping, creep; **plan de g., (cristallo.),** gliding plane.

glisser, to slip, to slump.

glisser par solifluxion, to flow, to creep.

globe terrestre, terrestrial globe.

Globigérine (pal.), Globigerine; **boue à Globigérines,** Globigerine ooze.

globulaire, spherulitic.

gloméro-blastique (structure), glomeroblastic structure.

gloméroporphyrique (pétro.), glomerophyric, glomero-porphyritic.

Glossoptéridées (paléobot.), Glossopteridales.

glycérine, glycerin, glycerol.

glyptogenèse, glyptogenesis, mechanical weathering or erosion, geomorphology.

Gnathostomes (pal.), Gnathostoma.

gneiss, gneiss; **g. d'injection,** composite g.; **g. du socle,** high grade g., fundamental g.; **g. en feuillets,** foliated g., leaf g.; **g. granitisé,** migmatite; **g. lité,** banded g.; **g. oeillé,** lenticular banded g., g. with discordal segregations; **ortho-g.,** orthogneiss; **parag.,** paragneiss.

gneissique, gneissic, gneissose, gneissoid; **structure g.,** gneissic structure.

gneissosité, gneissosity.

gnomonique (projection), gnomonic (projection).

godet (de drague), bucket; **chaîne à g.,** skip hoist; **élévateur à g.,** skip hoist.

goethite, goethite.

golfe, gulf, bay; **g. de corrosion,** etching pit.

gondolé (terrain), warped.

gondolement (de terrain), buckling, warping.

Gondwana (continent de), Gondwanaland.

gonflement (d'une couche), swelling up, bulging; **g. du mur (mine),** heave.

Goniatidés (pal.), Goniatitides.

Goniatite (pal.), Goniatite.

goniomètre (minéral.), goniometer.

gonothèque (pal.), gonotheca.

gore, clay parting, grit, clay weathered from granites; **g. blanc,** white clay (coal measures).

gorge (de rivière), gorge, pass, defile, gullet, gully, gulch, canyon.

Gothlandien, Gothlandian, Gotlandian (obsolete = Silurian sensu stricto).

goudron, tar; **g. bitumineux,** bituminous t.; **g. de houille,** coal t.; **g. de pétrole,** oil t.; **g. minéral,** mineral t.

goudronneux, tarry.

gouffre (karst), abyss, chasm, pit, sink-hole, cave, cave-in (collapse).

gouge (coup de, glaciol.), jumping gouge.

goule, swallow-hole.

goulet, gully, bottle-neck, narrow gorge, inlet.

goulot siphonal, siphonal neck.

gour (Auvergne), volcanic lake.

goutte de pluie, raindrop.

gouttelette, droplet.

gouttière fluvio-glaciaire, marginal channel.

graben, graben, fault trough, rift (valley).

gradient, gradient; **g. de gravité,** gravity g.; **g. de pression,** pressure g.; **g. de température,** temperature g.; **g. géothermique,** geothermal g.; **g. hydraulique,** hydraulic g.; **g. métamorphique,** metamorphic g.

gradin, step, scarp, rock step; **g. avec triage (périgl.),** sorted step; **g. de confluence glaciaire,** confluence step; **g. de faille,** fault step, fault scarp; **g. de plage,** beach terrace, beach; **g. droit (mine),** underhand stope, ridge; **g. renversé (mine),** overhand stope; **g. sans triage,** non-sorted step; **en gradins,** stepped.

graduer (un appareil), to calibrate, to graduate.

grahamite, grahamite.

grain, grain; **g. de quartz émoussé-luisant,** blunt-shining quartz g.; **g. fin,** fine g.; **g. grossier,** coarse g.; **à g. inégal,** unequigranular; **à gros g.,** coarse grained.

graine (du globe terrestre), inner core.

grammatite, grammatite (var. of tremolite).

gramme, gram, gramme (28,35 g = 1 ounce U.K.).

grande échelle (carte), large scale.

grande oolithe, bathonian « great oolite ».

grandeur (astro.), magnitude.

grandissement (d'un appareil), magnification.

granite, granite; **g. à augite,** augite g.; **g. à biotite,** normal g.; **g. à aegyrine,** aegirine g.; **g. à deux**

micas, binary g.; **g. alcalin**, alkali g.; **g. à hornblende**, hornblende g.; **g. à muscovite**, muscovite g.; **g. à riebeckite**, riebeckite g.; **g. calco-alcalin**, calk-alcaline g.; **g. d'anatrexie**, ultrametamorphic rock; **g. gneissique**, gneissoid g.; **g. graphique**, graphic g.; **g. intrusif**, intrusive g.; **g. orbiculaire**, orbicular g.; **g. plutonique**, plutonic g.; **g. porphyroïde**, porphyritic g.; «**petit g.**», crystalline crinoïdal carboniferous limestone.

granitique, granitic; **aplite g.**, g. aplite; **arène g.**, g. sand; **arkose g.**, granite wash; **couche g.**, g. layer; **greisen g.**, g. greisen; **pegmatite g.**, g. pegmatite.

granitisation, granitization, granitification.

granitisé, granitized.

granito-gneiss, granite gneiss.

granitoïde, granitoid.

granoblastique, granoblastic; **texture g.**, granoblastic texture.

grano-classement, graded bedding.

grano-classement normal, fining-up.

grano-classement inversé, coarsening-up.

grano-diorite, granodiorite.

grano-dioritique, granodioritic.

granogabbro, granogabbro.

granophyre (microgranite), granophyre.

granophyrique (à fine texture graphique), granophyric, graniphyric, graphophyric.

granulaire, granular; **désagrégation g.**, granular disintegration.

granularité, granularity.

granule, granule, grain, particle.

granuleux, granulose, granulous, granulated.

granulite (pétro.), 1. granulite (metamorphic rock), granofels; 2. muscovite granite (obsolete); **faciès à g.**, g. facies.

granulitique, granulitic; **structure g.**, g. structure.

granulométrie, granulometry, grain-size distribution; **g. fine (du sol)**, fine texture; **g. grossière (du sol)**, coarse texture.

granulométrique, granulometric; **analyse g.**, granulometric analysis, mechanic(al) analysis; **composition g.**, granulometric composition; **courbe g.**, granulometric curve.

granulosité, coarseness of grain.

graphique, 1. adj: graphic; 2. n: diagram, graph; **microgranite à texture g.**, granophyre; **pegmatite g.**, g. granite; **structure g.**, g. structure; **texture g.**, g. intergrowth.

graphite (carbone natif), graphite, black lead; **g. filonien**, vein g.; **g. naturel**, naturel g.

graphiteux, graphitic.

graphitique, graphitic.

graphitisation, graphitization.

Graptolite, (pal.), Graptolite; **schistes à g.**, g. shale.

Graptolithidés (pal.), Graptolithina.

gras (éclat du quartz), greasy.

grasse (houille), bituminous coal.

grattage (de terrains), scraping.

gratter (au bull-dozer), to scrap.

grattoir (préhist.), scraper.

grau (passe dans un cordon littoral), inlet, grau.

grauwacke (arénite), graywacke (detritic rock).

gravé (préhist.), carved.

graveleux, gravelly.

gravelle (séd.), gravel.

graveluche (Champagne), fine periglacial chalk scree.

graves (ponts et chaussée, travaux publ.), alluvial pebbles.

gravette (trav. publics), fine gravel.

gravier, gravel, gravel stone; **g. alluvial**, alluvial g., river g.; **g. aurifère**, wash g., pit g.; **g. marin**, beach g., marine g.; **carrière de g.**, gravel pit.

gravière, gravel pit.

gravillon, fine gravel.

gravillonneux, gritter.

gravimètre, gravimeter; **g. astatisé**, astatic g.

gravimétrie, gravimetry.

gravimétrique, gravimetric(al); **anomalie g.**, gravity anomaly; **balance g.**, g. balance; **carte g.**, g. map; **levé g.**, g. survey; **prospection g.**, g. exploration.

gravitation, gravitation; **constante de g.**, gravitational constant.

gravitationnel, gravitational.

gravite (séd.), conglomerate.

gravité, gravity; **centre de g.**, g. center; **circulation par g.**, gravitational flow; **différenciation par g.**, gravitational differentiation; **éboulis de g.**, g. scree.

graviter, to gravitate.

gravures sur roche (préhist.), carvings.

greenockite, greenockite.

greisen (pétro.), greisen.

greisénisation, greisening.

grêle (météo), hail.

grêlon (météo), hail stone.

grenat, garnet; **g. almandite**, almandite g.; **g. alumino-calcique**, calcium-aluminum g.; **g. alumino-magnésien**, magnesium-aluminum g.; **g. andradite**, andradite g.; **g. chromifère**, chromium g.; **g. de Bohême**, pyrope g.; **g. grossularite**, grossular g., grossularite, gooseberry stone; **g. hélicitique**, g. with helicitic inclusions; **g. magnésien**, magnesian g.; **g. pyrope**, pyrope g.; **g. spessartite**, spessartite g.

grenatifère, garnetiferous; **amphibolite g.**, g. amphibolite; **roche g.**, garnet rock; **schiste g.**, g. schist.

grenatite, garnet plagioclase gneiss.

grenu (pétro.), granular; **roche g.**, grained rock; **structure g.**, g. structure; **texture g.**, g. texture.

grepp (Garonne), ironpan.

grès, sandstone; **g. à ciment argileux**, argillaceous cemented s.; **g. à ciment calcaire**, calcareous cemented s.; **g. à ciment d'anhydrite**, anhydritic cemented s. (cf. gypcrete); **g. à ciment dolomitique**, dolomitic cemented s.; **g. à ciment d'opale**, opal cemented s.; **g. à ciment ferrugineux**, ferrugineous s., ferricrete; **g. à ciment siliceux**, siliceous cemented s. (cf. silcrete, quartzite); **g. argilleux**, argillaceous s.; **g. arkosique**, arkosic s.; **g. armoricain**, Ordovician s. of brittany; **g. bitumineux**, bituminous s., **g. bigarré**, 1. variegated s.; **2.** buntsandstein (trias), bunter; **g. calcaire,** calcareous s., calcarenite, calcrete; **g. coquiller,** shelly s., coquina; **g. de plage,** beach s., beachrock; **g. dolomitique,** dolomitic s.; **g. feldspathique,** feldspathic s., arhosic s.; **g. ferrugineux,** ferruginous s., ferricrete; **g. fin,** fine-grained s.; **g. glauconieux,** glauconitic s.; **g. grossier,** coarse s., grit; **g. lumachellique,** coquina s., shelly s.; **g. marneux,** marly s.; **g. phosphaté,** phosphatic s.; **g. psammite,** psammitic s.; **g. quartzeux,** quartzose s.; **g. quartzite,** quartzitic s.; **g. vosgien,** Lower Triassic s. (New Bed Sandstone); **Vieux Grès Rouge,** Old Red Sandstone (devonian).

gréseux, sandstone-like, gritty.

grésière, sandstone quarry.

grésil (météo), soft hail.

grève crayeuse (périgl.), chalk scree.

grève de galets (littoral), shingle beach.

grève littorale, beach, shore.

grève périglaciaire, periglacial chalk scree.

grèze litée (périgl.), colluvium, bedded rock-fragments, bedded periglacial scree, talus (with frost chips), stratified scree, stratified head (U.K.)

griffon (d'une source), exsurgence, seep.

grillage (de minerai), roasting, calcination, calcinating.

grillé (minerai), roasted, calcined.

griller (un minerai), to calcine, to roast.

grisou, fire damp, pit gas.

grisoumètre, fire damp detector, methanometer.

grisoumétrie, science of fire damp.

grisouteux, fiery, gassy, gaseous.

groin (var. d'épi littoral), groin.

groise, grouine (périgl.), see grèze.

gros grain, coarse grain.

grosse mer, heavy sea.

grossi (au microscope), enlarged, magnified.

grossier (grain), coarse (grain).

grossissement (opt.), enlargement, magnification.

grossulaire (grenat), grossularite.

grossularite (grenat), grossularite, grossular garnet, gooseberry stone.

grotte, cave, grotto; **g. préhistorique,** prehistoric cave.

grouine (périgl.), gelifluxion deposit at the foot of cuestas (Lorraine), grit.

groupe (pal.), group.

grumeau (pédo.), crumb, pellet.

grumeleux (pédo.), grumous, clotty, crumby; **sol g.,** grumous soil.

grunérite (amphibole), grunerite.

gué (d'une rivière), ford, shoal.

gueulard, vent.

guidon (mine), marker.

guirlande (de solifluxion périglaciaire), (solifluction) guirland.

guirlande insulaire, island arc.

gummite, gummite.

Günz (glaciation), Gunz (glaciation).

Guttenberg (discontinuité de), Guttenberg discontinuity.

guyot (volcan sous-marin), sea-mount, guyot.

Gymnospermes (paléobot.), Gymnosperms.

gypse, gypsum, plaster rock; **g. fer de lance,** swallow-tail twinned gypsum; **g. pied d'alouette,** larkspur gypsum, twinned gypsum crystal; **g. saccharoïde,** sugary grained gypsum; **carrière de g.,** gypsum quarry; **lame de g. (teinte sensible),** gypsum plate.

gypseux, gypseous; **croûte g.,** gypseous crust, gypcrete (cf. calcrete, ferricrete); **roche g.,** gypseous rock.

gypsifère, gypsiferous, gypsum bearing; **argile g.,** gypsiferous clay.

gypsite, gypsite.

gyroconite (Charophyte), gyroconite.

H

habitat (pal.), habitat.
habitus (minér.), habit, habitus.
hachure, hachure, hatching.
hadale (zone), hadal (deep-sea zone).
halde, dump, dump heap, waste dump, waste heap;
  h. de minerai, ore d.
halite, halite, rock salt.
halloysite, halloysite.
halmyrolyse (sédim.), halmyrolysis.
halo (d'altération), halo.
halocinèse, diapirism.
haloclastie, salt weathering, salt fretting, exsudation.
halogène, 1. adj: halogenous; 2. n: halogen.
halogénique, halogenic.
halogénure, halide.
halomorphe, halomorphic.
halophile (organisme), halophilic (organism).
halophyte (paléobot.), halophytes.
halotrichite, halotrichite, hair salt, feather alum.
hammada, hammada, rocky desert, gibber plain (Australia).
happant à la langue (minéral), sticking to the tongue.
harpon de repêchage (forage), spear.
harpon préhistorique en bois de renne, antler harpoon.
hartine (résine fossile), hartin.
harzburgite (pétro.), harzburgite.
hastingsite, hastingsite.
hatchettite, hatchettite.
hauban (de forage), guy cable, stay.
haubanner (forage), to stay, to guy.
hausmannite (minér.), hausmannite.
haut de plage, beach ridge.
hauts-piliers (gypse), upper bed of Paris gypsum.
hautes eaux, high water.
haute énergie, high energy.
haut fond, shoal, shallow.
haute mer, main sea.
haute teneur, high content, high grade.
haute terre, upland.
Hauterivien, Hauterivian (l. Cretaceous).
hauteur, height, elevation; **h. d'ascension capillaire**, capillary rising; **h. d'eau**, water level; **h. de chute d'eau (hydro.)**, fall difference; **h. de marée**, tidal range; **h. du chantier (mine)**, head room; **h. piézométrique**, piezometric head; **h. relative**, relative height; **h. topographique**, highland, eminence, hill top.
haüyne (feldspathoïde), hauynite.
havage (mine), hewing, cutting, undercutting, holing, underholing, shearing.
havé (mine), undercut, holed, underholed.

havée, cut, kerf, kerving, kirve; **profondeur de h.**, depth of cut.
haver (mine), to cut, to hole, to underhole, to undercut, to kerve.
haveur (ouvrier), cutter, holer.
haveuse (machine), cutting machine, holing machine, shearer.
havre, natural haven.
hawaïen, hawaiian; **éruption h.**, h. eruption; **volcan h.**, h. volcano.
hawaïte (andésite), hawaite.
hectare, hectare = $10\,000\,\mathrm{m}^2$ = 2,471 acres (1 acre = $4\,046{,}7\,\mathrm{m}^2$ = $43560\,\mathrm{ft}^2$).
hectogramme, hectogramme, hectogram (U.S., 100 g).
hectolitre, hectolitre, hectoliter (U.S., 100 l).
hectomètre, hectometer (U.S., 100 m).
hedenbergite (pyroxène), hedenbergite.
hélicitique, helicitic.
hélicoïdal (structure), helicoid, helical.
hélictite, helictite.
héliopause (astro.), heliopause.
héliosphère, heliosphere.
héliotrope (silice), heliotrope.
hélium, helium.
Helminthoïde, Helminthoid.
Helvétien , Helvetian (m. Miocene).
hématite (minerai), hematite, iron glance, specular iron; **h. rouge**, red iron ore.
hématitique, hematitic.
hématoconite, red ferruginous marble.
héméra (strati.), hemera.
hémi (préfixe), hemi.
hémiarctique, hemiarctic.
hémièdre, hemihedron.
hémiédrie, hemihedrism, hemisymmetry.
hémiédrique, hemihedral, hemihedric, hemisymmetric(al).
hémihyalin, hemihyaline.
hémimorphie, hemimorphism, hemimorphous.
hémimorphite, hemimorphite.
hémipélagique, hemipelagic.
hémipélagite, hemipelagite.
hémisphère, hemisphere.
hémisphérique, hemispheric(al).
hémitrope, hemitropic, twinned.
hémitropie (minérale), hemitropism, hemitropy, twinning; **plan d'h.**, twin plan.
Hercynides (tecto.), Hercynides.
Hercynien, Hercynian; 1. trend, of Harz mts; 2. time, orogenic; **plissement h.**, Hercynian Folding, Hercynian Orogenesis (also: variscan).
hercynite, hercynite.
héritage (géomorph.), inheritance, inherited features.

héritée (forme), relict (landform).
hermatypique (pal.), hermatypic.
herschage (mine), haulage, hauling.
hessonite (grenat), hessonite.
hétéroblastique, heteroblastic.
hétérochrone, heterochronous.
hétérochronisme, heterochronism.
hétérodonte (pal.), heterodont, heterondonta.
hétérogène (couche), heterogeneous.
hétérogénéité (du manteau), heterogeneity.
hétérométrie, heterometry.
hétérométrique, heterometric.
hétéromorphe, heteromorphic.
hétéromorphisme, heteromorphism.
hétéromorphite, heteromorphite, feather ore.
hétérophyllétique, heterophylletic.
hétéropique, heteropical.
hétérosporé (paléobot.), heterosporous.
hétérotaxique, heterotaxial, heterotactic.
hétérozygote, heterozygous.
Hettangien, Hettangian (l. Jurassic).
heulandite, heulandite.
hexa (préfixe), six-(sided, etc.).
Hexacoralliaire (pal.), Hexacoral, Hexacoralla.
Hexactinellides, Hexactinellid.
hexaèdre, hexahedron, cube.
hexaédrique, hexahedral.
hexagonal, hexagonal; **prisme h.**, hexagonal prism; **pyramide h.**, hexagonal pyramid; **système h.**, hexagonal system.
hexahédrite, hexahedrite.
heumite (pétro.), heumite.
hexaoctaèdre, hexaoctahedron.
hexatétraèdre, hexatetrahedron.
hiatus, hiatus, stratigraphic gap, stratigraphic lacuna.
hiddénite, hiddenite.
hircine (résine), hircite.
Hippurite (pal.), Hippurites.
histogramme, histogram.
historique (géologie), historical (geology).
histosol, histosol.
hodochrone, time-distance curve.
hodographe, hodograph.
holoaxe, holoaxial.
Holocène, Holocene.
holocristallin, holocrystalline.
holoèdre, holosymmetric.
holoédrie (minéralo.), holohedrism.
holoédrique, holohedral, holohedric.
holohyalin, holohyaline.
hololeucocrate, hololeucratic.
holomélanocrate, holomelanocratic.
holomorphe, holomorphic.
holomorphique, holomorphic.
holophyte, holophyte.
Holostéens (pal.), Holosteii.
holosidérite, holosiderite.
holostome, holostomatous.
holotype (pal.), holotype.
homéoblastique, homeoblastic, homoeoblastic.

homéomorphe (minér.), homeomorphous.
homéomorphie (pal.), homeomorphism, homeomorphy.
Hominidés (anthrop.), Hominoids.
homoclinal (tecto.), homoclinal.
homogène (couche), homogeneous, homogène.
homogénéité (d'une couche), homogeneity.
homogénétique (de même origine), homogenetic.
homogénéisation (du magma), homogenization.
homologue (même composition chimique), homologous.
homologue climatique, homoclime.
homométrie, homometry.
homométrique, homometric.
homonyme (pal.), homonym.
homophyllétique (pal.), homophylletic.
homopolaire (liaison), homopolar.
homoséiste, coseismal line.
homoséismique, homoseismal.
homotaxie, homotaxis, homotaxy.
homotaxique, homotaxial.
horizon, horizon, layer; **h. aquifère**, aquiferous horizon; **h. argileux compact**, claypan; **h. carbonaté**, lime pan, caliche; **h. concrétionné**, hardpan; **h. d'accumulation**, B horizon, accumulation horizon; **h. éluvial**, eluvial horizon; **h. ferrugineux cimenté**, iron pan; **h. ferro-humique**, iron humus pan; **h. géologique**, geologic(al) layer; **h. humifié**, humus layer; **h. illuvial**, illuvial horizon; **h. lessivé**, leached horizon; **h. podzolisé**, podzolic horizon; **h. salé**, salt pan; **h. silicifié**, silica pan.
horizontal, horizontal; **composante h. du rejet net**, h. slip; **diaclase h.**, h. joint; **rejet h.**, h. displacement; **recouvrement h. (d'une faille inverse)**, heave.
horizontalité (d'une couche), horizontality.
hornblende (amphibole), hornblende; **h. basaltique**, basaltic h.; **h. brune**, brown h.; **h. verte**, green h.; **basalte à h.**, h. basalt; **gabbro à h.**, h. gabbro; **monzonite à h.**, h. monzonite; **norite à h.**, h. norite; **schiste à h.**, h. schist; **syénite à h.**, h. syenite.
hornblendite (pétro.), hornblendite.
hornito (volcano.), hornito, spatter cone.
hors production (puits), off production.
horst, horst, uplifted block.
hoséret (nord Cameroun), inselberg.
hôte (mineral, roche), host, palasome.
houle, swell.
houille, coal; **h. blanche**, water power; **h. bleue**, tide power; **h. demi-grasse**, semi-bituminous c.; **h. flambante**, longflame c.; **h. grasse**, bituminous c.; **h. maigre**, semi-anthracite; **h. pyriteuse**, brassy c.; **h. schisteuse**, shaly c.; **h. verte**, stream c.
houiller, 1. coal-bearing; 2. carboniferous (*strati.*); **couches h.**, coal measures.
houillère, coal mine, colliery.
houilleux, coaly.
houillification, coalification, carbonization.
houillifier, to convert into coal.
houlomètre, wave gage.
howardite (météorite), howardite.

Hoxnien, Hoxnian (Mindel-Riss interglacial, U.K.).
Hudsonien (plissement), Hudsonian (orogeny, Precambrian. 1,6 + 1,8 bill yr).
huile, oil; **h. brute (pétrole),** crude o.; **h. asphaltique,** asphaltic base o.; **h. brute non sulfurée,** sweet crude o.; **h. brute sulfurée,** sour crude o.; **h. combustible,** fuel o.; **h. de schiste,** shale o., slate o.; **h. lourde,** fuel o., heavy o.; **h. minérale,** mineral o.; **h. paraffinique,** wax o.; **h. sulfurée,** sulphurized o.
Huître (pal.), Oyster.
hum (karst), hum, mogote.
humide (terrain), humid, moist, wet.
humidifier, s'humidifier, to moisten, to damp.
humidité, wetness, moisture, moistness, humidity.
humifère, humus bearing.
humification, humification.
humine, humin.
huminite, huminite.
humique, humic; **acide h.,** humic acid; **alios h.,** humic iron pan; **composé h.,** humate; **couche h.,** humic layer.
humite, humite, humolite.
hummock, hummock.
humod (podzol humique), humod.
humodite, humodite.
humogélite, humogelite.
humox (sol ferralitique humifère), humox.
humult (ultisol humifère), humult.
humus, humus; **h. acide,** sour h.; **h. actif,** active h.; **h. brun,** brown h.; **h. brut,** raw h., mor; **h. doux,** soft h., earth h., mull; **h. intermédiaire,** mild h., moder; **h. forestier,** forest humus; **h. tourbeux,** peat h.; **appauvrissement en h.,** h. impoverishing.
Huronien, Huronian (Précambrian: Proterozoic).
hyacinthe (minér.), hyacinth.
hyalin (vitreuse), hyaline; **roche h. (vitreuse),** hyaline rock.
hyalite, hyalite.
hyaloclastite, hyaloclastite.
hyalocristallin, hyalocrystalline.
hyalomélane (verre basaltique), hyalomelane (obsolete).
hyalopilitique, hyalopilitic.
hyalosidérite (péridot.), hyalosiderite.
Hyalosponges (pal.), Hyalospongea.
hyalotourmalite (pétro.), hyalotourmalite.
hybridation magmatique, contamination of magma, magmatic stoping.
hybride (pal.), hybrid.
hydatogenèse, hydatogenesis.
hydatogène (formé en milieu aqueux), hydatogenic, hydatogenous.
hydrargillite, hydrargillite.
hydratation, hydration.
hydrate, hydrate.
hydraté, hydrated, hydrous; **chaux h.,** calcium hydrate.
hydrater, s'hydrater, to hydrate.
hydraulicien, hydraulic engineer.
hydraulicité, hydraulicity.

hydraulique, **1.** adj: hydraulic; **2.** n: hydraulics; **abattage h.,** h. mining, hydraulicing; **abattre par la méthode h.,** to hydraulic; **carte h.,** h. map; **chaux h.,** h. lime; **ciment h.,** h. cement; **dune h.,** antidune; **extraction h.,** h. hoisting; **fracturation h.,** h. hydraulicing.
hydrique, hydric.
hydrobios, hydrobios.
hydrobiotite, hydrobiotite.
hydrocarboné, hydrocarbonous, hydrocarbonaceous.
hydrocarbure, hydrocarbon; **h. à chaîne linéaire,** straight chain h.; **h. aliphatique,** aliphatic h.; **h. aromatique,** aromatic h.; **h. benzénique,** benzenic h.; **h. cyclique,** cyclic h.; **h. naphténique,** naphtenic h.; **h. non saturé,** unsaturated h.; **h. paraffinique,** paraffin h.; **h. saturé,** saturated h.
hydrochimie, hydrochemistry.
hydrochimique, hydrochemical.
hydroclasseur, hydraulic classifier.
Hydrocoralliaires (pal.), Hydrocorallines.
hydrocraquage, hydrocracking.
hydrocyanite (minér.), hydrocyanite.
hydrodésulfuration, hydrodesulfurizing.
hydrodynamique, **1.** adj: hydrodynamic; **2.** n: hydrodynamics.
hydroélectrique, hydroelectric; **réservoir h.,** hydroelectric reservoir.
hydrogénation, hydrogenation.
hydrogène, hydrogen; **h. naissant,** active h.; **h. sulfuré,** h. sulphide.
hydrogéner, to hydrogenate.
hydrogéochimique, hydrogeochemical.
hydrogéologie, hydrogeology.
hydrogéologique, hydrogeological.
hydrogéologue, hydrogeologist.
hydroglaciaire, hydroglacial.
hydrogramme, hydrograph.
hydrographe, hydrographer.
hydrographie, hydrography.
hydrographique, hydrographic; **bassin h.,** watershed, h. basin; **carte h.,** h. map; **réseau h.,** river pattern.
hydrohématite, hydrohematite.
hydrolaccolite, hydrolaccolith.
hydrolithosphérique, hydrolithospheric.
hydrologie, hydrology.
hydrologie appliquée, applied hydrology.
hydrologique, hydrologic(al).
hydrologue, hydrologist.
hydrolysat, hydrolyzate.
hydrolyse, hydrolysis.
hydrolyser, to hydrolyse.
hydromagnésite, hydromagnesite.
hydrométamorphisme, hydrometamorphism.
hydromètre, thermometric hydrometer.
hydrométrie, hydrometry.
hydromorphe (sol), hydromorphic (soil).
hydromuscovite, hydromuscovite.
hydronéphéline, hydronepheline.
hydrophane, hydrophane (variety of opal).
hydrophile, hydrophilic, hydrophilous.

hydrophilite, hydrophilite.
hydrophobe, hydrophobic.
hydrophone, hydrophone.
hydropore (pal.), hydropore.
hydroraffinage, hydrorefining.
hydroscopie, dowsing.
hydrosilicate, hydrosilicate.
hydrosol, hydrosol.
hydrosome, hydrosome.
hydrosphère, hydrosphere.
hydrosphérique, hydrospheric.
hydrostatique, hydrostatic; **niveau h.**, hydrostatic level; **pression h.**, hydrostatic head or pressure.
hydrotamis, jig.
hydrothermal, hydrothermal; **altération h.**, h. alteration; **eau h.**, h. water; **gisement h.**, h. deposit; **minéralisation h.**, h. synthesis, h. mineralization; **stade h.**, h. stage.
hydrotimétrique (degré), degree of hardness of water.
hydroxyde, hydroxide; **h. d'aluminium**, aluminium h.; **h. de calcium**, calcium h.; **h. de sodium**, sodium h.
hydrozincite, hydrozincite.
Hydrozoaire (pal.), Hydrozoan.
hydrure, hydride.
hygromètre, hygrometer; **h. enregistreur**, hygrograph.
hygrométrie (météo.), hygrometry.

hygrométrique, hygrometric.
hygroscopicité, hygroscopicity.
hygroscopique, hygroscopic.
Hyolithes (pal.), Hyolithes.
hypabyssal, hypabyssal.
hypersalin, hypersalin.
hypersthène (pyroxene), hypersthene.
hypersthénite (pétro.), hypersthenite.
hypocentre, hypocenter.
hypidiomorphe (pétro.), hypautomorphic.
hypocristallin, hypocrystalline.
hypogé (adj.), underground.
hypogène, hypogene.
hypomagma, hypomagma.
hyporelief, whole of sedimentary casts and marks of the lower part of a strate.
hypostructure (volcanique), subvolcanic structure.
hypothermal, hypothermal.
hypotype, hypotype.
hypovolcanique, hypovolcanic.
hypozone, hypozone.
hypsographe, hypsograph.
hypsographique, hypsographic.
hypsométrique, hypsometric.
hypsometry, hypsometry.
Hystrichomorphes, Hystrichomorpha.
Hystrichosphère, Hystrichosphere.

I

iceberg, iceberg.
ichnofossile, ichnofossil, mark of fossils.
ichnologie, ichonology.
ichnologique, ichnologic.
ichtyologie (pal.), ichtyology.
Ichtyoptérigiens (pal.), Ichtyopterygia.
Ichtyosaures (pal.), Ichtyosauria.
iddingsite, iddingsite.
idioblastique (texture), idioblastic.
idiogène, idiogenous.
idiogéosynclinal, idiogeosyncline.
idiomorphe (automorphe), idiomorphic, idiomorphous, automorphous.
idiomorphisme, idiomorphism.
idocrase (vésuvianite), idocrase, vésuvianite.
igné, igneous; **roches i.,** igneous rocks.
ignimbrite (r. pyroclastique), ignimbrite.
ijolite (pétro.), ijolith, ijolite.
île, island; **i. continentale,** continental i.; **i. corallienne,** coral i.; **i. de boue,** mud lump; **i. rattachée (à la côte),** tied i.; **i. volcanique,** volcanic i.
îlet, islet.
illite (minér.), illite, hydromica.
illuminer un cristal, to flood with light.
illuvial (horizon), illuvial (horizon).
illuviation, illuviation; **horizon d'i.,** illuvial horizon.
ilménite (minér.), ilmenite, titanic ore, titanoferrite.
ilménitite, ilmenitite.
ilménorutile, ilmenorutile.
îlot, islet.
imagerie radar, radar imagery.
imandrite (pétro.), imandrite.
imbiber, s'imbiber (une roche), to imbibe, to soak up.
imbibé d'eau, water logged.
imbibition, imbibition, soaking; **eau d'i.,** water of imbibition.
imbriqué, imbricate, over lapping; **structure i.,** imbricate structure.
immerger (un appareil), to immerse, to immerge, to plunge.
immersion, submergence, immersion.
immiscible, immiscible.
impactite, impactite.
Imparidigités (pal.), Imparidigitate, Perissodactylate.
imperméabilisation, waterproofing.
imperméabiliser, to waterproof.
imperméabilité, impermeability, imperviousness.
imperméable, impervious, impermeable, tight; **i. à l'eau,** watertight.
impetus (sismol.), impulse, onset.
implosion, implosion.
importance (d'un gisement), size.
imprégnation (par un fluide), impregnation, permeation; **i. saline,** salinization.

imprégner, to impregnate, to permeate.
improductif (puits), unproductive.
impulsion (sismo.), impulse, pulse; **i. de départ (séisme),** original pulse.
impur (minerai), impure, mixed.
impureté, impurity, dirt, foulness.
inaccessible (gisement), inaccessible.
inaltérable (minéral), unalterable.
inaltéré, unweathered, unaltered, fresh.
Inarticulés (Brachiopodes), Inarticulata.
incandescence, incandescence, glow.
incandescent (magma), incandescent, glowing.
incendie (mine), fire.
incidence (angle d'), incidence (angle).
incident (rayon), incident (ray).
inclinaison, incline, gradient, slope; **i. de l'axe d'un pli,** plunge, pitching; **i. d'une aiguille aimantée,** dip; **i. d'une couche,** dip, dipping; **i. d'une orbite,** inclination; **i. magnétique,** magnetic inclination, magnetic dip.
inclinée (couche), dipping, inclined; **plan i.,** decline; **puits i.,** decline.
incliner, s'incliner, to dip, to slope, to slant, to tilt.
inclinomètre, inclinometer, dipmeter.
inclusion, inclusion; **i. aqueuse,** aqueous i.; **i. fluide,** fluid i.; **i. gazeuse,** gaseous i., gas i.; **i. magmatique,** magmatic i.; **i. minérale,** mineral i.; **i. solide,** solid i.; **i. vitreuse.** vitreous i.
incolore (minéral), colorless.
incombustibilité, incombustibility.
incombustible, incombustible, unburnable, fireproof.
incompétente (couche), incompetent (bed).
incompressibilité, incompressibility.
incondensable, noncondensable.
inconformité, discontinuity, unconformity, discordance.
incongruente (fusion), incongruent (melting).
inconsistance (du sol), inconsistency, looseness.
inconsistant (terrain), loose, soft, running.
inconstant (écoulement), unsteady.
incorporation (d'une substance), incorporation.
incorporer, to incorporate, to embed.
incrustante (algue), incrusting (alga).
incrustation (de sel), incrustation.
incruster, s'incruster, to incrust.
incurvation (de couches), bending, incurvation.
incurver, s'incurver, to incurvate, to incursive, to bend.
indécomposable, undecomposable.
indécomposé (minéral), undecomposed.
indentation (géomorpho.), indentation.
indicateur, indicator, gauge; **i. chimique,** chemical indicator; **i. de débit,** flow meter; **i. de grisou,** gas detector; **i. de niveau,** level gage; **i. de profondeur,** depth indicator.

indicatrice (cristallo.), index ellipsoïde.
indicatrice (géogph.), time-distance curve.
indice, index; **i. cristallographique,** crystallographie i.; **i. d'aplatissement,** i. of flatness; **i. de basicité,** base number; **i. de coordination,** coordination number; **i. d'octane,** octane number; **i. de dureté,** hardness number; **i. d'émoussé,** degree of roudness; **i. de pétrole,** oil show, oil seepage; **i. de réfraction,** refraction i.; **i. de surface (pétrole),** seepage; **i. de triage granulométrique,** refractive i., sorting i.; **i. d'octane,** octane n.; **i. d'hétérométrie,** sorting i.; **i. thermique positif,** thaw i.; **i. thermique négatif,** freeze i.
indicolite (tourmaline), indicolite.
indissolubilité, indissolubility.
indissoluble, indissoluble.
induction, induction.
inductolog, inductolog.
induration (de sédiments), induration, hardening.
induré, indurated, hardened.
industrie, industry; **i. minière,** mining i.; **i. pétrolière,** oil i.; **i. préhistorique,** prehistoric tool assemblage, culture.
inégal (terrain), uneven.
inégalité (de terrain), unevenness.
inépuisable (réserves), inexhaustible.
inéquivalence (pal.), inequivalence.
inéquivalve (pal.), inequivalve.
inexact (mesure), inaccurate, inexact.
inexploitable (gisement), unworkable, inexploitable.
inexploité (gisement), unworked.
inexploré (site), unexplored.
infantile (stade), infancy (stage).
inférieur, lower, under; **i. (au sens stratigraphique),** lower.
inféroflux, undertow, inferoflux.
infiltration (d'eau, etc.), infiltration, seepage.
infiltration d'eau salée (littoral), salt water intrusion.
infiltrer, s'infiltrer, to infiltrate, to percolate, to seep.
infiltromètre, infiltrometer.
inflammabilité (d'un corps), inflammability.
inflammable (gaz), inflammable, ignitible.
inflexion, inflection; **i. de rayons,** bending; **point d'i.,** point of inflection.
influence, influence, effect; **i. de la température,** temperature effect.
informatique, computer science, informatics.
Infracambrien, Infracambrian.
Infracrétacé, lower Cretaceous.
Infralias, lower Lias.
infralittoral, infralittoral.
infranchissable (cours d'eau), impassable.
infrarouge, infrared.
infrastructure, substructure.
infratidal (océano.), infratidal.
infusibilité (d'un corps), infusibility.
infusible (substance), infusible, non-melting.
infusoires (terre à), diatomite.

ingénierie (trav. publics), engineering.
ingénieur, engineer; **i. conseil,** consulting e.; **i. des mines,** mining e.; **i. géologue,** geologic e.
ingression, ingression, incursion.
initial, initial; **écoulement i.,** initial open flow; **production i.,** initial production.
injecter, to inject.
injection, injection; **i. de boue (forage),** mud grouting; **i. d'eau,** water i.; **i. de gaz,** gas i.; **i. lit par lit,** bed by bed i.; **puits d'i.,** i. well.
inlandsis, ice cap.
inondation, inundation, flood, flooding.
inondation en nappes, sheet flood, sheet wash; **plaine d'i.,** flood plain.
inonder (des terrains), to inundate, to flood; **i. (un puits),** to wash out.
inondite (sédim.), inundite.
inorganique (substance), inorganic.
inosilicate, inosilicate.
inoxydable, inoxidizable.
inquartation (de minerai), quartering.
insaturé (chimie), unsaturated.
Insectes (pal.), Insects.
Insectivores (pal.), Insectivore.
inselberg, inselberg.
insequent, insequent.
insolubilité (chimie), insolubility.
insoluble, insoluble; **i. dans l'eau,** water i.
instable (écoulement), unsteady; **i. (terrain),** unstable.
installation, plant, installation; **i. de broyage,** crushing plant; **i. d'extraction,** extraction plant, hoisting plant; **i. de forage,** rig; **i. de lavage (de minerai),** washing plant; **i. de triage,** separating plant.
instant, time, instant; **i. d'explosion (géophys.),** time break; **i. zéro,** time break.
institut géologique, geological institute, department.
instrument, instrument; **i. de mesure,** measuring i.; **i. de nivellement,** levelling i.
insulaire, insular; **arc i.,** island arc; **chaîne i.,** island chain; **guirlande i.,** island arc.
intarissable (source), inexhaustible.
intégré (réseau fluvial), integrated drainage.
intégripallié (pal.), integripalliate.
intensité, intensity; **i. de la pesanteur,** gravity, gravitation constant; **i. de rayonnement,** radiation rate; **i. du champ magnétique,** i. of magnetic field; **échelle d'i. (des séismes),** intensity scale.
interambulacraire (zone), interambulacral (area).
intercalation (de couches), interstratification, intercalation, break; **i. d'argile (dans le charbon),** clay parting, gore.
intercalé (strati.), intercalated, interbedded, interstratified.
intercaler, s'intercaler (strati.), to interstratify, to intercalate.
intercristallin, intercrystalline.
intercroissance (minér.), intergrowth.
interdigitation, interfingering.
interface, interface.

interfacial (angle), interfacial (angle).
interférence, interference; **figure d'i.**, interference figure.
interféromètre, interferometer.
interférométrie, interferometry.
interfluve, interfluve.
interfoliaire (cristallo.), interlayer.
interglaciaire, 1. adj: interglacial; 2. n: interglacial stage.
intergranulaire, intergranular.
intérieur, 1. adj: interior, inner; 2. inside, inland; **i. du pays**, inland; **dépression i.**, interior basin; **mer i.**, inland sea.
intermédiaire (foyer de profondeur), intermediate (focus earthquake).
intermittent (écoulement), intermittent (flowing).
Internides, Internides (tectonic zone formerly site of eugeosyncline).
internival, internival.
interpénétration (périgl.), injection.
interpluvial, interpluvial.
interprétation (de données), interpretation.
interprétateur de photographies aériennes, airviews interpreter, photointerpreter.
intersection (de filons), intersection.
intersertal, intersertal; **structure i.**, intersertal structure.
interstade (du Quaternaire), interstade.
interstadiaire, interstadial.
interstellaire (astro.), interstellar.
interstice (pétro.), void, interstice.
interstitiel, interstitial; **eau i.**, interstitial water; **solution solide i.**, interstitial solid solution.
interstratification, interstratification, interbedding.
interstratifié, interbedded, interstratified; **minéral i.**, mixed layer clay.
intertidale (zone), tidal zone, intertidal.
intervalle de température, temperature range.
intervalle entre deux piliers (mine), span.
interzonal (sol), interzonal (soil).
intraclaste, intraclast.
intracratonique, intracratonic.
intraformationnel (strati.), intraformational.
intrafosse (tecto.), intrathrough.
intraglaciaire, intraglacial.
intramagmatique, intramagmatic.
intraplaque, intraplate.
intratellurique, intratelluric.
intrazonal (sol), intrazonal (soil).
intrusif, intrusive; **filon i. redressé**, dyke; **filon i. subhorizontal**, sill; **granite i.**, intrusive granite; **massif i.**, stock.
intrusion, intrusion; **i. d'évaporite**, diapir, salt dome; **i. discordante**, discordant i.; **i. entre des couches**, concordant i.; sill; **i. de sel**, diapir, salt i.; **i. rubanée**, ribbon injection.
intumescence (tecto.), intumescence.
inverse, inverted, reverse; **faille i.**, reverse fault; **flanc i.**, inverted limb; **pli i.**, reverse fold.
inversion, reversal; **i. de relief**, inverted relief, inversion of relief; **i. magnétique**, magnetic r.

Invertébré (pal.), Invertebrate.
involute (coquille), involute (shell).
involution (périgl.), involution.
iodargyrite, iodargyrite.
iode, iodine.
iodure, iodide.
iodyrite, iodargyrite.
iolite, iolite.
ion, ion.
ionique (rayon), ionic (radius).
ionisation, ionisation.
ionosphère, ionosphere.
iridium, iridium.
iridosmine, iridosmine.
irisées (marnes), Keuper.
irradier (R.X.), to radiate, to irradiate.
irréguliers (Oursins), exocyclic (sea Urchin).
irréversible (réaction), irreversible.
irrigation, irrigation, flooding.
irriguer, to irrigate.
irruption, inrush, irruption.
iso (préfixe), iso (equal, uniform values).
isobar, isobar.
isobathe, isobath.
isochore, isochore.
isochrone, 1. adj: isochronous; 2. n: isochron.
isoclinal, isoclinal; **pli i.**, isoclinal fold.
isocline, isocline.
isodynamique, isodynamic line.
isogamme, isogamme.
isogone, isogonic line.
isograde, isograd.
isohyète, isohyet.
isohypse, isohypse.
isomère (chimie), isomer.
isomérie, isomerism.
isomérique, isomeric.
isomérisation, isomerization.
isométrique (pétro., minér.), isometric.
isomorphe, isomorph, isomorphous.
isomorphisme, isomorphism.
isomyaire (pal.), homomyarian.
isopaque, isopachyte.
isopique (strati.), isopic.
isoplète, isopleth, isoline.
isopièze, isopiestic line.
isoséiste, isoseismic line.
isostasie, isostasy.
isostatique, isostatic(al).
isotherme, isotherm.
isothermique, isothermal.
isotope, isotope.
isotopique, isotopic(al); **dilution i**, i. dilution; **effet i.**, i. effect; **fractionnement i.**, i. fractionation; **géochimie i.**, i. geology; **rapport i.**, i. ratio.
isotrope, isotropic(al).
isotropie, isotropy.
issue (d'une vallée), outlet.
isthme, isthmus.
itabirite, itabirite.
itacolumite (grès), itacolumite.

J

jacupirangite (pétro.), jacupirangite.

jade, jade, jade-stone.

jadéite (pyroxène), jadeite.

jaillir (pétrole, eau), to gush out, to spout.

jaillissant, gushing; **nappe j.,** artesian layer; **puits j.,** g. well, gusher.

jaillissement (de laves), spatter cone, driblet cone.

jais, jet.

jalon (d'arpenteur), stake.

jalon-mire, levelling rod.

jalonner (arpentage), to stake.

jalpaïte (minér.), jalpaite.

jamesonite (minér.), jamesonite, feather ore.

jardang, yardang.

jarosite (minér.), jarosite.

jaspe, jaspe, jaspeite; **j. noir (lydienne),** lydite, lydian stone; **j. opale,** jaspopale; **j. rubané,** ribbon jasper; **j. sanguin,** bloodstone, heliotrope; **contenant du j.,** jaspidian; **se transformer en j.,** to jasperize.

jaspilite (pétro.), jaspilite.

Jatulien, Jatulian (Precambrian: Scandinavia).

jauge, gage, gauge; **j. de profondeur,** depth gage.

jaugeable, gaging, gauging.

jaugeage, gauging.

jauger, to gage, to gauge, to calibrate.

jayet, jet, black lignite.

jet (de liquide), stream, jet; **j. coronal (astro.),** coronal stream; **j. de gaz,** gas jet; **j. de plasma (astro.),** core jet; **j. de rive,** uprush; **j. de sable,** sand blast; **j. de vapeur,** stream jet.

jeu (d'une faille), faulting.

jeunesse (stade de), youth stage.

joint (= diaclase), joint; **j. de cisaillement,** shear j.; **j. de contraction,** shrinkage j.; **j. de tension,** tensional j., tension j.; **j. tectonique,** tectonic j.; **j. transversal,** cross j.; **espacement des j.,** j. spacing; **système de j.,** system of j.

jonction (de fleuves), confluent; **flèche de j.,** connecting bar.

jordanite (minér.), jordanite.

Jotnien, Jotnian (Precambrian: Scandinavia).

joue (de Tribolite), cheek; **j. fixe,** fixed c.; **j. mobile,** free c.

jour (mine), surface; **poste de j. (mine),** day shift.

jour-degré de fonte (glacio), thaw degree day.

jour-degré de gel, freeze degree day.

journal de sonde, log book.

journalière (production), daily output.

joyau, jewel.

jura Blanc, Malm (u. Jurassic).

jura Brun, Dogger (m. Jurassic).

jura Noir, Lias (l. Jurassic).

jurassien (style tecto.), jurassian.

Jurassique, Jurassic; **J. inférieur,** lower J. (Lias); **J. moyen,** middle J. (Dogger); **J. supérieur,** Upper J. (Malm); **période J.,** J. period, J. system.

jusant, ebb, ebb tide; **courant de j.,** ebb current.

juvénile (gaz, eau), juvenile, magmatic.

K

kaïnite (minér.), kainite.
Kalévien, Kalevian (fennoscandian Precambrian).
kalinite, kalinite.
kaliophilite (feldspathoïde), kaliophilite.
kame, kame.
kaolin, kaolin, porcelain clay, China clay.
kaolinite, kaolinite.
kaolinisation, kaolinization.
kaolinisé, kaolinized.
kaolinique, kaolinic.
Karélien, Karelian (Fennoscandian Precambrian).
Karslsbad (mâcle de), Karlsbad twin.
karst, karst, limestone area, solution land form, solution texture; **k. barré**, confined karst; **k. couvert**, covered karst; **k. des algues**, phytokarst; **k. profond**, deep karst; **k. superficiel**, shallow karst.
karsténite, karstenite.
karstification, karstification.
karstique, karst (adj.), karstic (rare).
karstologie, karstology.
katagenèse, katagenesis.
katagénique, katagenic.
katamorphisme (moins employé que catamorphisme), katamorphisme.
katazone (moins employé que catazone), katazone.

Katmaïen, Katmaian (éruption, violent ash explosion).
Kazanien, Kazanian (Permian).
Keewatin, Keewatin (Précambrian: Canada).
kélyphite (pétro.), kelyphite.
kérabitume, kerabitumen.
kératophyre, keratophyre.
kératophyrique (pétro.), keratophyric.
kermésite (minér.), kermesite, red antimony.
kérogène, kerogen.
kérosène, kerosen, kerosine.
kersantite (lamprophyre), kersantite.
Keuper, Keuper (u. Triassic; w. Europe).
kieselguhr (diatomite), kieselgurh, diatomite.
kilocalorie, kilocalorie, great calorie.
kilogramme, kilogram (10^3 g; 2,2046 lb, U.S., U.K.).
kilomètre, kilometer (10^3 m; 0,62137 mi, statute mile U.S., U.K.); **k. carré**, square k. (10^{10} cm^2; 0,3681 square mile; 247,1 acres, U.S., U.K.).
kilométrique, kilometric(al).
kimberlite (péridotite), kimberlite.
Kimméridgien, Kimmeridgian (u. Jurassic).
klippe, klippe.
Kongourien, Kungarian (l/m Permian).
kunzite (minér.), kunzite.

L

laboratoire, laboratory; **essai en l.**, l. test; **verrerie de l.**, l. glassware.

labradorite, labradorite.

labre (pal.), labrum.

Labyrinthodontes (pal.), Labyrinthodont.

lac, lake; **l. à bourrelet glaciel**, l. with rramparts (winter-ice pressure ridges); **l. boraté**, bitter l.; **l. cratère**, crater l., maar; **l. de barrage**, barrier l., dammed l.; **l. de barrage glaciaire**, ice-dammed l., ice-ponded l.; **l. de barrage morainique**, morainic l.; **l. de barrage volcanique**, lava-flow ponded l.; **l. de cirque glaciaire**, cirque l.; **l. de cuvette éolienne**, deflation l.; **l. de delta**, delta l.; **l. de doline**, sink-hole l.; **l. de fonte**, thaw l.; **l. de front glaciaire**, proglacial l.; **l. de lave**, lava l.; **l. de retenue**, barrier l.; **l. de trop plein**, ponded l.; **l. dimictique**, dimictic l. (with two-season overturn); **l. glaciaire**, glacial l.; **l. karstique**, karst l.; **l. orienté**, oriented l.; **l. salé**, salt l., alkali l.; **l. souterrain**, underground l.; **l. temporaire**, playa l., ephemerals; **petit l.**, lakelet.

laccolite, laccolith, laccolite.

laccolithique, laccolithic, laccolitic.

lâche (meuble), loose.

lacis de bras fluviatiles, tangled channels.

lacune, 1. gap, hiatus, lacuna; 2. interstice, void; **l. d'érosion**, erosional gap; **l. de sédimentation**, hiatus, sedimentary break; **l. minéralogique**, vacant site; **l. stratigraphique**, stratigraphic gap, stratigraphic break.

lacustre, lacustrine; **bassin l.**, lake basin; **faciès l.**, lake facies; **gisement l.** lake-bed placer; **sédiment l.**, lake deposit; **terrasse l.**, lake terrace.

ladère (grès), Cuisian sandstone (lower Eocene of Paris basin).

Ladinien, Ladinian (u. Triassic).

Lagénidés (Foram.), Lagenidae.

lagon, lagon.

lagunaire (sédiment), lagunal (deposit), lagoonal.

lagune, lagoon.

lahar, lahar, mud-flow.

laisse d'algues (littoral), drift of seaweed.

laisse de basse mer, low water mark.

laisse de crue, flood marks.

laisse de haute mer, high water mark.

laisse de vague déferlante, swash-mark.

lait, milk; **l. de chaux**, lime water, white wash, lime m.; **l. de glacier**, glacier m.

laiteux (quartz), milky (quartz).

laitier, slag, cinder, scoria; **l. de fonderie**, foundry slag.

laiton, brass.

Lamarkisme, Lamarkism.

lambeau (de charriage), nappe outlier, thrust outlier, klippe.

Lambert (projection équivalente de), Lambert equal area map.

lame, plate, blade, flake; **l. à encoche (préhist.)**, worked flake; **l. d'argile orientée (séd.)**, oriented clay slide; **l. de fond**, ground sea; **l. évaporée**, amount of evaporation; **l. Levallois**, Levallois blade; **l. mince (pétro.)**, thin plate, thin section; **l. minéralogique auxiliaire**, slide; **l. moustérienne**, mousterian blade.

lamellaire (minéral), lamellar; **structure l.**, lamination.

lamelle, lamina; **l. couvre-objets**, cover glass; **l. de mâcle**, twinning lamella.

lamelleux, lamellated, lamellose, lamellous.

Lamellibranches, Lamellibranchiata, Pelecypoda, Bivalves.

laminage, crushing out.

laminaire, laminar; **écoulement l.**, l. flow; **structure l.**, lamination.

lamination, lamination.

lamine, lamina.

lamines basales, bottomset laminae.

laminée (roche), laminated.

laminite, laminite.

lamprophyre (pétro.), lamprophyre.

lamprophyrique, lamprophyrique.

lance (gypse fer de), arrow-tail twinned gypsum.

lancéolé (biface), lanceolate (handaxe).

lande (géomorph.), moor, barrenlands, barrens.

Landénien, Landenian (u. Paleocene, incl. both Thanetian and Sparnacian).

langbéinite, langbeinite.

langue, tongue; **l. de boue**, mudflow; **l. de gélifluction**, solifluction lobe; **l. de terre**, spit, isthmus; **l. glaciaire**, glacier t.

lanière (tecto.), pinch.

lanthanides, lanthanides.

lapiaz, lapies, solution rills, klint.

lapidaire (relatif aux pierres précieuses), lapidary.

lapidification, lapidification, lithogenesis.

lapidifier (peu employé), to lapidify, to petrify.

lapiez, lapiaz, lapié, lapiaz, lapies, solution rills, clint; **l. dégagé**, revealed lapiaz; **l. littoral**, littoral lapiaz; **l. sous-cutané**, subcutaneous lapiaz; **l. souterrain**, subterraneous lapiaz.

lapilli, lapilli.

lapis-lazulli (minér.), lapis-lazulli.

Laramienne (orogénèse), Laramian (orogeny: late Cretaceous).

lardite (minér.), lardite, lardstone.

large (au), offshore.

larme volcanique, volcanic drop.

larnite, larnite.

larve (pal.), larva.

larvaire (pal.), larval.

latéral, lateral; **cône l.**, adventive cone; **érosion l.**, l. erosion; **migration l.**, l. migration; **moraine l.**, l. moraine.

latérite, laterite, allite; **l. alumineuse**, bauxitic laterite; **l. détritique**, detrital laterite; **l. gravillonnaire**, concretionary laterite; **l. scoriacée**, flaggy laterite; **l. vacuolaire**, vesicular laterite.

latéritique, lateritic; **limon rouge l.**, l. red loam; **sol l.**, l. soil, latosol.

latéritisation, laterization.

latéritisé, lateritised.

latérolog (for.), laterolog.

latite (trachyandésite), latite.

latitude géographique, geographic latitude.

latosol, latosol (great soil group, developed under deep tropical weathering).

Lattorfien, Lattorfian.

Laue (diagramme de), X ray diffraction pattern.

laumontite, laumontite.

Laurasie, Laurasia.

laurvickite (syénite), larvickite.

lauze (Auvergne), volcanic roofing-slab flagstone.

lavage, washing; **l. à l'acide**, acid w.; **l. au crible**, jigging; **l. du minerai**, ore w.; **l. sur table oscillante**, rocking.

lave, lava; **l. basaltique**, basaltic l.; **l. chaotique**, block l., aa l.; **l. cordée**, ropy l.; **l. en blocs**, block l.; **l. en coussins**, pillow-l.; **l. en oreillers**, pillow-l.; **l. figée**, congealed l.; **l. prismée**, columnar l.; **l. torrentielle**, mud flow; **bouclier de l.**, l. shield; **boursouflure de l.**, l. blister; **champ de l.**, l. field; **chenal de l.**, l. channel; **cône de l.**, l. cone; **coulée de l.**, l. flow, l. stream; **culot de l.**, l. plug; **débit de l.**, l. discharge; **dôme de l.**, l. dome; **filon de l.**, l. streak; **fontaine de l.**, l. fountain; **jaillissement de l.**, spatter l., blister l.; **lac de l.**, l. lake; **nappe de l.**, l. sheet; **plaine de l.**, l. plain; **plateau de l.**, l. plateau; **tunnel de l.**, l. tube; **volcan de l.**, l. volcano.

laver (un minerai), to wash.

laveur (de minerai), ore washer.

lavogne (Causses), small pond.

Laxfordien, Laxfordian (Precambrian: Scotland).

lazulite, lazulite.

lectotype (pal.), lectotype.

Lédien (Éocène), Ledian.

ledmorite (pétro.), ledmorite.

légende (d'une carte), key, legend.

léger, light; **fraction l.**, light fraction; **isotope l.**, light isotope; **minéral l.**, light mineral.

lehm, loessic soil, loam, lehm, lixiviated loess, leached loess; **l. argileux**, clay loam; **l. limono-argileux**, silt loam; **l. sableux**, sandy loam.

lehmification, leaching of loess and transformation into lehm.

Lémuriens (pal.), Lemuriformes, Lemurs.

lenticulaire (strati.), lenticular, lens-shaped, lentoid; **amas l.**, lenticule, lenticle; **masse l.**, lenticle;

stratification l., lensing; **structure l.**, flaser structure.

lentille, lens; **l. de sable**, sand l.; **l. gravitationnelle (astro.)**, gravitational l.; **l. optique**, l.; **l. rocheuse**, lentil; **l. salifère**, salt pillow.

léonhartite, leonhartite.

léonite (minér.), leonite.

lépidoblastique (structure), lepidoblastic.

lépidocrocite (minér.), lepidocrocite.

Lépidodendron (paléobot.), Lepidodendron.

lépidolite (mica), lepidolite.

lépidomélane (mica), lepidomelane.

Lépidosauriens (pal.), Lepidosauria.

Lépospondyles (pal.), Lepospondyle.

leptite (cf. granulite), leptite, granulite.

leptochlorite (minér.), leptochlorite.

leptynite (cf. granulite), leptynite, leptite.

leptynolite (r. métam.), leptinolite.

lessivage (pédol.), leaching, eluviation.

lessivé (sol.), leached.

lessiver (pédol.), to leach, to lixiviate.

leucite (feldspathoïde), leucite; **basalte à l.**, leucite basalt; **phonolite à l.**, leucite phonolite, leucitophyre; **téphrite à l.**, leucite tephrite; **trachyte à l.**, leucite trachyte.

leucitique, leucitic.

leucitite (basalte), leucitite.

leucitoèdre, leucitohedron.

leucitophyre (pétro.), leucitophyre.

leuco (préfixe), leuco, light-colored.

leucocrate (minéral), leucocratic.

leucogranite, leucogranite.

leucogranitique, leucogranitic.

leucogranodiorite, leucogranodiorite.

leucopyrite, leucopyrite.

leucorhyolite, leucorhyolite.

leucotéphrite, leucotephrite.

leucoxène (agrégat, cryptocristallin), leucoxene.

Levallois (éclat), Levallois (flake).

Levalloisien, Levalloisian (late Pleistocene industry).

levée, levee; **l. de berge**, bank deposit; **l. de galets**, shingle ridge; **l. de plage**, beach ridge.

lever, levé (n.), survey, surveying; **l. à la boussole**, compass survey; **l. à la planchette**, plane table survey; **l. de reconnaissance**, reconnaissance survey; **l. géophysique**, surveying; **l. hydrographique**, hydrographic survey; **l. par cheminement**, traversing survey; **l. photogrammétrique**, aerial survey; **l. topographique**, topographic survey.

lever (des courbes de niveau), to contour.

lévogyre (cristal), left-handed crystal.

lèvre (de faille), limb, side wall, rim; **l. affaissée**, down side, dropped side, lower wall; **l. de cratère**, rim; **l. inférieure**, down side, lowered side; **l. soulevée**, upthrown side, uplifted wall; **l. supérieure**, upper side, upthrown side.

Léwisien, Lewisien (Precambrian: Scotland).

lézarde (d'une couche), crevice, crack, split.

lherzite (pétro.), lherzite.

lherzolite (pétro.), lherzolite.

liaison, bond, bonding, link; **l. atomique**, atomic bond; **l. chimique**, chemical linkage, chemical binding; **l. covalente**, covalent bond; **l. de coordination**, coordination bond; **l. de valence**, valency bond; **l. homopolaire**, homopolar bond; **l. polaire**, polar bond.

liant (trav. publics), binding agent, binder.

liards (pierre à), nummulitic limestone (Eocene).

Lias, Lias (l. Jurassic).

liasique, liasic (l. Jurassic).

libération (de gaz, etc.), release, escape, liberation.

libéro-ligneux (paléobot.), fibro-vascular.

libéthenite (minéral), libethenite.

lichénométrie (datation), lichenometry.

lido (Océano.), barrier.

Liesegang (anneaux de), Liesegang (rings).

lieue, league; **l. marine**, nautical l. = 5.5 km (3 nautical miles, 3° arc); **l. terrestre**, land l. = 4 km (3 statute miles).

liévrite (minéral), lievrite.

ligament (pal.), ligament.

ligamentaire (région), ligament (area).

Ligérien (Turonien inférieur), Ligerian.

ligne, line; **l. andésitique**, andesitic l.; **l. d'affleurement**, l. of bearing, outcrop l.; **l. de champ magnétique**, magnetic l.; **l. de chevauchement**, overthrust l.; **l. de coupe (cartog.)**, section l.; **l. de courant**, stream l.; **l. de crête**, crest l.; **l. de direction**, strike l.; **l. de faille**, fault l.; **l. de faîte**, crest l.; **l. de flexion**, plunge l.; **l. de fracture**, fracture l.; **l. d'andésite**, andesitic l.; **l. de niveau**, level l.; **l. de partage des eaux**, dividing l., water parting, water divide; **l. de pente**, l. of dip; **l. de tir**, lead wire; **l. de sondage**, sounding l.; **l. de visée**, l. of sight; **l. de rivage**, shore l.; **l. de tir**, shooting l.; **l. en tirets**, dashed l.; **l. en pointillés**, dotted l.; **l. de temps (sism.)**, timer; **l. homoséiste**, homoseismal l.; **l. isanomale**, isanomalic l.; **l. isogone**, isogonic l.; **l. isomagnétique**, isomagnetic l.; **l. isoséiste**, isoseismal l.

lignée détritique, detritic load.

lignée évolutive (pal.), lineage.

lignée magmatique, rock suite.

ligneux (paléobot.), ligneous, lignified.

lignine (paléobot.), lignine.

lignite, lignite, brown coal.

lignitifère, lignitiferous.

liman, liman.

limburgite (petro.), limburgite.

limite, limit; **l. chronostratigraphique**, time-line; **l. d'élasticité**, elastic l.; **l. d'endurance**, endurance l.; **l. de charge**, maximum load; **l. de couche**, boundary; **l. de fluage**, yield point; **l. de liquidité**, liquid l.; **l. de plasticité**, plastic l.; **l. de rupture**, breaking strenght; **l. des neiges**, snow l.; **l. élastique**, yield l.; **angle l.**, l. angle; **pente l.**, angle of repose.

limitrophe (couche), angle of repose.

limivore (pal.), limivorous.

limnigramme, limnigram.

limnigraphe, water level recorder.

limnigraphe d'un puits, borehole logger.

limnique, limnic, limnetic.

limnologie, limnology.

limon, loam, silt; **l. à doublets (loess)**, foliated loam; **l. alluvial**, alluvial loam; **l. argileux**, silty clay loam; **l. argilo-sableux**, sandy clay loam; **l. brun lessivé**, brown bleached loam; **l. de pente**, slope loam; **l. des plateaux**, table-land loam; **l. des vallées**, bottom loam; **l. graveleux**, gravelly loam; **l. grossier**, coarse silt; **l. humifère à gley**, melanized gley loam; **l. loessique**, loessic loam; **l. loessoïde**, loess-like loam; **l. panaché**, variegated loam; **l. rouge**, red loam; **l. rouge calcaire**, calcareous red loam; **l. sableux**, sandy loam.

limoneux, loamy, silty.

limonite, limonite, pea iron, swamp ore iron; **à limonite**, limonitic.

limpide (eau), limpid.

Limule (pal.), sea Louse.

linarite, linarite.

linéaire (structure), linear; **anomalie l.**, lineated anomaly.

linéament (tecto.), lineament.

linéation, lineation.

Lingules (dalle à), Lingula (flags).

Lingulidés (pal.), Lingulid.

linnéenne (classif.), linnean.

linnéite, linnaeite.

liparite (rhyolite), liparite.

liparitique, liparitic.

liquéfaction, liquefaction.

liquéfiable, liquefiable.

liquéfié, liquefied; **gaz naturel l.**, liquefied natural gas.

liquéfier, se liquéfier, to liquefy.

liqueur lourde, heavy liquid.

liquid, liquid; **l. lourd**, heavy l.; **l. inflammable**, flammable l.; **l. obturateur (forage)**, sealing l.; **l. surfondu**, supercooled l.; **hydrocarbure l.**, l. hydrocarbon; **inclusion l.**, l. inclusion.

liquidus, liquidus.

liquidité (limite de), liquidity (limit).

liseré, border, edge; **l. de Becke**, Becke line.

listrique (faille), listric.

lit, bed, layer; **l. alternant**, alternating b.; **l. apparent**, low water channel; **l. aquifère**, water bearing bed; **l. de fleuve**, stream bed; **l. filtrant**, filter bed; **l. imperméable**, impervious bed; **l. majeur (d'une rivière)**, flood plain, first bottom, overbank; **l. mineur**, mean water channel, low flow channel; **l. mobile**, moving bed; **injection l. par l.**, bed by bed injection.

litage, bedding, stratification; **l. convoluté**, convolute bedding; **l. en chevron**, herring-bone bedding; **l. en « flaser »**, flaser bedding; **l. en lentilles**, lenticular bedding; **l. en mammelons**, hummocky bedding; **l. oblique et entrecroisé**, cross bedding.

lité, bedded, layered, stratified.

litharge, litharge.

lithification, lithification diagenesis.

lithinifère (mica), lithium (mica); **l. (tourmaline)**, lithium mica.
lithionite, lithionite, lithia mica.
lithique, lithic.
lithium, lithium.
lithoclase (désuet), lithoclase, fissure.
lithoclast, lithoclast.
lithofaciès, lithofacies; **carte de l.**, lithofacies map.
lithogénèse, lithogenesis.
lithogénétique, lithogenic, lithogenetic; **séquence l.**, lithogenic sequence.
lithographique, lithographic; **calcaire l.**, lithographic limestone.
lithologie, lithology.
lithologique, lithologic(al).
lithomarge, lithomarge.
lithophage (mollusque), saxicavous (mollusc).
lithophagé, mollusc-bored (ex. by Pholads).
lithophile (élément), lithophilic, lithophile.
lithophyse, lithophysa.
lithosol, lithosol, lithosolic soil, lithogenic soil.
lithosphère, lithosphere.
lithosphérique, lithospheric; **plaque l.**, lithospheric plate.
lithostratigraphique (unité), lithostratigraphic (unit).
lithostratigraphie, lithostratigraphy.
Lithothamniées (pal.), Lithothamnion (encrusting calcareous algae).
lithotope, lithotope.
lithotype, lithotype.
lithozone, lithozone.
litière (pédol.), litter.
littoral, 1. adj: littoral, coastal; **2.** n: shoreline, coast; **courant de dérive l.**, longshore current; **dérive l.**, littoral drift; **dune l.**, littoral dune; **milieu l.**, littoral environment; **zone l.**, littoral zone.
litre (mesure), liter.
lixiviation, leaching.
lixivier, to leach, to lixiviate.
Llandeilien (Ordovicien moyen), Llandeilian.
Llandovérien, Llandoverian.
Llanvirnien, Llanvirnian.
lobe, lobe.
lobe de gélifluxion, solifluction lobe.
lobe glaciaire, glacial lobe.
lobe morainique, morainic lobe.
localiser (sur une carte), to locate.
loch, loch (lake, gulf: Scotland).
lodranite (météorite), lodranite.
loehm, brickearth (weathered and reworked loess).
loess, loess.
loessification, loessification.
log, log.
loge (pal.), cell, chamber.
loge d'habitation (pal.), living chamber.

loge initiale, protoconch.
loi, law, principle, rule; **l. de constance des angles des faces cristallines**, law of constancy of interfacial angles; **l. de continuité originelle des couches**, law of original continuity; **l. de superposition**, law of superposition.
löllingite, lollingite.
longévité (pal.), longevity.
longitude, longitude.
longitudinal, longitudinal; **coupe l.**, l. section; **dune l.**, l. dune, linear dune; **faille l.**, l. fault; **moraine l.**, l. moraine; **profil l.**, l. profile.
longueur, length; **l. d'onde**, wave l.; **l. de tiges de forage**, drill pipe l.; **l. focale (opt.)**, focal l.
lophophore (pal.), lophophore.
lopolite (pétro.), lopolith.
loupe, lens magnyfying glass; **l. binoculaire**, binocular l.; **l. de glissement**, bulge; **l. portative**, hand l., magnifying l.
lourd, heavy; **fraction l.**, h. fraction; **liqueur l.**, h. liquid; **minéral l.**, h. mineral.
Love (ondes de), Love waves.
loxodromie, loxodromic curve (rhumb line).
Ludien, Ludian (u. Eocene).
Ludlovien, Ludlovian (u. Silurian).
lugarite (pétro.), lugarite.
luisant (grain), shining, shiny, glossy.
lujaurite (pétro.), lujaurite.
lumachelle (séd.), coquina, shelly limestone.
lumière polarisée, polarized light.
lumière polarisée non analysée (L.N.P.A.), plane polarized light.
lunaire, lunar; **cratère l.**, l. crater; **géologie l.**, l. geology; **massif l.**, l. terra; **sol l.**, l. soil, regolith.
Lune, moon; **pierre de l.**, adular, adularia.
lunette (dune), lunette.
lunette d'approche, telescope.
lunule (pal.), lunule.
luscladite (pétro.), luscladite.
Lusitanien, Lusitanian (u. Jurassic).
lusitanite, lusitanite.
lustré (schistes), lustrous (shales).
luté (lame), sealed.
lutécite, lutecin, lutecite.
Lutétien, Lutetian (m. Eocene).
lutite, lutite.
luxullianite, luxullianite.
Lycopodiales (pal.), Lycopodiales.
lydienne, lydite, lydian stone.
lydite, lydite (black flint).
lyophile, lyophilic.
lysimètre, lysimeter.
lysocline, lysocline.
Lytocératidés (pal.), Lytoceratids.

M

maar, maar (low crater formed by explosive eruptions).

maccaluba, mud volcano.

mâchefer (*trav. publics*), slag, clinker.

machine, machine, engine; **m. à calculer,** computer, calculator; **m. à remblayer,** stowing machine; **m. à tamiser,** mechanic sieve; **m. d'extraction,** hoisting engine.

mâchoire, (*pal.*) jaw; (*techn.*) grip, jaw; **m. de suspension (forage),** tubing catcher.

macigno (*grès*), macigno (calcereous fine sandstone).

maclage, twinning.

macle, chiastolite, macle, twin; **m. de Baveno,** Baveno, twin; **m. de Carlsbad,** Carlsbad twin; **m. de croissance,** growth twinning; **m. de déformation mécanique,** mechanical twinning; **m. de l'albite,** albite twin; **m. de Manebach,** Manebach twin; **m. des spinelles,** spinel twin; **m. d'interpénétration,** penetration twin; **m. en chevron,** herring-bone twin; **m. en crête de coq,** coxcomb twin; **m. en croix,** cross-shaped twin; **m. en fer de lance,** swallowtail twin; **m. en genou,** knee-shaped twin; **m. en x,** x-shaped twin; **m. par accolement,** juxtaposition contact; **m. par pénétration,** interpenetrant twin, penetration twin; **m. polysynthétique,** polysynthetic twin, repeated twin.

maclé, twinned, macled, hemitropic.

macler, se macler, to twin.

maclifère (*schiste*), chiastolite slate.

maçonnage, maçonnerie, masonry, bricklaying, brickwork.

macroagrégat, macroaggregate.

macroclimat, macroclimate.

macrocristal, phenocrystal.

macrocristallin, macrocrystalline.

macrodétritique, macroclastic.

macrodôme, macrodome.

macrofaciès, macrofacies.

macrofaune, macrofauna.

macroflore, megaflora.

macrofossile, macrofossil, megafossil.

macrogélifraction, macrogelifraction, frost shattering.

macrographie, macrography.

macropinacoïde, macropinacoid.

macropolygonation, macropolygonation.

macropolygone, toundra polygon.

macropore, macropore.

macroporosité, macroporosity.

macroprisme, macroprism.

macropyramide, macropyramide.

macroscopique, macroscopique.

macroséisme, macroseism.

macrosismique, macroseismic.

macrosphère, macrosphere.

macrosphérique, macrosphérique, mégaspérique.

macrospore, macrospore.

macrostructure, macrostructure.

Madréporaires (*pal.*), Madreporaria (Corals).

Madrépore, Madrepore.

madréporique (*pal.*), madreporic; **plaque m.,** madreporite.

maërl, nullipore gravel.

Maestrichtien, Maastrichtian (u. Cretaceous).

mafique (*minéral*), mafic, femic.

mafite (*pétro.*), mafite.

Magdalénien, Magdalenian (late Pleistocene industry).

maghémite (*minér.*), maghemite.

magma, magma, melt; **m. éruptif,** eruptive magma; **m. palingénétique,** neomagma; **m. primaire,** parental magma; **m. résiduel,** residual magma.

magmatique, magmatic; **assimilation m.,** m. stopping; **chambre m.,** m. chamber; **differenciation m.,** m. differenciation; **émanation m.,** m. emanation; **intrusion m.,** m. intrusion; **intumescence m.,** m. blister; **réservoir m.,** m. chamber.

magmatogène, magmatogene.

magnésie, magnesia.

magnésien, magnesian; **anthophyllite m.,** m. antophyllite; **calcaire m.,** m. limestone; **diopside m.,** m. diopside; **grenat alumino-m.,** magnesium-aluminium garnet; **mica m.,** m. mica; **rendzine m.,** magnesium rendzine.

magnésiochromite, magnesiochromite.

magnésioferrite, magnesioferrite.

magnésite, magnesite, giobertite.

magnésium, magnesium.

magnétique, magnetic; **azimuth m.,** m. azimuth; **anomalie m.,** m. anomaly; **attraction m.,** m. attraction; **champ m.,** m. field; **concentrateur m.,** m. concentrator; **déclinaison m.,** m. declination; **direction m.,** m. bearing; **équateur m.,** m. equator; **flux m.,** m. flux; **force m.,** m. force; **inversion m.,** m. reversal; **orage m.,** m. storm; **pôle m.,** m. pole; **prospection m.,** m. survey; **susceptibility m.,** m. susceptibility.

magnétisation, magnetization; **m. inverse,** reversed m.; **m. rémanente,** remanent m.; **m. thermorémanente,** thermoremanent.

magnétiser, to magnetize.

magnétisme, magnetism; **diamagnétisme,** diamagnetism; **ferrimagnétisme,** ferrimagnetism; **paléomagnétisme,** paleomagnetism.

magnétite, magnetite, magnetic ore iron, lodestone.

magnétohydrodynamique, magnetohydrodynamics.

magnétoilménite, magnetoilmenite.

magnétomètre, magnetometer; **m. aéroporté,** airborne m.; **m. à protons,** proton m.; **m. astatique,** astatic m. ; **m. rotatif,** spinner m.

magnétométrique (prospection), magnetometric survey.

magnétosphère, magnetosphere.

magnétostratigraphie, magnetic stratigraphy.

magnétotellurique, magnetotelluric.

magnitude (sismique), magnitude.

maigre (minerai), lean, poor.

maillage, grid.

maille, mesh; **m. de sondage,** drilling pattern; **m. élémentaire,** unit cell; **m. métallique (d'un tamis),** metallic wire mesh.

maillechort, maillechort, nickel-silver.

maillon, link.

maître-sondeur, drilling foreman.

maîtresse-tige, drilling-stem.

majeur (lit fluvial), floodplain, overbank.

majeure (forme), major (feature).

mal cristallisé, dyscrystalline.

malachite, malachite, green copper.

malacolite, malacolite.

malacologie, malacology.

malacon (zircon), malacon.

Malacostracés (pal.), Malacostraca.

malaxage (d'argile), mixing, malaxation.

malaxer, to mix, to malaxate.

malaxeur, mixer; **m. de béton,** concrete mixer.

malchite (pétro.), malchite.

maldonite (minér.), maldonite.

malléabilité (d'un corps), malleability.

Malm, Malm, Upper Jurassic.

malthe, maltha, brea (soft asphalt).

malthène, malthene.

mamelon (d'Échinoderme), mamelon, tubercle; **m. (topographique),** hillock, knob.

mamelonnée (topographie glaciaire), mamelonated, mammilary, mammilated, with many hummocks, hillocks, hummocky.

Mammifère, Mammal, Mammalia (pl.).

manchon (de tubage), casing coupling; **m. protecteur,** pipe thread protector.

mandibule (pal.), mandible.

mandrin relève-tubes (forage), casing spear.

manganèse, manganese; **dendrite de m.,** m. dendrite; **hydrate de m.,** psilomelane; **nodule de m.,** m. nodule.

manganésien, manganesian.

manganésifère, manganesiferous; **almandite m.,** manganesiferous, manganalmandite; **amphibole m.,** rhodonite; **ankérite m.,** manganankerite; **apatite m.,** manganapatite; **blende m.,** alabandite; **chlorite m.,** manganiferous chlorite; **fayalite m.,** manganese fayalite; **grenat m.,** spessartite; **ilménite m.,** manganilmenite; **magnétite m.,** manganmagnetite.

manganeux, manganous.

manganite (minér.), manganite, acerdese.

manganocalcite, manganocalcite.

manganoferrite, manganoferrite, jacobsite.

manganolite, manganolite, rhodonite.

manganomélane, manganomelane.

manganophyllite, manganophyllite.

manganosite (minér.), manganosite.

manganosidérite, manganosiderite.

mangrove (paléobot.), mangrove.

manifestation volcanique, volcanic event.

manteau, mantle; **m. de débris,** waste m., regolith; **m. de lamellibranche (pal.),** pallium; **m. détritique,** hillside waste, regolith; **m. externe,** outer m.; **m. interne,** inner m.; **m. nival,** snow cover; **m. terrestre,** earth's m.; **fusion du m.,** mantle melting.

maquis, landscape and scrub of poor soil siliceous (medit., macchia, Ital.).

marais, swamp, marsh; **m. d'eau douce,** fresh-water m.; **m. endigué,** dyked m.; **m. haut,** tourbiere; **m. littoral,** tidal m.; **m. maritime,** tidal m.; **m. salant,** salt m.; **m. saumâtre,** salt-water m., brackish m.; **m. tourbeux,** peat-bog; **m. tremblant,** quaking bog, floating bog, floating marsh; **m. troué (périgl. Canada),** pitted tidal m.

marbre, marble; **m. coquillier,** shelly m.; **m. de Carrare,** Carrare m.; **m. serpentin,** serpentine m.; **m. veiné,** veined m.; **m. vert antique,** vert antique; **carrière de m.,** m. quarry; **transformer en m,** to marmorize.

marbré (pédol.), marbled, variegated, mottled.

marbrière (inusité), marbre quarry; **industrie m.,** marbre industry.

marbrure (pédol.), mottling.

marcasite, marcasite, hepatic pyrite, radiated pyrite.

mardelle (karstique), swallow-hole; **m. (périgl.),** periglacial pond.

mare, pond.

mare à encorbellement (littoral), rimmed.

mare intertidale (autour d'un delta), intertidal mud flat.

marécage, swamp; **m. tourbeux,** bog.

marécageux, swampy, marshy, boggy.

marée, tide; **m. basse,** low t.; **m. de mortes-eaux,** neap t.; **m. de tempête,** storm t.; **m. de vives eaux,** spring t.; **m. descendante,** falling t., ebb t.; **m. montante,** rising t., incoming t.; **m. terrestre,** body t.; **courant de m.,** tidal current; **raz de m.,** tidal wave, tsunami; **rivière à m.,** tidal river; **zone de balancement des m.,** tidal zone.

marégramme, maregram.

marégraphe, tide jauge.

marelle (estuaire du Saint-Laurent), shorre pit; **schorre à marelle,** pitted schorre.

margarite (mica), margarite.

marge continentale, continental margin.

marge glaciaire, ice margin.

margino-littoral, margino-littoral.

marin (adj.), marine; **couche m.,** m. layer; **érosion m.,** m. abrasion; **faciès m.,** m. denudation; **formation m.,** m. formation; **géologie m.,** m. geology; **sediment m.,** m. deposit; **terrasse m.,** m. terrace.

maritime (transport), maritime.

mariupolite (syénite alcaline), mariupolite.

markfieldite (diorite), markfieldite.

marmite de géant, pot hole, eddy hole, glacial kettle.
marmorisation, marmorosis.
marmorisé, marbled.
marnage (agriculture), marling, liming; **m. (de marées),** tidal range.
marne, marl; **m. à huîtres,** Oligocene m. (Paris basin); **m. argileuse,** clayey m.; **m. calcaire,** calcareous m.; **m. dolomitique,** dolomitic m.; **m. indurée,** marlstone, marlite; **m. irisées,** Keuper m.; **m. phosphatée,** phosphatic m.; **m. sableuse,** sandy m.; **m. supragypseuses,** upper Eocene and lower Oligocene m. Paris basin; **m. vertes,** Sannoisian m. (middle Oligocene, Paris basin).
marneux, marly, marlaceous.
marnière, marl pit.
marno-calcaire, marly calcareous.
marque, mark, stamp, sign; **m. d'arrachement glaciaire,** crescentic gouge; **m. de courant,** flow m.; **m. de fond de lit,** bed m.; **m. de retour de vague,** backswash m.; **m. de surcharge,** load ‛m., load cast; **m. de vague,** wave m.; **m. de vague déferlante,** swash m.; **m. glaciaire,** glacial m.
marqueur (horizon), layer, marker (sism.), tracer; **m. radioactif,** radioactive marker.
marteau, hammer; **m. de géologue,** geologic h.; **m. perforateur,** hand drill; **m. perforateur à air,** pneumatic drill, pneumatic h.; **m. piqueur,** pneumatic pick; **m. pneumatique,** pneumatic drill, rock drill.
marteler (une roche), to hammer.
martite (minér.), martite.
mascagnite (minér.), mascagnite.
mascaret, tidal bore, tidal wave.
masqué (affleurement), buried, concealed, hidden.
masse (de terre, etc.), mass; **m. (instrument),** sledge hammer; **m. atomique,** atomique m.; **m. de gypse,** gypsum bed; **m. moléculaire,** molecular m.; **m. soslifluée,** gelifluxion sheet, solifluction deposit; **m. spécifique,** density; **m. volumique,** density; **écoulement en m.,** flow mass.
massette (mine), sledge, sledge hammer.
massicot (minér.), massicot.
massif, 1. adj: massive, bulky, solid; 2. n: block, massif boss; **m. ancien,** old block; **m. concordant,** laccolith; **m. effondré,** graben, sunken block; **m. en coupole,** cupola; **m. en dôme,** batholith; **m. granitique,** granitic block; **m. intrusif,** boss, intrusive block; **m. lenticulaire (et grand),** lopolith; **m. plutonique,** pluton; **m. surélevé,** horst; **grand m.,** batholith; **petit m.,** stock, dome.
mat (grain), dull, mat.
mât (de forage), (drilling) mast.
matelas de stériles (mine), rock cushion.
matériau (séd.), deposit, detritic deposit, waste; **m. (techn.),** material; **m. d'altération,** weathering deposit; **m. de construction,** building materials; **m. d'empierrement,** road metal; **m. de remblayage,** fill; **m. de solifluxion,** soliflucted deposit; **m. fins (travaux publ.),** fines; **m. fluviatiles,** river deposits; **m. glaciaires,** glacial drift, till; **m. grossiers (travaux publ.),** coarse materials.

matière, matter, material; **m. dissoute,** dissolved material; **m. en suspension,** suspended matter; **m. humique,** humic matter; **m. inerte,** inert material; **m. organique,** organic matter; **m. réfractaire,** refractory material; **m. solide,** sediment; **m. volatile,** volatile matter.
matériel, 1. adj: material, physical; 2. equipment, appliance; **m. de forage,** drilling plant; **m. de mines,** mining outfit.
matrice, matrix, gangue.
maturité (géomorph.), maturity; **m. avancée,** late m.; **paysage au stade de m.,** mature landscape; ; **région au stade de m.,** mature land; **vallée au stade de m.,** mature valley.
mauvais fossile, fossil with a wide range in time.
mauvaise qualité, low grade, low content.
mauvaises terres, badlands.
maxillaire (pal.), maxilla.
mazout, fuel oil.
méandre, meander, loop; **m. abandonné,** deserted m., abandoned m.; **m. composé,** compound m.; **m. encaissé,** incised m., entrenched m., enclosed m. goosenek (Utah); **m. mort,** ox bow; **m. recoupé,** cut-off m.; **m. surimposé;** inherited m.; **concavité de m.** m. scar; **courbure de m.,** m. curvature; **décrire des m.,** to m.; **fleuve à m.,** meandering stream; **lobe de m.,** m. core; **pédoncule de m.,** m. neck; **vallée à m.,** m. valley.
mécanique des sols, soils mechanics; **désagrégation m.,** mechanical analysis, physical disintegration, mechanical weathering; **macle d'origine m.,** mechanical twinning; **pelle m.,** mechanical shovel.
mécanisée (exploitation), mechanized (mining).
mécanisme au foyer, focal mechanism.
mèche de détonateur, fuse cap; **m. pour explosif,** fuse; **m. pour forer,** drill.
médiane (moraine), medial (moraine).
médiane granulométrique, median particle diameter.
médio (préfixe), mid; **m. atlantique (chaîne),** mid-atlantic ridge; **m. océanique (crête),** mid-oceanic ridge; **m. océanique (dorsale),** mid-oceanic rise; **m. océanique (fossé),** mid-oceanic ridge rift.
méditerranéen, mediterranean; **sol rouge m.,** mediterranean red soil.
méga (préfixe), mega (= giant).
mégacyclothème, megacyclothem.
mégafaciès, megafacies.
mégalithe, megalith.
mégalithique, megalithic.
mégaphénocristal, megaphenocryst.
mégaride (séd.), megaripple.
mégasphère (forme à), megalospheric.
Mégathérium, Megatherium.
méïonite (minér.), méionite.
meizoséismique, meizoseismal.
mélabasalte, melabasalt.
mélange, mixing, mixture; **m. binaire,** two-component mixture; **m. eutectique,** eutectic mixture; **m. gazeux,** gaseous mixture; **m. tectonique,** mélange, tectonic block complex, chaos.

mélanite (grenat), melanite.
mélanocrate, melanocratic; **basalte m.**, melabasalt; **diorite m.**, meladiorite; **gabbro m.**, melagabbro.
mélantérite (minér.), melanterite.
mélaphyre (pétro.), melaphyre.
mélilite (minér.), melilite; **basalte à m.**, melilite basalt.
mellite (minér.), mellite.
Mélobésiées (paléobot.), Melobesiae.
melteïgite (pétro.), melteigite.
meneau (tecto.), mullion; **m. (clivage)**, cleavage mullion; **m. (pli)**, fold mullion.
menhir, menhir, dolmen (mid-holocene megalith).
ménilite (var. d'opale), menilite.
méphitique (gaz), mephitic.
méplat (topogr.), flat surface, ledge.
mer, sea; **m. abyssale**, deep s.; **m. à Littorines**, Littorina s. (m. Holocene: Baltic); **m. bordière**, adjacent s.; **m. de rochers**, block field; **m. de sable**, sand s.; **m. épicontinentale**, epeiric shelf s., epicontinental s.; **m. étale**, slack tide; **m. fermée**, inland s.; **m. intérieure**, inland s., enclosed s.; **m. libre**, open s.; **m. lunaire**, mare; **m. marginale**, adjacent s.; **bras de m.**, arm of the s.; **basse m.**, low tide; **haute m.**, high tide.
Mercalli (échelle de), Mercalli's scale.
Mercator (projection de), Mercator's projection.
mercure, mercury, quicksilver; **extraire le m. d'un minerai**, to mercurify; **minerai de m.**, m. ore, cinnabar; **sulfure de m.**, m. sulphide, cinnabar.
mercureux, mercurous.
mercurifère, mercuriferous.
mercurique, mercuric.
mère, mother; **eaux m.**, m. water; **filon m.**, m. lode, main lode; **roche m.**, m. rock, source rock.
méridien, meridian; **m. d'origine**, first m., standard m.; **m. magnétique**, magnetic m.; **m. principal**, principal m.
mériédrie, merohedrism.
mériédrique, merohedral, merohedric.
mérokarst, merokarst.
Mérostomes (pal.), Merostomata.
mésa, mesa, tableland, small plateau.
meseta, meseta, tableland.
mésocrate, mesocratic.
mésocristallin, mesocrystalline.
mésoderme (pal.), mesoderm.
Mésogée, Mesogea (Tethys).
mésogène, mesogene.
mésohalin, mesohaline.
mésolite, mesolite.
Mésolithique (préhist.), Mesolithic age (early Holocene industry).
mésophyle, mesophyle.
Mésosauriens (pal.), Mesosauria.
mésosidérite, mesosiderite.
mésosphère, mesosphere.
mésostase (pétro.), mesostase, ground mass.
mésothèque (pal.), mesotheca.
mesothermal, mesothermal.
mésotype (pal.), mesotype.

Mésozoïque, Mesozoic; **ere m.** Mesozoic era.
mésozonal (métamorphisme), mesozonal.
mésozone, mesozone.
mesure de la pesanteur, gravity measurement.
méta (préfixe), meta (altered, metamorphosed).
métabasalte, metabasalte.
métabasite (pétro.), metabasite.
métacolloïde, metacolloid.
métadiabase (pétro.), metadiabase.
métadiorite (pétro.), métadiorite.
métagabbro (pétro.), metagabbro.
métal, metal; **m. ferreux**, ferrous m.; **m. lourd**, heavy m.; **m. natif**, native m.; **m. non ferreux**, nonferrous m.; **m. précieux**, precious m.; **exploitation de m.**, m. mining.
métallifère, metal bearing, metalliferous; **filon m.**, metallic vein; **mine m.**, metal mine.
métallique, metallic; **éclat m.**, m. luster.
métallisation, metallization; **m. tubulaire**, ore pipe.
métallogénique, metallogenetic; **carte m.**, m. map; **époque m.**, m. epoch; **minéral m.**, m. mineral; **province m.**, m. province.
métallogénie, metallogeny.
metallographe, metallographer.
metallographique, metallographic.
métalloïde, metalloid.
métamérie, metamerism.
métamicte (minér.), metamict.
métamorphique, metamorphic, metamorphous; **auréole m.**, m. aureole; **argillite m.**, m. shale; **calcaire m.**, metalimestone; **complexe m.**, m. complex; **différenciation m.**, m. differentiation; **dolomie m.**, metadolomite, dolomite marble; **pélite m.**, metargillite; **quartzite m.**, metaquartzite; **roche m.**, m. rock; **schiste m.**, m. schist; **sédiment m.**, m. sediment.
métamorphisé, metamorphic; **roches volcaniques m.**, metavolcanics.
métamorphisme, metamorphism; **m. de choc**, shock m.; **m. de contact**, contact m.; **m. d'enfouissement**, regional m.; **m. de pression**, load m., pressure m.; **m. de profondeur**, load m., regional m.; **m. d'injection**, injection m.; **m. dynamique**, dynamometamorphism, dynamothermal m.; **m. exomorphe**, exomorphic m.; **m. géothermique**, geothermal m.; **m. général**, dynamothermal, load m.; **m. hydrothermal**, hydrothermal m.; **m. local**, local m.; **m. périphérique**, contact m.; **m. prograde**, prograde m.; **m. régional**, regional m.; **m. régressif**, retromorphosis, diaphthoresis; **m. rétrograde**, retromorphosis; **m. thermique**, thermal m.; **m. thermodynamique**, thermodynamic(al) m.; **m. topochimique**, isochemical m.; **auréole de m.**, m. aureole; **degré de m.**, metamorphic, m. grade; **faciès de m.**, m. facies; **autom019métamorphisme**, autometamorphism; **dynamométamorphisme**, dynamometamorphism, dislocation metamorphism; **polymétamorphisme**, polymetamorphism; **pyrométamorphisme**, pyrometamorphism; **rétrométamorphisme**, retrogressive metamorphism,

diaphoresis; **ultramétamorphisme,** kinetic metamorphism.

métarhyolite, metarhyolite.

métasilicate, métasilicate.

métasomatique, metasomatic.

métasomatose, metasomatism, metasomatosis.

métasome, guest mineral.

métastable, metastable.

métatexie, metatexis.

métatropie, metatropy.

métatype (pal.), metatype.

Métazoaires (pal.), Metazoa.

météroïde, meteroid.

météore, meteor; **cratère de m.,** meteor crater, astrobleme.

météorique, meteoric; **fer m.,** meteor iron.

météorisation (peu employé), weathering (rare).

météorite, meteorite, meteoric stone, aerolith; **m. ferreuse,** iron meteorite; **m. pierreuse,** stony meteorite.

météoritique, meteoritic.

météorologie, meteorology.

météorologique, meteorologic(al); **station m.,** meteorological station.

méthane, methane, marsh gas, fire damp.

méthanier, methane tanker.

méthode, method; **m. acoustique,** acoustic m.; **m. d'exploitation,** working m.; **m. de diagraphie par induction,** induction logging m.; **m. de diagraphie par rayons gamma,** gamma ray well logging; **m. de flottation,** flotation m.; **m. de polarisation spontanée,** spontaneous potential m.; **m. électrique,** electric m.; **m. géologique,** geologic m.; **m. gravimétrique,** gravimetric m.; **m. magnétique,** magnetic m.; **m. sismique,** seismic m.

mètre, meter (1000 millimeters; $=10936$ yards; $=3,2808$ feet); **m. carré,** square m. $(=10^4$ cm^2; $=10,764$ ft^2); **m. cube,** cubic meter $(=10^6$ cm^3; 35,315 ft^3).

métrique, metric(al); **carat m.,** metric carat (200 mg); **système m.,** metric system; **tonne m.,** metric ton (1 000 kg; 1 short ton, 2000 lb, U.S.$=0,90718$ metric ton; 1 long to 2240 lb, U.K.$=1,016047$ metric tons).

mettre, to put; **m. à découvert,** to uncover; **m. au point (opt.),** to focus, to focalize; **m. au rebut,** to reject; **m. en production un puits,** to bring into production; **m. en tas,** to heap; **m. en valeur (un gisement),** to develop.

meuble, loose, uncemented, running, unlithified.

meule, millstone, grinding wheel; **m. lapidaire,** face-wheel.

meuler (une roche), to grind.

meulière, siliceous limestone; **m. de Beauce,** Beauce siliceous limestone (Upper Oligocene Paris basin); **m. de Brie,** Brie cavernous, siliceous limestone, Lower Oligocene Paris basin; **m. de Montmorency,** Montmorency cavernous siliceous limestone, Upper Oligocene Paris basin.

miargyrite, miargyrite.

miarolithique, miarolitic (with cavities as in igneous rocks, ex. some granites); **cavité m.,** mariolitic cavity, vough.

mica, mica; **m. blanc,** white m., muscovite; **m. clivable,** m. book; **m. lithinifère,** lithium m.; **m. noir,** biotite; **m. phlogopite,** rhombic m.; **m. potassique,** potash m.; **m. quart d'onde (lame auxiliaire),** m.-plate; **m. séricite,** sericite; **altération en m.,** micalization; **lamelle de m.,** m. sheet; **paillette de m.,** m. flake.

micacé, micaceous; **grès m.,** micaceous sandstone, micaceous flagstone.

micadiorite, micadiorite.

micaschiste, micaschist, micaslate.

micaschisteux, micaschistous, micaschistose.

Micoquien (préhist.), Micoquian.

micrite, micrite, microcrystalline calcite ooze.

micritique, micritic.

micro (préfixe), micro.

microanalyse, microanalysis.

microbenthos, microbenthos.

microbrèche, microbreccia.

microchimie, microchemistry.

microchimique, microchemical.

microclimat, microclimate.

microcline (feldspath), microcline.

microconglomérat, microconglomerate.

microdécrochement, microfault.

microdésintégration, comminution.

microdétritique, microclastic.

microdiagraphie, micrologging.

microdiorite (pétro.), microdiorite.

microfaciès, microfacies (texture in sedimentary rocks).

microfalaise, microcliff.

microfaune, microfauna.

microfelsite (pétro.), microfelsite.

microfelsitique, microfelsitic.

microfissuration, microfissuration.

microfissure, microcrack.

microflore, microflora.

microfluidal, microfluidal.

microforage, slim-hole drilling.

microfossile, microfossil.

microgabbro (pétro.), microgabbro.

microgélifluxion, microgelifluction.

microgélifraction, microgelifraction.

microgranite (pétro.), microgranite.

microgranitique, microgranitic.

microgranitoïde, microgranitoid.

microgranodiorite (pétro.), microgranodiorite.

microgranulitique, microgranulitic.

micrographique, micrographic.

microgrenu (pétro.), microgranular.

microlite, microlith, microlite.

microlité, microlaminated.

microlithique (pétro.), microlithic.

microlog, microlog.

micromagnétomètre, micromagnetometer.

micromammifères, micromammals.

micromètre, micrometer.

micrométrie, micrometry.

micrométrique, micrometric.
micromodelés (géomorph.), microforms.
micromorphologie (pédo.), micromorphology.
micron, micron (0,001 mm).
micro-organisme, microorganism.
micropaléontologie, micropaleontology.
micropegmatite (pétro.), micropegmatite.
micropegmatitique, micropegmatitic.
microperthite (pétro.), microperthite.
microphone, microphone.
microphotographie, microphotography.
microplaque, microplate.
micropli, microfold.
microplissé, microfolded, crenulated.
micropolygonation, micropolygonation.
micropore, micropore.
microporosité, microporosity.
microporphyrique, microphyric, microporphyric, miniphyric.
microschistosité, microfoliation.
microscope, microscope; **m. à réflexion**, mineragraphic m., reflected light m.; **m. binoculaire**, binocular m.; **m. électronique**, electron m.; **m. électronique à balayage**, scanning m.; **m. métallurgique**, metallurgical m.; **m. optique**, light m., photonic m.; **m. pétrographique**, petrographic m.; **m. polarisant**, polarization m., petrologic m.
microscopie, microscopy.
microscopie photonique, photonic microscopy.
microscopique, microscopic(al).
microscopiquement, microscopically.
microséisme, microseim.
microséparateur, microsplitter.
microsismique, 1. adj: microseismic(al); 2. n: microseismics.
microsonde électronique, electron microprobe, e. microanalyser.
microsonde ionique, ion probe.
microsphère (pal.), microsphere.
microsphérique, microspheric.
microsphérolithique, microspherulitic.
microstratification, microbedding.
microstructure (pédol.), microfabric, microstructure; **m. à revêtements**, coated fabric; **m. polyédrique**, polyhedrous fabric; **m. prismatique**, prismatic microstructure.
microsyénite, microsyenite.
microtectonique, microtectonic.
microtexture, microtexture.
microtremblement de terre, microearthquake.
migmatite (pétro.), migmatite.
migmatisation, migmatisation.
migration, migration; **m. des lignes de partage des eaux**, m. of divides; **m. des pôles**, polar drift, polar wandering; **m. primaire (du pétrole)**, primary m.; **m. secondaire**, secondary m.; **m. verticale**, vertical m.
migrer (pal.), to migrate.
milarite (minér.), milarite.
Milazzien, Milazzian (Pleistocene).

milieu (environnant), environment, medium; **m. abyssal**, abyssal environment; **m. fluviatile**, fluvial environment; **m. glaciaire**, glacial environment; **m. lacustre**, lacustrine environment; **m. lagunaire**, lagoonal environment; **m. marin**, marine environment; **m. pélagique**, pelagic environment; **m. saumâtre**, brackish environment.
Miliolidés (pal.), Miliolidae, Miliolacea.
mille, mile; **m. marin**, nautical m. (1,85325 km; = 1 minute d'arc; 6080,14 ft); **m. terrestre**, statute m. (1,60935 km; = 5280 ft; = 1760 yd); **distance en m.**, mileage.
millérite, millerite, capillary pyrite.
milliampèremètre, milliammeter.
millibar, millibar (10^{-3} bar = 10^3 dynes/cm^2 = pressure of 0,75006 Hg).
millidarcy, millidarcy (0,001 darcy: permeability).
milligal, milligal/abr. mgal (0,001 gal; 10^{-5} m/sec^2).
milligramme, milligram (0,001 g; 1 gram = 15,432 grains).
millilitre, milliliter (0,001 l; 0,006 inch^3; 0,27 fluidram).
millimètre, millimeter (0,001 m; 0,03937 inch).
millipoise, millipoise.
mimétèse, *mimétite*, mimetite, mimetesite.
minage, blasting.
Mindel (glaciation de), Mindel (glaciation).
Mindélien, Mindelian (Pleistocene: glacial stage).
Mindel-Riss, Mindel-Riss (Pleistocene: interglacial).
mine, mine, pit; **m. à ciel ouvert**, open pit; **m. de fer**, iron mine; **m. de pierres précieuses**, gem mine; **m. de houille**, coal mine, colliery; **m. de sel gemme**, rock-salt mine; **m. de soufre**, sulphur pit; **m. épuisée**, exhausted mine; **m. grisouteuse**, gaseous mine, gassy mine; **m. improductive**, non-producing mine; **m. métallique**, ore mine; **barre à m.**, miner's bar; **bois de m.**, mine timber; **carreau de m.**, mine yard; **chambre de m.**, mine chamber; **contremaître de m.**, mine foreman; **galerie de m.**, mine level; **géologie des mines**, mining geology; **ingénieur des m.**, mine inspector; **puits de m.**, mine shaft; **service de m.**, mine inspection; **trou de m.**, blast hole; **wagonnet de m.**, mine car.
miner, to mine, to undermine, to sap.
mineur, 1. adj: minor, accessory; 2. n: miner.
mineure (forme) (géogr.), minor feature.
minerai, ore; **m. à faible teneur**, low grade o., base o.; **m. à haute teneur**, high grade o.; **m. abattu**, broken o.; **m. bocardé**, stamped o.; **m. brisé**, crushed o., milled o.; **m. brut**, raw o.; **m. classé**, sorted o.; **m. concassé**, broken o.; **m. concentré**, concentrated o.; **m. d'uranium**, uranium o.; **m. de fer**, iron o.; **m. de fer argileux**, clay ironstone; **m. de fer oolithique**, oolite iron o.; **m. de mercure**, quick silver o.; **m. de plomb**, lead o.; **m. de plomb argentifère**, argentiferous lead o.; **m. des lacs**, marsh o.; **m. de scheidage**, cobbled o.; **m. disséminé**, disseminated o.; **m. en cocarde**, cockade o.; **m. en filons**, lode o., vein o.; **m. en rognons**, kidney

o.; **m. exploitable**, workable o.; **m. extrait**, extracted o.; **m. fin**, fine o.; **m. grillé**, roasted o.; **m. oxydé**, oxidised o.; **m. pauvre**, lean o., low grade o.; **m. sulfuré**, sulfide o.; **m. terreux**, earthy o.; **m. tout venant**, unsorted o.; **m. traité**, dressed o.; **m. trié**, sorted o.; **pilier de m.**, o. pillar.

minéral, **1.** adj: mineral; **2.** n: mineral; **m. accessoire**, accessory m.; **m. authigène**, authigenic m.; **m. caractéristique**, index m.; **m. clair**, felsic m.; **m. de faciès**, facies m.; **m. de la gangue**, gangue m.; **m. essentiel**, essential m.; **m. felsique**, felsic m.; **m. ferro-magnésien**, ferro-magnesian m., mafic m.; **m. filonien**, vein m.; **m. hôte**, palasome, host (m., ore); **m. interstratifié**, m. layer clay; **m. léger**, light m.; **m. métallique**, metalliferous m.; **m. métasomatique**, metasomatic m.; **m. normatif**, standard m.; **m. opaque**, opaque m.; **m. originel**, original m.; **m. pneumatolytique**, pneumatolytic m.; **m. primaire**, original m., **m. repère (métam.)**, index m.; **m. secondaire**, secondary m.; **m. symptomatique**, index m.; **m. virtuel**, standard m., normative m.; **asphalte m.**, m. pitch; **cire m.**, m. wax; **faciès m.**, m. facies; **filon m.**, m. vein; **fraction m.**, m. fraction; **gisement m.**, m. deposit; **gîte m.**, m. deposit; **goudron m.**, m. tar; **inclusion m.**, m. inclusion; **naphte m.**, m. naphta.

minéralier (navire), ore carrier.

minéralisable, mineralizable.

minéralisateur, **1.** adj: mineralizing; **2.** mineralizer; **agent m.**, mineralizer; **fluide m.**, mineralizing fluid.

minéralisation, mineralization; **m. pneumatolytique**, pneumatolytic m.

minéralisé, mineralized, mineral bearing; **eau m.**, mineral water; **filon m.**, mineral vein; **province m.**, mineral province.

minéraliser, se minéraliser, to mineralize.

minéralogenèse, mineralogenesis.

minéralogie, mineralogy, mineragraphy.

minéralogique, mineralogic(al); **échantillon m.**, m. sample or crop; **collection m.**, m. collection.

minéralogiste, mineralogist.

minette, **1.** oolithic iron ore; **2.** minette (var. of lamprophyre).

mineur, miner, hewer, mine digger; **m. de charbon**, collier, coal miner; **m. d'or**, gold miner.

minier, mining; **bail m.**, m. lease; **code m.**, m. code; **concession m.**, mineral claim; **district m.**, mineral district; **droit m.**, mineral right; **exploration m.**, m. exploration; **gisement m.**, m. field; **région m.**, m. district; **règlement m.**, m. regulation; **technique m.**, m. engineering; **travaux m.**, m. works.

minière (exploitation peu profonde), surface working.

minium, minium.

minute (de carte), map drawing.

minutieux (levé), detailed (survey).

minvérite (pétro.), minverite.

Miocène, Miocene.

miogéosynclinal, miogeosyncline.

mi-pente (d'une colline), mid-slope.

mirabilite (minér.), mirabilite.

mire, pole, staff, levelling staff; **m. de nivellement**, levelling pole; **m. graduée**, levelling rule.

miroir (horizon), reflecting horizon, mirror; **m. de faille**, slickenside.

miscibilité (de fluides), miscibility, mixability.

miscible, miscible, mixable.

mise, setting; **m. à feu (mine) (géoph.)**, firing, blowing in; **m. à nu (d'un terrain)**, denudation; **m. au point (opt.)**, adjustment, focussing; **m. au zéro**, zero s.; **m. en ligne**, line up; **m. en phase (géoph.)**, line up.

mispickel, mispickel, arsenopyrite.

Mississipien, Mississipian (l. Carboniferous).

Missourien, Missourian (u. Pennsylvanian).

missourite (pétro.), missourite.

mixte, mixed, composite, heterogeneous; **volcan m.**, mixed volcano.

mobilisation (des matériaux), weathering, erosion, abrasion.

mobilité (tecto.), mobility.

modal, modal; **analyse m.**, m. analysis; **classe m.**, m. classification; **unimodal**, unimodal.

mode (stat.), mode.

mode de gisement, kind of deposit.

mode d'exploitation, working method.

modèle analogique, analog model.

modèle en relief, relief model.

modèle hydraulique, hydraulic model.

modelé (du terrain), form, relief.

modélisation informatique, modeling, computer model.

moder (var. d'humus), moder.

modification (de composition), change.

module, modulus; **m. d'allongement**, strecht m.; **m. de cisaillement**, shear m.; **m. de compression**, m. of compression, bulk m.; **m. d'élasticité**, elasticity m., Young m.; **m. de rigidité**, rigidity m.; **m. de rupture**, m. de rupture; **m. hydraulique**, mean discharge.

moellon (construct.), quarry stone, cobble.

mofette, mofetten damp.

Mohorovicic (discontinuité de), Mohorovicic discontinuity, M. layer.

Mohs (échelle de), Mohs' scale.

molaire (chimie), molar.

molasse, **1.** molasse, post-orogenic facies (of any age); **2.** Miocene (of alpine belt, esp. boulder conglomerates and marine soft green sandstone).

molassique, molassic.

moldavite (var. d'ozocérite), moldavite.

mole (chimie), mol, gram-molecule.

môle (host), uplift block.

moléculaire, molecular; **liaison m.**, molecular bond; **poids m.**, molecular weight; **spectroscopie m.**, molecular spectroscopy.

molécule gramme, gram molecule.

mollisol (périgl.), mollisol, active layer.

Mollusque (pal.), Mollusca; **M. Amphineures**, Amphineura m.; **m. Céphalopodes**, Cephalopoda m.; **m. Gastéropodes**, Gastropoda m.; **m. Lamelli-**

branches, Lamellibranchiata, Pelecypoda m.; **m. Scaphopodes,** Scaphopoda m.

Mollweide (projection de), Mollweide (projection).

molybdène, molybdenum.

molybdénite (minerai), molybdenite.

molybdite (minér.), molybdite.

moment, moment; **m. d'inertie,** inertia m.; **m. de flexion,** bending m.

monadnock, monadnock.

monazite (minér.), monazite.

monchiquite (pétro.), monchiquite.

monochromatique, monochromatic.

monoclinal, monocline, monoclinous, monoclinal, uniclinal; **crêt m.,** hogback; **flexure m.,** monoclinal flexure; **pli m.,** monoclinal fold; **rivière m.,** down-dip river.

monoclinique, monoclinic, monosymmetric.

Monocotylédones (paléobot.), Monocots.

monocristal, single crystal.

monocyclique (pal.), monocyclic.

monogénique, monogenic, monogenetic; **brèche m.,** monogenic breccia; **conglomérat m.,** monogenic conglomerate; **sol m.,** monogenic soil.

monogéosynclinal, monogeosyncline.

monolithe, monolith.

monominéral, monomineral(ic); **roche m.,** monomineral rock.

monomyaire (pal.), monomyarian.

monophasé, monophase.

monophylétique (pal.), monophyletic.

monoréfringence, monorefringence.

monoréfringent, monorefringent.

monotype (pal.), monotype.

monotypique, monotypical.

monovalence (chimie), monovalence.

monovalent, monovalent.

mont, mount, mountain anticlinal ridge.

mont sous-marin, seamount.

montage, mounting, setting; **m. en dérivation,** parallel connection; **m. en parallèle,** parallel connection; **m. en série,** series connection; **m. microscopique,** microscopic mounting.

montagne, mountain; **m. à faible relief,** subdued degraded m.; **m. plissée,** folded m.; **chaîne de m.,** m. range, system; **éboulis de m.,** m. waste; **pente de m.,** m. slope; **pédiment de m.,** m. pediment; **versant de m.,** mountainside.

montagneux, mountainous.

montant de derrick, derrick post.

montebrasite (minér.), montebrasite.

montée (topogr.), rising, rise, acclivity.

monter (un appareil), to fit on, to set, to assemble, to mount; **m. (un forage),** to rig up; **m. (une pente),** to climb, to rise.

monticellite (péridot.), monticellite.

monticule, hillock, monticle; **m. de terre (périgl.),** earth hummock; **m. polygonal (périgl.),** polygonal mound.

Montien, Montian (Paleocene, above Danian).

montmorillonite, montmorillonite.

montueux, hilly.

monture, mounting, setting; **m. d'une pierre précieuse,** mounting of a precious stone.

monzodiorite, monzodiorite.

monzogabbro, monzogabbro.

monzonite (pétro.), monzonite; **m. quartzique,** quartz m.

monzonitique, monzonitic.

moraine, 1. moraine (*geomorph.*); 2. glacial till (*sedimen.*), glacial drift; **m. altérée,** weathered moraine; **m. consolidée,** tillite; **m. d'écoulement,** flow till; **m. de fond,** non stratified till; **m. de fond,** ground moraine, subglacial moraine; **m. de placage,** lodgement till; **m. de poussée,** push moraine; **m. de retrait,** recessional moraine, retreatal moraine; **m. déposée,** deposited moraine; **m. externe,** outer moraine; **m. frontale,** frontal moraine, terminal moraine, end moraine; **m. inférieure,** basal moraine; **m. interne,** internal moraine; **m. interlobaire,** interlobal moraine; **m. intraglaciaire,** intraglacial moraine; **m. latérale,** lateral moraine, flank moraine; **m. longitudinal,** longitudinal moraine; **m. marginale,** border moraine; **m. médiane,** medial moraine; **m. subquatique,** waterlain till; **m. superficielle,** surface moraine, superficial moraine.

morainique, morainal, morainic, morainial; **lac m.,** morainal lake; **rempart m.,** arcuate wall, arcuate moraine.

morganite (minér.), morganite.

morion (quartz fumé noir), morion.

morphogenèse, morphogenesis, morphogeny.

morphogénique, morphogenic.

morphologie, morphology.

morphologique, morphologic(al); **type m. (pal.),** morphotype.

morphologiquement, morphologically.

morphométrie, morphometry.

morphométrique, morphometric; **indice m.,** morphometric index.

morphoscopie, morphoscopy.

morphoscopique, morphoscopic(al).

morphosculpture, morphogenesis.

morphostructural (relief), morphostructure.

morphotectonique, morphotectonics.

mort-terrain, dead ground, overburden, cover, soil cap; **m. de recouvrement (mine),** muck.

mortier, mortar; **m. de chaux,** lime m.; **m. hydraulique,** hydraulic m.

mosaïque (de photographies aériennes), mosaic.

mosaïque de failles, fault mosaic.

Mosasauriens (pal.), Mosasauridae.

Moscovien, Moscovian (middle u. Carboniferous; above Namurian).

motte (pédo.), clod, clump.

motteux (terrain), cloddy.

mou (terrain), soft.

moudre (un minerai), to grind, to mill.

mouille (d'un fleuve), pool, scour, trough.

mouillé (terrain), damp, moist, wet.

mouiller (un terrain), to damp, to moisten, to wet.

moulage (hyporelief), cast; **m. d'affouillement,** scour c.; **m. de choc,** prod c.; **m. de rebond,** bounce c.; **m. de drainage,** drag c.; **m. d'outil,** tool c.; **m. en flûte,** flute c.

moule (pal.), mold, mould, moulding; **m. externe,** external mold, e. cast; **m. interne,** internal mold, internal cast.

moulin à bocards, stamp mill.

moulin glaciaire, moulin, glacial mill.

mousson, monsoon.

Moustérien, Mousterian (Late Pleistocene industry).

moustéroïde, mousteroid.

moutonnée, ice-smoothed rock, glaciated knob.

mouvant, moving, shifting; **dune m.,** shifting dune, moving dune; **sable m.,** drifting sand, flying sand.

mouvement, movement; **m. de masse,** mass. m.; **m. de terrain,** ground failures; **m. épirogénique,** epirogenic m.; **m. orogénique,** orogenic m.

moyen (adj.), mean, middle.

moyenne pression (métam.), medium pressure; **de dimension m.,** middle sized; **diamètre m. de particules,** median particle diameter; **latitude m.,** middle latitude.

moyenne (n.), mean value.

mucron (pal.), mucron; **à mucron,** mucronate.

mugéarite (andésite), mugearite.

mull (humus doux), mull (humus-mineral mix); **m. calcique,** calcic m.

multicanal (géoph.), multichannel.

multicouches (système), multilayer system.

multigélation, multigelation.

multiple, multiple; **failles m.,** m. faults; **réflections m.,** m. reflections.

multiplication (d'échelle), exaggeration (of scale).

multispectral, multispectral; **détecteur m.,** multispectral scanner; **télédétection m.,** multispectral remote sensing.

multitrace (géoph.), multichannel.

Multituberculé (pal.), Multituberculate.

mur (d'une couche), bottom, floor, footwall, lying wall, ledger; **m. (de rimaye),** headwall.

muraille (apl.), spirotheca.

murchisonite (minér.), murchisonite.

Muschelkalk, Muschelkalk (m. Triassic).

muscle (pal.), muscle; **m. adducteur antérieur,** anterior adductor m.; **m. adducteur postérieur,** posterior adductor m.

muscovite, muscovite.

muskeg, muskeg.

mutation (pal.), mutation.

mylonite, mylonite.

mylonitique, mylonitic.

mylonitisation, mylonitization.

Myriapodes, Myriapoda.

myrmékite (pétro.), myrmekite.

Mytiloïdés, Mytiloida.

N

nacre (pal.), nacre.
nacré (coquillage), nacreous, pearly.
nacrite (minér.), nacrite (lowest point; opposite to zenith).
nadir, nadir.
nageoire (pal.), fin.
nagyagite (minér.), nagyagite (Carboniferous, above Visean), below Wesphalian.
naledj (périgl.), pingo.
Namurien, Namurian.
nannofossile (pal.), nannofossil.
nannoplancton, nannoplankton.
Nansen (bouteille de), Nansen bottle.
naphtabitume, naphtabitumen.
naphte, naphta; **n. brut**, crude n.; **n. de pétrole**, petroleum n.; **n. de schiste**, shale n.; **n. minéral**, rock oil, petroleum fossil oil.
naphtène, naphtene; **teneur en n.**, naphtenicity.
naphténique, naphtenic; **série n.**, naphtenic series.
napoléonite (diorite orbiculaire), napoleonite.
nappe alluviale, alluvial sheet; **n. de charriage**, thrust sheet; **n. de chevauchement**, overthrust; **n. d'eau**, water table; **n. d'eau captive**, confined aquifer; **n. d'eau libre**, free-water table; **n. phréatique**, phreatic water-table; **n. souterraine**, groundwater table; **n. superficielle**, surface-water table; **n. de recouvrement**, allochtonous sheet.
natif, native, original; **élément n.**, native element; **métal n.**, native metal; **or n.**, native gold.
natrolite, natrolite.
natron, natron.
natroné (lac), alkali (lake).
naturel, natural; **gaz n.**, n. gas; **sélection n.**, n. selection.
Nautile (pal.), Nautilus.
Nautilidé, Nautiloid.
navigable (rivière), navigable (river).
navite (pétro.), navite.
Nazca (plaque), Nazca (plate).
Neanderthal (Homme de), Neanderthal man.
néanderthalien, neanderthalian.
nebka, nebka.
nébuleuse, nebula.
nébulite (pétro.), nebulite.
neck, neck.
necton, nekton.
nectonique (pal.), nektonic.
Needien, Needian (U.K. Mindel-Riss interglacial stage).
négatif, negative; **cristal n.**, n. crystal; **anomalie n. (de gravité)**, n. gravity anomaly.
neige, snow; **n. fondante**, slush; **n. incohérente**, loose s.; **n. sèche**, dry s.; **n. poudreuse**, powdery s.; **champ de n.**, s. field; **dune de n.**, s. dune; **limite des n.**, s. line; **tache de n.**, s. patch.

nelsonite (pétro.), nelsonite.
nématoblastique, nematoblastic.
néo-autochtone, neoautochtonous.
Néocomien, Neocomian (l. Cretaceous).
néodarwinisme, neodarwinism.
néodyme, neodymium.
néoformation (minéral de), crystallization after early diagenesis.
néoformé (minér.), crystallized after settling and early diagenesis.
Néogène, Neogene.
néogenèse, crystallization after diagenesis.
Néolithique, Neolithic new stone age (mid. Holocene); **civilisation n.**, neolithic age; **industrie n.**, neolithic tools.
néotectonique, neotectonics.
néoténie (pal.), neoteny.
néotype, neotype.
néovolcanique, neovolcanic.
néphéline, nepheline, nephelite; **basalte à n.**, nepheline basalt; **syénite à n.**, nepheline syenite.
néphélinique (syénite), nephelite syenite.
néphélinite, nephelinite.
néphéloïde (couche), nepheloid layer.
néphrite, nephrite.
Neptunien, Neptunian (ex. dike, i.e. synsedimentary non-igneous); **théorie neptunienne**, neptunian hypothesis.
neptunisme, neptunism.
néritique, neritic (pertaining to continental shelf, sublittoral); **zone n.**, neritic zone.
nésosilicate, nesosilicate.
net, nette, clean; **cassure n.**, c. break; **contour minéral n.**, sharp contour; **image n.**, c. image; **vision n.**, c. view.
netteté (opt.), sharpness, clearness.
neutralisation (chimie), neutralization.
neutraliser (chimie), to neutralize.
neutre, neutral; **roche n.**, neutral rock, intermediate rock.
neutron, neutron; **diagraphie neutron-neutron**, neutron-neutron log; **sonde à neutron**, neutron soil-moisture meter.
névé, neve, firn; **glace de névé**, firn ice.
Newton (échelle de), Newton's scale.
nez (d'un anticlinal), nose.
niccolite (= nickéline), niccolite.
niche (géomorphol.), hollow; **n. de corrosion**, solution h.; **n. de décollement**, scar; **n. de nivation**, nivation niche.
nickel, nickel.
nickélifère, nickeliferous.
nickélite, niccolite.
nickelochre, nickelocher, annabergite.

nicol, nicol, nicol prism; **n. croisés,** crossed nicols.

nicopyrite, nicopyrite, pentlandite.

nid de minerai, ore bunch, pocket.

nife (noyau de la terre), nife (Ni-Fe part of globe, core).

nimbostratus (météo.), nimbostratus cloud.

niobium, niobium.

nitrate, nitrate; **n. d'argent,** silver n.; **n. de potassium (salpêtre),** niter; **n. de soude,** nitratite, chili salpeter.

nitre, niter, salpeter.

nitreux (chimie), nitrous.

nitrification, nitrification.

nitrique, nitric; **acide n.,** nitric acid.

nitrobarite (minéral), nitrobarite.

nitrocalcite (minéral), nitrocalcite.

nitroglycérine (explosif), nitroglycerine.

nival, nival (fauna, climate); **ruissellement n.,** snow melt.

nivation, nivation; **creux de n.,** nivation hollow.

niveau, level, layer; **n. à bulle,** bubble level, spirit level; **n. aquifère,** water bearing layer; **n. d'eau, 1.** water gauge; **2.** water level; **n. de base,** base level; **n. de fond (mine),** bottom level; **n. de la mer,** sea level; **n. de mine,** floor level; **n. hydrostatique,** water table; **n. induré,** hard-ground; **n. minéralisé,** ore bed; **n. moyen,** mean level; **n. piézométrique,** water table, ground water; **n. principal,** main level table; **n. repère,** marker bed; **n. supérieur du pergélisol,** permafrost table.

niveler (travaux publics), to level.

nivellement, levelling, levelling survey, land levelling; **n. barométrique,** barometric levelling; **n. tachéométrique,** tacheometric levelling.

nivéoéolien (périgl.), niveo-eolian, niveolian.

nivofluvial, nivofluvial.

nivomètre, snow gauge.

nocif (gaz), noxious, harmful.

nodal, nodal; **point n.,** n. point.

nodule, nodule, ball; **n. d'argile,** clay ball; **n. de manganèse,** manganese nodule; **n. de péridotite,** peridotite nodule; **n. phosphaté,** phosphatic nodule; **n. polymétallique,** polymetallic n.

nœud, **1.** knot, (abrev. kt, velocity: nautical mile/hour); **2.** orogenic node, convergence (geotectonic).

nombre atomique, atomic number.

nomenclature (pal.), nomenclature.

nomogramme, nomogram.

non, not (prefix); **n. broyé,** uncrushed; **n. calcaire,** noncalcic; **n. capillaire,** noncapillary; **n. combustible,** noncombustible; **n.-conformité,** nonconformity; **n. corrosif,** noncorrosive; **n. cristallin,** noncrystalline; **n. dilué,** undiluted; **n. exploité,** unworked; **n. ferreux,** unferrous; **n. filtré,** unfiltered; **n. fondu,** unmelted; **n. magnétique,** nonmagnetic; **n. miscible,** nonconsolute; **n. perforé,** imperforated; **n. poreux,** imporous; **n. récupérable,** nonrecoverable; **n. remblayé,** unfilled; **n. saturé,** unsaturated; **n. solidifié,** unsolidified; **n. stratifié,** non bedded, unstratified; **n. traité,** nonprocessed; **n. trié (séd.),** nonsorted; **n. usé (grain),** unworn, nonworn, angular.

nontronite (m. argileux), nontronite.

nord, north; **n. géographique,** true n.; **n. magnétique,** magnetic n.

nordmarkite (syénite), nordmarkite.

Norien, Norian (u. Triassic).

norite (gabbro), norite.

normal (habituel) (perpendiculaire), normal; **déplacement n.,** n. displacement; **distribution n.,** n. grain size distribution; **érosion n.,** n. erosion (i.e. by water); **faille n.,** n. fault, gravity fault; **pli n.,** n. fold; **position n. (d'une couche),** n. position; **zonation n. (d'un feldspath),** n. zoning.

normatif (minéral), standard index (mineral), normative.

norme, standard, norm; **calcul de la n. (minér.),** norm analysis.

noséane (feldspathoïde), noseane.

Nothosauriens (pal.), Nothosauria.

notice (carto.), leaflet.

nouméite, noumeite.

Nouveaux-grès-rouges, New Red Sandstone (Permian and Triassic: Rotliegende Bunter sandstone).

novaculite (quartzophyllade), novaculite.

noyage (d'un puits), flooding.

noyau, core, nucleus, ring; **n. benzénique,** benzene ring; **n. d'un pli,** core; **n. terrestre,** earth core.

noyer (un puits), to flood, to drown.

nucléaire, nuclear; **centrale n.,** n. power plant; **combustible n.,** n. fuel; **diagraphie n.,** n. log; **énergie n.,** n. energy.

nucléation (cristallo.), nucleation.

nuée, cloud.

nuée ardente, nuée ardente, glowing ash c., pyroclastic flow.

nuée débordante, surge.

numéro atomique, atomic number.

Nummulite (pal.), Nummulite; **calcaire à Nummulites,** Nummulitic limestone (Lutetian, middle Eocene, Paris basin).

Nummulitique, Nummulitic (middle Eocene).

Nummulitidés, Nummulitids, Nummulitidae.

nunatak, nunatak.

nutation (de la terre), nutation.

nutritif (élément) (pédol.), nutrient, nutriment.

O

oasis, oasis.
obduction, obduction.
objectif, lens; **o. à immersion**, immersion l.; **o. grand angle**, wide-angle l.; **o. téléobjectif**, tele-l., tele-photographic l.
objet (platine porte-), object slide.
oblique, oblique; **extinction o.**, o. extinction, inclined extinction; **faille o.**, o. fault; **forage o.**, slant drilling.
oblitérée (structure), obliterated.
obséquent, obsequent; **escarpement o.**, o. scarp; **vallée o.**, o. valley.
observatoire, observatory.
obsidianite, obsidianite.
obsidienne, obsidian, volcanic glass.
obstruer, s'obstruer (une cheminée volcanique), to obstruct, to dam, to choke up.
obturation glaciaire, ice dam.
obturer (un puits), to seal off.
occipital (lobe) (pal.), occipital (lobe).
occlusion (de gaz), occlusion.
océan, ocean.
océan lunaire, ocean basin.
océanique, oceanic; **bassin o.**, o. basin, ocean basin; **circulation o.**, o. circulation; **courant o.**, o. current; **croûte o.**, o. crust; **dorsale o.**, mid-o. ridge; **expansion des fonds o.**, ocean floor spreading; **fond o.**, o. bottom, o. floor; **fosse o.**, o. trench; **île o.**, o. island; **influence o.**, oceanicity; **plaque o.**, o. plate; **seuil o.**, o. threshold, o. sill; **socle o.**, o. basement.
océanite (olivine), oceanite.
océanographie, oceanography.
océanographique, oceanographic(al).
océanologie, oceanology (applied ocean sciences).
ocelles (pal.), ocellus.
ochrept (sol brun tempéré), ochrept.
ocre, ocher, ochre; **o. jaune**, yellow ocher, nickel ocher; **o. rouge**, red ocher.
octaèdre, octahedron.
octaédrique, octahedral.
octaédrite (minér.), octahedrite, anatase.
octane, octane; **indice d'o.**, o. number.
Octocoralliaires (pal.), Octocorallia.
octophyllite, octophyllite.
oculaire, eye-piece, ocular; **o. à réticule**, eye-piece with cross wires.
œil, eye; **o. de chat**, cat's e.; **o. de tigre**, tiger's e. (*miner.*).
œillé (gneiss), augen gneiss.
ogive (glaciaire), ogive.
oléfine, olefin.
oléoduc, oil pipe.
Olénékien (Trias), Olenekian.
oligiste, oligist; **fer o.**, oligist iron.

oligo (préfixe), oligo (little, small).
Oligocène, Oligocene.
oligoclase (feldspath), oligoclase.
oligoclasite (diorite), oligoclasite.
oligohalin, oligohaline.
oligotrophe, oligotrophic.
olistolithe, olistolith.
olitostrome, olitostrome.
olivine (minér.), olivine; **basalte à o.**, o. basalt; **diabase à o.**, o. diabase; **gabbro à o.**, o. gabbro; **nodule d'o.**, o. nodule.
olivinite (dunite), olivinite.
ollaire (pierre), steatite, talcshist.
ombilic (pal.), umbilic.
ombilic glaciaire, overdeepened glacial basin.
omphacite, omphacite.
oncoïde, pisolith.
oncolite, algal ball.
onctueux (minéral à), soapy, greasy, unctuous.
onde, wave; **o. acoustique**, acoustic w.; **o. de choc**, shock w.; **o. de cisaillement**, shear w., transverse w.; **o. de compression**, compression w.; **o. de crue**, flood w.; **o. de Love**, Love w.; **o. de Rayleigh**, Rayleigh w.; **o. directe**, direct w.; **o. élastique**, elastic w.; **o. longitudinale**, longitudinal w., P w.; **o. lumineuse**, light w.; **o. P**, compressional w.; **o. primaire**, primary w.; **o. réfléchie**, reflected w.; **o. réfractée**, refracted w.; **o. S**, distortional w.; **o. secondaire**, secondary w.; **o. sismique**, seismic w.; **o. superficielle**, L. w., long w.; surface w.; **o. transversale**, transverse w., S wave; **longueur d'o.**, wavelength.
ondulation, swell, corrugation, undulation.
ondulé (terrain), corrugated, wrinkled, wavy, rolling.
onduler, to undulate, to corrugate.
Ongulés (pal.), Ungulates.
onguligrades (pal.), unguligrades.
ontogenèse (pal.), ontogenesis, ontogeny.
ontogénétique, ontogenetic.
onyx, onyx.
oogone, oogonia.
oolithe, oolite, oolith, ooid, eggstone.
oolithique, oolitic, oolithic, oolite; **calcaire o.**, oolitic limestone; **fer o.**, oolitic iron stone.
oomicrite (pétro.), oomicrite.
oosparite, oosparite.
oosporange (Charophytes), oosporange.
opacimètre, turbidimeter.
opacité (de l'eau), opacity, turbidity.
opale, opal; **o. de feu**, fire o.; **o. jaspe**, jasper o.; **o. noble**, precious o.; **o. xyloïde**, wood o.
opalescence, opalescence.
opalescent, opalescent.
opalisé, opalized.

opaque (minéral), opaque (mineral).
opération de forage, drilling operations.
opérationnelle (recherche), operational (research).
opercule (pal.), operculum.
ophicalcite, serpentinous marble.
ophiolite, ophiolite.
ophiolitique, ophiolitic.
ophite, ophite.
ophitique (texture), ophitic.
Ophiurides (pal.), Ophiuridea.
Opisthobranches (pal.), Opistobranchia.
opistocèle, opistocoelous.
opistogyre (pal.), opistogyrate.
optique, 1. adj: optic(al); 2. n: optics; **angle o.**, optical angle; **axe o.**, optical axis; **constantes o.**, optical constants; **extinction o.**, o. extinction; **microscope o.**, photonic microscope; **plan o.**, optic plane; **polarisation o.**, o. polarization; **signe o.**, optic character; **spectre o.**, o. spectrum.
or, gold; **o. affiné**, refined g.; **o. alluvionnaire**, placer g., alluvial g.; **o. en pépites**, nuggety g.; **o. filonien**, vein g.; **o. fin**, fine g.; **o. natif**, native g.
orangite, orangite.
orbiculaire (texture), orbicular; **diorite o.**, orbicular diorite; **granite o.**, orbicular granite.
orbicule, spherolith.
orbite (astro.), orbit.
Orbitolinidés (pal.), Orbitolinidae.
ordanchite, ordanchite.
ordinaire (rayon), ordinary (ray).
ordinateur, computer.
Ordovicien, Ordovician.
ordre, order; **o. de cristallisation**, crystallization o.; **o. de superposition**, succession o.; **o. originel de superposition**, original o. of stratification.
oreiller (lave en), pillow-lava.
organique (matière), organic.
organisme (pal.), organism, creature (animal), plant (vegetal); **o. euryhalin**, euryhaline o.; **o. sténohalin**, stenohaline o.
organoclastique, bioclastic.
organogène, organogenic.
organogénique, organogenous.
orgue basaltique, columnar basalt, basalt columns.
orientation, orientation, bearing; **o. des couches**, bed strike; **o. d'une faille**, fault strike.
oriental (rubis), oriental (rubis).
orientale (agate), oriental (agate).
orienter, s'orienter, to orient, to orientate; **échantillon orienté**, oriented specimen.
orifice, orifice, outlet, opening, aperture *(pal.)*; **o. d'un puits (mine)**, pit mouth, shaft collar; **o. volcanique**, volcanic vent.
ornementation (pal.), ornamentation.
Ornitischiens (pal.), Ornitischia.
ornoïte, ornoite.
oroclinale (zone), orocline.
orocratique, orocratic.
orogène, orogen.
orogenèse, orogénie, orogeny, orogenesis.

orogénique, orogenic, orogenetic; **cycle o.**, orogenic cycle; **phase o.**, orogenic phase; **zone o.**, orogenic belt.
orogéosynclinal, orogeosyncline.
orographie, orography.
orographique, orographic.
orohydrographie, orohydrography.
orohydrographique, orohydrographic(al).
oromètre, orometer.
orométrie, orometry.
orométrique, orometric.
orpaillage, gold washing, alluvial digging.
orpailleur, gold washer.
orpiment, orpiment, yellow arsenic.
orthent (lithosol), orthent.
orthite, orthite.
ortho (préfixe), ortho (straight, true derived directly from igneous source).
orthoamphibolite, orthoamphibolite.
Orthocératidés (pal.), Orthoceratidae.
orthochimique (calcaire), orthochemical.
orthoclase (orthose), orthoclase.
orthoclasite, orthoclasite.
orthoclastique, orthoclastic.
orthod (sol podzolique), orthod.
orthodromie, orthodromy.
orthofelsite, orthofelsite.
orthoferrosilite, orthoferrosilite.
orthogenèse, orthogenesis.
orthogéosynclinal, orthogeosyncline.
orthogneiss, orthogneiss.
orthographique (projection), orthographic (projection).
orthomagmatique (stade), orthomagmatic (stage).
orthophyre, orthophyre.
orthophyrique, orthophyric.
orthopinacoide, orthopinacoid.
orthoprisme, orthoprism.
orthorhombique, orthorhombic; **amphibole o.**, orthamphibole; **pyroxène o.**, orthaugite, orthopyroxene.
orthose, orthose, orthoclase; **porphyre à o.**, orthophyre.
orthosilicate, orthosilicate.
orthosite (pétro.), orthosite.
orthotectite, orthotectic.
orthox (sol ferrallitique de climat humide), orthox.
ortlérite, ortlerite.
os, osar (pl.), esker.
osannite (minér.), osannite.
oscillation, oscillation, swinging; **o. climatique**, o. climatic fluctuation; **ride d'o.**, o. ripple; **vague d'o.**, o. wave.
oscillographe cathodique, cathode ray oscillograph.
oscule (pal.), osculum.
osmium, osmium.
ossements (pal.), bones.
osseux (restes), osseous.
Ostéichtiens (pal.), Osteichthyes.
ostéologie (pal.), osteology.
ostéométrie (pal.), osteometry.

ostiole de toundra, tundra ostiole, periglacial mud circle.

Ostracodes (pal.), Ostracoda, Ostacods.

Ostracodermes (pal.), Ostracodermi.

Ostréidés, Ostreidae.

otolithe (pal.), otolith.

ottajanite (pétro.), ottajanite.

ottrélite (minér.), ottrelite.

oued, ouady, dry river.

ouragan (météo.), hurricane.

Ouralien, Uralian (u. Carboniferous).

ouralite, uralite.

ouralisation, uralitization.

Oursin, sea-Urchin; **o. exocycle,** exocyclic s-u.; **o. irrégulier,** exocyclic s-u.; **o. régulier,** endocyclic s-u.

outil, tool; **o. de forage,** boring t.; **o. à couronne de diamant,** diamond drill; **o. préhistorique,** prehistoric implement.

outremer, lazurite, lapis-lazzuli, ultra-marine.

ouvala (karst), uvala, glade.

ouvarovite, ouvarovite.

ouverture de la taille (mine), working thickness.

ouvrage d'amenée d'eaux (géotechnie), intake.

ouvrier, workman; **o. à l'extraction,** hoistman; **o. carrier,** stone cutter; **o. de fond (mine),** underground w.; **o. du jour,** surfaceman; **o. des plates-formes,** derrick man.

ovipare (pal.), oviparous.

Oxfordien, Oxfordian (u. Jurassic); **argiles oxfordiennes,** Oxford clay.

oxydation, oxidizing, oxidation.

oxyde, oxide; **o. d'aluminium,** alumine; **o. de carbone,** carbonic o.; **o. de fer,** iron o.; **o. ferreux,** ferrous o.; **o. ferrique,** ferric o.; **o. sulfureux,** sulfur dioxide.

oxydé, oxidized; **chapeau de fer o.,** gossan.

oxyder, s'oxyder, to oxidize, to oxidate.

oxydo-réduction, oxidation-reduction.

oxygène, oxygen.

oxygéné (milieu), oxygenous, oxygenated.

oxygéner, to oxidize.

ozocérite, ozocerite.

ozone, ozone.

P

Pachyodonte (pal.), Pachyodont.
pagodite (minér.), pagodite.
pailleteur (orpailleur), gold-washer, digger.
paillette, flake; **p. de mica**, f. of mica; **p. d'or**, floating-gold.
pain de sucre, sugar-loaf (Brazil), inselberg.
palaffite, lake dwelling.
palagonite, palagonite.
palagonitique, palagonitic.
palagonitisation, palagonitization.
palasome, palasome, host mineral.
paléo (préfixe), paleo, palaeo (U.K.), ancient, former.
Paléoanthropiens, Palaeoanthropians.
paléobathymétrie, palaeobathymetry.
paléobotanique, paleobotany, paleobotany.
Paléocène, Palaeocene, Paleocene.
paléochaîne, paleochain, buried ridge.
paléochenal, paleochannel.
paléoclimat, paleoclimate.
paléoclimatologie, palaeoclimatology, paleoclimatology.
paléocourant, paleocurrent.
paléoécologie, palaeoecology, paleoecology.
paléoenvironnement, palaeoenvironment.
paléofaciès, paleofacies.
Paléogène, Palaeogene, lower Tertiary (Paleocene-Oligocene).
paléogéographie, paleogeography.
paléogéographique (carte), paleogeographic map.
paléokarst, paleokarst.
Paléolithique, Paleolithic.
paléomagnétique, paleomagnetic.
paléomagnétisme, paleomagnetism.
paléontologie, paleontology; **p. animale**, paleozoology; **p. végétale**, paleobotany.
paléontologique, paleontologic.
paléoplaine, paleoplain.
paléoprofondeur, paleodepth.
paléorelief, paleolandcape.
paléosol, paleosol, paleosoil, fossil soil, buried soil.
paléotectonique (adj.), paleotectonic.
paléotempérature, paleotemperature.
paléovolcanique, paleovolcanic, precainozoic volcanic.
Paléozoïque, Paleozoic.
palichnologie, paleoichnology.
palier (mine), level.
palingenèse, palingenesis.
palinspatique, palinspatic.
pallasite (météorite), pallasite.
palléal, pallial; **cavité p.**, p. cavity; **ligne p.**, p. line; **sinus p.**, p. sinus.
palse (périgl.), palsa, small hydrolaccolith in peat.
palustre, paludal, palustral, palustrine, swampy.

palygorskite, palygorskite.
palynologie, palynology.
palynomorphe, palynomorph.
pan (de rocher), slab, pane.
panabase, panabase, fahl ore.
panache (point chaud), plumes.
panaché (limon), variegated, streaked.
panachure (pédol.), streak.
Pangée (continent), Pangea.
panidiomorphique, panidiomorphic.
panneau (mine), panel.
Pannonien, Pannonian (u. Miocene; s.e. Europe).
pantellérite (rhyolite), pantellerite.
Pantothériens (pal.), Pantotheria.
papier calque (dessin), tracing paper.
papier filtre (chimie), filter paper.
papier millimétré (carto.), quadrille paper, plotting scale paper.
paquet (de terrains charriés), outlier.
para-autochtone, parautochtonous.
paraclase, paraclase, fault.
paraffine, paraffin, paraffin wax; **p. brute**, crude wax; **p. de schiste**, shale wax.
paraffinique, paraffinic.
paragenèse, paragenesis.
paragénétique, paragenetic.
parageosynclinal, parageosyncline, intracratonic geosyncline.
paragneiss, paragneiss.
paragonite, paragonite.
paraliagéosynclinal, paraliageosyncline.
paralique, paralic.
parallaxe (opt.), parallax.
paramagnétique, paramagnetic.
paramagnétisme, paramagnetism.
paramétamorphique, parametamorphic.
paramorphique, paramorphic.
paraschiste, paraschist.
parasismique, aseismic.
parasite (cône), parasitic cone.
paratype (pal.), paratype.
parcours (d'une onde), path.
parcours de temps minimum, minimum time path.
pargasite (minér.), pargasite.
pariétal (art préhistorique), parietal.
paroi, wall; **p. d'un puits**, side of a shaft; **p. d'une galerie**, w.; **p. inférieure**, foot w.; **p. supérieure**, roof.
paroxysme (volcanique), paroxysm, eruption.
partage (ligne de), watershed, divide.
particule, particle; **p. argileuse**, clay p.; **p. colloïdale**, colloidal p.
pas de tir (géoph.), shooting interval.
passage au crible, screening.
passage au tamis, sifting.

passage latéral (strati.), lateral shift of facies.
passe (dans un cordon littoral), inlet, channel.
passer à travers les mailles d'un tamis, to pass through the meshes of a sieve.
passer au tamis, to sift.
passive (marge), passive (margin).
pâte (de roches volcaniques), groundmass.
patine (d'une roche), patina.
patine désertique, desert varnish, tan.
patte ambulatoire (pal.), walking leg.
pavage désertique, desert pavement, lag gravel.
pays, country; **p. accidenté**, rolling c.; **p. découvert**, open c.; **avant p.**, foreland; **arrière p.**, backland; **bas p.**, lowland; **haut p.**, upland.
paysage (géomorpho.), landscape.
peau d'éléphant (roche en), shrinkage joints pattern, contraction joints (in weathering crusts).
pechblende, pitchblende.
pechkohle, pitch coal.
pechstein, pitchstone.
péchurane, pitchblende, uraninite.
Pectinidés (pal.), Pectinids.
pectolite, pectolite.
pédalfer (pédo.), pedalfer.
pédicellaire (pal.), pedicellaria.
pédoncule, peduncle, pedicle (*pal.*, Brachiop.).
pédoncule (de méandre), neck.
pédiment, pediment; **p. coalescent**, coalescing p.; **p. désertique**, desert p.; **p. emboîté**, inset p.; **p. rocheux**, rock p.
pédiplaine, pediplain, pediplane.
pédocal, pedocal.
pédogenèse, pedogenesis.
pédologie, pedology, soil science.
pédologue, edaphologist, soil scientist.
pédon (pédo.), pedon.
pegmatisation, pegmatization.
pegmatite (pétro.), pegmatite.
pegmatite graphique, graphic pegmatite.
pegmatitique, pegmatitic.
pegmatoïde, pegmatoid.
pélagique, pelagic; **vase p.**, pelagic ooze.
pélagite, pelagic deposit.
Pelé (cheveux de), Pele's hair.
péléen, pelean; **aiguille péléenne**, pelean spine.
pélite, pelite.
pélitique, pelitic, argillaceous.
pelle mécanique de découverte, stripper.
pelletage mécanique, power shoveling.
pelletée (de minerai), shovelful.
pelliculaire, pellicular.
pellicule d'eau, water film.
Pelmatozoaires (pal.), Pelmatozoa.
pelmicrite, pelmicrite.
pelote fécale, pellet.
pelsparite, pelsparite.
pencher (relatif à une couche), to incline, to bend, to lean, to slope.
pendage, dip; **p. apparent**, apparent d.; **p. général**, regional d.; **p. inverse**, reverse d.; **p. originel**, original d.; **p. périclinal**, centroclinal d.; **p. radial**, quaquaversal d.; **p. réel, vrai**, true d.; **amont p.**, up of the d.; **aval p.**, down the d.
pendagemètre, dipmeter.
pendagemétrie, dip logging, dipmeter logging.
pénéplaine, peneplain, peneplane, denudation plain; **p. embryonnaire**, incipient peneplain; **p. exhumée**, exhumed peneplain; **p. naissante**, incipient peneplain; **p. rajeunie**, rejuvenated peneplain.
pénéplanation, peneplanation (jamais "peneplaination"), planation.
pénétrabilité (d'un sol), penetrability.
pénétromètre (mec. sols), penetrometer.
pénétrométrie, penetration test.
péninsulaire, peninsular.
péninsule, peninsula.
pénitent (rocheux), rock pinnacle, earth pillar.
pennine, 1. pennine (nom d'une chaîne de montagne, U.K.); 2. orogenic complex, Penninikum zone (Swiss Alps).
pennite, penninite (chlorite), pennite, penninite.
Pennsylvanien, Pennsylvanian (u. Carboniferous).
pente, slope, grade, gradient; **p. continentale**, continental slope; **p. d'éboulis**, talus slope; **p. d'érosion**, erosion slope; **p. de solifluxion**, solifluction slope; **p. descendante**, down slope; **p. d'un cours d'eau**, grade; **p. limite**, profil d'équilibre, grade; **p. raide**, steep slope; **p. montante**, up slope; **p. naturelle**, natural slope, angle of repose.
pente (en) (pentu), sloping.
pentlandite, pentlandite.
pénurie (de minerai), shortage.
pépérino (tuff volcano-sédimentaire), peperino.
pépérite (Auvergne), basaltic tuff.
pépite, pepita, nugget.
peralcalin, peralkaline.
perçage (mine), boring, piercing, drilling.
percée (géomorph.), consequent valley, water-gap.
percement, piercing, drilling, boring; **p. de recoupes (mine)**, cross-driving; **p. en travers-banc (mine)**, cross-cutting.
percer, to bore, to pierce, to drill out (a well), to tunnel (a drift), to hole; **p. en direction (mine)**, to drift, to drive; **p. en montant (mine)**, to rise; **p. en travers-banc (mine)**, to cross-cut.
percer un tunnel (géotechnie), to tunnel.
perceuse, drilling machine, driller, drill.
perché, perched; **aquifère p.**, p. aquifer; **bloc p.**, p. bloc, p. boulder; **nappe phréatique p.**, p. water-table; **synclinal p.**, p. syncline; **vallée p.**, p. valley.
perçoir (préhist.), awl.
percolation (d'eau), percolation.
percoler, to percolate, to infiltrate.
percussion, percussion; **marques de p.**, percussion markings; **sondage par p.**, percussion boring.
percuteur (préhist.), striker.
perdre (de l'eau), to leak.
perdre (se) (rivière), to lose itself.
pérenne, perennial.
perforateur, 1. adj: drilling; 2. n: perforator, drill; **p. à diamant**, diamong drill; **p. de tubage**, casing p.

perforation (forage), perforation, perforating, holing, drilling, boring.
perforation hydraulique, hydraulic drilling.
perforatrice (mine), drill, driller, rock-drill, borer; **p. à air comprimé**, air drill; **p. à injection d'eau**, water-drill; **p. à percussion**, percussion drill; **p. à pointes de diamant**, diamond drill; **p. pneumatique**, air drill; **p. rotative**, rotary drill.
perforer, to perforate, to drill, to pierce.
pergélisol (périgl.), pergelisol, permafrost, perennially frozen ground; **p. actuel**, active permafrost; **p. discontinu**, discontinuous permafrost; **p. pérenne**, permafrost; **p. résiduel**, relict permafrost; **p. sec**, dry permafrost.
périanticlinal, perianticlinal.
périarctique, periarctic.
périastre (astro.), periastron.
périclase (miner.), periclase.
périclinal, centroclinal, periclinal.
péricline, pericline.
péridot (minér.), 1. peridot; 2. olivine.
péridotite (pétro.), peridotite.
périgée, perigee.
périglaciaire, periglacial; **p. pérenne**, perennial p. condition; **climat p.**, p. climate; **faciès p.**, p. facies; **indice p.**, p. index; **phénomène p.**, p. phenomenon; **processus p.**, p. process; **régime p.**, p. regime; **zone p.**, p. zone.
Périgordien (préhist.), Perigordian.
périhélie, perihelion.
périmagmatique, perimagmatic.
périmètre mouillé, wetted perimeter.
période, period; **p. crétacée**, cretaceous p.; **p. de demi-vie (radioactive)**, half-life p.; **p. de fonte (des neiges)**, thaw season; **p. de gel**, freezing season; **p. de glace**, ice p.; **p. d'englacement**, p. of floating ice formation; **p. d'oscillation**, p. of oscillation; **p. glaciaire**, glacial p., boulder p.; **p. inverse (paléomagnétique)**, reverse p.; **p. interglaciaire**, interglacial p.
périodite (sédim.), periodite.
périostracum, periostracum.
périprocte (pal.), periproct.
périrécifal, perireefal.
Périssodactyles (pal.), Perissodactyla.
péristome (pal.), peristome.
périsynclinal, basin.
perle, pearl, bead.
perle (essai au chalumeau), bead test.
perlite, perlite.
perlitique, pearlitic, perlitic.
permagel, permafrost.
perméabilité, permeability; **p. latérale**, lateral p.; **p. magnétique**, magnetic p.; **p. secondaire**, secondary p.
perméable, permeable, porous, pervious.
Permien, Permian.
permis de forage, drilling permit.
permis de recherche, prospecting license.
Permo-Carbonifère, Permo-Carboniferous.
pérovskite, perovskite.

perré, rip-rap, stone packing.
perrière (désuet), quarry.
perrier, scree.
Périssodactyles (pal.), Périssodactyle.
perspective aérienne, aerial perspective.
perte, loss, leak, leakage, disappearance; **p. au feu**, fire loss; **p. d'eau**, water loss; **p. de pression**, loss of pressure; **p. karstique**, interrupted stream.
perthite (pétro.), perthite.
perthitique (texture), perthitic texture.
pertuis (Charente), inlet.
perturbation atmosphérique, statics, atmospheric disturbance.
perturbation magnétique, magnetic perturbation.
pesanteur, gravity force, gravity.
pétalite (minér.), petalite.
petite baie, cove.
petit ruisseau, rill.
pétri, 1. moulded; 2. full of...
pétrifiant, petrifying, petrescent, incrusting; **eaux p.**, incrusting waters.
pétrification, petrification, lithification, incrustation.
pétrifié (bois), petrified (wood).
pétrifier, se pétrifier, to petrify.
pétrir (de l'argile), to knead, to work (clay).
pétrochimie, petrochemistry.
pétrochimique, petrochemical.
pétrofabrique, petrofabric analysis.
pétrogenèse, petrogenesis, petrogeny.
pétrogénétique, petrogenetic.
pétrographe, petrograph, petrographer, petrologist.
pétrographie, petrography, petrology; **p. sédimentaire**, sedimentary petrography.
pétrographique, petrographic(al); **province p.**, p. province.
pétrole, petroleum, crude oil, oil, mineral oil, rock oil; **p. brut**, crude oil; **p. brut asphaltique**, asphaltic base crude; **p. brut à base paraffinique**, paraffin base oil; **p. brut naphténique**, naphtene base crude; **p. brut non sulfuré**, sweet crude; **p. léger**, light oil; **p. lourd**, heavy oil.
pétrolier (bateau), tanker; **p. génie**, petroleum engineering; **p. (ouvrier)**, oilman.
pétrolifère, petroliferous, oil bearing.
pétrologie, petrology; **p. structurale**, petrofabrics.
pétrologique, petrologic.
pétrosilex (felsite), petrosilex.
pétrosiliceux, microfelsitic.
petzite (minér.), petzite.
phacoïde, phacoid.
phacolite, 1. phacolite (*minér.*); 2. phacolith (intrusive body).
phanéritique, phaneritic, phanerocrystalline, coarsely crystalline.
Phanérogames (paléobot.), Phanerogams.
Phanérozoïque, Phanerozoic.
pharmacolite (minér.), pharmacolite.
pharmacosidérite (minér.), pharmacosiderite.
phase, phase, stage; **p. de distension**, distensive phase, rifting; **p. liquide**, liquid phase; **p. orogénique**, orogenic phase.

phénacite (minér.), phenacite.
phengite (mica), phengite.
phénoblaste (pétrol.), metacryst, metacrystal.
phénocristal, phenocryst.
phénomène géologique, geological event.
Phéophycées (paléobot.), Phaeophyceae.
phlogopite (mica), phlogopite, rhombic mica.
phonolite (pétro.), phonolite, clinkstone.
phonolitique, phonolitic.
phosgénite, phosgenite.
phosphate, phosphate.
phosphaté, phosphatic, phosphated; **craie p.**, phosphatic chalk; **grès p.**, phosphatic sandstone; **nodule p.**, phosphatic nodule.
phosphatisation, phosphatization.
phosphatique, phosphatic.
phosphore, phosphorus, phosphor.
phosphoré, phosphorated.
phosphoreux, phosphorous.
phosphorique, phosphoric.
phosphorite, phosphorite.
phosphorocalcite, phosphorocalcite.
photique, photic.
photogéologie, photogeology.
photogrammétrique, photogrammetric.
photogrammétrie, photogrammetry.
photographie aérienne, aerial view, aerial photography.
photographies aériennes multispectrales, aerial multispectral views.
photointerprétateur, photointerpreter.
photo-interprétation, photointerpretation.
photométrie, photometry.
photomosaïque, mosaic.
photoplan, photomap.
photorestituteur, photographic plotter, stereoplotter.
photorestitution, photorestitution.
photosphère (astro.), photosphere.
photothèque, photographic library.
phragmocône, phragmocone.
phréatique, phreatic; **explosion p.**, phreatic eruption; **nappe p.**, saturation level, water table.
phtanite, siliceous shale, schist phtanite (rare).
phylétique (pal.), phyletic.
phyllade, phyllite.
phyllite, phyllite.
phylliteux, phyllitic.
Phyllocératidés (pal.), Phylloceratids.
phyllonite (pétro.), phyllonite (pétro.).
Phyllopodes (pal.), Phyllopode.
phyllosilicate, phyllosilicate.
phylogénie (pal.), phylogeny.
phylum (pal.), phylum.
physiographie (= géomorphologie), physiography.
physiographique, physiographic(al).
phytéral, phyteral (vegetal remain).
phytocénose (paléobot.), phytocoenose.
phytophage (pal.), phytophagous.
phytoplancton, phytoplankton.

pic (montagne), peak; **p. de diffractogramme**, p.; **p. (outil)**, pick, pickaxe; **p. (d'un diagramme)**, pick; **p. (de mineur)**, miner's pick.
picot (mine), wedge.
picotite (spinelle), picotite.
picrite (r. ultrabasique), picrite.
pied, 1. foot, base, bottom; 2. foot (meas.)= 0,3048 m; **p. à coulisse**, slide gauge, slide calipers; **p. carré**, square foot; **p. cube**, cubic foot; **p. cube par seconde**, cusec; **p. de glace**, ice foot; **p. de pente**, foot of a slope.
piédmont, piémont, piedmont, foothill; **glacier de p.**, piedmont glacier; **glacis de p.**, piedmont slope; **plaine de p.**, piedmont plain.
piège, trap; **p. à gaz**, gas t.; **p. à sable**, sand t.; **p. de faille**, fault t.; **p. de perméabilité**, permeability t. reservoir; **p. de pincement**, pinch out t. reservoir; **p. diapir**, piercement t. reservoir; **p. pétrolifère**, oil t.; **p. stratigraphique**, stratigraphic gap; **p. structural**, structural gap.
piémontite (épidote), piemontite.
pierraille, brokenstone, crushed stone, chippings.
pierre, stone; **p. à aiguiser**, oilstone; **p. à bâtir**, building s.; **p. à chaux**, limestone; **p. à facettes**, sandblasted pebble, faceted pebble; **p. à feu**, flint; **p. à fusil**, flint; **p. à foulon**, smectite, fuller's earth; **p. à liards**, nummulitic limestone; **p. à plâtre**, plaster s., gypsum; **p. branlante**, logan s., rocking s.; **p. d'aimant**, magnetite; **p. d'alun**, alunite; **p. de bornage**, boundary s.; **p. de Caen**, white, bajocian oolite; **p. de croix**, staurolite; **p. de lune**, moonstone; **p. de soude**, natrolite; **p. de taille**, building s.; **p. de touche**, touchstone; **p. façonnée par le vent**, ventifact; **p. fine**, semiprecious s.; **p. gravée (préhist.)**, carved rock; **p. levée**, menhir, standing s.; **p. lithographique**, lithographic s.; **p. meulière**, millstone; **p. polie**, polished s., neolithic; **p. ponce**, pumice; **p. précieuse**, gem, gemstone; **p. précieuse sans défaut**, flawless gem; **p. précieuse taillée**, cut gem; **p. taillée**, chipped s., palaeolithic cut s.; **cercle de p. (périgl.)**, s. circle; **coulée de p.**, s. river, block stream; **glacier de p.**, rock glacier; **guirlande de p. (périgl.)**, s. festoon; **polygone de p. (périgl.)**, s. polygon; **réseaux de p.**, s. nets.
pierrerie, gem, precious stone.
pierreux, stony, cobbly.
piézocristallisation, piezocrystallization, piezocrescence.
piézoélectrique (quartz), piezoelectric.
piézoélectricité (du quartz), piezoelectricity.
piézomètre, piezometer.
piézométrique, piezometric; **surface p.**, saturation level.
pigeonite (pyroxène), pigeonite.
pilage (broyage), grinding, crushing.
pilier (mine), pillar, post, stack; **p. d'érosion**, erosion column, earth pillar; **p. de soutènement (mine)**, supporting pillar.
pinacle, stack; **p. corallien**, reef knoll, coral head (growing).
pinacoïde, pinacoid.

pince (labo.), tongs; **p. brucelles,** tweezers; **p. à creuset,** crucible t.; **p. pour tubes,** tube t.

pincée (tecto.), pinch.

pincement, pinching, wedge out.

pingo (périgl.), pingo, ice cored mound, periglacial hydrolaccolith.

pinite, pinite.

Pinnipèdes (pal.), Pinnipedia.

piochage, picking.

pioche, pickaxe, pick, mattock.

piocher (le sol), to dig (with a pick), to pick.

pipeline, pipe line.

pipette graduée, graduate pipette.

pipkrake, needle ice, pipkrake.

piquage (mine), hewing, digging.

piquant (d'Oursin), radiole.

piquetage (mine), stacking.

piquet de jalonnement, stake.

plicatulation (tecto.), puckering, minute folding.

piqueter (mine), to stake out.

piqueur (mine), pikeman, cutter, hewer.

pisolite, pisolith.

pisolithique, pisolitique, pisolitic; **calcaire p.,** pisolitic limestone; **minerai de fer p.,** pisolitic iron.

pistacite, pistazite, pistazite, epidote.

piste, 1. track, trail (invertebrate), ichnofossil; 2. footprint (vertebrate).

Pithécanthrope (pal.), Pithecanthrope.

piton, peak, pinnacle.

pixel (photogrammétrie), pixel.

placage, veneer, coating, superficial deposit.

placentaires (Mammifères), Placentalia.

placer, placer, alluvial digging; **p. alluvial,** river placer, placer deposit; **p. aurifère,** gold placer; **p. stannifère,** tin placer.

Placodermes (pal.), Placoderms.

Placodontes (pal.), Placodontia.

plafond (d'une mine), roof.

plage, beach, shore; **avant-p.,** fore-shore; **arrière-p.,** back-shore; **bas de p.,** fore-shore; **croissant de p.,** beach cusp; **gradin de p.,** beach terrace; **haut de p.,** beach ridge; **levée de p.,** beach ridge; **sillon de p.,** beach furrow; **p. soulevée,** raised beach.

plagioclase, plagioclase.

plagioclasite, plagioclasite.

plaine, plain, lowland; **p. abyssale,** abyssal plain, deep-sea plain; **p. alluviale,** alluvial plain, valley floor, first bottom flat; **p. côtière,** coastal plain; **p. d'abrasion marine,** plain of marine erosion; **p. d'alluvion,** flood plain; **p. de dénudation,** denudation plain; **p. deltaïque,** delta plain; **p. d'érosion,** erosion plain; **p. de piédmont,** piedmont plain; **p. d'inondation,** flood plain; **p. fluvio-glaciaire,** outwash plain, sandur, sandr; **p. glaciaire,** glacial plain; **p. littorale,** coastal plain, littoral plain; **p. périglaciaire,** periglacial plain, geliplain, **p. podzolisée,** podzolized sand plain; **p. ravinée,** dissected plain.

Plaisancien, Plaisancian (u. Pliocene).

plan, 1. adj: plane, level, even, planar; 2. n: design, project, tracing plan, map; **p. axial,** axial plane;

p. d'eau, water level; **p. de charriage, de chevauchement,** thrust plane, overthrust fault; **p. de cisaillement,** shear surface; **p. de clivage,** cleavage plane; **p. de crête d'un pli,** crestal plane; **p. de décollement,** detachment plane; **p. de diaclase,** joint plane; **p. de disjonction,** divisional plane; **p. de discontinuité,** plane of unconformity; **p. de faille,** fault plane; **p. de fracture,** slip plane; **p. de frappe,** striking plane; **p. de glissement,** gliding plane; **p. de la nappe phréatique,** ground water table; **p. de mâcle (d'hémitropie),** twinning plane; **p. de niveau,** datum line; **p. de polarisation,** polarization plane; **p. de poussée,** thrust plane; **p. de référence,** datum plane; **p. de schistosité,** foliation plane; **p. de stratification,** bedding plane; **p. de symétrie,** plane of symmetry; **p. équatorial (astro.),** equatorial plane; **p. focal (opt.),** focal plane; **p. incliné,** headway, slope, incline; **p. réticulaire,** lattice plane.

plancher (d'une couche), floor, bottom, lower part.

plancher océanique, o. floor.

planctivore (pal.), planktivorous.

plancton, plankton, plancton.

planctonique, planktonic, planctic.

planétaire, planetary.

planète, planet.

planétoïde, planetoid.

planétologie, planetology.

planèze (Auvergne), lava plateau.

planimétrage, plotting.

planimètre, planimeter.

planimétrie, planimetry.

planimétrique, planimetric.

plante à feuilles caduques (paléobot.), deciduous plant.

plaque, plate; **p. brachiales,** brachials; **p. de roche,** slab; **p. mince,** thin section, thin p.; **p. porte-objet (micro.),** slide, slider p.; **p. radiales,** radials; **p. tectonique,** lithospheric p.; **p. tectonique chevauchante,** overthrust p.; **p. tectonique chevauchée,** underthrust p.; **p. tectonique en subduction,** subducting p.

plaquette (de roche), plate, flag; **calcaire en p.,** platy limestone; **débit en p.,** platy parting.

plasticité, plasticity.

limite de plasticité, plasticity index.

plastique, plactic; **argile p.,** p. clay; **déformation p.,** p. flow, p. strain; **écoulement p.,** p. flow.

plat (paysage), flat.

plateau (géogr.), plateau, table-land; **p. basaltique,** basalt plateau; **p. cardinal (pal.),** hinge-plate; **p. continental,** continental shelf; **p. désertique,** desert plateau

plate-forme, platform, floor; **p. continentale,** continental shelf, shelf zone; **p. corallienne,** coral platform; **p. d'abrasion,** rock bench, abrasion platform; **p. d'abrasion marine,** wave cut platform; **p. d'accumulation marine,** wave built platform; **p. d'accrochage,** hooking platform, rocking platform; **p. de forage,** derrick platform; **p. flottante,** floating platform; **p. littorale,** rock bench; **p. semi-submersible**, semi-submersible platform; **p. structurale**, structural plateau, structural platform.

platier corallien, coral reef flat.

platier rocheux, abrasion platform.

platière (forêt de Fontainebleau), sandstone (or limestone) flat hill, flat land.

platine, platine, platinum; **p. de microscope**, stage; **p. porte-objets**, slide, slider; **p. tournante**, revolving stage; **p. universelle**, universal stage.

plâtre, plaster.

plâtrière, gypsum quarry.

plature, table reef, coral reef.

plauénite (pétro.), plauenite.

playa, playa, ephemeral lake.

Pléistocène, Pleistocene Quaternary (incl. Holocene).

pléochroïque, pleochroic.

pléochroïsme, pleochroism.

pléonaste (minér.), pleonaste.

Plésiosaures (pal.), Plesiosauria.

pleural (pal.), pleural.

plèvre (pal.), pleura.

pli, fold; **p. anticlinal**, up-fold, anticline; **p. coffré**, box f.; **p. composé**, composite f.; **p. concentrique**, concentric f.; **p. conjugué**, conjugate f.; **p. couché**, recumbent f.; **p. cylindrique**, cylindrical f.; **p. d'entraînement**, drag f.; **p. d'étirement**, drag f.; **p. de charriage**, thrust f.; **p. de cisaillement**, shear f.; **p. de couverture**, epidermic f.; **p. de revêtement**, drape f.; **p. décalé**, offset f.; **p. déjeté**, inclined f., asymmetric f.; **p. déversé**, overturned f., overfold; **p. déversé-faillé**, faulted overfold; **p. disharmonique**, disharmonic f.; **p. dissymétrique**, asymmetric f.; **p. droit**, upright f., symmetric f.; **p. en auge**, box syncline; **p. en chevrons**, zig-zag f.; **p. en échelons**, échelon f.; **p. en éventail**, fan-shaped f.; **p. en genou**, knee f.; **p. en retour**, back f.; **p. faille**, broken f., disrupted f.; **p. faille couché**, overthrust f.; **p. faille inverse**, reverse f. fault; **p. fermé**, closed f.; **p. isoclinal**, isoclinal f.; **p. monoclinal**, monoclinal f., uniclinal; **p. normal**, normal f.; **p. oblique**, oblique f., inclined f.; **p. parallèle**, parallel f.; **p. par flexion et glissement**, flexural slip f.; **p. plongeant**, dipping f., pitching f.; **p. posthume**, posthumous f.; **p. renversé**, overturned f.; **p. replissé**, replissed f.; **p. secondaire**, minor f.; **p. semblable**, similar f.; **p. serré**, compressed f.; **p. symétrique**, symmetric(al) f.; **p. synclinal**, synclinal f., syncline, down f.

plication, plication.

Pliensbachien, Pliensbachian (l. Jurassic).

plier, se plier (une couche), to bend.

Plinien (volc.), Plinian.

Pliocène, Pliocene.

Plio-Quaternaire, Plio-Pleistocene.

plissé, folded, plicated, wrinkled.

plissement, fold, folding, corrugation; **p. anticlinal**, anticlinal fold; **p. en retour**, back folding; **p. synsédimentaire**, contorted bedding.

plisser, to fold, to corrugate, to wrinkle.

plissotement, crumpling, puckering, minute folding.

plissoter, to crumple.

ploiement (d'une couche), bending, flexing.

plomb, lead; **p. argentifère**, silver l.; **p. jaune**, wulfenite; **p. phosphaté**, pyromorphite; **p. provenant de l'uranium**, uranium l.; **p. rouge**, red l. ore, crocoite; **p. sulfaté**, anglesite; **p. sulfuré**, galena; **p. sulfuré antimonifère**, jamesonite.

plombagine, plumbago, graphite.

plombifère, lead-bearing, plumbiferous.

plongeant (pli), plunging, pitching, dipping.

plongée, dive; **soucoupe de p.**, diving saucer.

plongement, plunge; **p. de l'axe d'un pli**, p. of axis, pitching, dipping; **p. d'un objet (par immersion)**, immersion, plunging; **structure à p. divergent**, quaquaversal dip, fold.

plonger dans l'eau, to dive, to immerse; **p. dans le sol (pli)**, to dip, to pitch.

plongeur (personne), diver.

pluie, rain, precipitation; **p. de cendres**, ash rain, shower; **p. de poussières**, blood rain; **p. de sable**, sand storm; **p. fine**, drizzle.

plumasite (anorthoclasite à corindon), plumasite.

plumosite, fibrous jamesonite, plumosite.

pluriloculaire (pal.), multilocular.

pluton, pluton, stock.

plutonique, plutonic; **roche p.**, p. rock.

Plutonisme (théorie), Plutonism.

plutonite, plutonite.

plutonium, plutonium.

pluvial, pluvial; **période p.**, pluvial period.

pluviomètre, pluviometer, rain gauge.

pluviométrie, pluviometry.

pluviométrique, pluviometric.

pluviosité, rainfall, precipitation.

pneumatogène, pneumatogenic.

pneumatolyse, pneumatolysis.

pneumatolytique, pneumatolytic.

pneumatophore (pal.), pneumatophore.

poche, pocket; **p. d'eau**, water p.; **p. de dissolution**, washout, solution cave; **p. de gaz**, gas p.; **p. de grisou**, fire damp p.; **p. de minerai**, p., bunch, nest of ore; **p. de solifluxion**, solifluction p.; **p. en chaudron (périgl.)**, p., pot hole.

podzol, podsol, podzol, bleached earth; **p. à gley**, gley podzolic soil; **p. ferrugineux**, iron podzol; **p. ferrugineux hydromorphe**, hydromorphic iron podzol; **p. humo-ferrugineux**, iron-humic podzol; **p. sableux**, sandy podzol.

podzolique, podsolic, podzolic; **sol p.**, podsolic soil.

podzolisation, podsolization.

podzolisé, podsolized, podzolized.

podzoluvisol, podzoluvisol.

poecilitique, poikilitic, poecilitic.

poeciloblastique, poekiloblastic.

poids, weight, load; **p. atomique**, atomic weight; **p. moléculaire**, molecular weight; **p. spécifique**, specific weight, density.

point, point, spot, dot; **p. cardinal**, cardinal point; **p. chaud**, hot spot; **p. coté**, height spot; **p. d'ébullition**, boiling point; **p. de condensation**,

condensation point; **p. de congélation,** freezing point; **p. de Curie,** Curie's point; **p. de fusion,** melting point; **p. de rosée,** dew point; **p. de rupture,** break point; **p. de tir,** shot point; **p. de vaporisation,** vaporization point; **p. d'inflexion (d'une pente),** nickpoint; **p. eutectique,** eutectic point; **p. triple,** triple junction.

pointe (préhist.), point; **p. à cran (*préhist.*),** shoulderet p.; **p. de crue,** flood crest; **p. de flèche (*préhist.*),** arrow head; **p. de terre,** headland, promontory; **p. génale (*pal.*),** genal spine; **p. littorale,** bar, spit; **p. littorale à crochets,** recurved spit, hooked bar.

pointement (remontée d'une couche), outcrop; **p. diapirique,** diapir.

pointillé (trait), dotted line.

poise (unité de viscosité), poise (g/cm sec).

poisson (pal.), fish.

poix, pitch; **p. minérale,** bitumen.

polaire, polar.

polarisant, polarizing; **microscope p.,** polarization microscope.

polarisation, polarization; **p. spontanée,** self p. (s.p.).

polariser, to polarize, to bias.

polariseur, polarizer.

polarité, polarity; **p. apparente,** apparent p.; **p. inverse,** reversed p.; **p. normale,** standart p.; **p. opposée,** reversed p.; **p. sédimentaire,** graded bedding, gradation.

polder, polder.

pôle, pole; **p. géographique,** geographic p.; **p. magnétique,** magnetic p.; **p. Nord,** North p.; **p. Sud,** South p.

poli glaciaire, glacial polish.

polianite (minér.), polianite.

polie (âge de la pierre), Neolithic.

polir (une roche), to smooth, to polish.

polissage éolien, wind polishing.

poljé, polje.

pollen, pollen; **p. non sylvatique,** nonarboreal p. (N.A.P.).

pollinique (analyse), pollenanalysis.

polluant (environnement), pollutant.

polluer, to pollute.

pollution, pollution; **p. des cours d'eau,** stream p.

polybasite (minér.), polybasite.

polychroïsme, polychroism.

polycyclique, polycyclic, recycled; **vallée p.,** polycyclic valley.

polyèdre, polyhedron.

polyédrique, polyhedral.

polygénique, polygénétique, polygenic, polygenous; **surface d'érosion p.,** facetted peneplain, polygenic peneplain.

polygéosynclinal, polygeosyncline.

polygonal (sol), patterned ground.

polygonation, polygonation (tropical weathering), polygonal jointing.

polygone (périgl.), polygon; **p. avec bourrelet,** rim p.; **p. avec triage,** sorted p.; **p. boueux,** mud p.; **p. boueux de dessication,** mud-crack p.; **p. de fentes de gel,** ice-wedge p.; **p. de fissuration par le gel,** frost-crack p.; **p. de pierres,** stone p.; **p. de terre,** earth ring, nonsorted p.

polyhalite, polyhalite.

polymère, polymer.

polymérisation, polymerization.

polymériser (chimie), to polymerize.

polymétallique (nodule), polymetallic.

polymétamorphisme, polymetamorphism.

polymorphe, polymorphous, polymorphic.

polymorphisme, polymorphism.

polype (pal.), polyp.

polyphasé, polyphase.

polyphylétique (pal.), polyphyletic.

Polypier (pal.), Coral, Polyp; **colonie de P.,** polyparium; **squelette d'un P.,** corallite.

polysynthétique, polysynthetic.

pompage (mine), pumpage.

pompe de mine, shaft pump.

ponce (volcanique), pumice.

ponceux, pumiceous.

ponor (karst), sink hole.

pont continental, continental bridge.

Pontien, Pontian (u. Miocene).

pore, pore; **p. fin,** fine p.; **p. moyen,** middle p.

poreux, 1. porous (soil); 2. poriferous (*pal.*).

porion (mine), overman.

porosimètre, porosimeter.

porosité, porosity; **p. capillaire,** capillary p.; **p. de fracture,** fracture p.; **p. primaire,** primary p.

porphyre, porphyry; **p. quartzifère,** quartz p.

porphyrique, porphyritic.

porphyrite, porphyrite.

porphyroblaste, porphyroblast.

porphyroblastique, porphyroblastic.

porphyroïde, porphyraceous.

portance (d'un sol), bearing capacity.

Portlandien, Portlandian (u. Jurassic).

position, location, position; **p. d'extinction,** extinction position, dark position; **p. inverse ou renversée,** inversion, inverted order; **p. normale,** right side up.

post-glaciaire, post-glacial.

posthume (pli), posthumous (fold).

postorogénique, postorogenic.

post-tectonique, post-tectonic.

pot de fusion, melting pot.

potamologie, potamology.

potasse, potash.

potassique, potassic.

potassium, potassium, kalium.

potassium-argon (datation), potassium-argon (dating).

poteau (mine), stake, pillar, post.

potentiel, potential; **p. capillaire,** capillary p.; capillary pressure; **p. d'oxydo-réduction,** oxido-reduction p.; **p. naturel du sol,** natural earth p.; **p.**

maximal d'un puits, openflow p.; **p. spontané,** spontaneous p.

potentiomètre, potentiometer.

Potsdamien, Potsdamian (u. Cambrian, U.S.).

pouce, inch (2,54 cm); **p. carré,** square i. (6,452 cm²); **p. cube, cubique,** cubic i. (16,387 cm³).

poudingue, pudding stone, conglomerate.

poudre, 1. powder; 2. silt; **p. de mine,** blasting powder; **p. d'or,** gold dust, gold flour.

poudreuse (neige), powdery (snow).

poulier, bar; **p. d'entrée de baie,** baymouth b.; **p. intérieur,** midway b.

poupée de loess, loess doll, puppet, calcareous concretion, lime concretion.

pourpre (conglomérat), Cambrian conglomerate (Normandy).

pourrir, se pourrir (décomposition des roches), to rot, to decay, to be weathered, to be disaggregated.

pourriture (des roches), decay.

poussée, thrust; **p. d'eau,** water drive; **p. de gaz,** gas drive; **p. de gel,** ice push; **p. de gel horizontale,** frost shove; **p. de mollisol,** frost boil; **p. latérale,** side t.; **p. tectonique,** tectonic t.

pousser (tecto.), to thrust, to push

poussier, coal dust.

poussière, dust; **p. d'eau,** spray; **p. d'or,** gold d.; **p. de charbon,** coal d.; **p. de minerai,** ore d.

poussiéreux (terrain), dusty.

pouvoir, power; **p. absorbant,** absorbing p.; **p. agglutinant,** agglutining p.; **p. calorifique,** heating p., calorific value; **p. de dispersion,** dispersive p.; **p. dispersif,** dispersive p.; **p. de rétention en eau,** moisture holding capacity; **p. grossissant,** magnifying p.; **p. réflecteur (d'un charbon),** reflectance; **p. séparateur,** partition efficiency; **p. tampon (chimie),** buffer-p.

pouzollane, pozzolana.

prase (minér.), prase.

prasinite (pétro.), prasinite.

Précambrien, Precambrian.

précipice, precipice.

précipitable (chimie), precipitable.

précipitation, precipitation, settlement; **p. atmosphérique,** precipitation rainfall; **p. chimique,** chemical precipitation.

précipité (chimie), precipitate.

précipiter (un sel, une substance), to precipitate.

précontinent, continental margin, continental shelf.

précontraint (béton), prestressed (concrete).

précurseur (séisme), precursor.

prédateur (pal.), predator.

prédation (pal.), predation.

prédominant (vent), prevailing (wind).

préglaciaire, preglacial.

préhistorique, prehistoric; **industrie p.,** prehistoric tool assemblage.

prehnite (minér.), prehnite.

prélèvement, sample; **p. d'échantillons,** sampling.

prélever un échantillon, to take a sample.

premier, first, prime.

première arrivée (d'une onde), first arrival.

prendre en masse (se solidifier), to harden.

préorogénique, preorogenic.

préparation mécanique (d'un minerai), dressing, ore-dressing.

préreconnaissance, preliminary survey.

présence (d'un minerai), occurrence, presence.

presqu'île, peninsula.

presser (comprimer), to squeeze, to press.

pression, pressure; **p. atmosphérique,** atmospheric p.; **p. cryostatique,** cryostatic; **p. d'eau,** head of water; **p. d'exploitation,** working p.; **p. de boue,** mud p.; **p. de débit,** flowing p.; **p. de formation,** rock p.; **p. de gaz,** gas p.; **p. de gisement,** formation p., field p.; **p. de gisement en écoulement,** open flow p.; **p. en tête de puits,** wellhead p.; **p. géostatique,** geostatic p.; **p. hydrostatique,** hydrostatic p.; **p. interstitielle,** pore p.; **p. statique,** static head, static p.

prétectonique, prekinematic.

prévision météorologique, weather forecast.

Priabonien, Priabonian (u. Eocene).

Primaire (ère), Paleozoic (era).

primitif (temps), Precambrian era.

prise, intake; **p. avec congélation,** freeze up; **p. d'air (mine),** air intake; **p. d'échantillons,** sampling, field sampling; **p. lente (de ciment),** slow hardening, slow setting; **p. rapide,** quick hardening, quick setting.

prismatique (cristal), prismatic; **débit p.,** p. jointing, columnar jointing.

prisme, prism; **p. d'accrétion tectonique,** arc-tranch gap; **p. de Nicol,** Nicol p.; **p. orthorhombique,** orthorhombic p.; **p. sédimentaire,** sedimentary p., wedge of sediments, deposition wedge.

prismé, prismatic(al).

probabilité, probability.

Proboscidiens, Proboscidea.

procédé, process; **p. d'extraction,** extraction p.; **p. de craquage,** cracking p.; **p. de désulfuration,** desulfurization p.; **p. de récupération,** recovery p.; **p. par flottage,** flotation p.; **p. par voie humide,** wet p.; **p. par voie sèche,** dry p.

processus, process; **p. interne,** endogeneous p.

producteur, 1. adj: productive, producing, yielding; 2. n: producer.

productif, producing, productive, yielding.

production, production, output, yield; **p. annuelle,** yearly output; **p. d'un puits de pétrole,** oilwell yield; **p. éruptive (d'un puits),** flush production; **p. initiale,** initial flow; **p. journalière,** daily output; **p. stabilisée,** settled production.

productivité, productivity; **indice de p.,** productivity index (P.I.).

produire, to yield, to produce, to bear.

produit, product, produce; **p. chimiques,** chemicals; **p. léger (pétrole),** front end tail; **p. lourd,** heavy end tail; **p. national brut (P.N.B.),** gross national product; **p. pétrochimiques,** petrochemicals; **p. réfractaires,** refractories.

profil, profile; **p. de plage,** beach p.; **p. d'équilibre,** equilibrium p., grade; **p. d'un sondage,** bore p.; **p. fluviatile,** river p.; **p. longitudinal,** longitudinal p., long p.; **p. pédologique,** soil p.; **p. régularisé,** graded p.; **p. sismique,** seismic line, seismic p.; **p. stratigraphique,** stratigraphic column; **p. topographique,** topographic p.; **p. transversal,** cross p.; **p. tronqué (pédol.),** truncated p.
profilage (géoph.), profiling.
profond, deep; **peu p.,** shallow.
profondément, deeply.
profondeur, depth; **p. abyssales,** abyssal d.; **p. au foyer,** d. of focus; **p. de compensation,** d. of compensation; **p. d'un puits,** well d.; **p. océaniques,** ocean d.
proglaciaire, proglacial.
progradant, prograding.
progradants (lits), offlap beds.
progradation, progradation.
programme, program, schedule; **p. de production (d'un puits),** production schedule.
progression (moraine de), progression moraine.
progression glaciaire, glacial advance.
projection, projection; **p. azimuthale,** azimuthal p.; **p. cartographique,** map p.; **p. cylindrique,** cylindrical map p.; **p. conforme,** conformal map p.; **p. conique,** conical map p.; **p. équivalente,** equal area map p.; **p. Lambert,** Lambert map p.; **p. Mercator,** Mercator map p.; **p. polyconique,** polyconic map p.; **p. volcanique,** ejecta, ejectamenta.
projeter (volcan), to project, to eject.
proluculus, proloculus.
promontoire, head, headland, cape, promontory.
propagation d'un écoulement, rooting.
propagation d'une onde, wave propagation.
proportion de pierres dans un sol, stoniness.
proportion pétrole-eau, oil water ratio.
propriété, property, characteristic; **p. mécanique,** mechanical property; **p. optique,** optical property; **p. physique,** physical property.
propylite (pétro.), propylite.
propylitisation, propylitization.
Prosimiens (pal.), Prosimii.
Prosobranches (pal.), Prosobranchia.
prospecter, to prospect.
prospecteur, prospector.
prospection, prospection, surveying; **p. géophysique,** geophysical surveying; **p. magnétique,** magnetic surveying; **p. pétrolière,** oil prospecting; **p. sismique,** seismic surveying.
protégé (abrité), sheltered, enclosed.
Protérozoïdes (tecto.), Proterozoides.
Protérozoïque, Proterozoic.
Protiste (pal.), Protista.
protobastite (minér.), enstatite (var.).
protoclase, protoclase.
protoclastique, protoclastic.
protocoquille, protoconque, protoconch.
protogine (pétro.), protogine.
protomylonite, protomylonite.

proton, proton.
protopétrole, protopetroleum.
Protozoaire (pal.), Protozoa.
protrusion, protrusion.
protubérance, knob; **p. solaire,** solar prominence.
proustite, proustite.
province métallogénique, metallogenic province.
province pétrographique, petrographic province.
provincialité (pal.), provinciality.
psammite, psammite, micaceous flagstone.
pséphite (pétro. séd.), psephite.
pséphitique, psephitic.
pseudobrèche, pseudobreccia.
pseudo-clivage, pseudo cleavage.
pseudo-cristallin, pseudocrystalline.
pseudo-fossile, pseudo-fossil.
pseudo-gley, pseudo-gley.
pseudo-karst, pseudokarren.
pseudo-lapiez, pseudo-karren.
pseudomorphe, pseudomorph.
pseudomorphisme, pseudomorphism.
pseudomorphose, chemical substitution, mineral substitution.
pseudo-schistosité, pseudo-lamination.
pseudo-tachylite, (minér.), pseudotachylite.
psilomélane, (minér.), psilomelane.
Psilopsidés (pal.), Psilopsidae.
Ptéridophytes (pal.), Pteridophyta.
Ptéridospermées (pal.), Pteridospermae.
Ptérocérien, Pterocerian.
Ptérodactyle, Pterodactyl.
Ptéropode (pal.), Pteropod; **vase à Ptéropodes,** Pteropod ooze.
Ptérosauriens (pal.), Pterosauria.
ptygmatique, ptygmatic.
puisage (de l'eau, mine), drawing water.
puisard, pit; **1.** collecting pit, sink, hole, sump, solution cave; **2.** disposal well.
puiser (de l'eau), to scoop out, to draw.
puissance (épaisseur d'une couche), thickness; **p. (de transport fluviatile),** competence, carrying power.
puissant (épais), thick.
puits, **1.** well (of water, of oil); **2.** shaft, pit (*mine*); **p. à balancier,** beam w.; **p. absorbant (karst),** sink-hole; **p. artésien,** artesian w.; **p. d'aération,** air-shaft; **p. d'eau,** water-shaft; **p. d'exhaure (mine),** pumping shaft; **p. d'exploration,** test w.; **p. d'extraction,** drawing shaft, extraction shaft, hoisting shaft; **p. de gaz,** gas w., gasser; **p. d'injection,** injection w.; **p. de mine,** mine shaft; **p. de pétrole,** oil w.; **p. dévié,** directional w.; **p. épuisé,** exhausted w.; **p. éruptif,** gusher, flowing w.; **p. fou,** wild w.; **p. intermittent,** intermittent w.; **p. jaillissant,** gusher.; **p. naturel,** natural w.; **p. perdu,** disposal w.; **p. pompé,** pumping w.; **p. producteur,** producing w.; **p. sec, tari,** nonproducing w., dry hole.
pulaskite (var. syénite), pulaskite.
Pulmonés (pal.), Pulmonata.
pulsation, pulsation.
pulvérulent, powdery, dusty.

pulsar (astro.), pulsar.
pumpellyite (minér.), pumpellyite.
pur (minerai), pure, native.
Purbeckien, Purbeckian (u. Jurassic).
purification, purification, purifying; **p. à l'eau**, elutriation.
purifier (un minerai), to purify, to treat.
putride (eau), putrid.
puy (volcanique), puy, dome, cone, neck.
pygidium (pal.), pygidium.
pyralspite (grenat), pyralspite.
pyramide, pyramid; **p. d'érosion**, earth pillar, erosion column; **p. hexagonale**, hexagonal pyramid.
pyrargyrite, pyrargyrite, red silver ore.
pyrénéen (glacier), hanging glacier.
Pyrénéenne (orogenèse), Pyrenean Orogeny (Europe: u. Eocene).
pyriboles (minér.), pyriboles (pyroxenes + amphiboles).
pyrite, pyrite; **p. arsenicale**, mispickel; **p. blanche**, marcasite; **p. crêtée**, spear p.; **p. cuivreuse**, copper p., chalcopyrite; **p. ferreuse**, iron p.; **p. grillée**, roasted p.; **p. magnétique**, pyrrhotite.
pyriteux, pyritaceous, pyritic.
pyritisation, pyritization.

pyritoèdre, pyritohedron.
pyrobitume, pyrobitumen.
pyrochlore, pyrochlore.
pyroclastique, pyroclastic; **brèche p.**, p. breccia; **coulée p.**, p. surge; **couronne p.**, tuff ring.
pyrocristallin, pyrocrystalline.
pyrogène, pyrogénique, pyrogenous, pyrogenetic, igneous.
pyrogenèse, pyrogenesis.
pyrolite, pyrolite.
pyrolusite, pyrolusite.
pyrolyse, pyrolysis.
pyromaque (silex), flint.
pyrométamorphisme, pyrometamorphism.
pyrométasomatique, formed by contact metamorphism.
pyromorphite (minér.), pyromorphite.
pyrope (grenat), pyrope.
pyrophyllite (phyllosilicate), pyrophyllite, pencil stone.
pyroschiste, oil shale, bituminous shale.
pyrosphère, pyrospher.
pyroxène (minér.), pyroxene.
pyroxénite (pétro.), pyroxenite.
pyrrhotine (minér.), pyrrhotite, magnetic pyrite.

Q

quadratique (système), quadratic.

quadrillage (cartogr.), grid, graticule, squaring.

quadriller (le terrain), to checker.

quantitative (analyse), quantitative.

quartz, quartz; q. à inclusions de rutile, rutilated q.; q. fumé, smoky q., morion; q. jaune, citrine; q. laiteux, milky q.; q. violet, amethyst; mine de q., q. mine.

quartzeux, quartzose; grès q., q. sandstone.

quartzifère, quartziferous.

quartzique, quartz-rich, quartzose; diorite q., q. diorite; monzonite q., q. monzonite.

quartzite, quartzite; q. métamorphique, metaquartzite; q. sédimentaire, orthoquartzite; grès-q., quartzitic sandstone.

quartzitique, quartzitic.

quartzo-feldspathique, felsic.

Quaternaire, Pleistocene (obsolete usage); Quaternary.

queue (de distillation), end products, bottom.

queue de comète (astro.), trailing spit.

queue de cristallisation, pressure fringe.

R

rabattement de nappe (hydro.), drawdown, lowering of ground water level.

rabotage glaciaire, subglacial planning, subglacial polishing.

raccord de tiges de sonde, drill rod bushing.

racine (de méandre), neck; **r. (de nappe de charriage)**, root.

raclage (de morts terrains), scraping.

racler (trav. publics), to scrape.

radarastronomie, radarastronomy.

radargraphie, radargraphy.

radeau de glace, ice raft; **bloc transporté par r. de g.**, i. rafted block.

radial, radié, radial; **faille r.**, r. fault; **réseau hydrographique r.**, r. pattern; **structure radiée**, radiolitic structure.

radiation, radiation; **r. cosmique**, cosmic r.; **détecteur de r.**, r. meter; **diagraphie par r.**, r. logging.

radier (géotechnie), invert.

radioactif radioactive; **datation au carbone r.**, radiocarbon dating (jamais « datation »); **élément r.**, radiometric element; **isotope r.**, radiogenic isotope.

radioactivité, radioactivity; **diagraphie par r.**, r. logging.

radioastronomie, radioastronomy.

radiocarottage, gamma ray logging.

radiochronologie, radiodating.

radiographie, radiography.

radiographique, radiographic.

radiointerferomètre, radiointerferometer.

radioisotope, radioisotope.

Radiolaires (pal.), Radiolaria; **vase à R.**, radiolarian ooze.

radiolarite, radiolarite, radiolarian chert.

radiole (pal.), radiole.

radiométrique, radiometric.

radiophotographie, radiophotography.

radium, radium.

radon (gaz), radon.

radula (pal.), radula.

rafale (météo), gust of wind, blast.

raffinage, refining; **r. du pétrole**, oil r.; **r. en phase vapeur**, vapor phase r.

raffiner, to refine.

raffinerie, refinery, refining plant.

rafraichissement (météo.), cooling.

raide (pente), steep.

raie de spectre, spectral line.

rai sismique, sismic ray.

rainure (figure séd.), groove, furrow, cast.

rajeunir (le relief), to rejuvenate.

rajeunissement (du relief), rejuvenation.

ramasser (des échantillons), to collect, to gather.

ramification (d'un filon), ramification, ramifying, branching, offshoot, splitting.

ramifié (cours d'eau), divided.

ramifier, se ramifier, to ramify, to branch out, to divide.

ramollissement (du sol), softening.

randannite, randanite (var. of diatomite).

rang (d'un charbon), rank (coal).

ranker (pédo.), ranker.

rapide, 1. adj: quick, fast; 2. n: rapid (of a stream).

rapport, ratio; **r. d'aplatissement (sédim.)**, flatness index, flatness ratio; **r. d'émoussé**, roundness ratio; **r. gaz-huile (pétrole)**, gas-oil ratio (G.O.R.).

rares (terres), rare (earth).

rasa (géomorpho.), rasa.

ratelier (à tiges de forage), rack.

Rauracien, Rauracian (u. Jurassic; m. Lusitanian).

ravin, ravine, gulch, gully; **r. sous-marin**, submarine canyon.

ravinement, gullying, channeling, cut-and-fill ravine, rill.

raviner, to ravine, to gully.

rayé (par un glacier), scratched, striped, grooved.

Rayleigh (onde de), Rayleigh wave.

rayon, ray; **r. cathodique**, cathode r.; **r. de lumière**, light r.; **r. extraordinaire**, extraordinary r.; **r. gamma**, gamma r.; **r. hydraulique**, hydraulic radius; **r. infra-rouge**, infra-red r.; **r. ordinaire**, ordinary r.; **r. terrestre**, earth radius; **r. ultraviolet**, ultraviolet r.; **r. X**, X r.

rayonnement, radiation.

rayure (glaciaire), stripe, streak, scratch, groove.

raz de marée, « tidal wave » (popular), tsunami.

réacteur (nucléaire), (nuclear) reactor.

réactif (chimique), (chemical) reagent, chemical reactant.

réaction, reaction; **r. chimique**, chemical r.; **r. de coloration**, staining test; **r. en chaîne**, breeding r.; **r. équilibrée**, balanced r.; **r. exothermique**, exothermal r.; **r. réversible**, reversible r.

réactionnelle (auréole), reaction (rim).

réactivation (de déchets radioactifs), reactivation.

réactivation (d'une faille), rejuvenation.

réactivée (faille), rejuvenated.

réalgar (minér.), realgar, red arsenic.

réaménagement (environnement), reclamation.

rebondir (grain de sable), to rebound, to bounce.

rebord du talus continental, break line.

reboucher (un puits), to block up, to plug.

rebroussement (de couches), upturning.

rebut (de mine), waste, refuse, scrap.

Récente (époque), Holocene (epoch).

récepteur acoustique, acoustic receiver.

récepteur multicanal, multichannel receiver.

réception (bassin de), catchment area.

récession glaciaire, glacial retreat.

recette (mine), landing station.

recharge (des nappes), groundwater recharge.

réchauffement (climatique), warming (of climate, cf. «greenhouse effect»).

recherche research, prospecting; **r. fondamentale**, fundamental research; **r. pétrolière**, oil prospecting; **r. scientifique**, scientific research.

rechercher (sur le terrain), to search for, to prospect.

récessif (caractère), recessive (character).

récif, reef; **r. annulaire**, annular r.; **r. barrière**, barrier r.; **r. corallien**, coral r.; **r. externe**, forereef; **r. frangeant**, fringing r.; **r. interne**, backreef; **r. submergé**, drowned r., submerged r.; **platier du r.**, r. flat; **socle du r.**, r. basement.

récifal, reefal, biohermal; **brèche r.**, reef breccia.

reconnaissance (d'un secteur), prospecting, exploration, scouting; **carte de r.**, r. map; **étude de r.**, r. survey; **mission de r.**, r. survey.

reconstitution paléogéographique, paleogeographic reconstruction.

recoupe (mine), cross drift, cross cut.

recoupement (de méandre), cut-off.

recouper (un filon), to intersect, to recut; **r. (un méandre)**, to cut off.

recouvrement (de terrain), over-lap, covering, overlay; **r. horizontal (d'une faille inverse)**, heave; **lambeau de r.**, thrust outlier; **roches de r. (mortterrain)**, overburden; **terrain de r.**, hanging wall.

recouvrir (un terrain), to overlap.

recristallisation, recrystallization, rejuvenation; **r. postérieure à un plissement**, post-tectonic recrystallization.

recristalliser, to recrystallize.

recueillir (des échantillons), to collect, to gather.

recul, recess, retreat, recession; **r. glaciaire**, glacier retreat, recession; **r. marin**, regression; **rivage en r.**, retrograding shoreline.

reculée (Jura), precipitous blind valley, dead end, reculée.

récupérable (pétrole), recoverable, retrievable.

récupération, recovery; **r. de charbon**, reclamation; **r. primaire (pétrole)**, primary r.; **r. secondaire**, secondary r.; **taux de r.**, r. ratio.

récupérer (du pétrole), to recover.

récurrence, recurrence.

récurrente (faune), recurrent (fauna).

recuveler (un puits), to recase.

redéposer, se redéposer, to redeposit.

redevance (pétrolière), (oil) royalty.

redissolution, resolution.

redissoudre, to resolve.

redistiller, to rerun.

redressement (de photographies aériennes), rectification.

redresser (une photographie aérienne), to rectify; **r. (un forage)**, to straighten.

réducteur (agent), reducing (agent).

réduction (chimique), reduction, reducing; **r. de Bouguer**, Bouguer reduction; **r. par l'hydrogène**, hydrogenation.

réduire, to reduce.

réduit (adj.), reduced.

référence (plan de), reference (plane).

réfléchi (rayon), reflected.

réfléchir (la lumière), to reflect; **r. (penser)**, to think.

réflectance, reflectance.

réflecteur (sismique), reflector, mirror.

réflectivité, reflectivity.

réflexion, reflection; **r. multiples**, multiple r.; **r. totale**, total r.; **sismique r.**, r. shooting; **taux de r.**, reflectance.

refluer (mer), to ebb, to surge, to flow back.

reflux (de la marée), ebb.

refondre, to remelt.

refonte, remelting.

reforage, reaming, redrilling.

reforer, to redrill.

réfractaire, refractory, fire-resisting; **argile r.**, r. clay; **matériaux r.**, r. materials; **qualité r.**, refractoriness; **sable r.**, refractory sand.

réfractée (onde), refracted.

réfracter, se réfracter, to refract.

réfraction, refraction; **r. de la schistosité**, cleavage r.; **r. du clivage schisteux**, cleavage r.; **double r.**, double r.; **indice de r.**, refractive index; **sismique r.**, r. shooting.

réfringence (opt.), refractivity, refringence.

réfringent, refringent.

refroidie (lave), cooled.

refroidir (météo.), to cool, to grow cold.

refus (de tamisage), oversize, refuse, screenings.

reg, reg, desert gravel.

regard (d'une falaise), front (of a cliff).

regel, freezing again, regelation.

régime, regime, kind of flow; **r. climatique**, climatic r.; **r. fluviatile**, river flow; **r. fluviométrique**, rainfall r.; **r. laminaire**, sheet flood; **r. torrentiel**, torrential flow; **r. turbulent**, eddy flow.

région, country, area, region; **r. aride**, arid area; **r. désertique**, desert area; **r. minière**, mining district; **r. sismique**, seismic area.

règle des phases (minér.), phase rule.

règlements miniers, mining regulations.

régolithe, régolite, waste mantle, regolith.

régosol, regosol.

règne minéral, mineral kingdom.

regradation, regradation.

régressif, regressive; **érosion r.**, retrogressive erosion; **métamorphisme r.**, diaphthoresis.

régression, regression, offlap, regressive overlap, reliction.

régularisation (du profil), grading, water-course regulation.

régularisation (du littoral), coastal grading.

régulariser une pente, to grade.

régulier (écoulement), steady (flow).

régulier (Oursin), endocyclic (sea-Urchin).

réinjection de gaz (forage), reinjection of gas.

rejeu de faille, recurrent faulting, playback (of the fault).

rejet (tecton.), throw; **r. d'eau usée,** effluent; **r. de faille,** fault t.; **r. fractionné,** distributive faulting; **r. horizontal,** heave, fault heave, horizontal displacement; **r. net,** net slip; **r. stratigraphique,** stratigraphic t.; **r. vertical,** vertical t.

relâchement des contraintes, strain release.

relais de failles, echelon faults.

relatif, relative; **âge r.,** r. age; **chronologie r.,** r. chronology; **datation r.,** r. dating; **perméabilité r.,** r. permeability.

relevé de terrain, ground plotting.

relevé hydrothimétrique, river record; **r. nivométrique,** snow survey.

relèvement à la boussole, compass bearing; **r. axial (d'un pli),** updipping (of the axis of a fold); **r. géophysique,** geophysical survey.

relief, relief, topography; **r. accidenté,** hummocky topography; **r. conforme,** conformable relief; **r. escarpé,** steep topography; **r. faible,** faint topography; **r. jeune,** young relief; **r. d'un minerai en lame mince,** optical relief; **r. résiduel,** residual hill; **r. tabulaire,** table land; **carte en r.,** relief map; **inversion du r.,** inverted relief.

reliquat magmatique, magmatic residue.

relique (forme), relic, relict.

rémanent (magnétisme), remanent (magnetization).

remanié (fossile), reworked, rehandled.

remaniement, reworking.

remblai, packing, filling, fill, backfill.

remblaiement, remblayage, packing, filling, stowage, aggradation; **matériaux de r.,** backfilling material.

remblaiement hydraulique, hydraulic backfilling.

remblayer, to stow, to fill, to pack.

remblayeuse mécanique, stowing machine.

rembourrage (mine), stuffing, padding.

remontée, climb, rise, upraise; **r. d'eau profonde,** upwelling; **r. du minerai,** ore raising; **r. du train de tiges,** pull out; **r. océanique,** oceanic rise.

remonter du minerai, to hoist ore.

remous (océano.), eddy.

rempart morainique, boulder wall, arcuate wall.

remplacement (d'un minéral), replacement.

remplissage d'un filon, lode filling.

remplissage d'un terrier (fig. séd.), dwelling trace.

remplissage karstique, karstic filling.

rendzine, rendzina; **r. blanche,** white r.; **r. brune,** brown r.; **r. vraie,** true humus calcareous soil.

rendzinification, rendzinification.

renflement, bulge, bulging, swelling.

renouvellement (des fonds océaniques), sea-floor spreading.

rentrant (d'une falaise), recess, reentrant.

renversé (pli), overturned, overtilted, overthrown (fold).

renversement (de terrain), overturn, reversal.

renverser, to overturn, to reverse.

réouvrir (un puits de mine), to reopen.

répartition bathymétrique, depth range.

répartition granulométrique, grain-size distribution.

repêchage (forage), fishing.

repêcher des outils, to fish up tools.

repère (topographique), topographic landmark; **couche r.,** key bed.

replat (topo.), bench, flat.

repli, fold.

réplique (sismique), after-shock.

replissement, refolding.

reposer sur (strati.), to overlie.

réponse spectrale, reflectance.

représentation triangulaire, triangular diagram, triangular plotting.

reptation (des sols), creeping, creep.

Reptile (pal.), Reptile, Reptilia.

reptilien, reptilian.

reséquent, resequent.

réseau, network, lattice; **r. avec triage (périgl.),** sorted net; **r. cristallin,** crystal lattice; **r. de détection sismique,** phase-array station; **r. de failles,** fault network; **r. de fentes de gel,** ice-wedge polygon; **r. de filons,** vein network; **r. (de mesures),** network; **r. de pierres,** stone net; **r. fluviatile,** network; **r. fluviatile confluent,** contributive network; **r. fluviatile diffluent,** distributive network; **r. hydrographique,** drainage pattern; **r. sans triage,** non-sorted net; **r. confluent,** contributive network; **r. diffluent,** distributive network.

resédimentation, redeposition.

réserves, reserves; **r. d'eau,** water supply, storage; **r. de gaz (naturel),** gas r.; **r. de minerai,** ore r.; **r. de pétrole,** oil r.; **r. d'uranium,** uranium r.; **r. probables,** probable r.; **r. récupérables,** recoverable r.

réservoir (naturel), reservoir; **r. de gaz naturel,** gas r.; **r. de pétrole,** oil r.; **r. magmatique,** magmatic chamber; **étude de r.,** r. engineering; **roche r.,** r. rock.

résidu, residue, remnant; **r. d'altération,** residual deposit, weathering residue; **r. de distillation,** residue, tailings, bottom; **r. d'exploitation minière,** waste, spall; **r. de craquage,** cracking residuum; **r. de déflation,** lag gravel.

résiduel (minéral, etc.), relic, residual, residuary; **argile r.,** residual clay; **pergélisol r.,** relic permafrost; **sol r.,** residual soil; **structure r.,** residual structure.

résine, resin; **r. de pétrole,** petroleum r.; **r. échangeuse d'ions,** ion exchange r.; **r. fossile,** fossil r.; **r. minérale,** mineral r., natural r.; **r. végétale,** natural r.

résinite, resinite.

résistance, resistance, strength; **r. à l'écrasement,** crushing strength; **r. à l'érosion,** abrasion resistance; **r. à la rupture,** breaking strength; **r. à la traction,** tensile strength; **r. au cisaillement,** shearing strength; **r. au gel**, frost strength; **r. au glissement**, slide resistance; **r. du sol**, ground strength.

résistant, resistant; **r. à l'érosion**, non-erodible; **r. au gel**, frost proof; **r. aux séismes**, quake proof.

résistivité (d'une couche), resistivity.

résorption (de la croûte), melting (of the crust).

ressac, undertow, surf.

ressaut, nip, scarp, rock step.

resserré (vallon), narrow, confined.

ressources en eau, water resources, storage.

ressources minérales, mineral resources.

ressuyage (pédo.), soil-water movement.

restes fossiles, fossil remains.

restitution (photogrammétrique), restitution.

résurgence, resurgence, exit of underground stream.

résurgent, resurgent.

retard (géoph.), delay, time lag; **r. d'arrivée des ondes P**, P-wave delay; **r. de longueur d'onde**, optical retardation; **r. de phase**, phase lag; **r. des ondes S sur P**, S-P interval; **détonateur à r.**, delay blasting cap.

rétention (de l'eau), retention (of water).

retenue (barrage de), dam, reservoir.

réticulaire (structure), reticulate (structure).

réticule (d'un oculaire), cross-wires, reticle, cross-hair.

réticulé (sol), reticulated (soil).

rétinite, retinite, pitch stone.

retirer (mine), to withdraw, to extract; **r. un tubage d'un puits**, to strip out a well.

retombées (volc.), fallout.

retouche (préhist.), flaking, retouch.

retour de courant de vagues, undertow.

retour de vague, backwash.

retrait (contraction du sol), shrinkage, shrink; **r. glaciaire**, glacier retreat; **r. des vagues**, backswash; **fentes de r.**, sun cracks; **moraine de r.**, recessional moraine.

retraitement (de combustibles nucléaires), reprocessing.

rétrécissement (d'une couche), narrowing, pinching out.

rétrocharriage, backthrusting.

rétrochevauchement, backthrust.

rétrograde (métamorphisme), retrogressive metamorphism, diaphthoresis.

rétrométamorphisme, retrograde metamorphism.

rétromorphose, retromorphosis, diaphthoresis.

rétrosiphoné (pal.), retrosiphonate.

rétrotectonique, palinspatic.

revers de cuesta, back slope.

revêtement, coating; **r. argileux**, clay c.; **r. de puits de mine**, shaft lining; **r. de puits de pétrole**, well casing; **r. d'une fente d'un sol**, c.

révolution (astronomique), revolution.

révolution (orogénique), revolution, orogenesis.

rhabdosome (pal.), rhabdosome.

Rhétien, Rhaetian, Rhaetic (u. Triassic).

rhexistasie, erosion and weathering following an uplift.

Rhizopodes (pal.), Rhizopoda.

rhodocrosite (minér.), rhodocrosite.

rhodolite (cf. maërl), rhodolite.

rhodonite (minér.), rhodonite.

Rhodophycées (paléobot.), Red algae.

rhombique, rhombic; **dodécaèdre r.**, rhombic dodecahedron.

rhomboèdre, rhombohedron.

rhomboédrique, rhombohedral, rhomboidal.

rhomboïdal, rhombic.

rhombophyre, rhombenporphyry.

rhyncholite, rhyncolith.

Rhynchonellacés (pal.), Rhynchonelloid.

rhyodacite (pétro.), rhyodacite.

rhyodacitique, rhyodacitic.

rhyolite (pétro.), rhyolite.

ria, ria (narrow drowned valley, non-glacial).

richesse minérale, mineral wealth.

Richter (échelle), Richter (scale).

ride, ripple; **r. à crête rectiligne**, straight crested r.; **r. de courant**, current r.; **r. des mers lunaires**, wrinkle ridges; **r. de plage**, r. mark; **r. de sable**, sand r.; **r. de vague**, wave r.; **r. dissymétrique**, assymetric(al) r.; **r. d'oscillation**, oscillatory r.; **r. en croissant**, lunate r.; **r. éolienne**, wind r.; **r. géante**, giant r.; **r. grimpante**, climbing r.; **r. médio-océanique**, mid-oceanic ridge; **r. sinueuse**, undulatory r.; **r. symmétrique**, symmetric r.; **longueur d'onde des r.**, ripple length.

rideau (Picardie), step.

riden, megaripple.

riébeckite (minér.), riebeckite.

rigidité (des matériaux), stiffness, rigidity; **module de r.**, rigidity modulus, young modulus.

rigole, rill, small ravine, small gully; **r. de plage**, rill wash; **r. de ruissellement**, rain rill, gully.

rigoureux (climat), hard.

rimaye (glaciaire), bergschrund.

ripidolite (minér.), ripidolite.

risque naturel, natural hazard.

risque sismique, seismic hazard.

Riss (glaciation), Riss (Pleistocene: penultimate glacial stage).

rissien, rissian.

rivage, coast, shore, shoreline; **r. régularisé**, graded shoreline; **ligne de r.**, shoreline.

rive (d'un fleuve), bank; **r. concave**, outer b.; **r. concave érodée**, cut side; **r. convexe**, inner b., alluviated b.

rivière, river; **r. à marée**, tidal r., creek; **r. « boudinée »**, roved r.; **r. captée**, beheaded r.; **r. régularisée**, graded r.; **r. souterraine**, underground r.

roc, rock.

rocaille, rubble, rock debris.

rocailleux, rocky, stony, bouldery.

roche, rock; **r. abyssale**, abyssal r.; **r. acide**, acid r.; **r. arénacée**, sandy r.; **r. argileuse**, clayed r.; **r. asphaltique**, asphaltic r.; **r. autochtone**, autochthonous r.; **r. basaltique**, basaltic r.; **r. basique**, basic r.; **r. bitumineuse**, bituminous r.; **r. champignon**, mushroom r.; **r. couverture**, cap-r.; **r.

corallienne, coral r.; **r. cristalline,** crystalline r.; **r. cristallophylienne,** metamorphic r.; **r. d'épanchement,** effusive r., volcanic r.; **r. de demiprofondeur,** hypabyssal r.; **r. de profondeur,** plutonic r.; **r. du socle,** bed r., basement r.; **r. détritique,** clastic r.; **r. encaissante,** country r., enclosing r.; **r. endogène,** endogeneous r.; **r. éruptive,** eruptive r., igneous r.; **r. extraterrestre,** meteorite; **r. filonienne,** dyke r., dike r. (U.S.); **r. granitique,** granitic r.; **r. grenue,** granular r.; **r. ignée,** igneous r.; **r. intrusive,** intrusive r.; **r. leucocrate,** leucocrate r.; **r. magazin,** reservoir r.; **r. mère,** parent r., source r.; **r. mélanocrate,** melanocratic r.; **r. métamorphique,** metamorphic r.; **r. monominérale,** monomineralic r.; **r. monogénique,** monogeneous r.; **r. moutonnée,** icesmoothed r.; **r. néovolcanique,** cainozoic (cenozoic) volcanic r.; **r. neutre,** intermediate r.; **r. organogène,** biogenic r.; **r. paléovolcanique,** precainozoic (pre-cenozoic) volcanic r.; **r. pétrolifère,** oil bearing r.; **r. plutonique,** plutonic r.; **r. pourrie,** rotten r., decayed r.; **r. pyroclastique,** pyroclastic r.; **r. réservoir,** reservoir r.; **r. saine,** fresh r.; **r. saline,** saline r.; **r. sans quartz,** quartzless r.; **r. sédimentaire,** sedimentary r.; **r. siliceuse,** silicic (siliceous) r.; **r. silicifiée,** silicified r.; **r. stratifiée,** stratified r.; **r. stérile,** barren r.; **r. striée,** striated r.; **r. ultrabasique,** ultrabasic r.; **r. verte,** green r.; **r. vitreuse,** glassy r.; **r. volcanique,** volcanic r.

rocher, rock, stone; **r. branlant,** rocking stone; **galerie au r.,** drift.

Rocheuses (montagnes), Rocky mountains.

rocheux, rocky; **masse r.,** rock mass.

rodite (var. de météorite), rodite.

rognon de minerai, kidney; **r. de silex,** flint nodule.

rond-mat (grain), round-frosted (grain).

rongé (rocher), corroded.

ronger (une roche), to corrode, to etch, to wear away.

Rongeur (pal.), Rodent.

rose, 1. adj: pink; 2. n: rose; **r. des sables,** gypsum rosette; **r. des vents,** compass card.

rosée, dew; **point de r.,** d. point.

rostre (pal.), rostrum.

Rotalidés (pal.), Rotalideae.

rotation (faille de), rotational fault; **r. (table de),** rotary table.

rotatoire (dispersion), rotatory (dispersion).

rouge, red.

  argile rouge, red clay.

  boue rouge, red mud.

  formations rouges (Trias), red beds.

  hématite rouge, red hematite.

rougines provençales, badlands.

rouille, iron rust.

roulé (fossile), rolled, worn by rolling.

rouleaux, rods, roddings.

roulement (sur le fond d'un fleuve), rolling.

ru, ruz, brooklet, gully.

ruban de sable, sand ribbon.

rubanée (structure), banded, stripped; **roche r.,** ribbon rock; **structure r.,** ribbon structure.

rubéfaction (pédo.), rubefaction, reddening.

rubellite, rubellite (tourmaline).

rubicelle (minér.), rubicelle.

rubidium, rubidium.

rubis, ruby; **r. de bohême,** rose quartz; **r. du Brésil,** burnt topaz; **r. oriental,** oriental r.; **r. spinelle,** spinel r.

Rudistes (pal.), Rudistids.

rudite, rudite, rudyte, rudaceous rock.

ruiniforme (paysage), tower-like.

ruisseau, brook, runnel, rivulet.

ruisselant, trickling.

ruissellement, running off, trickling, rain wash, runoff; **r. concentré,** rill wash; **r. diffus,** unconcentrated wash, rainwash; **r. en nappe,** sheet wash; **r. nival,** snowmelt wash; **r. pluvial,** rainwash; **r. retardé,** subsurface runoff.

ruisseler, to run down, to stream.

ruisselet, rivulet, brooklet.

Rupélien, Rupelian (m. Oligocene: equiv. Stampian).

rupestre, parietal.

rupture de pente, nickpoint, break of slope.

rutile (minér.), rutile; **inclusion de r.,** r. inclusion.

rythme de sédimentation, rate (of layering).

rythmique (sédimentation), rhythmic (settling).

S

Saalienne (glaciation), Saalian (glaciation, equiv. Rissian, pre-Eemian).

sable, sand; **s. aquifère,** water-bearing s.; **s. argileux,** clayey s.; **s. asphaltique,** asphaltic s.; **s. aurifère,** gold-bearing s.; **s. bitumineux,** bituminous s.; **s. boulant,** quick s.; **s. colmaté,** tight s.; **s. consolidé,** grit; **s. coquillier,** shelly s.; **s. de couverture,** cover s.; **s. désertique,** desert s.; **s. dunaire,** dune s.; **s. éolien,** eolian s.; **s. fin,** fine-grained s.; **s. fluent,** quicks.; **s. glaiseux,** clayey s.; **s. glauconieux,** green s.; **s. gleyifié,** gleyed s.; **s. grossier,** coarse grained s.; **s. lacustre,** lacustrine; **s. limoneux,** loamy s.; **s. mouvant,** running s., quick s., drifting s.; **s. moyen,** medium grained s.; **s. nivéo-éolien,** niveo-eolian s.; **s. perméable,** open s.; **s. pétrolifère,** oil-bearing s.; **s. phosphaté,** phosphatic s.; **s. poreux,** open s.; **s. quartzeux,** quartzose s.; **s. vaseux,** muddy s.; **s. volcanique,** volcanic ash; **bain de s. (labo.),** s. bath; **banc de s.,** s. bank; **barre de s.,** s. bar; **dune de s.,** s. dune; **flèche de s.,** s. spit; **lentille de s.,** s. lens; **tempête de s.,** s. storm; **tubulure de s.,** s. gall; **vent de s.,** s. drift.

sableux, sandy, sabuline, arenaceous; **argile s.,** clay; **calcaire s.,** s. limestone; **limon s.,** s. loam; **marne s.,** s. marl; **sol s.,** s. soil; **vase s.,** s. mud.

sablier (structure en), hourglass structure.

sablière, sand pit.

sablon, very fine sand.

sablonneux, finely sandy, sabulous.

sablonnière, sand pit.

saccharoïde, sugary-grained, saccharoid(al); **cassure s.,** saccharoidal fracture; **marbre s.,** saccharoidal marble; **structure s.,** saccharoidal texture.

sagénite (var. de rutile), sagenite.

sahlite (minér.), sahlite.

saignée, 1. kerf, kerving (mines); 2. trench, ditch.

saillant anticlinal, anticlinal bulge.

saillie (rocheuse), spur, outcrop.

saisie des données (inform.), encoding.

saison de gel, frost season.

saisonnier (écoulement), seasonal (run-off).

Sakmarien, Sakmarian (l. Permian).

salant (marais), salt (swamp).

salbande, salband, selvage, self-edge, vein wall, clay gouge.

salé, salted, saline; **eau s.,** saline water; **lac s.,** saline lake; **pré s.,** saline pasture; **source s.,** saline spring.

salifère, saliferous, salt bearing. **bassin salifère,** salt bottom.

salin, saline, briny; **roche s.,** saline rock; **sol s. acide,** salt earth podzol, soloth.

saline, salt works, salina, saltern.

salinelle, salse, mud-volcano.

salinifère, saliniferous.

salinisation, salinization.

salinité, salinity, saltness.

salite (minér.), salite.

salmiac (minér.), salmiac.

salpêtre, niter, saltpetre; **s. du Chili,** soda niter.

salpêtreux, salpetrous.

salpêtrière, salpeter works.

salse, salse, mud-volcano.

saltation, saltation (of a grain).

salure, salinity, salinization.

samarskite (minér.), samarskite.

sanidine (minér.), .sanidine, rhyacolite.

sanidinite (r. métamorphique), sanidinite.

Sannoisien (Oligocène inférieur), Sannoisian.

Santonien, Santonian.

sans, less, free of, without; **s. feldspaths,** feldspar-free; **s. quartz,** quartz-less.

sapement, undermining, undercutting, sapping.

saper, to sap, to undermine.

saphir, sapphire; **s. d'eau,** water s.; **s. de Ceylan,** salamstone s.; **s. oriental,** blue s.

saponite (minér.), saponite.

saprolite, saprolite, rotted rock in situ (cf. regolith).

sapropèle, sapropel.

sapropélique, sapropelic.

sapropélite, sapropelite.

sardoine, sar, sardonyx.

Sarmatien, Sarmatian (u. Miocene).

satellite, satellite.

saturant, saturating, saturant.

saturation, saturation; **s. en eau,** water s., waterlogging; **degré de s.,** s. degree; **facteur de s.,** s. factor; **pression de s.,** s. pressure; **taux de s.,** s. indice; **zone de s.,** s. zone.

saturé, saturated; **s. en eau,** waterlogged; **hydrocarbure s.,** s. hydrocarbon.

saturer, to sature.

saumâtre, brackish, briny; **lac s.,** salt pan.

saumure, brine, salt brine.

saunerie, salt refinery, salt works.

saupoudrage (volcanique), dusting, blanket.

Saurischiens (pal.), Saurischia.

Sauroptérygiens (pal.), Sauropterygia.

saussurite (pétro.), saussurite.

saussuritisation, saussuritization.

sauter (faire) (mine), to blow up, to blast.

savane (géomorpho.), savanna.

Saxonien (Permien), 1. Saxonian (m. Permian); 2. orogenic type (n. Germany); 3. polymetamorphic facies (Saxony).

saxonite (pétro.), saxonite.

scalénoèdre, scalenohedron.

Scaphopode (Mollusque), Scaphopoda.

scapolite, scapolite, scapolith.

scellement, sealing.

scènes (landsat), scenes, views.
scheelite, scheelite.
scheidage (mine), sorting, cobbing, bucking.
scheider, to sort, to cob.
scheideur, ore-sorter, cobber.
schillerisation, schillerization.
schillerspath, schiller spar.
schiste, schist, slate (metam.), shale (compact), fissile clay; **s. à chlorite**, chloritic schist; **s. à grenats**, garnetiferous schist; **s. à hornblende**, hornschist; **s. à séricite**, sericite schist; **s. alunifère**, alum shale; **s. ampéliteux**, ampelitic shale; **s. ardoisier**, slate; **s. argileux**, shale, mudstone; **s. bitumineux**, oil shale, bituminous shale; **s. bleuté**, blue schist; **s. carton**, paper schist; **s. charbonneux**, carbonaceous shale; **s. chloriteux**, chlorite schist; **s. cristallins**, metamorphic rock, metasediments; **s. cuprifère**, copper schist; **s. graphitique**, graphitic schist; **s. houiller**, carboniferous shale; **s. kérobitumineux**, oil shale, **s. métamorphique**, metamorphic schist; **s. micacé**, micaceous schist; **s. noduleux**, knotted schist; **s. pyrobitumineux**, pyrobituminous schist; **s. talqueux**, talcschist; **s. tacheté**, spotted schist; **s. vert**, greenstone schist.
schisteux, schistose, schistous, slaty, shaly; **roche schisteuse**, schistic rock, metamorphic rock.
schistosité, schistosity; **s. de fracture**, fracture cleavage; **s. de flux**, foliation; **s. de plan axial**, axial plane cleavage; **s. en éventail**, cleavage fan.
Schizodonte (Mollusque), Schizodont.
schorl (minér.), schorl, black tourmaline.
schorlacé, schorlaceous.
schorlifère, schorliferous.
scie (à pierres), (stone) saw; **s. diamantée**, diamond s.
scier (une roche), to saw.
scintillation (compteur à), scintillation (counter).
scintillomètre, scintillometer.
scolécodonte (pal.), scolecodont.
scoriacé, scoriated, scoriaceous, scorious, slaggy.
scorie industrielle, slag, clinker; **s. volcanique**, scoria.
scorification (industr.), slagging, scorification.
scorodite, scorodite.
Scyphozoaires, Scyphozoa.
Scythien, Scythian (l. Triassic).
séchage à l'étuve, stove drying.
sécher, assécher (un fleuve), to dry.
sécheresse, aridity, drought.
séchoir à minerai, ore-drying.
sebhka, sebhka (n. Africa); sabkha (Arabia).
Secondaire (ère), Mesozoic (era).
secouer (sur un crible), to shake.
secousse, shock, shaking, jog; **s. d'explosion de mine**, rock burst; **s. sismique**, shock, earth tremor.
section mouillée (hydro.), wetted area.
section polie, polished section.
section transversale, cross section.
séculaire (variation), secular (variation).
sédigraphe (séd.), sedigraph.
sédiment, sediment, deposit; **s. classé**, graded deposit; **s. continental**, land deposit; **s. détritique**, detrital deposit, clastic deposit; **s. éolien**, eolian deposit; **s. euxinique**, euxinic deposit; **s. fluviatile**, alluvial deposit; **s. marin**, marine deposit; **s. stratifié**, layered deposit, bedded deposit, stratified deposit.
sédimentaire, sedimentary; **bassin s.**, s. basin; **cycle s.**, s. cycle; **manteau s.**, s. mantle; **piège s.**, s. trap; **roche s.**, s. rock; **structure s.**, s. structure.
sédimentation, sedimentation, settling; **s. abyssale**, abyssal s.; **s. marine**, marine s.; **s. marine négative**, sequence coarsening upward; **s. marine positive**, sequence fining upward; **s. terrigène**, terrigenous s.; **balance à s.**, s. balance; **courbe de s.**, s. curve; **essai de s.**, s. test; **lacune de s.**, s. break; **taux de s.**, s. rate.
Sedimentologie, Sedimentology.
segment (pal.), segment.
ségrégation, segregation; **s. de glace**, ground ice s.; **s. magmatique**, magmatic s.; **filon de s.**, segregated vein.
seiche (géogr.), seiche.
seif (dune), seif, sif, longitudinal dune.
seislog, sieslog.
séisme, seism, earthquake, seismic event; **s. profond**, deep seism; **s. sous-marin**, submarine seism, seaquake; **s. superficiel**, shallow seism, surface earthquake.
séismicité, seismicity, seismism.
séismique, seismic(al), seismal; **activité s.**, seismic activity; **bruit s.**, seismic noise; **carte s.**, seismic map; **détecteur s.**, seismic detector; **discontinuité s.**, seismic discontinuity; **onde s.**, seismic wave; **vitesse s.**, seismic velocity; **zone s.**, seismic zone.
séismogramme, seismographic record, seismogram.
séismographe, seismograph; **s. électromagnétique**, electromagnetic s.; **s. vertical**, vertical s.
séismographie, seismography.
séismographique, seismographic.
séismologie, seismology.
séismologique, seismologic(al).
séismologue, seismologist.
séismomètre, seismometer.
sel, salt; **s. de Glauber**, mirabilite; **s. gemme**, sodium chloride, rock s., fossil s., halite; **s. marin**, sea s., marine s.; **croûte de s.**, s. crust; **culot de s.**, s. diapir; **dôme de s.**, s. diapir; **intrusion de s.**, s. intrusion; **mine de s.**, s. mine.
sélection (naturelle), selection.
sélénieux, selenious.
sélénifère, seleniferous.
sélénium, selenium.
séléniure, selenide.
sélénite (var. de gypse), selenite; **s. fibreuse**, satin spar.
séléniteux, selenitic.
sélénologie (étude de la Lune), selenology.
selle (anticlinale), (anticlinal) saddle.
selle (élément de la suture des Ammonites), saddle.
semelle (tectonique), sole thrust.
semelle (d'une couche), bedsole.
semi (préfixe), semi; **s. aride**, s.-arid; **s. cristallin**, s.-crystalline; **s. fluid**, s.-fluid; **s. marécageux (sol)**,

s.-swamp (soil); **s. planosol,** s.-planosol; **s. précieux,** s.-precious; **s. rigide,** s.-rigid; **s. transparent,** subtranslucent.

sénestre (faille), sinistral (fault).

sénile (géogr.), senile.

sénilité (géomorpho.), senility.

Sénonien, Senonian (v. Cretaceous).

sensible (gypse teinte), sensitive teint.

séparateur, separator; **s. de minerai,** ore s.; **s. hydraulique,** elutriator; **s. magnétique,** magnetic s.

séparation, separation, separating, parting; **s. centrifuge,** centrifugal separation; **s. des continents,** rifting; **s. électrolytique,** electrolytic parting; **s. magnétique,** magnetic separation; **s. par liqueurs lourdes,** heavy liquor separation; **s. par tamisage,** sieving, sizing process; **entonnoir à s.,** sorting funnel.

séparer (des fractions minérales), to sort, to segregate, to divide, to separate.

sépiolite, sepiolite.

septal (pal.), septal.

septaria, septaria.

septum, septa (pl.), septum, septa.

Séquanien, Sequanian.

séquence, sequence, succession, series; **s. climatique,** climosequence; **s. glaciaire,** glacial sequence; **s. lithologique,** lithosequence, lithogenic sequence; **s. négative,** offlap; **s. pédologique,** soil series; **s. pollinique,** palynologic s.; **s. positive,** onlap; **s. sédimentaire,** rhythmic sedimentation.

séquentiel (stade), sequential (stage).

sérac, serac.

séricite (mica), sericite.

sériciteux, sericitic.

séricitique, sericitic.

séricitisation, sericitization.

séricitoschiste (pétro.), sericite schist.

série, series, suite, set; **s. chevauchante,** overlapping series; **s. chevauchée,** overlapped series; **s. de failles,** fault set; **s. inverse (tecto.),** reversed series; **s. naphténique,** naphtene series; **s. pacifique (pétro.),** pacific suite; **s. paraffinique,** paraffin series; **s. sédimentaire,** series, sequence; **s. sédimentaire inverse,** downgraded sequence; **s. sédimentaire normale,** upgraded sequence.

serpentine (minér.), serpentine.

serpentineux, serpentinous.

serpentinisation, serpentinization.

serpentinisé, serpentinous.

Serpule (pal.), Serpula.

service géologique, geological survey.

sessile (pal.), sessile.

seuil, limit, sill, threshold; **s. de congélation,** freezing point; **s. de déformation,** yield limit; **s. du 0 °C,** zero °C curtain; **s. maritime,** shoal.

shale, shale.

sial, sial.

siallitique, siallitic; **sol s.,** siallitic soil.

siallitisation, siallitization.

sicula (pal.), sicula.

sidérite (météorite), siderite.

sidérolithique (argile), cainozoic clayey residue of weathering limestone.

sidéronatrite, sideronatrite.

sidérophile (élément), siderophile.

siderophyllite, siderophyllite.

sidérose, siderite.

sidérurgie, siderurgy.

Siegénien, Siegenian (l. Devonian).

Sienne (terre de), Sienna.

sierra, sierra.

signal (géoph.), signal, pulse.

signature spectrale (astro.), spectral reflectance feature, spectral fingerprint.

signe (optique), optical character.

silex, flint, flintstone, chert, silex; **s. de la craie,** flint (concretions, nodules); **s. pyromaque,** flintstone, gun flint; **argile à s.,** clay with flints; **nodule de s.,** flint nodule.

silexite (pétro.), silexite.

silicarénite (pétro.), silicarenite.

silicate, silicate; **s. en chaîne,** inosilicate; **s. en feuillets,** phyllosilicate; **nésosilicate,** nesosilicate; **phyllosilicate,** phyllosilicate; **tectosilicate,** tectosilicate.

silicaté, silicated.

silicatisation, silicatization.

silice, silica.

siliceux, siliceous, silicious; **boue s.,** siliceous ooze; **calcaire s.,** siliceous limestone; **concrétion s.,** siliceous c.; **éponges s.,** silicospongiae; **roche s.,** siliceous rock; **sable s.,** siliceous sand; **sol s.,** siliceous soil.

silicicalcaire, silicicalcareous.

siliciclastique, siliciclastic.

silicification, silicification, chertification.

silicifié, silicified wood; **bois s.,** silicified wood; **horizon s. induré,** silicified pan, silcrete.

silicifier, to silicify.

Silicoflagellés (pal.), Silicoflagellata.

silicose, silicosis.

sillimanite (minér.), sillimanite.

sillon, furrow, groove, trough; **s. d'érosion,** furrow cast; **s. d'estran,** runnel; **s. houiller,** coal belt; **s. pré-littoral,** off-shore trough; **s. tardiorogénique,** foredeep.

silt (terme anglais utilisé en français), silt.

silteux, silty.

Silurien, Silurian.

sima, sima.

simoun (vent), simoon.

simulation d'exploitation minière, mining simulation.

Sinémurien, Sinemurian.

sinopite (minér.), sinopite.

sinueux (cours), sinuate, sinuous.

sinuosité, sinuosity, meandering.

sinupallié, sinupalliate.

sinus palléal (pal.), pallial sinus.

siphon (pal.), siphon, siphuncle; **s. exhalant,** excurrent siphon; **s. inhalant,** incurrent siphon.

siphonal (canal), siphonal funnel.

sismique, seismic(al), seismal; **s. marine,** off-shore shooting; **s. réflexion,** reflexion shooting; **s. réfraction,** refraction shooting; **s. terrestre,** land shooting; **bruit s.,** seismic noise; **enregistrement s.,** seismic record; **prospection s.,** seismic prospecting, seismic survey; **rayons s.,** sismic rays; **tir s.,** seismic shooting.

sismogramme, seismogram.

sismographe, seismograph, geophone.

sismologie, seismology.

sismologue, seismologist.

sismomètre, seismometer.

sismosédimentologie, seismic stratigraphy.

sismosondage, well shooting.

sismotectonique, sismotectonics.

skarn, scarn, skarn.

smaltine, smaltite, smaltite, grey cobalt.

smectique (argile), fuller's earth.

smectite (m. argileux), smectite.

smithsonite (minér.), smithsonite.

socle, basal complex, shield, craton, base, basement, bedrock, bottom, floor; **s. acoustique,** seismic basement; **s. granitique,** granitic basement; **s. océanique,** oceanic basement.

sodalite (minér.), sodalite.

sodé, soda.

sodique, soda, sodium.

sodium, sodium; **chlorure de s.,** halite, salt rock; **nitrate de s.,** chili nitrate, niter.

sol, soil, ground; **s. à alcalis,** alkali soil; **s. à croûte,** crust soil; **s. à croûte gypseuse,** gypsum crust soil, gypcrete; **s. à cuirasse ferrugineuse,** iron crust soil; ferricrete; **s. à fentes de froid,** frozen crack soil; **s. à festons (périgl.),** soil with involutions; **s. à figures géométriques,** patterned ground; **s. à gley,** gley soil; **s. à gley profond,** deep gley soil; **s. à gley superficiel,** shallow gley soil; **s. alcalin,** alkaline soil; **s. allochtone,** allochtonal soil; **s. alluvial,** alluvial soil, fluvent; **s. alluvial à gley,** alluvial gley soil, aquent; **s. alluvial de prairie,** meadow soil; **s. anthropique,** anthropic soil; **s. à pseudogley,** pseudogley soil; **s. arctique,** arctic soil; **s. à réseaux de pierres,** polygonal soil; **s. argileux,** clayey soil; **s. argileux à gley,** gley clay soil; **s. aride,** aridosol; **s. autochtone,** autochtonous soil; **s. azonal,** azonal soil; **s. brun,** brown forest soil (G.B.), gray brown podzolic soil (U.S.); **s. brun alluvial,** alluvial brown soil; **s. brun calcaire,** calcareous brown soil; **s. brun forestier,** brown forest soil; **s. brun fortement lessivé,** brown podzolic soil; **s. calcaire,** calcareous soil; **s. calcaire de rendzine,** rendzina soil, rendoll; **s. carbonaté humique,** rendzinic soil; **s. cendreux (décoloré),** podzol; **s. châtain,** chestnut coloured soil, xeroll; **s. colluvial,** colluvial soil; **s. complexe,** polygenetic soil; **s. d'altération,** weathering soil; **s. décalcifié,** decalcified soil; **s. décoloré,** bleached soil; **s. de marais,** bog soil; **s. de prairie,** prairie soil; **s. désertique,** desert soil; **s. de toundra,** toundra soil; **s. d'inondation,** flooding soil; **s. ferralitique,** ferralitic soil; **s. ferrugineux rouge,** ferrimorphic soil; **s. fossile,** fossil soil; **s. gelé,** permafrost, frozen soil; **s. glaciaire (sur moraines argileuses),** gumbotil; **s. gris désertique,** gray desert soil; **s. humifère à gley,** humic gley soil, aquoll; **s. hydromorphe,** hydromorphic soil, aquent; **s. intrazonal,** intrazonal soil; **s. latéritique,** lateritic soil, latosol; **s. lessivé,** leached soil, alfisol; **s. limoneux,** loamy soil; **s. loessique,** loessic soil; **s. marécageux noir,** meadow bog soil; **s. minéral,** mineral soil; **s. mouvant,** shifting soil; **s. noir steppique,** chernozem, steppe black soil; **s. pierreux,** stony soil; **s. podzolique,** podzolic soil, spodosol; **s. polygonal,** polygonal ground; **s. résiduel,** residual soil; **s. réticulé,** patterned ground; **s. rouge,** red soil; **s. rouge latéritique,** iron lateritic soil, latosol; **s. rouge lessivé,** red podzolic soil; **s. rouge méditerranéen,** mediterranean soil, terra rossa; **s. sableux,** sandy soil; **s. sablo-limoneux,** loamy soil; **s. salin,** saline soil; **s. salin blanc non structuré,** solonchak; **s. salin lessivé acide,** steppe bleached earth; **s. salin podzolisé,** solod; **s. steppique,** steppe soil; **s. steppique gris,** grey earth; **s. strié (périgl.),** striped ground, soil stripes; **s. structuré,** patterned ground; **s. tourbeux,** peaty soil; **s. tronqué,** truncated soil; **s. zonal,** zonal soil; **analyse de s.,** soil analysis; **carte des s.,** soil map; **compactage de s.,** soil compaction, soil densification; **échantillon de s.,** soil sample; **érosion du s.,** soil erosion; **glissement de s.,** soil creeping; **horizon de s.,** soil horizon; **lessivage des s.,** soil leaching; **mécanique des s.,** soil mechanics; **type de s.,** soil type.

solaire, solar; **centrale s.,** s. plant; **collecteur solaire,** s. collector; **énergie s.,** s. energy; **radiation s.,** s. radiation.

sole (mine), floor.

solfatare, solfatare.

solide, **1.** adj.: solid, sound, tight; **2.** n: solid; **charge s.,** solid load; **combustible s.,** solid fuel; **état s.,** solid state; **inclusion s.,** solid inclusion; **phase s.,** solid phase.

solidification, solidification; **point de s.,** solidifying point.

solidifié (magma, etc.), solidified.

solidifier, se solidifier, to solidify.

solidus, solidus.

soliflué, soliflucted.

solifluxion, solifluction, solifluction; **bourrelet de s.,** s. wrinkle, buckle; **coulée de s.,** s. flow, s. stream; **dépôt de s.,** s. deposit; **guirlande de s.,** s. festoon; **manteau de s.,** s. sheet; **poche de s.,** s. pocket; **s. périglaciaire,** gelifluxion (rare).

solodisation, solodization.

solodisé, solodized, soloth-like.

solonchak, solonchak.

solonetz, solonetz (saline black earth).

solonisation (formation d'un solonetz), solonization.

solstice (astro.), solstice.

solubilisation, solvency.

solubiliser, to solubilize.

solubilité, solubility; **courbe de s.,** solubility curve.

soluble, soluble.

soluté, solute.

solution, solution; **s. acide,** acid s.; **s. aqueuse,** aqueous s.; **s. colloïdale,** colloidal s.; **s. concentrée,** concentrated s.; **s. diluée,** dilute s.; **s. étendue,** weak s.; **s. normale,** normal s.; **s. saturée,** saturated s.; **s. solide,** exsolution; **s. sursaturée,** supersaturated s.

Solutréen (préhist.), Solutrean (Pleistocene industry).

solvus, solvus.

sondage, bore, bore hole, boring, well boring; **s. acoustique,** sonic sounding; **s. à injection,** flush drilling; **s. à la corde,** rope drilling; **s. au diamant,** diamond drilling; **s. carotté,** core drilling; **s. d'exploration,** wildcat drilling, exploration borehole; **s. dévié,** deflected well, slanted well; **s. électrique,** electric logging; **s. géologique,** structural test hole; **s. non tubé,** open hole; **s. océanographique,** sounding; **s. par battage,** boring by percussion; **s. rotatif,** rotary drilling; **s. sismique continu,** continuous seismic profiling; **s. sismique profond,** deep seismic sounding; **s. stérile,** barren boring; **s. tubé,** cased boring; **carotte de s.,** drill core; **echo-s.,** echo-sounding.

sonde, sound; **s. acoustique,** sonoprobe; **s. à neutrons,** neutron probe; **s. d'induction,** induction s.; **s. électronique,** electronic microprobe; **s. (forage),** drilling rig, borer; **s. nautique,** sounding line; **s. pédologique,** earth borer; **ballon-s.,** sounding balloon; **écho-s.,** sonic altimeter.

sonder (océano.), to probe; **s. (pétrole),** to bore, to hill.

sondeur (appareil), sonic depth finder; **s. à multifaisceaux,** sea-beam; **s. (ouvrier),** drill man, driller, borer; **s. sismique,** sub-bottom profiler.

sondeur de vase, mud penetrator.

sondeuse, drill, drilling machine.

sorosilicate, sorosilicate.

sortie (de galerie), outlet, exit.

sotch, solution pit.

soubassement, basement, bedrock, underlying rock.

soudage (de tubes de gazoduc), welding.

soude, soda; **s. carbonatée,** natron; **s. caustique,** caustic.

soudure (de plaques lithosphériques), suturing.

souflard, blow hole, blower, suffione.

soufflés (sables), blown out sands, wind-borne sands.

soufre, sulphur, sulfur; **s. natif,** native s.; **s. précipité,** precipitated s.

soufré, sulphuretted.

soufrière, sulphur pit, solfatara.

souiller (une nappe phréatique), to pollute, to contaminate.

soulevé (bloc), uplifted.

soulevée (plage), raised beach.

soulèvement, uplift, upheaval; **s. glacio-isostatique,** glacio-isostatic rise; **s. intermittent,** intermittent uplift; **s. par le gel,** frost heaving; **s. structural,** structural uplift.

soulever, to lift, to uplift.

source, spring; **s. artésienne,** artesian s.; **s. ascendante,** ascending s.; **s. chaude,** hot s.; **s. d'affleurement,** outcrop s.; **s. d'eau salée,** salt s.; **s. de fracture,** fracture s.; **s. d'infiltration,** filtration s.; **s. ferrugineuse,** ferruginous s.; **s. hydrothermale,** hydrothermal s.; **s. hydrothermale chaude (dorsale),** hydrothermal hot event; **s. incrustante,** incrustating s.; **s. intarissable,** perennial s.; **s. intermittente,** intermittent s.; **s. jaillissante,** spouting s.; **s. juvénile,** juvenile s.; **s. karstique,** karstic s.; **s. minérale,** mineral s.; **s. suintante,** seepage s.; **s. structurale,** structural s.; **s. sulfureuse,** sulphurous s.; **s. thermale,** thermal s.; **s. vauclusienne,** exsurgence.

sourcier, dowser, water-diviner, "water-witch".

sourdre, to ooze, to spring, to well up.

sous, under; **s. affleurement,** subcrop; **s. cavage,** undermining, undercutting; **s. caver,** to undermine; **s. charriage,** underthrust; **s. classe,** subclass; **s. delta,** pro-delta; **s. écoulement,** underflow; **s. espèce,** subspecies; **s. étage,** substage, subage; **s. faciès,** subfacies; **s. famille,** subfamily; **s. fluvial,** subfluvial; **s. genre,** subgenus; **s. glaciaire,** subglacial; **s. groupe,** subgroup; **s. jacence,** subterposition; **s. jacent,** underlying; **s. lacustre,** sublacustrine; **s. le vent,** leeward; **s.-marin,** submarine; **s. ordre,** suborder; **s. platine,** substage; **s. produit,** by-product; **s. règne,** subkingdom; **s. saturé,** undersaturated; **s. saturation,** undersaturation; **s. sol,** undersoil, subsoil; **s. solage,** subsoiling, substilling.

soutènement (mine), support(ing), roof support.

soutenir (étayer), to support, to prop, to stay.

souterrain (adj.), subterranean, subterraneous, underground.

soutirage (hydro.), withdrawing, draw off, tapping; **point de s.,** draw point.

soutirer (de l'eau, etc.), to withdraw.

Sparnacien, Sparnacian (l. Eocene).

sparagmite (pétro.), sparagmite (feldspathic sandstone: Precambrian, Scandinavia).

sparite (pétro.), sparite.

spath, spar; **s. adamantin,** adamantine s.; **s. calcaire,** calc s.; **s. d'Islande,** Iceland s.; **s. fluor,** fluorite; **s. pesant,** barytine, heavy s.

spathique, spathic, sparry, spathose.

spéciation (pal.), speciation.

spécifique (densité), specific (gravity).

spécimen, specimen, sample.

spectral, spectral; **analyse s.,** spectral analysis; **canal s.,** spectral channel; **diagraphie s. par rayons gamma,** spectral gamma ray log; **domaine s.,** spectrum area.

spectre, spectrum; **s. d'absorption,** absorption s.; **s. de diffraction,** diffraction s.; **s. d'émission,** emission s.; **s. de flamme,** flame s.; **s. de masse,** mass s.; **s. magnétique,** magnetic s.; **s. pollinique,** pollen s.

spectrogramme, spectrogram.

spectrographe, spectrograph; **s. de masse,** mass s.

spectographie, spectography.

spectrographique, spectrographic.
spectromètre, spectrometer; **s. à réseau,** grating s.; **s. de masse,** mass s.
spectrométrie, spectrometry.
spectrométrique, spectrometry.
spectrophotométrique, spectrophotometric.
spectroscope, spectroscope.
spectroscopique, spectroscopic.
spéculaire, specular; **fer s.,** specular iron ore, specularite.
spéléologie, speleology.
spéléologique, speleological.
spéléologue, speleologist.
spéléothème, speleothem.
spessartite (minér.), spessartite.
sphalérite, spalerite, blende.
Sphénopsidés (pal.), Sphenopsida.
sphéroïdale (désagrégation), spheroidal weathering.
sphérolite, spherolite.
sphérolitique (structure), spheroidal texture, spherulitic texture.
sphérosidérite, spherosiderite.
sphérulite, spherolite.
spicule (pal.), spicule.
spilite (pétro.), spilite.
spilitique, spilitic.
spinelle (minér.), spinel; **mâcle des s.,** s. twin.
spiracle (pal.), spiracle.
spire (de coquille), spire, coil.
Spiriféridés (pal.), Spiriferid.
spodique (pédo.), spodic.
spodosol (var. de podzol), spodosol.
spodumène, spodumene.
spondylium, spondylium.
Spongiaires (pal.), Spongiae, Porifera; **s. siliceux,** Silicospongiae.
spongolite (gaize), sponge-spicule deposit, gaize, spongolite.
spontanée (polarisation), spontaneous (polarization), self potential.
spore (paléobot.), spore; **charbon de s.,** spore coal.
sporifère, sporiferous.
sporomorphe, sporomorph.
squelette (pal.), skeleton.
squelette corallien, limy case.
squelettique (sol), skeletal (soil).
stabilisation, stabilization; **s. des dunes,** dunes s.; **s. des talus,** slope s.; **s. du sol,** soil s.
stabiliser (géotechnie), to stabilize.
stabilité (chimique), chemical stability.
stable (chimie, géomorpho.), stable, steady.
stade, stage, substage; **s. de jeunesse (géogr.),** youth stage; **s. de maturité (géogr.),** maturity stage; **s. final de cristallisation,** late stage of crystallization; **s. glaciaire (Quat.),** glacial stage; **s. interglaciaire,** interglacial stage; **s. pneumatolytique,** pneumatolytic stage; **s. successifs de stationnement d'un niveau d'eau,** water levels marks.
stadiaire, stadial.
stagnantes (eaux), still, stagnant (waters).

stagnation, stagnation; **s. de l'eau,** water s.; **s. d'un glacier,** glacier s.
stalactite, stalactite.
stalactitique, stalactitic.
stalagmite, stalagmite.
Stampien, Stampian (m. Oligocene: Paris Basin).
stanneux, stannous.
stannifère, stanniferous, tin bearing.
stannique, stannic.
stannite (minér.), stannite.
statif (de microscope), stand.
station, station; **s. de pompage,** pumping plant; **s. de recompression,** recompression plant; **s. météorologique,** weather s.; **s. océanographique,** marine research center.
stationnaire (onde), stationary (wave).
statique (pression), static pressure.
staurotide, staurolite (minér.), staurotide, staurolite.
stéatite (talc), steatite, soapstone.
stéatiteux, steatitic.
Stégocéphales (pal.), Stegocephalia.
Stégosauriens (pal.), Stegosauria.
sténohalin, stenohaline.
Stéphanien, Stephanian.
stéphanite, stephanite, black silver.
stéréocomparateur, stereocomparator.
stéréogramme, stereogram, block diagram.
stéréographique, stereographic; **projection s.,** s. projection, stereonet.
stéréophotogrammétrie, stereophotogrammetry.
stéréorestituteur, stereoplotter.
stéréorestitution, stereocompilation.
stéréoscope, stereoscope; **s. à balayage,** scanning s.
stéréoscopique, stereoscopic; **plaquettes s.,** stereoscopic stereoscope; **vision s.,** stereoscopic vision.
steppe, steppe; **s. à thufur,** steppe with grassy hummocks.
stérile, sterile, barren, unproductive; **couche s.,** barren stratum; **puits s.,** unproductive well; **roche s.,** barren rock.
stibine (minér.), stibnite, sulphide of antimony.
stilbite (minér.), stilbite.
stilpnomélane (minér.), stilpnomelane.
stockage, storage; **s. de gaz,** gas s.; **s. de pétrole,** oil s.; **s. souterrain,** underground s.
stocker (du gaz), to stock.
Stockes (loi de), Stockes' law.
stolon (pal.), stolon.
strate, stratum, layer, bed.
stratification, stratification, bedding; **s. concordante,** conformable bedding; **s. discordante,** unconformable bedding; **s. entrecroisée,** cross bedding, cross lamination, cross stratification; **s. horizontale,** horizontal stratification; **s. lenticulaire,** lensoid stratification; **s. oblique,** oblique stratification; **s. thermique,** thermal s.; **plan de s.,** stratification plane.
stratifié, stratified, bedded, layered, laminated; **finement s.,** straticulate; **moraine s.,** stratified drift.
stratifier (se), to stratify.

stratiforme (gisement), stratiform deposit, strata-bound deposit.

stratigraphe, stratigrapher.

stratigraphie, stratigraphy.

stratigraphique, stratigraphic; **classification s.**, s. classification; **colonne s.**, s. column; **corrélation s.**, s. correlation; **lacune s.**, s. lacune; **limite s.**, s. boundary; **paléontologie s.**, s. paleontology; **piège s.**, s. trap; **répartition s.**, s. range; **rejet s.**, s. throw; **unité s.**, s. unit.

stratocône, stratocone.

stratosphère, stratosphere.

stratotype, stratotype.

stratotypique, stratotypic.

stratovolvan, stratovolcano.

stratus (météo.), stratus.

striage (glaciaire), scratching.

strie (glaciaire), scratch, striation, groove; **s. d'accroissement**, growth ring; **s. de faille**, fault striate.

strié, streaked, striated.

strier, to scratch, to striate.

stromatolite (pal.), stromatolite.

Stromatoporoïdés (pal.), Stromatoporoids.

strombolien, strombolian.

strontianite (minér.), strontianite.

strontium, strontium.

structural, structural; **analyse s.**, s. analysis; **carte s.**, s. map; **cuvette s.**, s. basin; **diagramme s.**, s. diagram; **fermeture s.**, s. fault closure; **gîtologie s.**, s. control; **pénéplaine s.**, s. plain; **piège s.**, s. trap; **plate-forme s.**, s. plateform, s. surface; **relief s.**, s. relief; **terrasse s.**, s. rock terrace.

structure, structure, texture; **s. alvéolaire**, honeycomb s.; **s. amygdalaire**, amydaloid s.; **s. annulaire**, ring s.; **s. anticlinale**, anticlinal s.; **s. bréchique**, brecciation; **s. cataclastique**, cataclastic s.; **s. d'effondrement**, collapse s.; **s. en boules et en coussins**, ball and pillow s.; **s. en chapelet**, bedded s.; **s. en cônes emboîtés**, cone in cone s.; **s. en cocarde**, cockade s.; **s. en cuvette**, dish s.; **s. en écailles**, imbricate s.; **s. en mortier**, mortar s.; **s. en mosaïque**, mosaic s.; **s. en plaquettes**, platy s.; **s. en sablier**, hour glass s.; **s. faillée**, faulted s.; **s. feuilletée**, leaflike s.; **s. fibreuse**, fibrous s.; **s. fluidale**, flow s., fluidal s.; **s. géologique**, geological s.; **s. géotrope**, geopetal s.; **s. graphique**, graphic texture; **s. granulaire**, granular s.; **s. grenue**, holocrystalline and granular s.; **s. grumeleuse (du sol)**, crumbly s.; **s. hélicitique**, helicitic tecture; **s. homéoblastique**, homeoblastic tecture; **s. imbriquée**, imbricated s.; **s. lamellaire**, lamellar s.; **s. lépidoblastique**, lepidoblastic texture; **s. litée**, layered s.; **s. microcristalline**, microcrystalline tecture; **s. microgrenue**, microgranular texture, fine-grained holocrystalline tecture; **s. microlithic**, microlitic tecture, volcanic texture; **s. monoclinale**, monoclinal s.; **s. orbiculaire**, orbicular s.; **s. pétrolifère**, oil-bearing s.; **s. plissée**, folded s.; **s. prismatique**, prismatic s., columnar s.; **s. polyédrique**, angular blocky s.; **s. prismée**, columnar s.; **s. résiduelle**, relic s.; **s. réticulée**, network s.; **s.**

rubanée, ribbon s.; **s. sphérolitique**, spherulitic s.; **s. superficielle**, surface s.; **s. superposée**, superimposed s.; **s. synclinale**, synclinal s.; **s. tabulaire**, table-like s.; **s. vitreuse**, holohyaline texture; **s. zonée**, zonal s.

structurologie, tectonics.

Strunien, Strunian (u. Devonian, equiv. Etroeungtian).

style jurassien (tecto.), jurassian tectonics.

stylet (d'un seismographe), pen.

stylolite, stylolith, suture joint, cone in cone structure.

stylolitique, stylolitic.

subaérien, subaerial, superterranean.

subalcalin, subalkaline.

subanguleux, subangular.

subarctique, subarctic.

subaride, subarid.

subarrondi, subrounded.

subautochtone, subautochtonous.

subcrustal, subcrustal.

subduction (tecto.), subduction.

subduite (plaque), subducted.

sublimation (d'un solide), sublimation.

sublimer (se), to sublimate, to substilize.

sublithographique, sublithographic.

sublittoral, sublittoral.

submergé, submerged, drowned.

submerger, to submerge, to drown, to flood.

submersion, submergence, submersion; **côte de s.**, submerged shoreline.

subpergélisol, subgelisol.

subpolaire, subpolar.

subséquent, subsequent; **faille s.**, s. fault; **vallée s.**, s. valley.

subsidence, subsidence.

subsident, subsiding.

substitution (chimique), substitution, replacement.

substitution (filon de), substitution vein.

substrat, substratum, substratum, bedrock, bottom, subterrane.

subsurface, subsurface.

subtidal, subtidal.

subtropical, subtropical.

subtrusion, subtrusion.

subvitreux, subvitreous, subglassy.

suc phonolitique, phonolitic plug.

succession de couches, succession of strata.

succin, amber.

succinite, succinite.

succion (autour d'un forage), coning.

Sudète (phase), Sudetan (orogenesis).

suffosion, suffosion (groundwater eruptions, pits, mudvolcanoes, tundra craters).

suintant (roche), seeping, oozing.

suintement, seepage, seeping; **s. de pétrole**, oil seepage.

suinter, to ooze, to seep, to exude.

suite réactionnelle, reaction series.

suivre (un affleurement), to follow, to trace, to strike.

sulfatation, sulfation, sulfatation, sulfating.

sulfate, sulfate; **s. de cuivre**, copper s., blue vitriol; **s. de fer**, iron s., green copperas; **s. de magnésium**, magnesium s.; **s. ferreux**, ferrous s.

sulfite, sulfite.

sulfohalite, sulfohalite.

sulfosel, sulfosalt.

sulfure, sulfide; **s. de fer**, iron s.

sulfuré, sulfidic; **antimoine s.**, antimony sulfide; **minéral s.**, sulfide ore; **minerai s.**, sulfurous ore.

sulfureux, sulfurous ore; **anhydride s.**, sulfur dioxide; **boue s.**, sulfur mud; **eau s.**, sulfur water; **source s.**, sulfuric spring; **vapeur s.**, sulfur fume.

sulfurique, sulfuric.

superficiel, superficial; **dépôt s.**, s. deposit; **écoulement s.**, surface runoff; **encroûtement s.**, surface crust; **érosion s.**, surface erosion; **moraine s.**, surface moraine; **onde s.**, surface wave; **pression s.**, surface pression; **tectonique s.**, superstructure.

supergène (minéral), secondary (mineral), supergene.

supérieur (strati.), upper.

superposé (strati.), overlying, superincumbent, superjacent; **pli s.**, superposed fold; **être s. à**, to overlie.

superposer (se), to superimpose, to superpose.

superposition, superposition; **s. inverse**, anormal s.; **s. normale**, original s.

superstructure (de derrick), headgear.

Supracrétacé, Upper Cretaceous.

supracrustal, supracrustal.

supraglaciaire, superglacial.

supralittoral, supralittoral.

suprapergélisol, supragelisol.

supratidal, supratidal.

surcharge, overload(ing), overstressing.

surcharger, to overload.

surchauffe (magmatique), overheating.

surcreusement (géomorpho.), overdeepening.

surélévation, uplift.

surélever (relief), to raise.

surface, surface, area; **s. d'aplanissement**, planation surface; **s. de charriage**, overthrust plane; **s. de compensation du CO_3Ca**, compensation level; **s. de couche**, bed surface; **s. de discontinuité**, discontinuity surface; **s. de discordance**, surface of unconformity; **s. de faille**, fault surface; **s. de glissement**, sliding surface, slip plane; **s. d'érosion**, erosional surface; **s. d'érosion sous-marine**, wash out, erosional truncation; **s. de séparation**, parting surface, boundary surface, interface; **s. d'onde**, wave front; **s. du sol**, land surface; **s. gauchie**, warped surface; **s. limite**, boundary surface; **s. limite eau-pétrole**, oil-water contact; **s. mamelonnée**, mamillated contact; **s. ondulée**, corrugated contact; **s. perforée**, mollusc bored surface; **s. piézométrique**, water-level contact, piezometric contact; **s. polie**, polished contact; **s. structurale**, back slope; **affaissement de la s.**, surface break, collapse; **installation de s.**, surface plant; **tir en s.**, surface shooting.

surfondre, to supercool, to surfuse.

surfondu, supercooled.

surfusion, supercooling, surfusion.

surhaussement (tecto.), uplift.

surimposé (géomorpho.), superposed, superimposed, epigenetic.

surimposition, superimposition.

surnageant (minéral), supernatant, floating.

surplatine (micro.), superstage.

surplomb, overhanging.

surplomber (une vallée), to overhang.

surpression, overpressure.

surrection, uplift.

sursaturation (cristallo.), oversaturation, supersaturation.

sursaturé, supersaturated, oversaturated.

sursaturer, to supersaturate.

susceptibilité magnétique, magnetic susceptibility.

sus-jacent, overlying; **couche sus-jacente**, superstratum.

suspendu (glacier), hanging (glacier).

suspendue (vallée), hanging (valley), perched (valley).

suspension, suspension.

suspension (en), suspended; **charge en s.**, suspended load; **sédiment en s.**, suspended deposit.

suture (ligne de) (pal.), suture (line).

suture faciale (pal.), facial suture.

syénite, syenite; **s. à feldspathoides**, syenoid; **s. néphélinique**, nephelite s.; **s. quartzique**, quartz s.

syénitique, syenitic; **aplite s.**, syenitic aplite; **pegmatite s.**, syenitic pegmatite.

sylvanite (minér.), sylvanite.

sylvinite, sylvinite.

sylvite (minér.), sylvite.

symbiose (pal.), symbiosis.

symbiotique (pal.), symbiotic.

symétrie, symmetry; **s. bilatérale**, bilateral s.; **s. de type cinq**, pentamerous s.; **s. radiale**, radial s.; **axe de s.**, s. axis; **plan de s.**, s. plane.

symétrique pli, symmetrical fold.

symplectique (association minéralogique), symplektic (intergrowth).

synchronisme (d'un événement tectonique), synchroneity.

synclase (désuet), synclase.

synclinal, 1. adj: synclinal; 2. n; syncline; **s. fermé**, closed syncline; **s. perché**, upstanding syncline; **axe du s.**, synclinal axis; **charnière s.**, synclinal bend; **cuvette s.**, syncline; **dépression s.**, synclinal trough; **flanc d'un s.**, limb; **pli s.**, syncline.

synclinorium, synclinorium, syncline.

syndiagenèse, syndiagenesis, penecontemporaneous.

synéclise, syneclise, sediment alteration.

synérèse, syneresis.

synforme (tecto.), synform.

syngénèse, syngenesis.

syngénétique, syngenetic.

synglaciaire, synglacial.

synsédimentaire, synsedimentary, penecontemporaneous.

syntaxie, syntaxy.
syntectonique, synkinematic, synorogenic, syntectonic.
syntectique, syntectic.
synthétique (pétrole), synthetic (crude).
syntype (pal.), syntype.
système, system, method; **s. Carbonifère,** Carboniferous system; **s. cubique,** cubic system, isometric system; **s. d'écoulement,** pattern of drainage; **s. d'exploitation,** working method; **s. de cristallisation,** crystallization system; **s. de diaclases,** set of joints; **s. hexagonal,** hexagonal system; **s. métrique,** metric system; **s. monoclinique,** monoclinic system; **s. orthorhombique,** orthorhombic system; **s. quadratique,** quadratic system; **s. rhomboédrique,** rhombohedral system; **s. triclinique,** triclinic system.
syzygie, syzygy.

T

tabétisol (périgl.), tabetisol.
table, table; t. à secousse, concentrating t.; t. basaltique, basaltic t., trapp; t. de concentration, concentrating t.; t. de lavage, washing t.
tabulaire, tabular, table-like.
Tabulés (pal.), Tabulata.
tache solaire (astro.), sunspot.
tacheté (schiste), spotted (schist).
tachygénèse (pal.), tachygenesis.
tachylite (verre basaltique), tachylite.
Taconique (phase), Taconic (orogeny).
taconite (pétro.), taconite.
tactite (pétro.), tactite.
taffoni, tafoni, taffoni, honeycomb weathering, cavities.
taïga (géomorpho.), taiga.
taillant (mine), bit; t. amovible, detachable b.; t. en croix, cross b.
taille, cutting; t. de la pierre, stone c.; t. (dimension), size; t. (mine), c., dressing; front de t., face; largeur de t., face width.
taillée (pierre précieuse), cut (gemstone).
taillée (époque de la pierre), Paleolithic.
tailler (mine), to cut, to hew; t. par éclats, to chip.
tailleur (de pierres), cutter, hewer, dresser.
talc, talc, soapstone, talcum.
talcochloritique, talcochloritic.
talcomicacé, talcomicaceous.
talcschiste, talcshist.
talqueux, talcous, talcose, talcy.
talus, slope, escarpment; t. continental, continental slope; t. d'éboulis, scree, talus, slope of debris; t. de déferlement, swash ramp, beach face slope; t. d'équilibre, angle of repose, gravity slope, terrace scarp.
talutage (géotechnie), sloping.
taluter, to slope.
tambour d'extraction (mine), hoist drum.
tambour de treuil, hoisting drum.
tamis, sieve, sifter, screen; t. à mailles, mesh sieve; t. à secousses, shaking sieve; t. métallique, wire-cloth sieve; t. oscillant, swinging sieve; t. vibreur, vibrating sieve.
tamisage, sieving, sifting, screening.
tamisat, sieved particles.
tamiser, to sift.
tamiseuse, shaking sieve.
tamponné (chimie), buffered.
tangentiel (charriage), tangential (thrust).
tangentielle (poussée), tangential (stress).
tangue (mt Saint-Michel), calcareous sandy shelly mud, slikke, ooze (shallow water).
tantale, tantalum.
tantalite (minér.), tantalite.
taphonomie, taphonomy.

taphrogénèse, taphrogenesis, block-faulting orogeny.
taphrogénie, taphrogeny.
taphrogéosynclinal, taphrogeosyncline.
taphronomique, taphronomic.
tapiolite (minér.), tapiolite.
tapis algaire (littoral), algal mat.
tardiglaciaire (quat.), lateglacial.
tarditectonique, latertectonics, laterorogenic events.
tarière, auger, earth auger; t. à main, hand auger; t. à vis, screw auger; t. de pédologue, earth auger, surface auger, clay auger; t. rubanée, auger worm.
tarir, se tarir, to dry up, to exhaust.
tarissement d'une source, drying up.
tarissement d'un gisement, exhaustion, depletion.
tas (de minerai), heap, pile.
tassé, compressed, packed, compact, tight.
tassement (compression), packing, compression, compaction; t. (de terrain), collapsing, sinking.
tasser, se tasser, to compact, to pack, to ram, to squeeze.
taurite (pétro.), taurite.
taux, rate, ratio; t. d'évaporation, evaporation rate; t. d'expansion (des fonds océaniques), rate of sea-floor spreading; t. de ruissellement, flow rate.
taxite, taxite.
taxitique (structure), taxitic (structure).
taxodonte (Mollusque), taxodonte.
taxon (pal.), taxon.
taxonomie, taxonomy.
taxonomique, taxonomic.
tchernozem, chernozem soil.
technique minière, mining engineering.
tectite (météorite), tektite.
tectofaciès, tectofacies.
tectogénèse, tectogenesis.
tectomorphique, tectomorphic.
tectonique, 1. adj: tectonic; 2. n: tectonics; t. active, active tectonics; t. cassante, faulting tectonics; t. de couverture, fold tectonics; t. de fond, deep-seated structures; t. de glissement, sliding tectonics; t. de socle, basement tectonics; t. des plaques, plate tectonics; t. globale, plate tectonics; t. salifère, salt tectonics, diapirism; t. tangentielle, compression tectonics; t. souple, folding tectonics; fossé t., rift, graben.
tectonisation, occuring of tectonic events.
tectonite, tectonite.
tectonophysique, tectonophysics.
tectonosphère, tectonosphere.
tectosilicate, tectosilicate, frame-work silicate.
téguline (argile), tile-clay (Albian).
teinte sensible (gypse), sensitive tint.
télédétection, remote sensing; t. aéroportée, aircraft sensing; t. infrarouge, infrared sensing; t. multi-spectrale, multispectral sensing.

téléimages, remote-sensing images.

télémètre, telemeter.

télémétrie, telemetry.

téléobjectif, telephotographic lens.

Téléostéens (pal.), Teleostei.

télescope (astro.), telescope.

télésismique (signal), teleseismic signal.

tellurique, telluric (pertaining to earth, ex. electrical currents); **courant t.,** telluric current; **méthode t.,** telluric method.

tellurite (minér.), tellurite.

tellurifère, telluriferous.

telluromètre, tellurometer.

témoin (butte), outlier.

température, temperature; **t. absolue,** absolute t.; **t. atmosphérique,** atmospheric t.; **t. d'ébullition,** boiling point; **t. de congélation,** freezing point; **t. de fusion,** melting point; **t. moyenne,** average t.; **chute de t.,** t. drop; **correction de t.,** t. correction; **enregistreur de t.,** t. recorder.

tempéré, temperate; **climat t.,** temperate climate; **glacier t.,** temperate glacier.

tempestite (sédim.), tempestite storm deposit.

tempête, storm; **t. de neige,** snow s.; **t. de poussière,** dust s.; **t. de sable,** sandstorm; **t. magnétique,** magnetic s.

temps (météo.), weather; **t. (durée),** time; **t. de forage,** drilling time; **t. de propagation,** travel time; **t. froid,** cold w.; **t. glacial,** frosty w.; **t. pluvieux,** rainy w.; **t. zéro (géophys.),** shot break.

teneur, content, amount, grade, ratio; **t. en carbone,** carbone content; **t. en eau,** water content, moisture content; **t. en fer,** iron content; **t. en minerai,** ore content; **t. en sel,** salinity content; **t. faible,** low grade (of an ore); **t. moyenne,** average grade; **t. volumique,** volumic content.

tennantite (minér.), tennantite.

ténorite (minér.), tenorite.

tension de cisaillement, shearing stress.

tension de flexion, bending stress.

tension de polarisation, bias.

tension de vapeur, stream pressure, vapour pressure.

tension superficielle, capillary content, superficial tension.

tensiomètre, tensiometer.

tentacule (pal.), tentacle.

téphra, tephra, volcanic ash layer.

téphrite (pétro.), tephrite.

téphrochronologie, tephrochronology.

tératologie, teratology.

Térébratulidés (pal.), Terebratulidae.

terme (d'une série stratigraphique), member (of a series).

terminaison anticlinale, anticlinal ending; **t. synclinale,** terminal curvature.

terminaison synclinale tronquée, synclinal troncation.

terminer en biseau (se) (Strati.), to thin out, to peter out, to taper.

terrain, ground, land, earth, rocks, strata, formations, terrane (U.S.), terrain (U.K.); **t. aquifère,** water-bearing bed; **t. aurifère,** gold-bearing ground; **t. caillouteux,** stony land; **t. de couverture,** overburden, capping, cap; **t. encaissant,** country rock; **t. ébouleux,** loose ground; **t. ferme,** solid ground; **t. houiller,** coal measures; **t. marécageux,** swampy ground; **t. pétrolifère,** oil land; **t. sableux,** sandy land, barren ground; **t. tourbeux,** peaty ground; **étude de t., travail de t.,** field work, studies in the field.

terra rossa, red earth, soil, terra rossa.

terrasse, terrace; **t. alluviale,** alluvial t.; **t. alluviale construite,** aggradational t.; **t. climatique,** climatic t.; **t. couplées,** matched t.; **t. d'accumulation,** fill t.; **t. d'érosion,** bedrock t.; **t. emboîtée,** inner t., fill and fill t.; **t. en gradins,** stepped t., bench t.; **t. étagée,** stepped t.; **t. eustatique,** eustatic t.; **t. fluviatile,** river t., stream t.; **t. fluvio-glaciaire,** fluvio-glacial t.; **t. littorale,** shore t.; **t. rocheuse,** rock t.; **t. tectonique,** tectonic t., structural t.; **en t.,** terraced; **niveau de t.,** t. level; **talus de t.,** t. scarp.

terrassette (périgl.), terracette, step, sheep or cattle track.

terrassement (travaux de), earthworks, excavation works.

terrasser (géotechnie), to bank up, to embank.

terre, earth, ground, soil; **t. à briques,** brick earth, lehm; **t. à diatomées,** diatomite; **t. à foulon,** fuller's earth; **t. argileuse,** clayey earth; **t. calcaire,** calcareous earth; **t. d'infusoires,** infusorial earth, diatomite; **t. de bruyère,** heather soil; **t. de remblais,** back fill; **t. en friche,** fallow land; **t. ferme,** mainland; **t. forte,** loam; **t. glaise,** clayey loam; **t. lourde,** heavy earth; **t. meuble,** loose ground; **t. rare,** rare earth element; **t. réfractaire,** clay; **t. tourbeuse,** peaty earth.

terrestre, terrestrial, terrene, non-marine, continental.

terreux, earthy.

terrigène, terrigenous, terrigene.

terril, terris, waste dump, refuse dump, heap, dumping ground.

territoire (pal.), territory.

Tertiaire (ère), Tertiary, Cainozoic, Cenozoic (era), Kainozoic.

tertre, hillock, mound.

teschénite (gabbro alcalin), teschenite.

test, **1.** test, testing, try, trial; **2.** shell *(pal.).*

tétartoèdre, tetartohedron.

tête, head; **t. de bassin (hydro.),** headwaters; **t. de chat,** siliceous or calcareous concretion, Eocene; **t. de colonne de production,** tubing h.; **t. de comète,** h. of a comet; **t. de distillation,** first running; **t. de nappe,** nappe front; **t. de puits,** well h.; **t. de tubage,** casing h.; **t. de vallée,** valley h.; **t. plongeante (tecto.),** dipping fold.

Téthys, Tethyan ocean, Tethys.

Tétrabranchiaux (pal.), Tetrabranchiata.

Tétracoralliaires (pal.), Tetracorallia, Rugosa; **t. isolé,** solitary Tetracorallia; **t. colonial,** compound Tetracorallia.

tétradymite (minér.), tetradymite.

tétraèdre, tetrahedron.

tétraédrique, tetrahedral.

tétrahédrite (minér.), tetrahedrite.

tétrahexaèdre, tetrahexahedron.

Tétrapodes (pal.), Tetrapoda.

tétravalence, tetravalency.

tétravalent (chimie), tetravalent.

textural, textural.

texture (voir aussi structure), texture; **t. aplitique,** aplitic t.; **t. diablastique,** diablastic t.; **t. en mosaïque,** mosaic t.; **t. fenêtrée,** lattice t.; **t. fibreuse,** fibrous t.; **t. foliacée,** foliated t.; **t. granoblastique,** granoblastic t.; **t. granophyrique,** granophyric t.; **t. graphique,** graphic t.; **t. hyaline,** hyaline t.; **t. hyalopilitique,** hyalopilitic t.; **t. intersertale,** intersertal t.; **t. microgrenue,** microgranular (holocrystalline and fine-grained) t.; **t. microlithique,** microlitic t., volcanic t.; **t. mylonitique,** mylonitic t.; **t. nématoblastique,** nematoblastic t.; **t. ophitique,** ophitic t.; **t. pilotaxitique,** pilotaxitic t.; **t. poécilitique,** poikilitic t.; **t. porphyrique,** porphyritic t.; **t. porphyroblastique,** porphyroblastic t.; **t. sphérolithique,** sperulitic t.; **t. vitreuse,** holohyaline t.

thalassique, thalassic (pertaining to ocean).

thalassocratique, thalassocratic (paleogeography dominated by ocean).

thalassocraton, thalassocraton.

thalassoid, bassin lunaire.

Thallophytes (pal.), Thallophyta.

thalweg, thalweg.

thanatocoenose (pal.), thanatocoenosis, death association (cf. brocoenosis).

Thanétien, Thanetian (l. Eocene: Paris basin).

thénardite (minér.), thenardite.

théodolite, theodolite; **t. à boussole,** transit compass; **t. pour mines,** mining transit.

théorie, theory; **t. de la dérive des continents,** continental d. theory; **t. de la tectonique des plaques,** plate tectonics.

thèque (pal.), theca.

théralite (gabbro alcalin), theralite.

Thérapsidés (pal.), therapsida.

thermal, thermal; **eau t.,** thermal water; **hydrothermal,** hydrothermal; **source t.,** thermae.

thermalité, thermality.

thermique, thermic(al), thermal; **analyse t.,** t. analysis; **désaimantation t.,** t. demagnetization; **diagraphie t.,** t. logging; **dilatation t.,** t. expansion; **gradient t.,** t. gradient; **pollution t.,** t. pollution; **rendement t.,** t. efficiency; **unité t.,** t. unit.

thermoclastie, thermal weathering.

thermocline, thermocline.

thermoforage, jet piercing.

thermogramme, thermogram.

thermographie, thermography; **t. aérienne,** aerial t.

thermokarst, thermokarst; **trou de t.,** thermokarst hole.

thermokarstique, thermokarstic.

thermoluminescence, thermoluminescence.

thermoluminescent, thermoluminescent.

thermolyse, thermolysis.

thermomagnétique, thermomagnetic.

thermométamorphisme, thermal metamorphism.

thermométrie, thermometry.

thermométrique, thermometric(al).

thermominéral, thermomineral.

thermonatrite (minér.), thermonatrite.

thermophile, thermophile.

thermorémanence, thermoremanence.

thermostable, thermostable.

Théropodes (pal.), Theropoda.

Thétis, Thétys, Tethys, Tethyan ocean.

thixotropie, thixotropy.

thixotropique, thixotropic.

tholéiite (pétro.), tholeiite.

tholéiitique, tholeiitic; **basalte t.,** tholeiitic basalt.

thomsonite (minér.), thomsonite.

thorianite (minér.), thorianite.

thorite (minér.), thorite.

thorium, thorium.

thulite (minér.), thulite.

Thuringien, Thuringian (u. Permian: Europe).

thuringite (minér.), thuringite.

tidal (littoral), tidal.

tidalite, tidalite.

tige, **1.** rod; **2.** stem (palaeobot.); **t. de fleuret,** drill rod; **t. de forage,** drill pipe, drill rod; **t. de pompage,** pumping rod; **t. de production,** tubing; **t. de sonde,** drilling rod.

tigre (œil, crocidolite), tiger's eye.

Tiglien, Tiglian (u. Pliocene, Netherlands).

tillite, tillite (palaeozoic indurated till).

tir, shooting, shot, blasting; **t. à l'air comprimé,** air blasting; **t. de mine,** firing; **t. de profondeur,** depth shooting; **t. de réflexion,** reflection shooting; **t. de réfraction,** refraction shooting; **t. électrique,** electric blasting; **t. en amont pendage,** up-dip blasting; **t. en arc,** arc shooting; **t. en éventail,** fan shooting; **t. en parallèle,** parallel shooting; **t. sismique,** seismic shooting; **ligne de t.,** blasting cable; **zone de t.,** blast area.

tirant (d'eau), draught, draft.

tirer (retirer), to draw off, to extract; **t. (mine),** to shoot, to blast; **t. (sismique),** to fire off, to shoot.

titane, titanium.

titane oxydé, titanic oxide.

titané, titanic, titanitic; **fer t.,** titanoferrite, titanitic iron ore.

titaneux, titanous.

titanifère, titaniferous; **augite t.,** titanaugite; **grenat t.,** titangarnet; **hornblende t.,** titanhornblende; **mica t.,** titanmica; **tourmaline t.,** titantourmaline.

titanite (sphène), titanite.

titanium, titanium.

titanomagnétite, titanomagnetite.

Tithonique, Tithonian (u. Jurassic, l. Cretaceous transition: Alps).

titrage (chimie), titration.

titre (de l'or), title (of gold); **t. (d'un métal),** title, grade, fineness.

titrer, to titrate.

tjäle (périgl.), tjale, permafrost.

Toarcien, Toarcian (l. Jurassic).

toit (d'une couche), hanging wall, roof, top, top wall; **coup de t. (mine),** rock burst; **éboulement du t.,** roof fall.

tombolo, tombolo.

tonalite (pétro.), tonalite.

Tongrien, Tongrian (l. Oligocene).

tonnage (de minerai), tonnage.

tonne, ton; **t. courte (américaine),** short ton = 907, 185 kg (U.S., 2000 lb); **t. forte (avoir-du-poids, G.B.),** long ton, gross ton = 1016,05 kg (U.K., 2 240 lb); **t. métrique = 1 000 kg,** metric ton.

tonstein (var. d'argile), tonstein.

topaze, topaz.

topazolite (andradite jaune), topazolite.

topochimique (métamorphisme), metamorphism occuring without chemical change.

topographe, topographer.

topographie, topography.

topographique, topographic(al); **carte t., t.** map; **correction t., t.** correction; **dépression t., t.** low; **discordance t., t.** unconformity; **levé t., t.** survey; **planchette t., t.** sheet.

topotype (pal.), topotype.

torbernite (minér.), torbernite.

torchère (pétrole), flare pit.

tornade (météo.), tornado.

torpiller un puits (pétrole), to torpedo.

torrent, torrent, mountain stream.

torrentiel, torrential.

torride (météo.), torrid.

torsion, twisting, torsion; **balance de t.,** t. balance; **cisaillement par t.,** t. shearing; **coefficient de t.,** t. coefficient; **essai de t.,** torsional test; **fil de t.,** torsional wire; **résistance à la t.,** torsional strain.

Tortonien, Tortonian (u. Miocene).

totale (réflexion), total (reflection).

toundra, tundra; **t. à monticules,** hillocky t.; **t. sèche,** dry t.; **tertre de t.,** hydrolaccolith.

tour, tower; **t. de fractionnement,** fractionnating t.; **t. de forage,** derrick, well rig; **t. de sondage,** derrick, boring t.; **t. karstique,** karst t., hill mogote, hum.

tourbe, peat; **t. acide,** acid p.; **t. de bruyère,** heather p.; **t. de carex,** sedge p.; **t. de mousses,** moss p.; **t. de sphaignes,** sphagnum p.; **t. lacustre,** limnic p.; **t. ligneuse,** wood p.; **t. limoneuse,** peaty loam; **t. vaseuse,** muddy p.

tourbeux, peaty, turfy.

tourbière, peat bog, bog; **t. exploitée,** turf pit; **t. basse,** low moor; **t. bombée,** raised bog; **t. haute,** high moor.

tourbification, peat formation.

tourbillon, eddy, whirl, swirl; **t. à axe horizontal,** roller; **t. de vent,** whirlwind.

tourbillonnaire, vortical.

tourbillonner, to whirl, to swirl.

tourmaline, tourmalin(e); **t. bleue,** blue t.; **t. brune,** brown t.; **t. lithinifère,** lithium t.; **t. noire,** black t.; **t. rouge,** red t.

granite à tourmaline, tourmaline granite.

tourmalinite (pétro.), schorl rock.

« tout venant » (mine), run of mine.

trace, trace, trail; **t. d'alimentation,** feeding traces; **t. de courants,** ripple marks, rillmarks; **t. de faille,** fault trace; **t. de roulement,** roll-mark; **t. de traînage,** groove cast; **t. d'outil,** tool mark; **t. d'un enregistrement géophysique,** trace; **t. de fossile,** trail, print.

trace (élément), trace element.

tracé, plotting, drawing; **t. cartographique,** plotting; **t. de faille,** fault trace; **t. d'un fleuve,** river course; **t. photogrammétrique,** plotting.

tracer (une courbe), to draw, to plot (a curve).

traceur chimique, chemical tracer.

traceur radioactif, radioactive tracer.

trachée (pal.), trachiae.

trachyandésite (pétro.), trachyandesite.

trachybasalte (pétro.), trachybasalt.

trachydolérite (pétro.), trachydolerite.

trachyte (pétro.), trachyte.

trachytique, trachytic.

trachytoïde, trachytoid.

traction (courant de), tractive (current).

train d'ondes, wave front.

train de tiges de forage, drill pipe string.

traînage (mine), haulage.

traînée (périgl.), stripe; **t. avec triage,** sorted s.; **t. de blocs,** block s.; **t. de gélifluction,** gelifluction s.

trait de côte, shoreline, coastline.

trait pédologique, pedologic feature.

traitement, treatment, treating; **t. acide,** acid t.; **t. alcalin,** alkaline t.; **t. chimique,** chemical t.; **t. de l'eau,** water t.; **t. du minerai,** ore dressing; **t. géophysique,** processing; **t. thermique,** thermal t.

traiter, to treat, to work, to process.

trajectoire (d'une onde), (wave) path.

tranchée (géotechnie), ditch, digging.

trancher (une roche), to slice, to cut.

trancheuse (machine), ditcher, trencher.

transfluent (glacier), transection (glacier).

transformante (faille), transform (fault).

transformisme, transformism.

transgressif, transgressive.

transgression, transgression, transgressive overlap.

transgressivité (parallèle), paraunconformity.

transition (couche de), transition (bed).

translation (vague de), translation (wave).

translucide (minéral), translucent.

transmettre (la lumière), to transmit.

transmission thermique, thermal transmission.

transparent (minéral), transparent.

transport (géomorpho.), transportation; **t. éolien,** eolian t.; **t. fluviatile,** river t.; **t. glaciaire,** glacial t.; **puissance de t. fluviatile,** carrying power, competency.

transporteur (mine), conveyor; **t. à bande,** belt c.; **t. à chaîne,** chain c.; **t. à godets,** bucket c.; **t. mécanique,** c.

transvaser (un fluide), to transvase.

transversal, transverse; **coupe t.,** t. section; **crevasse t.,** t. crevice; **dune t.,** t. dune; **faille t.,** t. fault; **onde t.,** t. wave; **vallée t.,** t. valley, cluse.

trapézoèdre, trapezohedron, leucitohedron.

travaux, workings, works; **t. d'exploitation,** mining works; **t. d'exploration,** exploratory works; **t. miniers,** mining works; **t. publics,** public works; **t. de terrassement,** earthworks.

travée (mine), lift, stage.

travers-banc (mine), cross-cut.

travers (en), cross-wise.

travertin, travertine, calcareous tufa (mineral spring or lake); soil travertine; calcrete; **t. de Sézanne,** lacustrine calcareous Thanetian travertine (Paris basin).

trébuchet (labo.), assay balance.

treillis, lattice; **réseau hydrographique en t.,** lattice-like river pattern.

Trémadoc(ien), Tremadocian (l. Ordovician).

tremblement de terre, earthquake, seism, earth tremor; **t. de forte magnitude,** megaseism; **t. sous-marin,** sea-quake.

trémie, funnel, hopper; **t. à graviers,** gravel hopper; **t. à minerai,** ore bin; **t. à sable,** sand hopper; **t. de chargement (mine),** loading pocket; **t. de sel,** salt pan; **cristal en t.,** hopper shaped crystal.

trémolite (minér.), tremolite.

trempage (par un liquide), soaking.

trempe (métall.), hardening, tempering.

tremper (métall.), to harden, to temper, to treat.

trépan, bit, drill bit; **t. à cônes,** cone rock bit; **t. à couronne,** crown bit; **t. à disque,** rotary disc bit; **t. à lames,** blade bit; **t. à molettes,** rock bit; **t. carottier,** rock bit; **t. en croix,** cross bit; **t. tricône,** tricone rock bit.

treuil, winch; **t. à câble,** rope w.; **t. d'extraction,** extracting w.

triage (de minerai), sorting, bucking; **t. à l'eau,** wet sorting; **t. à la main,** hand sorting.

triangulaire (diagramme), triangular (diagram), ternary diagram.

triangulation, triangulation; **t. aérienne,** stereo-triangulation.

trianguler (géodésie), to triangulate.

Trias, Trias.

Triasique, Triassic.

tributaire (cours d'eau), tributary.

triclinique, triclinic; **système t.,** triclinic system.

tricone, rotary bit.

tridymite (minér.), tridymite.

trier (un minerai), to sort, to classify, to separate.

trieur (ouvrier), picker, sorter.

Trilobitidés (pal.), Trilobita.

trioctaédrique, trioctahedral.

triphane (spodumène), triphane.

triphyllite (minér.), triphyllite.

triplite (minér.), triplite.

tripoli, tripolite, diatomite.

trituration (mine), trituration, grinding.

trivalent, trivalent.

troctolite (gabbro), troctolite.

troglodytique, troglodytic(al).

troïlite (minér.), troilite.

trombe d'eau, cloud burst, waterspout.

trommel (mine), trommel; **passer au t.,** to trommel.

troncature (minér.), truncation.

tronqué, truncated.

tronquer, to truncate.

troostite (minér.), troostite.

trop-plein (d'un barrage), overflow, weir, waste.

tropopause, tropopause.

troposphère, troposphere.

trottoir, intertidal organic corniche or ledge (encrusting algae, vermids, etc.)

trou, hole; **t. de mine,** drill h., blast h.; **t. de sonde,** drill h.; **t. de tir,** shot h.; **t. (karst),** sink-h.; **t. souffleur,** spouting h.

trouble (de l'eau), turbidity.

troubler (se) (eau), to become muddy.

tsunami (océano.), tsunami; tidal wave (popular).

tubage, casing; **t. de protection,** protection c.; **t. de puits,** well c.; **t. perforé,** perforated c.

tube, tube, pipe; **t. à essai,** test tube; **t. à rayons X,** X ray tube; **t. à sédimentation,** settling tube; **t. carottier,** coring barrel.

tubé (puits), cased (well).

tuber (for.), to case.

tubercule (pal.), knob.

tubulure (karst), pipe.

tuf, 1. tuff (volcanic ash, fine volcano-clastic); 2. tufa (spring deposit, calcareous, siliceous); **t. calcaire,** calcareous tufa, travertine; **t. siliceux,** siliceous sinter; **t. soudé,** welded tuff (ignimbritic tuff), ignimbrite; **t. volcanique,** volcanic tuff, tuff.

tufacé, tuffaceous.

tuffeau, sandy chalk (u. Cretaceous, Touraine, France).

tuffite (pétro.), tuffite.

tumulus (préhist.), tumulus.

tungstène, tungstene, tungstenium.

tungsténite (minér.), tungstenite.

tunnel (géotechnie), subway.

turbidites, turbidites.

turbidité (courant de), turbidity (current).

turbocarottage, turbocoring.

turboforage, turbodrilling.

turbulence (hydro.), turbulence.

turbulent, turbulent.

Turonien, Turonian (u. Cretaceous).

turquoise, turquoise.

Turritellidés (pal.), Turritellidae.

typhon, typhon.

typologie, typology.

typomorphique, typomorphic.

Tyrrhénien, Tyrrhenian (u. Pleistocene).

U

ubac, northern slope of a moutain (in northern hemisphere and vice versa); cf. adret.
udalf (sol lessivé de climat humide), udalf.
udoll (brunizem), udoll.
udomètre, pluviometer.
udométrie, pluviometry.
udométrique, pluviometric.
udult (ultisol de climat humide), udult.
ulexite (minér.), ulexite.
ullmannite (minér.), ulmannite.
ultisol (pédo.), ultisol.
ultrabasique, ultrabasic, ultramafic; **roche u.,** ultramafite.
ultrabasite (pétro.), ultramafite, ultramafitolite.
ultramétamorphique, ultrametamorphic.
ultramétamorphisme, anatexis metamorphism, ultrametamorphism.
ultrason, ultrasound.
ultraviolet, ultraviolet.
ultrazone, zone of anatexis.
umbo (pal.), beak.
uni (préfixe), uni-(prefixe), single, unique.
uni (terrain), even, smooth, flat.
uniaxe, uniaxial.
unicellulaire (pal.), unicellular.
uniforme (terrain), uniform, even.

uniformitarisme, uniformitarism.
uniloculaire (pal.), unilocular.
unisérié (pal.), uniseriate.
univalve (pal.), univalve.
univers, universe.
universelle platine, universal stage.
uranifère, uraniferous.
uraninite (minér.), uraninite.
uranite (autunite), uranite, uran mica.
uranium, uranium.
uranothorite (minér.), uranothorite.
uranotile (minér.), uranotil.
Urgonien, Urgonian (l. Cretaceous: facies, France).
urtite (syénite alcaline), urtite.
usé, worn; **u. par l'eau,** water w.; **u. par le vent,** wind w., wind facetted; **non-usé,** non-worn, angular.
ustalf (sol fersiallitique), ustalf.
ustert (vertisol de climat-chaud), ustert.
ustoll (chernozem méridional), ustoll.
ustox (sol ferralitique de climat chaud), ustox.
ustult (ultisol de climat chaud), ustult.
usure (géomorpho.), wear, wearing.
uvala (karst), uvala.
uvarovite (minér.), uvarovite.

V

vacuolaire (roche), vesicular, vuggy.
vacuole (petite cavité), vesicle.
vadose (eau), vadose (water).
va-et-vient (des vagues), swash.
vague, wave; **v. déferlante**, breaker; **v. de courant**, current w.; **v. de fond**, ground swell; **v. de froid**, cold w.; **v. de sable**, megaripple, sand w.; **v. de tempête**, storm w.; **v. de translation**, translatory w.; **v. d'oscillation**, oscillatory w.; **v. sismique**, seaquake w.; **v. stationnaire**, standing w.; **creux de v.**, w. trough; **hauteur de v.**, w. height.
valence (chimie), valency.
Valanginien, Valanginian (l. Cretaceous).
valentinite (minér.), valentinite.
valeur (sans-) (minéral), value-less (mineral).
vallée, valley, vale; **v. absorbante**, absorbent v.; **v. à fond plat**, flat floored v.; **v. antécédente**, antecedent v.; **v. aveugle**, blind v.; **v. conséquente**, consequent v.; **v. emboîtée**, inner v.; **v. en auge**, trough v.; **v. en U**, U-shaped v.; **v. en V**, V-shaped v.; **v. encaissée**, enclosed v.; **v. évasée**, wide v.; **v. fermée**, bolson; **v. glaciaire**, glacial v., glacial trough; **v. inadaptée**, underfit, misfit v.; **v. inondée**, ria; **v. monoclinale**, monoclinal v.; **v. morte**, abandoned v.; **v. obséquente**, obsequent v.; **v. sèche**, dry v., dell; **v. sous-marine**, submarine v.; **v. submergée**, drowned v.; **v. subséquente**, subsequent v.; **v. surimposée**, epigenetic v., superposed v.; **v. suspendue**, hanging v., suspended v.; **v. tectonique**, rift v.; **v. transversale**, transverse v., water gap, cluse; **fond de v.**, v. bottom, v. flat; **versant de v.**, v. side.
valleuse, hanging valley above shoreline.
vallon, dale, dell, glen (Écosse); **v. périglaciaire**, periglacial dry valley, dell, gelivation valley.
vallonné, undulating.
vallonnement, undulation.
valve (pal.), valve; **v. dorsale (Brachiopode)**, brachial v.; **v. ventrale (Brachiopode)**, pedicle v.
valvé (pal.), valvate.
vanadifère, vanadiferous.
vanadinite (minér.), vanadinite, vanadic ocher.
vanadium, vanadium.
vannage (éolien), eolian winnowing, eolian sorting, deflation.
vanner (les minerais), to van.
vapeur, steam, vapour; **v. d'eau**, water vapor; **v. de pétrole**, oil vapor; **v. sulfureuse**, sulphur fumes; **pression de v.**, vapor pressure.
vaporisable, vaporizable.
vaporisation, vaporization, evaporation.
vaporiser (se), to vaporize.
variation, variation, change; **v. barométrique**, barometric changes; **v. glaciaire**, glacial fluctuation; **v. magnétique**, magnetic variation, polarity transition; **v. séculaire géomagnétique**, geomagnetic secular variation.
variographe à vanne de flux, fluxgate variometer.
variolite (pétro.), variolite.
variolitique (structure), variolitic (structure).
variomètre, magnetometer.
variscite (minér.), variscite.
Varisque (orogénèse), Variscan, Varoscian (orogeny); Hercynian (orogeny).
varve, varve; **argile à v.**, varved clay; **disposition en v.**, varvity.
varvé, varved.
vasculaire (plante) (paléobot.), vascular (plant).
vase, mud, ooze, slime; **v. à Globigérines**, Globigerinid ooze; **v. d'étang**, pond mud; **v. rouge**, red ooze; **v. sableuse**, sandy mud; **v. putride**, sapropel; **v. tourbeuse**, peaty mud.
vaseux, muddy, silty.
vasière (d'estran), slikke.
vasière maritime, slikke, mud flat.
vaste, large, wide (rarely "vast").
vauclusienne (source), vauclusian (spring).
végétal (paléobot.), vegetal; **couverture v.**, v. cover.
végétal (fossile), vegetal fossil.
végétation, vegetation.
veine (mine), vein, lode, seam; **v. aurifère**, gold-bearing vein; **v. de charbon**, coal seam; **v. interstratifiée (éruptive)**, sill; **v. intrusive et oblique**, dyke.
veinule (mine), veinlet, veinule, stringlet.
vêlage (glaciol.), calving.
vêler (glacier), to calve.
vent, wind; **v. alizé**, trade w.; **v. contralizé**, antitrade w.; **v. de mer**, onshore w.; **v. de terre**, offshore w.; **v. de sable**, sand storm, sand w.; **sous le v.**, leeward.
ventifact, ventifact.
ventilateur (mine), ventilator, fan.
ventral (pal.), ventral.
venue d'eau (mine), water inflow, water inrush.
verglas (météo.), glazed frost.
vérins (banc à), lutetian calcareous bed with molds of *Cerithium giganteum* (Paris basin).
vermiculé (caillou), vermiculated (pebble).
vermiculite (minér.), vermiculite.
vernaculaire (pal.), vernacular.
vernis (désertique), (desert) varnish, patina.
verre, glass; **v. basaltique**, tachylite, basaltic g.; **v. volcanique**, volcanic g.
verrou glaciaire, rock bar, cross cliff, rock sill.
vers, towards; **v. l'amont**, upwards, upstream; **v. l'aval**, downwards, downstream; **v. le continent**, landwards.
versant, side, limb; **v. abrupt**, steep valley side; **v. de montagne**, mountain side; **v. d'un pli**, limb; **v.**

d'une vallée, valley side, valley wall; **bassin v.,** catchment area.

verser (un liquide), to pour down.

« vertes » (argiles), Brie clays, lower Oligocene, Paris basin.

vertes (roches), greenstones.

Vertébré (pal.), Vertebrate.

vertical, vertical, upright; **coupe v.,** vertical section; **faille v.,** vertical fault; **forage v.,** vertical drilling; **pendage v.,** vertical dip; **photographies aériennes v.,** vertical airviews, verticals; **rejet v.,** vertical separation, vertical displacement.

verticalité (d'une couche), verticality.

vertisol, vertisol.

vésicule (petite cavité), vesicle.

vésiculeux, vésiculaire, vesicular.

vestige (pal.), remain.

vestigial (pal.), vestigial.

vésuvianite (idocrase), vesuvianite.

vibrant (tamis), vibrating (screen).

vibration sonore, sound vibration.

vibratoire (mouvement), vibrational.

vibroseis, vibroseis.

vibro-séparateur, vibro-classifier.

vidange (d'un réservoir), emptying.

vider (un gisement), to exhaust.

vider (se) (s'écouler), to flow out.

vierge (minerai), native, pure; **v. (région),** unexplored.

Vieux Grès Rouge, Old Red Sandstone (Devonian: continental facies).

vieux travaux, waste.

vif argent, mercury.

Villafranchien, Villafranchian (u. Pliocene − l. Pleistocene: continental facies).

Vindobonien, Vindobonian (l. Miocene, obs).

virgation, virgation.

Virglorien (Trias), Virglorian (m. Triassic, equiv. Anisian).

virgula (pal.), virgula.

virtuelle (composition minéralogique), norm; **minéral v.,** normative mineral.

viscosité, viscosity; **coefficient de v.,** v. coefficient; **indice de v.,** v. index.

viscoélasticité, viscoelasticity.

visée, viewing, sighting.

Viséen, Visean (l. Carboniferous).

viseur (photog.), view-finder.

visibilité, visibility.

visible, visible.

visqueux, viscous.

Vistule (glaciation), Vistulian (glaciation) (u. Pleistocene).

vitesse, velocity, speed; **v. de forage,** drilling speed; **v. de balayage,** scanning speed; **v. de la lumière,** light velocity; **v. de précipitation,** settlement rate; **v. de propagation,** travel velocity; **v. de réaction,** reaction velocity; **v. de rotation (forage),** rotating speed; **v. de sédimentation,** settling rate; **v. limite**

de chute, terminal velocity; **discontinuité de v. sismique,** velocity discontinuity.

vitrain, vitrain.

vitreux, glassy, hyaline, vitreous, vitric; **roche v.,** holohyaline rock; **texture v.,** holohyaline texture.

vitrine (pour échantillons), glass-walled showcase.

vitrinite (cf. charbon), vitrinite.

vitriol, vitriol; **v. blanc,** zinc sulphate; **v. bleu,** copper sulphate; **v. vert,** iron sulphate.

Vitrollien, Vitrollian (lacustrine facies, Paleocene S.E. France).

vitrophyre, vitrophyre.

vitrophyrique, vitrophyric.

vivante (faille), active (fault).

vivianite (minér.), vivianite.

Vocontienne (fosse), Vocontian (trough; l. Cretaceous, S.E. France).

voie, 1. way; 2. channel (géoph.); **v. d'eau,** water way; **v. de fond (mine),** deep level; **v. de roulage (mine),** haulway.

volatil (élément), volatile.

volatilisation, volatilization.

volatiliser (se), to volatilize.

volatilité, volatility.

volcan, volcano, volcanoe; **v. actif,** active v.; **v. assoupi,** dormant v.; **v. bouclier,** shield v.; **v. de boue,** mud v., salse; **v. de sable,** pit and mound structure; **v. embryonnaire,** embryonic v.; **v. éteint,** extinct v.; **v. hawaïen,** hawaiian v., shield v.; **v. mixte,** mixed v.; **v. monogénique,** monogenic v.; **v. péléen,** pelean v.; **v. punctiforme,** central v.; **v. secondaire,** adventice v.; **v. sous-marin,** submarine v., sea-mount; **v. stratifié,** stratovolcano; **v. strombolien,** strombolian v.; **v. vulcanien,** vulcanian v.

volcanicité, volcanicity.

volcanique, volcanic; **aiguille v.,** v. spine; **bombe v.,** v. bomb; **cendre v.,** v. ash; **cheminée v.,** v. pipe, diatreme; **cône v.,** v. cone; **culot v.,** v. plug, neck; **dôme v.,** v. dome; **éruption v.,** v. eruption; **orifice v.,** v. vent; **projections v.,** v. ejectamenta; **sable v.,** v. sand; **tuff v.,** v. tuff, cinerite.

volcanisme, vulcanism, volcanism.

volcanite, pyroclastic rock.

volcanokarst, volcanokarst (tuff corrosion, solution, taffoni cavernous weathering).

volcanologie, volcanology.

volcanologue, volcanologist, vulcanologist.

volcanosédimentaire, volcanoclastic.

volcanotectonique, volcanotectonic.

volume spécifique, specific volume.

Vraconien, Vraconian (l. Cretaceous, u. Cretaceous transition: Europe).

voussure (tecto.), structural salient.

voûte anticlinale, upfold.

vue, view; **v. de face,** front v.; **v. en coupe,** sectional v.; **v. latérale,** side v.

vulcanien (volcanisme), vulcanian.

vulcanologie, vulcanology.

vulcanorium, vulcanorium.

W

wad (minér.), wad, bog manganese (cf. silomelane).
wawellite (minér.), wawellite.
Wealdien, Wealdian (l. Cretaceous, U.K., lacustrine-paludal facies).
Weichsélien, Weichselian (u. Pleistocene: glacial stage, equiv. Würmian, Wisconsinian).
Wentworth (échelle de), Wentworth (scale).
Werfénien, Werfenian (l. Triassic) alpine facies: equiv. Scythian).
Westphalien, Westphalian (u. Carboniferous: between Namurian and Stephanian).
wildflysh, wildflysch (u. Cretaceous: alpine, gravitational slump facies).

willémite (minér.), willemite.
williamsite (minér.), williamsite.
withérite (minér.), witherite.
wolfram, wolfram; **w. ocre (tungstite)**, wolfram-ocher.
wolframite, wolframite.
wollastonite (minér.), wollastonite.
wulfénite (minér.), wulfenite.
Wulff (canevas de), Wulff (net).
Wurm (glaciation de), Wurm.
Wurmien, Wurmian (u. Pleistocene: glacial stage; equiv: Weichselian, Wisconsinian).
Wurm-Riss, Wurm-Riss.

X

xanthophyllite (minér.), xanthophyllite.
xénoblaste, xenoblast.
xénoblastique, xenoblastic.
xénocristal, xenocryst.

xénolite, xenolith.
xénomorphe (minéral), xenomorph.
xénothermique (pal.), xenothermal.
xénotime (minér.), xenotime.

Y

yardang, yardang.
Young (module de), Young (modulus).

Yprésien, Ypresian (l. Eocene Paris basin).
yttrium, yttrium.

Z

Zechstein, Zechstein (u. Permian: facies in Germany; equiv. approx. Thuringian).

zénith, zenith.

zéolite, zeolite.

zéolitique, zeolitic.

zéolitisation, zeolitization.

zinc, zinc.

zincifère, zinciferous, zincky.

zincite (minér.), zincite.

zinckénite (minér.), zincenite.

zingueux, zincous.

zinnwaldite (minér.), zinnwaldite.

zircon (minér.), zircon.

zirconifère, zircon-bearing.

zirconium, zirconium.

Zoanthaires (pal.), Zoantharia.

zoécie (pal.), zoecium.

zoïsite (minér.), zoisite.

zone, zone, belt; **z. abyssale,** abyssal zone; **z. aquifère,** water-bearing zone; **z. d'altération,** weathering zone; **z. de brisants,** breaker zone; **z. de broyage,** shattered zone; **z. de déferlement,** surf zone; **z. de dislocation,** shear zone; **z. de lessivage,** leached zone; **z. de pergélisol continu,** continuous pergélisol zone; **z. de plissement,** folding zone; **z. désertique,** desert zone; **z. de subsidence,** subsiding area; **z. fracturée,** fractured zone; **z. inondable,** floodable zone; **z. intertidale,** tidal zone, intertidal zone; **z. littorale,** coastal zone; **z. minéralisée,** mineralized zone; **z. orogénique,** orogenic zone; **z. prélittorale,** nearshore zone.

zoné (cristal), banded, zoned (crystal).

zoogene (pal.), zoogene.

zoologie, zoology.

Répertoire d'abréviations scientifiques anglo-américaines

AAS	atomic absorption spectrometry	spectrométrie d'absorption atomique
ABW	antarctic bottom water	eau de fond antarctique
AIW	antarctic intermediate water	eau antarctique intermédiaire
APF	absolute pollen frequency	fréquence pollinique absolue
API	american petroleum institute	institut américain du pétrole
APT	automatic picture transmission	transmission automatique des images par satellite
ARC	chemical remanent aimantation	aimantation chimique rémanente
ARD	detritic remanent aimantation	aimantation rémanente chimique
ARN	natural remanent magnetization	aimantation rémanente naturelle
ARV	viscous remanent magnetization	aimantation rémanente visqueuse
ASL	above sea level	au-dessus du niveau de la mer
ATR	thermoremanent magnetization	aimantation thermo-rémanente
BHT	bottomhole temperature	température du fond de trou
BIF	banded iron formation	formation ferrifère zonée
BOD	biochemical oxygen demand	besoin chimique en oxygène
BP	before present	avant l'époque actuelle
BPD	barrels per day	barrils de pétrole par jour
BPG	blende-pyrite-galen mineral association	association minérale à blende-pyrite-galène
BPGC	blende-pyrite-galen-chalcopyrite association	association minérale à blende-pyrite-galène-chalcopyrite
BRT	below rotary table	sous la table de rotation
CCD	calcite compensation depth	profondeur de compensation des carbonates
CEC	cation exchange capacity	capacité d'échange cationique
CI	contour interval	équidistance des courbes de niveau
COD	chemical oxygen demand	demande chimique en oxygène
CRM	chemical remanent magnetization	magnétisme chimique rémanent
CSP	continuous seismic profiling	profilage (sondage sismique continu)
CVL	continuous velocity log	programme continu de vitesse
CW	continuous wave	onde entretenue
D and A	dry and abandoned	puits sec et abandonné
DD	drilling deeper	forage profond
D/H ratio		rapport deutériumH2/hydrogène H1
DDPS	dynamically positioned drillship	navire de forage positionné dynamiquement
DRM	detrital remanent magnetization	aimantation rémanente détritique
DSDP	deep sea drilling programme	programme de forages océaniques profonds
DST	drillstem system	système de forage
DSL	deep scattering layer	couche dispersive profonde (océano.)
EEZ	exclusive economic zone	zone économique exclusive
ELF	extremely low frequence	très basse fréquence
EOR	enhanced oil recovery	récupération assistée des hydrocarbures
FAD	first appearance datum	première date d'apparition
FAL	formation analysis log	diagraphie d'analyse des formations
FAMOUS	franco-american mid-ocean undersea study	site d'étude sous-marine de la dorsale médio-atlantique nord (expédition franco-américaine)
FDL	formation-density log	diagraphie de densité des formations
FP	freezing point	point de congélation
GARP	global atmospheric research programme	programme de recherche globale atmosphérique
GCM	gas-cut mud	boue de forage émulsionnée de gaz
GEOS	geodynamic experimental oceanic satellite	satellite océanique expérimental géodynamique
GL	ground level	niveau du sol
GLR	gas/liquid ratio	proportion gaz liquide
GMT	Greenwich mean time	heure moyenne de Greenwich
GOR	gas/oil ratio	proportion gaz/pétrole
GPS	global positioning system	système global de positionnement

HDR	hot dry rock	roche chaude et sèche (énergie radioactive)
HFU	heat-flow unit	unité de flux thermique
HHW	high high water	niveau maximum de l'eau
HP-BT	high pression-low temperature	métamorphisme régional-haute pression-basse température
HREE	heavy rare earth element	élément terrestre lourd et rare
HRIR	high resolution infrared radiometer	radiomètre infrarouge à haute résolution
ICSU	international council of scientific unions	conseil international des unions scientifiques
ID	inside diameter	diamètre intérieur d'un tube de forage
IGC	international geologic congress	congrès géologique international
IP	primary waves impulse	impetus des ondes
IPOD	international programme of ocean drilling	programme international de forage océanique
ITCZ	intertropical convergence zone	zone de couvergence intertropicale des courants
IUGS	international union of geological sciences	union internationale des sciences géologiques
IX	ion exchange	échange d'ions
JOIDES	joint oceanographic institutions (deep earth sounding)	sondages profonds dirigés par les instituts océano-graphiques
GPS	global positioning system	système de positionnement par satellite
GPH	gallon per hour	débit d'un puits en gallons/heure
GFU	geothermic flux unit	unité de flux géothermique
HRV	high resolution visible	haute résolution dans le visible
LAP	last apparition period	dernière période d'apparition
LCM	lost-circulation material	colmatant pour boue de forage
LVZ	low-velocity zone	zone à faible vitesse de propagation sous l'océan
LWD	logging while drilling	diagraphie en cours de forage
MEP	mean effective pressure	pression effective moyenne
MD	measured depth	profondeur mesurée
Mev	million electronvolt	millions d'électrons volt
Mg/d	million of gallons by day	millions de gallons par jour
Mu	millimicron	millimicron (10^{-6} mm)
MHD	magnetohydrodynamic	magnétohydrodynamique
ML	microlog	microlog (diagraphie)
MLL	microlaterolog	microlatérolog
MORB	mid-oceanic ridge basalt	basalte des crêtes médio-océaniques
MSL	mean sea level	niveau moyen de la mer
MWD	measurements while drilling	mesures en cours de forage
NADW	north atlantic deep water	eau profonde de l'océan nord-atlantique
NASA	national aeronautics and space administration	administration nationale aéronautique et de l'espace
NGL	natural gas liquids	gaz naturels liquéfiés
NNSS	navy navigation satellite system	système de navigation par satellite
NOAA	national oceanographic and atmospheric administration	direction nationale des océans et de l'atmosphère
NRM	natural remanent magnetization	magnétisme naturel rémanent
NSF	national science foundation	fondation nationale des sciences
O and G	oil and gas	huile et gaz
OCS	outer continental shelf	plateau continental externe
OD	outside diameter	diamètre extérieur
OIB	ocean-island basalt	basalte d'île océanique
OIH	oil in hole	huile en place
P and A	plugged and abandoned	puits bouché et fermé
PI	productive index	indice de productivité
PIOCW	pacific and indian ocean common water	eaux communes aux océans indien et pacifique
PNL	pulsed neutron logging	diagraphie de neutrons pulsés
POOH	pulled-out of hole	remontée du puits
PV	pore volume	volume des pores
QHM	quartz horizontal magnetometer	magnétomètre horizontal à quartz
REE	rare earth element	élément terrestre rare
RIH	run in hole	descendu dans le puits
ROP	rate of penetration	taux de pénétration
RPF	relative pollen frequency	fréquence pollinique relative

RR	rerun	redescente d'un outil dans le puits
RRR	triple junction of three ridges	point triple
RS	remote sensing	télédétection
RSS	regional stratigraphic scale	échelle stratigraphique régionale
RT	rotary table	table de rotation
SEM	scanning electron microscope	microscope électronique à balayage
SP	1. shot point	1. point de tir
	2. self potential	2. polarisation spontanée
SPS	satellite positioning system	système de positionnement par satellite
SSS	standart stratigraphic scale	échelle stratigraphique standard
SSTT	subsea test tree	tête de colonne pour essai de puits sous-marin
TCR	total core recovery	récupération de la totalité de la carotte
TD	1. time-depth curve	1. courbe temps-profondeur
	2. time-distance curve	2. dromochronique
	3. total depth of a hole	3. profondeur totale d'un puits
TDM	thermo-dynamic metamorphism	thermo-dynamo-métamorphisme
TP	tubing pressure	pression statique
TVD	true vertical depth	profondeur verticale réelle
T-X curve	time-distance curve	hodochrone
UPS	universal polarstereographic projection	projection stéréographique universelle polaire
UTS	unified stratigraphic time scale	échelle stratigraphique universelle
UVS	spectrophotometry by U.V. rays	spectrophotométrie aux rayons U.V.
VGP	virtual geomagnetic pole location	emplacement du pôle géomagnétique virtuel
VLBI	very large base interferometry	interférométrie à très grande base
VLF	very low frequency	très basse fréquence
VHF	very high frequency	très haute fréquence
VRM	viscous remanent magnetism	magnétisme rémanent visqueux
VSP	vertical seismic profile	profil sismique vertical
WOR	water oil ratio	rapport eau/huile
WIP	within plate basalt	basalte intraplaque
WNSS	world network seismograph system	réseau mondial standardisé de sismographes
YP	yeald point	seuil de plasticité, de cisaillement.

Conversion d'unités
Unit conversion

Multiplier les valeurs américaines	Par	Pour obtenir les valeurs décennales (système international)
Multiply inch-pound unit	By	To obtain SI unit
acre	4,047	square meter
	.4047	hectare
cubic yard	.7646	cubic meter
cubic feet per second	28.32	liters per second
	.02832	cubic meters per second
cubic mile	4.166	cubic kilometer
foot	.3048	meter
gallon	.003785	cubic meter
	3.785	liter
gallons per minute	.06309	liters per second
inch	2.540	centimeter
mile	1.609	kilometer
miles per hour	1.609	kilometers per hour
square mile	2.59	square kilometer
ton, short (2,000 pounds)	.9072	
	907.1848	kilograms

OUVRAGES CONSULTÉS/SHORT BIBLIOGRAPHY

BAULIG H., 1966. — *Vocabulaire franco-anglo-allemand de Géomorphologie*, 229 p., Public. Fac. Lettres de Strasbourg.

BERRY L. G., MASSON B. et DIETRICH R. V., 1983. — *Mineralogy*, 561 p., 300 fig., Éd. Freeman, Oxford G.B.

BEST M. G., 1982. — *Igneous and metamorphic petrology*, 630 p., 468 fig., Éd. Freeman, Oxford G.B.

BLANC J. J., 1982. — *Sédimentation des marges continentales actuelles et anciennes*, 159 p., 99 fig., Éd. Masson. Paris.

BLATT Harvey, 1982. — *Sedimentary petrology*, 564 p., 343 fig., Éd. Freeman, Oxford.

BODELLE J. et MARGAT J., 1980. — *L'eau souterraine en France*, 208 p., 77 fig., Éd. Masson, Paris.

BOILLOT G., 1983. — *Géologie des marges continentales*, 139 p., Éd. Masson, Paris.

BOILLOT G., MONTADERT L., LEMOINE M., BIJU-DUVAL, 1984. — *Les marges continentales actuelles et fossiles autour de la France*, 352 p., 204 fig., Éd. Masson.

CAILLEUX A., 1969. — *La Science de la Terre*, 799 p., 777 fig., Éd. Bordas.

CAMPY M. et MACAIRE J. J., 1989. — *Géologie des formations superficielles :* Géodynamique, faciès, utilisation, 433 p., Éd. Masson, Paris.

DECKER R. et DECKER B., 1980. — *Volcanoes*, 244 p., 117 fig.

DERRUAU M., 1988. — *Précis de Géomorphologie*, 7e éd., 533 p., 172 fig., Éd. Masson, Paris.

DUCHAUFOUR Ph., 1983. — *Pédologie, t. I*, Pédogenèse et classification, 491 p., 101 fig., Éd. Masson.

FAIRBRIDGE R. W., 1966. — *The Encyclopedia of Oceanography*, 1021 p., Éd. Reinhold, New York.

FAIRBRIDGE R. W., 1967. — *The Encyclopedia of Atmospheric Sciences and Astrogeology*, 1200 p., Éd. Reinhold, New York.

FAIRBRIDGE R. W., 1968. — *The Encyclopedia of Geomorphology*, 1295 p., Éd. Reinhold, New York.

FAIRBRIDGE R. W., 1972. — *The Encyclopedia of Geochemistry and Environmental Sciences*. 1321 p., Éd. Reinhold, New York.

FISCHER R. V., SCHMINCKE H. U., 1984. — *Pyroclastic Rocks*, 472 p., 339 fig., Éd. Springer-Verlag, Berlin.

FOUCAULT A. et RAOULT J. F., 1988. — *Dictionnaire de Géologie*, 3e éd., 351 p., Éd. Masson.

LLIBOUTRY L., 1982. — *Tectonophysique et Géodynamique*, 339 p., Éd. Masson, Paris.

MARSILY G. (de), 1981 — *Hydrogéologie quantitative*, 214 p., Éd. Masson, Paris.

NICOLAS A., 1989. — *Principes de Tectonique*, 2e éd., 223 p., Éd. Masson, Paris.

PAUL S., ALOUGES A., BONNEVAL H., PONTIER L., 1982. — *Dictionnaire de Télédétection aérospatiale*, 236 p., Éd. Masson.

PRESS F. et SIEVER R., 1982 — *Earth*, 3e éd., 613 p., Éd. Freeman, Oxford.

REINECK-SINGH, 1975. — *Depositional Sedimentary Environments*, 439 p., 579 fig., Éd. Springer-Verlag, Berlin.

ROCHE M. F., 1986 — *Dictionnaire français d'hydrologie de surface*, 288 p., Éd. Masson, Paris.

TARDY Y., 1986 — *Le cycle de l'eau*, 335 p., 134 fig., Éd. Masson, Paris.

MASSON Éditeur,
120, Boulevard Saint-Germain,
75280 Paris Cedex 06
Dépôt légal : mai 1992

Imprimerie Nouvelle
45800 Saint-Jean-de-Braye
N° d'impression : 12448
Dépôt légal : avril 1992